Methods in Enzymology

Volume 320
CUMULATIVE SUBJECT INDEX
Volumes 290–319

METHODS IN ENZYMOLOGY

EDITORS-IN-CHIEF

John N. Abelson   Melvin I. Simon

DIVISION OF BIOLOGY
CALIFORNIA INSTITUTE OF TECHNOLOGY
PASADENA, CALIFORNIA

FOUNDING EDITORS

Sidney P. Colowick and Nathan O. Kaplan

Methods in Enzymology

Volume 320

Cumulative Subject Index

Volumes 290–319

ACADEMIC PRESS
San Diego   London   Boston   New York   Sydney   Tokyo   Toronto

Academic Press
A Harcourt Science and Technology Company
525 B Street, Suite 1900, San Diego, California 92101-4495, USA

http://www.academicpress.com

Academic Press Limited
32 Jamestown Road, London NW1 7BY, UK

International Standard Book Number: 0-12-182221-4

PRINTED IN THE UNITED STATES OF AMERICA
00 01 02 03 04 05 06 MM 9 8 7 6 5 4 3 2 1

Table of Contents

Preface

The need for a cumulative subject index was recognized by the founding editors of *Methods in Enzymology* who prepared one for Volumes I through VI by weeding and interpolating from the entries that had been indexed in the individual volumes. As the series developed in number and complexity, different individuals with different backgrounds served as volume indexers. Subsequently, the series was fortunate in having Dr. Martha G. Dennis and Dr. Edward A. Dennis accept the challenge of digitizing the data available from these indexes, and this effort resulted in Volumes 33, 75, and 95, which cover Volumes 1 through 80.

Although each of these three books produced with the aid of digitization provided an appropriate cumulative index, major problems were encountered. One was time, both expensive computer time and lag time before such efforts resulted in publication. The most important difficulty was that the compilers were hampered by the lack of uniformity in the indexing of the individual volumes, resulting in the need for much hand editing to achieve a reasonable collation. The products were very decent, if uneven, indexes that also contributed to the methodology of computerized indexing, albeit with much delay and great expense.

This cumulative index was produced by Keith Shostak. Following a set of established guidelines, Dr. Shostak collected and compiled the entries from the individual volume indexes to produce the cumulative index. Since there is a tendency to identify specific topics and methods with particular individuals, a contributor index is included as well as the complete table of contents of each of the volumes indexed. In addition to Volume 320, the cumulative indexes in this series include Volumes 33, 75, 95, 120, 140, 175, 199, 229, 265, and 285.

Contents of Volumes 290–319

VOLUME 290
MOLECULAR CHAPERONES
George H. Lorimer and Thomas O. Baldwin

1. Protein Folding and Assembly in a Cell-Free Expression System — ALEXEY N. FEDOROV AND THOMAS O. BALDWIN — 1

2. Preparation and Application of Chaperone-Deficient *Escherichia coli* Cell-Free Translation Systems — GISELA KRAMER, TONG ZHANG, WIESLAW KUDLICKI, AND BOYD HARDESTY — 18

3. Protein Disulfide Isomerase — H. F. GILBERT — 26

4. Thermophilic Fungal Protein Disulfide Isomerase — TSUTOMU KAJINO, CHIE MIYAZAKI, OSAMU ASAMI, MASANA HIRAI, YUKIO YAMADA, AND SHIGEZO UDAKA — 50

5. Disulfide Bond Catalysts in *Escherichia coli* — THOMAS ZANDER, NIKHIL D. PHADKE, AND JAMES C. A. BARDWELL — 59

6. Yeast Immunophilins: Purification and Assay of Yeast FKBP12 — GREGORY WIEDERRECHT AND JOHN J. SIEKIERKA — 74

7. Peptidylprolyl *cis–trans*-Isomerases from Plant Organelles — AUTAR K. MATTOO — 84

8. Purification of GroEL with Low Fluorescence Background — A. CLAY CLARK, RAGULAN RAMANATHAN, AND CARL FRIEDEN — 100

9. Overexpression, Purification, and Properties of GroES from *Escherichia coli* — EDWARD EISENSTEIN, PRASAD REDDY, AND MARK T. FISHER — 119

10. Criteria for Assessing the Purity and Quality of GroEL — MATTHEW J. TODD AND GEORGE H. LORIMER — 135

11. Construction of Single-Ring and Two-Ring Hybrid Versions of Bacterial Chaperonin GroEL — ARTHUR L. HORWICH, STEVEN G. BURSTON, HAYS S. RYE, JONATHAN S. WEISSMAN, AND WAYNE A. FENTON — 141

12. Chaperonin 60₁₄ and Co-Chaperonin 10₇ from *Chromatium vinosum* — JOSE A. TORRES-RUIZ AND BRUCE A. MCFADDEN   147

13. Chaperonins of the Purple Nonsulfur Bacterium *Rhodobacter sphaeroides* — W. THEODORE LEE, GREGORY M. F. WATSON, AND F. ROBERT TABITA   154

14. Chaperonins from *Thermoanaerobacter* Species — ROBERT K. SCOPES AND KAYE TRUSCOTT   161

15. Chaperonin from Thermophile *Thermus thermophilus* — HIDEKI TAGUCHI AND MASASUKE YOSHIDA   169

16. Insect Chaperonin 60: Symbionin — MIZUE MORIOKA AND HAJIME ISHIKAWA   181

17. Purification of Yeast Mitochondrial Chaperonin 60 and Co-Chaperonin 10 — YVES DUBAQUIÉ, GOTTFRIED SCHATZ, AND SABINE ROSPERT   193

18. Purification of Mammalian Mitochondrial Chaperonin 60 through *in Vitro* Reconstitution of Active Oligomers — PAUL V. VIITANEN, GEORGE LORIMER, WOLFGANG BERGMEIER, CELESTE WEISS, MARTIN KESSEL, AND PIERRE GOLOUBINOFF   203

19. Purification of Recombinant Plant and Animal GroES Homologs: Chloroplast and Mitochondrial Chaperonin 10 — PAUL V. VIITANEN, KAREN BACOT, RAMONA DICKSON, AND TOM WEBB   218

20. Mammalian Cytosolic Chaperonin — NICHOLAS J. COWAN   230

21. Electron Microscopy of Chaperonins — S. CHEN, A. M. ROSEMAN, AND H. R. SAIBIL   242

22. Structural Analysis of GroE Chaperonin Complexes Using Chemical Cross-Linking — ABDUSSALAM AZEM, CELESTE WEISS, AND PIERRE GOLOUBINOFF   253

23. Molecular Chaperones and Their Interactions Investigated by Analytical Ultracentrifugation and Other Methodologies — HANS-JOACHIM SCHÖNFELD AND JOACHIM BEHLKE   269

24. Probing Conformations of GroEL-Bound Substrate Proteins by Mass Spectrometry — CAROL V. ROBINSON, MICHAEL GROSS, AND SHEENA E. RADFORD   296

25. Fluorescence Anisotropy Method for Investigation of GroEL–GroES Interaction — BORIS M. GOROVITS AND PAUL M. HOROWITZ   313

26. Photoincorporation of Fluorescent Probe into GroEL: Defining Site of Interaction — JEFFREY W. SEALE, BILL T. BRAZIL, AND PAUL M. HOROWITZ   318

27. Analysis of Chaperone Function Using Citrate Syn- JOHANNES BUCHNER,
    thase as Nonnative Substrate Protein HOLGER GRALLERT, AND
                                           URSULA JAKOB   323

28. Purification and Characterization of Small Heat JOHANNES BUCHNER,
    Shock Proteins MONIKA EHRNSPERGER,
                    MATTHIAS GAESTEL, AND
                    STEFAN WALKE   339

29. Expression, Purification, and Molecular Chaperone GARRETT J. LEE AND
    Activity of Plant Recombinant Small Heat ELIZABETH VIERLING   350
    Shock Proteins

30. Lens α-Crystallin: Chaperone-Like Properties JOSEPH HORWITZ,
                                                   QING-LING HUANG,
                                                   LINLIN DING, AND
                                                   MICHAEL P. BOVA   365

31. Purification and Properties of BiP MATHIEU CHEVALIER,
                                         LASHAUNDA KING, AND
                                         SYLVIE BLOND   384

32. Purification and Characterization of Prokaryotic JOHANNES BUCHNER,
    and Eukaryotic Hsp90 SUCHIRA BOSE,
                          CHRISTIAN MAYR, AND
                          URSULA JAKOB   409

33. Purification of Hsp90 Partner Proteins Hop/p60, JOHANNES BUCHNER,
    p23, and FKBP52 TINA WEIKL,
                     HANS BÜGL,
                     FRANZISKA PIRKL, AND
                     SUCHIRA BOSE   418

34. Purification and Properties of Hsp104 from Yeast ERIC C. SCHIRMER AND
                                                       SUSAN LINDQUIST   430

35. SecB: A Chaperone from *Escherichia coli* LINDA L. RANDALL,
                                                TRACI B. TOPPING,
                                                VIRGINIA F. SMITH,
                                                DEBORAH L. DIAMOND, AND
                                                SIMON J. S. HARDY   444

VOLUME 291
CAGED COMPOUNDS
Gerard Marriott

1. New Photoprotecting Groups: Desyl and *p*-Hy- RICHARD S. GIVENS,
   droxyphenacyl Phosphate and Carboxylate JÖRG F. W. WEBER,
   Esters ANDREAS H. JUNG, AND
           CHAN-HO PARK   1

2. Synthesis, Photochemistry, and Biological Charac- KYLE R. GEE,
terization of Photolabile Protecting Groups for BARRY K. CARPENTER, AND
Carboxylic Acids and Neurotransmitters GEORGE P. HESS 30

3. New Caged Groups: 7-Substituted Coumarinyl- TOSHIAKI FURUTA AND
methyl Phosphate Esters MICHIKO IWAMURA 50

4. Caged Fluorescent Probes T. J. MITCHISON,
K. E. SAWIN,
J. A. THERIOT,
K. GEE, AND
A. MALLAVARAPU 63

5. Biologically Active Peptides Caged on Tyrosine R. SREEKUMAR,
MITSUO IKEBE,
FREDRIC S. FAY, AND
JEFFERY W. WALKER 78

6. Light-Directed Activation of Protein Activity from GERARD MARRIOTT,
Caged Protein Conjugates JOHANNES OTTL,
MANFRED HEIDECKER, AND
DANIELA GABRIEL 95

7. Caged Peptides and Proteins by Targeted Chemi- HAGAN BAYLEY,
cal Modification CHUNG-YU CHANG,
W. TODD MILLER,
BRETT NIBLACK, AND
PENG PAN 117

8. Photocleavable Affinity Tags for Isolation and De- JERZY OLEJNIK,
tection of Biomolecules EDYTA KRZYMAŃSKA-
OLEJNIK, AND
KENNETH J. ROTHSCHILD 135

9. Synthesis and Applications of Heterobifunctional GERARD MARRIOTT AND
Photocleavable Cross-Linking Reagents JOHANNES OTTL 155

10. Use of Lasers for One- and Two-Photon Pho- JAMES A. MCCRAY 175
tolysis of Caged Compounds

11. Flash Lamp-Based Irradiation of Caged Com- GERT RAPP 202
pounds

12. Fourier Transform Infrared Photolysis Studies of VALENTIN CEPUS,
Caged Compounds CAROLA ULBRICH,
CHRISTOPH ALLIN,
AGNES TROULLIER, AND
KLAUS GERWERT 223

13. Use of Caged Compounds in Studies of Bioelec- C. W. WHARTON AND
tronic Imaging and Pattern Recognition R. S. CHITTOCK 245

14. Use of Caged Nucleotides to Characterize Unstable AXEL SCHEIDIG,
Intermediates by X-Ray Crystallography CHRISTOPH BURMESTER, AND
ROGER S. GOODY 251

15. Photoregulation of Cholinesterase Activities with Caged Cholinergic Ligands — LING PENG AND MAURICE GOELDNER — 265

16. Caged Substrates for Measuring Enzymatic Activity *in Vivo:* Photoactivated Caged Glucose 6-Phosphate — ROBERT R. SWEZEY AND DAVID EPEL — 278

17. Investigation of Charge Translocation by Ion Pumps and Carriers Using Caged Substrates — K. FENDLER, K. HARTUNG, G. NAGEL, AND E. BAMBERG — 289

18. Studies of Molecular Motors Using Caged Compounds — JODY A. DANTZIG, HIDEO HIGUCHI, AND YALE E. GOLDMAN — 307

19. Application of Caged Fluorescein-Labeled Tubulin to Studies of Microtubule Dynamics and Transport of Tubulin Molecules in Axons — TAKESHI FUNAKOSHI AND NOBUTAKA HIROKAWA — 348

20. Two-Photon Activation of Caged Calcium with Submicron, Submillisecond Resolution — EDWARD B. BROWN AND WATT W. WEBB — 356

21. Caged Inositol 1,4,5-Trisphosphate for Studying Release of Ca^{2+} from Intracellular Stores — NICK CALLAMARAS AND IAN PARKER — 380

22. Characterization and Application of Photogeneration of Calcium Mobilizers cADP-Ribose and Nicotinic Acid Adenine Dinucleotide Phosphate from Caged Analogs — KYLE R. GEE AND HON CHEUNG LEE — 403

23. Applications of Caged Compounds of Hydrolysis-Resistant Analogs of cAMP and cGMP — U. BENJAMIN KAUPP, CLAUDIA DZEJA, STEPHAN FRINGS, JÜRGEN BENDIG, AND VOLKER HAGEN — 415

24. Caged Probes for Studying Cellular Physiology: Application of *o*-Nitromandelyloxycarbonyl (Nmoc) Caging Method to Glutamate and a Ca^{2+}-ATPase Inhibitor — FRANCIS M. ROSSI, MICHAEL MARGULIS, ROBERT E. HOESCH, CHA-MIN TANG, AND JOSEPH P. Y. KAO — 431

25. Development and Application of Caged Ligands for Neurotransmitter Receptors in Transient Kinetic and Neuronal Circuit Mapping Studies — GEORGE P. HESS AND CHRISTOF GREWER — 443

26. Caged Plant Growth Regulators — ANDREW C. ALLAN, JANE L. WARD, MICHAEL H. BEALE, AND ANTHONY J. TREWAVAS — 474

27. Use of Caged Compounds in Studies of the Kinetics of DNA Repair — R. A. MELDRUM, R. S. CHITTOCK, AND C. W. WHARTON — 483

VOLUME 292
ABC TRANSPORTERS: BIOCHEMICAL, CELLULAR, AND MOLECULAR ASPECTS
Suresh V. Ambudkar and Michael M. Gottesman

Section I. Prokaryotic ABC Transporters

1. Overview of Bacterial ABC Transporters — HIROSHI NIKAIDO AND JASON A. HALL — 3

2. Preparation and Reconstitution of Membrane-Associated Maltose Transporter Complex of *Escherichia coli* — JASON A. HALL, AMY L. DAVIDSON, AND HIROSHI NIKAIDO — 20

3. Maltose Transport in *Escherichia coli:* Mutations That Uncouple ATP Hydrolysis from Transport — CYNTHIA H. PANAGIOTIDIS AND HOWARD A. SHUMAN — 30

4. Binding Protein-Dependent ABC Transport System for Glycerol 3-Phosphate of *Escherichia coli* — WINFRIED BOOS — 40

5. Structure–Function Analysis of Hemolysin B — FANG ZHANG, JONATHAN A. SHEPS, AND VICTOR LING — 51

6. *Erwinia* Metalloprotease Permease: Aspects of Secretion Pathway and Secretion Functions — PHILIPPE DELEPELAIRE — 67

7. Arsenical Pumps in Prokaryotes and Eukaryotes — MASAYUKI KURODA, HIRANMOY BHATTACHARJEE, AND BARRY P. ROSEN — 82

Section II. Eukaryotic ABC Transporters

8. Evolutionary Relationships among ABC Transporters — JAMES M. CROOP — 101

9. Cloning of Novel ABC Transporter Genes — RANDO ALLIKMETS AND MICHAEL DEAN — 116

10. *Saccharomyces cerevisiae* ABC Proteins and Their Relevance to Human Health and Disease — DANIEL TAGLICHT AND SUSAN MICHAELIS — 130

A. Nonmammalian ABC Transport Systems

11. *Arabidopsis MDR* Genes: Molecular Cloning and Protein Chemical Aspects — ROBERT DUDLER AND MICHAEL SIDLER — 162

12. Functional Analysis of *pfmdr1* Gene of *Plasmodium falciparum* — SARAH VOLKMAN AND DYANN WIRTH — 174

13. Amplification of ABC Transporter Gene *pgpA* and   MARC OUELLETTE,
    of Other Heavy Metal Resistance Genes in *Leish-*   ANASS HAIMEUR,
    mania tarentolae and Their Study by Gene Trans-   KATHERINE GRONDIN,
    fection and Gene Disruption   DANIELLE LÉGARÉ, AND
                                                         BARBARA PAPADOPOULOU   182

14. Functional Assay for Analysis of Yeast *ste6* Mutants GABY L. NIJBROEK AND
                                                           SUSAN MICHAELIS   193

15. ABC Transporters Involved in Transport of Eye   GARY D. EWART AND
    Pigment Precursors in *Drosophila melanogaster* ANTHONY J. HOWELLS   213

B. Mammalian P-Glycoproteins

16. Isolation of Altered-Function Mutants and Genetic IGOR B. RONINSON,
    Suppressor Elements of Multidrug Transporter   DONALD ZUHN,
    P-Glycoprotein by Expression Selection from   ADAM RUTH, AND
    Retroviral Libraries   DAVID DE GRAAF   225

17. Selection and Maintenance of Multidrug-Resis- MICHAEL M. GOTTESMAN,
    tant Cells   CAROL CARDARELLI,
                                                   SARAH GOLDENBERG,
                                                   THOMAS LICHT, AND
                                                   IRA PASTAN   248

18. Monoclonal Antibodies to P-Glycoprotein: Prepa- MIKIHIKO NAITO AND
    ration and Applications to Basic and Clinical Re- TAKASHI TSURUO   258
    search

19. Topology of P-Glycoproteins   WILLIAM R. SKACH   265

20. Use of Cell-Free Systems to Determine P-Glycopro- JIAN-TING ZHANG   279
    tein Transmembrane Topology

21. Photoaffinity Labels for Characterizing Drug Inter- AHMAD R. SAFA   289
    action Sites of P-Glycoprotein

22. Identification of Drug Interaction Sites in P-Glyco- LEE M. GREENBERGER   307
    protein

23. Photoaffinity Labeling of Human P-Glycoprotein: SAIBAL DEY,
    Effect of Modulator Interaction and ATP Hydro-   MURALIDHARA
    lysis on Substrate Binding   RAMACHANDRA,
                                                      IRA PASTAN,
                                                      MICHAEL M. GOTTESMAN, AND
                                                      SURESH V. AMBUDKAR   318

24. Identification of Phosphorylation Sites in Human TIMOTHY C. CHAMBERS   328
    MDR1 P-Glycoprotein

25. Identification of *in Vivo* Phosphorylation Sites for J. S. GLAVY, M. WOLFSON,
    Basic-Directed Kinases in Murine *mdr1b* P-Gly-   E. NIEVES, E.-K. HAN,
    coprotein by Combination of Mass Spectrometry   C.-P. H. YANG,
    and Site-Directed Mutagenesis   S. B. HORWITZ, AND
                                                      G. A. ORR   342

26. P-Glycoprotein and Swelling-Activated Chloride Channels — TAMARA D. BOND, CHRISTOPHER F. HIGGINS, AND MIGUEL A. VALVERDE — 359

27. Functional Expression of *mdr* and *mdr*-like cDNAs in *Escherichia coli* — EITAN BIBI, ROTEM EDGAR, AND ODED BÉJÀ — 370

28. Yeast Secretory Vesicle System for Expression and Functional Characterization of P-Glycoproteins — STEPHAN RUETZ — 382

29. High-Level Expression of Mouse Mdr3 P-Glycoprotein in Yeast *Pichia pastoris* and Characterization of ATPase Activity — LUCILLE BEAUDET, INA L. URBATSCH, AND PHILIPPE GROS — 397

30. Mutational Analysis of P-Glycoprotein in Yeast *Saccharomyces cerevisiae* — LUCILLE BEAUDET AND PHILIPPE GROS — 414

31. Baculovirus-Mediated Expression of Human Multidrug Resistance cDNA in Insect Cells and Functional Analysis of Recombinant P-Glycoprotein — URSULA A. GERMANN — 427

32. Recombinant Vaccinia Virus Vectors for Functional Expression of P-Glycoprotein in Mammalian Cells — MURALIDHARA RAMACHANDRA, MICHAEL M. GOTTESMAN, AND IRA PASTAN — 441

33. Functional Expression of Human P-Glycoprotein from Plasmids Using Vaccinia Virus–Bacteriophage T7 RNA Polymerase System — CHRISTINE A. HRYCYNA, MURALIDHARA RAMACHANDRA, IRA PASTAN, AND MICHAEL M. GOTTESMAN — 456

34. pHaMDR-DHFR Bicistronic Expression System for Mutation Analysis of P-Glycoprotein — SHUDONG ZHANG, YOSHIKAZU SUGIMOTO, TZIPORA SHOSHANI, IRA PASTAN, AND MICHAEL M. GOTTESMAN — 474

35. Mutational Analysis of Human P-Glycoprotein — TIP W. LOO AND DAVID M. CLARKE — 480

36. Purification and Reconstitution of Human P-Glycoprotein — SURESH V. AMBUDKAR, ISABELLE H. LELONG, JIAPING ZHANG, AND CAROL CARDARELLI — 492

37. Drug-Stimulatable ATPase Activity in Crude Membranes of Human *MDR1*-Transfected Mammalian Cells — SURESH V. AMBUDKAR — 504

38. ATPase Activity of Chinese Hamster P-Glycoprotein — ALAN E. SENIOR, MARWAN K. AL-SHAWI, AND INA L. URBATSCH — 514

39. Construction of *MDR1* Vectors for Gene Therapy   YOSHIKAZU SUGIMOTO,
                                                      MICHAEL M. GOTTESMAN,
                                                      IRA PASTAN, AND
                                                      TAKASHI TSURUO   523

40. Construction of *MDR1* Adeno-Associated Virus   MARION BAUDARD   538
    Vectors for Gene Therapy

41. Retroviral Transfer of Multidrug Transporter to THOMAS LICHT,
    Murine Hematopoietic Stem Cells   MICHAEL M. GOTTESMAN, AND
                                                      IRA PASTAN   546

42. Retroviral Transfer of Human *MDR1* Gene into   CAROLINE G. L. LEE,
    Human T Lymphocytes   IRA PASTAN, AND
                                                      MICHAEL M. GOTTESMAN   557

43. Construction and Analysis of Multidrug Resistance GREGORY L. EVANS   572
    Transgenic Mice

C. Multidrug Resistance Associated Protein

44. Cloning, Transfer, and Characterization of Multi- CAROLINE E. GRANT,
    drug Resistance Protein   GABU BHARDWAJ,
                                                      SUSAN P. C. COLE, AND
                                                      ROGER G. DEELEY   594

45. Transport Function and Substrate Specificity of DIETRICH KEPPLER,
    Multidrug Resistance Protein   GABRIELE JEDLITSCHKY, AND
                                                      INKA LEIER   607

D. Cystic Fibrosis Transmembrane Conductance Regulator

46. Heterologous Expression Systems for Study of   XIU-BAO CHANG,
    Cystic Fibrosis Transmembrane Conductance   NORBERT KARTNER,
    Regulator   FABIAN S. SEIBERT,
                                                      ANDREI A. ALEKSANDROV,
                                                      ANDREW W. KLOSER,
                                                      GRETCHEN L. KISER, AND
                                                      JOHN R. RIORDAN   616

47. Characterization of Polyclonal and Monoclonal An- NORBERT KARTNER AND
    tibodies to Cystic Fibrosis Transmembrane Con-   JOHN R. RIORDAN   629
    ductance Regulator

48. Identification of Cystic Fibrosis Transmembrane   ISABELLE T. CRAWFORD AND
    Conductance Regulator in Renal Endosomes   PETER C. MALONEY   652

49. Assays of Dynamics, Mechanisms, and Regulation   ERIK M. SCHWIEBERT,
    of ATP Transport and Release: Implications for   MARIE E. EGAN, AND
    Study of ABC Transporter Function   WILLIAM B. GUGGINO   664

50. Overexpression, Purification, and Function of First Nucleotide-Binding Fold of Cystic Fibrosis Transmembrane Conductance Regulator — YOUNG HEE KO AND PETER L. PEDERSEN — 675

51. Recombinant Synthesis of Cystic Fibrosis Transmembrane Conductance Regulator and Functional Nucleotide-Binding Domains — SCOTT A. KING AND ERIC J. SORSCHER — 686

52. Cationic Lipid Formulations for Intracellular Gene Delivery of Cystic Fibrosis Transmembrane Conductance Regulator to Airway Epithelia — SENG HING CHENG, JOHN MARSHALL, RONALD K. SCHEULE, AND ALAN E. SMITH — 697

53. Adeno-Associated Virus Vectors for Gene Therapy of Cystic Fibrosis — TERENCE R. FLOTTE AND BARRIE J. CARTER — 717

E. Sulfonylurea Receptor

54. Sulfonylurea Receptors and ATP-Sensitive Potassium Ion Channels — LYDIA AGUILAR-BRYAN, JOHN P. CLEMENT IV, AND DANIEL A. NELSON — 732

F. Intracellular ABC Transporters

55. Peptide Transport Assay for TAP Function — YE WANG, DAVID S. GUTTOH, AND MATTHEW J. ANDROLEWICZ — 745

56. Peroxisomal ABC Transporters — NOAM SHANI AND DAVID VALLE — 753

57. Mitochondrial ABC Transporters — JONATHAN LEIGHTON — 776

VOLUME 293
ION CHANNELS (PART B)
P. Michael Conn

Section I. Assembly

1. Methods Used to Study Subunit Assembly of Potassium Channels — JIA XU AND MIN LI — 3

2. Assembly of Ion Channels — ZUFANG SHENG AND CAROL DEUTSCH — 17

3. Analysis of K$^+$ Channel Biosynthesis and Assembly in Transfected Mammalian Cells   JAMES S. TRIMMER   32

Section II. Genetics

4. Site-Directed Mutagenesis   TAKAHIRO M. ISHII, PATRICIA ZERR, XIAO-MING XIA, CHRIS T. BOND, JAMES MAYLIE, AND JOHN P. ADELMAN   53

5. Molecular Physiology of Human Cardiovascular Ion Channels: From Electrophysiology to Molecular Genetics   SYLVAIN RICHARD, PHILIPPE LORY, EMMANUEL BOURINET, AND JOËL NARGEOT   71

6. Studying Ion Channels Using Yeast Genetics   ROBERT L. NAKAMURA AND RICHARD F. GABER   89

7. Identification of Ion Channel-Associated Proteins Using the Yeast Two-Hybrid System   MARTIN NIETHAMMER AND MORGAN SHENG   104

8. Substituted-Cysteine Accessibility Method   ARTHUR KARLIN AND MYLES H. AKABAS   123

9. Explorations of Voltage-Dependent Conformational Changes Using Cysteine Scanning   RICHARD HORN   145

10. Assessment of Distribution of Cloned Ion Channels in Neuronal Tissues   KOHJI SATO AND MASAYA TOHYAMA   155

Section III. Electrophysiology

11. Patch-Clamp Studies of Cystic Fibrosis Transmembrane Conductance Regulator Chloride Channel   JOHN W. HANRAHAN, ZIE KONE, CERI J. MATHEWS, JIEXIN LUO, YANLIN JIA, AND PAUL LINSDELL   169

12. Identification of Ion Channel Regulating Proteins by Patch-Clamp Analysis   THOMAS J. NELSON, PAVEL A. GUSEV, AND DANIEL L. ALKON   194

13. Tight-Seal Whole-Cell Patch Clamping of *Caenorhabditis elegans* Neurons   SHAWN R. LOCKERY AND M. B. GOODMAN   201

14. Low-Noise Patch-Clamp Techniques   RICHARD A. LEVIS AND JAMES L. RAE   218

15. Giant Membrane Patches: Improvements and Applications   DONALD W. HILGEMANN AND CHIN-CHIH LU   267

16. Electrophysiologic Recordings from *Xenopus* Oocytes — WALTER STÜHMER — 280

17. Cut-Open Oocyte Voltage-Clamp Technique — ENRICO STEFANI AND FRANCISCO BEZANILLA — 300

18. Cut-Open Recording Techniques — SHUJI KANEKO, AKINORI AKAIKE, AND MASAMICHI SATOH — 319

19. Gating Currents — FRANCISCO BEZANILLA AND ENRICO STEFANI — 331

20. Calcium Influx during an Action Potential — J. GERARD G. BORST AND FRITJOF HELMCHEN — 352

21. Combined Whole-Cell and Single-Channel Current Measurement with Quantitative Ca^{2+} Injection or Fura-2 Measurement of Ca^{2+} — L. DONALD PARTRIDGE, HANNS ULRICH ZEILHOFER, AND DIETER SWANDULLA — 371

22. Determining Ion Channel Permeation Properties — TED BEGENISICH — 383

23. Voltage-Dependent Ion Channels: Analysis of Non-ideal Macroscopic Current Data — RÜDIGER STEFFAN, CHRISTIAN HENNESTHAL, AND STEFAN H. HEINEMANN — 391

24. Signal Processing Techniques for Channel Current Analysis Based on Hidden Markov Models — SHIN-HO CHUNG AND PETER W. GAGE — 420

25. Investigating Single-Channel Gating Mechanisms through Analysis of Two-Dimensional Dwell-Time Distributions — BRAD S. ROTHBERG AND KARL L. MAGLEBY — 437

Section IV. Expression Systems

26. Expression of Ligand-Gated Ion Channels Using Semliki Forest Virus and Baculovirus — KATHRYN RADFORD AND GARY BUELL — 459

27. Recombinant Adenovirus-Mediated Expression in Nervous System of Genes Coding for Ion Channels and Other Molecules Involved in Synaptic Function — MARKUS U. EHRENGRUBER, MARKUS LANZREIN, YOUFENG XU, MARK C. JASEK, DAVID B. KANTOR, ERIN M. SCHUMAN, HENRY A. LESTER, AND NORMAN DAVIDSON — 483

28. *In Vivo* Incorporation of Unnatural Amino Acids into Ion Channels in *Xenopus* Oocyte Expression System — MARK W. NOWAK, JUSTIN P. GALLIVAN, SCOTT K. SILVERMAN, CESAR G. LABARCA, DENNIS A. DOUGHERTY, AND HENRY A. LESTER — 504

29. High-Level Expression and Detection of Ion Chan- THEODORE M. SHIH,
nels in *Xenopus* Oocytes RAYMOND D. SMITH,
LIGIA TORO, AND
ALAN L. GOLDIN 529

30. Unidirectional Fluxes through Ion Channels Ex- PER STAMPE AND
pressed in *Xenopus* Oocytes TED BEGENISICH 556

31. Transient Expression of Heteromeric Ion Channels ALISON L. EERTMOED,
YOLANDA F. VALLEJO, AND
WILLIAM N. GREEN 564

Section V. Model Simulations

32. Molecular Modeling of Ligand-Gated Ion Channels MICHAEL J. SUTCLIFFE,
ALLISTER H. SMEETON,
Z. GALEN WO, AND
ROBERT E. OSWALD 589

33. Use of Homology Modeling to Predict Residues SEAN-PATRICK SCOTT AND
Involved in Ligand Recognition JACQUELINE C. TANAKA 620

34. Ion Channels: Molecular Modeling and Simula- MARK S. P. SANSOM 647
tion Studies

35. Computer Simulations and Modeling of Ion MICHAEL E. GREEN 694
Channels

36. Kinetic Models and Simulation: Practical Ap- VLADIMIR AVDONIN AND
proaches and Implementation Notes TOSHINORI HOSHI 724

VOLUME 294
ION CHANNELS (PART C)
P. Michael Conn

Section I. Physical Methods

1. Optical Measurements of Calcium Signals in Mam- PETER SAGGAU,
malian Presynaptic Terminals RICHARD GRAY, AND
JOHN A. DANI 3

2. Fluorescent Techniques for Measuring Ion Chan- GÖNÜL VELIÇELEBI,
nel Activity KENNETH A. STAUDERMAN,
MARK A. VARNEY,
MICHAEL AKONG,
STEPHEN D. HESS, AND
EDWIN C. JOHNSON 20

3. Tagging Potassium Ion Channels with Green Fluo-   EDWIN S. LEVITAN   47
   rescent Protein to Study Mobility and Interactions
   with Other Proteins

4. Spin Label Electron Spin Resonance and Fourier   DEREK MARSH   59
   Transform Spectroscopy for Structural/Dynamic
   Measurements on Ion Channels

5. Nuclear Magnetic Resonance Spectroscopy of Pep-   MICHAEL D. REILY,
   tide Ion Channel Ligands: Cloning and Expres-   ANNE M. BOKMAN,
   sion as Aid to Evaluation of Structural and Dy-   JAMES OFFORD, AND
   namic Properties   PATRICK MCCONNELL   92

6. Ligand Binding Methods for Analysis of Ion Chan- STEEN E. PEDERSEN,
   nel Structure and Function   MONICA M. LURTZ, AND
                                                     RAO V. L. PAPINENI   117

7. Three-Dimensional Structure of Membrane Pro-   MARK YEAGER,
   teins Determined by Two-Dimensional Crystalli-   VINZENZ M. UNGER, AND
   zation, Electron Cryomicroscopy, and Image   ALOK K. MITRA   135
   Analysis

8. Drug-Dependent Ion Channel Gating by Applica-   FRANK BRETSCHNEIDER AND
   tion of Concentration Jumps Using U-Tube Tech-   FRITZ MARKWARDT   180
   nique

9. Voltage Clamp Biosensors for Capillary Electro-   OWE ORWAR,
   phoresis   KENT JARDEMARK,
                                                     CECILIA FARRE,
                                                     INGEMAR JACOBSON,
                                                     ALEXANDER MOSCHO,
                                                     JASON B. SHEAR,
                                                     HARVEY A. FISHMAN,
                                                     SHERI J. LILLARD, AND
                                                     RICHARD N. ZARE   189

10. Ion Channels as Tools to Monitor Lipid Bilayer– O. S. ANDERSEN,
    Membrane Protein Interactions: Gramicidin   C. NIELSEN,
    Channels as Molecular Force Transducers   A. M. MAER,
                                                     J. A. LUNDBÆK,
                                                     M. GOULIAN, AND
                                                     R. E. KOEPPE II   208

Section II. Purification and Reconstitution

11. Purification and Reconstitution of Epithelial Chlo- MOHABIR RAMJEESINGH,
    ride Channel Cystic Fibrosis Transmembrane   ELIZABETH GARAMI,
    Conductance Regulator   LING-JUN HUAN,
                                                        KEVIN GALLEY,
                                                        CANHUI LI,
                                                        YANCHUN WANG, AND
                                                        CHRISTINE E. BEAR   227

Section III. Second Messengers and Biochemical Approaches

Section IV. Special Channels

24. Mechanosensitive Channels of Bacteria   PAUL BLOUNT,
SERGEI I. SUKHAREV,
PAUL C. MOE,
BORIS MARTINAC, AND
CHING KUNG   458

25. Simplified Fast Pressure-Clamp Technique for   DON W. MCBRIDE, JR., AND
Studying Mechanically Gated Channels   OWEN P. HAMILL   482

26. Virus Ion Channels   DAVID C. OGDEN,
I. V. CHIZHMAKOV,
F. M. GERAGHTY, AND
A. J. HAY   490

27. Ion Channels in Microbes   YOSHIRO SAIMI,
STEPHEN H. LOUKIN,
XIN-LIANG ZHOU,
BORIS MARTINAC, AND
CHING KUNG   507

28. Design and Characterization of Gramicidin   DENISE V. GREATHOUSE,
Channels   ROGER E. KOEPPE II,
LYNDON L. PROVIDENCE,
SUNDARAM SHOBANA, AND
OLAF S. ANDERSEN   525

29. Functional Analyses of Aquaporin Water Chan-   PETER AGRE,
nel Proteins   JOHN C. MATHAI,
BARBARA L. SMITH, AND
GREGORY M. PRESTON   550

Section V. Toxins and Other Membrane Active Compounds

30. Pore-Blocking Toxins as Probes of Voltage-Depen-   ROBERT J. FRENCH AND
dent Channels   SAMUEL C. DUDLEY, JR.   575

31. Conus Peptides as Probes for Ion Channels   J. MICHAEL MCINTOSH,
BALDOMERO M. OLIVERA, AND
LOURDES J. CRUZ   605

32. Scorpion Toxins as Tools for Studying Potassium   MARIA L. GARCIA,
Channels   MARKUS HANNER,
HANS-GÜNTHER KNAUS,
ROBERT SLAUGHTER, AND
GREGORY J. KACZOROWSKI   624

33. Potassium Ion Channel Inactivation Peptides   RUTH D. MURRELL-LAGNADO   640

34. Interactions of Snake Dendrotoxins with Potas-   WILLIAM F. HOPKINS,
sium Channels   MARGARET ALLEN, AND
BRUCE L. TEMPEL   649

35. Use of Planar Lipid Bilayer Membranes for Rapid   TAJIB A. MIRZABEKOV,
    Screening of Membrane Active Compounds   ANATOLY Y. SILBERSTEIN, AND
                                                      BRUCE L. KAGAN   661

Section VI. Reagent and Information Sources

36. Antibodies to Ion Channels   ANGELA C. VINCENT,
                                                      IAN K. HART,
                                                      ASHWIN PINTO, AND
                                                      F. ANNE STEPHENSON   677

37. Internet Information on Ion Channels: Issues of   EDWARD C. CONLEY   704
    Access and Organization

VOLUME 295
ENERGETICS OF BIOLOGICAL MACROMOLECULES (PART B)
Gary K. Ackers and Michael L. Johnson

1. Deciphering Rules of Helix Stability in Peptides   CAROL A. ROHL AND
                                                      ROBERT L. BALDWIN   1

2. Modified Amino Acids as Probes of Helix Stability NEVILLE R. KALLENBACH AND
                                                      ERIK J. SPEK   26

3. Measuring Hydration Changes of Proteins in Solu-   VINCE J. LICATA AND
   tion: Applications of Osmotic Stress and Struc-   NORMA M. ALLEWELL   42
   ture-Based Calculations

4. Protein Folding in Membranes: Determining   STEPHEN H. WHITE,
   Energetics of Peptide–Bilayer Interactions   WILLIAM C. WIMLEY,
                                                      ALEXEY S. LADOKHIN, AND
                                                      KALINA HRISTOVA   62

5. Tight Ligand Binding Affinities Determined from   MICHAEL L. DOYLE AND
   Thermodynamic Linkage to Temperature by   PRESTON HENSLEY   88
   Titration Calorimetry

6. Structure-Based Prediction of Binding Affinities   IRENE LUQUE AND
   and Molecular Design of Peptide Ligands   ERNESTO FREIRE   100

7. Molecular Crowding: Analysis of Effects of High   ALLEN P. MINTON   127
   Concentrations of Inert Cosolutes on Biochemi-
   cal Equilibria and Rates in Terms of Volume Exclusion

full

8. Application of Automated Methods for Determination of Protein Conformational Stability — JEFFERY M. SCHWEHM AND WESLEY E. STITES — 150

9. Calculations of Proton-Binding Thermodynamics in Proteins — PAUL BEROZA AND DAVID A. CASE — 170

10. Analysis of Spectra from Multiwavelength Oxygen-Binding Studies of Mixed Metal Hybrid Hemoglobins — ALEXANDRA L. KLINGER AND GARY K. ACKERS — 190

11. Study of the Bohr Effect in Hemoglobin Intermediates — LOUISE BENAZZI, ROSARIA RUSSO, MARILENA RIPAMONTI, AND MICHELE PERRELLA — 208

12. Application of Planck–Benzinger Relationships to Biology — PAUL W. CHUN — 227

13. Kinetic Analysis of Macromolecular Interactions Using Surface Plasmon Resonance Biosensors — THOMAS A. MORTON AND DAVID G. MYSZKA — 268

14. Prediction of Binding Energetics from Structure Using Empirical Parameterization — BRIAN M. BAKER AND KENNETH P. MURPHY — 294

15. Photoacoustic Calorimetry of Proteins — MARK A. MCLEAN, CARMELO DI PRIMO, ERIC DEPREZ, GASTON HUI BON HOA, AND STEPHEN G. SLIGAR — 316

16. Isoergonic Cooperativity: A Novel Form of Allostery — HARVEY F. FISHER AND JON TALLY — 331

17. Theoretical Aspects of Isothermal Titration Calorimetry — LAWRENCE INDYK AND HARVEY F. FISHER — 350

18. Use of Poisson–Boltzmann Equation to Analyze Ion Binding to DNA — VICTOR A. BLOOMFIELD AND IOULIA ROUZINA — 364

19. Following the Folding of RNA with Time-Resolved Synchrotron X-Ray Footprinting — BIANCA SCLAVI, SARAH WOODSON, MICHAEL SULLIVAN, MARK CHANCE, AND MICHAEL BRENOWITZ — 379

20. Analysis of Interactions between CytR and CRP at CytR-Regulated Promoters — DONALD F. SENEAR, LAURA T. PERINI, AND STACEY A. GAVIGAN — 403

21. Energetic Methods to Study Bifunctional Biotin Operon Repressor — DOROTHY BECKETT — 424

22. Thermal Melting Properties of C-Terminal Domain Mutants of Bacteriophage λ cI Repressor — EDDINE K. MERABET, DAVID S. BURZ, AND GARY K. ACKERS — 450

23. Investigation of Phase Transitions in Bilayer Mem-   JEFFREY T. MASON   468
    branes

24. Membrane-Confined Analytical Electrophoresis   THOMAS M. LAUE,
                                                   HARVEY K. SHEPARD,
                                                   THERESA M. RIDGEWAY,
                                                   THOMAS P. MOODY, AND
                                                   TIMOTHY J. WILSON   494

VOLUME 296
NEUROTRANSMITTER TRANSPORTERS
Susan G. Amara

Section I. Purification of Transporter Proteins and cDNAs

A. Plasma Membrane Carriers

1. Purification, Hydrodynamic Properties, and Glyco-   CARMEN ARAGÓN AND
   sylation Analysis of Glycine Transporters   BEATRIZ LÓPEZ-CORCUERA   3

2. Expression Cloning Using *Xenopus laevis* Oocytes   MICHAEL F. ROMERO,
                                                       YOSHIKATSU KANAI,
                                                       HIROMI GUNSHIN, AND
                                                       MATTHIAS A. HEDIGER   17

3. Cloning of Genes or cDNAs Encoding Neurotrans-   NATHAN NELSON AND
   mitter Transporters and Their Localization by   QING-RONG LIU   52
   Immunocytochemistry

B. Vesicular Carriers

4. Purification of Vesicular Monoamine Transporters: RODRIGO YELIN AND
   From Classical Techniques to Histidine Tags   SHIMON SCHULDINER   64

5. Noncovalent and Covalent Labeling of Vesicular   JEAN-PIERRE HENRY,
   Monoamine Transporter with Tetrabenazine and   CORINNE SAGNÉ,
   Ketanserin Derivatives; Purification of Photola-   MARIE-FRANÇOISE ISAMBERT,
   beled Protein   AND BRUNO GASNIER   73

6. Functional Identification of Vesicular Monoamine   HELENE VAROQUI AND
   and Acetylcholine Transporters   JEFFREY D. ERICKSON   84

7. Photoaffinity Labeling of Vesicular Acetylcholine   STANLEY M. PARSONS,
   Transporter from Electric Organ of *Torpedo*   GARY A. ROGERS, AND
                                                       LAWRENCE M. GRACZ   99

8. Bioenergetic Characterization of γ-Aminobutyric   JOHANNES W. HELL AND
   Acid Transporter of Synaptic Vesicles   REINHARD JAHN   116

9. Solubilization and Reconstitution of Synaptic Vesi- SCOTT M. LEWIS AND
   cle Glutamate Transport System   TETSUFUMI UEDA   125

10. Analysis of Neurotransmitter Transport into Secre- J. PATRICK FINN III,
    tory Vesicles   ANDREW MERICKEL, AND
                                                        ROBERT H. EDWARDS   144

Section II. Pharmacological Approaches and Binding Studies

11. Inhibitors of γ-Aminobutyric Acid Transport as Ex- POVL KROGSGAARD-LARSEN,
    perimental Tools and Therapeutic Agents   BENTE F. FRØLUND, AND
                                                       ERIK FALCH   165

12. Design and Synthesis of Conformationally Con-   A. RICHARD CHAMBERLIN,
    strained Inhibitors of High-Affinity, Sodium-   HANS P. KOCH, AND
    Dependent Glutamate Transporters   RICHARD J. BRIDGES   175

13. Examination of Glutamate Transporter Heteroge-   MICHAEL B. ROBINSON   189
    neity Using Synaptosomal Preparations

14. Specificity and Ion Dependence of Binding of   AMADOU T. CORERA,
    GBR Analogs   JEAN-CLAUDE DO RÉGO, AND
                                                       JEAN-JACQUES BONNET   203

15. Cocaine and GBR Photoaffinity Labels as Probes   ROXANNE A. VAUGHAN   219
    of Dopamine Transporter Structure

Section III. Transport Assays and Kinetic Analyses

16. Ion-Coupled Neurotransmitter Transport: Thermo-   GARY RUDNICK   233
    dynamic vs. Kinetic Determinations of Stoichi-
    ometry

17. Inhibition of [³H]Dopamine Translocation and   MAARTEN E. A. REITH,
    [³H]Cocaine Analog Binding: A Potential Screen-   CEN XU, F. IVY CARROLL, AND
    ing Device for Cocaine Antagonists   NIAN-HANG CHEN   248

18. Transport and Drug Binding Kinetics in Membrane   HEINZ BÖNISCH   259
    Vesicle Preparation

19. Use of Human Placenta in Studies of Monoamine   VADIVEL GANAPATHY,
    Transporters   PUTTUR D. PRASAD, AND
                                                       FREDERICK H. LEIBACH   278

Section IV. Biochemical Approaches for Structure–Function Analyses

20. Analysis of Transporter Topology Using Deletion and Epitope Tagging — JANET A. CLARK — 293

21. Selective Labeling of Neurotransmitter Transporters at the Cell Surface — GWYNN M. DANIELS AND SUSAN G. AMARA — 307

22. Transmembrane Topology Mapping Using Biotin-Containing Sulfhydryl Reagents — REBECCA P. SEAL, BARBARA H. LEIGHTON, AND SUSAN G. AMARA — 318

23. Probing Structure of Neurotransmitter Transporters by Substituted-Cysteine Accessibility Method — JONATHAN A. JAVITCH — 331

24. Biosynthesis, N-Glycosylation, and Surface Trafficking of Biogenic Amine Transporter Proteins — SAMMANDA RAMAMOORTHY, HALEY E. MELIKIAN, YAN QIAN, AND RANDY D. BLAKELY — 347

25. Expression of Neurotransmitter Transport Systems in Polarized Cells — JINHI AHN, GRAZIA PIETRINI, THEODORE R. MUTH, AND MICHAEL J. CAPLAN — 370

26. Localization of Transporters Using Transporter-Specific Antibodies — N. C. DANBOLT, K. P. LEHRE, Y. DEHNES, F. A. CHAUDHRY, AND L. M. LEVY — 388

27. Generation of Transporter-Specific Antibodies — GARY W. MILLER, MICHELLE L. GILMOR, AND ALLAN I. LEVEY — 407

Section V. Expression Systems and Molecular Genetic Approaches

28. Homologies and Family Relationships among Na^+/Cl^- Neurotransmitter Transporters — HOLGER LILL AND NATHAN NELSON — 425

29. Vaccinia Virus–T7 RNA Polymerase Expression System for Neurotransmitter Transporters — SUE L. POVLOCK AND SUSAN G. AMARA — 436

30. Baculovirus-Mediated Expression of Neurotransmitter Transporters — CHRISTOPHER G. TATE — 443

31. Dopamine Transporter Mutants, Small Molecules, and Approaches to Cocaine Antagonist/Dopamine Transporter Disinhibitor Development — GEORGE UHL, ZHICHENG LIN, THOMAS METZGER, AND DALIT E. DAR — 456

32. *In Vivo* Generation of Chimeras — KARI JOHNSON BUCK AND SUSAN G. AMARA — 466

33. Structural Determinants of Neurotransmitter   ERIC L. BARKER AND
    Transport Using Cross-Species Chimeras: Studies   RANDY D. BLAKELY   475
    on Serotonin Transporter

34. Molecular Cloning of Neurotransmitter Trans-   DAVID J. VANDENBERGH   498
    porter Genes: Beyond Coding Region of cDNA

35. Use of Antisense Oligodeoxynucleotides to Inhibit   LYNN A. BRISTOL AND
    Expression of Glutamate Transporter Subtypes   JEFFREY D. ROTHSTEIN   514

36. Using *Caenorhabditis elegans* to Study Vesicular   JAMES B. RAND,
    Transport   JANET S. DUERR, AND
    DENNIS L. FRISBY   529

Section VI. Application of Electrophysiological Techniques to Neurotransmitter Carriers

37. Measurement of Transient Currents from Neuro-   SELA MAGER,
    transmitter Transporters Expressed in *Xenopus*   YONGWEI CAO, AND
    Oocytes   HENRY A. LESTER   551

38. Fluorescence Techniques for Studying Cloned   ALBERT CHA,
    Channels and Transporters Expressed in *Xeno-*   NOA ZERANGUE,
    pus Oocytes   MICHAEL KAVANAUGH, AND
    FRANCISCO BEZANILLA   566

39. Fluctuation Analysis of Norepinephrine and Seroto-   LOUIS J. DEFELICE AND
    nin Transporter Currents   AURELIO GALLI   578

40. Serotonin Transport in Cultured Leech Neurons   DIETER BRUNS   593

41. Glutamate Uptake in Purkinje Cells in Rat Cerebel-   MICHIKO TAKAHASHI,
    lar Slices   MONIQUE SARANTIS, AND
    DAVID ATTWELL   608

42. Patch-Clamp, Ion-Sensing, and Glutamate-Sensing   BRIAN BILLUPS,
    Techniques to Study Glutamate Transport in Iso-   MAREK SZATKOWSKI,
    lated Retinal Glial Cells   DAVID ROSSI, AND
    DAVID ATTWELL   617

43. Measurement of Glial Transport Currents in Micro-   STEVEN MENNERICK AND
    cultures: Application to Excitatory Neurotrans-   CHARLES F. ZORUMSKI   632
    mission

Section VII. Microdialysis and Electrochemical Measurements

44. Voltammetric Studies on Kinetics of Uptake and   KRISSTINA DANEK AND
    Efflux at Catecholamine Transporters   JAY B. JUSTICE, JR.   649

45. Resolution of Biogenic Amine Transporter Kinetics   CYNTHIA EARLES,
    by Rotating Disk Electrode Voltammetry: Meth-   HOLLIE WAYMENT,
    odology and Mechanistic Interpretations   MITCHELL GREEN, AND
    JAMES O. SCHENK   660

46. Electrochemical Detection of Reverse Transport   BRIAN B. ANDERSON,
    from *Planorbis* Giant Dopamine Neuron   ANDREW G. EWING, AND
                                                     DAVID SULZER   675

47. Measuring Uptake Rates in Intact Tissue   MELISSA A. BUNIN AND
                                                     R. MARK WIGHTMAN   689

48. High-Speed Chronoamperometric Electrochemical   NANCY R. ZAHNISER,
    Measurements of Dopamine Clearance   SHELLY D. DICKINSON, AND
                                                     GREG A. GERHARDT   708

49. *In Vivo* Microdialysis for Measurement of Extracel- NIAN-HANG CHEN AND
    lular Monoamine Levels Following Inhibition of   MAARTEN E. A. REITH   719
    Monoamine Transporters

VOLUME 297
PHOTOSYNTHESIS: MOLECULAR BIOLOGY OF ENERGY CAPTURE
Lee McIntosh

Section I. Genetic Approaches to Dissect Complex Functions

1. Transposon Mutagenesis of Heterocyst-Forming   MICHAEL F. COHEN,
   Filamentous Cyanobacteria   JOHN C. MEEKS,
                                                    YUPING A. CAI, AND
                                                    C. PETER WOLK   3

2. Isolation and Genetic Characterization of Pseudore- JIANPING YU AND
   vertants from Site-Directed PSI Mutants in *Sy-*   LEE MCINTOSH   18
   nechocystis 6803

3. High-Frequency Transformation of *Chlamydomo-*   KAREN L. KINDLE   27
   nas reinhardtii

4. Approaches to Investigating Nuclear Genes That   ALICE BARKAN   38
   Function in Chloroplast Biogenesis in Land
   Plants

5. Obtaining Crystal Structures from Bacterial Photo- GÜNTER FRITZSCH   57
   synthetic Reaction Centers

Section II. Photosynthetic Complexes: Function/Structure

6. Using Genetics to Explore Cytochrome Function   HANS-GEORG KOCH,
   and Structure in *Rhodobacter*   HANNU MYLLYKALLIO, AND
                                                    FEVZI DALDAL   81

7. Comparison of *in Vitro* and *in Vivo* Mutants of   JOHN H. GOLBECK   95
   Photosystem I: Protocols for Mutagenesis and
   Techniques for Analysis

8. Isolation and Functional Study of Photosystem I   JUN SUN, AN KE, PING JIN,
   Subunits in the Cyanobacterium *Synechocystis* sp.   VAISHALI P. CHITNIS, AND
   PCC 6803   PARAG R. CHITNIS   124

9. Delineation of Critical Regions of the ε Subunit of   RICHARD E. MCCARTY AND
   Chloroplast ATP Synthase through a Combina-   JEFFREY A. CRUZ   139
   tion of Biochemical and Site-Directed Mutagene-
   sis Approaches

Section III. Gene Expression of Photosynthetic Components

10. Transcriptional Regulation of Photosynthesis Op-   JILL H. ZEILSTRA-RYALLS,
    erons in *Rhodobacter sphaeroides* 2.4.1   MARK GOMELSKY,
                                                        ALEXEI A. YELISEEV,
                                                        JESUS M. ERASO, AND
                                                        SAMUEL KAPLAN   151

11. Use of *Synechocystis* 6803 to Study Expression of   CHRISTER JANSSON,
    a *psbA* Gene Family   GAZA SALIH,
                                                        JAN ERIKSSON,
                                                        RONNEY WIKLUND, AND
                                                        HAILE GHEBRAMEDHIN   166

12. Analysis of Light-Regulated Gene Expression   PAUL P. DIJKWEL,
                                                        FRED ROOK, AND
                                                        SJEF C. M. SMEEKENS   182

13. Translational Regulation of Chloroplast Gene Ex-   AMYBETH COHEN,
    pression in *Chlamydomonas reinhardtii*   CHRISTOPHER B. YOHN,
                                                        RICHARD K. BRUICK, AND
                                                        STEPHEN P. MAYFIELD   192

14. Use of Antisense Mutants to Study Regulation of   STEVEN RODERMEL   209
    Photosynthesis during Leaf Development

15. Assessing the Potential for Chloroplast Redox   DION G. DURNFORD,
    Regulation of Nuclear Gene Expression   ONDREJ PRASIL,
                                                      JEAN-MICHEL ESCOUBAS, AND
                                                      PAUL G. FALKOWSKI   220

Section IV. Biogenesis and Adaptation of Photosynthetic Components

16. Molecular and Genetic Analysis of Light-Depen-   GREGORY ARMSTRONG AND
    dent Chlorophyll Biosynthesis   KLAUS APEL   237

17. Molecular Biology of Carotenoid Biosynthesis in   MARK HARKER AND
    Photosynthetic Organisms   JOSEPH HIRSCHBERG   244

18. Copper-Responsive Gene Expression during Adap-   JEANETTE M. QUINN AND
    tation to Copper Deficiency   SABEEHA MERCHANT   263

19. Using Molecular Genetics to Investigate Comple-   DAVID M. KEHOE AND
    mentary Chromatic Adaptations: Advances in   ARTHUR R. GROSSMAN   279
    Transformation and Complementation

Section V. Photosynthetic Mutants: Construction and
Biological/Biochemical/Biophysical Analyses

20. Gene Modification and Mutation Mapping to Study   WIM F. J. VERMAAS   293
    the Function of Photosystem II

21. Specific Mutagenesis of Reaction Center Proteins HYEONMOO LEE,
    by Chloroplast Transformation of Chlamydomo-   SCOTT E. BINGHAM, AND
    nas reinhardtii   ANDREW N. WEBBER   310

22. Directed Mutagenesis in Photosystem II: Analysis TERRY M. BRICKER,
    of the CP 47 Protein   CINDY PUTNAM-EVANS, AND
                                                      JITUO WU   320

23. Application of Spectroscopic Techniques to the   BRUCE A. DINER   337
    Study of Mutations Engineered in Synechocystis
    and Chlamydomonas

VOLUME 298
MOLECULAR MOTORS AND THE CYTOSKELETON (PART B)
Richard B. Vallee

Section I. Analysis of Actomyosin-Related Systems

1. Purification and Assay of Myosin V   RICHARD E. CHENEY   3

2. Use of Latrunculin-A, an Actin Monomer-Bind-   KATHRYN AYSCOUGH   18
   ing Drug

3. Use of the F-Actin Binding Drugs, Misakinolide A   MICHAEL R. BUBB AND
   and Swinholide A   ILAN SPECTOR   26

4. F-Actin Blot Overlays   ELIZABETH J. LUNA   32

5. Purification and Assay of the Arp2/3 Complex from JOSEPH F. KELLEHER,
   Acanthamoeba castellani   R. DYCHE MULLINS, AND
                                                      THOMAS D. POLLARD   42

6. Purification and Assay of the Platelet Arp2/3   MATTHEW D. WELCH AND
   Complex   TIMOTHY J. MITCHISON   52

7. Purification and Assay of Zyxin   KAREN L. SCHMEICHEL,
                                          BETH E. STRONACH, AND
                                          MARY C. BECKERLE   62

8. Purification and Assays for Paxillin   CHRISTOPHER E. TURNER AND
                                          MICHAEL C. BROWN   77

9. Assay and Purification of Focal Adhesion Kinase   JUDITH LACOSTE,
                                          AMY MA, AND
                                          J. THOMAS PARSONS   89

10. Purification and Assays of VASP   THOMAS JARCHAU,
                                          THOMAS MUND,
                                          MATTHIAS REINHARD, AND
                                          ULRICH WALTER   103

11. *Listeria monocytogenes*-Based Assays for Actin As-   JULIE A. THERIOT AND
    sembly Factors   DAVID C. FUNG   114

Section II. Analysis of Microtubule-Related Systems

12. Preparation and Characterization of Caged Fluores-   ARSHAD DESAI AND
    cein Tubulin   TIMOTHY J. MITCHISON   125

13. Purification of Novel Kinesins from Embryonic   DAVID MEYER,
    Systems   DANIEL R. RINES,
                                          ANNA KASHINA, AND
                                          JONATHAN M. SCHOLEY   133

14. Assaying Processive Movement of Kinesin by Fluo-   DANIEL W. PIERCE AND
    rescence Microscopy   RONALD D. VALE   154

15. Purification of Dynactin and Dynein from Brain   JAMES B. BINGHAM,
    Tissue   STEPHEN J. KING, AND
                                          TRINA A. SCHROER   171

16. Isolation and Characterization of Kinectin   JANARDAN KUMAR AND
                                          MICHAEL P. SHEETZ   185

17. Purification and Assay of CLIP-170   GEORGIOS S.
                                          DIAMANTOPOULOS,
                                          THOMAS E. KREIS, AND
                                          JANET E. RICKARD   197

18. Purification and Assay of the Microtubule-Severing   FRANK MCNALLY   206
    Protein, Katanin

19. Purification and Assay of γ Tubulin Ring Complex   YIXIAN ZHENG,
                                          MEI LIE WONG,
                                          BRUCE ALBERTS, AND
                                          TIMOTHY J. MITCHISON   218

20. Rapid Isolation of Centrosomes   MAUREEN
                                          BLOMBERG-WIRSCHELL AND
                                          STEPHEN J. DOXSEY   228

21. Photoaffinity Labeling Approach to Map Taxol-   GEORGE A. ORR,
    Binding Site on the Microtubule   SRINIVASA RAO,
                                                       CHARLES S. SWINDELL,
                                                       DAVID G. I. KINGSTON, AND
                                                       SUSAN BAND HORWITZ   238

22. Use of Drugs to Study Role of Microtubule Assem- MARY ANN JORDAN AND
    bly Dynamics in Living Cells   LESLIE WILSON   252

Section III. Other Cytoskeletal Systems

23. Purification and Assay of a Septin Complex from   KAREN OEGEMA,
    Drosophila Embryos   ARSHAD DESAI,
                                                       MEI LIE WONG,
                                                       TIMOTHY J. MITCHISON, AND
                                                       CHRISTINE M. FIELD   279

24. Purification, Assembly, and Localization of FtsZ AMIT MUKHERJEE AND
                                                       JOE LUTKENHAUS   296

25. Purification and Assembly of FtsZ   CHUNLIN LU AND
                                                       HAROLD P. ERICKSON   305

Section IV. Cell-Free and Genetic Systems

26. High Resolution Video and Digital Enhanced Dif-   E. D. SALMON, ELAINE YEH,
    ferential Interference Contrast Light Microscopy   SID SHAW,
    of Cell Division in Budding Yeast   BOB SKIBBENS, AND
                                                       KERRY BLOOM   317

27. Production of M-Phase and I-Phase Extracts from   DUANE A. COMPTON   331
    Mammalian Cells

28. Organelle Motility and Membrane Network Forma-   VIKI J. ALLAN   339
    tion in Metaphase and Interphase Cell-Free Ex-
    tracts

29. Organelle Motility and Membrane Network Forma-   CAROL MCGOLDRICK AND
    tion from Cultured Mammalian Cells   MICHAEL P. SHEETZ   353

30. An *in Vitro* Motility Assay for Melanophore Pig- STEPHEN L. ROGERS,
    ment Organelles   IRINA S. TINT, AND
                                                       VLADIMIR I. GELFAND   361

31. Purification of Phagosomes and Assays for Microtu- JANIS K. BURKHARDT   373
    bule Binding

32. Magnetic Bead Assay for Characterization of Mi-   JOCHEN SCHEEL AND
    crotubule–Membrane Interactions   THOMAS E. KREIS   381

33. Reactivation of Vesicle Transport in Lysed Tele- LEAH HAIMO   389
    ost Melanophores

34. Reactivatable Cell Models of Giant Amoeba *Reticu-* MICHAEL P. KOONCE,
    lomyxa   URSULA EUTENEUER, AND
                                                           MANFRED SCHLIWA   400

Section V. Force Production Assays

35. Design and Use of the Centrifuge Microscope to RONALD J. BASKIN   413
    Assay Force Production

36. Motility Assays on Molluscan Native Thick Fila- YUNG JIN HAN AND
    ments   JAMES R. SELLERS   427

37. Use of Optical Traps in Single Molecule Study on AMIT D. MEHTA,
    Nonprocessive Biological Motors   JEFFREY T. FINER, AND
                                                       JAMES A. SPUDICH   436

38. Versatile Optical Traps with Feedback Control   KOEN VISSCHER AND
                                                     STEVEN M. BLOCK   460

39. Preparation of a Flexible, Porous Polyacrylamide YU-LI WANG AND
    Substrate for Mechanical Studies of Cultured   ROBERT J. PELHAM, JR.   489
    Cells

40. Design and Use of Substrata to Measure Traction TIM OLIVER,
    Forces Exerted by Cultured Cells   KEN JACOBSON, AND
                                                      MICAH DEMBO   497

Section VI. General Methods

41. Purification of Cytoskeletal Proteins Using Pep- CHRISTINE M. FIELD,
    tide Antibodies   KAREN OEGEMA,
                                                       YIXIAN ZHENG,
                                                       TIMOTHY J. MITCHISON, AND
                                                       CLAIRE E. WALCZAK   525

42. Strategies to Assess Phosphoprotein Phosphatase JOHN E. ERIKSSON,
    and Protein Kinase-Mediated Regulation of the   DIANA M. TOIVOLA,
    Cytoskeleton   CECILIA SAHLGREN,
                                                     ANDREY MIKHAILOV, AND
                                                     ANN-SOFI HÄRMÄLÄ-BRASKÉN   542

43. Correlative Light and Electron Microscopy of the TATYANA M. SVITKINA AND
    Cytoskeleton of Cultured Cells   GARY G. BORISY   570

VOLUME 299
OXIDANTS AND ANTIOXIDANTS (PART A)
Lester Packer

Section I. Total Antioxidant Activity

1. Total Antioxidant Activity Measured by Chemilu- HANNU ALHO AND
   minescence Methods   JANNE LEINONEN   3

2. Ferric Reducing/Antioxidant Power Assay: Direct   IRIS F. F. BENZIE AND
   Measure of Total Antioxidant Activity of Biologi-   J. J. STRAIN   15
   cal Fluids and Modified Version for Simultaneous
   Measurement of Total Antioxidant Power and
   Ascorbic Acid Concentration

3. Automated Electron Spin Resonance Free Radical   YASUKO NODA,
   Detector Assays for Antioxidant Activity in Nat-   MASAHIRO KOHNO,
   ural Extracts   AKITANE MORI, AND
                                                        LESTER PACKER   28

4. Measurement of Oxidizability of Blood Plasma   ANATOL KONTUSH AND
                                                        ULRIKE BEISIEGEL   35

5. Measurement of Oxygen Radical Absorbance   GUOHUA CAO AND
   Capacity in Biological Samples   RONALD L. PRIOR   50

Section II. Vitamin C

6. Analysis of Ascorbic and Dehydroascorbic Acid in   MARK LEVINE,
   Biological Samples   YAOHUI WANG, AND
                                                        STEVEN C. RUMSEY   65

7. Analysis of Water-Soluble Antioxidants by High-   ANN M. BODE AND
   Performance Liquid Chromatography with Elec-   RICHARD C. ROSE   77
   trochemical Detection

8. Ascorbic Acid Recycling in Rat Hepatocytes as   JENS LYKKESFELDT AND
   Measurement of Antioxidant Capacity: Decline   BRUCE N. AMES   83
   with Age

Section III. Polyphenols and Flavonoids

9. Detecting and Measuring Bioavailability of Pheno-   LOUISE C. BOURNE AND
   lics and Flavonoids in Humans: Pharmacokinetics   CATHERINE A. RICE-EVANS   91
   of Urinary Excretion of Dietary Ferulic Acid

10. Simultaneous Analysis of Individual Catechins and   TETSUHISA GOTO AND
    Caffeine in Green Tea   YUKO YOSHIDA   107

11. Reversed-Phase High-Performance Liquid Chro-   ANDREW L. WATERHOUSE,
    matography Methods for Analysis of Wine Poly-   STEVEN F. PRICE, AND
    phenols   JEFFREY D. McCORD   113

12. Analysis of Antioxidant Wine Phenols by High-   DAVID M. GOLDBERG AND
    Performance Liquid Chromatography   GEORGE J. SOLEAS   122

13. Analysis of Antioxidant Wine Polyphenols by Gas   GEORGE J. SOLEAS AND
    Chromatography–Mass Spectrometry   DAVID M. GOLDBERG   137

14. Analysis of Total Phenols and Other Oxidation Substrates and Antioxidants by Means of Folin–Ciocalteu Reagent   VERNON L. SINGLETON, RUDOLF ORTHOFER, AND ROSA M. LAMUELA-RAVENTÓS   152

15. Size Separation of Condensed Tannins by Normal-Phase High-Performance Liquid Chromatography   VÉRONIQUE CHEYNIER, JEAN-MARC SOUQUET, ERWAN LE ROUX, SYLVAIN GUYOT, AND JACQUES RIGAUD   178

16. Resveratrol and Piceid in Wine   ROSA M. LAMUELA-RAVENTÓS AND ANDREW L. WATERHOUSE   184

17. Antioxidant Activity by Headspace Gas Chromatography of Volatile Oxidation Products of ω-6 and ω-3 Polyunsaturated Lipids   E. N. FRANKEL   190

18. Determination of Tea Catechins by Reversed-Phase High-Performance Liquid Chromatography   PETER C. H. HOLLMAN, DINI P. VENEMA, SANTOSH KHOKHAR, AND ILJA C. W. ARTS   202

19. Flavonoids as Peroxynitrite Scavengers in Vitro   ANANTH SEKHER PANNALA, SURINDER SINGH, AND CATHERINE RICE-EVANS   207

Section IV. Thiols

20. Determination of Oxidized and Reduced Lipoic Acid Using High-Performance Liquid Chromatography and Coulometric Detection   CHANDAN K. SEN, SASHWATI ROY, SAVITA KHANNA, AND LESTER PACKER   239

21. Flow Cytometric Determination of Cellular Thiols   CHANDAN K. SEN, SASHWATI ROY, AND LESTER PACKER   247

22. Measurement of Glutathione, Glutathione Disulfide, and Other Thiols in Mammalian Cell and Tissue Homogenates Using High-Performance Liquid Chromatography Separation of N-(1-Pyrenyl)Maleimide Derivatives   LISA A. RIDNOUR, ROGER A. WINTERS, NURAN ERCAL, AND DOUGLAS R. SPITZ   258

23. Ratio of Reduced to Oxidized Glutathione as Indicator of Oxidative Stress Status and DNA Damage   MIGUEL ASENSI, JUAN SASTRE, FEDERICO V. PALLARDÓ, ANA LLORET, MARTIN LEHNER, JOSE GARCIA-DE-LA ASUNCION, AND JOSÉ VIÑA   267

24. Nonenzymatic Colorimetric Assay of Glutathione in Presence of Other Mercaptans   JEAN CHAUDIÈRE, NADIA AGUINI, AND JEAN-CLAUDE YADAN   276

25. Protozoological Method for Assaying Lipoate in Human Biologic Fluids and Tissue   HERMAN BAKER, BARBARA DEANGELIS, ELLIOTT R. BAKER, AND SEYMOUR H. HUTNER   287

26. Antioxidant Activity of Amidothionophosphates   OREN TIROSH, YEHOSHUA KATZHENDLER, YECHEZKEL BARENHOLZ, AND RON KOHEN   293

27. Quantitation of Anethole Dithiolthione Using High-Performance Liquid Chromatography with Electrochemical Detection   KATRINA TRABER AND LESTER PACKER   300

Section V. Vitamin E and Coenzyme Q_{10}

28. Gas Chromatography–Mass Spectrometry Analysis of Vitamin E and Its Oxidation Products   DANIEL C. LIEBLER, JEANNE A. BURR, AND AMY J. L. HAM   309

29. Determination of Vitamin E Forms in Tissues and Diets by High-Performance Liquid Chromatography Using Normal-Phase Diol Column   JOHN K. G. KRAMER, ROBERT C. FOUCHARD, AND KRISHNA M. R. KALLURY   318

30. Sensitive High-Performance Liquid Chromatography Techniques for Simultaneous Determination of Tocopherols, Tocotrienols, Ubiquinols, and Ubiquinones in Biological Samples   MAURIZIO PODDA, CHRISTINE WEBER, MARET G. TRABER, RAINER MILBRADT, AND LESTER PACKER   330

31. High-Performance Liquid Chromatography–Coulometric Electrochemical Detection of Ubiquinol 10, Ubiquinone 10, Carotenoids, and Tocopherols in Neonatal Plasma   BARBARA FINCKH, ANATOL KONTUSH, JENS COMMENTZ, CHRISTOPH HÜBNER, MARTIN BURDELSKI, AND ALFRIED KOHLSCHÜTTER   341

32. Simultaneous Determination of Retinol, Tocopherols, Carotene, Lycopene, and Xanthophylls in Plasma by Means of Reversed-Phase High-Performance Liquid Chromatography   CLAUDE-PIERRE AEBISCHER, JOSEPH SCHIERLE, AND WILLY SCHÜEP   348

33. Assessment of Prooxidant Activity of Vitamin E in Human Low Density Lipoprotein and Plasma   PAUL K. WITTING, DETLEF MOHR, AND ROLAND STOCKER   362

Section VI. Carotenoids and Retinoids

34. Screening of Dietary Carotenoids and Carotenoid-Rich Fruit Extracts for Antioxidant Activities Applying 2,2'-Azinobis(3-ethylbenzothiazoline-6-sulfonic acid) Radical Cation Decolorization Assay — NICOLETTA PELLEGRINI, ROBERTA RE, MIN YANG, AND CATHERINE RICE-EVANS ... 379

35. Matrix-Assisted Laser Desorption Ionization–Postsource Decay-Mass Spectrometry in Carotenoid Analysis — THOMAS WINGERATH, DIETER KIRSCH, RAIMUND KAUFMANN, WILHELM STAHL, AND HELMUT SIES ... 390

36. Carotenoid Photobleaching — ALAN MORTENSEN AND LEIF H. SKIBSTED ... 408

37. Interactions between Vitamin A and Vitamin E in Liposomes and in Biological Contexts — MARIA A. LIVREA AND LUISA TESORIERE ... 421

38. On-Line Solid-Phase Extraction and Isocratic Separation of Retinoic Acid Isomers in Microbore Column Switching System — THOMAS E. GUNDERSEN AND RUNE BLOMHOFF ... 430

39. Purification and Characterization of Cellular Carotenoid-Binding Protein from Mammalian Liver — M. R. LAKSHMAN AND MANJUNATH N. RAO ... 441

40. Analysis of Zeaxanthin Stereoisomer Distribution within Individual Human Retinas — JOHN T. LANDRUM, RICHARD A. BONE, LINDA L. MOORE, AND CHRISTINA M. GOMEZ ... 457

VOLUME 300
OXIDANTS AND ANTIOXIDANTS (PART B)
Lester Packer

Section I. Oxidative Damage to Lipids, Proteins, and Nucleic Acids

A. Lipids

1. Mass Spectrometric Quantification of F_2-Isoprostanes in Biological Fluids and Tissues as Measure of Oxidant Stress — JASON D. MORROW AND L. JACKSON ROBERTS II ... 3

2. Gas Chromatography–Mass Spectrometry Assay   JAFFAR NOUROOZ-ZADEH   13
   for Measurement of Plasma Isoprostanes

3. Preparation of Lipid Hydroperoxide-Free Low   GLENN T. SHWAERY,
   Density Lipoproteins   HIRO-OMI MOWRI,
                                                 JOHN F. KEANEY, JR., AND
                                                 BALZ FREI   17

4. Lipid Hydroperoxide Analysis by High-Perfor- WITOLD KORYTOWSKI,
   mance Liquid Chromatography with Mercury   PETER G. GEIGER, AND
   Cathode Electrochemical Detection   ALBERT W. GIROTTI   23

5. Measurement of Lipid Hydroperoxides in Human ANTONIO M. PASTORINO,
   Plasma and Lipoproteins by Kinetic Analysis of ADRIANA ZAMBURLINI,
   Photon Emission   LUCIO ZENNARO,
                                                 MATILDE MAIORINO, AND
                                                 FULVIO URSINI   33

6. High-Performance Liquid Chromatography Analy- GARRY J. HANDELMAN   43
   sis of Cholesterol Linoleate Hydroperoxide in
   Oxidized Low Density Lipoproteins: Calibration
   by Conjugated Diene Internal Standard

7. Determination of Phospholipid Oxidation in Cul- GLENN T. SHWAERY,
   tured Cells   JASON M. SAMII,
                                                 BALZ FREI, AND
                                                 JOHN F. KEANEY, JR.   51

8. Ferrous Ion Oxidation in Presence of Xylenol Or- JAFFAR NOUROOZ-ZADEH   58
   ange for Detection of Lipid Hydroperoxides in
   Plasma

9. Purification and Characterization of Phospholipid TAMMY R. DUGAS AND
   for Use in Lipid Oxidation Studies   DANIEL F. CHURCH   63

10. Malondialdehyde Assay in Biological Fluids by Gas HELEN C. YEO, JIANKANG LIU,
    Chromatography–Mass Spectrometry   HAROLD J. HELBOCK, AND
                                                 BRUCE N. AMES   70

11. Assessment of Physiological Interaction between PETER A. GLASCOTT, JR., AND
    Vitamin E and Vitamin C   JOHN L. FARBER   78

B. Proteins and Nucleic Acids

12. Modification of Proteins and Lipids by Myeloperox- STANLEY L. HAZEN,
    idase   FONG F. HSU, JOSEPH P. GAUT,
                                                  JAN R. CROWLEY, AND
                                                  JAY W. HEINECKE   88

13. Protein Carbonyl Measurement by Enzyme-Linked CHRISTINE C. WINTERBOURN
    Immunosorbent Assay   AND I. HENDRIKJE BUSS   106

14. Detection of 3-Chlorotyrosine in Proteins Exposed ANTHONY J. KETTLE   111
    to Neutrophil Oxidants

15. Determination of 2-Oxohistidine by Amino Acid Analysis — SANDRA A. LEWISCH AND RODNEY L. LEVINE — 120

16. Detecting Oxidative Modifications of Biomolecules with Isotope Dilution Mass Spectrometry: Sensitive and Quantitative Assays for Oxidized Amino Acids in Proteins and Tissues — JAY W. HEINECKE, FONG F. HSU, JAN R. CROWLEY, STANLEY L. HAZEN, CHRISTIAAN LEEUWENBURGH, DIANNE M. MUELLER, JANE E. RASMUSSEN, AND JOHN TURK — 124

17. Measurement of Protein Carbonyls in Human Brain Tissue — PATRICIA EVANS, LEONIDAS LYRAS, AND BARRY HALLIWELL — 145

18. 8-Hydroxydeoxyguanosine and 8-Hydroxyguanine as Biomarkers of Oxidative DNA Damage — HAROLD J. HELBOCK, KENNETH B. BECKMAN, AND BRUCE N. AMES — 156

19. Markers of Oxidative Damage to DNA Antioxidants and Molecular Damage — STEFFEN LOFT AND HENRIK ENGHUSEN POULSEN — 166

20. Comet Assay for Nuclear DNA Damage — STYLIANOS M. PIPERAKIS, EVANGELOS-E. VISVARDIS, AND ASPASIA M. TASSIOU — 184

21. Generation of Hydroxyl Radical by Photolysis of Mercaptopyridine N-Oxides: Application to Redox Chemistry of Purines — ABEL J. S. C. VIEIRA, JOÃO P. TELO, AND RUI M. B. DIAS — 194

22. Trace Determination of Hydroxyl Radical Using Fluorescence Detection — BEIBEI LI, PETER L. GUTIERREZ, AND NEIL V. BLOUGH — 202

Section II. Assays in Cells, Body Fluids, and Tissues

23. Isoforms of Mammalian Peroxiredoxin That Reduce Peroxides in Presence of Thioredoxin — HO ZOON CHAE, SANG WON KANG, AND SUE GOO RHEE — 219

24. Preparation and Assay of Mammalian Thioredoxin and Thioredoxin Reductase — ELIAS S. J. ARNÉR, LIANGWEI ZHONG, AND ARNE HOLMGREN — 226

25. Methionine Sulfoxide Reductase in Antioxidant Defense — JACKOB MOSKOVITZ, BARBARA S. BERLETT, J. MICHAEL POSTON, AND EARL R. STADTMAN — 239

26. Determination of Tissue Susceptibility to Oxidative Stress by Enhanced Luminescence Technique — P. VENDITTI, T. DE LEO, AND S. DI MEO — 245

27. Measurement of Hydroxyl Radical by Salicylate in Striatum of Intact Brain — MIDORI HIRAMATSU AND MAKIKO KOMATSU — 252

28. Analytical and Numerical Techniques for Evaluation of Free Radical Damage in Cultured Cells Using Imaging Cytometry and Fluorescent Indicators — STEPHEN E. BUXSER, GERI SAWADA, AND THOMAS J. RAUB — 256

29. *In Vivo* Measurement of Hydrogen Peroxide by Microelectrodes — HIDEKATSU YOKOYAMA — 275

30. Overall Low Molecular Weight Antioxidant Activity of Biological Fluids and Tissues by Cyclic Voltammetry — RON KOHEN, ELIE BEIT-YANNAI, ELLIOT M. BERRY, AND OREN TIROSH — 285

31. Estimation of Hydroxyl Free Radical Levels *in Vivo* Based on Liquid Chromatography with Electrochemical Detection — IAN N. ACWORTH, MIKHAIL B. BOGDANOV, DOUGLAS R. McCABE, AND M. FLINT BEAL — 297

32. Mechanism of Interaction of *in Situ* Produced Nitroimidazole Reduction Derivatives with DNA Using Electrochemical DNA Biosensor — ANA MARIA OLIVEIRA BRETT, SILVIA H. P. SERRANO, MAURO A. LA-SCALEA, IVANO G. R. GUTZ, AND MARIA L. CRUZ — 314

33. Heme Oxygenase Activity Determination by High-Performance Liquid Chromatography — STEFAN RYTER, EGIL KVAM, AND REX M. TYRRELL — 322

Section III. Oxidant and Redox-Sensitive Steps in Signal Transduction and Gene Expression

34. Assay for Redox-Sensitive Kinases — ASHOK KUMAR AND BHARAT B. AGGARWAL — 339

35. Inhibition of NF-κB Activation *in Vitro* and *in Vivo*: Role of 26S Proteasome — MATTHEW B. GRISHAM, VITO J. PALOMBELLA, PETER ELLIOTT, ELAINE M. CONNER, STEPHEN BRAND, HENRY WONG, CHRISTINE PIEN, AND ANTONIO DESTREE — 345

36. Nuclear Factor κB Activity in Response to Oxidants and Antioxidants — YVONNE M. W. JANSSEN AND CHANDAN K. SEN — 363

37. Assessing Induction of IκB by Nitric Oxide — MARTIN SPIECKER AND JAMES K. LIAO — 374

38. Nitrosative Stress — ALFRED HAUSLADEN AND JONATHAN S. STAMLER — 389

39. Determination of Cell–Cell Adhesion in Response   SASHWATI ROY,
    to Oxidants and Antioxidants   CHANDAN K. SEN, AND
                                                     LESTER PACKER   395

40. Antioxidant Regulation of Gene Expression: Anal- KISHORCHANDRA GOHIL,
    ysis of Differentially Expressed mRNAs   SASHWATI ROY,
                                                      LESTER PACKER, AND
                                                      CHANDAN K. SEN   402

Section IV. Noninvasive Methods

41. Noninvasive Measurement of α-Tocopherol Gradi- JENS J. THIELE AND
    ents in Human Stratum Corneum by High-Perfor-   LESTER PACKER   413
    mance Liquid Chromatography Analysis of Se-
    quential Tape Strippings

42. Ultraweak Photon Emission of Human Skin *in Vivo:* GERHARD SAUERMANN,
    Influence of Topically Applied Antioxidants on   WEI PING MEI, UDO HOPPE,
    Human Skin   AND FRANZ STÄB   419

43. Noninvasive *in Vivo* Evaluation of Skin Antioxidant RON KOHEN,
    Activity and Oxidation Status   DAVID FANBERSTEIN,
                                                     ABRAHAM ZELKOWICZ,
                                                     OREN TIROSH, AND
                                                     SHARON FARFOURI   428

44. Antioxidative Homeostasis: Characterization by   IGOR POPOV AND
    Means of Chemiluminescent Technique   GUDRUN LEWIN   437

45. Clinical Use of Photoionization Gas Chromatogra- EMILE R. MOHLER III AND
    phy for Detection of Lipid Peroxidation   DAVID R. HATHAWAY   456

46. Intravital Fluorescence Microscopy for Study of   HANS-ANTON LEHR,
    Leukocyte Interaction with Platelets and Endo-   BRIGITTE VOLLMAR,
    thelial Cells   PETER VAJKOCZY, AND
                                                     MICHAEL D. MENGER   462

VOLUME 301
NITRIC OXIDE: (PART C: BIOLOGICAL AND ANTIOXIDANT ACTIVITIES)
Lester Packer

Section I. Biological Activity

1. Quantifying Role of Nitric Oxide in Endothelial   J. STEVEN ALEXANDER AND
   Barrier Regulation   NAOTSUKA OKAYAMA   3

2. *In Vivo* Measurements of Endothelial Cell Adhesion Molecule Expression — MICHAEL J. EPPIHIMER AND D. NEIL GRANGER — 14

3. Ruthenium Red as Measure of Mast Cell Activation — PAUL KUBES — 22

4. Quantitation of Protein-Bound 3-Nitrotyrosine by High-Performance Liquid Chromatography with Electrochemical Detection — MARK K. SHIGENAGA — 27

5. Quantitative Estimation of Endogenous N-Nitrosation in Humans by Monitoring *N*-Nitrosoproline in Urine — HIROSHI OHSHIMA AND HELMUT BARTSCH — 40

6. Measuring Nitric Oxide Production in Human Clinical Studies — DONALD L. GRANGER, NICHOLAS M. ANSTEY, WILLIAM C. MILLER, AND J. BRICE WEINBERG — 49

7. Nitric Oxide and Superoxide Detection in Human Platelets — JANE E. FREEDMAN AND JOHN F. KEANEY, JR. — 61

8. Assay of Isoforms of *Escherichia coli*-Expressed Nitric Oxide Synthase — PAVEL MARTÁSEK, R. TIMOTHY MILLER, LINDA J. ROMAN, THOMAS SHEA, AND BETTY SUE S. MASTERS — 70

9. Identification of Carrier Systems in Plasma Membrane of Mammalian Cells Involved in Transport of L-Arginine — ELLEN I. CLOSS AND GIOVANNI E. MANN — 78

10. Measurement of Nitric Oxide Synthesis in Humans Using L-[^{15}N$_2$]Arginine — PABLO FORTE, LORNA M. SMITH, ERIC MILNE, AND NIGEL BENJAMIN — 92

11. Concerted Nitric Oxide/Oxygen Delivery by Hemoglobin — TIMOTHY J. McMAHON AND JONATHAN S. STAMLER — 99

12. Scintillation Proximity Assay to Measure Nitroarginine and Tetrahydrobiopterin Binding to Heme Domain of Neuronal Nitric Oxide Synthase — WENDY K. ALDERTON AND PETER N. LOWE — 114

13. Measurement of Nitric Oxide-Mediated Effects on Zinc Homeostasis and Zinc Finger Transcription Factors — K.-D. KRÖNCKE AND V. KOLB-BACHOFEN — 126

14. Immunoprecipitation of Nitrotyrosine-Containing Proteins — LEE ANN MacMILLAN-CROW AND JOHN A. THOMPSON — 135

15. Gel Electrofocusing Method for Studying Protein S-Nitrosylation — YANBIN JI, THEODORUS P. M. AKERBOOM, HELMUT SIES, AND JAMES A. THOMAS — 145

16. Measurement and Significance of Free and Pro-   JOHN P. CROW   151
    tein-Bound 3-Nitrotyrosine, Free 3-Nitro-4-Hy-
    droxyphenylacetic Acid, and 3-Chlorotyrosine
    in Biologic Samples: A High-Performance Liq-
    uid Chromatography Method Using Electro-
    chemical Detection

17. Determination of Nitric Oxide Saturated Solution STEFAN MESÁROŠ   160
    by Amperometry on Modified Microelectrode

18. Electron Spin Resonance Spin-Trapping Detection   JEANETTE VÁSQUEZ-VIVAR,
    of Superoxide Generated by Neuronal Nitric Ox-   PAVEL MARTÁSEK,
    ide Synthase   NEIL HOGG,
                                                       HAKIM KAROUI,
                                                       BETTIE SUE SILER MASTERS,
                                                       KIRKWOOD A. PRITCHARD, JR.,
                                                       AND B. KALYANARAMAN   169

19. Direct and Simultaneous Ultraviolet Second-Deriv- RACHEL M. BOLZAN,
    ative Spectrophotometric Determination of   RAFAEL CUETO,
    Nitrite and Nitrate in Preparations of Peroxy-   GIUSEPPE L. SQUADRITO,
    nitrite   RAO M. UPPU, AND
                                                       WILLIAM A. PRYOR   178

20. Regulation of Mitochondrial Respiration by Aden-   ALBERTO BOVERIS,
    osine Diphosphate, Oxygen, and Nitric Oxide   LIDIA E. COSTA,
                                                       ENRIQUE CADENAS, AND
                                                       JUAN J. PODEROSO   188

Section II. Nitric Oxide Donors: Nitrosothiols and Nitroxyls

21. Detection of S-Nitrosothiols by Fluorometric and   DAVID A. WINK,
    Colorimetric Methods   SUNGMEE KIM,
                                                       DEBORAH COFFIN,
                                                       JOHN C. COOK,
                                                       YORAM VODOVOTZ,
                                                       DANAE CHRISTODOULOU,
                                                       DAVID JOURD'HEUIL, AND
                                                       MATTHEW B. GRISHAM   201

22. Chemical Approaches toward Generation of Ni-   S. BRUCE KING AND
    troxyl   HERBERT T. NAGASAWA   211

23. Stability of S-Nitrosothiols in Presence of Copper, DAVID JOURD'HEUIL,
    Zinc-Superoxide Dismutase   F. STEPHEN LAROUX,
                                                       DAVID KANG,
                                                       ALLEN M. MILES,
                                                       DAVID A. WINK, AND
                                                       MATTHEW B. GRISHAM   220

24. Fluorimetric Detection of S-Nitrosothiols   PETER KOSTKA AND
                                                       JULIET K. J. PARK   227

25. Redox-Sensitive Nitric Oxide Donors: Nitric Oxide   ANDREA D. McGILL,
    Generation through Electrolysis   YIFAN YANG,
                                                          JIANQIANG WANG,
                                                          LUIS ECHEGOYEN, AND
                                                          PENG GEORGE WANG   235

26. Glyco-S-Nitrosothiols: Sugar–SNAP, a New Type   YONGCHUN HOU,
    of Nitric Oxide Donor   JIANQIANG WANG,
                                                          JOHNNY RAMIREZ, AND
                                                          PENG GEORGE WANG   242

27. Protein S-Nitrosating Agents   ZHENGMAO GUO,
                                                          NEIL MIRANDA, AND
                                                          PENG GEORGE WANG   249

28. Measurement of Nitrite and Nitrate by High-Perfor-   MICHAEL R. L. STRATFORD   259
    mance Ion Chromatography

29. Iron Diethyldithiocarbamate as Spin Trap for Nitric   A. F. VANIN   269
    Oxide Detection

30. Synthesis, Chemistry, and Applications of Nitroxyl   MARTIN N. HUGHES AND
    Ion Releasers Sodium Trioxodinitrate or Angeli's   RICHARD CAMMACK   279
    Salt and Piloty's Acid

31. Nitric Oxide Donor Generation from Reactions   C. ROGER WHITE,
    of Peroxynitrite   RAKESH P. PATEL, AND
                                                          VICTOR DARLEY-USMAR   288

Section III. Peroxynitrite

32. Defenses against Peroxynitrite   KARLIS BRIVIBA,
                                                          LARS-OLIVER KLOTZ, AND
                                                          HELMUT SIES   301

33. Use of Repair Endonucleases to Assess DNA Dam-   DANIEL BALLMAIER,
    age by Peroxynitrite   KARLIS BRIVIBA,
                                                          HELMUT SIES, AND
                                                          BERND EPE   311

34. Interaction of Carotenoids and Tocopherols with   ANANTH SEKHER PANNALA,
    Peroxynitrite   SURINDER SINGH, AND
                                                          CATHERINE RICE-EVANS   319

35. Assessment of Peroxynitrite Scavengers in Vitro   BARRY HALLIWELL,
                                                          PAT EVANS, AND   333
                                                          MATTHEW WHITEMAN

36. Peroxynitrite Studied by Stopped-Flow Spec-   REINHARD KISSNER,
    troscopy   JOSEPH S. BECKMAN, AND
                                                          WILLEM H. KOPPENOL   342

37. Peroxynitrite Reactions with Carbon Dioxide–   RAFAEL RADI,
    Bicarbonate   ANA DENICOLA, AND
                                                          BRUCE A. FREEMAN   353

38. Detection of Reactive Nitrogen Species Using 2,7-Dichlorodihydrofluorescein and Dihydrorhodamine 123   HARRY ISCHIROPOULOS, ANDREW GOW, STEPHEN R. THOM, NEIL W. KOOY, JAMES A. ROYALL, AND JOHN P. CROW   367

39. Immunohistochemical Methods to Detect Nitrotyrosine   LILIANA VIERA, YAO ZU YE, ALVARO G. ESTÉVEZ, AND JOSEPH S. BECKMAN   373

40. Mitochondria, Nitric Oxide, and Peroxynitrite   CHRISTOPH RICHTER, MATTHIAS SCHWEIZER, AND PEDRAM GHAFOURIFAR   381

41. Examining Apoptosis in Cultured Cells after Exposure to Nitric Oxide and Peroxynitrite   ALVARO G. ESTÉVEZ, NATHAN SPEAR, HUGO PELLUFFO, ANDRES KAMAID, LUIS BARBEITO, AND JOSEPH S. BECKMAN   393

42. Application of Authentic Peroxynitrite to Biological Materials   HIDEYUKI ISHIDA, CHOKOH GENKA, AND HIROE NAKAZAWA   402

Section IV. Oxidant and Antioxidant Action

43. Antioxidant Effects of Nitric Oxide   DAVID WINK, YORAM VODOVOTZ, MATTHEW B. GRISHAM, WILLIAM DEGRAFF, JOHN C. COOK, ROBERTO PACELLI, MURALI KRISHNA, AND JAMES B. MITCHELL   413

44. Assay of Antioxidant and Anti-inflammatory Activity of Nitric Oxide *in Vivo*   HAJIME HIGUCHI, D. NEIL GRANGER, HIRIHISA SAITO, AND IWAO KUROSE   424

45. Effect of Nitric Oxide on Iron or Hemoprotein-Catalyzed Oxidative Reactions   DAVID JOURD'HEUIL, ALLEN M. MILES, AND MATTHEW B. GRISHAM   437

46. Antioxidant Effects of Nitric Oxide and Nitric Oxide Donor Compounds on Low-Density Lipoprotein Oxidation   STEVEN P. A. GOSS, B. KALYANARAMAN, AND NEIL HOGG   444

47. Nitration of Unsaturated Fatty Acids by Nitric Ox-   VALERIE B. O'DONNELL,
    ide-Derived Reactive Species   JASON P. EISERICH,
                                                          ALLISON BLOODSWORTH,
                                                          PHILLIP H. CHUMLEY,
                                                          MARION KIRK,
                                                          STEPHEN BARNES,
                                                          VICTOR M. DARLEY-USMAR,
                                                          AND BRUCE A. FREEMAN   454

48. Analysis of Aromatic Nitration, Chlorination, and   ALBERT VAN DER VLIET,
    Hydroxylation by Gas Chromatography–Mass   ANDREW JENNER,
    Spectrometry   JASON P. EISERICH,
                                                          CARROLL E. CROSS, AND
                                                          BARRY HALLIWELL   471

49. Products from Reaction of Peroxynitrite with γ-   NIEL C. HOGLEN AND
    Tocopherol   DANIEL C. LIEBLER   483

50. Nitric Oxide Radical Scavenging of Flavonoids   GUIDO R. M. M. HAENEN AND
                                                          AALT BAST   490

51. Assay of Inducible Form of Nitric Oxide Synthase   HIROTSUGU KOBUCHI,
    Activity: Effect of Flavonoids and Plant Extracts   FABIO VIRGILI, AND
                                                          LESTER PACKER   504

52. In Vitro System to Study Role of Blood Flow on   YOUNG-MI GO,
    Nitric Oxide Production and Cell Signaling in En-   HEONYONG PARK,
    dothelial Cells   MATHEW C. MALAND, AND
                                                          HANJOONG JO   513

53. Pharmacological Approaches of Endothelial Nitric   RAMAROSON
    Oxide-Dependent Vasorelaxation Induced by   ANDRIANTSITOHAINA,
    Polyphenols from Plant Extract   EMILE ANDRIAMBELOSON, AND
                                                          JEAN CLAUDE STOCLET   522

54. Fluorescence Detection of Nitric Oxide Based on   MICHAEL BÄTZ,
    Cheletropic Spin Traps   HANS-GERT KORTH,
                                                          PETRA MEINEKE, AND
                                                          REINER SUSTMANN   532

VOLUME 302
GREEN FLUORESCENT PROTEIN
P. Michael Conn

Section I. Monitoring of Physiological Processes

1. Membrane Trafficking   SABINE KUPZIG,
                                                          SAN SAN LEE, AND
                                                          GEORGE BANTING   3

2. Monitoring of Protein Secretion with Green Fluo- CHRISTOPH KAETHER AND
   rescent Protein   HANS-HERMANN GERDES   11

3. Green Fluorescent Protein to Visualize Cancer Pro- ROBERT M. HOFFMAN   20
   gression and Metastasis

4. Comparison of Enhanced Green Fluorescent Pro- XIAONING ZHAO,
   tein and Its Destabilized Form as Transcription TOMMY DUONG,
   Reporters   CHIAO-CHIAN HUANG,
                                                   STEVEN R. KAIN, AND
                                                   XIANQIANG LI   32

5. Early Detection of Apoptosis with Annexin STEVEN R. KAIN AND
   V–Enhanced Green Fluorescent Protein   JING-TYAN MA   38

6. Green Fluorescent Protein in the Visualization of MARKUS MANIAK   43
   Particle Uptake and Fluid-Phase Endocytosis

7. Monitoring Intracellular Shuttling of Histidine-Rich FRANK HANAKAM AND
   pH-Sensor Proteins Tagged with Green Fluores- GÜNTHER GERISCH   51
   cent Protein

8. Measuring Protein Degradation with Green Fluo- STEPHEN R. CRONIN AND
   rescent Protein   RANDOLPH Y. HAMPTON   58

9. Studying Nuclear Receptors with Green Fluores- GORDON L. HAGER   73
   cent Protein Fusions

Section II. Localization of Molecules

10. Localization of Calmodulin in Budding Yeast and MARK R. FLORY AND
    Fission Yeast Using Green Fluorescent Protein TRISHA N. DAVIS   87

11. Analysis of Microtubule Organization and Dynam- KEITH R. OLSON AND
    ics in Living Cells Using Green Fluorescent Pro- J. B. OLMSTED   103
    tein–Microtubule-Associated Protein 4 Chimeras

12. Trafficking of the Androgen Receptor   VIRGINIE GEORGET,
                                                  BÉATRICE TEROUANNE,
                                                  JEAN-CLAUDE NICOLAS, AND
                                                  CHARLES SULTAN   121

13. Use of Green Fluorescent Protein for Visualization CHRIS D. WEBB AND
    of Cell-Specific Gene Expression and Subcellular ORNA RESNEKOV   136
    Protein Localization in Bacillus subtilis

14. Signaling, Desensitization, and Trafficking of G Pro- LARRY S. BARAK,
    tein-Coupled Receptors Revealed by Green Fluo- JIE ZHANG,
    rescent Protein Conjugates   STEPHEN S. G. FERGUSON,
                                                   STEPHANE A. LAPORTE, AND
                                                   MARC G. CARON   153

15. Use of Green Fluorescent Proteins Linked to Cy- GUISSOU A. DABIRI,
    toskeletal Proteins to Analyze Myofibrillogenesis JOSEPH C. AYOOB,
    in Living Cells KENAN K. TURNACIOGLU,
    JEAN M. SANGER, AND
    JOSEPH W. SANGER   171

Section III. Special Uses

16. Fluorescent Proteins in Single- and Multicolor LONNIE LYBARGER AND
    Flow Cytometry ROBERT CHERVENAK   189

17. Flow Cytometric Analysis of Transcription: Use of LONNIE LYBARGER AND
    Green Fluorescent Protein Variants to Control ROBERT CHERVENAK   199
    Transfection Efficiency

18. Use of Coexpressed Enhanced Green Fluorescent YU FANG,
    Protein as a Marker for Identifying Transfected CHIAO-CHIAN HUANG,
    Cells STEVEN R. KAIN, AND
    XIANQIANG LI   207

19. The Jellyfish Green Fluorescent Protein: A Tool for L. A. C. BLAIR,
    Studying Ion Channels and Second-Messenger K. K. BENCE, AND
    Signaling in Neurons J. MARSHALL   213

20. Expression of Green Fluorescent Protein and Inosi- ATSUSHI MIYAWAKI,
    tol 1,4,5-Triphosphate Receptor in *Xenopus* JULIE M. MATHESON,
    laevis Oocytes LEE G. SAYERS,
    AKIRA MUTO,
    TAKAYUKI MICHIKAWA,
    TEIICHI FURUICHI, AND
    KATSUHIKO MIKOSHIBA   225

21. Expression of Green Fluorescent Protein in TATSUYUKI TAKADA,
    Transgenic Mice KENICHI YOSHIDA,
    KENJI NAKAMURA,
    KAZUKI NAKAO,
    GOZOH TSUJIMOTO,
    MOTOYA KATSUKI, AND
    SUMIO SUGANO   233

22. Green Fluorescent Protein as a Probe to Study In- ALAN S. VERKMAN   250
    tracellular Solute Diffusion

23. Application of Green Fluorescent Protein–Protein TAKASHI AOKI,
    A Fusion Protein to Western Blotting KATHERINE S. KOCH,
    HYAM L. LEFFERT, AND
    HIROYUKI WATABE   264

24. Green Fluorescent Protein as a Reporter for Pro- P. PRABHAKARA REDDI,
    moter Analysis of Testis-Specific Genes in MARKO KALLIO, AND
    Transgenic Mice JOHN C. HERR   272

25. Green Fluorescent Protein in the Measurement of   LUIZ E. BERMUDEZ,
    Bacteria–Host Interactions   FELIX J. SANGARI, AND
                                                       AMY PARKER   285

26. Flow Cytometric Analysis of Transgene Expression   DAVID W. GALBRAITH,
    in Higher Plants: Green Fluorescent Protein   LEONARD A. HERZENBERG,
                                                       AND MICHAEL T. ANDERSON   296

27. Continual Green Fluorescent Protein Monitoring   PETER E. URWIN,
    of Promoter Activity in Plants   SIMON G. MØLLER,
                                                       JENNIFER K. BLUMSOM, AND
                                                       HOWARD J. ATKINSON   316

28. Retroviral Expression of Green Fluorescent Protein ILYA A. MAZO,
                                                       JOHN P. LEVY,
                                                       REBECCA R. MULDOON,
                                                       CHARLES J. LINK, JR., AND
                                                       STEVEN R. KAIN   329

29. Confocal Imaging of Ca^{2+}, pH, Electrical Potential, JOHN J. LEMASTERS,
    and Membrane Permeability in Single Living   DONNA R. TROLLINGER,
    Cells   TING QIAN,
                                                       WAYNE E. CASCIO, AND
                                                       HISAYUKI OHATA   341

30. *In Vivo* Retroviral Transduction and Expression of JOHN P. LEVY,
    Green Fluorescent Protein   REBECCA R. MULDOON,
                                                       ILYA A. MAZO,
                                                       STEVEN R. KAIN, AND
                                                       CHARLES J. LINK, JR.   358

Section IV. Mutants and Variants of Green Fluorescent Protein

31. S147P Green Fluorescent Protein: A Less Thermo-   YUKIO KIMATA,
    sensitive Green Fluorescent Protein Variant   CHUN REN LIM, AND
                                                       KENJI KOHNO   373

32. Spectral Variants of Green Fluorescent Protein   GOTTFRIED J. PALM AND
                                                       ALEXANDER WLODAWER   378

33. Expression of Green Fluorescent Protein Using   LINDA A. KING,
    Baculovirus Vectors   CAROLE J. THOMAS,
                                                       NICOLA WILKINSON, AND
                                                       ROBERT D. POSSEE   394

34. Green Fluorescent Protein Forms for Energy   ROGER HEIM   408
    Transfer

35. Use of Codon-Modified, Red-Shifted Variants of   CHARLES J. LINK, JR.,
    Green Fluorescent Protein Genes to Study Virus-   SUMING WANG,
    Mediated Gene Transfer   REBECCA R. MULDOON,
                                                       TATIANA SEREGINA, AND
                                                       JOHN P. LEVY   424

36. Generation of a Destabilized Form of Enhanced   XIAONING ZHAO,
    Green Fluorescent Protein   XIN JIANG,
                                                     CHIAO-CHIAN HUANG,
                                                     STEVEN R. KAIN, AND
                                                     XIANQIANG LI   438

37. Spectral   Properties   of   Green   Fluorescent   SATOSHI INOUYE,
    Protein–S65A   KAZUHIKO UMESONO, AND
                                                        FREDERICK I. TSUJI   444

VOLUME 303
cDNA PREPARATION AND CHARACTERIZATION
Sherman M. Weissman

Section I. cDNA Preparation

1. Preparation of cDNA from Single Cells and Sub-   JANET ESTEE KACHARMINA,
   cellular Regions   PETER B. CRINO, AND
                                                      JAMES EBERWINE   3

2. High-Efficiency Full-Length cDNA Cloning   PIERO CARNINCI AND
                                                      YOSHIHIDE HAYASHIZAKI   19

3. Preparation and Analysis of cDNA from a Small   MENG LIU,
   Number of Hematopoietic Cells   Y. V. B. K. SUBRAMANYAM,
                                                      AND NAMADEV BASKARAN   45

4. Optical Mapping: An Approach for Fine Mapping   CHRISTOPHER ASTON,
                                                      CATHARINA HIORT, AND
                                                      DAVID C. SCHWARTZ   55

Section II. Gene Identification

5. Current Status of Computational Gene Finding: RICHARD J. MURAL   77
   A Perspective

6. Gene Identification by Exon Amplification   DEANNA M. CHURCH AND
                                                      ALAN J. BUCKLER   83

7. Cosmid-Based Exon Trapping   JOHAN T. DEN DUNNEN   100

8. Direct cDNA Selection Using Large Genomic   ANDREW D. SIMMONS AND
   DNA Targets   MICHAEL LOVETT   111

9. cDNA Selection: An Approach for Isolation of   SATISH PARIMOO AND
   Chromosome-Specific cDNAs   SHERMAN M. WEISSMAN   127

10. Saturation Identification of Coding Sequences in   KATHELEEN GARDINER   144
    Genomic DNA

Section III. Patterns of mRNA Expression

11. cDNA Screening by Array Hybridization   RADOJE DRMANAC AND
                                                     SNEZANA DRMANAC   165

12. DNA Arrays for Analysis of Gene Expression   MICHAEL B. EISEN AND
                                                     PATRICK O. BROWN   179

13. Construction and Analysis of Arrayed cDNA   MATTHEW D. CLARK,
    Libraries   GEORGIA D. PANOPOULOU,
                                                     DOLORES J. CAHILL,
                                                     KONRAD BÜSSOW, AND
                                                     HANS LEHRACH   205

14. Principles of Differential Display   KATHERINE J. MARTIN AND
                                                     ARTHUR B. PARDEE   234

15. READS: A Method for Display of 3′-End Frag-   YATINDRA PRASHAR AND
    ments of Restriction Enzyme-Digested cDNAs   SHERMAN M. WEISSMAN   258
    for Analysis of Differential Gene Expression

16. A Modified Method for the Display of 3′-End Re- Y. V. B. K. SUBRAMANYAM,
    striction Fragments of cDNAs: Molecular Profil- NAMADEV BASKARAN,
    ing of Gene Expression in Neutrophils   PETER E. NEWBURGER, AND
                                                     SHERMAN M. WEISSMAN   272

17. Fluorescent Differential Display: A Fast and Reli- TAKASHI ITO AND
    able Method for Message Display Polymerase   YOSHIYUKI SAKAKI   298
    Chain Reaction

18. Identification of Differentially Expressed Genes FRANÇOISE MATHIEU-DAUDÉ,
    Using RNA Fingerprinting by Arbitrarily Primed   THOMAS TRENKLE,
    Polymerase Chain Reaction   JOHN WELSH,
                                                     BARBARA JUNG,
                                                     THOMAS VOGT, AND
                                                     MICHAEL MCCLELLAND   309

19. cDNA Representational Difference Analysis: A   MICHAEL HUBANK AND
    Sensitive and Flexible Method for the Identifica- DAVID G. SCHATZ   325
    tion of Differentially Expressed Genes

20. Suppression Subtractive Hybridization: A Versatile LUDA DIATCHENKO,
    Method for Identifying Differentially Expressed   SERGEY LUKYANOV,
    Genes   YUN-FAI CHRIS LAU, AND
                                                     PAUL D. SIEBERT   349

21. Reduced Complexity Probes for DNA Arrays   THOMAS TRENKLE,
                                                     FRANÇOISE MATHIEU-DAUDÉ,
                                                     JOHN WELSH, AND
                                                     MICHAEL MCCLELLAND   380

22. Targeted Display: A New Technique for the Analy- ALASTAIR J. H. BROWN,
    sis of Differential Gene Expression   CATHERINE HUTCHINGS,
                                                   JULIAN F. BURKE, AND
                                                   LYNNE V. MAYNE   392

Section IV. Functional Relationship among cDNA Translation Products

23. RNA Polymerase III–Based Two-Hybrid System   MARIE-CLAUDE MARSOLIER
                                                 AND ANDRÉ SENTENAC   411

24. Screening for Protein–Protein Interactions   F. JOSEPH GERMINO AND
                                                 NEAL K. MOSKOWITZ   422

25. Using the Lac Repressor System to Identify Inter- NICOLE L. STRICKER,
    acting Proteins   PETER SCHATZ, AND
                                                   MIN LI   451

26. A Genetic Selection for Isolating cDNA Clones KENNETH A. JACOBS,
    That Encode Signal Peptides   LISA A. COLLINS-RACIE,
                                                  MAUREEN COLBERT,
                                                  MCKEOUGH DUCKETT,
                                                  CHERYL EVANS,
                                                  MARGARET GOLDEN-FLEET,
                                                  KERRY KELLEHER,
                                                  RONALD KRIZ,
                                                  EDWARD R. LAVALLIE,
                                                  DAVID MERBERG,
                                                  VIKKI SPAULDING,
                                                  JEN STOVER,
                                                  MARK J. WILLIAMSON, AND
                                                  JOHN M. MCCOY   468

27. The Signal Sequence Trap Method   KEI TASHIRO,
                                                 TOMOYUKI NAKAMURA, AND
                                                 TASUKU HONJO   479

28. Solid-Phase Differential Display and Bacterial Ex- STEFAN STÅHL,
    pression Systems in Selection and Functional JACOB ODEBERG,
    Analysis of cDNAs   MAGNUS LARSSON,
                                                  ØYSTEIN RØSOK,
                                                  ANNE HANSEN REE, AND
                                                  JOAKIM LUNDEBERG   495

29. Transposon Mutagenesis for the Analysis of Protein PETRA ROSS-MACDONALD,
    Production, Function, and Localization   AMY SHEEHAN,
                                                  CARL FRIDDLE,
                                                  G. SHIRLEEN ROEDER, AND
                                                  MICHAEL SNYDER   512

VOLUME 304
CHROMATIN
Paul M. Wasserman and Alan P. Wolffe

Section I. Assembly of Nucleosomes, Chromatin, and Nuclei

1. Preparation of Nucleosome Core Particle from Recombinant Histones — KAROLIN LUGER, THOMAS J. RECHSTEINER, AND TIMOTHY J. RICHMOND — 3

2. Assembly of Defined Nucleosomal and Chromatin Arrays from Pure Components — LENNY M. CARRUTHERS, CHRISTIN TSE, KERFOOT P. WALKER III, AND JEFFREY C. HANSEN — 19

3. Isolation of Minichromosomes from Yeast Cells — JENNIFER A. ALFIERI AND DAVID J. CLARK — 35

4. Chromatin Assembly in *Xenopus* Extracts — DAVID JOHN TREMETHICK — 50

5. *In Vitro* Reconstitution of Nuclei for Replication and Transcription — ALISON J. CROWE AND MICHELLE CRAIG BARTON — 63

6. Preparation/Analysis of Chromatin Replicated *in Vivo* and in Isolated Nuclei — LOUISE CHANG, COLLEEN A. RYAN, CAROLYN A. SCHNEIDER, AND ANTHONY T. ANNUNZIATO — 76

7. Interactions of High Mobility Group Box Proteins with DNA and Chromatin — MAIR E. A. CHURCHILL, ANITA CHANGELA, LINDA K. DOW, AND ADAM J. KRIEG — 99

8. Reconstitution of High Mobility Group 14/17 Proteins into Nucleosomes and Chromatin — YURI V. POSTNIKOV AND MICHAEL BUSTIN — 133

9. Purification and Assays for High Mobility Group HMG-I(Y) Protein Function — RAYMOND REEVES AND MARK S. NISSEN — 155

Section II. Assays for Nucleosome Structure and Function *in Vitro*

10. Cryoelectron Microscopic Analysis of Nucleosomes and Chromatin — JAN BEDNAR AND CHRISTOPHER L. WOODCOCK — 191

11. Electron Microscopy of DNA–Protein Complexes and Chromatin — JACK GRIFFITH, SUSAN MICHALOWSKI, AND ALEXANDER M. MAKHOV — 214

12. Targeted Cross-Linking and DNA Cleavage within Model Chromatin Complexes — KYU-MIN LEE, DAVID R. CHAFIN, AND JEFFREY J. HAYES — 231

13. Base-Pair Resolution Mapping of Nucleosome Posi-   ANDREW FLAUS AND
    tions Using Site-Directed Hydroxy Radicals   TIMOTHY J. RICHMOND   251

14. Localization of Specific Histone Regions on Nucleo- SERGEI I. USACHENKO AND
    somal DNA by Cross-Linking   E. MORTON BRADBURY   264

15. Restriction Enzymes as Probes of Nucleosome Sta-   K. J. POLACH AND
    bility and Dynamics   J. WIDOM   278

16. Nucleoprotein Gel Assays for Nucleosome Position- SARI PENNINGS   298
    ing and Mobility

17. Assays for Interaction of Transcription Factor   QIAO LI,
    with Nucleosome   ULLA BJÖRK, AND
                                                       ÖRJAN WRANGE   313

18. Assays for Chromatin Remodeling during DNA   JONATHAN G. MOGGS AND
    Repair   GENEVIÈVE ALMOUZNI   333

19. Nuclear Run-On Assays: Assessing Transcription   KAZUNORI HIRAYOSHI AND
    by Measuring Density of Engaged RNA Poly-   JOHN T. LIS   351
    merases

Section III. Assays for Chromatin Structure and Function *in Vivo*

20. Mapping Chromatin Structure in Yeast   PHILIP D. GREGORY AND
                                                      WOLFRAM HÖRZ   365

21. Assays for Nucleosome Positioning in Yeast   MICHAEL P. RYAN,
                                                      GRACE A. STAFFORD,
                                                      LIUNING YU,
                                                      KELLIE B. CUMMINGS, AND
                                                      RANDALL H. MORSE   376

22. Mapping DNA Interaction Sites of Chromosomal   ANDREAS HECHT AND
    Proteins Using Immunoprecipitation and Poly-   MICHAEL GRUNSTEIN
    merase Chain Reaction   399

23. Immunological Analysis of Yeast Chromatin   PAMELA B. MELUH AND
                                                      JAMES R. BROACH   414

24. DNA Methyltransferases as Probes of Chromatin   MICHAEL P. KLADDE,
    Structure *in Vivo*   M. XU, AND
                                                      ROBERT T. SIMPSON   431

25. Mapping Cyclobutane–Pyrimidine Dimers in DNA   BERNHARD SUTER,
    and Using DNA–Repair by Photolyase for Chro-   MAGDALENA LIVINGSTONE-
    matin Analysis in Yeast   ZATCHEJ, AND
                                                      FRITZ THOMA   447

26. Analysis of *Drosophila* Chromatin Structure *in Vivo* IAIN L. CARTWRIGHT,
DIANE E. CRYDERMAN,
DAVID S. GILMOUR,
LORI A. PILE,
LORI L. WALLRATH,
JANET A. WEBER, AND
SARAH C. R. ELGIN   462

27. Ultraviolet Cross-Linking Assay to Measure Se- MARK D. BIGGIN   496
quence-Specific DNA Binding *in Vivo*

28. DNA–Protein Cross-Linking Applications for DMITRY PRUSS,
Chromatin Studies *in Vitro* and *in Vivo*   IGOR M. GAVIN,
SVETLANA MELNIK, AND
SERGEI G. BAVYKIN   516

29. Chromatin Immunoprecipitation Assays in Acetyla- COLYN CRANE-ROBINSON,
tion Mapping of Higher Eukaryotes   FIONA A. MYERS,
TIM R. HEBBES,
ALISON L. CLAYTON, AND
ALAN W. THORNE   533

30. Chromatin Structure Analysis by Ligation-Medi- GERD P. PFEIFER,
ated and Terminal Transferase-Mediated Poly- HSIU-HUA CHEN,
merase Chain Reaction   JUON-ICHIRO KOMURA, AND
ARTHUR D. RIGGS   548

31. Guanine–Adenine Ligation-Mediated Polymerase ERICH C. STRAUSS AND
Chain Reaction *in Vivo* Footprinting   STUART H. ORKIN   572

32. Exonuclease III as Probe of Chromatin Structure TREVOR K. ARCHER AND
in Vivo   ANDREA R. RICCI   584

33. Protein Image Hybridization: Mapping and Posi- S. BELIKOV,
tioning DNA–Protein Contacts along DNA   O. PREOBRAZHENSKAYA, AND
V. KARPOV   600

34. *In Vivo* Analysis of Chromatin Structure   KENNETH S. ZARET   612

35. Analysis of Nucleosome Positioning in Mamma- GORDON L. HAGER AND
lian Cells   GILBERTO FRAGOSO   626

36. Measurement of Localized DNA Supercoiling and PHILLIP R. KRAMER,
Topological Domain Size in Eukaryotic Cells   OLGA BAT, AND
RICHARD R. SINDEN   639

37. Fluorescent *in Situ* Hybridization Analysis of Chro- WENDY BICKMORE   650
mosome and Chromatin Structure

38. Analysis of Nuclear Organization in *Saccharo-* MONICA GOTTA,
myces cerevisiae   THIERRY LAROCHE, AND
SUSAN M. GASSER   663

Section IV. Chromatin Remodeling Complexes

39. Histone Acetyltransferases: Preparation of Sub- CRAIG A. MIZZEN,
    strates and Assay Procedures   JAMES E. BROWNELL,
                                                   RICHARD G. COOK, AND
                                                   C. DAVID ALLIS   675

40. Analysis of Activity and Regulation of hGcn5, Hu- NICKOLAI A. BARLEV,
    man Histone Acetyltransferase   JERRY L. WORKMAN, AND
                                                      SHELLEY L. BERGER   696

41. Purification of a Histone Deacetylase Complex from PAUL A. WADE,
    Xenopus laevis: Preparation of Substrates and PETER L. JONES,
    Assay Procedures   DANIELLE VERMAAK, AND
                                                      ALAN P. WOLFFE   715

42. Purification and Biochemical Properties of Yeast COLIN LOGIE AND
    SWI/SNF Complex   CRAIG L. PETERSON   726

43. Analysis of Modulators of Chromatin Structure in PATRICK D. VARGA-WEISZ,
    Drosophila   EDGAR J. BONTE, AND
                                                     PETER B. BECKER   742

44. Purification of *Drosophila* Nucleosome Remodel- RAPHAEL SANDALTZOPOULOS,
    ing Factor   VINCENT OSSIPOW,
                                                     DAVID A. GDULA,
                                                     TOSHIO TSUKIYAMA, AND
                                                     CARL WU   757

VOLUME 305
BIOLUMINESCENCE AND CHEMILUMINESCENCE (PART C)
Miriam M. Ziegler and Thomas O. Baldwin

Section I. Chemiluminescence and Bioluminescence: Overviews

1. Chemical Generation of Excited States: The Basis FRANK MCCAPRA   3
   of Chemiluminescence and Bioluminescence

2. Some Brief Notes on Nomenclature and Units and PHILIP E. STANLEY   47
   Standards Used in Bioluminescence and Chemi-
   luminescence

Section II. Instrumentation

3. Physics of Low Light Detectors   KENNETH J. VOSS   53

4. Luminometer Design and Low Light Detection   FRITZ BERTHOLD,
KLAUS HERICK, AND
RUTH M. SIEWE   62

5. Absolute Calibration of Luminometers with Low-   DENNIS J. O'KANE AND
Level Light Standards   JOHN LEE   87

6. Commercially Available Luminometers and Low-   PHILIP E. STANLEY   96
Level Light Imaging Devices

7. Automated Recordings of Bioluminescence with   TILL ROENNEBERG AND
Special Reference to the Analysis of Circadian   WALTER TAYLOR   104
Rhythms

8. Chemiluminescence Imaging Systems for the Anal-   ALDO RODA,
ysis of Macrosamples: Microtiter Format, Blot   PATRIZIA PASINI,
Membrane, and Whole Organs   MONICA MUSIANI, AND
MARIO BARALDINI   120

Section III. Luciferases, Luminescence Accessory Proteins, and Substrates

9. Overexpression of Bacterial Luciferase and Purifi-   THOMAS O. BALDWIN,
cation from Recombinant Sources   MIRIAM M. ZIEGLER,
VICKI A. GREEN, AND
MICHAEL D. THOMAS   135

10. Purification of Luciferase Subunits from Recombi-   JAMES F. SINCLAIR   152
nant Sources

11. A Rapid Chromatographic Method to Separate   A. CLAY CLARK,
the Subunits of Bacterial Luciferase in Urea-   BRIAN W. NOLAND, AND
Containing Buffer   THOMAS O. BALDWIN   157

12. Purification and Ligand Exchange Protocols for An-   VALENTIN N. PETUSHKOV,
tenna Proteins from Bioluminescent Bacteria   BRUCE G. GIBSON,
ANTONIE J. W. G. VISSER, AND
JOHN LEE   164

13. Purification of Firefly Luciferase from Recombi-   THOMAS O. BALDWIN AND
nant Sources   VICKI A. GREEN   180

14. Chemical Synthesis of Firefly Luciferin Analogs   BRUCE R. BRANCHINI   188
and Inhibitors

15. Structural Basis for Understanding Spectral Varia-   S. JAMES REMINGTON   196
tions in Green Fluorescent Protein

16. Large-Scale Purification of Recombinant GFP from   DANIEL G. GONZÁLEZ AND
Escherichia coli   WILLIAM W. WARD   212

17. Recombinant Obelin: Cloning and Expression of cDNA, Purification, and Characterization as a Calcium Indicator
BORIS A. ILLARIONOV, LUDMILA A. FRANK, VICTORIA A. ILLARIONOVA, VLADIMIR S. BONDAR, EUGENE S. VYSOTSKI, AND JOHN R. BLINKS   223

18. *Gonyaulax* Luciferase: Gene Structure, Protein Expression, and Purification from Recombinant Sources
LIMING LI   249

19. Dinoflagellate Luciferin-Binding Protein
DAVID MORSE AND MARIA MITTAG   258

Section IV. Bacterial Autoinduction System and Its Applications

20. Assay of Autoinducer Activity with Luminescent *Escherichia coli* Sensor Strains Harboring a Modified *Vibrio fischeri lux* Regulon
JERRY H. DEVINE AND GERALD S. SHADEL   279

21. Detection, Purification, and Structural Elucidation of the Acylhomoserine Lactone Inducer of *Vibrio fischeri* Luminescence and Other Related Molecules
AMY L. SCHAEFER, BRIAN L. HANZELKA, MATTHEW R. PARSEK, AND E. PETER GREENBERG   288

22. Chemical Synthesis of Bacterial Autoinducers and Analogs
ANATOL EBERHARD AND JEFFREY B. SCHINELLER   301

23. Overexpression of Foreign Proteins Using the *Vibrio fischeri lux* Control System
MICHAEL D. THOMAS AND ANITA VAN TILBURG   315

Section V. Luminescence-Based Assays *in Vitro*

24. Application of Bioluminescence and Chemiluminescence in the Biomedical Sciences
LARRY J. KRICKA   333

25. Use of Firefly Luciferase in ATP-Related Assays of Biomass, Enzymes, and Metabolites
ARNE LUNDIN   346

26. Chemiluminescent Methods for Detecting and Quantitating Enzyme Activity
LARRY J. KRICKA, JOHN C. VOYTA, AND IRENA BRONSTEIN   370

27. Chemiluminescence Assay of Serum Alkaline Phosphatase and Phosphoprotein Phosphatases
BRENDA J. HALLAWAY AND DENNIS J. O'KANE   390

28. Chemiluminescence Screening Assays for Erythrocytes and Leukocytes in Urine
VALERIE J. BUSH, BRENDA J. HALLAWAY, THOMAS A. EBERT, DAVID M. WILSON, AND DENNIS J. O'KANE   402

29. Immunoassay Protocol for Quantitation of Protein   JENNIFER MOSIER,
    Kinase Activities   CORINNE E. M. OLESEN,
                                                        JOHN C. VOYTA, AND
                                                        IRENA BRONSTEIN   410

30. Chemiluminescent Immunodetection Protocols   CORINNE E. M. OLESEN,
    with 1,2-Dioxetane Substrates   JENNIFER MOSIER,
                                                        JOHN C. VOYTA, AND
                                                        IRENA BRONSTEIN   417

31. Chemiluminescent Reporter Gene Assays with 1,2-   CORINNE E. M. OLESEN,
    Dioxetane Substrates   CHRIS S. MARTIN,
                                                        JENNIFER MOSIER,
                                                        BETTY LIU,
                                                        JOHN C. VOYTA, AND
                                                        IRENA BRONSTEIN   428

32. Clinical Application of Southern Blot Hybridization   KATHLEEN S. TENNER AND
    with Chemiluminescence Detection   DENNIS J. O'KANE   450

33. Quantitative Polymerase Chain Reaction and Solid-   CHRIS S. MARTIN,
    Phase Capture Nucleic Acid Detection   JOHN C. VOYTA, AND
                                                        IRENA BRONSTEIN   466

Section VI. Luminescence Monitoring *in Vivo*

34. Targeted Bioluminescent Indicators in Living Cells   GRACIELA B. SALA-NEWBY,
                                                        MICHAEL N. BADMINTON,
                                                        W. HOWARD EVANS,
                                                        CHRISTOPHER H. GEORGE,
                                                        HELEN E. JONES,
                                                        JONATHAN M. KENDALL,
                                                        ANGELA R. RIBEIRO, AND
                                                        ANTHONY K. CAMPBELL   479

35. Green Fluorescent Protein as a Reporter of Tran-   DEBORAH A. SIEGELE,
    scriptional Activity in a Prokaryotic System   LISA CAMPBELL, AND
                                                        JAMES C. HU   499

36. Bacterial *lux* Genes as Reporters in Cyanobacteria   F. FERNÁNDEZ-PIÑAS,
                                                        F. LEGANÉS, AND
                                                        C. PETER WOLK   513

37. Application of Bioluminescence to the Study of   CAROL R. ANDERSSON,
    Circadian Rhythms in Cyanobacteria   NICHOLAS F. TSINOREMAS,
                                                        JEFFREY SHELTON,
                                                        NADYA V. LEBEDEVA,
                                                        JUSTIN YARROW,
                                                        HONGTAO MIN, AND
                                                        SUSAN S. GOLDEN   527

38. Construction of *lux* Bacteriophages and the Determination of Specific Bacteria and Their Antibiotic Sensitivities   S. ULITZUR AND J. KUHN   543

39. Luciferase Gene as Reporter: Comparison with the CAT Gene and Use in Transfection and Microinjection of Mammalian Cells   S. GELMINI, P. PINZANI, AND M. PAZZAGLI   557

40. *In Situ* Hybridization and Immunohistochemistry with Enzyme-Triggered Chemiluminescent Probes   ALDO RODA, MONICA MUSIANI, PATRIZIA PASINI, MARIO BARALDINI, AND JEAN E. CRABTREE   577

41. Blood Phagocyte Luminescence: Gauging Systemic Immune Activation   ROBERT C. ALLEN, DAVID C. DALE, AND FLETCHER B. TAYLOR, JR.   591

Section VII. Bioluminescence as an Education Tool

42. Demonstrations of Chemiluminescence   FRANK MCCAPRA   633

43. Bioluminescence as a Classroom Tool for Scientist Volunteers   MARA HAMMER AND JOSEPH D. ANDRADE   660

44. Green Fluorescent Protein in Biotechnology Education   WILLIAM W. WARD, GAVIN C. SWIATEK, AND DANIEL G. GONZÁLEZ   672

VOLUME 306
EXPRESSION OF RECOMBINANT GENES IN EUKARYOTIC SYSTEMS
Joseph C. Glorioso and Martin C. Schmidt

Section I. Analysis of Gene Expression

1. Fluorescence-Based Expression Monitoring Using Microarrays   ELIZABETH A. WINZELER, MARK SCHENA, AND RONALD W. DAVIS   3

2. Protein Expression in Mammalian and Insect Cell Systems   SABINE GEISSE AND HANS P. KOCHER   19

3. Analysis of Mammalian Cis-regulatory DNA Elements by Homologous Recombination   STEVEN FIERING, M. A. BENDER, AND MARK GROUDINE   42

4. Genetically Directed Representational Difference MARK W. SORNSON AND
   Analysis-Based Positional Cloning and Snorthern MICHAEL G. ROSENFELD   67
   Analysis of Candidate Genes

Section II. Gene Expression Systems for Functional Analysis

5. Yeast Three-Hybrid System to Detect and Analyze BEILIN ZHANG,
   Interactions between RNA and Protein   BRIAN KRAEMER,
                                                        DHRUBA SENGUPTA,
                                                        STANLEY FIELDS, AND
                                                        MARVIN WICKENS   93

6. Estimating Subunit Dissociation Constants in Natu- LOIS R. MANNING,
   ral and Recombinant Proteins   ANTOINE DUMOULIN,
                                                         W. TERRY JENKINS,
                                                         ROBERT M. WINSLOW, AND
                                                         JAMES M. MANNING   113

7. Ectopic Expression Systems in *Drosophila*   PIER PAOLO D'AVINO AND
                                                        CARL S. THUMMEL   129

Section III. Yeast Expression Systems

8. Copper Ion Inducible and Repressible Promoter SIMON LABBÉ AND
   Systems in Yeast   DENNIS J. THIELE   145

9. Use of *Pichia pastoris* for Expression of Recombi- STUART A. ROSENFELD   154
   nant Proteins
10. Use of Yeast *sec6* Mutant for Purification of Vesicles LARRY A. COURY,
    Containing Recombinant Membrane Proteins   MARK L. ZEIDEL, AND
                                                           JEFFREY L. BRODSKY 169

11. Use of Yeast Artificial Chromosomes to Express KENNETH R. PETERSON   186
    Genes in Transgenic Mice

Section IV. RNA-Based Control of Recombinant Gene Expression

12. Use of Ribozymes to Inhibit Gene Expression MARTIN KRÜGER,
                                                  CARMELA BEGER, AND
                                                  FLOSSIE WONG-STAAL   207

13. Translational Control of Gene Expression Using JENNIFER L. DAVIS,
    Retroviral Vectors with Iron Response Elements PAUL R. GROSS, AND
                                                     OLIVIER DANOS   225

14. Strategies to Express Structural and Catalytic JAMES D. THOMPSON   241
    RNAs in Mammalian Cells

Section V. Small Molecule Control of Recombinant Gene Expression

15. Regulation of Gene Expression with Synthetic   ROY POLLOCK AND
    Dimerizers   VICTOR M. RIVERA   263

16. Antiprogestin Regulable Gene Switch for Induction   YAOLIN WANG,
    of Gene Expression *in Vivo*   SOPHIA Y. TSAI, AND
                                                         BERT W. O'MALLEY   281

Section VI. Viral Systems for Recombinant Gene Expression

17. Expression of SV40 Large T Antigen in Baculovirus   PAUL CANTALUPO,
    Systems and Purification by Immunoaffinity   M. TERESA SÁENZ-ROBLES,
    Chromatography   AND JAMES M. PIPAS   297

18. Applications of *ori*P Plasmids and Their Mode of   DAVID MACKEY AND
    Replication   BILL SUGDEN   308

19. Expression of E1 Protein of Human Papillomavi-   SALEEM A. KHAN,
    ruses in Eukaryotic Cells   SAIFUDDIN SHEIKH,
                                                       LAURA SHEAHAN,
                                                       GERALD VAN HORN,
                                                       VIDYA GOPALAKRISHNAN, AND
                                                       FRANCIS SVERDRUP   328

20. Genetic Manipulation of Herpes Simplex Virus   BRIAN C. HORSBURGH,
    Using Bacterial Artificial Chromosomes   MARIA M. HUBINETTE, AND
                                                    FRANK TUFARO   337

21. Protein and RNA Affinity Selection Techniques   MITCHELL E. GARBER,
    Using HIV-1 Tat   PING WEI,
                                                     GEORG CADERAS, AND
                                                     KATHERINE A. JONES   352

VOLUME 307
CONFOCAL MICROSCOPY
P. Michael Conn

Section I. Theory and Practical Considerations

1. Theoretical Basis of Confocal Microscopy   ROBERT H. WEBB   3

2. Practical Considerations in Acquiring Biological   S. K. Kong,
   Signals from Confocal Microscope   Samuel Ko,
                                                       C. Y. Lee, and
                                                       P. Y. Lui   20

Section II. General Techniques

3. Equipment for Mass Storage and Processing of Data   Guy Cox   29

4. Antifading Agents for Confocal Fluorescence Mi-   Miguel Berrios,
   croscopy   Kimberly A. Conlon, and
                                                       David E. Colflesh   55

5. Mounting Techniques for Confocal Microscopy   Manabu Kagayama and
                                                       Yasuyuki Sasano   79

6. Preparation of Whole Mounts and Thick Sections   Hisashi Hashimoto,
   for Confocal Microscopy   Hiroshi Ishikawa, and
                                                       Moriaki Kusakabe   84

7. Cytotomography   Torsten Porwol,
                                                       Anja-Rose Strohmaier,
                                                       and Eberhard Spiess   108

8. Use of Confocal Microscopy to Investigate Cell   Ghassan Bkaily,
   Structure and Function   Danielle Jacques, and
                                                       Pierre Pothier   119

9. Combining Laser Scanning Confocal Microscopy   Xuejun Sun   135
   and Electron Microscopy
10. Construction of Line-Scan Confocal Microscope for   Nick Callamaras and
    Physiological Recording   Ian Parker   152

Section III. Measurement of Subcellular Relations and Volume Determinations

11. Three-Dimensional Visualization of Cytoskeleton   Werner Baschong,
    by Confocal Laser Scanning Microscopy   Markus Dürrenberger,
                                                       Anna Mandinova, and
                                                       Rosmarie Suetterlin   173

12. Chromosome Spread for Confocal Microscopy   Nadir M. Maraldi,
                                                       Silvano Capitani,
                                                       Caterina Cinti,
                                                       Luca M. Neri,
                                                       Spartaco Santi,
                                                       Stefano Squarzoni,
                                                       Liborio Stuppia, and
                                                       Francesco A. Manzoli   190

13. Visualization of Nuclear Pore Complex and Its Dis- Ulrich Kubitscheck,
    tribution by Confocal Laser Scanning Microscopy   Thorsten Kues, and
                                                       Reiner Peters   207

14. Measurement of Tissue Thickness Using Confocal Microscopy — JAMES V. JESTER, W. MATTHEW PETROLL, AND H. DWIGHT CAVANAGH — 230

15. Measurements of Vascular Remodeling by Confocal Microscopy — SILVIA M. ARRIBAS, CRAIG J. DALY, AND JOHN C. MCGRATH — 246

16. Cell Volume Measurements by Fluorescence Confocal Microscopy: Theoretical and Practical Aspects — LOTHAR A. BLATTER — 274

17. Volume Measurements in Confocal Microscopy — GARY C. SIECK, CARLOS B. MANTILLA, AND Y. S. PRAKASH — 296

Section IV. Imaging of Processes

18. Quantitation of Phagocytosis by Confocal Microscopy — GEORGE F. BABCOCK — 319

19. Measurement of Secretion in Confocal Microscopy — AKIHISA SEGAWA — 328

20. Receptor–Ligand Internalization — GUIDO ORLANDINI, NICOLETTA RONDA, RITA GATTI, GIAN CARLO GAZZOLA, AND ALBERICO BORGHETTI — 340

21. Quantitative Imaging of Metabolism by Two-Photon Excitation Microscopy — DAVID W. PISTON AND SUSAN M. KNOBEL — 351

22. Localization of Proteases and Peptide Receptors by Confocal Microscopy — CELIA J. SNYMAN, DESHANDRA M. RAIDOO, AND KANTI D. BHOOLA — 368

23. Observation of Microcirculatory Kinetics by Real-Time Confocal Laser Scanning Microscopy — KAZUHIRO YAMAGUCHI — 394

Section V. Imaging of Ions

24. Confocal Imaging Analysis of Intracellular Ions in Mixed Cellular Systems or *in Situ* Using Two Types of Confocal Microscopic Systems — HISAYUKI OHATA, MASAYUKI YAMAMOTO, YOSUKE UJIKE, GOUSEI RIE, AND KAZUTAKA MOMOSE — 425

25. Confocal Ca^{2+} Imaging of Organelles, Cells, Tissues, and Organs — DAVID ALAN WILLIAMS, DAVID N. BOWSER, AND STEVEN PETROU — 441

26. Using Confocal Microscopy and Fluorescent Indicator, 6-Methoxy-N-ethylquinolinium Iodide, to Measure Changes in Intracellular Chloride — JON R. INGLEFIELD AND ROCHELLE D. SCHWARTZ-BLOOM — 469

Section VI. Imaging of Specialized Tissues

Section VII. Imaging Viruses and Fungi

VOLUME 308
ENZYME KINETICS AND MECHANISM (PART E: ENERGETICS OF ENZYME CATALYSIS)
Vern L. Schramm and Daniel L. Purich

Section I. Energetic Coupling in Enzymatic Reactions

2. Energy Coupling Through Molecular Discrimina-   CHARLES TIMMIS GRUBMEYER,
   tion: Nicotinate Phosphoribosyltransferase   JEFFREY W. GROSS, AND
                                                   MATHUMATHI RAJAVEL   28

3. On the Advantages of Imperfect Energetic Linkage   THOMAS S. LEYH   48

4. Reaction Dynamics of G-Protein Catalyzed Hydro-   DAVID E. COLEMAN AND
   lysis of GFP as Viewed by X-Ray Crystallographic   STEPHEN R. SPRANG   70
   Snapshots of $G_{i\alpha1}$

5. Energetics of Nucleotide Hydrolysis in Polymer   DANIEL L. PURICH AND
   Assembly/Disassembly: The Cases of Actin and   FREDERICK S. SOUTHWICK 93
   Tubulin

6. Fundamental Mechanisms of Substrate Channeling   KAREN S. ANDERSON   111

Section II. Intermediates and Complexes in Catalysis

7. Intermediates and Energetics in Pyruvate Phos-   DEBRA DUNAWAY-MARIANO   149
   phate Dikinase

8. Raman Spectroscopic Studies of the Structures,   HUA DENG AND
   Energetics, and Bond Distortions of Substrates   ROBERT CALLENDER   176
   Bound to Enzymes

9. Crystallographic Analysis of Solvent-Trapped In-   GREGORY K. FARBER   201
   termediates of Chymotrypsin

Section III. Detection and Properties of Low-Barrier Hydrogen Bonds

10. Nuclear Magnetic Resonance Methods for the De-   ALBERT S. MILDVAN,
    tection and Study of Low-Barrier Hydrogen   THOMAS K. HARRIS, AND
    Bonds of Enzymes   CHITRANANDA
                                                     ABEYGUNAWARDANA   219

11. Hydrogen Bonding in Enzymatic Catalysis: Analysis   SHU-OU SHAN AND
    of Energetic Contributions   DANIEL HERSCHLAG   246

12. Application of Marcus Rate Theory to Proton   A. JERRY KRESGE AND
    Transfer in Enzyme-Catalyzed Reactions   DAVID N. SILVERMAN   276

Section IV. Transition State Determination and Inhibitors

13. Enzymatic Transition-State Analysis and Transi-   VERN L. SCHRAMM   301
    tion-State Analogues

14. Determining Transition States from Kinetic Iso-   PAUL J. BERTI   355
    tope Effects

15. Computational Methods for Transition State and   BENJAMIN B. BRAUNHEIM AND
    Inhibitor Recognition   STEVEN D. SCHWARTZ   398

VOLUME 309
AMYLOID, PRIONS, AND OTHER PROTEIN AGGREGATES
Ronald Wetzel

Section I. Characterization of *in Vivo* Protein Deposition

A. Identification and Isolation of Aggregates

1. Staining Methods for Identification of Amyloid in   GUNILLA T. WESTERMARK,
   Tissue   KENNETH H. JOHNSON, AND
                                                        PER WESTERMARK   3

2. Isolation and Characterization of Amyloid Fibrils   GLENYS A. TENNENT   26
   from Tissue

3. Isolating Inclusion Bodies from Bacteria   GEORGE GEORGIOU AND
                                                        PASCAL VALAX   48

4. Isolation of Amyloid Deposits from Brain   ALEX EUGENE ROHER AND
                                                        YU-MIN KUO   58

B. Isolation and Characterization of Protein Deposit Components

5. Microextraction and Purification Techniques Appli- BATIA KAPLAN,
   cable to Chemical Characterization of Amyloid   RUDI HRNCIC,
   Proteins in Minute Amounts of Tissue   CHARLES L. MURPHY,
                                                        GLORIA GALLO,
                                                        DEBORAH T. WEISS, AND
                                                        ALAN SOLOMON   69

6. Purification of Paired Helical Filament Tau and   VIRGINIA M.-Y. LEE,
   Normal Tau from Human Brain Tissue   JUN WANG, AND
                                                        JOHN TROJANOWSKI   81

7. Chemical Modifications of Deposited Amyloid-β JONATHAN D. LOWENSON,
   Peptides   STEVEN CLARKE, AND
                                                        ALEX EUGENE ROHER   89

C. Characterization of Aggregates *in Situ* and *in Vitro*

8. Monoclonal Antibodies Specific for the Native, Disease-Associated Isoform of Prion Protein — CARSTEN KORTH, PETER STREIT, AND BRUNO OESCH — 106

9. Assays of Protease-Resistant Prion Protein and Its Formation — BRYON CAUGHEY, MOTOHIRO HORIUCHI, RÉMI DEMAIMAY, AND GREGORY RAYMOND — 122

10. *In Situ* Methods for Detection and Localization of Markers of Oxidative Stress: Application in Neurodegenerative Disorders — LAWRENCE M. SAYRE, GEORGE PERRY, AND MARK A. SMITH — 133

11. Advanced Glycation End Products: Detection and Reversal — YOUSEF AL-ABED, APHRODITE KAPURNIOTU, AND RICHARD BUCALA — 152

12. Analysis of Transglutaminase-Catalyzed Isopeptide Bonds in Paired Helical Filaments and Neurofibrillary Tangles from Alzheimer's Disease — BRIAN J. BALIN, ARIEL G. LOEWY, AND DENAH M. APPELT — 172

Section II. Characterization of *in Vitro* Protein Deposition

A. Managing the Aggregation Process

13. Methodological and Chemical Factors Affecting Amyloid-β Peptide Amyloidogenicity — MICHAEL G. ZAGORSKI, JING YANG, HAIYAN SHAO, KAN MA, HONG ZENG, AND ANITA HONG — 189

14. *In Vitro* Immunoglobulin Light Chain Fibrillogenesis — JONATHAN WALL, CHARLES L. MURPHY, AND ALAN SOLOMON — 204

15. Inhibition of Aggregation Side Reactions during *in Vitro* Protein Folding — ELIANA DE BERNARDEZ CLARK, ELISABETH SCHWARZ, AND RAINER RUDOLPH — 217

16. Inhibition of Stress-Induced Aggregation of Protein Therapeutics — JOHN F. CARPENTER, BRENT S. KENDRICK, BYEONG S. CHANG, MARK C. MANNING, AND THEODORE W. RANDOLPH — 236

B. Aggregation Theory

17. Analysis of Protein Aggregation Kinetics — FRANK FERRONE — 256

C. Monitoring Aggregate Growth by Dye Binding

18. Quantification of β-Sheet Amyloid Fibril Structures with Thioflavin T   HARRY LEVINE III   274

19. Quantifying Amyloid by Congo Red Spectral Assay   WILLIAM E. KLUNK, ROBERT F. JACOB, AND R. PRESTON MASON   285

20. Kinetic Analysis of Amyloid Fibril Formation   HIRONOBU NAIKI AND FUMITAKE GEJYO   305

D. Measurement and Characterization of Assembly Intermediates

21. Small-Zone, High-Speed Gel Filtration Chromatography to Detect Protein Aggregation Associated with Light Chain Pathologies   ROSEMARIE RAFFEN AND FRED J. STEVENS   318

22. Detection of Early Aggregation Intermediates by Native Gel Electrophoresis and Native Western Blotting   SCOTT D. BETTS, MARGARET SPEED, AND JONATHAN KING   333

E. Monitoring Aggregate Growth by Measuring Solid-Phase Accumulation

23. Deposition of Soluble Amyloid-β onto Amyloid Templates: Identification of Amyloid Fibril Extension Inhibitors   WILLIAM P. ESLER, EVELYN R. STIMSON, PATRICK W. MANTYH, AND JOHN E. MAGGIO   350

24. Membrane Filter Assay for Detection of Amyloid-like Polyglutamine-Containing Protein Aggregates   ERICH E. WANKER, EBERHARD SCHERZINGER, VOLKER HEISER, ANNIE SITTLER, HOLGER EICKHOF, AND HANS LEHRACH   375

25. Analysis of Fibril Elongation Using Surface Plasmon Resonance Biosensors   DAVID G. MYSZKA, STEPHEN J. WOOD, AND ANJA LEONA BIERE   386

26. Methods for Studying Protein Adsorption   VLADIMIR HLADY, JOS BUIJS, AND HERBERT P. JENNISSEN   402

F. Monitoring Aggregate Growth and Structure Using Light Scattering

27. Monitoring Protein Assembly Using Quasielastic Light Scattering Spectroscopy   ALEKSEY LOMAKIN, GEORGE B. BENEDEK, AND DAVID B. TEPLOW   429

28. Flow Cytometric Characterization of Amyloid Fi-   JONATHAN WALL AND
    brils   ALAN SOLOMON   460

G. Aggregation Inhibitors

29. Screening for Pharmacologic Inhibitors of Amyloid   HARRY LEVINE III AND
    Fibril Formation   JEFFREY D. SCHOLTEN   467

30. Design and Testing of Inhibitors of Fibril Formation   MARK A. FINDEIS AND
                                                       SUSAN M. MOLINEAUX   476

Section III. Aggregate and Precursor Protein Structure

A. Aggregate Morphology

31. Electron Microscopy of Prefibrillar Structures and   ELLEN HOLM NIELSEN,
    Amyloid Fibrils   MADS NYBO, AND
                                                       SVEN-ERIK SVEHAG   491

32. *In Situ* Electron Microscopy of Amyloid Deposits   SADAYUKI INOUE AND
    in Tissues   ROBERT KISILEVSKY   496

33. Analysis of Amyloid-β Assemblies Using Tapping   TOMAS T. DING AND
    Mode Atomic Force Microscopy under Ambi-   JAMES D. HARPER   510
    ent Conditions

B. Molecular Level Aggregate Structure

34. X-Ray Fiber Diffraction of Amyloid Fibrils   LOUISE C. SERPELL,
                                                       PAUL E. FRASER, AND
                                                       MARGARET SUNDE   526

35. Solid State Nuclear Magnetic Resonance of Pro-   DAVID E. WEMMER   536
    tein Deposits

36. Fourier Transform Infrared Spectroscopy in Analy-   SANGITA SESHADRI,
    sis of Protein Deposits   RITU KHURANA, AND
                                                       ANTHONY L. FINK   559

37. Stable Isotope-Labeled Peptides in Study of Pro-   MICHAEL A. BALDWIN   576
    tein Aggregation

38. Mapping Protein Conformations in Fibril Structures   ERIK LUNDGREN,
    Using Monoclonal Antibodies   HAKAN PERSSON,
                                                       KARIN ANDERSSON,
                                                       ANDERS OLOFSSON,
                                                       INGRID DACKLIN, AND
                                                       GUNDARS GOLDSTEINS   591

C. Characterization of Precursor Protein Structure

39. Analysis of Protein Structure by Solution Optical   WILFREDO COLÓN   605
    Spectroscopy

40. Probing Conformations of Amyloidogenic Proteins   EWAN J. NETTLETON AND
    by Hydrogen Exchange and Mass Spectroscopy   CAROL V. ROBINSON   633

Section IV. Cellular and Organismic Consequences of Protein Deposition

A. Microbial Model Systems

41. Yeast Prion [X+] and Its Determinant, Sup35p   TRICIA R. SERIO,
                                                   ANIL G. CASHIKAR,
                                                   JAHAN J. MOSLEHI,
                                                   ANTHONY S. KOWAL, AND
                                                   SUSAN L. LINDQUIST   649

B. Animal Models of Protein Deposition Diseases

42. The Senescence-Accelerated Mouse   KEIICHI HIGUCHI,
                                                   MASANORI HOSOKAWA, AND
                                                   TOSHIO TAKEDA   674

43. Detection of Polyglutamine Aggregation in   STEPHEN W. DAVIES,
    Mouse Models   KIRUPA SATHASIVAM,
                                                   CARL HOBBS,
                                                   PATRICK DOHERTY,
                                                   LAURA MANGIARINI,
                                                   EBERHARD SCHERZINGER,
                                                   ERICH E. WANKER, AND
                                                   GILLIAM P. BATES   687

44. A Mouse Model for Serum Amyloid A Amyloidosis   MARK S. KINDY AND
                                                   FREDERICK C. DE BEER   701

C. Cell Studies on Protein Aggregate Cytotoxicity

45. Toxicity of Protein Aggregates in PC12 Cells: MARK S. SHEARMAN   716
    3-(4,5-Dimethylthiazol-2-yl)-2,5-diphenyl-
    tetrazolium Bromide Assay

46. Inflammatory Responses to Amyloid Fibrils   STEPHEN L. YATES,
                                                   JUNE KOCSIS-ANGLE,
                                                   PAULA EMBURY, AND
                                                   KURT R. BRUNDEN   723

47. Impairment of Membrane Transport and Signal   MARK P. MATTSON   733
    Transduction Systems by Amyloidogenic Pro-
    teins

48. Amyloid β-Peptide-Associated Free Radical Oxida- D. ALLAN BUTTERFIELD,
    tive Stress, Neurotoxicity, and Alzheimer's Dis- SERVET M. YATIN,
    ease   SRIDHAR VARADARAJAN, AND
                                                   TANUJA KOPPAL   746

VOLUME 310
BIOFILMS
Ron J. Doyle

Section I. Molecular Biology of Biofilm Bacteria

1. Reporter Systems for Microscopic Analysis of Microbial Biofilms — DARREN R. KORBER, GIDEON M. WOLFAARDT, VOLKER S. BRÖZEL, RAYNARD MACDONALD, AND TANJA NIEPEL — 3

2. Molecular Tools for Study of Biofilm Physiology — BJARKE BAK CHRISTENSEN, CLAUS STERNBERG, JENS BO ANDERSEN, ROBERT J. PALMER, JR., ALEX TOFTGAARD NIELSEN, MICHAEL GIVSKOV, AND SØREN MOLIN — 20

3. Quorum Sensing Signals in Development of *Pseudomonas aeruginosa* Biofilms — MATTHEW R. PARSEK AND E. PETER GREENBERG — 43

4. Monitoring Gene Expression in Biofilms — CLAIRE PRIGENT-COMBARET AND PHILIPPE LEJEUNE — 56

5. *In Situ* Analysis of Microbial Biofilms by rRNA-Targeted Oligonucleotide Probing — WERNER MANZ — 79

6. Genetic Approaches to Study of Biofilms — GEORGE A. O'TOOLE, LESLIE A. PRATT, PAULA I. WATNICK, DIANNE K. NEWMAN, VALERIE B. WEAVER, AND ROBERTO KOLTER — 91

7. Identifying *in Vivo* Expressed Streptococcal Genes in Endocarditis — LIN TAO AND MARK C. HERZBERG — 109

8. Quorum Sensing, Gene Expression, and *Pseudomonas* Biofilms — TERESA R. DE KIEVIT AND BARBARA H. IGLEWSKI — 117

Section II. Microscopic Methods of Biofilm Formation and Physiology

9. Confocal Laser Scanning Microscopy for Analysis of Microbial Biofilms — JOHN R. LAWRENCE AND THOMAS R. NEU — 131

10. Lectin-Binding Analysis in Biofilm Systems — THOMAS R. NEU AND JOHN R. LAWRENCE — 145

11. Spatially Resolved, Quantitative Determination of Luciferase Activity by Photon-Counting Microscopy — ROBERT J. PALMER, JR., AND DAVID C. WHITE — 152

12. Microscopy Flow Cells: Perfusion Chambers for Real-Time Study of Biofilms — ROBERT J. PALMER, JR. — 160

13. Fluorescent Probes Applied to Physiological and Structural Characterization of Bacterial Biofilms — JOHN T. LISLE, PHILIP S. STEWART, AND GORDON A. McFETERS — 166

14. Deconvolution Fluorescence Microscopy for Observation and Analysis of Membrane Biofilm Architecture — DONALD PHIPPS, GRISEL RODRIGUEZ, AND HARRY F. RIDGWAY — 178

15. Bacterial Biofilms: Strategies for Preparing Glycocalyx for Electron Microscopy — THERESA A. FASSEL AND CHARLES E. EDMISTON, JR. — 194

Section III. Flow and Steady-State Methods

16. Robbins Device in Biofilm Research — ARSALAN KHARAZMI, BIRGIT GIWERCMAN, AND NIELS HØIBY — 207

17. Controlled Environment Model for Accumulation of Biofilms of Oral Bacteria — GEORGE H. W. BOWDEN — 216

18. Laminar Flow Chamber for Continuous Monitoring of Biofilm Formation and Succession — MANFRED S. ZINN, ROBIN D. KIRKEGAARD, ROBERT J. PALMER, JR., AND DAVID C. WHITE — 224

19. Perfused Biofilm Fermenters — DAVID G. ALLISON AND PETER GILBERT — 232

20. Laboratory Techniques for Studying Biofilm Growth, Physiology, and Gene Expression in Flowing Systems and Porous Media — ROBERT J. C. McLEAN, MARVIN WHITELEY, BRIAN C. HOSKINS, PAUL D. MAJORS, AND MUKUL M. SHARMA — 248

21. Use of Constant Depth Fermenter in Studies of Biofilms of Oral Bacteria — MICHAEL WILSON — 264

22. Use of Continuous Flow Techniques in Modeling Dental Plaque Biofilms — DAVID J. BRADSHAW AND PHILIP D. MARSH — 279

23. Steady-State Biofilm: Practical and Theoretical Models — GEORGE H. DIBDIN AND JULIAN WIMPENNY — 296

24. Spatial Organization of Oral Bacteria in Biofilms — PAUL E. KOLENBRANDER, ROXANNA N. ANDERSEN, KAREN KAZMERZAK, ROSEMARY WU, AND ROBERT J. PALMER, JR. — 322

Section IV. Biofilms in Archaea

25. Biofilm Formation in Hyperthermophilic Archaea   PATRICIA L. HARTZELL,
                                                       JACK H. MILLSTEIN, AND
                                                       CHRISTOPHER LAPAGLIA   335

Section V. Physical Methods

26. Use of Conductance Measurements for Determina-   CHARLOTTE JOHANSEN,
    tion of Enzymatic Degradation of Microbial   BERIT K. BREDTVED, AND
    Biofilm   SØREN MØLLER   353

27. Evaluation and Quantification of Bacterial Attach-   WOLFGANG SAND AND
    ment, Microbial Activity, and Biocide Efficacy by   HENRY VON RÈGE   361
    Microcalorimetry

28. Surface Analyses by X-Ray Photoelectron Spectros-   YVES F. DUFRÊNE,
    copy in Study of Bioadhesion and Biofilms   CHRISTOPHE J. P. BOONAERT,
                                                        AND PAUL G. ROUXHET   375

Section VI. Physiology of Biofilm-Associated Microorganisms

29. Evaluating Biofilm Activity in Response to Mass   RYAN N. JORDAN   393
    Transfer-Limited Bioavailability of Sorbed Nu-
    trients

30. Bacterial Survival in Biofilms: Probes for Exopoly-   RONALD WEINER,
    saccharide and Its Hydrolysis, and Measurements   E. SEAGREN,
    of Intra- and Interphase Mass Fluxes   C. ARNOSTI, AND
                                                         E. QUINTERO   403

31. Biosurfactants Produced by *Lactobacillus*   GREGOR REID,
                                                      CHRISTINE HEINEMANN,
                                                      MARTINE VELRAEDS, AND
                                                      HENK J. BUSSCHER   426

32. Proteome Analysis of Biofilms: Growth of *Bacillus*   BRIAN S. MILLER AND
    subtilis on Solid Medium as Model   MARIA R. DIAZ-TORRES   433

33. Physiologic Homeostasis and Stress Responses of   ROBERT A. BURNE,
    Oral Biofilms   ROBERT G. QUIVEY, JR., AND
                                                        ROBERT E. MARQUIS   441

Section VII. Substrate for Biofilm Development

34. Biofouling of Membranes: Membrane Preparation;   HARRY F. RIDGWAY,
    Characterization and Analysis of Bacterial Ad-   KENNETH ISHIDA,
    hesion   GRISEL RODRIGUEZ,
                                                       JANA SAFARIK,
                                                       TOM KNOELL, AND
                                                       RICHARD BOLD   463

35. Simple Physical Model to Study Formation and   DAVID J. STICKLER,
    Physiology of Biofilms on Urethral Catheters   NICOLA S. MORRIS, AND
                                                    CAROLE WINTERS   494

36. Studying Initial Phase of Biofilm Formation: Molec-   ATSUO AMANO,
    ular Interaction of Host Matrix Proteins and Bac-   ICHIRO NAKAGAWA, AND
    terial Surface Components   SHIGEYUKI HAMADA   501

37. Studying Biofilm Formation of Mutans Streptococci   SHIGETADA KAWABATA AND
                                                         SHIGEYUKI HAMADA   513

38. Models for Studying Initial Adhesion and Surface   BOVRT GOTTENBOS,
    Growth in Biofilm Formation on Surfaces   HENNY C. VAN DER MEI, AND
                                                        HENK J. BUSSCHER   523

39. Recovery and Characterization of Biofilm Bacteria   MARC W. MITTELMAN   534
    Associated with Medical Devices

40. Primary Adhesion of *Pseudomonas aeruginosa* to   DONALD G. AHEARN,
    Inanimate Surfaces Including Biomaterials   ROYA N. BORAZJANI,
                                                      ROBERT B. SIMMONS, AND
                                                      MANAL M. GABRIEL   551

41. Biofilm and Biofilm-Related Encrustation of Uri-   MICHAEL M. TUNNEY,
    nary Tract Devices   DAVID S. JONES, AND
                                                       SEAN P. GORMAN   558

42. Detection of Prosthetic Joint Biofilm Infection Us-   MICHAEL M. TUNNEY,
    ing Immunological and Molecular Techniques   SHEILA PATRICK,
                                                          MARTIN D. CURRAN,
                                                          GORDON RAMAGE,
                                                          NEIL ANDERSON,
                                                          RICHARD I. DAVIS,
                                                          SEAN P. GORMAN, AND
                                                          JAMES R. NIXON   566

43. *In Vitro* and *in Vivo* Models of Bacterial Biofilms   HIROSHI YASUDA,
                                                            TETSUFUMI KOGA, AND
                                                            TAKASHI FUKUOKA   577

Section VIII. Antifouling Methods

44. *Pseudomonas aeruginosa* Biofilm Sensitivity to Bio-   DANIEL J. HASSETT,
    cides: Use of Hydrogen Peroxide as Model Anti-   JAMES G. ELKINS,
    microbial Agent for Examining Resistance Mech-   JU-FANG MA, AND
    anisms   TIMOTHY R. MCDERMOTT   599

45. Measuring Antimicrobial Effects on Biofilm Bacte-   NICK ZELVER,
    ria: From Laboratory to Field   MARTY HAMILTON,
                                                        BETSEY PITTS,
                                                        DARLA GOERES,
                                                        DIANE WALKER,
                                                        PAUL STURMAN, AND
                                                        JOANNA HEERSINK   608

46. Quantitative Assessment of Biocide Control of Bio-   JAMES T. WALKER,
    films Including *Legionella pneumophila* Using   A. D. G. ROBERTS,
    Total Viable Counts, Fluorescence Microscopy,   V. J. LUCAS,
    and Image Analysis   M. A. ROPER, AND
                                                        R. G. BROWN   629

47. Quantifying Effects of Antifouling Paints on Micro- JOSEPH J. COONEY AND
    bial Biofilm Formation   RUEY-JING TANG   637

48. *Candida* Biofilms and Their Susceptibility to Anti- GEORGE S. BAILLIE AND
    fungal Agents   L. JULIA DOUGLAS
                                                                             644

49. Enhanced Bacterial Biofilm Control Using Electro-   BRUCE R. MCLEOD,
    magnetic Fields in Combination with Antibiotics   SUSANA FORTUN,
                                                        J. WILLIAM COSTERTON, AND
                                                        PHILIP S. STEWART   656

VOLUME 311
SPHINGOLIPID METABOLISM AND CELL SIGNALING (PART A)
Alfred H. Merrill, Jr. and Yusuf A. Hannun

Section I. Sphingolipid Metabolism

A. Biosynthesis

1. Serine Palmitoyltransferase   ROBERT C. DICKSON,
                                                        ROBERT L. LESTER, AND
                                                        M. MAREK NAGIEC   3

2. Assay of the *Saccharomyces cerevisiae* Dihydro-   MICHELLE M. GRILLEY AND
   sphingosine C-4 Hydroxylase   JON Y. TAKEMOTO   9

3. Ceramide Synthase   ELAINE WANG AND
                                                        ALFRED H. MERRILL, JR. 15

4. Dihydroceramide Desaturase   HEIKE SCHULZE,
                                                        CHRISTOPH MICHEL, AND
                                                        GERHILD VAN ECHTEN-
                                                        DECKERT   22

5. Assays for the Biosynthesis of Sphingomyelin and   MARIANA NIKOLOVA-
   Ceramide-Phosphoethanolamine   KARAKASHIAN   31

6. Glucosylceramide Synthase: Assay and Properties   JAMES A. SHAYMAN AND
                                                      AKIRA ABE   42

7. Methods for Studying Glucosylceramide Synthase   DAVID L. MARKS,
                                                      PASCAL PAUL,
                                                      YASUSHI KAMISAKA, AND
                                                      RICHARD E. PAGANO   50

8. Analysis of Galactolipids and UDP-Galactose: HEIN SPRONG,
   Ceramide Galactosyltransferase   GERRIT VAN MEER, AND
                                                      PETER VAN DER SLUIJS   59

9. Assay of Lactosylceramide Synthase and Comments   SUBROTO CHATTERJEE   73
   on Its Potential Role in Signal Transduction

10. *In Vitro* Assays for Enzymes of Ganglioside Syn- GOTTFRIED POHLENTZ,
    thesis   CHRISTOPH KAES, AND
                                                      KONRAD SANDHOFF   82

11. Analyses of Sulfatide and Enzymes of Sulfatide Me- FIROZE B. JUNGALWALA,
    tabolism   PRASSON CHATTURVEDI,
                                                      DAVID S. NEWBURG, AND
                                                      MARVIN R. NATOWICZ   94

12. 1-*O*-Acylceramide Synthase   JAMES A. SHAYMAN AND
                                                      AKIRA ABE   105

13. *N*-Acetylation of Sphingosine by Platelet-Activat- TEN-CHING LEE   117
    ing Factor: Sphingosine Transacetylase

14. Inositolphosphoryl Ceramide Synthase from Yeast ANTHONY S. FISCHL,
                                                      YONGSHENG LIU,
                                                      ANGELEAH BROWDY, AND
                                                      AIDA E. CREMESTI   123

15. Enzymes of Sphingolipid Metabolism in Plants   DANIEL V. LYNCH   130

B. Turnover

16. Purification and Characterization of Recombinant STEFANIE LANSMANN,
    Human Acid Sphingomyelinase Expressed in In-   OLIVER BARTELSEN, AND
    sect *Sf* 21 Cells   KONRAD SANDHOFF   149

17. Purification of Rat Brain Membrane Neutral Sphin- BIN LIU AND
    gomyelinase   YUSUF A. HANNUN   156

18. Sphingomyelinase Assay Using Radiolabeled Sub-   BIN LIU AND
    strate   YUSUF A. HANNUN   164

19. Robotic Assay of Sphingomyelinase for High-   ARMINDA G. BARBONE,
    Throughput Screening   ALISA C. JACKSON,
                                                      DAVID M. RITCHIE, AND
                                                      DENNIS C. ARGENTIERI   168

20. A High Throughput Sphingomyelinase Assay   DANIEL F. HASSLER, RONALD M. LAETHEM, AND GARY K. SMITH   176

21. Analyses of Sphingomyelin Hydrolysis in Caveolar Membranes   RICK T. DOBROWSKY AND VALESWARA-RAO GAZULA   184

22. Ceramidases   MARIANA NIKOLOVA-KARAKASHIAN AND ALFRED H. MERRILL, JR.   194

23. Purification of Acid Ceramidase from Human Placenta   THOMES LINKE, STEFANIE LANSMANN, AND KONRAD SANDHOFF   201

24. Ceramide Kinase   SANDRA BAJJALIEH AND ROBERT BATCHELOR   207

25. Assaying Sphingosine Kinase Activity   ANA OLIVERA, KEITH D. BARLOW, AND SARAH SPIEGEL   215

26. Yeast Sphingosine-1-Phosphate Phosphatases: Assay, Expression, Deletion, Purification, and Cellular Localization by GFP Tagging   CUNGUI MAO AND LINA M. OBEID   223

27. Analysis of Ceramide 1-Phosphate and Sphingosine-1-Phosphate Phosphatase Activities   DAVID N. BRINDLEY, JIM XU, RENATA JASINSKA, AND DAVID W. WAGGONER   233

28. Sphingosine-1-Phosphate Lyase   PAUL P. VAN VELDHOVEN   244

29. Sphingolipid Hydrolases and Activator Proteins   UWE BIERFREUND, THOMAS KOLTER, AND KONRAD SANDHOFF   255

30. Sphingolipid Hydrolyzing Enzymes in the Gastrointestinal Tract   RUI-DONG DUAN AND ÅKE NILSSON   276

31. Properties of Animal Ceramide Glycanases   MANJU BASU, PATRICK KELLY, MARK GIRZADAS, ZHIXIONG LI, AND SUBHASH C. BASU   287

C. Genetic Approaches

32. Enzymatic N-Deacylation of Sphingolipids   MAKOTO ITO, KATSUHIRO KITA, TOYOHISA KURITA, NORIYUKI SUEYOSHI, AND HIROYUKI IZU   297

33. Genetic Approaches for Studies of Glycolipid Syn-   SHINICHI ICHIKAWA AND
    thetic Enzymes   YOSHIO HIRABAYASHI   303

34. Use of Yeast as a Model System for Studies of   NAMJIN CHUNG AND
    Sphingolipid Metabolism and Signaling   LINA M. OBEID   319

Section II. Inhibitors of Sphingolipid Biosynthesis

35. Isolation and Characterization of Novel Inhibitors   SUZANNE M. MANDALA AND
    of Sphingolipid Biosynthesis: Australifungin, Vir-   GUY H. HARRIS   335
    idiofungins, Rustmicin, and Khafrefungin

36. Fermentation, Partial Purification, and Use of Ser- RONALD T. RILEY AND
    ine Palmitoyltransferase Inhibitors from *Isaria*   RONALD D. PLATTNER   348
    (= *Cordyceps*) *sinclairii*

37. Isolation and Characterization of Fumonisins   FILMORE I. MEREDITH   361

38. Inhibitors of Glucosylceramide Synthase   JAMES A. SHAYMAN,
                                                         LIHSUEH LEE,
                                                         AKIRA ABE, AND
                                                         LIMING SHU   373

Section III. Chemical and Enzymatic Synthesis

39. Synthesis of Sphingosine and Sphingoid Bases   CHRISTOPHER CURFMAN AND
                                                         DENNIS C. LIOTTA   391

40. Synthesis of Sphingosine, Radiolabeled Sphingo-   KARL-HEINZ JUNG AND
    sine, 4-Methyl-*cis*-Sphingosine, and 1-Amino De-   RICHARD R. SCHMIDT   441
    rivatives of Sphingosine via Their Azido Deriva-
    tives

41. Total Synthesis of Sphingosine and Its Analogs   PÄIVI M. KOSKINEN AND
                                                         ARI M. P. KOSKINEN   458

42. Radiolabeling of the Sphingolipid Backbone   ALICJA BIELAWSKA,
                                                         YUSUF A. HANNUN, AND
                                                         ZDZISLAW SZULC   480

43. Preparation of Radiolabeled Ceramides and Phos-   ALICJA BIELAWSKA AND
    phosphingolipids   YUSUF A. HANNUN   499

44. Synthesis of Key Precursors of Radiolabeled Sphin-   ALICJA BIELAWSKA,
    golipids   ZDZISLAW SZULC, AND
                                                         YUSUF A. HANNUN   518

45. Practical Synthesis of *N*-Palmitoylsphingomyelin   ANATOLIY S. BUSHNEV AND
    and *N*-Palmitoyldihydrosphingomyelin   DENNIS C. LIOTTA   535

46. Synthesis and Biological Activity of Glycolipids,   TAKAO IKAMI,
    with a Focus on Gangliosides and Sulfatide An-   HIDEHARU ISHIDA, AND
    alogs   MAKOTO KISO   547

47. Sphingolipid Photoaffinity Labels   FRIEDERIKE KNOLL,
                                                     THOMAS KOLTER, AND
                                                     KONRAD SANDHOFF   568

48. Synthesis and Characterization of Metabolically GÜNTER SCHWARZMANN   601
    Stable Sphingolipids

49. Synthetic Soluble Analogs of Glycolipids for Studies JACQUES FANTINI   626
    of Virus–Glycolipid Interactions

50. Preparation of Radioactive Gangliosides ^{3}H or ^{14}C SANDRO SONNINO,
    Isotopically Labeled at the Oligosaccharide or VANNA CHIGORNO, AND
    Ceramide Moieties   GUIDO TETTAMANTI   639

51. Estimating Sphingolipid Metabolism and Traffick- LAURA RIBONI,
    ing in Cultured Cells Using Radiolabeled Com-   PAOLA VIANI, AND
    pounds   GUIDO TETTAMANTI   656

52. Enzymatic Synthesis of [^{14}C]Ceramide, [^{14}C]Glyco- MAKOTO ITO,
    sphingolipids, and ω-Aminoceramide   SUSUMU MITSUTAKE,
                                                     MOTOHIRO TANI, AND
                                                     KATSUHIRO KITA   682

VOLUME 312
SPHINGOLIPID METABOLISM AND CELL SIGNALING (PART B)
Alfred H. Merrill, Jr. and Yusuf A. Hannun

Section I. Methods for Analyzing Sphingolipids

1. Analysis of Sphingoid Bases and Sphingoid Base 1- ALFRED H. MERRILL, JR.,
   Phosphates by High-Performance Liquid Chro-   THOMAS B. CALIGAN,
   matography   ELAINE WANG,
                                                    KATHERINE PETERS, AND
                                                    JOYCE OU   3

2. Enzymatic Method for Measurement of Sphingo-   LISA EDSALL,
   sine 1-Phosphate   LEWIS VANN,
                                                    SHELDON MILSTIEN, AND
                                                    SARAH SPIEGEL   9

3. Ceramide Mass Analysis by Normal-Phase High-   ANNEMIEK D. TEPPER AND
   Performance Liquid Chromatography   WIM J. VAN BLITTERSWIJK   16

4. Quantitative Determination of Ceramide Using Di- DAVID K. PERRY,
   glyceride Kinase   ALICJA E. BIELAWSKA, AND
                                                    YUSUF A. HANNUN   22

5. Analysis of Sphingomyelin, Glucosylceramide, Ceramide, Sphingosine, and Sphingosine 1-Phosphate by Tandem Mass Spectrometry — M. CAMERON SULLARDS — 32

6. Analyses of Glycosphingolipids by High-Performance Liquid Chromatography — JOHANNES MÜTHING — 45

7. Sphingolipid Extraction and Analysis by Thin-Layer Chromatography — GERHILD VAN ECHTEN-DECKERT — 64

8. Extraction and Analysis of Multiple Sphingolipids from a Single Sample — TAMOTSU TAKETOMI AND EIKO SUGIYAMA — 80

9. Purification of Sphingolipid Classes by Solid-Phase Extraction with Aminopropyl and Weak Cation Exchanger Cartridges — JACQUES BODENNEC, GÉRARD BRICHON, GEORGES ZWINGELSTEIN, AND JACQUES PORTOUKALIAN — 101

10. Ganglioside Analysis by High-Performance Thin-Layer Chromatography — TOSHIO ARIGA AND ROBERT K. YU — 115

11. Purification and Analysis of Gangliosides — STEPHAN LADISCH AND RUIXIANG LI — 135

12. Thin-Layer Chromatography Blotting Using Polyvinylidene Difluoride Membrane (Far-Eastern Blotting) and Its Applications — DAI ISHIKAWA AND TAKAO TAKI — 145

13. Thin-Layer Chromatography Immunostaining — DAI ISHIKAWA AND TAKAO TAKI — 157

14. Monoclonal Anti-Glycosphingolipid Antibodies — REIJI KANNAGI — 160

15. Immunolocalization of Gangliosides by Light Microscopy Using Anti-Ganglioside Antibodies — ANDREAS SCHWARZ AND ANTHONY H. FUTERMAN — 179

16. Cloud-Point Extraction of Gangliosides Using Nonionic Detergent C14EO6 — GEORG C. TERSTAPPEN, ANTHONY H. FUTERMAN, AND ANDREAS SCHWARZ — 187

17. Analyses of Glycosphingolipids Using Clam, *Mercenaria mercenaria,* Ceramide Glycanase — SARA DASTGHEIB, SHIB S. BASU, ZHIXIONG LI, MANJU BASU, AND SUBHASH BASU — 196

18. Quantitative Analyses of Binding Affinity and Specificity for Glycolipid Receptors by Surface Plasmon Resonance — C. ROGER MacKENZIE AND TOMOKO HIRAMA — 205

19. Use of Circular Dichroism for Assigning Stereochemistry of Sphingosine and Other Long-Chain Bases — AKIRA KAWAMURA, KOJI NAKANISHI, AND NINA BEROVA — 217

20. Infrared Determination of Conformational Order and Phase Behavior in Ceramides and Stratum Corneum Models — RICHARD MENDELSOHN AND DAVID J. MOORE — 228

21. Use of Nuclear Magnetic Resonance Spectroscopy   DOMENICO ACQUOTTI AND
    in Evaluation of Ganglioside Structure, Confor-   SANDRO SONNINO   247
    mation, and Dynamics

22. Fluorescence Quenching Assay of Sphingolipid/   ERWIN LONDON,
    Phospholipid Phase Separation in Model Mem-   DEBORAH A. BROWN, AND
    branes   XIAOLIAN XU   272

Section II. Methods for Analyzing Aspects of Sphingolipid Metabolism in Intact Cells

23. Synthesis of Fluorescent Substrates and Their Ap- ARIE DAGAN,
    plication to Study of Sphingolipid Metabolism in VERED AGMON,
    Vitro and in Intact Cells   SHIMON GATT, AND
                                                      TAMA DINUR   293

24. Selection of Mammalian Cell Mutants in Sphingo-   KENTARO HANADA AND
    lipid Biosynthesis   MASAHIRO NISHIJIMA   304

25. Selection of Yeast Mutants in Sphingolipid Metab- TERESA M. DUNN,
    olism   KEN GABLE,
                                                      ERIN MONAGHAN, AND
                                                      DAGMAR BACIKOVA   317

26. Fluorescence-Based Selection of Gene-Corrected   EDWARD H. SCHUCHMAN,
    Hematopoietic Stem and Progenitor Cells Based   SHAI ERLICH,
    on Acid Sphingomyelinase Expression   SILVIA R. P. MIRANDA,
                                                      TAMA DINUR,
                                                      ARIE DAGAN, AND
                                                      SHIMON GATT   330

27. Mammalian Ganglioside Sialidases: Preparation   ERHARD BIEBERICH,
    and Activity Assays   SEAN S. LIOUR, AND
                                                      ROBERT K. YU   339

Section III. Sphingolipid–Protein Interactions and Cellular Targets

28. Effects of Sphingosine and Other Sphingolipids on E. R. SMITH,
    Protein Kinase C   ALFRED H. MERRILL, JR.,
                                                      LINA M. OBEID, AND
                                                      YUSUF A. HANNUN   361

29. Kinetic Analysis of Sphingoid Base Inhibition of WEN-I. WU AND
    Yeast Phosphatidate Phosphatase   GEORGE M. CARMAN   373

30. Assays of Sphingosine-Dependent Kinase for 14-3- TAMAR MEGIDISH,
    3 Protein   AKIKAZU HAMAGUCHI,
                                                      KAZUHISA IWABUCHI, AND
                                                      SEN-ITIROH HAKOMORI   381

31. Synthesis and Use of Caged Sphingolipids   RODERICK H. SCOTT,
JAMIE POLLOCK,
AHMET AYAR,
NICOLA M. THATCHER, AND
URI ZEHAVI   387

32. Binding of Sphingosine 1-Phosphate to Cell Sur-   JAMES R. VAN BROCKLYN AND
face Receptors   SARAH SPIEGEL   401

33. Use of Short-Chain Ceramides   CHIARA LUBERTO AND
YUSUF A. HANNUN   407

34. Analysis of Ceramide-Activated Protein Phospha-   CHARLES E. CHALFANT,
tases   KATSUYA KISHIKAWA,
ALICJA BIELAWSKA, AND
YUSUF A. HANNUN   420

35. Use of Affinity Chromatography and TID-Cer-   STEFAN SCHÜTZE,
amide Photoaffinity Labeling for Detection of   MARC WICKEL,
Ceramide-Binding Proteins   MICHAEL HEINRICH,
SUPANDI WINOTO-MORBACH,
THOMAS WEBER,
JOSEF BRUNNER, AND
MARTIN KRÖNKE   429

36. Lectin-Mediated Cell Adhesion to Immobilized   BRIAN E. COLLINS,
Glycosphingolipids   LYNDA J.-S. YANG, AND
RONALD L. SCHNAAR   438

37. Analysis of Glycolipid-Dependent Cell Adhesion   KAZUKO HANDA,
Based on Carbohydrate–Carbohydrate Inter-   NAOYA KOJIMA, AND
action   SEN-ITIROH HAKOMORI   447

38. Analysis of Interactions between Glycosphingo-   CLIFFORD A. LINGWOOD,
lipids and Microbial Toxins   BETH BOYD, AND
ANITA NUTIKKA   459

39. Oxidation of Aglycone of Glycosphingolipids: Ser-   MURUGESAPILLAI
ine and Ceramide Acid Precursors for Soluble   MYLVAGANAM AND
Glycoconjugates   CLIFFORD A. LINGWOOD   473

40. Separation of Glycosphingolipid-Enriched Micro-   KAZUHISA IWABUCHI,
domains from Caveolar Membrane Character-   KAZUKO HANDA, AND
ized by Presence of Caveolin   SEN-ITIROH HAKOMORI   488

41. Reconstitution of Sphingolipid–Cholesterol Plasma   DJILAI HAMMACHE,
Membrane Microdomains for Studies of Virus–   GÉRARD PIÉRONI,
Glycolipid Interactions   MARC MARESCA,
SERGE IVALDI,
NOUARA YAHI, AND
JACQUES FANTINI   495

42. Analysis of Ceramides Present in Glycosylphospha-   ISABELLE GUILLAS,
tidylinositol Anchored Proteins of Saccharo-   MARTINE PFEFFERLI, AND
myces cerevisiae   ANDREAS CONZELMANN   506

43. Preparation of Functionalized Lipid Tubules for Electron Crystallography of Macromolecules — ELIZABETH M. WILSON-KUBALEK — 515

Section IV. Sphingolipid Transport and Trafficking

44. Applications of BIODIPY–Sphingolipid Analogs to Study Lipid Traffic and Metabolism in Cells — RICHARD E. PAGANO, RIKIO WATANABE, CHRISTINE WHEATLEY, AND MICHEL DOMINGUEZ — 523

45. Using Biotinylated Gangliosides to Study Their Distribution and Traffic in Cells by Immunoelectron Microscopy — GÜNTER SCHWARZMANN, ALEXANDER VON COBURG, AND WIEBKE MÖBIUS — 534

46. Assays for Transmembrane Movement of Sphingolipids — DAN J. SILLENCE, RENÉ J. RAGGERS, AND GERRIT VAN MEER — 562

Section V. Other Methods

47. Compilation of Methods Published in Previous Volumes of *Methods in Enzymology* — ALFRED H. MERRILL, JR. — 583

VOLUME 313
ANTISENSE TECHNOLOGY (PART A: GENERAL METHODS, METHODS OF DELIVERY, AND RNA STUDIES)
M. Ian Phillips

Section I. General Methods

1. Progress in Antisense Technology: The End of the Beginning — STANLEY T. CROOKE — 3

2. Basic Principles of Using Antisense Oligonucleotides *in Vivo* — M. IAN PHILLIPS AND Y. CLARE ZHANG — 46

3. Polyethyleneimine-Mediated Transfection to Improve Antisense Activity of 3′-Capped Phosphodiester Oligonucleotides — SONIA DHEUR AND E. TULA SAISON-BEHMOARAS — 56

4. Design of Antisense and Triplex-Forming Oligonucleotides — JEAN-CHRISTOPHE FRANÇOIS, JÉROME LACOSTE, LAURENT LACROIX, AND JEAN-LOUIS MERGNY — 74

5. Chimeric Oligodeoxynucleotide Analogs: Chemical Synthesis, Purification, and Molecular and Cellular Biology Protocols — RICHARD V. GILES, DAVID G. SPILLER, AND DAVID M. TIDD — 95

6. Evaluation of Antisense Mechanisms of Action — CHANDRAMALLIKA GHOSH, DAVID STEIN, DWIGHT WELLER, AND PATRICK IVERSEN — 135

7. Physiology and Molecular Biology Brought to Single-Cell Level — PIERRE-MARIE LLEDO, DAVID DESMAISONS, ALAN CARLETON, AND JEAN-DIDIER VINCENT — 143

8. Antisense Properties of Peptide Nucleic Acid — PETER E. NIELSEN — 156

9. Synthesis of Oligonucleotide Conjugates in Anhydrous Dimethyl Sulfoxide — DAVID MILESI, IGOR KUTYAVIN, EUGENE A. LUKHTANOV, VLADIMIR V. GORN, AND MICHAEL W. REED — 164

10. Gene Switching: Analyzing Broad Range of Mutations Using Steric Block Antisense Oligonucleotides — PAUL A. MORCOS — 174

11. Vesicular Stomatitis Virus as Model System for Studies of Antisense Oligonucleotide Translation Arrest — IAN ROBBINS AND BERNARD LEBLEU — 189

12. Purification of Antisense Oligonucleotides — RANJIT R. DESHMUKH, DOUGLAS L. COLE, AND YOGESH S. SANGHVI — 203

13. Boranophosphate Backbone: A Mimic of Phosphodiesters, Phosphorothioates, and Methyl Phosphonates — BARBARA RAMSAY SHAW, DMITRI SERGUEEV, KAIZHANG HE, KEN PORTER, JACK SUMMERS, ZINAIDA SERGUEEVA, AND VLADIMIR RAIT — 226

14. Intracellular Distribution of Digoxigenin-Labeled Phosphorothiate Oligonucleotides — GEMMA TARRASÓN, DAVID BELLIDO, RAMON ERITJA, SENÉN VILARÓ, AND JAUME PIULATS — 257

15. Use of Minimally Modified Antisense Oligonucleotides for Specific Inhibition of Gene Expression — EUGEN UHLMANN, ANUSCH PEYMAN, ANTONINA RYTE, ANNETTE SCHMIDT, AND ECKHART BUDDECKE — 268

Section II. Methods of Delivery

16. Determination of Cellular Internalization of Fluo- LYUBA BENIMETSKAYA,
    rescent Oligonucleotides   JOHN TONKINSON, AND
                                                     C. A. STEIN   287

17. Intrabody Tissue-Specific Delivery of Antisense ROBERT J. DUFF,
    Conjugates in Animals: Ligand–Linker–Anti-   SCOTT F. DEAMOND,
    sense Oligomer Conjugates   CLINTON ROBY,
                                                    YUANZHONG ZHOU, AND
                                                    PAUL O. P. TS'O   297

18. Lipid-Based Formulations of Antisense Oligonucle- SEAN C. SEMPLE,
    otides for Systemic Delivery Applications   SANDRA K. KLIMUK,
                                                     TROY O. HARASYM, AND
                                                     MICHAEL J. HOPE   322

19. In Vitro Transport and Delivery of Antisense Oligo- JEFF HUGHES,
    nucleotides   ANNA ASTRIAB,
                                                      HOON YOO,
                                                      SURESH ALAHARI,
                                                      EARVIN LIANG,
                                                      DMITRI SERGUEEV,
                                                      BARBARA RAMSAY SHAW, AND
                                                      R. L. JULIANO   342

20. In Vitro and in Vivo Delivery of Antisense Oligo- AKIRA MATSUNO,
    deoxynucleotides Using Lipofection: Application TADASHI NAGASHIMA,
    of Antisense Technique to Growth Suppression   HARUKO KATAYAMA, AND
    of Experimental Glioma   AKIRA TAMURA   359

21. Preparation and Application of Liposome-Incorpo- ANA M. TARI   372
    rated Oligodeoxynucleotides

22. Cell-Specific Optimization of Phosphorothioate SHIRLEY A. WILLIAMS AND
    Antisense Oligodeoxynucleotide Delivery by JEFFREY S. BUZBY   388
    Cationic Lipids

Section III. RNA Studies

23. Intracellular Expression and Function of Antisense DANIELA CASTANOTTO,
    Catalytic RNAs   MICHAELA SCHERR, AND
                                                      JOHN J. ROSSI   401

24. Selective Degradation of Targeted mRNAs Using JOHN M. DAGLE AND
    Partially Modified Oligonucleotides   DANIEL L. WEEKS   420

25. Reversible Depletion of Specific RNAs by Anti- GEORGE L. ELICEIRI   436
    sense Oligodeoxynucleotide-Targeted Degrada-
    tion in Frog Oocytes

26. Inactivation of Gene Expression Using Ribo- CECILIA GUERRIER-TAKADA
    nuclease P and External Guide Sequences   AND SIDNEY ALTMAN   442

27. Disruption of mRNA-RNP Formation and Sorting   EMMA R. JAKOI AND
to Dendritic Synapses by Antisense Oligonucleo-   W. L. SEVERT   456
tides

28. Antisense RNA and DNA in *Escherichia coli*   OLEG MIROCHNITCHENKO AND
MASAYORI INOUYE   467

29. Utilization of Properties of Natural Catalytic RNA   LEONIDAS A. PHYLACTOU,
to Design and Synthesize Functional Ribozymes   CHARLOTTE DARRAH,
LOUISE EVERATT,
DESPINA MANIOTIS, AND
MICHAEL W. KILPATRICK   485

30. Antisense Oligonucleotides and RNAs as Modula-   HALINA SIERAKOWSKA,
tors of pre-mRNA Splicing   LINDA GORMAN,
SHIN-HONG KANG, AND
RYSZARD KOLE   506

31. Selective RNA Cleavage by Isolated RNase L Acti-   ROBERT H. SILVERMAN,
vated with 2–5A Antisense Chimeric Oligonucle-   BEIHUA DONG,
otides   RATAN K. MAITRA,
MARK R. PLAYER, AND
PAUL F. TORRENCE   522

VOLUME 314
ANTISENSE TECHNOLOGY (PART B: APPLICATIONS)
M. Ian Phillips

Section I. Antisense Receptor Targets

1. *In Vivo* Modulation of G Proteins and Opioid Re-   JAVIER GARZÓN,
ceptor Function by Antisense Oligodeoxynucleo-   ISABEL DE ANTONIO, AND
tides   PILAR SÁNCHEZ-BLÁZQUEZ   3

2. Targeting Brain GABA$_A$ Receptors with Antisense   JESPER KARLE AND
Oligonucleotides: Implications for Epilepsy   MOGENS NIELSEN   20

3. Delivery of Antisense DNA by Vectors for Pro-   DAGMARA MOHUCZY,
longed Effects *in Vitro* and *in Vivo*   XIAOPING TANG, AND
M. IAN PHILLIPS   32

4. Antisense Mapping: Assessing Functional Signifi-   GAVRIL W. PASTERNAK AND
cance of Genes and Splice Variants   YING-XIAN PAN   51

5. *In Vitro* and *in Vivo* Effects of Antisense on   EMMA S. J. ROBINSON AND
α_2-Adrenoceptor Expression   ALAN L. HUDSON   61

6. Design and Efficacy of a Serotonin-2A Receptor   JOANNE M. SCALZITTI AND
   Antisense Oligodeoxynucleotide   JULIE G. HENSLER   76

7. Reduction of Neurotransmitter Receptor and G-   KELLY M. STANDIFER   90
   Protein Expression *in Vivo* and *in Vitro* by Anti-
   sense Oligodeoxynucleotide Treatment

8. Use of Expression of Antisense mRNA for Con-   DAESETY VISHNUVARDHAN
   vertases 1 and 2 in Prohormone Processing   AND MARGERY C. BEINFELD   103

Section II. Antisense Neuroscience Targets

9. Strategies for the Design and Delivery of Antisense WILLIAM C. BROADDUS,
   Oligonucleotides in Central Nervous System   SUJIT S. PRABHU,
                                                        SUSANNA WU-PONG,
                                                        GEORGE T. GILLIES, AND
                                                        HELEN L. FILLMORE   121

10. Use of Antisense Expression Plasmids to Attenuate   NOEL J. BUCKLEY,
    G-Protein Expression in Primary Neurons   FE C. ABOGADIE,
                                                        DAVID A. BROWN,
                                                        MARIZA DAYRELL,
                                                        MALCOLM P. CAULFIELD,
                                                        PATRICK DELMAS, AND
                                                        JANE E. HALEY   136

11. Gene Functional Analysis in Nervous System   FINN HALLBÖÖK,
                                                        MARINA CATSICAS,
                                                        JULIE K. STAPLE,
                                                        GABRIELE GRENNINGLOH, AND
                                                        STEFAN CATSICAS   148

12. RNA Mapping: Selection of Potent Oligonucleotide   SIEW PENG HO,
    Sequences for Antisense Experiments   DUSTIN H. O. BRITTON,
                                                        YIJIA BAO, AND
                                                        MICHAEL S. SCULLY   168

13. Effects of Centrally Administered Antisense Oligo-   PUSHPA S. KALRA,
    deoxynucleotides on Feeding Behavior and Hor-   MICHAEL G. DUBE, AND
    mone Secretion   SATYA P. KALRA   184

14. Blockade of Neuropathic Pain by Antisense Tar-   JOSEPHINE LAI,
    geting of Tetrodotoxin-Resistant Sodium Chan-   JOHN C. HUNTER,
    nels in Sensory Neurons   MICHAEL H. OSSIPOV, AND
                                                        FRANK PORRECA   201

15. Antisense Approach for Study of Cell Adhesion   RADMILA MILEUSNIC   213
    Molecules in Central Nervous System

16. Sequence Design and Practical Implementation of   INGA D. NEUMANN AND
    Antisense Oligonucleotides in Neuroendocri-   NICOLA TOSCHI   223
    nology

17. Localization of Oligonucleotides in Brain by *in Situ* Hybridization — KATSUMI YUFU, YUTAKA YAIDA, AND THADDEUS S. NOWAK, JR.   238

18. Use of Antisense Oligonucleotides in Human Neuronal and Astrocytic Cultures — VALERIA SOGOS, MONICA CURTO, MARIELLA SETZU, ISABELLA MUSSINI, MARIA GRAZIA ENNAS, AND FULVIA GREMO   247

19. Pharmacokinetic Properties of Oligonucleotides in Brain — WOLFGANG SOMMER, MATTHEW O. HEBB, AND MARKUS HEILIG   261

20. Antisense Oligonucleotides: Preparation and *in Vivo* Application to Rat Brain — WOLFGANG TISCHMEYER   275

21. Application of Antisense Techniques to Characterize Neuronal Ion Channels *in Vitro* — STEPHEN G. VOLSEN, REGIS C. LAMBERT, YVES MAULET, MICHEL dE WAARD, SAMANTHA GILLARD, PETER J. CRAIG, RUTH BEATTIE, AND ANNE FELTZ   290

Section III. Antisense in Nonneuronal Tissues

22. Antisense Inhibition of Sodium–Calcium Exchanger — MARTIN K. SLODZINSKI AND MAGDALENA JUHASZOVA   313

23. Targeted Delivery of Antisense Oligonucleotides to Parenchymal Liver Cell *in Vivo* — E. A. L. BIESSEN, H. VIETSCH, E. T. RUMP, K. FLUITER, M. K. BIJSTERBOSCH, AND T. J. C. VAN BERKEL   324

24. Antisense Methods for Discrimination of Phenotypic Properties of Closely Related Gene Products: Jun Kinase Family — FRÉDÉRIC BOST, ROBERT MCKAY, NICHOLAS M. DEAN, O. POTAPOVA, AND DAN MERCOLA   342

25. Evaluation of Biological Role of c-Jun N-Terminal Kinase Using an Antisense Approach — CATHERINE L. CIOFFI AND BRETT P. MONIA   363

26. Role of Antisense in Kidney Cells — MICHAEL B. GANZ   378

27. Use of Antisense Techniques in Rat Renal Medulla — DAVID L. MATTSON   389

28. Antisense Approaches to *in Vitro* Organ Culture   TAKAHIRO OCHIYA AND MASAAKI TERADA   401

29. *In Vivo* and *in Vitro* Antiproliferative Effects of Antisense Interleukin 10 Oligonucleotides   GEORGE A. PARKER, BIHAI PENG, MAI HE, SUSAN GOULD-FOGERITE, CHUAN-CHU CHOU, AND ELIZABETH S. RAVECHÉ   411

30. Inhibition of *c*-ABL Expression in Hematopoietic Progenitor Cells Using Antisense Oligodeoxynucleotides   VITTORIO ROSTI, CLAUDIA LUCOTTI, MARIO CAZZOLA, AND GAETANO BERGAMASCHI   429

31. Cellular Pharmacology of Antisense Oligodeoxynucleotides   CHRISTOPH SCHUMACHER   440

32. Application of Antisense Oligodeoxynucleotides for Suppression of Na^+/Ca^{2+} Exchange   BEAT SCHWALLER, MARCEL EGGER, PETER LIPP, AND ERNST NIGGLI   454

33. Optimizing Efficacy of Antisense Oligodeoxynucleotides Targeting Inhibitors of Apoptosis   ANNEMARIE ZIEGLER, A. PAULA SIMÕES-WUEST, AND UWE ZANGEMEISTER-WITTKE   477

Section IV. Antisense in Therapy

34. *In Vitro* and *in Vivo* Modulation of Transforming Growth Factor β_1 Gene Expression by Antisense Oligomer   HUN-TAEG CHUNG, DONG-HWAN SOHN, BYUNG-MIN CHOI, JI-CHANG YOO, HYUN-OCK PAE, AND CHANG-DUK JUN   493

35. Analysis of Cancer Gene Functions through Gene Inhibition with Antisense Oligonucleotides   CHERYL ROBINSON-BENION, ROY A. JENSEN, AND JEFFREY T. HOLT   499

36. Dosimetry and Optimization of *in Vivo* Targeting with Radiolabeled Antisense Oligodeoxynucleotides: Oligonucleotide Radiotherapy   KALEVI J. A. KAIREMO, ANTTI P. JEKUNEN, AND MIKKO TENHUNEN   506

37. Antisense Oligonucleotide Therapy of Hepadnavirus Infection   WOLF-BERNHARD OFFENSPERGER, C. THOMA, D. MORADPOUR, F. VON WEIZSÄCKER, S. OFFENSPERGER, AND H. E. BLUM   524

38. Preclinical Antisense DNA Therapy of Cancer in Mice   JANET B. SMITH AND ERIC WICKSTROM   537

39. Retrovirally Mediated Delivery of Angiotensin II   HONGWEI WANG,
    Type 1 Receptor Antisense *in Vitro* and *in Vivo*   DI LU,
                                                    PHYLLIS Y. REAVES,
                                                    MICHAEL J. KATOVICH, AND
                                                    MOHAN K. RAIZADA   581

VOLUME 315
VERTEBRATE PHOTOTRANSDUCTION AND THE VISUAL CYCLE (PART A)
Krzysztof Palczewski

Section I. Expression and Isolation of Opsins

1. Heterologous Expression of Bovine Opsin in   NAJMOUTIN G. ABDULAEV AND
   Pichia pastoris   KEVIN D. RIDGE   3

2. Baculovirus Expression System for Expression and   CORNÉ H. W. KLAASSEN AND
   Characterization of Functional Recombinant Vi-   WILLEM J. DEGRIP   12
   sual Pigments

3. Development of Stable Cell Lines Expressing High   JACK M. SULLIVAN AND
   Levels of Point Mutants of Human Opsin for Bio-   MICHAEL F. SATCHWELL   30
   chemical and Biophysical Studies

4. Folding and Assembly of Rhodopsin from Ex-   KEVIN D. RIDGE AND
   pressed Fragments   NAJMOUTIN G. ABDULAEV   59

5. Isolation of Isoelectric Species of Phosphorylated   J. HUGH MCDOWELL,
   Rhodopsin   JOSEPH P. NAWROCKI, AND
                                                PAUL A. HARGRAVE   70

6. Rhodopsin Trafficking in Photoreceptors Using   DUSANKA DERETIC   77
   Retinal Cell-Free System

Section II. Characterization of Opsins

7. Preparation and Analysis of Two-Dimensional   GEBHARD F. X. SCHERTLER
   Crystals of Rhodopsin   AND PAUL A. HARGRAVE   91

8. Domain Approach to Three-Dimensional Structure   ARLENE D. ALBERT AND
   of Rhodopsin Using High Resolution Nuclear   PHILIP L. YEAGLE   107
   Magnetic Resonance

9. Analysis of Functional Microdomains in Rhodopsin   STEVEN W. LIN, MAY HAN,
                                                AND THOMAS P. SAKMAR   116

10. Mapping Tertiary Contacts between Amino Acid   MARY STRUTHERS AND
    Residues within Rhodopsin   DANIEL D. OPRIAN   130

11. Energetics of Rhodopsin Photobleaching: Photoca- ROBERT R. BIRGE AND
    lorimetric Studies of Energy Storage in Early and BRYAN W. VOUGHT   143
    Late Intermediates

12. Absorption Spectroscopy in Studies of Visual Pig- JAMES W. LEWIS AND
    ments: Spectral and Kinetic Characterization of DAVID S. KLIGER   164
    Intermediates

13. Structural Determinants of Active State Conforma- KARIM FAHMY,
    tion of Rhodopsin: Molecular Biophysics Ap-   THOMAS P. SAKMAR, AND
    proaches   FRIEDRICH SIEBERT   178

14. pK_a of Protonated Schiff Base of Visual Pigments THOMAS G. EBREY   196

15. Assays for Detection of Constitutively Active   PHYLLIS R. ROBINSON   207
    Opsins

16. Synthetic Retinals: Convenient Probes of Rhodop- JIHONG LOU, QIANG TAN,
    sin and Visual Transduction Process   ELENA KARNAUKHOVA,
                                                    NINA BEROVA,
                                                    KOJI NAKANISHI, AND
                                                    ROSALIE K. CROUCH   219

17. Assays for Activation of Opsin by all-*trans*-Retinal KRISTINA SACHS,
                                                    DIETER MARETZKI, AND
                                                    KLAUS PETER HOFMANN   238

18. Assays for Activation of Recombinant Expressed   MAY HAN AND
    Opsin by all-*trans*-Retinal   THOMAS P. SAKMAR   251

19. Electrical Approach to Study Rhodopsin Activation JACK M. SULLIVAN,
    in Single Cells with Early Receptor Current Assay LIOUBOV BRUEGGEMANN, AND
                                                    PRAGATI SHUKLA   268

20. Analysis of Amino Acid Residues in Rhodopsin   HIROO IMAI,
    and Cone Visual Pigments That Determine Their   AKIHISA TERAKITA, AND
    Molecular Properties   YOSHINORI SHICHIDA   293

21. Phylogenetic Analysis and Experimental Ap-   SHOZO YOKOYAMA   312
    proaches to Study Color Vision in Vertebrates

Section III. Proteins That Interact with Rhodopsin

22. Light Scattering Methods to Monitor Interaction MARTIN HECK,
    between Rhodopsin-Containing Membranes and   ALEXANDER PULVERMÜLLER,
    Soluble Proteins   AND KLAUS PETER HOFMANN   329

23. Heterogeneity of Rhodopsin Intermediate State In- YOSHINORI SHICHIDA,
    teracting with Transducin   SHUJI TACHIBANAKI,
                                                    HIROO IMAI, AND
                                                    AKIHISA TERAKITA   347

24. Limited Proteolytic Digestion Studies of G Protein–   MARIA R. MAZZONI AND
     Receptor Interaction   HEIDI E. HAMM   363

25. Monitoring Proton Uptake from Aqueous Phase   CHRISTOPH K. MEYER AND
     during Rhodopsin Activation   KLAUS PETER HOFMANN   377

26. Use of Peptides-on-Plasmids Combinatorial Library   ANNETTE GILCHRIST, ANLI LI,
     to Identify High Affinity Peptides That Bind Rho-   AND HEIDI E. HAMM   388
     dopsin

27. Purification of Rhodopsin Kinase by Recoverin Af-   CHING-KANG CHEN AND
     finity Chromatography   JAMES B. HURLEY   404

28. Heterologous Expression and Reconstitution of   SHOJI OSAWA,
     Rhodopsin with Rhodopsin Kinase and Arrestin   DAYANIDHI RAMAN, AND
                                                            ELLEN R. WEISS   411

29. Arrestin: Mutagenesis, Expression, Purification,   VSEVOLOD V. GUREVICH AND
     and Functional Characterization   JEFFREY L. BENOVIC   422

30. Mapping Interaction Sites between Rhodopsin and   W. CLAY SMITH AND
     Arrestin by Phage Display and Synthetic Peptides   PAUL A. HARGRAVE   437

31. Characterization of RanBP2-Associated Molecular   PAULO A. FERREIRA   455
     Components in Neuroretina

Section IV. Transducin and Regulators of G-Protein Signaling

32. Intrinsic Biophysical Monitors of Transducin Acti-   OLIVER P. ERNST,
     vation: Fluorescence, UV/Visible Spectroscopy,   CHRISTOPH BIERI,
     Light Scattering, and Evanescent Field Tech-   HORST VOGEL, AND
     niques   KLAUS PETER HOFMANN   471

33. Fluorescent Probes as Indicators of Conformation   CHII-SHEN YANG,
     Changes in Transducin on Activation   NIKOLAI P. SKIBA,
                                                            MARIA R. MAZZONI,
                                                            TARITA O. THOMAS, AND
                                                            HEIDI E. HAMM   490

34. $G\alpha_t/G\alpha_{i1}$ Chimeras Used to Define Structural Basis   NIKOLAI P. SKIBA,
     of Specific Functions of $G\alpha_t$ TARITA O. THOMAS, AND
                                                            HEIDI E. HAMM   502

35. Enzymology of GTPase Acceleration in Phototrans-   CHRISTOPHER W. COWAN,
     duction   THEODORE G. WENSEL, AND
                                                            VADIM Y. ARSHAVSKY   524

36. Mutational Analysis of Effectors and Regulators of   MICHAEL NATOCHIN AND
     G-Protein Signaling Interfaces of Transducin   NIKOLAI O. ARTEMYEV   539

Section V. Photoreceptor Protein Phosphatases

37. Isolation and Properties of Protein Phosphatase 2A   MUHAMMAD AKHTAR,
     in Photoreceptors   ALASTAIR J. KING, AND
                                                            NINA E. M. MCCARTHY   557

38. Purification and Characterization of Protein Phosphate Type 2C in Photoreceptors — SUSANNE KLUMPP AND DAGMAR SELKE — 570

39. Photoreceptor Serine/Threonine Protein Phosphatase Type 7: Cloning, Expression, and Functional Analysis — XIZHONG HUANG, MARK R. SWINGLE, AND RICHARD E. HONKANEN — 579

Section VI. Photoreceptor Phosphodiesterase and Guanylyl Cyclase

40. Purification and Assay of Bovine Type 6 Photoreceptor Phosphodiesterase and Its Subunits — TERRY A. COOK AND JOSEPH A. BEAVO — 597

41. Transcriptional Regulation of cGMP Phosphodiesterase β-Subunit Gene — LEONID E. LERNER AND DEBORA B. FARBER — 617

42. Inhibition of Photoreceptor cGMP Phosphodiesterase by Its γ Subunit — ALEXEY E. GRANOVSKY, KHAKIM G. MURADOV, AND NIKOLAI O. ARTEMYEV — 635

43. Kinetics and Regulation of cGMP Binding to Noncatalytic Binding Sites on Photoreceptor Phosphodiesterase — RICK H. COTE — 646

44. Purification and Autophosphorylation of Retinal Guanylate Cyclase — JEFFREY P. JOHNSTON, JENNIFER G. APARICIO, AND MEREDITHE L. APPLEBURY — 673

45. Use of Nucleotide α-Phosphorothioates in Studies of Photoreceptor Guanylyl Cyclase: Purification of Guanylyl Cyclase Activating Proteins — WOJCIECH A. GORCZYCA — 689

46. Heterologous Expression and Assays for Photoreceptor Guanylyl Cyclases and Guanylyl Cyclase Activating Proteins — JAMES B. HURLEY AND ALEXANDER M. DIZHOOR — 708

47. Spectrophotometric Determination of Retinal Rod Guanylyl Cyclase — GREGOR WOLBRING AND PAUL P. M. SCHNETKAMP — 718

48. Calcium-Dependent Activation of Membrane Guanylate Cyclase by S100 Proteins — ARI SITARAMAYYA, NIKOLAY POZDNYAKOV, ALEXANDER MARGULIS, AND AKIKO YOSHIDA — 730

49. Characterization of Guanylyl Cyclase and Phosphodiesterase Activities in Single Rod Outer Segments — YIANNIS KOUTALOS AND KING-WAI YAU — 742

Section VII. Cyclic Nucleotide-Gated Channels

50. Covalent Tethering of Ligands to Retinal Rod Cyclic Nucleotide-Gated Channels: Binding Site Structure and Allosteric Mechanism — JEFFREY W. KARPEN, MARIALUISA RUIZ, AND R. LANE BROWN — 755

51. Using State-Specific Modifiers to Study Rod cGMP- SHARONA E. GORDON   772
    Activated Ion Channels Expressed in *Xenopus*
    Oocytes

52. Identification and Characterization of Calmodulin KARL-WILHELM KOCH   785
    Binding Sites in cGMP-Gated Channel Using Sur-
    face Plasmon Resonance Spectroscopy

53. Determination of Fractional Calcium Ion Current STEPHAN FRINGS,
    in Cyclic Nucleotide-Gated Channels   DAVID H. HACKOS,
                                                    CLAUDIA DZEJA,
                                                    TSUYOSHI OHYAMA,
                                                    VOLKER HAGEN,
                                                    U. BENJAMIN KAUPP, AND
                                                    JUAN I. KORENBROT   797

54. Modulation of Rod cGMP-Gated Cation Channel MARIA E. GRUNWALD AND
    by Calmodulin   KING-WAI YAU   817

Section VIII. Na$^+$/Ca^{2+}-K$^+$ Exchanger and ABCR Transporter

55. Purification and Biochemical Analysis of cGMP- ROBERT S. MOLDAY,
    Gated Channel and Na$^+$/Ca^{2+}-K$^+$ Exchanger of RENÉ WARREN, AND
    Rod Photoreceptors   TOM S. Y. KIM   831

56. Spectrofluorometric Detection of Na$^+$/Ca^{2+}-K$^+$ Ex- CONAN B. COOPER,
    change   ROBERT T. SZERENCSEI, AND
                                                    PAUL P. M. SCHNETKAMP   847

57. Purification and Characterization of ABCR from JINHI AHN AND
    Bovine Rod Outer Segments   ROBERT S. MOLDAY   864

58. ABCR: Rod Photoreceptor-Specific ABC Trans- HUI SUN AND
    porter Responsible for Stargardt Disease   JEREMY NATHANS   879

VOLUME 316
VERTEBRATE PHOTOTRANSDUCTION AND THE VISUAL CYCLE
(PART B)
Krzysztof Palczewski

Section I. Photoreceptor Proteins

1. Inhibition of Rhodopsin Phosphorylation by S- SATORU KAWAMURA   3
   Modulins: Purification, Reconstitution, and
   Assays

2. Analysis of Protein–Protein Interactions in Photo-transduction Cascade Using Surface Plasmon Resonance — DAULET K. SATPAEV AND VLADLEN Z. SLEPAK — 20

3. Expression of Phototransduction Proteins in *Xenopus* Oocytes — Barry E. KNOX, ROBERT B. BARLOW, DEBRA A. THOMPSON, RICHARD SWANSON, AND ENRICO NASI — 41

4. *Xenopus* Rod Photoreceptor: Model for Expression of Retinal Genes — SUCHITRA BATNI, SHOBANA S. MANI, CHARISSE SCHLUETER, MING JI, AND BARRY E. KNOX — 50

5. Fusion between Retinal Rod Outer Segment Membranes and Model Membranes: Functional Assays and Role for Peripherin/rds — KATHLEEN BOESZE-BATTAGLIA — 65

6. Bovine Retinal Nucleoside Diphosphate Kinase: Biochemistry and Molecular Cloning — NAJMOUTIN G. ABDULAEV, DMITRI L. KAKUEV, AND KEVIN D. RIDGE — 87

Section II. Calcium-Binding Proteins and Calcium Measurements in Photoreceptor Cells

7. Calcium-Binding Proteins and Their Assessment in Ocular Diseases — ARTHUR S. POLANS, RICARDO L. GEE, TERESA M. WALKER, AND PAUL R. VAN GINKEL — 103

8. Molecular Structure of Membrane-Targeting Calcium Sensors in Vision: Recoverin and Guanylate Cyclase-Activating Protein 2 — JAMES B. AMES, MITSUHIKO IKURA, AND LUBERT STRYER — 121

9. Measurement of Light-Evoked Changes in Cytoplasmic Calcium in Functionally Intact Isolated Rod Outer Segments — PETER B. DETWILER AND MARK P. GRAY-KELLER — 133

10. Laser Spot Confocal Technique to Measure Cytoplasmic Calcium Concentration in Photoreceptors — HUGH R. MATTHEWS AND GORDON L. FAIN — 146

Section III. Phototransduction

11. Functional Study of Rhodopsin Phosphorylation *in Vivo* — ANA MENDEZ, NATALIJA V. KRASNOPEROVA, JANIS LEM, AND JEANNIE CHEN — 167

12. Mechanisms of Single-Photon Detection in Rod Photoreceptors — FRED RIEKE — 186

13. Electroretinographic Determination of Human Rod   DAVID R. PEPPERBERG,
Flash Response *in Vivo*   DAVID G. BIRCH, AND
    DONALD C. HOOD   202

14. Electrophysiological Methods for Measurement of   M. CARTER CORNWALL,
Activation of Phototransduction by Bleached Vi-   G. J. JONES,
sual Pigment in Salamander Photoreceptors   V. J. KEFALOV,
    G. L. FAIN, AND
    H. R. MATTHEWS   224

15. Exploring Kinetics of Visual Transduction with   T. MINH VUONG   253
Time-Resolved Microcalorimetry

16. Use of α-Toxin-Permeabilized Photoreceptors in *in*   ANNIE E. OTTO-BRUC,
Vitro Phototransduction Studies   J. PRESTON VAN HOOSER, AND
    ROBERT N. FARISS   269

17. Photoreceptors in Pineal Gland and Brain: Cloning,   TOSHIYUKI OKANO AND
Localization, and Overexpression   YOSHITAKA FUKADA   278

18. Cultured Amphibian Melanophores: A Model Sys-   MARK D. ROLLAG,
tem to Study Melanopsin Photobiology   IGNACIO PROVENCIO,
    DAVID SUGDEN, AND
    CARLA B. GREEN   291

Section IV. Enzymes of the Visual Cycle

19. High-Performance Liquid Chromatography Analy-   GREGORY G. GARWIN AND
sis of Visual Cycle Retinoids   JOHN C. SAARI   313

20. Quantitative Measurements of Isomerohydrolase   ANETTE WINSTON AND
Activity   ROBERT R. RANDO   324

21. Multienzyme Analysis of Visual Cycle   HARTMUT STECHER AND
    KRZYSZTOF PALCZEWSKI   330

22. Analyzing Membrane Topology of 11-*cis*-Retinol   ANDRAS SIMON,
Dehydrogenase   ANNA ROMERT, AND
    ULF ERIKSSON   344

23. Phase Partition and High-Performance Liquid   JOHN C. SAARI,
Chromatography Assays of Retinoid Dehydroge-   GREGORY G. GARWIN,
nases   FRANÇOISE HAESELEER,
    GEENG-FU JANG, AND
    KRZYSZTOF PALCZEWSKI   359

24. Short-Chain Dehydrogenases/Reductases in Retina   FRANÇOISE HAESELEER AND
    KRZYSZTOF PALCZEWSKI   372

25. Substrate Specificities of Retinyl Ester Hydrolases   ANDREW T. C. TSIN,
in Retinal Pigment Epithelium   NATHAN L. MATA,
    JENNIFER A. RAY, AND
    ELIA T. VILLAZANA   384

26. Molecular Characterization of Lecithin-Retinol   ALBERTO RUIZ AND
    Acyltransferase   DEAN BOK   400

27. Analysis of Chromophore of RGR: Retinal G-   WENSHAN HAO,
    Protein–Coupled Receptor from Pigment Epi-   PU CHEN, AND
    thelium   HENRY K. W. FONG   413

Section V. Posttranslational and Chemical Modifications

28. Low Temperature Photoaffinity Labeling of Rho-   MARIA L. SOUTO,
    dopsin and Intermediates along Transduction   BABAK BORHAN, AND
    Path   KOJI NAKANISHI   423

29. Structural Analysis of Protein Prenyl Groups and   MARK E. WHITTEN,
    Associated C-Terminal Modifications   KOHEI YOKOYAMA,
                                                        DAVID SCHIELTZ,
                                                        FARIDEH GHOMASHCHI,
                                                        DEREK LAM,
                                                        JOHN R. YATES III,
                                                        KRZYSZTOF PALCZEWSKI, AND
                                                        MICHAEL H. GELB   436

30. Isoprenylation/Methylation and Transducin Func-   CRAIG A. PARISH AND
    tion   ROBERT R. RANDO   451

31. Functional Analysis of Farnesylation and Methyla- TAKAHIKO MATSUDA AND
    tion of Transducin   YOSHITAKA FUKADA   465

32. Identification of Phosphorylation Sites within Ver- HIROSHI OHGURO   482
    tebrate and Invertebrate Rhodopsin

33. Ocular Proteomics: Cataloging Photoreceptor Pro-   HIROYUKI MATSUMOTO AND
    teins by Two-Dimensional Gel Electrophoresis   NAOKA KOMORI   492
    and Mass Spectrometry

Section VI. Analysis of Animal Models of Retinal Diseases

34. Animal Models for Inherited Retinal Diseases   JEANNE FREDERICK,
                                                        J. DARIN BRONSON, AND
                                                        WOLFGANG BAEHR   515

35. Avian Models of Inherited Retinal Disease   SUSAN L. SEMPLE-ROWLAND
                                                        AND NANCY R. LEE   526

36. Biochemical Analysis of Phototransduction and Vi- MICHAEL R. TAYLOR,
    sual Cycle in Zebrafish Larvae   HEATHER A. VAN EPPS,
                                                        MATTHEW J. KENNEDY,
                                                        JOHN C. SAARI,
                                                        JAMES B. HURLEY, AND
                                                        SUSAN E. BROCKERHOFF   536

37. Genetic Models to Study Guanylyl Cyclase   SUSAN W. ROBINSON AND
     Function   DAVID L. GARBERS   558

38. Analysis of Visual Cycle in Normal and   J. PRESTON VAN HOOSER,
     Transgenic Mice   GREGORY G. GARWIN, AND
     JOHN C. SAARI   565

Section VII. Inherited Retinal Disease: From the Defective Gene to Its Function and Repair

39. *In Situ* Hybridization Studies of Retinal Neurons   LINDA K. BARTHEL AND
     PAMELA A. RAYMOND   579

40. Cloning and Characterization of Retinal Transcrip-   SHIMING CHEN AND
     tion Factors, Using Target Site–Based Method-   DONALD J. ZACK   590
     ology

41. *In Vivo* Assessment of Photoreceptor Function in   ARTUR V. CIDECIYAN   611
     Human Diseases Caused by Photoreceptor-Spe-
     cific Gene Mutations

42. Spectral Sensitivities of Human Cone Visual Pig-   ANDREW STOCKMAN,
     ments Determined *in Vivo* and *in Vitro*   LINDSAY T. SHARPE,
     SHANNATH MERBS, AND
     JEREMY NATHANS   626

43. Molecular Analysis of Human Red/Green Visual   SAMIR S. DEEB,
     Pigment Gene Locus: Relationship to Color   TAKAAKI HAYASHI,
     Vision   JORIS WINDERICKX, AND
     TOMOHIKO YAMAGUCHI   651

44. Expression and Characterization of Peripherin/rds-   ANDREW F. X. GOLDBERG AND
     rom-1 Complexes and Mutants Implicated in Ret-   ROBERT S. MOLDAY   671
     inal Degenerative Diseases

45. Isolation of Retinal Proteins That Interact with Ret-   RONALD ROEPMAN,
     initis Pigmentosa GTPase Regulator by Interac-   DIANA SCHICK, AND
     tion Trap Screen in Yeast   PAULO A. FERREIRA   688

46. Genetic Analysis of RPE65: From Human Disease   T. MICHAEL REDMOND AND
     to Mouse Model   CHRISTIAN P. HAMEL   705

47. Construction of Encapsidated Gutted Adenovirus   RAJENDRA KUMAR-SINGH,
     Minichromosomes and Their Application to Res-   CLYDE K. YAMASHITA,
     cue of Photoreceptor Degeneration   KEN TRAN, AND
     DEBORA B. FARBER   724

48. Production and Purification of Recombinant Ad-   WILLIAM W. HAUSWIRTH,
     eno-Associated Virus   ALFRED S. LEWIN,
     SERGEI ZOLOTUKHIN, AND
     NICHOLAS MUZYCZKA   743

49. Ribozymes in Treatment of Inherited Retinal   LYNN C. SHAW,
    Disease   PATRICK O. WHALEN,
                                                   KIMBERLY A. DRENSER,
                                                   WEIMING YAN,
                                                   WILLIAM W. HAUSWIRTH, AND
                                                   ALFRED S. LEWIN   761

50. Cross-Species Comparison of *in Vivo* Reporter JEAN BENNETT,
    Gene Expression after Recombinant Adeno-Asso- VIBHA ANAND,
    ciated Virus–Mediated Retinal Transduction   GREGORY M. ACLAND, AND
                                                   ALBERT M. MAGUIRE   777

VOLUME 317
RNA–LIGAND INTERACTIONS (PART A: STRUCTURAL
BIOLOGY METHODS)
Daniel W. Celander and John N. Abelson

Section I. Semisynthetic Methodologies

A. RNA Synthetic Methods

1. Advanced 5'-Silyl-2'-Orthoester Approach to RNA STEPHEN A. SCARINGE   3
   Oligonucleotide Synthesis

2. Preparation of Specifically ^{2}H- and ^{13}C-Labeled LINCOLN G. SCOTT,
   Ribonucleotides   THOMAS J. TOLBERT, AND
                                                    JAMES R. WILLIAMSON   18

3. Base-Modified Phosphoramidite Analogs of Pyrimi- LEONID BEIGELMAN,
   dine Ribonucleosides for RNA Structure–   JASENKA MATULIC-ADAMIC,
   Activity Studies   ALEXANDER KARPEISKY,
                                                    PETER HAEBERLI, AND
                                                    DAVID SWEEDLER   39

B. Derivatization of RNA

4. Use of T7 RNA Polymerase and Its Mutants for   RUI SOUSA   65
   Incorporation of Nucleoside Analogs into RNA

5. Phosphorothioate Modification of RNA for Stereo- L. CLAUS S. VÖRTLER AND
   chemical and Interference Analyses   FRITZ ECKSTEIN   74

6. Chemical Probing of RNA by Nucleotide Analog   SEAN P. RYDER,
   Interference Mapping   LORI ORTOLEVA-DONNELLY,
                                                   ANNE B. KOSEK, AND
                                                   SCOTT A. STROBEL   92

7. Joining of RNAs by Splinted Ligation — MELISSA J. MOORE AND CHARLES C. QUERY — 109

8. Heavy Atom Derivatives of RNA — BARBARA L. GOLDEN — 124

9. Site-Specific Cleavage of Transcript RNA — JON LAPHAM AND DONALD M. CROTHERS — 132

10. Using DNAzymes to Cut, Process, and Map RNA Molecules for Structural Studies or Modification — ANNA MARIE PYLE, VI T. CHU, ECKHARD JANKOWSKY, AND MARC BOUDVILLAIN — 140

Section II. RNA Structure Determination

A. X-Ray Crystallography

11. Purification, Crystallization, and X-Ray Diffraction Analysis of Small Ribozymes — JOSEPH E. WEDEKIND AND DAVID B. MCKAY — 149

12. Solving Large RNA Structures by X-Ray Crystallography — JAMIE H. CATE AND JENNIFER A. DOUDNA — 169

13. Conventional and Time-Resolved Ribozyme X-Ray Crystallography — WILLIAM G. SCOTT AND JAMES B. MURRAY — 180

B. Nuclear Magnetic Resonance Spectroscopy

14. Nuclear Magnetic Resonance Methods to Study RNA–Protein Complexes — PETER BAYER, LUCA VARANI, AND GABRIELE VARANI — 198

15. Filamentous Bacteriophage for Aligning RNA, DNA, and Proteins for Measurement of Nuclear Magnetic Resonance Dipolar Coupling Interactions — MARK R. HANSEN, PAUL HANSON, AND ARTHUR PARDI — 220

16. Biochemical and Nuclear Magnetic Resonance Studies of Aminoglycoside–RNA Complexes — STEPHEN R. LYNCH, MICHAEL I. RECHT, AND JOSEPH D. PUGLISI — 240

C. Electron Microscopy

17. Experimental Prerequisites for Determination of tRNA Binding to Ribosomes from *Escherichia coli* — FRANCISCO J. TRIANA-ALONSO, CHRISTIAN M. T. SPAHN, NILS BURKHARDT, BEATRIX RÖHRDANZ, AND KNUD H. NIERHAUS — 261

18. Three-Dimensional Cryoelectron Microscopy of Ribosomes — JOACHIM FRANK, PAWEL PENCZEK, RAJENDRA K. AGRAWAL, ROBERT A. GRASSUCCI, AND AMY B. HEAGLE — 276

19. Preparation of Functional Ribosomal Complexes   GREGOR BLAHA,
    and Effect of Buffer Conditions on tRNA Posi-   ULRICH STELZL,
    tions Observed by Cryoelectron Microscopy   CHRISTIAN M. T. SPAHN,
                                                     RAJENDRA K. AGRAWAL,
                                                     JOACHIM FRANK, AND
                                                     KNUD H. NIERHAUS   292

Section III. Techniques for Monitoring RNA Conformation and Dynamics

A. Solution Methods

20. Probing RNA Structure and Function by Circular   TAO PAN   313
    Permutation

21. Kinetic Oligonucleotide Hybridization for Monitor-   DANIEL K. TREIBER AND
    ing Kinetic Folding of Large RNAs   JAMES R. WILLIAMSON   330

22. Time-Resolved Synchrotron X-Ray Footprinting   CORIE Y. RALSTON,
    and Its Application to RNA Folding   BIANCA SCLAVI,
                                                     MICHAEL SULLIVAN,
                                                     MICHAEL L. DERAS,
                                                     SARAH A. WOODSON,
                                                     MARK R. CHANCE, AND
                                                     MICHAEL BRENOWITZ   353

B. Electrophoretic and Spectroscopic Methods

23. Analysis of Global Conformation of Branched   DAVID M. J. LILLEY   368
    RNA Species Using Electrophoresis and Fluores-
    cence

24. Application of Circular Dichroism to Study RNA   TOBIN R. SOSNICK,
    Folding Transitions   XINGWANG FANG, AND
                                                     VALERIE M. SHELTON   393

25. Fluorescence Assays to Study Structure, Dynamics,   NILS G. WALTER AND
    and Function of RNA and RNA–Ligand Com-   JOHN M. BURKE   409
    plexes

26. Transient Electric Birefringence for Determining   PAUL J. HAGERMAN   440
    Global Conformations of Nonhelix Elements and
    Protein-Induced Bends in RNA

Section IV. Modeling Tertiary Structure

27. Structure–Function Relationships of RNA: A Mod-   FABRICE LECLERC,
    eling Approach   BELSIS LLORENTE, AND
                                                     ROBERT CEDERGREN   457

28. Computational Modeling of Structural Experimen-   MICHAEL A. BADA AND
    tal Data   RUSS B. ALTMAN   470

29. Modeling RNA Tertiary Structure from Patterns of   FRANÇOIS MICHEL,
    Sequence Variation   MARIA COSTA,
                                              CHRISTIAN MASSIRE, AND
                                              ERIC WESTHOF   491

VOLUME 318
RNA–LIGAND INTERACTIONS (PART B: MOLECULAR BIOLOGY METHODS)
Daniel W. Celander and John N. Abelson

Section I. Solution Probe Methods

1. Probing RNA Structure and RNA–Ligand Com-   CHRISTINE BRUNEL AND
   plexes with Chemical Probes   PASCALE ROMBY   3

2. Methylene Blue-Mediated Cross-Linking of Pro-   ZHI-REN LIU,
   teins to Double-Stranded RNA   BRUNO SARGUEIL, AND
                                            CHRISTOPHER W. J. SMITH   22

3. Structure and Distance Determination in RNA with   THOMAS HERMANN AND
   Copper Phenanthroline Probing   HERMANN HEUMANN   33

4. Applications of Uranyl Cleavage Mapping of   NIELS ERIK MØLLEGAARD AND
   RNA Structure   PETER E. NIELSEN   43

5. Identification of Messenger RNAs That Contain   DANIEL P. MORSE   48
   Inosine

Section II. Tethered-Probe Methodologies

A. Photochemical Reagents

6. Site-Specific 4-Thiouridine Incorporation into   YI-TAO YU   71
   RNA Molecules

7. Nucleoprotein Photo-Cross-Linking Using Halo-   KRISTEN M. MEISENHEIMER,
   pyrimidine-Substituted RNAs   PONCHO L. MEISENHEIMER,
                                            AND TAD H. KOCH   88

8. Structure Determination by Directed Photo-Cross-   DMITRI MUNDIS AND
   Linking in Large RNA Molecules with Site-   PAUL WOLLENZIEN   104
   Specific Psoralen

9. Photolabile Derivatives of Oligonucleotides as   BARRY S. COOPERMAN,
   Probes of Ribosomal Structure   REBECCA W. ALEXANDER,
                                                      YURI BUKHTIYAROV,
                                                      SERGUEI N. VLADIMIROV,
                                                      ZHANNA DRUZINA,
                                                      RUO WANG, AND
                                                      NORA ZUÑO   118

10. Photoaffinity Cross-Linking and RNA Structure   BRIAN C. THOMAS,
    Analysis   ALEXEI V. KAZANTSEV,
                                                      JIUNN-LIANG CHEN, AND
                                                      NORMAN R. PACE   136

B. Chemical Reagents

11. Cleavage of RNA with Synthetic Ribonuclease   RICHARD GIEGÉ,
    Mimics   BRICE FELDEN,
                                                    MARINA A. ZENKOVA,
                                                    VLADIMIR N. SIL'NIKOV, AND
                                                    VALENTIN V. VLASSOV   147

12. Site-Specific Sulfhydryl Groups for Study of RNA   SNORRI TH. SIGURDSSON   165
    Conformation via Disulfide Cross-Linking

13. Directed Hydroxyl Radical Probing Using Iron(II)   SIMPSON JOSEPH AND
    Tethered to RNA   HARRY F. NOLLER   175

Section III. *In Vitro* Affinity Selection Methodologies

14. *In Vitro* Selection of RNA Aptamers   KRISTIN A. MARSHALL AND
                                                    ANDREW D. ELLINGTON   193

15. *In Vitro* Selection and Characterization of RNAs   SCOT T. WALLACE AND
    with High Affinity to Antibiotics   RENÉE SCHROEDER   214

16. Selection for RNA:Peptide Recognition through   MATT WECKER AND
    Sulfur Alkylation Chemistry   DREW SMITH   229

17. *In Vitro* Selection of RNA Substrates for Ribo-   FENYONG LIU,
    nuclease P and Its Catalytic RNA   JUN WANG, AND
                                                        PHONG TRANG   238

18. RNA–Protein Interactions in Ribosomes: *In Vitro*   ULRICH STELZL,
    Selection from Randomly Fragmented rRNA   CHRISTIAN M. T. SPAHN, AND
                                                         KNUD H. NIERHAUS   251

19. Optimized Synthesis of RNA–Protein Fusions for   RIHE LIU,
    in Vitro Protein Selection   JEFFREY E. BARRICK,
                                                      JACK W. SZOSTAK, AND
                                                      RICHARD W. ROBERTS   268

Section IV. Genetic Methodologies for Detecting RNA–Protein Interactions

A. Bacterial Systems

20. Screening RNA–Binding Libraries by Transcriptional Antitermination in Bacteria — HADAS PELED-ZEHAVI, COLIN A. SMITH, KAZUO HARADA, AND ALAN D. FRANKEL ... 297

21. Rapid Genetic Analysis of RNA–Protein Interactions by Translational Repression in *Escherichia coli* — CHAITANYA JAIN AND JOEL G. BELASCO ... 309

22. RNA Challenge Phages as Genetic Tools for Study of RNA–Ligand Interactions — DANIEL W. CELANDER, KRISTINE A. BENNETT, DERRICK E. FOUTS, ERICA A. STEITZ, AND HEATHER L. TRUE ... 332

B. Eukaryotic Systems

23. Screening RNA-Binding Libraries Using Tat-Fusion System in Mammalian Cells — STEPHEN G. LANDT, RUOYING TAN, AND ALAN D. FRANKEL ... 350

24. Frameshifting Assay to Characterize RNA–Protein Interactions in Eukaryotic Cells — HEIKE KOLLMUS AND HANSJÖRG HAUSER ... 363

25. Translational Repression Assay Procedure: A Method to Study RNA–Protein Interactions in Yeast — EFROSYNI PARASKEVA AND MATTHIAS W. HENTZE ... 374

26. Analysis of Low-Abundance Ribonucleoprotein Particles from Yeast by Affinity Chromatography and Mass Spectrometry Microsequencing — SCOTT W. STEVENS ... 385

27. Yeast Three-Hybrid System to Detect and Analyze RNA–Protein Interactions — BEILIN ZHANG, BRIAN KRAEMER, DHRUBA SENGUPTA, STANLEY FIELD, AND MARVIN WICKENS ... 399

Section V. Protein Engineering Methodologies Useful for RNA–Protein Interaction Studies

28. Using Peptides as Models of RNA–Protein Interactions — COLIN A. SMITH, LILY CHEN, AND ALAN D. FRANKEL ... 423

29. Analysis of RNA–Protein Cross-Link Sites by   BERND THIEDE AND
    Matrix-Assisted Laser Desorption/Ionization   BRIGITTE WITTMANN-LIEBOLD   438
    Mass Spectrometry and N-Terminal Micro-
    sequencing

30. *In Vitro* Reconstitution of 30S Ribosomal Subunits   GLORIA M. CULVER AND
    Using Complete Set of Recombinant Proteins   HARRY F. NOLLER   446

31. Directed Hydroxyl Radical Probing of RNA from   GLORIA M. CULVER AND
    Iron(II) Tethered to Proteins in Ribonucleopro-   HARRY F. NOLLER   461
    tein Complexes

Section VI. Cell Biology Methods

32. Use of Dimethyl Sulfate to Probe RNA Structure   SANDRA E. WELLS,
    in Vivo   JOHN M. X. HUGHES,
    A. HALLER IGEL, AND
    MANUEL ARES, JR.   479

33. Sensitive and High-Resolution Detection of RNA   PASCAL CHARTRAND,
    in Situ   EDOUARD BERTRAND,
    ROBERT H. SINGER, AND
    ROY M. LONG   493

VOLUME 319
SINGLET OXYGEN, UV-A, AND OZONE
Lester Packer and Helmut Sies

Section I. Singlet Oxygen

A. Generation and Detection: Chemical Systems

1. Naphthalene Endoperoxides as Generators of Sin-   CHRISTEL PIERLOT,
   glet Oxygen in Biological Media   JEAN-MARIE AUBRY,
   KARLIS BRIVIBA,
   HELMUT SIES, AND
   PAOLO DI MASCIO   3

2. Photosensitized Production of Singlet Oxygen   IRENE E. KOCHEVAR AND
   ROBERT W. REDMOND   20

3. Singlet Oxygen from Irradiated Titanium Dioxide   YORIHIRO YAMAMOTO,
   and Zinc Oxide   NORITAKA IMAI,
   RYUICHI MASHIMA,
   RYUSEI KONAKA,
   MASAYASU INOUE, AND
   WALTER C. DUNLAP   29

4. Time-Resolved Singlet Oxygen Detection   SANTI NONELL AND
                                                SILVIA E. BRASLAVSKY   37

5. Measurement of Photogenerated Singlet Oxygen in VÉRONIQUE NARDELLO AND
   Aqueous Media   JEAN-MARIE AUBRY   50

B. Generation and Detection: Biological Systems

6. Assay for Singlet-Oxygen Generation by Peroxi- JEFFREY R. KANOFSKY   59
   dases Using 1270-nm Chemiluminescence

7. Formation of Electronically Excited States during ENRIQUE CADENAS AND
   the Oxidation of Arachidonic Acid by Prostaglan- HELMUT SIES   67
   din Endoperoxide Synthase

8. Singlet Oxygen Detection with Sterically Hindered ÉVA HIDEG,
   Amine Derivatives in Plants under Light Stress IMRE VASS,
                                                TAMÁS KÁLAI, AND
                                                KÁLMÁN HIDEG   77

9. Cholesterol as a Singlet Oxygen Detector in Biologi- ALBERT W. GIROTTI AND
   cal Systems   WITOLD KORYTOWSKI   85

10. Singlet Oxygen Scavenging in Phospholipid Mem- KENJI FUKUZAWA   101
    branes

11. Catalase Modification as a Marker for Singlet FERNANDO LLEDIAS AND
    Oxygen   WILHELM HANSBERG   110

C. Biological Effects: Role in Signaling

12. Nuclear Factor-κB Activation by Singlet Oxygen JEAN-YVES MATROULE AND
    Produced during Photosensitization   JACQUES PIETTE   119

13. Mitogen-Activated Protein Kinase Activation by LARS-OLIVER KLOTZ,
    Singlet Oxygen and Ultraviolet A   KARLIS BRIVIBA, AND
                                                HELMUT SIES   130

D. Toxic Effects

14. Singlet Oxygen DNA Damage Products: Formation JEAN CADET,
    and Measurement   THIERRY DOUKI,
                                                JEAN-PIERRE POUGET, AND
                                                JEAN-LUC RAVANET   143

15. Ultraviolet A- and Singlet Oxygen-Induced Muta- ANNE STARY AND
    tion Spectra   ALAIN SARASIN   153

16. Damage to DNA by Long-Range Charge Transport MEGAN E. NÚÑEZ,
                                                SCOTT R. RAJSKI, AND
                                                JACQUELINE K. BARTON   165

17. Cholesterol Photodynamic Oxidation by Ultraviolet KYOICHI OSADA AND
    Irradiation and Cholesterol Ozonization by ALEX SEVANIAN   188
    Ozone Exposure

18. Bactericidal and Virucidal Activities of Singlet Oxygen Generated by Thermolysis of Naphthalene Endoperoxides — CORINNE PELLIEUX, ANNY DEWILDE, CHRISTEL PIERLOT, AND JEAN-MARIE AUBRY   197

19. Inactivation of Viruses in Human Plasma — Harald MOHR   207

20. 3-(4'-Methyl-1'-naphthyl)propionic Acid, 1',4'-Endoperoxide for Dioxygenation of Squalene and for Bacterial Killing — MINORU NAKANO, YASUHIRO KAMBAYASHI, AND HIDETAKA TATSUZAWA   216

E. Protection: Singlet Oxygen Quenchers

21. Biological Singlet Oxygen Quenchers Assessed by Monomol Light Emission — KARLIS BRIVIBA AND HELMUT SIES   222

22. Synthetic Singlet Oxygen Quenchers — STEFAN BEUTNER, BRITTA BLOEDORN, THOMAS HOFFMANN, AND HANS-DIETER MARTIN   226

Section II. Ultraviolet A

A. Dosimetry

23. Dosimetry of Ultraviolet A Radiation — BRIAN L. DIFFEY   245

B. Biological Responses: Signaling

24. Radiation-Induced Signal Transduction — AXEL KNEBEL, FRANK D. BÖHMER, AND PETER HERRLICH   255

25. Signaling Pathways Leading to Nuclear Factor-κB Activation — NANXIN LI AND MICHAEL KARIN   273

26. Gene Regulation by Ultraviolet A Radiation and Singlet Oxygen — SUSANNE GRETHER-BECK AND JEAN KRUTMANN   280

27. Role for Singlet Oxygen in Biological Effects of Ultraviolet A Radiation — REX M. TYRRELL   290

28. Ultraviolet A-1 Irradiation as a Tool to Study the Pathogenesis of Atopic Dermatitis — JEAN KRUTMANN   296

29. Ultraviolet A Radiation-Induced Apoptosis — AKIMICHI MORITA AND JEAN KRUTMANN   302

30. Singlet Oxygen-Triggered Immediate Preprogrammed Apoptosis — DIANNE E. GODAR   309

31. Determination of DNA Damage, Peroxide Generation, Mitochondrial Membrane Potential, and Caspase-3 Activity during Ultraviolet A-Induced Apoptosis — SAEKO TADA-OIKAWA, SHINJI OIKAWA, AND SHOSUKE KAWANISHI   331

32. Mechanism of Photodynamic Therapy-Induced Cell Death — NIHAL AHMAD AND HASAN MUKHTAR — 342

C. Biological Responses: Photocarcinogenesis, Photoaging, and Photoallergic Reactions

33. Photocarcinogenesis: UVA vs UVB — FRANK R. DE GRUIJL — 359

34. Photoaging-Associated Large-Scale Deletions of Mitochondrial DNA — MARK BERNEBURG AND JEAN KRUTMANN — 366

35. Role of Activated Oxygen Species in Photodynamic Therapy — WESLEY M. SHARMAN, CYNTHIA M. ALLEN, AND JOHAN E. VAN LIER — 376

36. Gas Chromatography–Mass Spectrometry Analysis of DNA: Optimization of Protocols for Isolation and Analysis of DNA from Human Blood — ALMAS REHMAN, ANDREW JENNER, AND BARRY HALLIWELL — 401

D. Oxidative Damage Markers

37. Sequence Specificity of Ultraviolet A-Induced DNA Damage in the Presence of Photosensitizer — KIMIKO ITO AND SHOSUKE KAWANISHI — 417

38. Protein Oxidative Damage — EMILY SHACTER — 428

39. DNA Damage Induced by Ultraviolet and Visible Light and Its Wavelength Dependence — CHRISTOPHER KIELBASSA AND BERND EPE — 436

40. Photoprotection of Skin against Ultraviolet A Damage — HANS SCHAEFER, ALAIN CHARDON, AND DOMINIQUE MOYAL — 445

E. Protection

41. Topically Applied Antioxidants in Skin Protection — FRANZ STÄB, RAINER WOLBER, THOMAS BLATT, REZA KEYHANI, AND GERHARD SAUERMANN — 465

42. Erythropoietic Protoporphyria: Treatment with Antioxidants and Potential Cure with Gene Therapy — MICHELINE M. MATHEWS-ROTH — 479

43. Porphyrias: Photosensitivity and Phototherapy — MAUREEN B. POH-FITZPATRICK — 485

44. Carotenoids in Human Skin: Noninvasive Measurement and Identification of Dermal Carotenoids and Carotenol Esters — WILHELM STAHL, ULRIKE HEINRICH, HOLGER JUNGMANN, HAGEN TRONNIER, AND HELMUT SIES — 494

Section III. Ozone

45. Reactive Absorption of Ozone: An Assay for Reaction Rates of Ozone with Sulfhydryl Compounds and with Other Biological Molecules — JEFFREY R. KANOFSKY AND PAUL D. SIMA   505

46. Assay for Singlet Oxygen Generation by Plant Leaves Exposed to Ozone — JEFFERY R. KANOFSKY AND PAUL D. SIMA   512

47. Ozone Effects on Plant Defense — CHRISTIAN LANGEBARTELS, DIETER ERNST, JAAKKO KANGASJÄRVI, AND HEINRICH SANDERMANN, JR.   520

48. High-Pressure Liquid Chromatography Analysis of Ozone-Induced Depletion of Hydrophilic and Lipophilic Antioxidants in Murine Skin — STEFAN U. WEBER, SUMANA JOTHI, AND JENS J. THIELE   536

49. Reactions of Vitamin E with Ozone — DANIEL C. LIEBLER   546

50. Induction of Nuclear Factor-κB by Exposure to Ozone and Inhibition by Glucocorticoids — Kian Fan CHUNG AND Ian M. ADCOCK   551

51. Detection of 4-Hydroxy-2-nonenol Adducts Following Lipid Peroxidation from Ozone Exposure — LUKE I. SZWEDA, PAMELA A. SZWEDA, AND ANDRIJ HOLIAN   562

52. Synthesis of Inflammatory Signal Transduction Species Formed during Ozonation and/or Peroxidation of Tissue Lipids — GIUSEPPE L. SQUADRITO, MARIA G. SALGO, FRANK R. FRONCZEK, AND WILLIAM A. PRYOR   570

Section IV. General Methods

53. Assay for Redox-Sensitive Transcription Factor — MADAN M. CHATURVEDI, ASOK MUKHOPADHYAY, AND BHARAT B. AGGARWAL   585

54. Fluorescent Fatty Acid to Monitor Reactive Oxygen in Single Cells — Eward H. W. PAP, G. P. C. DRUMMEN, J. A. POST, P. J. RIJKEN, AND K. W. A. WIRTZ   603

55. Noninvasive Techniques for Measuring Oxidation Products on the Surface of Human Skin — DANIEL MAES, TOM MOMMONE, MARYANN MCKEEVER, ED PELLE, CHRISTINA FTHENAKIS, LIEVE DECLERCQ, PAOLO U. GIACOMONI, AND KEN MARENUS   612

METHODS IN ENZYMOLOGY

VOLUME I. Preparation and Assay of Enzymes
Edited by SIDNEY P. COLOWICK AND NATHAN O. KAPLAN

VOLUME II. Preparation and Assay of Enzymes
Edited by SIDNEY P. COLOWICK AND NATHAN O. KAPLAN

VOLUME III. Preparation and Assay of Substrates
Edited by SIDNEY P. COLOWICK AND NATHAN O. KAPLAN

VOLUME IV. Special Techniques for the Enzymologist
Edited by SIDNEY P. COLOWICK AND NATHAN O. KAPLAN

VOLUME V. Preparation and Assay of Enzymes
Edited by SIDNEY P. COLOWICK AND NATHAN O. KAPLAN

VOLUME VI. Preparation and Assay of Enzymes (*Continued*)
Preparation and Assay of Substrates
Special Techniques
Edited by SIDNEY P. COLOWICK AND NATHAN O. KAPLAN

VOLUME VII. Cumulative Subject Index
Edited by SIDNEY P. COLOWICK AND NATHAN O. KAPLAN

VOLUME VIII. Complex Carbohydrates
Edited by ELIZABETH F. NEUFELD AND VICTOR GINSBURG

VOLUME IX. Carbohydrate Metabolism
Edited by WILLIS A. WOOD

VOLUME X. Oxidation and Phosphorylation
Edited by RONALD W. ESTABROOK AND MAYNARD E. PULLMAN

VOLUME XI. Enzyme Structure
Edited by C. H. W. HIRS

VOLUME XII. Nucleic Acids (Parts A and B)
Edited by LAWRENCE GROSSMAN AND KIVIE MOLDAVE

VOLUME XIII. Citric Acid Cycle
Edited by J. M. LOWENSTEIN

VOLUME XIV. Lipids
Edited by J. M. LOWENSTEIN

VOLUME XV. Steroids and Terpenoids
Edited by RAYMOND B. CLAYTON

VOLUME XVI. Fast Reactions
Edited by KENNETH KUSTIN

VOLUME XVII. Metabolism of Amino Acids and Amines (Parts A and B)
Edited by HERBERT TABOR AND CELIA WHITE TABOR

VOLUME XVIII. Vitamins and Coenzymes (Parts A, B, and C)
Edited by DONALD B. MCCORMICK AND LEMUEL D. WRIGHT

VOLUME XIX. Proteolytic Enzymes
Edited by GERTRUDE E. PERLMANN AND LASZLO LORAND

VOLUME XX. Nucleic Acids and Protein Synthesis (Part C)
Edited by KIVIE MOLDAVE AND LAWRENCE GROSSMAN

VOLUME XXI. Nucleic Acids (Part D)
Edited by LAWRENCE GROSSMAN AND KIVIE MOLDAVE

VOLUME XXII. Enzyme Purification and Related Techniques
Edited by WILLIAM B. JAKOBY

VOLUME XXIII. Photosynthesis (Part A)
Edited by ANTHONY SAN PIETRO

VOLUME XXIV. Photosynthesis and Nitrogen Fixation (Part B)
Edited by ANTHONY SAN PIETRO

VOLUME XXV. Enzyme Structure (Part B)
Edited by C. H. W. HIRS AND SERGE N. TIMASHEFF

VOLUME XXVI. Enzyme Structure (Part C)
Edited by C. H. W. HIRS AND SERGE N. TIMASHEFF

VOLUME XXVII. Enzyme Structure (Part D)
Edited by C. H. W. HIRS AND SERGE N. TIMASHEFF

VOLUME XXVIII. Complex Carbohydrates (Part B)
Edited by VICTOR GINSBURG

VOLUME XXIX. Nucleic Acids and Protein Synthesis (Part E)
Edited by LAWRENCE GROSSMAN AND KIVIE MOLDAVE

VOLUME XXX. Nucleic Acids and Protein Synthesis (Part F)
Edited by KIVIE MOLDAVE AND LAWRENCE GROSSMAN

VOLUME XXXI. Biomembranes (Part A)
Edited by SIDNEY FLEISCHER AND LESTER PACKER

VOLUME XXXII. Biomembranes (Part B)
Edited by SIDNEY FLEISCHER AND LESTER PACKER

VOLUME XXXIII. Cumulative Subject Index Volumes I–XXX
Edited by MARTHA G. DENNIS AND EDWARD A. DENNIS

VOLUME XXXIV. Affinity Techniques (Enzyme Purification: Part B)
Edited by WILLIAM B. JAKOBY AND MEIR WILCHEK

VOLUME XXXV. Lipids (Part B)
Edited by JOHN M. LOWENSTEIN

VOLUME XXXVI. Hormone Action (Part A: Steroid Hormones)
Edited by BERT W. O'MALLEY AND JOEL G. HARDMAN

VOLUME XXXVII. Hormone Action (Part B: Peptide Hormones)
Edited by BERT W. O'MALLEY AND JOEL G. HARDMAN

VOLUME XXXVIII. Hormone Action (Part C: Cyclic Nucleotides)
Edited by JOEL G. HARDMAN AND BERT W. O'MALLEY

VOLUME XXXIX. Hormone Action (Part D: Isolated Cells, Tissues, and Organ
Systems)
Edited by JOEL G. HARDMAN AND BERT W. O'MALLEY

VOLUME XL. Hormone Action (Part E: Nuclear Structure and Function)
Edited by BERT W. O'MALLEY AND JOEL G. HARDMAN

VOLUME XLI. Carbohydrate Metabolism (Part B)
Edited by W. A. WOOD

VOLUME XLII. Carbohydrate Metabolism (Part C)
Edited by W. A. WOOD

VOLUME XLIII. Antibiotics
Edited by JOHN H. HASH

VOLUME XLIV. Immobilized Enzymes
Edited by KLAUS MOSBACH

VOLUME XLV. Proteolytic Enzymes (Part B)
Edited by LASZLO LORAND

VOLUME XLVI. Affinity Labeling
Edited by WILLIAM B. JAKOBY AND MEIR WILCHEK

VOLUME XLVII. Enzyme Structure (Part E)
Edited by C. H. W. HIRS AND SERGE N. TIMASHEFF

VOLUME XLVIII. Enzyme Structure (Part F)
Edited by C. H. W. HIRS AND SERGE N. TIMASHEFF

VOLUME XLIX. Enzyme Structure (Part G)
Edited by C. H. W. HIRS AND SERGE N. TIMASHEFF

VOLUME L. Complex Carbohydrates (Part C)
Edited by VICTOR GINSBURG

VOLUME LI. Purine and Pyrimidine Nucleotide Metabolism
Edited by PATRICIA A. HOFFEE AND MARY ELLEN JONES

VOLUME LII. Biomembranes (Part C: Biological Oxidations)
Edited by SIDNEY FLEISCHER AND LESTER PACKER

VOLUME LIII. Biomembranes (Part D: Biological Oxidations)
Edited by SIDNEY FLEISCHER AND LESTER PACKER

VOLUME LIV. Biomembranes (Part E: Biological Oxidations)
Edited by SIDNEY FLEISCHER AND LESTER PACKER

VOLUME 73. Immunochemical Techniques (Part B)
Edited by JOHN J. LANGONE AND HELEN VAN VUNAKIS

VOLUME 74. Immunochemical Techniques (Part C)
Edited by JOHN J. LANGONE AND HELEN VAN VUNAKIS

VOLUME 75. Cumulative Subject Index Volumes XXXI, XXXII, XXXIV–LX
Edited by EDWARD A. DENNIS AND MARTHA G. DENNIS

VOLUME 76. Hemoglobins
Edited by ERALDO ANTONINI, LUIGI ROSSI-BERNARDI, AND EMILIA CHIANCONE

VOLUME 77. Detoxication and Drug Metabolism
Edited by WILLIAM B. JAKOBY

VOLUME 78. Interferons (Part A)
Edited by SIDNEY PESTKA

VOLUME 79. Interferons (Part B)
Edited by SIDNEY PESTKA

VOLUME 80. Proteolytic Enzymes (Part C)
Edited by LASZLO LORAND

VOLUME 81. Biomembranes (Part H: Visual Pigments and Purple Membranes, I)
Edited by LESTER PACKER

VOLUME 82. Structural and Contractile Proteins (Part A: Extracellular Matrix)
Edited by LEON W. CUNNINGHAM AND DIXIE W. FREDERIKSEN

VOLUME 83. Complex Carbohydrates (Part D)
Edited by VICTOR GINSBURG

VOLUME 84. Immunochemical Techniques (Part D: Selected Immunoassays)
Edited by JOHN J. LANGONE AND HELEN VAN VUNAKIS

VOLUME 85. Structural and Contractile Proteins (Part B: The Contractile Apparatus and the Cytoskeleton)
Edited by DIXIE W. FREDERIKSEN AND LEON W. CUNNINGHAM

VOLUME 86. Prostaglandins and Arachidonate Metabolites
Edited by WILLIAM E. M. LANDS AND WILLIAM L. SMITH

VOLUME 87. Enzyme Kinetics and Mechanism (Part C: Intermediates, Stereochemistry, and Rate Studies)
Edited by DANIEL L. PURICH

VOLUME 88. Biomembranes (Part I: Visual Pigments and Purple Membranes, II)
Edited by LESTER PACKER

VOLUME 89. Carbohydrate Metabolism (Part D)
Edited by WILLIS A. WOOD

VOLUME 90. Carbohydrate Metabolism (Part E)
Edited by WILLIS A. WOOD

VOLUME 91. Enzyme Structure (Part I)
Edited by C. H. W. HIRS AND SERGE N. TIMASHEFF

VOLUME 92. Immunochemical Techniques (Part E: Monoclonal Antibodies and General Immunoassay Methods)
Edited by JOHN J. LANGONE AND HELEN VAN VUNAKIS

VOLUME 93. Immunochemical Techniques (Part F: Conventional Antibodies, Fc Receptors, and Cytotoxicity)
Edited by JOHN J. LANGONE AND HELEN VAN VUNAKIS

VOLUME 94. Polyamines
Edited by HERBERT TABOR AND CELIA WHITE TABOR

VOLUME 95. Cumulative Subject Index Volumes 61–74, 76–80
Edited by EDWARD A. DENNIS AND MARTHA G. DENNIS

VOLUME 96. Biomembranes [Part J: Membrane Biogenesis: Assembly and Targeting (General Methods; Eukaryotes)]
Edited by SIDNEY FLEISCHER AND BECCA FLEISCHER

VOLUME 97. Biomembranes [Part K: Membrane Biogenesis: Assembly and Targeting (Prokaryotes, Mitochondria, and Chloroplasts)]
Edited by SIDNEY FLEISCHER AND BECCA FLEISCHER

VOLUME 98. Biomembranes (Part L: Membrane Biogenesis: Processing and Recycling)
Edited by SIDNEY FLEISCHER AND BECCA FLEISCHER

VOLUME 99. Hormone Action (Part F: Protein Kinases)
Edited by JACKIE D. CORBIN AND JOEL G. HARDMAN

VOLUME 100. Recombinant DNA (Part B)
Edited by RAY WU, LAWRENCE GROSSMAN, AND KIVIE MOLDAVE

VOLUME 101. Recombinant DNA (Part C)
Edited by RAY WU, LAWRENCE GROSSMAN, AND KIVIE MOLDAVE

VOLUME 102. Hormone Action (Part G: Calmodulin and Calcium-Binding Proteins)
Edited by ANTHONY R. MEANS AND BERT W. O'MALLEY

VOLUME 103. Hormone Action (Part H: Neuroendocrine Peptides)
Edited by P. MICHAEL CONN

VOLUME 104. Enzyme Purification and Related Techniques (Part C)
Edited by WILLIAM B. JAKOBY

VOLUME 105. Oxygen Radicals in Biological Systems
Edited by LESTER PACKER

VOLUME 106. Posttranslational Modifications (Part A)
Edited by FINN WOLD AND KIVIE MOLDAVE

VOLUME 107. Posttranslational Modifications (Part B)
Edited by FINN WOLD AND KIVIE MOLDAVE

VOLUME 108. Immunochemical Techniques (Part G: Separation and Characterization of Lymphoid Cells)
Edited by GIOVANNI DI SABATO, JOHN J. LANGONE, AND HELEN VAN VUNAKIS

VOLUME 109. Hormone Action (Part I: Peptide Hormones)
Edited by LUTZ BIRNBAUMER AND BERT W. O'MALLEY

VOLUME 110. Steroids and Isoprenoids (Part A)
Edited by JOHN H. LAW AND HANS C. RILLING

VOLUME 111. Steroids and Isoprenoids (Part B)
Edited by JOHN H. LAW AND HANS C. RILLING

VOLUME 112. Drug and Enzyme Targeting (Part A)
Edited by KENNETH J. WIDDER AND RALPH GREEN

VOLUME 113. Glutamate, Glutamine, Glutathione, and Related Compounds
Edited by ALTON MEISTER

VOLUME 114. Diffraction Methods for Biological Macromolecules (Part A)
Edited by HAROLD W. WYCKOFF, C. H. W. HIRS, AND SERGE N. TIMASHEFF

VOLUME 115. Diffraction Methods for Biological Macromolecules (Part B)
Edited by HAROLD W. WYCKOFF, C. H. W. HIRS, AND SERGE N. TIMASHEFF

VOLUME 116. Immunochemical Techniques (Part H: Effectors and Mediators of Lymphoid Cell Functions)
Edited by GIOVANNI DI SABATO, JOHN J. LANGONE, AND HELEN VAN VUNAKIS

VOLUME 117. Enzyme Structure (Part J)
Edited by C. H. W. HIRS AND SERGE N. TIMASHEFF

VOLUME 118. Plant Molecular Biology
Edited by ARTHUR WEISSBACH AND HERBERT WEISSBACH

VOLUME 119. Interferons (Part C)
Edited by SIDNEY PESTKA

VOLUME 120. Cumulative Subject Index Volumes 81–94, 96–101

VOLUME 121. Immunochemical Techniques (Part I: Hybridoma Technology and Monoclonal Antibodies)
Edited by JOHN J. LANGONE AND HELEN VAN VUNAKIS

VOLUME 122. Vitamins and Coenzymes (Part G)
Edited by FRANK CHYTIL AND DONALD B. MCCORMICK

VOLUME 123. Vitamins and Coenzymes (Part H)
Edited by FRANK CHYTIL AND DONALD B. MCCORMICK

VOLUME 124. Hormone Action (Part J: Neuroendocrine Peptides)
Edited by P. MICHAEL CONN

VOLUME 125. Biomembranes (Part M: Transport in Bacteria, Mitochondria, and Chloroplasts: General Approaches and Transport Systems)
Edited by SIDNEY FLEISCHER AND BECCA FLEISCHER

VOLUME 126. Biomembranes (Part N: Transport in Bacteria, Mitochondria, and Chloroplasts: Protonmotive Force)
Edited by SIDNEY FLEISCHER AND BECCA FLEISCHER

VOLUME 127. Biomembranes (Part O: Protons and Water: Structure and Translocation)
Edited by LESTER PACKER

VOLUME 128. Plasma Lipoproteins (Part A: Preparation, Structure, and Molecular Biology)
Edited by JERE P. SEGREST AND JOHN J. ALBERS

VOLUME 129. Plasma Lipoproteins (Part B: Characterization, Cell Biology, and Metabolism)
Edited by JOHN J. ALBERS AND JERE P. SEGREST

VOLUME 130. Enzyme Structure (Part K)
Edited by C. H. W. HIRS AND SERGE N. TIMASHEFF

VOLUME 131. Enzyme Structure (Part L)
Edited by C. H. W. HIRS AND SERGE N. TIMASHEFF

VOLUME 132. Immunochemical Techniques (Part J: Phagocytosis and Cell-Mediated Cytotoxicity)
Edited by GIOVANNI DI SABATO AND JOHANNES EVERSE

VOLUME 133. Bioluminescence and Chemiluminescence (Part B)
Edited by MARLENE DELUCA AND WILLIAM D. MCELROY

VOLUME 134. Structural and Contractile Proteins (Part C: The Contractile Apparatus and the Cytoskeleton)
Edited by RICHARD B. VALLEE

VOLUME 135. Immobilized Enzymes and Cells (Part B)
Edited by KLAUS MOSBACH

VOLUME 136. Immobilized Enzymes and Cells (Part C)
Edited by KLAUS MOSBACH

VOLUME 137. Immobilized Enzymes and Cells (Part D)
Edited by KLAUS MOSBACH

VOLUME 138. Complex Carbohydrates (Part E)
Edited by VICTOR GINSBURG

VOLUME 139. Cellular Regulators (Part A: Calcium- and Calmodulin-Binding Proteins)
Edited by ANTHONY R. MEANS AND P. MICHAEL CONN

VOLUME 140. Cumulative Subject Index Volumes 102–119, 121–134

VOLUME 141. Cellular Regulators (Part B: Calcium and Lipids)
Edited by P. MICHAEL CONN AND ANTHONY R. MEANS

VOLUME 142. Metabolism of Aromatic Amino Acids and Amines
Edited by SEYMOUR KAUFMAN

VOLUME 143. Sulfur and Sulfur Amino Acids
Edited by WILLIAM B. JAKOBY AND OWEN GRIFFITH

VOLUME 144. Structural and Contractile Proteins (Part D: Extracellular Matrix)
Edited by LEON W. CUNNINGHAM

VOLUME 145. Structural and Contractile Proteins (Part E: Extracellular Matrix)
Edited by LEON W. CUNNINGHAM

VOLUME 146. Peptide Growth Factors (Part A)
Edited by DAVID BARNES AND DAVID A. SIRBASKU

VOLUME 147. Peptide Growth Factors (Part B)
Edited by DAVID BARNES AND DAVID A. SIRBASKU

VOLUME 148. Plant Cell Membranes
Edited by LESTER PACKER AND ROLAND DOUCE

VOLUME 149. Drug and Enzyme Targeting (Part B)
Edited by RALPH GREEN AND KENNETH J. WIDDER

VOLUME 150. Immunochemical Techniques (Part K: *In Vitro* Models of B and T Cell Functions and Lymphoid Cell Receptors)
Edited by GIOVANNI DI SABATO

VOLUME 151. Molecular Genetics of Mammalian Cells
Edited by MICHAEL M. GOTTESMAN

VOLUME 152. Guide to Molecular Cloning Techniques
Edited by SHELBY L. BERGER AND ALAN R. KIMMEL

VOLUME 153. Recombinant DNA (Part D)
Edited by RAY WU AND LAWRENCE GROSSMAN

VOLUME 154. Recombinant DNA (Part E)
Edited by RAY WU AND LAWRENCE GROSSMAN

VOLUME 155. Recombinant DNA (Part F)
Edited by RAY WU

VOLUME 156. Biomembranes (Part P: ATP-Driven Pumps and Related Transport: The Na,K-Pump)
Edited by SIDNEY FLEISCHER AND BECCA FLEISCHER

VOLUME 157. Biomembranes (Part Q: ATP-Driven Pumps and Related Transport: Calcium, Proton, and Potassium Pumps)
Edited by SIDNEY FLEISCHER AND BECCA FLEISCHER

VOLUME 158. Metalloproteins (Part A)
Edited by JAMES F. RIORDAN AND BERT L. VALLEE

VOLUME 159. Initiation and Termination of Cyclic Nucleotide Action
Edited by JACKIE D. CORBIN AND ROGER A. JOHNSON

VOLUME 160. Biomass (Part A: Cellulose and Hemicellulose)
Edited by WILLIS A. WOOD AND SCOTT T. KELLOGG

VOLUME 161. Biomass (Part B: Lignin, Pectin, and Chitin)
Edited by WILLIS A. WOOD AND SCOTT T. KELLOGG

VOLUME 162. Immunochemical Techniques (Part L: Chemotaxis and Inflammation)
Edited by GIOVANNI DI SABATO

VOLUME 163. Immunochemical Techniques (Part M: Chemotaxis and Inflammation)
Edited by GIOVANNI DI SABATO

VOLUME 164. Ribosomes
Edited by HARRY F. NOLLER, JR., AND KIVIE MOLDAVE

VOLUME 165. Microbial Toxins: Tools for Enzymology
Edited by SIDNEY HARSHMAN

VOLUME 166. Branched-Chain Amino Acids
Edited by ROBERT HARRIS AND JOHN R. SOKATCH

VOLUME 167. Cyanobacteria
Edited by LESTER PACKER AND ALEXANDER N. GLAZER

VOLUME 168. Hormone Action (Part K: Neuroendocrine Peptides)
Edited by P. MICHAEL CONN

VOLUME 169. Platelets: Receptors, Adhesion, Secretion (Part A)
Edited by JACEK HAWIGER

VOLUME 170. Nucleosomes
Edited by PAUL M. WASSARMAN AND ROGER D. KORNBERG

VOLUME 171. Biomembranes (Part R: Transport Theory: Cells and Model Membranes)
Edited by SIDNEY FLEISCHER AND BECCA FLEISCHER

VOLUME 172. Biomembranes (Part S: Transport: Membrane Isolation and Characterization)
Edited by SIDNEY FLEISCHER AND BECCA FLEISCHER

VOLUME 173. Biomembranes [Part T: Cellular and Subcellular Transport: Eukaryotic (Nonepithelial) Cells]
Edited by SIDNEY FLEISCHER AND BECCA FLEISCHER

VOLUME 174. Biomembranes [Part U: Cellular and Subcellular Transport: Eukaryotic (Nonepithelial) Cells]
Edited by SIDNEY FLEISCHER AND BECCA FLEISCHER

VOLUME 175. Cumulative Subject Index Volumes 135–139, 141–167

VOLUME 176. Nuclear Magnetic Resonance (Part A: Spectral Techniques and Dynamics)
Edited by NORMAN J. OPPENHEIMER AND THOMAS L. JAMES

VOLUME 177. Nuclear Magnetic Resonance (Part B: Structure and Mechanism)
Edited by NORMAN J. OPPENHEIMER AND THOMAS L. JAMES

VOLUME 178. Antibodies, Antigens, and Molecular Mimicry
Edited by JOHN J. LANGONE

VOLUME 179. Complex Carbohydrates (Part F)
Edited by VICTOR GINSBURG

VOLUME 180. RNA Processing (Part A: General Methods)
Edited by JAMES E. DAHLBERG AND JOHN N. ABELSON

VOLUME 181. RNA Processing (Part B: Specific Methods)
Edited by JAMES E. DAHLBERG AND JOHN N. ABELSON

VOLUME 182. Guide to Protein Purification
Edited by MURRAY P. DEUTSCHER

VOLUME 183. Molecular Evolution: Computer Analysis of Protein and Nucleic Acid Sequences
Edited by RUSSELL F. DOOLITTLE

VOLUME 184. Avidin–Biotin Technology
Edited by MEIR WILCHEK AND EDWARD A. BAYER

VOLUME 185. Gene Expression Technology
Edited by DAVID V. GOEDDEL

VOLUME 186. Oxygen Radicals in Biological Systems (Part B: Oxygen Radicals and Antioxidants)
Edited by LESTER PACKER AND ALEXANDER N. GLAZER

VOLUME 187. Arachidonate Related Lipid Mediators
Edited by ROBERT C. MURPHY AND FRANK A. FITZPATRICK

VOLUME 188. Hydrocarbons and Methylotrophy
Edited by MARY E. LIDSTROM

VOLUME 189. Retinoids (Part A: Molecular and Metabolic Aspects)
Edited by LESTER PACKER

VOLUME 190. Retinoids (Part B: Cell Differentiation and Clinical Applications)
Edited by LESTER PACKER

VOLUME 191. Biomembranes (Part V: Cellular and Subcellular Transport: Epithelial Cells)
Edited by SIDNEY FLEISCHER AND BECCA FLEISCHER

VOLUME 192. Biomembranes (Part W: Cellular and Subcellular Transport: Epithelial Cells)
Edited by SIDNEY FLEISCHER AND BECCA FLEISCHER

VOLUME 193. Mass Spectrometry
Edited by JAMES A. MCCLOSKEY

VOLUME 194. Guide to Yeast Genetics and Molecular Biology
Edited by CHRISTINE GUTHRIE AND GERALD R. FINK

VOLUME 195. Adenylyl Cyclase, G Proteins, and Guanylyl Cyclase
Edited by ROGER A. JOHNSON AND JACKIE D. CORBIN

VOLUME 196. Molecular Motors and the Cytoskeleton
Edited by RICHARD B. VALLEE

VOLUME 197. Phospholipases
Edited by EDWARD A. DENNIS

VOLUME 198. Peptide Growth Factors (Part C)
Edited by DAVID BARNES, J. P. MATHER, AND GORDON H. SATO

VOLUME 199. Cumulative Subject Index Volumes 168–174, 176–194

VOLUME 200. Protein Phosphorylation (Part A: Protein Kinases: Assays, Purification, Antibodies, Functional Analysis, Cloning, and Expression)
Edited by TONY HUNTER AND BARTHOLOMEW M. SEFTON

VOLUME 201. Protein Phosphorylation (Part B: Analysis of Protein Phosphorylation, Protein Kinase Inhibitors, and Protein Phosphatases)
Edited by TONY HUNTER AND BARTHOLOMEW M. SEFTON

VOLUME 202. Molecular Design and Modeling: Concepts and Applications (Part A: Proteins, Peptides, and Enzymes)
Edited by JOHN J. LANGONE

VOLUME 203. Molecular Design and Modeling: Concepts and Applications (Part B: Antibodies and Antigens, Nucleic Acids, Polysaccharides, and Drugs)
Edited by JOHN J. LANGONE

VOLUME 204. Bacterial Genetic Systems
Edited by JEFFREY H. MILLER

VOLUME 205. Metallobiochemistry (Part B: Metallothionein and Related Molecules)
Edited by JAMES F. RIORDAN AND BERT L. VALLEE

VOLUME 206. Cytochrome P450
Edited by MICHAEL R. WATERMAN AND ERIC F. JOHNSON

VOLUME 207. Ion Channels
Edited by BERNARDO RUDY AND LINDA E. IVERSON

VOLUME 208. Protein–DNA Interactions
Edited by ROBERT T. SAUER

VOLUME 209. Phospholipid Biosynthesis
Edited by EDWARD A. DENNIS AND DENNIS E. VANCE

VOLUME 210. Numerical Computer Methods
Edited by LUDWIG BRAND AND MICHAEL L. JOHNSON

VOLUME 211. DNA Structures (Part A: Synthesis and Physical Analysis of DNA)
Edited by DAVID M. J. LILLEY AND JAMES E. DAHLBERG

VOLUME 212. DNA Structures (Part B: Chemical and Electrophoretic Analysis of DNA)
Edited by DAVID M. J. LILLEY AND JAMES E. DAHLBERG

VOLUME 213. Carotenoids (Part A: Chemistry, Separation, Quantitation, and Antioxidation)
Edited by LESTER PACKER

VOLUME 214. Carotenoids (Part B: Metabolism, Genetics, and Biosynthesis)
Edited by LESTER PACKER

VOLUME 215. Platelets: Receptors, Adhesion, Secretion (Part B)
Edited by JACEK J. HAWIGER

VOLUME 216. Recombinant DNA (Part G)
Edited by RAY WU

VOLUME 217. Recombinant DNA (Part H)
Edited by RAY WU

VOLUME 218. Recombinant DNA (Part I)
Edited by RAY WU

VOLUME 219. Reconstitution of Intracellular Transport
Edited by JAMES E. ROTHMAN

VOLUME 220. Membrane Fusion Techniques (Part A)
Edited by NEJAT DÜZGÜNEŞ

VOLUME 221. Membrane Fusion Techniques (Part B)
Edited by NEJAT DÜZGÜNEŞ

VOLUME 222. Proteolytic Enzymes in Coagulation, Fibrinolysis, and Complement Activation (Part A: Mammalian Blood Coagulation Factors and Inhibitors)
Edited by LASZLO LORAND AND KENNETH G. MANN

VOLUME 223. Proteolytic Enzymes in Coagulation, Fibrinolysis, and Complement Activation (Part B: Complement Activation, Fibrinolysis, and Nonmammalian Blood Coagulation Factors)
Edited by LASZLO LORAND AND KENNETH G. MANN

VOLUME 224. Molecular Evolution: Producing the Biochemical Data
Edited by ELIZABETH ANNE ZIMMER, THOMAS J. WHITE, REBECCA L. CANN, AND ALLAN C. WILSON

VOLUME 225. Guide to Techniques in Mouse Development
Edited by PAUL M. WASSARMAN AND MELVIN L. DEPAMPHILIS

VOLUME 226. Metallobiochemistry (Part C: Spectroscopic and Physical Methods for Probing Metal Ion Environments in Metalloenzymes and Metalloproteins)
Edited by JAMES F. RIORDAN AND BERT L. VALLEE

VOLUME 227. Metallobiochemistry (Part D: Physical and Spectroscopic Methods for Probing Metal Ion Environments in Metalloproteins)
Edited by JAMES F. RIORDAN AND BERT L. VALLEE

VOLUME 228. Aqueous Two-Phase Systems
Edited by HARRY WALTER AND GÖTE JOHANSSON

VOLUME 229. Cumulative Subject Index Volumes 195–198, 200–227

VOLUME 230. Guide to Techniques in Glycobiology
Edited by WILLIAM J. LENNARZ AND GERALD W. HART

VOLUME 231. Hemoglobins (Part B: Biochemical and Analytical Methods)
Edited by JOHANNES EVERSE, KIM D. VANDEGRIFF, AND ROBERT M. WINSLOW

VOLUME 232. Hemoglobins (Part C: Biophysical Methods)
Edited by JOHANNES EVERSE, KIM D. VANDEGRIFF, AND ROBERT M. WINSLOW

VOLUME 233. Oxygen Radicals in Biological Systems (Part C)
Edited by LESTER PACKER

VOLUME 234. Oxygen Radicals in Biological Systems (Part D)
Edited by LESTER PACKER

VOLUME 235. Bacterial Pathogenesis (Part A: Identification and Regulation of Virulence Factors)
Edited by VIRGINIA L. CLARK AND PATRIK M. BAVOIL

VOLUME 236. Bacterial Pathogenesis (Part B: Integration of Pathogenic Bacteria with Host Cells)
Edited by VIRGINIA L. CLARK AND PATRIK M. BAVOIL

VOLUME 237. Heterotrimeric G Proteins
Edited by RAVI IYENGAR

VOLUME 238. Heterotrimeric G-Protein Effectors
Edited by RAVI IYENGAR

VOLUME 239. Nuclear Magnetic Resonance (Part C)
Edited by THOMAS L. JAMES AND NORMAN J. OPPENHEIMER

VOLUME 240. Numerical Computer Methods (Part B)
Edited by MICHAEL L. JOHNSON AND LUDWIG BRAND

VOLUME 241. Retroviral Proteases
Edited by LAWRENCE C. KUO AND JULES A. SHAFER

VOLUME 242. Neoglycoconjugates (Part A)
Edited by Y. C. LEE AND REIKO T. LEE

VOLUME 243. Inorganic Microbial Sulfur Metabolism
Edited by HARRY D. PECK, JR., AND JEAN LEGALL

VOLUME 244. Proteolytic Enzymes: Serine and Cysteine Peptidases
Edited by ALAN J. BARRETT

VOLUME 245. Extracellular Matrix Components
Edited by E. RUOSLAHTI AND E. ENGVALL

VOLUME 246. Biochemical Spectroscopy
Edited by KENNETH SAUER

VOLUME 247. Neoglycoconjugates (Part B: Biomedical Applications)
Edited by Y. C. LEE AND REIKO T. LEE

VOLUME 248. Proteolytic Enzymes: Aspartic and Metallo Peptidases
Edited by ALAN J. BARRETT

VOLUME 249. Enzyme Kinetics and Mechanism (Part D: Developments in Enzyme Dynamics)
Edited by DANIEL L. PURICH

VOLUME 250. Lipid Modifications of Proteins
Edited by PATRICK J. CASEY AND JANICE E. BUSS

VOLUME 251. Biothiols (Part A: Monothiols and Dithiols, Protein Thiols, and Thiyl Radicals)
Edited by LESTER PACKER

VOLUME 252. Biothiols (Part B: Glutathione and Thioredoxin; Thiols in Signal Transduction and Gene Regulation)
Edited by LESTER PACKER

VOLUME 253. Adhesion of Microbial Pathogens
Edited by RON J. DOYLE AND ITZHAK OFEK

VOLUME 254. Oncogene Techniques
Edited by PETER K. VOGT AND INDER M. VERMA

VOLUME 255. Small GTPases and Their Regulators (Part A: Ras Family)
Edited by W. E. BALCH, CHANNING J. DER, AND ALAN HALL

VOLUME 256. Small GTPases and Their Regulators (Part B: Rho Family)
Edited by W. E. BALCH, CHANNING J. DER, AND ALAN HALL

VOLUME 257. Small GTPases and Their Regulators (Part C: Proteins Involved in Transport)
Edited by W. E. BALCH, CHANNING J. DER, AND ALAN HALL

VOLUME 258. Redox-Active Amino Acids in Biology
Edited by JUDITH P. KLINMAN

VOLUME 259. Energetics of Biological Macromolecules
Edited by MICHAEL L. JOHNSON AND GARY K. ACKERS

VOLUME 260. Mitochondrial Biogenesis and Genetics (Part A)
Edited by GIUSEPPE M. ATTARDI AND ANNE CHOMYN

VOLUME 261. Nuclear Magnetic Resonance and Nucleic Acids
Edited by THOMAS L. JAMES

VOLUME 262. DNA Replication
Edited by JUDITH L. CAMPBELL

VOLUME 282. Vitamins and Coenzymes (Part L)
Edited by DONALD B. MCCORMICK, JOHN W. SUTTIE, AND CONRAD WAGNER

VOLUME 283. Cell Cycle Control
Edited by WILLIAM G. DUNPHY

VOLUME 284. Lipases (Part A: Biotechnology)
Edited by BYRON RUBIN AND EDWARD A. DENNIS

VOLUME 285. Cumulative Subject Index Volumes 263, 264, 266–284, 286–289

VOLUME 286. Lipases (Part B: Enzyme Characterization and Utilization)
Edited by BYRON RUBIN AND EDWARD A. DENNIS

VOLUME 287. Chemokines
Edited by RICHARD HORUK

VOLUME 288. Chemokine Receptors
Edited by RICHARD HORUK

VOLUME 289. Solid Phase Peptide Synthesis
Edited by GREGG B. FIELDS

VOLUME 290. Molecular Chaperones
Edited by GEORGE H. LORIMER AND THOMAS BALDWIN

VOLUME 291. Caged Compounds
Edited by GERARD MARRIOTT

VOLUME 292. ABC Transporters: Biochemical, Cellular, and Molecular Aspects
Edited by SURESH V. AMBUDKAR AND MICHAEL M. GOTTESMAN

VOLUME 293. Ion Channels (Part B)
Edited by P. MICHAEL CONN

VOLUME 294. Ion Channels (Part C)
Edited by P. MICHAEL CONN

VOLUME 295. Energetics of Biological Macromolecules (Part B)
Edited by GARY K. ACKERS AND MICHAEL L. JOHNSON

VOLUME 296. Neurotransmitter Transporters
Edited by SUSAN G. AMARA

VOLUME 297. Photosynthesis: Molecular Biology of Energy Capture
Edited by LEE MCINTOSH

VOLUME 298. Molecular Motors and the Cytoskeleton (Part B)
Edited by RICHARD B. VALLEE

VOLUME 299. Oxidants and Antioxidants (Part A)
Edited by LESTER PACKER

VOLUME 300. Oxidants and Antioxidants (Part B)
Edited by LESTER PACKER

VOLUME 301. Nitric Oxide: Biological and Antioxidant Activities (Part C)
Edited by LESTER PACKER

VOLUME 320. Cumulative Subject Index Volumes 290–319 (in preparation)

VOLUME 321. Numerical Computer Methods (Part C)
Edited by MICHAEL L. JOHNSON AND LUDWIG BRAND

VOLUME 322. Apoptosis
Edited by JOHN C. REED

VOLUME 323. Energetics of Biological Macromolecules (Part C)
Edited by MICHAEL L. JOHNSON AND GARY K. ACKERS

VOLUME 324. Branched-Chain Amino Acids (Part B)
Edited by ROBERT A. HARRIS AND JOHN R. SOKATCH

VOLUME 325. Regulators and Effectors of Small GTPases (Part D: Rho Family)
Edited by W. E. BALCH, CHANNING J. DER, AND ALAN HALL

VOLUME 326. Applications of Chimeric Genes and Hybrid Proteins (Part A: Gene Expression and Protein Purification)
Edited by JEREMY THORNER, SCOTT D. EMR, AND JOHN N. ABELSON

VOLUME 327. Applications of Chimeric Genes and Hybrid Proteins (Part B: Cell Biology and Physiology)
Edited by JEREMY THORNER, SCOTT D. EMR, AND JOHN N. ABELSON

VOLUME 328. Applications of Chimeric Genes and Hybrid Proteins (Part C: Protein–Protein Interactions and Genomics)
Edited by JEREMY THORNER, SCOTT D. EMR, AND JOHN N. ABELSON

VOLUME 329. Regulators and Effectors of Small GTPases (Part E: GTPases Involved in Vesicular Traffic) (in preparation)
Edited by W. E. BALCH, CHANNING J. DER, AND ALAN HALL

VOLUME 330. Hyperthermophilic Enzymes (Part A) (in preparation)
Edited by MICHAEL W. W. ADAMS AND ROBERT M. KELLY

VOLUME 331. Hyperthermophilic Enzymes (Part B) (in preparation)
Edited by MICHAEL W. W. ADAMS AND ROBERT M. KELLY

VOLUME 332. Regulators and Effectors of Small GTPases (Part F: Ras Family I) (in preparation)
Edited by W. E. BALCH, CHANNING J. DER, AND ALAN HALL

VOLUME 333. Regulators and Effectors of Small GTPases (Part G: Ras Family II) (in preparation)
Edited by W. E. BALCH, CHANNING J. DER, AND ALAN HALL

VOLUME 334. Hyperthermophilic Enzymes (Part C) (in preparation)
Edited by MICHAEL W. W. ADAMS AND ROBERT M. KELLY

VOLUME 335. Flavonoids and Other Polyphenols (in preparation)
Edited by LESTER PACKER

VOLUME 336. Microbial Growth in Biofilms (Part A: Developmental and Molecular Biological Aspects) (in preparation)
Edited by RONALD J. DOYLE

VOLUME 337. Microbial Growth in Biofilms (Part B: Special Environments and Physicochemical Aspects) (in preparation)
Edited by RONALD J. DOYLE

Subject Index

Boldface numerals indicate volume number.

A

A23187, 1-(4,5-dimethoxy-2-nitrophenyl)-ethyl carboxylate synthesis, **291:**39–40

AAC, *see* ADP, ATP carrier

AAPH, *see* 2,2′-Azobis[2-amidinopro-pane]hydrochloride

AApoAII, *see* Murine senile amyloid fibril

AAV, *see* Adeno-associated virus

ABC transporter, *see* ATP-binding cassette transporter

ABCR

  ATPase activity

    activators and inhibitors, overview, **315:**871–872, 887–888

    all-*trans*-retinal activation

      dose response, **315:**888–889

      factors affecting activation, **315:**889–890

      mechanisms, **315:**894–895

      synergistic activation with amioda-rone, **315:**891–892, 894

    amiodarone activation

      dose response, **315:**888

      synergistic activation with all-*trans*-retinal, **315:**891–892, 894

    charcoal-binding assay, **315:**882

    kinetic parameters, **315:**871

    nucleotide-binding domains expressed as maltose binding protein fusion proteins, **315:**878

    reconstituted protein, **315:**874, 886

    thin-layer chromatography assay

      incubation conditions, **315:**869–870

      linearity, **315:**870

      product separation, **315:**870

      sensitivity, **315:**878–879

      solutions, **315:**869

  ATP-binding cassette, **315:**865

  gene cloning, **315:**864

  localization in eye

    immunostaining, **315:**883–884

    in situ hybridization, **315:**883–884

  mutation in disease, **315:**864, 879–880

  nucleotide-binding domain expression as maltose binding protein fusion pro-teins

    affinity chromatography, **315:**877–878

    expression vector construction, **315:**876–877

    induction of *Escherichia coli*, **315:**877

    solubility, **315:**875

  purification from bovine retina

    gel electrophoresis and Western blot analysis, **315:**868–869

    immunoaffinity chromatography

      antibody preparation, **315:**866

      binding and elution, **315:**867–868, 881, 885

      buffers, **315:**866

      column preparation, **315:**881

      sample preparation and loading, **315:**866–867, 884–885

    rod outer segment isolation, **315:**865

    stability, **315:**870–871

    yield, **315:**872–873

  reconstitution

    porcine brain lipids, **315:**882, 885–886

    soybean phospholipid vesicles, **315:**873–874

  retinal flipping function, **315:**895

  Stargardt disease pathogenesis, **315:**896–897

  structure, **315:**865

c-ABL

  activation in chronic myeloid leukemia, **314:**429–430, 440

  antisense knockdown in CD34$^+$ hemato-poietic progenitor cells

    cell separations, **314:**432–433

    clonogenic assays, **314:**436–437

    flow cytometry analysis of cell cycle

and bromodeoxyuridine uptake,
314:437–439
incubation conditions, **314:**434–435
oligodeoxynucleotide design, **314:**434
reverse transcription-polymerase chain
reaction analysis, **314:**435–436
S phase assessment by cytosine arabino-
side suicide, **314:**437, 439
cell cycle regulation, **314:**431–432,
438–439
developmental expression, **314:**431
protein structure and function,
314:430–431
retrovirus homologs, **314:**429
RNA splicing, **314:**430
[³H]Azidoacetylcholine
photoaffinity labeling of vesicular acetylcholine transporter from *Torpedo*
vesicles
denaturing gel electrophoresis and
autofluorography, **296:**113
photolysis conditions, **296:**110–113
reversibility of binding,
296:107–109
structure, **296:**101
synthesis, **296:**101–102
[³H]Azidoaminobenzovesamicol
photoaffinity labeling of vesicular acetylcholine transporter from *Torpedo*
vesicles
denaturing gel electrophoresis and
autofluorography, **296:**114–116
photolysis conditions, **296:**113–114
reversibility of binding, **296:**109
structure, **296:**101
synthesis, **296:**102–104
Ab initio modeling, *see* Three-dimensional
structure, molecular modeling
Abscisic acid, caged
biological function after photolysis,
291:480–483
calcium response measurement with fluorescent probes, **291:**478, 481–482
injection into plant cells, **291:**477–478
photolysis conditions, **291:**478–480
structure, **291:**475
synthesis, **291:**475–477
Absorption spectroscopy, *see* Absorption
spectroscopy, rhodopsin; Ultraviolet/vis-
ible spectroscopy

Absorption spectroscopy, rhodopsin
advantages and applications, **315:**164
all-*trans*-retinal activation of opsin,
315:244–245
bovine opsin expressed in *Pichia pastoris*,
315:8
constitutively active opsin mutants,
315:216–217
fragment reconstitution studies,
315:66–68
kinetic range of photoreactions, **315:**166
meta-rhodopsin intermediates, **315:**302–
303, 305–308
primary photolysis
laser sources, **315:**165
signal/noise ratio optimization, **315:**165
Schiff base
bleach measurement, **315:**171
flash lamps as light sources,
315:166–167
flow cells, **315:**168–169
global fitting of time-dependent spectra, **315:**171–173
indicator dye monitoring, **315:**177–
178
isospectral intermediate analysis,
315:177
kinetic resolution of activation steps,
315:176
kinetic scheme determination, **315:**173,
175
pK_a determination from absorbance
spectra with model chromophores,
315:201, 203
probe beam polarization, **315:**167–168
protonated form, **315:**196–198
signal averaging, **315:**168–169, 171
time-resolved low-temperature absorption spectroscopy of rhodopsin intermediates interacting with transducin
absorption maxima of intermediates,
315:361–362
applications and overview, **315:**350–351
collected spectra, **315:**353, 355
experimental setup, **315:**351–352
GTP analog studies, **315:**352, 359, 361,
363
intermediates interacting with transducin, **315:**348–349, 358–359,
361–363

photoconverted pigment, quantification, **315:**352–353
rhodopsin purification, **315:**349
spectral analysis, **315:**355, 357–358
transducin purification, **315:**349–350
transducin activation assays
extra metarhodopsin-II, **315:**476, 478
photoregeneration from metarhodopsin-II state, **315:**479
tryptophans in transmembrane segment
data collection, **315:**119–120
difference spectroscopy analysis of mutants, **315:**120, 122
chromophore photoisomerization sensing, **315:**123–124
wavelengths of maximal absorption, identification of critical amino acids
ancestral pigment sequence elucidation, **315:**320–321
phylogenetic tree construction
neighbor-joining method, **315:**315,
319–320
programs, **315:**313–314
unweighted pair-group method with
arithmetic mean, **315:**314–315
site-directed mutagenesis in functional
testing
absorbance spectroscopy, **315:**324
applications, **315:**325
bovine A292S, **315:**324
cloning of genes, **315:**321–322
mutagenesis, **315:**322–323
purification, **315:**324
regeneration, **315:**323–324
RNA isolation, **315:**321
wild-type pigment absorption, **315:**312,
323–324
ABTS, *see* 2,2′-Azinobis(3-ethylbenzothiazo-
line-6-sulfonic acid)
3-Acetamido-4,4-dimethylthietan-2-one, synthesis, **301:**244
Acetic acid, caged compound
pH changes using photolysis, **291:**246,
248–249
synthesis, **291:**247–248
Acetylation mapping, *see* Histone
Acetylcholine receptor
molecular dynamics simulated annealing
with restrained molecular dynamics,
293:656, 658–659

substituted-cysteine accesibility method
analysis, **293:**124, 130, 133, 135–139,
143–144
transient expression
assembly efficiency assays, **293:**579–
581, 585
α-bungarotoxin binding assays in detection
cell-surface binding assay,
293:569–570
fluorescence microscopy,
293:570–572
calcium phosphate transfection,
293:568–569
cell line selection, **293:**568, 572–573
cell-surface distribution, effects of expression efficiency, **293:**581–582,
584–585
heteromeric receptor expression, complications, **293:**576–579
homomeric versus heteromeric expression levels, **293:**574–576, 581, 584
metabolic labeling and immunoprecipitation in detection, **293:**572
overview of gene delivery methods,
293:565–566
receptor types, **293:**567
Acetylcholinesterase
caged ligands
carbamylcholine, **291:**266–267, 269,
273–276
choline, **291:**266–267, 269, 273–276
criteria for ideal compounds, **291:**273
inhibition analysis, **291:**272–274
kinetics of photolysis, **291:**273
noracetylcholine, **291:**266–267, 269
norbutyrylcholine, **291:**267, 269
photochemical properties,
291:269–272
photolysis conditions, **291:**277–278
recovery of activity following photolysis, **291:**275–276, 278
inhibitor evaluation with chemiluminescent imaging of microtiter plates,
305:123–124, 126, 128
physiological function, **291:**265
Acetylcholine transporter, *see* Vesicular acetylcholine transporter
N-Acetylcysteine, oxidative stress inhibition, **319:**260–261

N-Acetylgalactosaminyltransferase
    assay
        incubation conditions, **311:**86–87
        materials, **311:**85
        product identification and structural
            analysis, **311:**89–91
        product separations, **311:**88,
            93–94
        quantitative analysis, **311:**88–89
        stock solutions, **311:**86
    knockout mice, **311:**82
    reaction catalyzed, **311:**83
β-N-Acetylhexosaminidase, sphingolipid
    identification, **311:**668–669
N-Acetylneuramine lactitol, ganglioside si-
    alidase assay, **312:**353–354
N-Acetylsphingosine, tritiated compound
    synthesis, **311:**112
AChE, *see* Acetylcholinesterase
AChR, *see* Acetylcholine receptor
Acid ceramidase
    activator proteins, **311:**201–202, 206
    assay
        alkaline hydrolysis, **311:**203
        incubation conditions, **311:**202–203
        quantification of liberated sphingosine,
            311:203–204
        reagents, **311:**202
    properties of human placenta enzyme,
        311:207
    purification from human placenta
        ammonium sulfate precipitation,
            311:204–205
        anion-exchange chromatography,
            311:206
        concanavalin A affinity chromatogra-
            phy, **311:**205
        dye affinity chromatography,
            311:205–206
        homogenization, **311:**204
        hydrophobic affinity chromatography,
            311:205
        sphingolipid activator protein D separa-
            tion, **311:**206
Acid sphingomyelinase, *see also* Sphingomy-
    elinase
    assays
        incubation conditions, **311:**150–151, 166
        precipitation assay for high
            throughput, **311:**183

principle, **311:**165
radioactivity quantification,
    311:166–167
reagents, **311:**165–166
sample preparation, **311:**165
substrate/detergent-mixed micelle prep-
    aration, **311:**166
caveolae, biological significance,
    311:185–186
deficiency in Niemann-Pick disease,
    311:149
human enzyme properties
    glycosylation, **311:**155–156
    proteolytic processing, **311:**155–156
    size, **311:**155–156
human enzyme purification
    placental enzyme, **311:**155–156
    recombinant protein from baculovirus/
        insect cell system
        ammonium sulfate precipitation,
            311:152–153
        anion-exchange chromatography,
            311:154
        cell growth, **311:**151
        concanavalin A affinity chromatogra-
            phy, **311:**153–154
        dye affinity chromatography,
            311:153–154
        expression, **311:**151
        hydrophobic affinity chromatogra-
            phy, **311:**153–154
        modified protocol, **311:**154–155
        overview of purification scheme,
            311:152
    urine enzyme, **311:**149
posttranslational processing, **311:**150
reaction catalyzed, **311:**149
Acousto-optic deflector, rapid confocal mi-
    croscope scanning, **307:**400–402
Acridan luminescence
    educational demonstrations, **305:**651–652
    enzyme substrate incorporation and
        assay, **305:**381, 383
    theory, **305:**32–35
Acridine luminescence, *see also* Lucigenin
    educational applications
        demonstrations, **305:**649–650
        mechanism of luminescence, **305:**650
        overview, **305:**647–648
    synthesis

acridine-9-carbonyl chloride, **305:**649
acridine-9-carboxylic acid,
    305:648–649
phenyl acridine-9-carboxylate,
    305:649
phenyl 10-methylacridinium-9-car-
    boxylate methosulfate, **305:**649
safety, **305:**648
emission characteristics, **305:**74
theory, **305:**32–35
Acridine orange
cell counting in biofilms, **310:**545–548
general bacterial activity assessment in
    biofilms, **310:**172
Actin, *see also* Actin polymerization; Cy-
    toskeleton, confocal microscopy;
    F-actin; G-actin; Muscle fiber; Myocyte
cross-linking with photocleavable re-
    agents
    caged G-actin complex preparation,
        291:173–175
    cross-linking with 4-bromomethyl-3-
        nitrobenzoic acid succinimide es-
        ter, **291:**159–162
    labeling with tetramethylrhodamine
        iodoacetamide, **291:**157–158
    labeling with thiol-reactive caged rhoda-
        mine reagent, **291:**170–171, 173
G-actin, 6-nitroveratryloxycarbonyl chlo-
    ride caged conjugate
    fluorescence photoactivation micros-
        copy, **291:**104–105
    functional analysis of caged and de-
        caged proteins, **291:**105–107
    labeling ratio, **291:**101–102
    photoactivation conditions, **291:**104
    polymerization competence,
        291:102–103
    preparation, **291:**101
    tetramethylrhodamine labeling of
        caged protein, **291:**103
labeling with fluorescent caged com-
    pounds, **291:**72–74
α-Actinin, *see* Myocyte
Actin polymerization
minimal mechanism, **308:**93
nucleotide exchange promotion by F,
    308:93
self-assembly models
    ATP role

actin depolymerizing factor control,
    308:103, 105
bidirectional polymerization with hy-
    drolysis at time of monomer ad-
    dition, **308:**99–101
delayed hydrolysis after monomer ad-
    dition, **308:**101–102
difficulty of study, **308:**110–111
overview, **308:**94
timing control of assembly and disas-
    sembly, **308:**102–103, 110
unidirectional polymerization with
    hydrolysis at time of monomer
    addition, **308:**97–99
condensation equilibrium model,
    308:94–96
elongation without nucleotide hydroly-
    sis, **308:**96–97
Action potential, calcium influx mea-
    surement
action potential waveform voltage clamp
    applications, **293:**353–354
    isolation of calcium currents, **293:**354
    quality of voltage clamp, **293:**354–356
    signal-to-noise ratio, **293:**364–366, 371
    voltage template selection, **293:**356–358
comparison of voltage clamp and fluores-
    cence measurements, **293:**363–370
fluorometric measurement
    calibration, **293:**361–362
    detection systems, **293:**362–363,
        368–369
    dye overload, **293:**358–361
    dye types, **293:**359
    quantification of fluorescence change,
        293:358, 360–362
    signal-to-noise ratio, **293:**366–368
overview, **293:**352
simultaneous measurement of voltage
    clamp and fluorescence, **293:**369–370
Activation energy
energy for catalysis, **308:**20–24
rate constant relationship, **308:**4–5
rate acceleration calculation
    bimolecular enzymatic reactions,
        308:9–10
    unimolecular enzymatic reactions,
        308:6
temperature effects, **308:**13–15
Acute myelogenous leukemia, cerebrospi-

nal fluid analysis of gangliosides, **312**:137–138, 143

1-*O*-Acylceramide
discovery, **311**:106
synthetic pathway characterization
1-*O*-acyl-*N*-acetylsphingosine identification, **311**:107
acyl donor identification, **311**:108–110
cell homogenate studies, **311**:107
transacylase specificity for sphingolipid acceptors, **311**:110

Acyl-CoA:sphinganine *N*-acyltransferase
assay, **311**:135, 138–139
biological significance in plants, **311**:140
substrate specificity of plant enzyme, **311**:139–140

N-Acyl-3,6-dihydro-1,2-oxazines, nitroxyl synthesis by thermal decomposition, **301**:219–220

Acylhomoserine lactone autoinducers
bioassays for detection
analog assays, **305**:313–315
butanoylhomoserine lactone assay, **305**:293
Chromobacterium violaceum assay, **305**:294
controls, **305**:290
hexanoylhomoserine lactone assay, **305**:293
homoserine lactone assay for side chain lengths of 8–14 carbons, **305**:293
3-hydroxybutanoylhomoserine lactone assay, **305**:293
3-oxohexanoylhomoserine lactone assay, **305**:292–293
3-oxooctanoylhomoserine lactone assay, **305**:294
principles, **305**:289–290
reporter strains, **305**:290–291
sample preparation, **305**:290
standard curves, **305**:290, 292
unsubstitututed homoserine lactone assay with longer side chain lengths, **305**:294
chemical synthesis
chloroformate method for simple acyl-homoserine lactone synthesis, **305**:307–308

dicyclohexylcarbodiimide coupling, **305**:308–309
large-scale synthesis
analogs, **305**:310
3-oxohexanoylhomoserine lactone, **305**:309–310
overview, **305**:300–303
3-oxoacylhomoserine lactones
coupling with 1-ethyl-3-(3-dimethyl-aminopropyl)carbodiimide, **305**:305–307
Meldrum's acid synthesis, **305**:303–305
storage of analogs, **305**:303
yields, **305**:302
infrared spectroscopy, **305**:310
nuclear magnetic resonance spectra
carbon-13 spectroscopy, **305**:312
proton spectroscopy, **305**:311–312
overview, **305**:279–281, 288
purification
cell growth, **305**:295
extraction, **305**:295
reverse-phase chromatography, **305**:295–296
stability, **305**:295, 300–301
structure
comparisons, **305**:288–289, 302
elucidation
mass spectrometry, **305**:296–297, 312
nuclear magnetic resonance of 7,8-*cis*-tetradecenoylhomoserine lactone, **305**:297, 299
thin-layer chromatography identification, **305**:300

N-Acylsphingosine aminohydrolase, *see* Acid ceramidase

N-Acylsphingosine deacylase, *see* Ceramidase

AD, *see* Alzheimer's disease

Adenine phosphoribosyltransferase
assay, **317**:25–27
purification of recombinant enzyme from *Escherichia coli*, **317**:23–25

Adeno-associated virus
antisense delivery
angiotensinogen targeting, **314**:45
angiotensin II type 1 receptor targeting effects, **314**:45–46, 50–51

plasmid preparation
  construction and design, **314:**45–46
  large-scale preparation, **314:**42,
    44–45
  recombinant virus preparation,
    314:47–49
  tissue-specific expression, **314:**51
  titer assay, **314:**49–50
chromosomal integration, **314:**40–41
cystic fibrosis transmembrane regulator
  gene transfer
  in vitro transfer, **292:**728–729
  in vivo transfer
    clinical trials, **292:**731–732
    macaque, **292:**729, 731
    rabbit, **292:**729
gene therapy applications, **314:**41–42
genomic features of virus, **292:**725–726,
  314:40
helper plasmid, **316:**745
infectious center assay, **316:**756–757,
  759–760
insertion locus, **292:**539, 725, 728
integration of *rep* negative vectors,
  292:544–545
inverted terminal repeat sequences,
  292:539, 544
latency, **292:**725
MDR1
  recombinant virion production,
    292:540–543
  titration of vector, **292:**543
  transduction and selection of express-
    ing cells, **292:**543, 545
packaging size and site, **292:**539–540, 545,
  727, **314:**41
proliferating cell transfection efficiency,
  292:728
purification of recombinant virus
  concentrating with filters, **316:**755–756
  containment facility, **316:**746
  cotransfection of viral and helper plas-
    mids, **316:**746–747
  heparin affinity chromatography,
    316:753–756
  host cells, **316:**746
  iodixanol density step gradient purifi-
    cation
    advantages over cesium chloride,
      316:748, 760–761

aggregation prevention, **316:**747–748,
  760
density, **316:**748
efficiency and resolving capacity,
  316:749, 751, 753
gradient formation and centrifugation, **316:**753
iodixanol properties, **316:**748
lysis of cells, **316:**747
overview, **316:**744–746, 760–761
plasmid vector, **316:**746
technique effects on gene transfer efficiency, **316:**760
recombinant gene transfer applications,
  316:743–744, 777
replication, **314:**40
retinal transduction
  animal species comparisons
    humoral immune response, **316:**785,
      787, 789
    time of onset for expression,
      316:784–785
  green fluorescent protein reporter
    advantages, **316:**777
    histological analysis, **316:**785
    ophthalmoscopy, **316:**783–784
  intraocular administration in animals,
    316:779–783
  promoter, **316:**777
  recombinant virus preparation,
    316:778–779
safety as vector, **314:**41
titering by quantitative polymerase chain
  reaction, **316:**756–757
transfer efficiencies in various cells,
  292:727–728
Adenosine A$_1$ receptor, antisense targeting,
  313:29–30
Adenosine deaminase
  molecular surface point similarity mea-
    sures in binding energy analysis,
    308:404–405, 407, 409
  substrate identification, *see* Inosine-con-
    taining messenger RNA
S-Adenosylmethionine synthetase, kinetic
  isotope effect modeling of transition-
  state, **308:**343
Adenovirus
  antisense amylin delivery, **314:**38
  cell infection

hippocampal slices, **293:**499–500, 502–503

pathway, **293:**484

toxicity problems, **293:**494–495, 502–503

gene therapy applications, **314:**39

G-protein-coupled inward rectifier potassium channel expression, **293:**496–497, 499

inducible systems, **293:**503

recombinant virus preparation

cotransfection into human embryonic kidney cells, **293:**489

DNA purification, **293:**492

expression cassette, **293:**485, 487, 493–494

large-scale amplification, **293:**491

overview, **293:**484–485

plaque purification, **293:**489–490

right arm viral DNA preparation, **293:**488

screening, **293:**490–491

virus purification with cesium chloride gradients, **293:**491–492

replication, **314:**38–39

ribozyme gene delivery, **313:**503–506

safety, **293:**488

Shaker-type potassium channel expression

electrophysiologic studies, **293:**496–497

transfection tests using green fluorescent protein as reporter, **293:**495–496

titration of virus

cell lysis assay, **293:**493

plaque assay, **293:**492

Adenovirus minichromosome, *see* Encapsidated adenovirus minichromosome

Adhesion, *see* Cell–cell adhesion assay

ADP, *p*-hydroxyphenacyl caged compound synthesis, **291:**22–24

ADP1, functions in *Saccharomyces cerevisiae*, **292:**152

ADP, ATP carrier

electrophysiological measurements with caged ATP

bilayer system, **291:**298–299

time-resolved measurements, **291:**300–301

transport modes, **291:**300

physiological function, **291:**298

Adrenoleukodystrophy

ALD gene

homology with PMP70, **292:**759–761

locus, **292:**759

mutation in disease, **292:**111–112, 118, 134, 757, 759

ALD-related protein, **292:**762

phenotypes, **292:**757, 759

α_2-Adrenoreceptor

antisense oligonucleotide targeting

functional inhibition versus actual expression inhibition, **314:**70

intracerebroventricular infusion

autoradiography of knockdown, **314:**73

controls, **314:**72

distribution of oligonucleotides, **314:**71

dose, **314:**71

infusion system, **314:**72

nonsecific toxicity, **314:**74–76

overview of studies, **314:**68

primary cortical neuron system

AraC-induced inhibition of DNA synthesis, **314:**64, 66–68

cell culture, **314:**62–63

controls, **314:**64

culture condition effects on receptor expression, **314:**66–67

oligonucleotide uptake, **314:**67–68

overview of studies, **314:**61–62

radioligand binding experiments, **314:**63–64

sequences of probes, **314:**63

specificity, **314:**61

subtypes, **314:**61

turnover rates, **314:**68, 70–71

β_2-Adrenoreceptor, *see* G protein-coupled receptor

ADRP, *see* Autosomal dominant retinitis pigmentosa

Adsorption, proteins

adsorbent chacterization, **309:**406–407

autoradiography, **309:**411, 422–424

flat surface adsorption experiments

advantages, **309:**421

autoradiography and adsorption from flowing solution, **309:**422–424

radiolabeled protein adsorption onto
    silica surface, **309:**421–422
fluorescein isothiocyanate labeling of pro-
    teins, **309:**413–414
high surface area adsorption experiments
    batch depletion, **309:**418–420
    batch repletion, **309:**420–421
    limiting sample-load column,
        309:417–418
    saturating sample-load column,
        309:416–417
interfacial activity, **309:**402–403
isotherms
    axes, **309:**403
    bulk ligand binding function,
        309:404–405
    equilibrium and heterogeneity,
        309:404
    lattice site binding function,
        309:404–405
    two-dimensional lattice of binding
        sites, **309:**404–405
kinetics, **309:**405–406, 427–428
optical techniques, **309:**409
radioiodination of proteins
    chloramine T labeling, **309:**413
    iodine monochloride labeling,
        309:412–413
    isotope features, **309:**411–412
silica surfaces, chemical modification
    cleaning, **309:**414–415
    ligand immobilization, **309:**415–416
    silanization, **309:**415
single versus multiprotein adsorption,
    309:406
solution depletion technique
    adsorption media, **309:**407–408
    batch methods, **309:**408, 418–420
    principle, **309:**407
spectroscopic techniques
    fluorescence
        calibration of total internal reflection
            fluorescence, **309:**426–427
        emission spectra analysis, **309:**428
        intrinsic protein fluorescence, total in-
            ternal reflection fluorescence
            spectroscopy, **309:**424–426
        kinetic analysis with total internal re-
            flection fluorescence,
            309:427–428

overview, **309:**409–410
        quenching studies, **309:**428–429
    infrared spectroscopy, **309:**410–411
    microscopy, **309:**411
    surface plasmon resonance, *see* Surface
        plasmon resonance
    water role in binding, **309:**406–407
ADT, *see* Anethole dithiolthione
Advanced glycation end products
    Alzheimer's disease pathogenesis role,
        309:153
    antibody production and sources,
        309:156–158, 162–163
    enzyme-linked immunosorbent assay
        hemoglobin assay
            clinical application, **309:**166
            hemolysate preparation from fresh
                blood, **309:**166–167
            hemolysate preparation from frozen
                and packed red blood cells,
                309:167
            incubations and washes, **309:**167–168
        incubations and washes, **309:**164
        low-density lipoprotein assay, **309:**168
        plate preparation, **309:**163–164
        serum assay
            clinical application, **309:**164–165
            proteinase K digestion of samples,
                309:165
    formation, overview, **309:**135, 154
    high-performance liquid chromatography,
        309:158
    immunochemical assay
        applications, **309:**157–158
        epitopes, **309:**156–157
        nervous tissue samples, **309:**144–145,
            149
        slide preparation and staining, **309:**169
    mass spectrometry, **309:**158, 170
    preparation of modified proteins
        albumin modification, **309:**162
        buffer selection, **309:**161–162
        glucose as modifying sugar, **309:**161
    reversal of cross-links, **309:**159–161,
        170–172
    structures, **309:**154–156
Advanced lipoxidation end products
    formation, overview, **309:**135
    immunochemical assay, **309:**144–146,
        149–150

AEAETS, *see* 2-Aminoethyl-2-aminoethane thiosulfonate
AEC, *see* 3-Amino-9-ethylcarbazole
AEF, *see* Amyloid enhancing factor
AEMTS, *see* 2-Aminoethylmethane thiosulfonate
Aequorin
  calcium indicator compared with obelin
    calcium concentration–effect curves, **305:**243–245
    fractional rate of discharge, expression of results, **305:**240–241
    magnesium effects, **305:**245–248
    overview, **305:**224–225, 248–249
    rapid kinetics measured by stopped-flow, **305:**245–248
  emission characteristics, **305:**74
  fusion protein targeting to subcellular organelles for calcium measurements
    activation of recombinant protein, **305:**490
    calibration of calcium-dependent luminescence, **305:**491–492
    construction of fusion proteins
      cloning vectors, **305:**483–484
      ligation of complementary DNAs, **305:**492
      oligonucleotides, **305:**484–485
      plasmids and bacterial strains, **305:**482–483
      polymerase chain reaction, **305:**486, 488–490
      reagents, **305:**486
      solutions, **305:**485–486
      transformation of *Escherichia coli*, **305:**492
    endoplasmic reticulum targeting, **305:**481
    expression in living bacteria, **305:**492–493
    imaging, **305:**486, 495–498
    immunolocalization, **305:**495
    nucleus targeting, **305:**482
    plasma membrane targeting with luciferase fusion protein, **305:**482, 494
    rationale, **305:**479–480
    selective permeabilization of plasma membrane, **305:**495
    specific activity determination for re-

combinant proteins, **305:**490–491
  stability in cells, **305:**497–498
  subplasma membrane calcium and ATP measurements with luciferase fusion protein, **305:**494, 498
  recombinant protein preparation, **305:**239–240
  stability, **305:**248
a-factor
  export by Ste6p, **292:**156, 193
  halo assays of export
    patch halo assays, **292:**201
    principle, **292:**198
    sensitivity, **292:**198–199
    spot assays, **292:**199–201
  immunoprecipitation, **292:**201–202
  posttranslational processing, **292:**194
Affinity biopanning, *see* Phage display library
Affinity chromatography, *see* ATP-affinity chromatography; Dye affinity chromatography; Lectin affinity chromatography; Nickel affinity chromatography; Small nuclear ribonucleoprotein; Zinc affinity chromatography
Affinity labeling, *see* Biotin affinity labeling; Photocleavable affinity tags
AFM, *see* Atomic force microscopy
AFP, *see* α-Fetoprotein
ω-Agatoxin, immunohistochemical analysis, **314:**302–303
ω-Agatoxin-IVA, antibody, **294:**677, 693
ω-Agatoxin-IVB
  biological assays, **294:**108–109
  expression and purification in *Escherichia coli*
    expression vectors, **294:**101
    gene cloning, **294:**101
    inclusion body isolation, **294:**103
    large-scale expression, **294:**103–104
    leader peptide cleavage, **294:**108
    matrix-assisted laser desorption mass spectrometry, **294:**109
    nickel affinity chromatography, **294:**104–105
    recombinant protein design, **294:**100–101
    refolding conditions, **294:**106–107

reversed-phase high-performance liquid chromatography, **294:**105–106

small-scale expression, **294:**103

structure determination by nuclear magnetic resonance

buffers, **294:**111

confirmation of properly-refolded proteins, **294:**109–110

dynamic analysis with relaxation studies, **294:**112–116

heteronuclear-based determinations, **294:**97–98, 111–112

homonuclear proton-based determination, **294:**96–97, 109–110

isotope labeling, **294:**99–100, 116–117

relaxation times, senstivity to correlation time, **294:**98

rigidity in solution, **294:**93–95

suitability for nuclear magnetic resonance, **294:**92–93

1-*O*-Acylceramide synthase

assay, **311:**112–113

ceramide synthesis inhibitor studies, **311:**105–106

inhibitors, **311:**117, 385–386

pH optimum, **311:**117

phospholipase activity, **311:**117

purification

ammonium sulfate precipitation, **311:**113

anion-exchange chromatography, **311:**113–114

cation-exchange chromatography, **311:**114

concanavalin A affinity chromatography, **311:**114–116

gel filtration, **311:**114

heparin affinity chromatography, **311:**116

hydrophobic interaction chromatography, **311:**114

postmitochondrial supernatant, **311:**112–113

solutions, **311:**111

reaction catalyzed, **311:**105, 379–380

size, **311:**117

storage and stability, **311:**116–117

substrate specificity, **311:**108–110

Age-related macular degeneration, protection by macular pigments, **299:**459

AGEs, *see* Advanced glycation end products

Aggregation kinetics

addition reaction, **309:**256–257

agitation-stimulated fibrillogenesis, **309:**207–209

amyloid fibril formation, kinetic analysis with thioflavin T fluorescence

assumptions, **309:**311

binding conditions, **309:**310–311

first-order kinetic model of extension, **309:**313, 315–317

initial rate measurement, **309:**316

polymerization reaction

composition, **309:**312–313

initiation and termination, **309:**313

reaction temperature, **309:**313

wavelengths for detection, **309:**310

competition between folding and aggregation, **309:**224, 235–236

conformational change and association rate, **309:**271

Fourier transform infrared spectroscopy, **309:**569

initiation modeling

concentration dependence of initial rate constant, **309:**262–264, 270, 274

dimensionless variable method, **309:**272–273

droplet condensation/crystallization approach, **309:**272

growth curve shape, **309:**273–274

lag time, **309:**266–267, 273–274, 305–306

near-nuclear species effects, **309:**267–269

nucleation barriers, **309:**269–270

nucleation rate constant, **309:**271

nucleus elongation rate, **309:**259–260, 263, 270

overview, **309:**256

rate constant variation from later aggregation, **309:**257

thermodynamic nucleus, **309:**258–259, 262–263

polymer elongation rate, **309:**257, 259–261, 263, 270

secondary pathway modeling

branching, **309:**264–265

equations, **309:**265
fragmentation, **309:**264
heterogeneous nucleation, **309:**265
standard state relation to Gibbs free energy, **309:**258
Agitation-stimulated fibrillogenesis, *see also* Amyloid-β fibril
aggregation conditions, **309:**207
cuvette conditions, **309:**215–216
kinetic analysis, **309:**207–209
orbital shaker conditions, **309:**214–215
peptide synthesis, **309:**206–207
protein quantification, **309:**209
recombinant V_L proteins, **309:**212–214
thioflavin T monitoring
advantages, **309:**209–210, 212
controls, **309:**210
fluorescence detection, **309:**211–212
sampling versus *in situ* monitoring, **309:**216
stock solution preparation, **309:**210
AGM1
nomenclature, **311:**639
radiolabeled compound synthesis
materials, **311:**640, 642
overview, **311:**639–641
stability, **311:**655–656
tritium labeling at position 6 of terminal galactose and *N*-acetylgalactosamine residues, **311:**642–645
AGR, *see* α-Glucosyl rutin
Airy disk, *see* Point-spread function
Akt1, heterologous expression in yeast, **293:**94, 103
ALA, *see* 5-Aminolevulinic acid
Alamethicin
pore size, **293:**661
sequence, **293:**659–660
three-dimensional structure and molecular dynamics simulation, **293:**660–663, 689–692
voltage dependence of channel formation, **293:**660
β-Alanine
γ-aminobutyric acid transporter inhibition, **296:**167–168
2-methoxy-5-nitrophenyl carboxylate caged compound
kinetics of photolysis, **291:**40–41

quantum yield determination, **291:**43, 45
synthesis, **291:**38–39
receptor, *see* Neurotransmitter receptor
Alcian blue
amyloid staining, **309:**13–14
glycocalyx preservation for electron microscopy
glutaraldehyde–lysine fixation, **310:**196, 203
paraformaldehyde–glutaraldehyde–lysine fixation, **310:**196, 198–199, 203
structure, **310:**195
transmission electron microscopy support treatment, **304:**221–222
ALD, see Adrenoleukodystrophy
ALEs, *see* Advanced lipoxidation end products
Alkaline phosphatase
activity assay in phosphate-starved biofilms, **310:**175–176
amyloid immunohistochemistry detection
blocking of endogenous enzyme, **309:**19
primary versus secondary detection systems, **309:**21–22
substrates, **309:**23–24
chemiluminescence assays
challenges, **305:**392
CSPD, dioxetane substrate assay
colorimetric assay comparison, **305:**395
fluorescein isothiocyanate-labeled serum preparation, **305:**393
incubation conditions, **305:**393
interferences, **305:**394
linearity and precision, **305:**394
matrix preparation, **305:**393
pH effects, **305:**394
reagents, **305:**393–393
recovery, **305:**394–395
substrate preparation, **305:**393
temperature effects, **305:**394
dioxetane-based substrates, **305:**335–336, 392–395
NADH assay, **305:**388
4-nitrophenol assay, **305:**389
prosthetogenesis assay, **305:**389

conjugated secondary antibody detection
in Western blot
dual detection, **305:**424–425
single detection, **305:**421
disulfide bond formation defect in *Esche-richia coli*, detection, **290:**69–71
function, **305:**391
immunohistochemistry detection, **305:**578–579, 588
levels in disease, **305:**391
reporter gene assay with chemilumines-cence
cell extract preparation, **305:**443
controls, **305:**444
CSPD dioxetane substrate, **305:**440–441
instrumentation, **305:**449–450
microplate luminometer detection
nonsecreted enzyme, **305:**443–444
secreted enzyme, **305:**442–443
overview, **305:**428–429, 440–441
reagents, **305:**441
sensitivity, **305:**450
tube luminometer detection, **305:**442
in situ hybridization detection, **305:**578–579, 583–584
Alkaline sphingomyelinase, *see also* Sphin-gomyelinase
assay, **311:**278
bile salt dependence, **311:**281
biological significance, **311:**284–286
divalent cation effects, **311:**281
expression in disease, **311:**286
inhibitors, **311:**281–282
intestinal tract distribution, **311:**277–278
isolation
human bile enzyme
acetone precipitation, **311:**280
anion-exchange chromatography, **311:**280
dextran sulfate precipitation, **311:**279–280
hydrophobic interaction chromatogra-phy, **311:**280
sphingosylphosphocoline affinity chromatography, **311:**280
rat intestinal mucosa
acetone precipitation, **311:**279
anion-exchange chromatography, **311:**279

elution from brush border by bile salt, **311:**278–279
gel filtration chromatography, **311:**279
hydrophobic interaction chromatogra-phy, **311:**279
pH optimum, **311:**276, 280–281
stability, **311:**281
substrate specificity, **311:**281
tissue distribution, **311:**276–277
Alkanal
gas chromatography–negative-ion chemi-cal ionization mass spectrometry assay, **300:**76–78
lipid peroxidation marker, **300:**76
All-*trans*-retinal
ABCR interactions
ATPase activation
dose response, **315:**888–889
factors affecting activation, **315:**889–890
mechanisms, **315:**894–895
synergistic activation with amioda-rone, **315:**891–892, 894
flipping function, **315:**895
activation of constitutively active opsin mutants
basal activity correlation with *trans*-reti-nal activation, **315:**258–260
immunoaffinity chromatography, **315:**254
membrane preparation, **315:**253–254
modeling of activation, **315:**258–260
retinal analog studies of activation, **315:**264, 266
Schiff base formation in mutants, **315:**260–263
transducin activation
filter-binding assay, **315:**255–256
intrinsic tryptophan fluorescence assay, **315:**256
transient transfection in COS-1 cells, **315:**253
wild-type activation, **315:**257
activation of opsin
absorption spectroscopy, **315:**244–245
activity level, **315:**238, 249, 252, 257–258
arrestin binding, **315:**244
mechanism of activation, **315:**245

opsin preparations
  membrane-bound preparation,
    315:239–240
  methylation of membranes, **315:**240
  phosphorylated opsin, **315:**240
  pH dependence, **315:**249–250, 257
  physiological role, **315:**238–239,
    266–267
  rhodopsin kinase binding, **315:**244,
    249–250
  transducin activation assay by intrinsic
    tryptophan fluorescence, **315:**241,
    243–245, 247–249
  visual desensitization implications,
    315:266–267
agonist functions, **315:**257–260, 264
reduction by retinol dehydrogenase,
  315:238
All-*trans* retinol dehydrogenase
  high-performance liquid chromatography
    assay, **316:**365, 368, 370–371
  phase partion assay
    all-*trans* retinol tritiation, **316:**362
    endogenous substrate assay, **316:**363
    enzyme preparation, **316:**363
    exogenous substrate assay, **316:**363–364
    materials, **316:**361
    membrane preparation, **316:**363
    nicotinamide adenine dinucleotide triti-
      ation, **316:**361
    pH, **316:**360
    principle, **316:**360
  visual cycle role, **316:**359
Alpha-helix, *see* Helix–coil transition
Alpha blending, rendering of confocal images, **307:**47, 49
Alzheimer's disease, *see also* Amyloid-β
  advanced glycation end products, role,
    309:153
  amyloid-β fibril
    deposition, **309:**58–59, 89–90, 152, 350,
      386, 450, 467–468, 476–477, 511–
      512, 716–717, 733, 749
    insertion into planar lipid membranes,
      294:670–672
    spontaneous chemical modifications,
      309:89–90, 105
  epidemiology, **309:**748–749
  oxidative stress markers and role,
    309:149–150, 747–750, 767–768

paired helical filament-tau in pathogenesis, **309:**81–83
transglutaminase, isopeptide bond formation
  activity assay, **309:**174–175
  colocalization with insoluble protein
    complexes, **309:**178–179
  cross-linking of proteins, **309:**172–173
  detection of isopeptide bonds
    immunohistochemistry, **309:**176
    overview, **309:**174
  immunoelectron microscopy with
    paired helical filaments and tan-
    gles, **309:**182–183
  immunohistochemistry, **309:**176,
    178–179
  isopeptidase cleavage of bonds
    assay, **309:**173
    digestion of paired helical filaments
      and tangles, **309:**184–185
    mechanism of cleavage in assay sys-
      tem, **309:**183–184
  levels in brain, **309:**173–175
  substrate specificity, **309:**174
  Western blot analysis, **309:**176–178
AMD, *see* Age-related macular degeneration
Amidothionophosphates
  antioxidant activity mechanism,
    299:294–295
  peroxyl radical reactivity, **299:**299
  reactivity
    hydrogen peroxide decomposition,
      299:296
    lipid hydroperoxide reduction,
      299:297–300
    sodium hypochlorite, **299:**296–297
  synthesis
    2-hydroxyethylamidoethyl thionophos-
      phate, **299:**295
    N,N',N''-tripropylamidothionophos-
      phate, **299:**296
Aminoacylation, *see* Transfer RNA
γ-Aminobutyric acid
  α-Carboxy-*o*-nitrobenzyl carboxylate
    caged compound
    kinetics of photolysis, **291:**43
    synthesis, **291:**35–37
  desyl caged compound, **291:**10–11

prospects, **317:**261
resonance assignments
    antibiotic, **317:**246–248
    exchangeable protons, **317:**245–246
    RNA protons, **317:**245
structure
    calculations, **317:**255–256
    comparison with chemical modification
        data, **317:**256–257
    comparison with mutational data,
        317:257–258
    titration of antibiotic into RNA,
        317:249–250
ω-Aminoceramide, enzymatic synthesis and
    purification, **311:**687–688
cis-1-Aminocyclobutane 1,3-dicarboxylate,
    glutamate transporter inhibition and
    transport, **296:**176, 180, 187–188
cis-3-Aminocyclohexanecarboxylic acid, γ-
    aminobutyric acid transporter inhibi-
    tion, **296:**168
2-Aminoethyl-2-aminoethane thiosulfonate,
    synthesis, **293:**126–127
3-Amino-9-ethylcarbazole, peroxidase sub-
    strate in amyloid immunohistochemis-
    try, **309:**23
2-Aminoethylmethane thiosulfonate, thiol
    trapping, **290:**63
α-Amino-3-hydroxy-5-methylisoxazole-4-
    propioninc acid receptor, *see also* Excit-
    atory amino acid receptors
    antibodies, **294:**688–689, 701–702
    domain labeling with radioligands,
        294:394
    drug screening with fluorescence assay of
        calcium
        activity of various drugs, **294:**36
        advantages of assay, **294:**46–47
        limitations of assay, **294:**45–46
        pharmacologic validation, **294:**26–27
    radioactive filter assays
        ligand selection, **294:**388
        reaction mixtures, **294:**388–389
        synaptic membrane preparation from
            rat brain, **294:**386–387
        Triton X-100 treatment of crude frac-
            tions, **294:**387–388
    stable cell line expression of recombinant
        channels, **294:**25–26
5-Aminolevulinic acid

Arabidopsis mutants in feedback loop
    controlling synthesis, **297:**243–244
limitation of precursor for chlorophyll
    biosynthesis, **297:**237, 241–242
mitogen-activated protein kinase induc-
    tion in photodynamic therapy,
    319:141–142
photodynamic therapy principles,
    319:381–382
singlet oxygen generation, **319:**137
trans-4-Aminopent-2-enoic acid, γ-aminobu-
    tyric acid transporter inhibition,
    296:168
Aminopropyl-bonded silica gel, *see* Solid-
    phase extraction
3-Aminopropyltriethoxysilane
    coating of microslides, **293:**156
    coverslip derivatization for optical map-
        ping, **303:**63–64
    distillation, **303:**63
Aminopyropheophorbide
    colon carcinoma cell photosensitization
        deuterium oxide effects, **319:**123–124
        irradiation conditions, **319:**122–123
        nuclear factor-κB activation
            antioxidant effects, **319:**128
            deuterium oxide effects, **319:**127, 129
            electrophoretic mobility shift assay,
                319:124–125, 127
            irradiation conditions, **319:**124
        solution handling, **319:**122
        subcellular localization, **319:**128
Aminosilane, preparation for mounting of
    amyloid immunohistochemistry sam-
    ples, **309:**17
Amiodarone, ABCR ATPase activation
    dose response, **315:**888
    synergistic activation with all-*trans*-reti-
        nal, **315:**891–892, 894
AML, *see* Acute myelogenous leukemia
Ammonium molybdate, chromatin staining
    for transmission electron microscopy,
    304:229
AMP, isotopic labeling for kinetic isotope
    effect studies, **308:**309
AMPA receptor, *see* α-Amino-3-hydroxy-5-
    methylisoxazole-4-propioninc acid re-
    ceptor
AMP deaminase
    catalytic mechanism, **308:**333, 335

inhibitor design, **308:**348–349, 394
isotope effects and transition-state modeling, **308:**335–336
molecular electrostatic potential surface
of transition-state, **308:**347–348
AMP nucleosidase, molecular surface point
similarity measures in binding energy
analysis, **308:**405, 407, 409
Amylamine, transmission electron microscopy support treatment, **304:**221
Amylase, chemiluminescence assay,
305:388–389
Amylin, *see also* Islet amyloid polypeptide
insertion into planar lipid membranes,
294:672
Amyloid A, *see* Experimental murine AA
amyloid
Amyloid enhancing factor, serum A amyloidosis induction in mice, **309:**502–503,
708–710
Amyloid-β fibril, *see also* Alzheimer's
disease
acetonitrile, protein aggregation properties, **309:**190
aggregation assays, overview,
309:352–353
aggregation versus deposition in formation, **309:**350, 352
Alzheimer's disease pathogenesis,
309:58–59, 89–90, 152, 350, 386, 450,
467–468, 476–477, 511–512, 716–717,
733, 749
aspartyl isomerization assay
dialysis of samples, **309:**90
D-isoaspartyl *O*-methyltransferase
assay
limitations, **309:**98, 100
principle, **309:**94–95
reaction conditions, **309:**95
sources of enzyme, **309:**97–98
validation, **309:**95–97
pathogenesis role, **309:**93
prevalence in Alzheimer's disease,
309:105
trypsin digestion and reversed-phase
high-performance liquid chromatography, **309:**90–91, 93–94
atomic force microscopy, *see* Atomic
force microscopy

circular dichroism analysis of secondary
structure, **309:**312
conformational conversion factors,
309:189–190
Congo red staining, *see* Congo red
deposition assays
advantages and disadvantages of assay
types, **309:**474–476
autoradiography using Alzheimer's disease tissue sections, **309:**367–370
brain homogenate deposition assay,
309:367, 370–372
detection method selection, **309:**361
overview, **309:**475
radioactivity quantification,
309:366–367
radiolabeled amyloid-β tracer
adsorption prevention, **309:**363, 367
aggregation prevention in synthesis,
309:362
iodination, **309:**361–363, 470–471
purification by high-performance liquid chromatography, **309:**362,
365
quality control, **309:**366
separation of unbound isotope,
309:362–363, 365
separation of bound and free material,
309:360
synthetic deposition assays
overview, **309:**368–369
plates, **309:**372–373
slides, **309:**373–374
templates
agitation in preparation, **309:**359
brain homogenate preparation,
309:370–371
characterization, **309:**359–360
controls, **309:**358, 367
immobilization, **309:**360–361
plate preparation, **309:**372–373
selection, **309:**353, 355
slide preparation, **309:**373–374
synthetic templates, **309:**358–359,
372–374
tissue characterization, **309:**355, 358
types, **309:**353
electron microscopy, *see* Electron microscopy

elongation assay with BIACORE surface
  plasmon resonance system
  advantages, **309:**387, 401–402
  association phase time, varying,
    309:397–398
  association time, **309:**392
  background binding, **309:**390–391
  binding responses, assessment,
    309:394
  contact time and peptide concentra-
    tions, varying, **309:**398–399
  data analysis, **309:**400–401
  dissociation phase, normalizing,
    309:398
  experimental conditions, varying,
    309:400
  fibril decay monitoring, **309:**395–396
  fibril preparation, **309:**390
  flat surface data, **309:**399–400
  immobilization of fibrils, **309:**391–392
  limitations, **309:**388
  model of fibril elongation, overview,
    309:395
  principle, **309:**388, 390
  reference surfaces, **309:**392
  reproducibility of binding experiments,
    309:392, 394
fibrillogenesis mechanism, **309:**395, 450–
    451, 469, 478, 511–512
formation from synthetic amyloid β-pep-
    tides
  peptide preparation, **309:**308
  reaction conditions, **309:**307–308
inhibitors of fibrillogenesis, identification
  4'-deoxy-4-iododoxorubicin, **309:**478
  extension assay, thioflavin T fluores-
    cence, **309:**481–482, 487–488
  nucleation assay using turbidity or
    thioflavin T fluorescence, **309:**480–
    481, 486–488
  peptide analogs
    design strategy, **309:**479–480
    secondary structure modeling,
      309:482–483, 485
    synthesis, **309:**485
  rationale, **309:**369, 467–468
  screening criteria, **309:**470
  seeded fibril assay, **309:**473–474
  spontaneous fibril formation assay,
    309:472–473

structure–activity relationship analysis,
    309:470
isolation
  cerebral cortex diffuse amyloid,
    309:65–67
  cortical vessel amyloid isolation,
    309:61–63
  leptomeningeal vessel amyloid isola-
    tion, **309:**63–64
  senile plaque amyloid cores
    cerebral cortex dissection and ho-
      mogenization, **309:**59–60
    fast protein liquid chromatography,
      309:61
    materials, **309:**59
    sucrose density gradient centrifuga-
      tion, **309:**60
light scattering, correction in Congo red
    assay, **309:**289–291, 294–296
peptide, *see* Amyloid-β peptide
N-phenacylthiazolium bromide, reversal
    of advanced glycation end product
    cross-links, **309:**160–161, 172
polarized light microscopy, **309:**312
pyroglutamyl residue characterization,
    309:104–105
quasielastic light-scattering spectroscopy,
    see Quasielastic light-scattering spec-
    troscopy
reproducibility of studies, **309:**189–191,
    203, 472
thioflavin T assays, *see* Thioflavin T
X-ray fiber diffraction of senile plaques,
    309:528–529
Amyloidosis, *see also* Amyloid-β; Dialysis-
    related amyloid fibril; Experimental
    murine AA amyloid; Immunoglobulin
    light chain amyloidosis; Islet amyloid
    polypeptide; Murine senile amyloid fi-
    bril; Serum amyloid P component;
    Transthyretin amyloid
  criteria for identification, **309:**205
  immunohistochemistry
    antibodies
      absorption, **309:**24–25
      dilution, **309:**19–20
      primary antibody incubation, **309:**20,
        38
      purification, **309:**20
      types, **309:**14–15, 37

antigenic epitope retrieval, **309:**18
controls, **309:**24, 39
coupling enzymes
    blocking of endogenous enzymes,
        309:19, 37–38
    types and substrates, **309:**22–24
detection systems, **309:**20–22
fixation, **309:**15–16
mounting, **309:**16–17, 24
sectioning, **309:**16
wash buffers, **309:**17–18, 38
isolation of fibrils from tissue, *see also*
    Amyloid-β fibril
differential centrifugation and sucrose
    gradient centrifugation, **309:**26–27
homogenization of tissue in water
    buffer preparation, **309:**30–31
    overview, **309:**27–29
    salt precipitation for amyloid recov-
        ery in suspension, **309:**35–36
    salt removal for fibril recovery,
        309:34
    serum amyloid P component re-
        moval, **309:**27–28, 32–34
    tissue homogenization, **309:**31–35
microextraction
    amyloid content of biopsies, **309:**70–
        71, 79
    clinical value, **309:**68–69, 78–81
    extraction, **309:**69–72, 74, 76–77
    formalin-fixed tissues, **309:**76–80
    fresh tissue, **309:**71–72, 74, 76
    high-performance liquid chromatog-
        raphy for purification, **309:**72, 77
    microsequencing, **309:**69–70, 74,
        76 78
    preparative gel electrophoresis,
        309:72, 74
    validation of technique, **309:**71
optical density in quantification,
    309:40–41
physical separation and extraction with
    denaturing solvents, **309:**29
safety, **309:**29–30
storage of fibrils, **309:**46–47
precursor proteins, **309:**26, 28, 67–68, 204
sequencing of fibrils, **309:**46
staining of deposits
    Alcian blue, **309:**13–14
    Congo red, *see* Congo red

cotton dye development, **309:**10–11
iodine, **309:**3
silver staining, **309:**13
Sirius red, **309:**9–10
thioflavins, *see* Thioflavin S; Thioflavin
    T
Western blot analysis of isolated fibrils,
    309:42–44
Amyloid-β peptide
3-(4,5-dimethylthiazol-2-yl)-2,5-diphenyl-
    tetrazolium bromide cytotoxicity
    assay
    amyloid-β treatment results, **309:**719,
        721–723
    5-(3-carboxymethoxyphenyl)-2-(4,5-
        dimethylthiazolyl)-3-(4-sulfophe-
        nyl)tetrazolium cell viability assay,
        309:719
    cell culture, **309:**717
    materials, **309:**717
    peptide preparation, **309:**718
    reduction assay, **309:**718–719, 761
    sensitivity of cell lines, **309:**722
excitotoxicity in cell death mechanism,
    309:739
fibril, *see* Amyloid-β fibril
glucose transporter inhibition
    assay, **309:**741, 743
    mechanism, **309:**740
glutamate transporter inhibition
    assay, **309:**741, 743
    overview, **309:**740–741
G protein-coupled receptor signaling im-
    pairment
    GTPase assay, **309:**744–746
    muscarinic receptor, **309:**743–744
    thrombin receptor, **309:**744
inflammatory response in human mono-
    cytes
    cytokine enzyme-linked immunosor-
        bent assay
        interleukin-1β, **309:**731–732
        materials, **309:**730
        principle, **309:**729–730
        solutions, **309:**730–731
    cytokine induction by fibrils, mecha-
        nism, **309:**724
    lipopolysaccharide-treated cells,
        309:724, 732
    overview, **309:**723–724

peptide characterization for studies
  principle, **309:**725–726
  stock peptide preparation, **309:**726
  thioflavin T assay of fibrils, **309:**726
  THP-1 cell culture
    lysate preparation, **309:**729
    maintenance, **309:**728
    materials and solutions, **309:**727–728
    plating for cytokine assays,
      309:728–729
    principle, **309:**727
ion-motive ATPase inhibition
  assay, **309:**741
  overview, **309:**739
methionine-35 oxidation
  aggregation effects, **309:**478
  homogeneity of peptide samples,
    309:195–196
oxidative stress
  Alzheimer's disease, oxidative stress
    markers and pathogenesis,
    309:149–150, 747–750, 767–768
  carbonyl determination in proteins
    2,4-dinitrophenylhydrazine assay,
      309:762–763
    histofluorescence, **309:**764–765
    immunochemistry, **309:**763–764
  cortical synaptosome membrane oxida-
    tion model, **309:**752–754
  creatine kinase inhibition and assay,
    309:748, 760
  electron paramagnetic resonance assay
    advantages, **309:**754–755
    spectral acquisition, **309:**757
    spin labels and labeling, **309:**755,
      757–758
    spin trapping, **309:**758
    stearic acid spin probes and lipid per-
      oxidation evaluation,
      309:765–766
  free fatty acid analysis, **309:**766–767
  glutamine synthetase inhibition and
    assay, **309:**748, 752, 759–760
  glutathione protection, **309:**737
  hippocampal neuronal culture oxida-
    tion, **309:**754
  4-hydroxynonenal–protein conjugates
    glucose transporter inhibition,
      309:740
    immunohistochemistry, **309:**738

    immunoprecipitation, **309:**737–738
    induction, **309:**734–735
    Western blot analysis, **309:**738, 766
  malondialdehyde assay, **309:**737
  neurotoxicity assays
    mitochondrial function assay,
      309:761
    morphometric cell counting assay,
      309:761
    overview, **309:**760–761
    Trypan blue exclusion assay, **309:**761
  overview of lipid peroxidation,
    309:734–735
  peptide assessment for study,
    309:751–752
  reactive oxygen species detection with
    fluorescent probes, **309:**762
purity of peptide in formation studies
  effects on aggregation behavior,
    309:191–192, 203, 472
  high-performance liquid chromatogra-
    phy analysis, **309:**192–193, 486
  mass spectrometry analysis, **309:**192,
    472
  nuclear magnetic resonance analysis,
    309:193–196
racemization
  aggregation rate effects of aspartyl ra-
    cemization, **309:**202–203
  aspartyl racemization assay, **309:**100–
    101, 105, 201, 204
  nicotine effects on racemic peptide ag-
    gregation, **309:**200–203
  seryl racemization assay, **309:**103–105
salt and metal effects on peptide aggregation, **309:**199, 204
size and deposition sites, **309:**89
solid-state nuclear magnetic resonance of
  peptide fragments, **309:**557
solvent activation restructuring process,
  309:752
solvents for peptide storage and starting
  aggregation state, **309:**196–198,
  203–204
time-dependent aggregation of peptides,
  minimization of variability,
  309:199–200
β-Amyloid precursor protein
  antisense studies in chick learning model
    behavioral effects, **314:**221

chick model overview, **314:**216

intracerebral injection, **314:**216–217, 222

oligodeoxynucleotide design and purification, **314:**217–218, 221

specificity of effects, **314:**222–223

functions, **314:**215

isoforms, **314:**214

processing, **314:**214–215

Analytical ultracentrifugation, *see* Sedimentation equilibrium; Sedimentation velocity; Sucrose density gradient centrifugation

Androgen receptor

domains, **302:**121

green fluorescent protein fusion protein

androgen-binding assay, **302:**124

applications

androgen insensitivity syndrome, functional characterization of mutations, **302:**132–135

antiandrogen screening, **302:**131–132

confocal scanning microscopy, **302:**127–128

epifluorescence microscopy imaging, **302:**128–131

immunodetection

comparison to intrinsic fluorescence localization, **302:**126–127

materials, **302:**125–126

permeabilization conditions, **302:**126–127

laser scanning cytometry, **302:**128

nuclear translocation, **302:**77–78, 130–131

plasmid construction, **302:**122–123

quantification of dynamics, **302:**129–131

transactivation assay, **302:**124–125

transfection, **302:**123

Western blot analysis, **302:**123–124

immunostaining for subcellular localization, **302:**121–122

Anethole dithiolthione

effects on nuclear factor-κB, **299:**300–301

high-performance liquid chromatography with electrochemical detection

cell culture, **299:**302

chromatography conditions, **299:**302–303

extraction, **299:**301–306

instrumentation, **299:**301

Angeli's salt

effects on oxidative stress, **301:**415–418

synthesis and nitroxyl release, **301:**214–216, 282–285, 287

Angelman syndrome, Southern blot analysis with chemiluminescence detection

background on syndrome, **305:**460–462

blotting, **305:**463

dot blots, **305:**462–463

hybridization and washes, **305:**463

principle, **305:**462

probe generation, **305:**462

reagents and solutions, **305:**462

signal generation and detection, **305:**463

Angiotensinogen, antisense targeting, **313:**55–56

Angiotensin II receptor, confocal microscopy, **307:**132–133

Angiotensin II type 1 receptor, antisense targeting, **313:**29

adeno-associated virus vectors

hypertension effects, **314:**45–46, 50–51

plasmid preparation

construction and design, **314:**45–46

large-scale preparation, **314:**42, 44–45

recombinant virus preparation, **314:**47–49

tissue-specific expression, **314:**51

titer assay, **314:**49–50

multiple oligonucleotide targeting to single sites, **314:**182

rationale, **314:**581–582, 590

retroviral vectors

complementary DNA preparation, **314:**582–583

concentrating of vectors, **314:**588

infection of target cells, **314:**588

titration of virus, **314:**587–588

transduction, *in vivo*, **314:**588, 590

viral particle preparation, **314:**583–585, 587

RNA mapping for antisense sequence selection, **314:**182–183

tissue distribution, **314:**581

8-Anilino-1-naphthalenesulfonic acid

fluorescence enhancement in SecB binding of ligands, **290:**456

Sup35p NM fragment assembly assay, **309**:667

Anion-exchange chromatography
  BiP, **290**:389, 392
  chaperonin 10, **290**:165, 171–172, 200, 202
  cytosolic chaperonin, **290**:232–234
  FKBP52, **290**: 423–424
  GroEL, **290**:104, 107–110, 118, 137–139
  GroES, **290**:125–126
  heat shock protein 18.1, **290**:357–358
  heat shock protein 60, **290**:197–198
  heat shock protein 90, **290**:417
  heat shock protein 104, **290**:435–436, 440–441
  Hop/p60, **290**:427
  HtpG, **290**:411–412
  mammalian mitochondrial chaperonin 60, **290**:206–208
  p23, **290**:425
  protein disulfide isomerase, **290**:30–31, 53
  SecB, **290**:447–449

Annexin V–green fluorescent protein fusion protein
  apoptosis assay
    flow cytometry detection, **302**:42–43
    fluorescence microscopy detection, **302**:43
    incubation of cells, **302**:42
    principle, **302**:40–41
  phosphatidylserine binding, **302**:40

ANS, *see* 8-Anilino-1-naphthalenesulfonic acid

Anthocyanins, *see* Phenolic antioxidants

9,10-Anthracenedipropanoate, singlet oxygen trapping, **319**:50–51

Antibiotic, *see also* Aminoglycoside
  detection and sensitivity assays using *lux*-containing phage, **305**:555–557
  RNA complex, *see* Antibiotic–RNA complex

Antibiotic–RNA complex
  examples of aptamers, **318**:215
  mechanism of antibiotic action, **318**:214
  recognition motifs, **318**:215
  RNA aptamer characterization
    boundary mapping
      lead(II) acetate cleavage, **318**:223, 225–226
      partial hydrolysis, **318**:222–223
    dimethyl sulfate modification

hybridization, **318**:227–228
modification reaction, **318**:226–229
polyacrylamide gel electrophoresis analysis, **318**:228
primer extension, **318**:228
renaturation, **318**:226
dissociation constant determination, **318**:222
sequence analysis, **318**:221
structure characterization by nuclear magnetic resonance, **318**:229
in vitro selection
  affinity chromatography
    antibiotic coupling to Sepharose, **318**:218–220
    buffers, **318**:219
    counterselection, **318**:220
  initial DNA pool design, **318**:217–218
  overview, **318**:215, 217
  polymerase chain reaction, **318**:221
  reverse transcription, **318**:220–221
  transcription, **318**:219, 221

Antibody, *see also* Enzyme-linked immunosorbent assay; Immunoaffinity chromatography; Immunochemical assay; Immunocytochemistry; Immunofluorescence; Immunoprecipitation; Monoclonal antibody; Peptide antibody; Western blot analysis
  advanced glycation end products, production and sources of antibodies, **309**:156–158, 162–163
  affinity-tagged antigen generation from mammalian complementary DNAs
    advantages, **303**:504–505
    affinity tagging
      ABP tag, **303**:506
      requirements of tag, **303**:505, 507
      ZZ tag, **303**:505
    expression in *Escherichia coli*, **303**:507–508
    purification, **303**:506–508
    Western blot analysis, **303**:508–509
  antiaffinity-tagged antibody generation
    affinity enrichment of antisera, **303**:509–510
    immunization of rabbits, **303**:509
    subcellular localization of target proteins, **303**:510–511
  fluorochrome-conjugated

secondary antibody probing, **307:**62, 65
vendors, **307:**65–66
folding enhancement of inclusion body
     proteins, **309:**231–232
ion channel antigens, *see* Antibody, ion
     channels
prion protein conformational mapping
     with monoclonal antibodies
   fusion, **309:**110
   immunization and peripheral immune
        response, **309:**110, 113, 116
   immunogen preparation, recombinant
        bovine prion protein, **309:**108–110
   immunoprecipitation, **309:**112–113
   nitrocellulose freezing of protein con-
        formation, **309:**122
   peptide libraries, **309:**112
   screening of hybridomas
      enzyme-linked immunofiltration
           assay, **309:**110–112
      enzyme-linked immunosorbent assay,
           309:110–111
      Western blot analysis, **309:**112
   specificity for native and disease prions
      historical overview, **309:**106–108
      PrP^C, **309:**116, 118
      PrP^{Sc}, **309:**118–119
transthyretin amyloid, monoclonal anti-
     body probing of conformation
   antibody production
      immunization of mice, **309:**596
      screening of hybridomas,
           309:596–597
   applications, **309:**604–605
   binding to extracted fibrils, **309:**603
   enzyme-linked immunosorbent assay
      cryptic epitope demonstration,
           309:600–601
      direct assay, **309:**595
      sandwich-type assay, **309:**596
   immunoprecipitation, **309:**602–603
   mutants of protein for aggregation,
        309:593–595
   phage display of epitopes
      DNase I digestion, **309:**598
      fractionation of DNA fragments,
           309:598–599
      fusion gene cloning, **309:**599
      overview, **309:**595
      screening, **309:**599–600

precautions, **309:**604
rationale, **309:**592–593
strategy overview, **309:**593
Antibody, ion channels
   α-amino-3-hydroxy-5-methylisoxazole-4-
        propioninc acid receptors, **294:**688–
        689, 701–702
   γ-aminobutyric acid-A receptors,
        294:685–686, 701
   autoimmune antibodies in humans,
        294:678, 700–703
   calcium channels, voltage-gated
      autoimmunity, **294:**702
      channel antibodies, **294:**690–692, 702
      neurotoxin antibodies, **294:**693, 702
   commercial availability, **294:**703–704
   cyclic nucleotide-gated channels, **294:**699,
        703
   cystic fibrosis transmembrane conduc-
        tance regulator, **294:**698, 703
   GluR1 phosphorylation, site-specific anti-
        bodies, **294:**364–366
   glycine receptor, **294:**685, 701
   N-methyl-D-aspartate receptors, **294:**687,
        701–702
   nicotinic acetylcholine receptors
      autoimmunity, **294:**701
      muscle receptors
         mouse monoclonal antibodies,
              294:681, 700–701
         rabbit polyclonal antibodies, **294:**682,
              701
         rat monoclonal antibodies, **294:**679–
              680, 700
      neuronal receptors
         monoclonal antibodies, **294:**683
         rabbit polyclonal antibodies raised
              against extracellular domain pep-
              tides, **294:**684
   potassium channels, voltage-gated
      autoimmunity, **294:**703
      mouse monoclonal channel antibodies,
           294:694, 702
      rabbit polyclonal antibodies against syn-
           thetic peptides, **294:**695–697, 702
   sodium channels, voltage-gated, **294:**703
Antifading agents
   commercial agents, **307:**67–68, 134
   effectiveness with immunofluorescence
        specimens, **307:**68–69, 73

mounting of immunofluorescence specimens, **307:**65–67, 80
vascular casts with fluorochrome-labeled
gelatin, **307:**99–100
Antimicrobial resistance, biofilms
assays for biofilms, *see also Candida* biofilm; *Legionella pneumophila* biofilm;
Pseudomonas aeruginosa biofilm; Rotating disk reactor
criteria
repeatability, **310:**611
reproducibility, **310:**611–612
unbiased assays, **310:**610–611
development of standard methods,
310:612–613
federal regulatory agency acceptance,
310:609–610
bioelectric effect, *see* Electromagnetic
field
mechanisms in biofilms, **310:**249
paints, *see* Copper; Tributyltin
Antimony
pumps, *see* Arsenite pump; PgpA
therapy in *Leishmania* infection,
292:182–183, 192–193
Antioxidative homeostasis
antioxidant capacity, measurement in
model systems, **300:**439–441
compartmentalization of defenses,
300:438–439
hydrogen peroxide-stimulated chemiluminescence, measurement of oxidative
damage
apparatus, **300:**451
blood ultraviolet irradiation effects,
300:451
correlation with antioxidative parameters, **300:**452–453
smoking effects, **300:**453
overview of protection mechanisms,
300:437–438
photoinduced chemiluminescence assay
apparatus, **300:**441
applications, **300:**455–456
blood ultraviolet irradiation effects,
300:446, 448
calibration, **300:**443
comparison to other assays, **300:**444,
446, 448–450
disease effects, **300:**449–450

lag phase, **300:**442
lipid-soluble antioxidants, measurement in plasma, **300:**442–443, 456
precision, **300:**443–444
principle, **300:**441–442
stress effects, **300:**444, 446
rate of oxidative cell injury, **300:**454–455
α_1-Antiproteinase, peroxynitrite scavenger
identification by inactivation protection
controls, **301:**340
interpretation, **301:**341–342
mechanism, **301:**338
reaction conditions, **301:**338, 340
Antisense oligonucleotide
acute versus long-term effects,
314:230–231
advantages and disadvantages in protein
function analysis, **296:**515
advantages over knockout mice,
314:51–52
biological effects, factors affecting
kinetics of effects, **313:**6–7
oligonucleotide
control effects, **313:**6
nonspecific binding, **313:**5–6
purity, **313:**3–4, 204, 361
structure, **313:**4
uptake and distribution, **313:**4–5
RNA structure, **313:**4
terminating mechanism, **313:**6
colocalization with imunolabeled proteins
in brain, **314:**274–275
control sequences, **296:**518
delivery, *see also* Microinjection; Patch-
clamp loading, antisense oligonucleotides; Transfection, antisense oligonucleotides
biodegradable pH-sensitive surfactants
to facilitate endosome release,
313:351–352
dendrimers, **313:**352–354
receptor targeting with ligand atta-
chement
asialoglycoprotein receptor targeting,
see Asialoglycoprotein receptor
overview, **313:**297–300
transport-enhancing peptides, **313:**354–
355, 357–358
design, *see also* RNA mapping
array technology, **314:**292

backbone modifications for stability,
    314:81, 91, 94, 123, 229, 249, 293,
    314, 362, 480, 507–508, 539–541
base content, **313**:52
controls, **313**:48, 52, 130–131, 362,
    427–428
criteria, **313**:47, 144, 287, 360–361, 370
G–C content, **314**:79–80
homology analysis, **313**:49, 147, 361,
    314:80–81, 122–123, 292, 315, 484
length, **313**:47, 54, 145, 361, **314**:79–80,
    248, 313, 478–480
overview, **314**:121–122
receptor ligand attachment, **314**:325–
    337, 339–342
secondary structure analysis, **314**:123,
    225–229, 315, 483
target selection on messenger RNA,
    313:48–50, **314**:81, 90–91, 122,
    225–228, 291, 314, 480–481, 483
dosing and delivery, **296**:522–523
5′-end radiolabeling, **314**:240–241, 271–
    272, 283–284
experimental design recommendations,
    313:7–8, 28
fetal brain culture studies, *see* Fetal
    brain, human culture antisense
    studies
gene switching, *see* Gene switching
glutamate uptake assay for glutamate
    transporter analysis, **296**:527–529
inhibition efficiency, Western blot analy-
    sis, **296**:523–526
intrahippocampal injection in rat
    animal care, **314**:287
    distribution of oligodeoxynucleotides,
        314:289–290
    injection, **314**:288–289
    materials, **314**:287
    surgery, **314**:287–288
intraparenchymal delivery
    blood–brain barrier disruption,
        314:128–129
    diffusion of oligonucleotides, **314**:130
    distribution of phosphorothioate oligo-
        nucleotides
        autoradiography, **314**:132–132
        infusion, **314**:131–132
        nucleic acid extraction and analysis

        from brain and cerebrospinal
            fluid, **314**:133–134
        oligonucleotide preparation, **314**:131
        rat preparation, **314**:131
        volume of distribution and brain tis-
            sue content, **314**:134–135
    positive pressure controlled-rate infu-
        sion, **314**:130–131
knockdown studies, *see specific proteins*
local infusion in brain, **314**:231–232, 237
mechanisms, *see also* Triplex-forming oli-
    gonucleotide
    5′-capping inhibition, **313**:11
    double-strand ribonuclease, cleavage of
        targets, **313**:16
    end sequence targeting, **313**:12–13, 137
    initiation codon binding, **313**:9–10, 137
    3′-polyadenylation inhibition,
        313:11–12
    protein synthesis inhibition, **296**:516,
        314:76–77, 314, 379
    ribonuclease activation, *see* Ribonucle-
        ase H; Ribonuclease L
    secondary structure disruption in
        RNA, **313**:10–11
    splicing, **313**:8–9, 506–510
    steric inhibition of translation, **313**:423
model system selection, **314**:92
modifications, *see also* Boranophosphate
    oligonucleotides; Methylphosphono-
    diester oligodeoxynucleotide; Minor
    groove binder–oligodeoxynucleotide
    conjugate; Peptide nucleic acid;
    P-ethoxy oligonucleotide; Phosphor-
    amidate oligonucleotide; Phospho-
    rothioate oligodeoxynucleotide
    cell entry enhancement, **296**:520
    conjugate preparation, overview,
        313:164–165
    effects *in vivo*
        backbone modifications, **313**:44–45
        end modifications, **313**:42
        sugar modifications, **313**:42–44
    end modifications and nuclease stabil-
        ity, **313**:40–41, 57
    morpholino oligonucleotides for gene
        switching, **313**:175–176, 182–185
    purine modifications, **313**:39–40
    pyrimidine modifications, **313**:38–39,
        77, 79

RNA cleaving group incorporation,
  313:42
selection experiment recommendations,
  313:73
stabilization against nucleases, **296:**516,
  518, 520
triplex-forming oligonucleotides,
  313:76–79
uptake-enhancing modifications,
  313:41–42, 148–149, 258
neuroendocrine experiment design,
  314:224–225
nonantisense effects, **313:**27, **314:**77–79,
  123, 314–315, 389, 456, 557–558
optimization of conditions for receptor
  studies
  in vitro studies, **314:**93–97
  in vitro studies, **314:**100–101
pharmacokinetics, *see* Pharmacokinetics,
  antisense oligonucleotides
preparation for chronic intraventricular
  fusion, **296:**520–522
radiotherapy, *see* Oligonucleotide radiotherapy
receptor targeting, *see* α_2-Adrenorecep-
  tor; γ-Aminobutyric acid receptor;
  Angiotensin II type 1 receptor; Asialoglycoprotein receptor; Opioid receptor; Serotonin 2A receptor
reporter gene assays in HeLa cells,
  313:140–141, 143, 343–345
ribonucleoprotein disruption, *see* Ribonucleoprotein
RNA, *see* Antisense RNA, *Escherichia
  coli*; Ribozyme
screening of binding sites, overview,
  313:17
sequence selection, **296:**517–518
specificity of inhibition, **313:**155–156,
  314:90
stability
  central nervous system, **314:**53, 224,
    262–266
  oligodeoxynucleotides, **313:**52–53, 96
synthesis and purification of oligodeoxynucleotides
  automated synthesis, **314:**277–278
  cleavage and deprotection, **314:**278–279
  denaturing polyacrylamide gel electro-
    phoresis

electrophoresis and detection,
  314:285–286
gel preparation, **314:**285
materials, **314:**284–285
purification of radiolabeled probes,
  314:286–287
desalting, **314:**282
detritylation and precipitation,
  314:281–282
ether extraction, **314:**282–283
purification by reversed-phase high-performance liquid chromatography,
  314:279–281
yield determination, **314:**283
toxicity
  CG immunostimulation, **314:**167, 558–
    560, 567–568, 573, 576–578
  phosphorothioate analogs, **314:**75, 161–
    162, 165–166, 194–195, 231, 262,
    555–557
translational arrest assays, *see* Reticulocyte lysate; Vesicular stomatitis virus
uptake by cells, *see also* Flow cytometry
  adsorptive endocytosis, **313:**288–289
  efflux, **313:**53–54, 291, 293–294
  fluid-phase pinocytosis, **313:**289–290
  heparin-binding proteins as receptors,
    313:289, 342
  length effects, **313:**53
  neurons and glial cells, **314:**67–68, 97,
    151–152, 158, 223, 266–268
  nuclear delivery, **313:**342–343
Antisense RNA
  biological function in plants, **297:**209
  Escherichia coli, *see* Antisense RNA,
    Escherichia coli
  gene identification, **306:**259
  gene therapy, **306:**244
  mechanisms of gene expression downregu-
    lation, **297:**209
  mutant generation
    growth medium for selection, **297:**212
    promoters, **297:**211–212
    Rubisco mutants
      developmental changes in expres-
        sion, **297:**217–219
      feedback regulation analysis,
        297:219–220
      photosynthesis rate control studies,
        297:216–217, 219–220

pulse-labeling and posttranscriptional control, **297:**213
transcription analysis, **297:**212–214
Western blotting, **297:**212–213
target gene considerations, **297:**210–211
target sequence selection, **297:**211
Antisense RNA, *Escherichia coli*
artificial genes
construction, **313:**469
design, **313:**468–469
expression, **313:**469, 471
target RNA analysis, **313:**472
Western blot of target protein inhibition, **313:**472–473
bacteriophage protection
applications in research, **313:**473
overview, **313:**473
SP replication inhibition
assays, **313:**475–476
targets, **313:**474–477
vectors, **313:**474
target size, **313:**473
classification, **313:**467
control processes, **313:**467
DNA, *see* Multicopy single-stranded DNA
expression characteristics, **313:**467
Northern blot analysis, **313:**471–472
secondary structure, **313:**468–469
Antitermination, RNA-binding library screening
components of λ N antitermination complex, **318:**297–299
doped libraries, **318:**308
β-galactosidase assays for scoring interactions
colony color assay
overview, **318:**300–301
plating and color development, **318:**301
reagents, **318:**301
screening capacity, **318:**304–305
solution assay, **318:**301–302
modifications of system, **318:**308
N–RNA-binding protein fusion, **318:**299, 302–304
RNA-binding domain library generation
ligation, **318:**307
oligonucleotide extraction, **318:**306
overview, **318:**305–306

primer extension, **318:**306
selection of RNA binders from libraries
false-positive elimination, **318:**307
nonspecific-positive elimination, **318:**307–308
primary screen, **318:**307
two-plasmid reporter system, principles, **318:**298–299
α_1-Antitrypsin, native gel electrophoresis of aggregation intermediates, **309:**341
AOD, *see* Acousto-optic deflector
AP-2
skin end point for ultraviolet A exposure, **319:**464
ultraviolet A-induced gene expression in keratinocytes, **319:**280, 288, 290
AP1510, dimerizer-regulated gene expression
advantages over FK1012, 264–265
modifications to prevent FKBP binding, **306:**265
AP1889, dimerizer-regulated gene expression
FKBP mutant dimerization, **306:**267, 279
stable cell line characteristics
decay of gene expression following dimerizer removal, **306:**278–279
dose response, **306:**277
inhibitor response of gene expression, **306:**278
kinetics of expression, **306:**278
storage and handling, **306:**272
APD, *see* Avalanche photodiode
Apolipoprotein A-I, photoaffinity labeling, **311:**587
Apolipoprotein A-II amyloid, *see* Murine senile amyloid fibril
Apolipoprotein E, photoaffinity labeling, **311:**587
Apoptosis, *see also* c-ABL
annexin V–green fluorescent protein fusion protein assay
flow cytometry detection, **302:**42–43
fluorescence microscopy detection, **302:**43
incubation of cells, **302:**42
principle, **302:**40–41
Bcl-2 and Bcl-xL, antisense knockdown in cancer cells
clinical application, **314:**478

controls, **314**:483–484
lipofection of cultured cells
  A549 cells, **314**:488–489
  parameters, **314**:487
  principle, **314**:485, 487
  SW2 cells, **314**:487–488
  ZR-75-1 cells, **314**:488–489
oligodeoxynucleotide design
  backbone modification, **314**:480
  homology searching, **314**:484
  length, **314**:478–480
  secondary structure analysis, **314**:481,
    483
  target region selection, **314**:480–481,
    483
phosphorothioate oligodeoxynucleotide
  purification, **314**:484–485
secondary test systems, **314**:489–490
C11-BODIPY studies in fibroblasts,
  319:607–608
caspase-3 role, **319**:332
definition, **319**:309–310
flow cytometry assay
  cell requirements, **319**:324–325
  forward light scattering, **319**:325, 329
  instrumentation, **319**:325
  mitochondria transmembrane potential
    analysis, **319**:326–328, 330
  side light-scattering, **319**:325, 329
HL-60 cell studies of ultraviolet A radiation
  caspase-3 assay, **319**:340–341
  cell culture, **319**:332
  DNA analysis
    gel electrophoresis of fragments,
      319:333–334
    ladder detection, **319**:334–336
    8-oxodeoxyguanosine determination,
      319:333
  flow cytometry assays
    mitochondrial transmembrane poten-
      tial, **319**:336–340
    peroxide detection, **319**:336
  irradiation conditions, **319**:332–333
  materials, **319**:332
  mechanisms of apoptosis, **319**:341–342
irradiation of cells for studies
  adherent cells, **319**:315
  buffers, **319**:312–313
  postexposure treatment

adherent cells, **319**:315–316
suspended cells, **319**:314–315
sources
  ultraviolet, **319**:313–314
  visible light, **319**:314
  X-ray, **319**:314
suspended cells, **319**:314–315
temperature control, **319**:313–314
vessels, **319**:313
c-Jun N-terminal kinase antisense knockdown effects in kidney cells
  cell culture, **314**:366
  fluorescence microscopy assay of
    apoptosis, **314**:370–371
  JNK isoform specificity, **314**:372, 374
  knockdown efficacy, **314**:371–372
  oligonucleotide treatment, **314**:366–367
  rationale for study, **314**:371
mitochondrial membrane potential role,
  319:332
morphological features, **319**:309
necrosis following apoptosis, **319**:310–311
nitric oxide protection
  exposure of cells, **301**:394–395
  guanylate cyclase stimulation, **301**:393,
    395–396
overview of regulators, **314**:477
p53 modulation in tumors, **314**:523
peroxynitrite induction
  cell exposure, **301**:396–399
  generation from 3-morpholinosyd-
    nonimine, **301**:400–401
  signal transduction, **301**:401–402
  toxic effects versus apoptosis,
    301:399–400
phosphatidylserine externalization by
  cells, **302**:38
photodynamic therapy apoptosis
  bcl2 role, **319**:348–349
  calcium signaling, **319**:346–348
  caspase activation
    Bap31 fragments, **319**:350–351
    inhibitor studies, **319**:349–350
  cell cycle mediation, **319**:351–353
  ceramide generation, **319**:351
  early event in tumor shrinkage,
    319:345–346
  mechanisms, **319**:346
  nitric oxide role, **319**:353
  p21 role, **319**:352, 354

p53 role, **319:**349, 352
phospholipase C activation,
    319:346–348
retinoblastoma protein role,
    319:352–353
serine/threonine dephosphorylation,
    319:348
preprogrammed cell death versus pro-
    grammed cell death, assessment by
    translation inhibition with cyclohex-
    imide, **319:**323–324
reactive oxygen species
induction, overview, **319:**310, 316
identification of types in apoptosis in-
    duction
    cyclosporin A inhibition of singlet ox-
        ygen, **319:**316, 321–322, 328–329
    deuterium oxide, **319:**317–318
    DNA damage induction, **319:**320
    hydroxyl radical generators,
        319:319
    sham controls, **319:**317–318
    singlet oxygen generators,
        319:318–320
    singlet oxygen quenchers, **319:**317
    superoxide generators, **319:**319–320
recovery and handling of singlet oxygen-
    damaged cells
    centrifugation, **319:**311
    pipetting, **319:**312
    suspending, **319:**312
    vortexing, **319:**312
    washing, **319:**312
T cells in atopic dermatitis ultraviolet A
    therapy
    cell isolation, **319:**303, 305
    deuterium oxide effects, **319:**304
    DNA laddering assay, **319:**306
    double-staining technique for detec-
        tion, **319:**308
    immunofluorescence microscopy,
        319:301
    irradiation device and dosimetry,
        319:304
    irradiation studies, *in vitro*,
        319:303–309
    mitochondria transmembrane potential
        changes, **319:**307
    morphology, **319:**300–301
    phosphatidylserine assay, **319:**307

singlet oxygen generator and quencher
    effects. **319:**304–305
TUNEL assay, **319:**306–308
ultraviolet A-1 radiation therapy,
    319:300–302
ultraviolet B effects, **319:**302
APP, *see* Aminopyropheophorbide; β-Amy-
    loid precursor protein
Appetite, *see* Galanin; Neuropeptide Y
APS, *see* ATP sulfurylase
Aptamer, *see In Vitro* selection, RNA ap-
    tamers
Aquaporin, *see also* Aquaporin I
    activation energy for water transport,
        306:180–181
    sec6-4 mutant, expression of recombinant
        protein
    promoters, **306:**172
    proteoliposome reconstitution,
        306:180–181
    vesicle characteristics, **306:**177
    yield and plasmid loss, **306:**183
    shrinkage assay of function, **306:**177–180
    water permeability calculation,
        306:180–181
Aquaporin I
    discovery, **294:**551
    expression in *Xenopus* oocytes
        complementary DNA expression con-
            struct preparation, **294:**568–569
        complementary RNA
            injection, **294:**570
            preparation, **294:**569–570
        materials, **294:**568
        osmotic water permeability measure-
            ment, **294:**570
        Western blot analysis, **294:**570, 572
    expression in yeast
        materials, **294:**555
        plasmid construction, **294:**555
        reconstitution into proteoliposomes,
            294:558–559
        spheroplast preparation and secretory
            vesicle isolation, **294:**555–556
    homology cloning by degenerate oligonu-
        cleotide polymerase chain reaction
        amplification reaction, **294:**565, 567
        gel electrophoresis analysis,
            294:565–566
        materials, **294:**562–565

overview, **294**:562
secondary amplification and DNA puri-
    fication, **294**:566–567
purification from red cells
    anion-exchange chromatography,
        294:553–554
    materials, **294**:552
    membrane vesicle preparation,
        294:552–553
    reconstitution into proteoliposomes,
        294:556–558
two-dimensional crystallization by *in
    vitro* reconstitution
    dialysis, **294**:148–150
    electron cryomicroscopy, image pro-
        cessing, **294**:165–167, 169, 171,
        173–175
    lipids for reconstitution, **294**:147–148
water permeability assay
    activation energy, **294**:561
    calculations, **294**:559–561
    carboxyfluorescein fluorescence quench-
        ing, **294**:560
    materials, **294**:560
    stopped-flow analysis, **294**:561
    tetramer density determination,
        294:561
AR, *see* Androgen receptor
AraC, inhibition of DNA synthesis in anti-
    sense studies, **314**:64, 66–68
Arbitrarily primed polymerase chain reac-
    tion, *see* RNA-arbitrarily primed poly-
    merase chain reaction
Archaeoglobus fulgidus biofilm
    Archaea biofilms, overview, **310**:335–336
    autofluorescence, **310**:342
    carbohydrate content assays, **310**:347
    culture
        inoculum, **310**:339, 341
        media preparation, **310**:337–339
        temperature and pH induction of bio-
            films, **310**:341–342
    electron microscopy, **310**:342–343
    environmental response, **310**:335–336
    gene induction, **310**:336–337
    lipid assay, **310**:347–349
    metal effects on formation
        induction conditions, **310**:343
        atomic absorption spectrometry,
            310:345

chelation mechanisms, **310**:345–346
viability assays, **310**:343–345
nucleic acid content assays, **310**:346–347
phase contrast microscopy, **310**:342
protein assay, **310**:348
D-Arginine, folding enhancement of inclu-
    sion body proteins, **309**:227, 231
L-Arginine transport
assays
    adherent cells cultured and superfused
        on microcarriers, **301**:83–84
    adherent cells grown on monolayers,
        301:82–83
    cell suspensions, **301**:84–85
    influx versus efflux, **301**:82
    membrane vesicles, **301**:85–86
    tissues, **301**:85
$b^{0,+}$, **301**:81
$B^{0,+}$, **301**:81
b_1^+, **301**:80–81
b_2^+, **301**:80–81
cationic amino acid transporters
    expression patterns, **301**:87
    ribonuclease protection assays,
        301:88–89
    types, **301**:86–87
    Western blot analysis, **301**:89–91
regulation of nitric oxide synthesis,
    301:78–79, 91
sodium dependence, **301**:80
y^+, **301**:80
y^+L, **301**:81
Argon ion laser, *see* Confocal microscopy
Arp2/3
assay by actin polymerization on *Listeria
    monocytogenes*
    fluorescence detection, **298**:61
    incubation conditions, **298**:61
    killed fluorescence-labeled bacteria
        preparation, **298**:59–60
enzyme-linked immunosorbent assay,
    298:44–45
functions, **298**:42, 52–53
purification from *Acanthamoeba
    castellani*
    anion-exchange chromatography,
        298:46, 49, 51
    buffer preparation, **298**:45–46
    cation-exchange chromatography,
        298:47, 50–51

cell growth and extraction, **298:**48–49
hydroxylapatite chromatography, **298:**47–48
phosphocellulose chromatography, **298:**47, 49–51
poly-L-proline affinity chromatography, **298:**46–47, 49
yield, **298:**51
purification from platelets
  anion-exchange chromatography, **298:**57–58
  cation-exchange chromatography, **298:**58
  cytoskeleton extract preparation, **298:**55–56
  gel filtration chromatography, **298:**56, 58
  overview, **298:**53
  Triton-insoluble cytoskeleton preparation, **298:**53–55
  yield, **298:**59
subunit structure, **298:**52
Western blot analysis, **298:**43–44
Array, *see* DNA array; DNA microarray; Protein array; Sequencing by hybridization
Arrestin, *see also* G protein-coupled receptor
binding of opsin/all-*trans*-retinal complex, **315:**244
direct binding assay for rhodopsin
  dissociation constant determination, **315:**435–436
  incubation conditions, **315:**434–435
  radiolabeled arrestin preparation, **315:**433–434
  troubleshooting, **315:**435–436
  variations of assay, **315:**436
expression and purification of recombinant protein from *Escherichia coli*
  ammonium sulfate precipitation, **315:**430
  anion-exchange chromatography, **315:**431
  cells
    growth, **315:**429
    lysis, **315:**429–430
  expression vector construction, **315:**428–429
  gel electrophoresis, **315:**428

heparin-Sepharose chromatography, **315:**430–431
  troubleshooting, **315:**432–433
  Western blot analysis, **315:**431–432
expression in cell-free translation system
  gel electrophoresis and fluorography, **315:**427
  radiolabeling, **315:**423–424
  stock solutions, **315:**425
  transcription, **315:**423–425
  translation reaction, **315:**424–426
  troubleshooting, **315:**427–428
  yield calculation, **315:**426–429
gel electrophoresis assay of rhodopsin binding
  arrestin preparation and metabolic radiolabeling, **315:**417, 419
  binding reaction, **315:**419, 421
  dissociation constant determination, **315:**421–422
  gel electrophoresis, **315:**421
  rhodopsin preparation and phosphorylation, **315:**417
kinetic light scattering of rhodopsin complexes
  arrestin preparation, **315:**335
  modeling of binding reaction, **315:**342
peptide competition assay for rhodopsin ligands, **315:**453–455
phage display of rhodopsin ligands
  advantages in ligand screening, **315:**439–440
  data analysis, **315:**448–451
  data confirmation, **315:**451–452
  insert preparation, **315:**440–442
  library evaluation, **315:**442–443
  M13 coat protein utilization, **315:**438–439
  panning
    controls, **315:**445–446, 448
    DNA preparation and sequencing, **315:**447–448
    elution of bound phage, **315:**444–445, 447
    phage culture, **315:**446–447
    rounds, **315:**443–444, 447
    substrate coating on dishes, **315:**444
  vectors, **315:**440
promoter analysis in transiently transfected *Xenopus* embryos, **316:**57

purification, **315**:240
rhodopsin binding
  overview, **315**:70–71, 238, 411, 422–
    423, 437
  phosphorylation requirement, **315**:422–
    423, 433–435
  selectivity profile, **315**:433–434
ARRP, *see* Autosomal recessive retinitis
  pigmentosa
Arsenite pump, *see also* PgpA
  ars operon and prokaryotic proteins,
    292:82–84
  ArsA ATPase
    allosteric activation by antimonite or
      arsenite, **292**:90–91
    ArsB interactions, assay, **292**:91–93
    assay, **292**:89–90
    purification, **292**:88–89
  assays in eukaryotes
    Chinese hamster cells, **292**:94–95
    Leishmania, *see also* PgpA
      plasma membrane-enriched vesicle
        assays, **292**:95–96
      promastigote assays, **292**:95
  assays in prokaryotes
    arsenite extrusion from intact cells,
      292:84–85
    arsenite uptake in intact cells, **292**:86
    ATP-dependent arsenite uptake in
      everted membrane vesicles,
      292:86–87
    NADH-dependent arsenite uptake in
      everted membrane vesicles, **292**:88
  evolution in prokaryotes and eukaryotes,
    292:82, 96–97, 102
Arylazides, *see* RNA–RNA cross-linking
Arylsulfatase A
  assay
    incubation conditions, **311**:102
    principle, **311**:101
    quantification, **311**:102
    radiolabeled sulfatide preparation,
      311:101
    reagents, **311**:101
  deficiency in metachromic leukodystro-
    phy, **311**:95–96
  gene, **311**:102
  leukocyte enzyme properties, **311**:102
ASA, *see* Arylsulfatase A
Ascorbic acid, *see also* Ferric reducing/anti-

oxidant power and ascorbic acid concentration assay
C11-BODIPY, multiwell screening of topical effectiveness, **319**:608
free radical neutralization, **300**:78
high-performance liquid chromatography,
  analysis of ozone depletion in skin
  equipment, **319**:539
  operation, **319**:540
  reagents, **319**:539
  sample preparation, **319**:540
  standards, **319**:540
high-performance liquid chromatography
  with electrochemical detection
  amperometric detection, **299**:66
  chromatography conditions, **299**:68,
    70–72
  coulometric detection, **299**:66–67, 72
  hydrodynamic voltammetry plot,
    299:79, 82
  instrumentation, **299**:67–68, 78–79
  materials, **299**:68, 78
  passivation, **299**:72
  sample preparation, **299**:67, 72–73,
    79–82
  standards, **299**:72
hydroxyl and superoxide anion radical
  scavenging, **299**:28, 32, 34
ozone-induced singlet oxygen in plants,
  chemiluminescence assay
  ascorbic acid reaction, **319**:512–513,
    518–519
  assay in leaves
    carrier gas effects on emission inten-
      sity, **319**:517–518, 520
    cell wall emission, **319**:516–517
    gas volume per unit surface area ef-
      fects, **319**:519–520
    instrumentation, **319**:513–514
    intensity of gas phase emission,
      319:516
    intercellular extract measurements,
      319:515
    kinetics, **319**:519
    principle, **319**:513
    spectral analysis, **319**:517
  reactants in generation, **319**:514–515
radiolabeled compound detection,
  299:75–76
recycling from dehydroascorbic acid

measurement from various systems, **299:**83–84

rat hepatocyte assay
  aging effects on results, **299:**87–88
  hepatocyte isolation, **299:**84–86
  high-performance liquid chromatography with electrochemical detection, **299:**86–87
  incubation conditions, **299:**86
  materials, **299:**84
  pH, **299:**85–86

regeneration of vitamin E
  hepatocyte assay system from ascorbate-deficient rats
    animal maintenance, **300:**80–81
    ascorbate determinations, **300:**83–84
    tert-butyl hydroperoxide protection studies, **300:**85, 88
    isolation and culture of hepatocytes, **300:**81–82
    tocopherol determinations, **300:**83
    vitamin content of cells, **300:**79–80
    vitamin supplementation, **300:**82–83
  mechanism, **300:**79

stability in solution, **299:**65, 79–80

ASF, *see* Agitation-stimulated fibrillogenesis

Asialoglycoprotein receptor, targeting for oligodeoxynucleotide uptake
  cell distribution of receptors, **313:**315
  glycopeptide L₃G₄-modified antisense oligodeoxynucleotide for liver parenchymal cell uptake
    asialoglycoprotein receptor binding analysis, **314:**330–333, 335–336, 341
    conjugation, **314:**326–328
    pH-dependence of receptor binding, **314:**335–336
    phosphorous-32 labeling of oligodeoxynucleotide, **314:**328–330
    stability studies, **314:**330, 332
    synthesis of L₃G₄, **314:**326
    uptake studies, **314:**333–335, 341
    in vivo fate in rats following intravenous injection, **314:**336–337, 339
  glycopeptide YEE(ahGalNAc)₃-modified antisense oligodeoxynucleotide targeting
    advantages of system, **313:**300–301

affinity for receptor, **313:**300
animals, **313:**302
cell culture, **313:**304
clearance, **313:**306, 315–316, 321
equipment, **313:**302
hepatitis B virus inhibition
  cell culture, **313:**308
  controls, **313:**320
  DNA analysis, **313:**308–309, 319
  dose–response, **313:**319
  stability of neoglycoconjugate, **313:**309
  surface antigen assay, **313:**308, 319–320
  target sequences, **313:**307–308
  toxicity analysis, **313:**209, 320–321
metabolite analysis, **313:**306–307, 316–318, 321
reagents and buffers, **313:**301–302
structure, **313:**299
synthesis, **313:**303–304, 310
tissue distribution in mice, **313:**306, 313, 315
uptake
  cell-type specificity, **313:**311
  efficacy, **313:**310–311, 313
  kinetics in liver, **313:**315–316
  radioassay, **313:**304–306
hepatocellular carcinoma uptake of oligodeoxynucleotides, **313:**298
rationale, **313:**297–298, **314:**325–326, 339–340, 342

Aspartate transcarbamylase
  conformational probing with osmotic stress measurement
    allosteric regulation by nucleotide triphosphates, **295:**57–58
    assumptions, **295:**56
    mutant studies and physiological significance, **295:**61
    T to R transition, **295:**56–57
  hydration changes in protein, measurement
    controls
      molecular crowding effects, **295:**51, 53
      protein dissociation or aggregation, **295:**54–55
      solute size dependence, **295:**48–51

solute types and protein interactions, **295**:47–48

freezing point depression osmometry, **295**:45–46

overview, **295**:42–43

solute-induced osmotic pressure in buffer versus water, **295**:46–47, 53–54

structure-based calculations of hydration change, **295**:60–61

vapor pressure osmometry, **295**:45

structure, **295**:44

D-Aspartyl O-methyltransferase, *see* D-Isoaspartyl O-methyltransferase

Assembly, ion channels

expression systems in analysis, **293**:466

membrane dependence, **293**:24–25

potassium channel assembly

functional tagging studies, **293**:31

overview, **293**:33

yeast two-hybrid analysis, *see* Yeast two-hybrid system

ATCase, *see* Aspartate transcarbamylase

Atherosclerosis, *see also* Low-density lipoprotein; Shear stress

ceramide lactosyltransferase role

cell adhesion, **311**:81

cell proliferation, **311**:78–81

overview, **311**:73–75

signal transduction, **311**:78

Atm1p

epitope tagging, **292**:780–782

functions in yeast, **292**:157–158, 785–786

immunofluorescence

antibody treatment, **292**:783

fixing cells, **292**:782

treatment of slides, **292**:782

submitochondrial localization, **292**:783–785

Atomic absorption spectroscopy

Archaeoglobus fulgidus biofilms, **310**:345

copper determination, **297**:277–278

urinary tract device biofilms, **310**:308, 501, 562

Atomic force microscopy

advantages in fibril studies, **309**:510

amyloid-β fibrils

image interpretation, **309**:517–518

morphology classifications

Alzheimer's disease morphologies, **309**:523

4 day samples, **309**:518

small species, **309**:523–525

variability in time course of fibrillogenesis, **309**:523

3 week samples, **309**:518, 520

4 week samples, **309**:520

5 week samples, **309**:520

scanning, **309**:515

seed-free aggregation solution preparation, **309**:513

specimen preparation, **309**:514–515

tapping mode scan parameters, optimization, **309**:516

topographical data acquisition, **309**:517, 525

resolution, **309**:510

surface property characterization of membranes, **310**:477–478

X-ray photoelectron microscopy samples, **310**:387, 389

ATP

actin polymerization role, *see* Actin polymerization

agonist behaviors, **292**:664, 674

ATPase classification, **308**:28

free energy of hydrolysis, **308**:28

intracellular concentration and gradient, **292**:665

isotopic labeling, glycolysis and pentose phosphate pathway enzymes

enzyme preparation and assays, **317**:23–28

monitoring of reaction mixtures, **317**:32

overview, **317**:19–23

preparation from glucose, **317**:29–32

purification from reaction pools, **317**:33–35

isotopic labeling for kinetic isotope effect studies, **308**:308, 310, 312

molecular discrimination in hydrolysis, *see also* ATP sulfurylase; DNA gyrase; Nicotinate phosphoribosyltransferase

clock mechanism, **308**:34–35

conformational coupling, **308**:28–29, 50–52

conformer affinities for substrates and products, **308**:31–34, 36

examples of enzymes, **308:**29, 49
ion transporter energetics, **308:**30–36
leaks, origin, **308:**52
necessary imperfection, **308:**49, 52, 70
optimization of ATP use, **308:**32–34
RecA, **308:**47–48
thermodynamic modeling, **308:**30–31,
35
phosphorothioate-tagged nucleotide triphosphate analogs for nucleotide analog interference mapping,
317:104–106
photoaffinity labeling, *see* 8-Azidoadeno-
sine-5′-triphosphate
purinergic receptors, **292:**664
release
assays
bioluminescence assay, **292:**668–673
cell culture, **292:**666–667
patch-clamp analysis, **292:**673–674
radioactive trapping assay, **292:**667–
668, 672–673
cystic fibrosis transmembrane regulator
role, **292:**665
ATP, caged
ADP, ATP carrier, electrophysiological
measurements with caged ATP
bilayer system, **291:**298–299
time-resolved measurements,
291:300–301
transport modes, **291:**300
calcium affinity, **291:**321, 324, 346–347
dideoxy ATP in DNA repair kinetic
analysis
analytical high-performance liquid chromatography, **291:**487–488
apparatus design for studies,
291:491–492
DNA damage and analysis, **291:**490,
494
panic response, **291:**494–495
photolysis conditions, **291:**488
purification, **291:**487
stability in mammalian cells,
291:489–490
synthesis of high specific activity α-^{32}P
compound, **291:**486–487
efficiency of photoconversion,
291:341–342
p-hydroxyphenacyl caged compound

photolysis conditions, **291:**28–29
properties and applications, **291:**8
stability testing, **291:**29
luciferase activation for bioelectronic imaging, **291:**246–248, 250
magnesium affinity, **291:**321, 324,
346–347
muscle fiber photolysis studies
actomyosin studies
ATPase assays, **291:**332–333
ATP binding, **291:**330–332
caged calcium studies, **291:**332
cross-bridge detachment, **291:**330,
334, 344–346
relaxation of tension, **291:**330, 334
muscle shortening velocity, inhibition
before photolysis, **291:**343–346
photochemical conversion, quantification,
291:338–341
sodium, potassium-ATPase, electrophysiological measurements with caged
ATP
bilayer experiments, **291:**291–295
binding of caged compounds, **291:**293
patch clamp studies, **291:**295–298
photolysis conditions, **291:**291, 293
time-resolved measurements, **291:**293–
294, 297–298
ATP, firefly luciferase assays
biomass assays
antibiotic sensitivity testing, **305:**361
bacteriuria assays, **305:**360–361
cytotoxicity and cell proliferation
assays, **305:**361–362
extraction of intracellular ATP,
305:358–359
hygiene monitoring, **305:**362–363
neutralization of extractants,
305:359–360
overview, **305:**356–357
sample collection and pretreatment,
305:357
sensitivity, **305:**360
sterility testing, **305:**360
calibration with ATP standard,
305:354–356
classification of reagents
flash reagents, **305:**350, 369
selection by application, **305:**369–370
slow decay reagents, **305:**349–350

stable light-emitting reagents, **305:**349, 369

enzyme assays

ATP/ADP/AMP production, simultaneous measurement, **305:**364

ATP/phosphocreatine production, simultaneous measurement, **305:**364–365

creatine kinase isoenzymes, **305:**366–367

glycerol kinase quantification of glycerol, **305:**365–366

overview, **305:**363–364

oxidative phosphorylation assay, **305:**368–369

photophosphorylation assay, **305:**368–369

pyrophosphate quantification with ATP sulfurylase, **305:**367–368

urease quantification of urea, **305:**366

kinetics of luciferase reaction

ATP response curve, **305:**353

calculations, **305:**350–351

peak intensity extrapolation, **305:**352–353

total light emitted, **305:**352–353

principle, **305:**348–349

ATP-affinity chromatography

BiP, **290:**388–389, 392

cytosolic chaperonin, **290:**234

ATPase, ATP-binding cassette proteins

ArsA ATPase

allosteric activation by antimonite or arsenite, **292:**90–91

ArsB interactions, assay, **292:**91–93

assay, **292:**89–90

purification, **292:**88–89

Chinese hamster P-glycoprotein ATPase assays

lactate dehydrogenase coupled assay, **292:**520

malachite green assay, **292:**519–520

kinetic parameters, **292:**520–521

modification of essential residues, **292:**521

overexpression in CR1R12 cells, **292:**515

plasma membrane preparation, **292:**516–517

purification of P-glycoprotein

dye affinity chromatography, **292:**518–519

reconstitution, **292:**519

solubilization, **292:**517–518

stimulators, **292:**521

suspension culture of CR1R12 cells, **292:**516

vanadate trapping of nucleotide, **292:**521–523

ligand transporters, **292:**17–18

malachite green assay, **292:**38–39, 519–520

maltose transporter

activity and assays, **292:**9, 17–18, 27–29, 37–39

mutants with uncoupled ATPase activity

ATP hydrolysis assays, **292:**37–39

selection for restored transport in *malE* strain, **292:**33–34

transport property characterization, **292:**34, 36–37

MDR1 ATPase

assay of reconstituted protein, **292:**489–490, 492

assays, **292:**438–441, 467–469, 489–490, 499–500, 507

baculovirus–insect cell protein characterization, **292:**438–441

characterization of expressed protein from rodent cultured cells

activity dependence on ATPase expression level, **292:**511–512

contaminating ATPase inhibition, **292:**510–511, 513

crude membrane preparation, **292:**505–506

inhibitor analysis, **292:**509–511

inorganic phosphate colorimetric assay, **292:**507

phosphorylation-deficient mutants, **292:**512–513

kinetic parameters, **292:**397–398

purified or reconstituted proteins, **292:**499–500

site-directed mutant characterization, **292:**490, 492, 505

Mdr3 ATPase

assay, **292:**402–403

contaminating ATPase inhibition,
292:411–412
kinetic parameters, **292:**410–411
pH optimum, **292:**406–407
valinomycin stimulation, **292:**409
vanadate inhibition, **292:**410–411
verapamil stimulation, **292:**408–409
vinblastine stimulation, **292:**409
PrtD ATPase
activity in various preparations,
292:79–81
assay, **292:**79
inhibitors, **292:**81
reconstition, **292:**81
ATPase, chaperone molecules
BiP
activity, **290:**384
assays
calculations, **290:**399–400
phosphomolybdate complex extraction assay, **290:**398–399
thin-layer chromatography,
290:398–399
expression of recombinant protein in
Escherichia coli, **290:**396–397
resolubilization from inclusion bodies,
290:397
chaperonin 60 assay, **290:**152–153,
157–158
GroEL assay, **290:**117
GroEL–GroES complex, dimerization effects on activity, **290:**129–130, 292
heat shock protein 104, **290:**431, 442–443
ATP-binding cassette transporter, *see also
specific proteins*
ATP-binding proteins
ATPase activity, **292:**9
clustering of domains in *Saccharomyces cerevisiae,* **292:**139, 145
membrane topology, **292:**8–9
sequence alignment in *Saccharomyces cerevisiae,* **292:**146–150
sequence homology in eukaryotes,
292:101–102
sequence homology with other proteins, **292:**3, 7–8
structure, overview, **292:**135–136
three-dimensional structure, **292:**9–10
classification
eukaryotes, **292:**103, 105, 109–110

humans, **292:**129–130
prokaryotes, **292:**3–4
Saccharomyces cerevisiae, **292:**137, 139,
145, 148–159
clinical importance, overview, **292:**132–133, 161–162
domain organization
eukaryotes, **292:**102, 111–112, 116,
130–131, 136
prokaryotes, **292:**5–6, 67–68, 113
functions in bacteria, **292:**30–31, 67, 112
gene cloning strategies
antibody screening, **292:**127
criteria for ATP-binding cassette transporter gene definition, **292:**118
database searching, **292:**122–127, 130,
137–138
degenerate polymerase chain reaction,
292:121–122
function/positional cloning, **292:**117–118, 120
hybridization, **292:**120–121
yeast two-hybrid system, **292:**127
half-molecules, **292:**111–112, 145,
768–769
heterologous expression in yeast, **292:**159
ligand transport
assay systems
accumulation in intact cells,
292:13–14
ATP addition to inside-out vesicles,
292:16
ATP hydrolysis, **292:**17–18
binding protein addition to cells
from external medium, **292:**14
binding protein addition to right-side-out vesicles, **292:**15
reconstitution into proteoliposomes,
292:16–17
ligand-binding protein function and mutation studies, **292:**11–13, 31
mechanism, **292:**18–20
ligand-binding proteins, **292:**10–11
membrane channel proteins, **292:**6–7,
136–137, 145, 148–149
multicomponent exporters, **292:**114–115,
131
phylogenetic analysis, **292:**52–53, 102,
104, 113–114, 144–145
ATPGP1

antibody generation, **292:**167

degenerate polymerase chain reaction in gene cloning

 probe design, **292:**164

 screening of libraries, **292:**165–166

epitope tagging

 c-MYC fusion protein generation, **292:**170–171

 expression and purification of fusion protein, **292:**169–170

 plasmid construction, **292:**167–169

sequence homology with other P-glyco-proteins, **292:**166–167

Western blot analysis

 extraction of protein, **292:**171

 gel electrophoresis and blotting, **292:**171

 immunostaining, **292:**171–172

 transgenic plant expression analysis, **292:**173

 wild-type plant expression analysis, **292:**172–173

ATPGP2

degenerate polymerase chain reaction in gene cloning

 probe design, **292:**164

 screening of libraries, **292:**165–166

sequence homology with other P-glyco-proteins, **292:**166–167

ATP-sensitive potassium channel

components in assembly, **294:**445–443

expression in *Escherichia coli*, **294:**521

expression in tissue culture cells

 maintenance of cells, **294:**451

 membrane isolation, **294:**453

 transfection, **294:**451–452

rubidium efflux assay, **294:**452–453

solubilization

 3-[(3-cholamidopropyl)dimethyl-ammonio]-1-propane sulfonate, **294:**455

 digitonin, **294:**454–455

 Triton X-100, 455

sucrose gradient centrifugation, **294:**456–457

SUR1

 binding assays

  gibenclamide, **294:**457–458

  P1075, **294:**458

concentration of receptor, **294:**456

dye affinity chromatography, **294:**456

lectin affinity chromatography, **294:**455

nickel affinity chromatography, **294:**455–456

phenylboronate-10 affinity chromatography, **294:**456

photolabeling

 live cells, **294:**454

 membranes, **294:**453

radioligand synthesis

 azidoiodoglibenclamide, **294:**449–451

 iodoglibenclamide, **294:**446–448

types, **294:**442–443

ATP sulfurylase

AMP formation, **308:**57

bond breaking, interdependence, **308:**56

central complex

 chemistry, **308:**55

 formation, **308:**54–55

coupled versus uncoupled mechanisms, **308:**54

functions of activated sulfates, **308:**53–54

phosphate plugging of coupling efficiency leak, **308:**60–63

reaction cycle and subunits, **308:**52–53

stoichiometry and energy transfer, **308:**57

structural modeling of coupling step, **308:**56–57

transient product synthesis, **308:**57–58, 60

ATP synthase, *see* Chloroplast ATP synthase

ATR–FTIR, *see* Attenuated total reflection–Fourier transform infrared spectrometry

Attenuated total reflection–Fourier transform infrared spectrometry, *see also* Fourier transform infrared spectroscopy; Infrared spectroscopy

bacterial adhesion assay

 advantages and disadvantages, **310:**491

 bands for monitoring, **310:**491–492

 data acquisition, **310:**492–494

 internal reflection element, polymer coating, **310:**473–475, 492

surface property characterization of membranes, **310:**475–477

Aureobasidin, inositol phosphoceramide synthase inhibition, **311:**335, 348

Aureobasidium pullulans, green fluorescent protein expression
  confocal microscopy
    image acquisition, **307:**613
    instrumentation, **307:**612–613
    propidium iodide multicolor staining, **307:**615–616
    rhodamine-conjugated antibody, actin staining in multicolor detection, **307:**617–618
    rhodamine-conjugated phalloidin, actin staining in multicolor detection, **307:**616–617
  epifluorescence microscopy and digital deconvolution, **307:**618–620, 622
  fixation, **307:**612
  imaging in living samples, **307:**111
  mutants and detection wavelengths, **307:**611–612
  quantitative analysis on leaves, **307:**620–621
  transcriptional control, **307:**610
  transformation, **307:**610
  vectors, **307:**611
Aurothioglucose, chromatin staining for transmission electron microscopy, **304:**229
Australifungin
  bioassay for sphingolipid inhibition in *Candida albicans*, **311:**337–339
  ceramide synthase inhibition, **311:**335, 347
  keto-enamine complexes, **311:**347
  preparation
    fermentation, **311:**341–342
    isolation, **311:**342–343
  structure, **311:**337
Autofluorescence
  minimization in immunofluorescence specimens
    buffer selection, **307:**60
    chambers and sealing, **307:**61
    reducing agent, **307:**60–62, 185
  quenching in biofilm microscopy, **310:**170–171
Autoinduction, *see* Acylhomoserine lactone autoinducers; *lux* control system, *Vibrio*
Autosomal dominant retinitis pigmentosa, *see* Retinitis pigmentosa

Autosomal recessive retinitis pigmentosa, cyclic GMP phosphodiesterase mutations, **315:**617
Avalanche photodiode
  luminometer prospects, **305:**63
  overview of features, **305:**53, 89
  principle of detection, **305:**61
Avidin–biotin, amyloid immunohistochemistry system, **309:**22
Azid-1, calcium uncaging efficiency by two-photon photolysis, **291:**367–369, 379
Azides, *see* RNA–RNA cross-linking
8-Azidoadenosine-5'-triphosphate, photolabeling
  Chinese hamster P-glycoprotein, **292:**521
  MDR1 expresed in vaccinia virus–T7 RNA polymerase system, **292:**464–465
2,2'-Azinobis(3-ethylbenzothiazoline-6-sulfonic acid)
  radical cation spectrum, **299:**379–380
  Trolox equivalent antioxidant capacity assay
    decolorization assay, **299:**384, 389
    extraction and analysis of tomato, **299:**383–385, 387, 389
    materials, **299:**381
    preparation of preformed radical cation, **299:**381
    principle, **299:**380
    standards, preparation, **299:**381–383
Azidoiodoglibenclamide, radioligand synthesis, **294:**449–451
7-Azido[8 ^{125}I]iodoketanserin, photoaffinity labeling of vesicular monoamine transporter
  purification of labeled protein, **296:**79–81, 83
  quantitative analysis, **296:**77–79
  reaction conditions, **296:**77, 79
  synthesis, **296:**77
8-*p*-Azidophenacylthio-cyclic GMP, *see* Photoaffinity labeling
Azidopine, photoaffinity labeling of P-glycoprotein, **292:**294, 302, 304, 309–311, 437–438
2,2'-Azobis[2-amidinopropane]hydrochloride

decomposition and antioxidant consumption, **299**:4

plasma oxidizability assay, **299**:38–40

AZIK, *see* 7-azido[8-^{125}I]iodoketanserin

Azure A, ganglioside detection on high-performance thin-layer chromatography plates, **312**:123

B

B$^{0,+}$, *see* L-Arginine transport

b$^{0,+}$, *see* L-Arginine transport

b$_1^+$, *see* L-Arginine transport

b$_2^+$, *see* L-Arginine transport

B800-850, *Rhodobacter sphaeroides* 2.4.1

components, **297**:151–152

mutagenesis techniques, **297**:155–156

regulatory gene isolation

complementation of regulatory mutants, **297**:156

heterologous host gene expression, **297**:157

mapping and cloning, **297**:158

reverse genetics, **297**:159

suppressor isolation, **297**:156–157

regulatory mutant isolation

cis-acting mutations, **297**:152–153

decreased photosynthesis gene isolation using transcriptional fusions to *sacB*, **297**:154–155

enrichment using *lacZ* fusions, **297**:153–154

increased photosynthesis gene isolation using transcriptional fusions to *aph*, **297**:153

spontaneous mutant isolation by pigmentation and growth analysis, **297**:155

trans-acting mutations, **297**:153

transcriptional regulatory factors

biochemical characterization, **297**:161–162

interactions between regulatory factors, **297**:162

phenotypic characterization, **297**:159–160

protein–DNA interactions, **297**:162–163

quantitation of photosynthetic expression, **297**:160–161

sequence analysis, **297**:159

structure–function analysis, **297**:161

types and gene targets, **297**:163–165

B875, *Rhodobacter sphaeroides* 2.4.1

components, **297**:151–152

mutagenesis techniques, **297**:155–156

regulatory gene isolation

complementation of regulatory mutants, **297**:156

heterologous host gene expression, **297**:157

mapping and cloning, **297**:158

reverse genetics, **297**:159

suppressor isolation, **297**:156–157

regulatory mutant isolation

cis-acting mutations, **297**:152–153

decreased photosynthesis gene isolation using transcriptional fusions to *sacB*, **297**:154–155

enrichment using *lacZ* fusions, **297**:153–154

increased photosynthesis gene isolation using transcriptional fusions to *aph*, **297**:153

spontaneous mutant isolation by pigmentation and growth analysis, **297**:155

trans-acting mutations, **297**:153

transcriptional regulatory factors

biochemical characterization, **297**:161–162

interactions between regulatory factors, **297**:162

phenotypic characterization, **297**:159–160

protein–DNA interactions, **297**:162–163

quantitation of photosynthetic expression, **297**:160–161

sequence analysis, **297**:159

structure–function analysis, **297**:161

types and gene targets, **297**:163–165

BABE, *see* 1(*p*-Bromoacetamidobenzyl)-EDTA

BAC, *see* Bacterial artificial chromosome

Bacillus subtilis biofilm

microcalorimetry, adhesion analysis of river biofilms, **310**:366–368

proteome analysis
    metabolic radiolabeling of proteins,
        310:434–435
    phenotypic variability, 310:434
    two-dimensional polyacrylamide gel
        electrophoresis
        autoradiogram analysis, 310:435–436
        database of maps, 310:441
        identification of proteins, 310:436,
            438
        overview, 310:433–434
        running conditions, 310:435
        sample preparation, 310:435
        staining, 310:435–436
bcr-abl, antisense targeting, 313:30, 287
Bacteria–host interaction, see Host–
    pathogen interaction
Bacterial artificial chromosome
    complementary DNA selection
        biotinylation of DNA, 306:84–85
        digestion of DNA, 306:84
        preparation of DNA, 306:84
    herpes simplex virus mutagenesis
        DNA purification, 306:341
        electrocompetent cell preparation,
            306:341
        electroporation, 306:341
        gene replacement
            day-to-day protocol, 306:347
            screening and characterization,
                306:348
            vector construction, 306:344–347
        infectious viral DNA preparation,
            306:349–350
        media and growth conditions, 306:340
        overview, 306:337, 339–340
        plasmid
            construction, 306:342–344
            transfection into mammalian cells,
                306:348
        rescue and growth of virus,
            306:348–349
        strains of bacteria, 306:340
        vhs gene deletion, 306:350, 352
Bacterial bioluminescence
    antenna proteins, see also Blue fluores-
        cence protein; Lumazine proteins;
        Yellow fluorescence protein
        absorption spectroscopy, 305:166, 174
        bioluminescence activity, 305:180

    fluorescence spectra, 305:176–179
    overview, 305:164–165
    wavelength shifting for emission,
        305:164
bacteriophage incorporation of lux genes
    aldehyde addition, 305:546
    antibiotic detection and sensitivity
        assays, 305:555–557
    applications, 305:543, 555–557
    bacterial strains, 305:544
    cloning, insertion, and recombination
        in construction
        Felix 01 lux, 305:554–555
        λ lux, 305:554
        overview, 305:553–554
    conjugation, 305:547
    direct cloning, 305:550
    instrumentation for luminescence detec-
        tion, 305:545–546
    media, 305:543–544
    phage
        specificity, 305:549
        stocks, 305:545
        strains, 305:544
    plasmids, 305:544–545
    transduction, 305:549
    transformation, 305:547–548
    transposition of genes into phage,
        305:550–553
cyanobacteria incorporation of lux genes
    aldehyde supply
        endogenous, 305:520
        exogenous, 305:518, 520
    biosensing application, 305:516–517
    circadian rhythm studies
        hit-and-run allele replacement,
            305:533–536
        overview, 305:516, 527–528
        random insertion of luxAB,
            305:541–542
        transposon mutagenesis of Synecho-
            coccus sp. strain PCC 9742 using
            Escherichia coli conjugation,
            305:536–537, 539, 541
        vector, 305:528–529, 531, 533
    constructs, 305:518–519
    luciferase assays in vivo
        colonies, 305:523–524
        comparison with other reporters,
            305:526–527

heterocyst considerations,
305:525–526
imaging of single cells, **305:**524–525
suspended cells, **305:**521, 523
reporters of gene expression during differentiation, **305:**514–516
luciferase
acidity of subunits, **305:**135
aldehyde inhibition, **305:**149–150
assays
dithionite assay, **305:**151–152
flavin injection assay, **305:**150–151
flavin mononucleotide reduction for assay, **305:**146–147
instrumentation, **305:**147
materials, **305:**145–146
oxidoreductase-coupled assay, **305:**152
assembly, **305:**156–158
catalytic efficiency, **305:**135
fused subunit construct as reporter in mammalian cells compared with firefly luciferase and chloramphenicol acetyltransferase
applications, **305:**571–572
calibration curves, **305:**559
expression levels, **305:**560–563
materials, **305:**558–559
promoter for studies, **305:**558
rationale for fusion, **305:**558
sensitivity, **305:**559–560
stable transfection effects on assay reproducibility, **305:**564–567
thermal stability, **305:**563–564
transient transfected cell splitting, effects on assay reproducibility, **305:**567, 569–571
kinetic mechanism, **305:**147–149
rate constants and equilibrium constants, **305:**149
recombinant *Vibrio harveyi* enzyme expression and purification in *Escherichia coli*
ammonium sulfate fractionation, **305:**144–145
anion-exchange chromatography, **305:**145
buffers and solution for purification, **305:**142–143
cell growth, **305:**137–141

cell lysis, **305:**141–142
expression levels, **305:**143
glycerol stocks of cells, **305:**136–137
overview, **305:**135–136
plasmid maintenance, **305:**136–137
recombinant *Vibrio harveyi* subunit expression and purification in *Escherichia coli*
anion-exchange chromatography, **305:**155, 158
anion-exchange chromatography of denatured subunits, **305:**158–162
bioluminescence activity, **305:**156
cell growth, **305:**153–154
cell lysis, **305:**154
extinction coefficients, **305:**156
fluorescence, **305:**163–164
heterodimer formation, **305:**157
homodimer formation, **305:**156, 158
plasmid construction, **305:**153
refolding, **305:**162–163
storage, **305:**155–156
subunit types and structure, **305:**152–153, 157
lux, see lux control system, *Vibrio*
mechanisms, **305:**42–46
Bacteriochlorophyll, biosynthesis, **297:**237
Bacteriophage
antisense inhibition, *see* Antisense RNA, *Escherichia coli*
ncorporation of *lux* genes, *see* Bacterial bioluminescence
Bacteriophage P22, *see* P22; RNA challenge phage system
Bacteriorhodopsin
crystallization for X-ray diffraction, **297:**58–59
energy storage studies with laser-induced optoacoustic spectroscopy, **315:**147–148
photocycle energetics, **315:**154, 156
Baculovirus–insect cell expression system
affinity tagging of proteins, **293:**463–464
baculovirus production
amplification, **293:**473–474, **296:**449
overview, **296:**445–446
plaque purification, **296:**447–448
recombinant virus construction, **296:**447
titration, **296:**448–449

cell culture, **306:**33–34
cell infection, **306:**35
cell line maintenance, **296:**446
comparison with other expression sys-
    tems, **306:**36–41
cyclic GMP phosphodiesterase δ subunit,
    expression and purification,
    315:609–610
cystic fibrosis transmembrane conduc-
    tance regulator, expression and puri-
    fication
  gel filtration chromatography,
    294:233–235
  hydroxyapatatite chromatography,
    294:233, 235
  nickel affinity chromatography,
    294:236–237
  overview, **294:**228–230
  solubilization in detergents
    sodium dodecyl sulfate, **294:**232–233
    sodium pentadecafluorooctanoic
      acid, **294:**235–236
  transfer vector construction,
    294:230–231
cystic fibrosis transmembrane regulator–
    glutathione S-transferase fusion
    protein
  cell purification, **292:**692–693
  fluorescence-activated cell sorting in se-
    lection, **292:**692
  glutathione affinity chromatography,
    292:693
  transfection, **292:**691–692
  vector construction, **292:**691–692
  verification of expression, **292:**693
E1
  expression, **306:**334–335
  purification from Sf21 cells, **306:**335
G$_t$α expression, **315:**539
glycine receptor, **294:**273
green fluorescent protein expression
  confocal microscopy, **302:**406–407
  fixation of infected cells, **302:**407–408
  fusion proteins, **302:**396
  insect cell expression, **302:**399–400
  native protein, **302:**395–396
  promoter selection, **302:**394–395
  recombinant baculovirus
    amplification of stocks, **302:**403–404
    isolation, **302:**396, 398

  preparation, **302:**400–401
  separation from parental baculovi-
    rus, **302:**401–403
  titer determination, **302:**396, 404–406
host cell line
  culture, **293:**471–472
  selection, **293:**471
infection, **296:**449–450
inhibitor binding assays, **296:**453–455
insect cell lines, **315:**14–15
ion channel expression
  electrophysiologic recording,
    293:479–480
  glycosylation analysis, **293:**481–483
  heteromeric receptor expression and
    detection, **293:**480–481
  types expressed, **293:**459–460
marker genes, **306:**32
MDR1
  advantages of expression system,
    292:428–429
  ATPase characterization, **292:**438–441
  cell infection and culture, **292:**434–435
  histidine-tagged protein
    expression system, **292:**319–320
    insect cell membrane preparation,
      292:320–321
    metal affinity chromatography,
      292:321–322
  membrane isolation, **292:**436–437
  recombinant baculovirus construction
    amplification of virus, **292:**433–434
    calcium phosphate tranfection,
      292:432
    DNA preparation, **292:**431–432
    purification of virus, **292:**433
  site-directed mutants, **292:**483–484
  vector construction for expression,
    292:429–431
membrane preparation, **296:**450
optimization of expression, **293:**464–465,
    477–478
posttranslational modification advantages,
    293:465–466, **315:**13–14
principle, **296:**444–445
promoter selection, **293:**463, **306:**33,
    315:15
protein expression level, **293:**459
protein phosphatase type 7, expression of
    human protein

complementary DNA production, **315:**581
identification and cloning of gene, **315:**581–582, 584
transfection and expression in Sf21 cells, **315:**586–587
vector construction, **315:**584–586
receptor assembly analysis, **293:**466
recombinant baculovirus production
  approaches, **293:**468–469
  homologous recombination, **315:**15–16
  linearized DNA transfection, **293:**469–470
  pFastBAC transfection, **293:**470–471
  transposon-mediated exchange, **315:**16–20
rhodopsin expression
  bovine protein, optimization of expression, **315:**25
  functional characterization, overview of assays, **315:**29
  infection duration optimization, **315:**23–24
  insect cell adaptation to serum-free and protein-free culture conditions, **315:**22–23
  large-scale production, **315:**25
  media, **315:**21–22
  multiplicity of infection
    effects on yield, **315:**20–21
    optimization, **315:**23–24
  purification
    immunoaffinity chromatography, **315:**26
    lectin affinity chromatography, **315:**26–27
    nickel affinity chromatography of histidine-tagged protein, **315:**27
    reconstitution into proteoliposomes, **315:**28
    regeneration of pigment, **315:**24–25
    stable-isotope labeling, **315:**25–26
    yield relationship to functional protein, **315:**20–21
rhodopsin kinase expression and recoverin affinity purification, **315:**409–410, 409–410
Sf9 cell culture, **315:**16
simian virus 40 large T antigen expression for immunoaffinity chromatography
  cell lines, **306:**299–300
  growth media, **306:**300
  infection of cells, **306:**303
  lysate preparation, **306:**304, 306
titration of virus
  end-point dilution assay, **293:**475–476
  forms of virus, **293:**474
  MTT assay, **293:**476
  plaque assay, **293:**475
transfection, **306:**34
uptake assays, **296:**452–453
virus
  features, **306:**32
  high-titer stock establishment, **306:**35
  recombinant virus analysis, **306:**34–35
Western blot analysis of neurotransmitter transporter expression, **296:**450–452
Badger–Bauer relationship, *see* Raman spectroscopy
Base pairing, geometry inference from base variation patterns in RNA comparative sequence analysis, **317:**500–502
Bax, Bcl-2 interaction analysis with green fluorescent protein mutants, **302:**422
BBB, *see* Blood–brain barrier
Bcl2, photodynamic therapy, apoptosis role, **319:**348–349
BCECF, *see* 2′,7′-Bis(carboxyethyl)carboxyfluorescein
B-1 cell, CD5$^+$
  cell culture, **314:**419
  chronic lymphocytic leukemia in humans, **314:**415–416
  developmental changes, **314:**414
  flow cytometry, **314:**413
  neoplasia in mice
    antisense knockdown of interleukin-10
    clinical implications and costs, **314:**426–429
    flow cytometry analysis, **314:**421–422
    histologic analysis, **314:**422–424
    infusion pump delivery, **314:**420–421
    lipofection with cochleates, **314:**418–421, 423–425
    oligonucleotide design, **314:**419
    survival effects, **314:**422, 425–426
    toxicity, **314:**428

chronic lymphocytic leukemia similari-
ties, **314:**411, 414
interleukin-10 as growth factor,
314:411, 415
model establishment, **314:**419–420
phenotype, **314:**412
tissue distribution, **314:**413
Bcl-2
antisense knockdown in cancer cells
clinical application, **314:**478
controls, **314:**483–484
lipofection of cultured cells
A549 cells, **314:**488–489
parameters, **314:**487
principle, **314:**485, 487
SW2 cells, **314:**487–488
ZR-75-1 cells, **314:**488–489
oligodeoxynucleotide design
backbone modification, **314:**480
homology searching, **314:**484
length, **314:**478–480
secondary structure analysis, **314:**481,
483
target region selection, **314:**480–481,
483
phosphorothioate oligodeoxynucleotide
purification, **314:**484–485
secondary test systems, **314:**489–490
Bax interaction analysis with green fluorescent protein mutants, **302:**422
functions, **314:**477
inhibition with P-ethoxy antisense oligonucleotides, **313:**378–379
Beam radius, Gaussian beam optics,
291:187
BEBOVA, *see* Bond energy/bond order vibrational analysis
BEC, *see* Biofilm elimination concentration
Beer–Lambert law, **309:**607
Bence Jones proteins, *see also* Immunoglobulin light chain amyloidosis
agitation-stimulated fibrillogenesis
aggregation conditions, **309:**207
cuvette conditions, **309:**215–216
kinetic analysis, **309:**207–209
orbital shaker conditions, **309:**214–215
peptide synthesis, **309:**206–207
protein quantification, **309:**209
thioflavin T monitoring, **309:**209–212,
216

small zone chromatography, **309:**329–330,
332
types, **309:**329
Benzoin phosphate
limitations in photochemical applications,
291:7
photolysis properties, **291:**6
2-(4-Benzoylphenyl)thiazole-4-carboxylic
acid, synthesis, **305:**195
2′,7′-Bis(carboxyethyl)carboxyfluorescein
vascular remodeling studies, **307:**257
intracellular pH change measurement during glutamate transport
acidification and glutamate transport,
296:623
loading of cells, **296:**622
ratio fluorescence, **296:**622–623
Beta cell, glucose response analysis of
NAD(P)H with two-photon excitation
microscopy
immunostaining of enzymes and hormones, **307:**362–365
islet cell studies, **307:**360–362
principle, **307:**360
Betaine, polymerase chain reaction improvement of efficiency with dimethyl sulfoxide, **303:**45–46, 50
Beta-sheet, propensity of formation,
295:27
BFP, *see* Blue fluorescence protein
BIACORE, *see* Surface plasmon resonance
Bilayer, *see* Membrane bilayer
Bile salt-stimulated lipase
assay of ceramidase activity, **311:**282–283
bile salt effects on ceramidase activity,
311:283
Binding, Gibbs energy, *see* Gibbs energy of
binding
Binding isotherm, analysis
competitive inhibition data analysis,
294:130–131
curve fitting, **294:**129–130
subtraction of nonspecific binding,
294:128–129
Bioelectric effect, *see* Electromagnetic field
Biofilm
adhesion forces, **310:**375–376, 501–502,
526
antimicrobial resistance, *see* Antimicrobial resistance, biofilms

capsular polysaccharide, *see* Capsular polysaccharide, biofilms

conductance, *see* Conductance, assays for enzymatic degradation of biofilms

culture apparatus, *see* Biofilm fermenter; Chemostat, biofilm culture; Constant depth film fermentor; Flow cell; Robbins device; Rotating disk reactor; Rototorque

culture techniques, overview, **310:**250–253, 609

definition, **310:**56, 91, 95, 146, 232, 594

eradication techniques, **310:**525–527

examples, **310:**523–524

extracellular polysaccharides, *see* Exopolysaccharide, biofilms

formation steps, overview, **310:**375–376, 524–525

fouling, *see* Biofouling, membranes

growth rate assay for isolated bacteria, **310:**97–98

lectin staining, *see* Lectin staining, biofilms

mass transfer, *see* Mass transfer-limited bioavailability, biofilm nutrients

medical devices, *see* Bladder catheter biofilm; *Candida* biofilm; Medical device implant; Prosthetic joint biofilm

metal working biofilm, *see* Metal working biofilm

microscopy, *see* Atomic force microscopy; Attenuated total reflection–Fourier transform infrared spectrometry; Confocal laser scanning microscopy; Deconvolution fluorescence microscopy; Electron microscopy; Epifluorescence microscopy; Immunofluorescence microscopy; Microscope flow cell; Phase-contrast microscopy; Photon-counting microscopy; X-ray photoelectron microscopy

microtiter plate assay for formation
applications, **310:**92–93
growth conditions, **310:**93–94
inoculation, **310:**93
mutant screening, **310:**96–97
quantitative analysis, **310:**95–96
staining, **310:**93, 95

nuclear magnetic resonance, *see* Nuclear magnetic resonance, opaque biofilms

oral, *see* Oral biofilm

penetration of antimicrobials, **310:**608

rescue of mutants by growth condition changes, **310:**100

species-specific films, *see* *Archaeoglobus fulgidus* biofilm; *Candida* biofilm; *Escherichia coli* biofilm; *Legionella pneumophila* biofilm ; *Pseudomonas aeruginosa* biofilm; *Staphylococcus epidermidis*; *Streptococcus* biofilm

steady-state model
acid/base equilibrium, **310:**317–320
charge coupling modeling, **310:**316–317
concentration change calculations, **310:**315–316
diffusion, **310:**314–315, 318
ideal requirements for computer model, **310:**320–321
reaction rates, **310:**319
requirements, **310:**312–313

structural classification, **310:**297

substrata, *see* Substrata

surface variety attachment assay, **310:**99–100

suspension culture comparison, **310:**20–21, 57

viability assessment, **310:**177–178

in vivo growth models, **310:**589–595

Biofilm elimination concentration, assay, **310:**543–544

Biofilm fermenter
advantages over closed systems, **310:**233
biological system comparison, **310:**233–234
perfused biofilm fermenter
applications, **310:**235, 239–240
biomass generation, **310:**234–235
doubling time of cells on membrane, **310:**239
establishment of biofilms, **310:**235, 238–239
growth rate calculation, **310:**239
harvesting, **310:**239
overview, **310:**234–235
Sorbarod biofilm fermenter
applications, **310:**243, 247–248
design, **310:**243

establishment of biofilms, **310:**244–247
  growth rates, **310:**245–247
  harvesting, **310:**247
  yield, **310:**247
Swinnex biofilm fermenter
  applications, **310:**242–243
  design, **310:**240–241
  establishment of biofilms, **310:**241–242
  harvesting, **310:**242
Biofouling, membranes
  adhesion assays
    attenuated total reflection–Fourier
      transform infrared spectrometry
      advantages and disadvantages,
        310:491
      bands for monitoring, **310:**491–492
      data acquisition, **310:**492–494
      internal reflection element, polymer
        coating, **310:**473–475, 492
    autoradiographic scanning, **310:**494
    bacteria selection, **310:**482
    overview, **310:**482–483
    radiometric disk assay
      cell detachment assay, **310:**487–488
      cell growth, **310:**483
      cell pretreatment assay, **310:**487
      controls, **310:**484, 486
      defective membranes, **310:**488–489
      edge effects, **310:**488
      incubation conditions, **310:**486
      inhibitor screening, **310:**491
      liquid scintillation counting,
        310:486–487
      membrane pretreatment assay,
        310:487
      polyamide membranes, **310:**490
      tube preparation, **310:**483–484
  biofilm formation, **310:**463–465
  membrane module retrieval for sampling
    archiving of samples, **310:**469
    dissection, **310:**467
    draining, **310:**467
    scraping, **310:**467–468
    storage and transport, **310:**467
    types of samples, **310:**468–469
  microscopy approaches for study, **310:**636
  substrata preparation for studies
    cellulose acetate thin films on internal
      reflection elements and glass cou-
      pons, **310:**473–475, 492

cellulose acetate, **310:**469–471
    polyamide thin-film composite, **310:**471,
      473
    rationale, **310:**469
  surface property characterization of mem-
    branes
    atomic force microscopy,
      310:477–478
    attenuated total reflection–Fourier
      transform infrared spectrometry,
      310:475–477
    captive bubble contact angle measure-
      ment of hydrophobicity
      advantages, **310:**479
      apparatus, **310:**479–480
      contact angle calculations, **310:**481
      rationale, **310:**478
    rationale, **310:**475
    uranyl cation-binding assay of mem-
      brane surface charge, **310:**481–482
  symptoms, **310:**466
Biology Workbench project, *see* Internet re-
  sources, ion channels
Bioluminescence, *see* Luminescence
Biopanning, *see* Phage display library
Biosensor, *see* Capillary electrophoresis;
  Gramicidin; Sniffer-patch detector
Biosensor, *see* Surface plasma resonance
  biosensor
Biosurfactant
  adhesion assays
    flow chamber assay, **310:**429–430
    polystyrene assay, **310:**428–429
    vaginal epithelial cell assay, **310:**429–
      430, 432
  axisymmetric drop shape analysis by pen-
    dant drop method, **310:**426, 428, 430
  composition, **310:**426–427
  isolation
    cell growth, **310:**427
    dialysis, **310:**427
    Lactobacillus strains, **310:**427, 432
  surlactin functions in lactobacilli, **310:**426,
    432–433
Biotin affinity labeling, *see also* Photocleavable affinity tags
  applications, **291:**135–136, 152
  avidin affinity, **291:**136
  biotinylation of macromolecules, **291:**136
  release of target molecule, **291:**136–137

Biotin operon repressor
  assembly, overview, **295:**429–430
  BirA
    biotinyl-5′-AMP, binding model,
      295:430–431, 433–437
    covalent linkage of biotin to carrier
      protein, **295:**426–427
    DNA binding, **295:**427
    footprinting and thermodynamic analy-
      sis of holoBiRA binding to biotin
      operator, **295:**441–450
    ligand-induced conformational change,
      295:439–440
    monomer assembly, **295:**440–441
    three-dimensional structure, **295:**428
  components of regulation, **295:**426
  corepressor binding to repressor, thermo-
    dynamic analysis, **295:**437–439
  operon organization, **295:**427–428
Biotinylation, *see also* Cell surface biotinyla-
    tion, neurotransmitter transporters;
    Transmembrane topology
  complementary DNA full-length cloning
    with high efficiency
    cap trapper in selection, **303:**19, 21
    capture reaction, **303:**35
    complementary DNA biotinylation re-
      action, **303:**33–34
    complementary DNA release from
      beads, **303:**36
    streptavidin-coated magnetic bed prepa-
      ration, **303:**34–35
    washing of beads, **303:**35
  direct complementary DNA selection,
    303:115–116
  gangliosides, *see* GM1
BiP
  ATPase
    activity, **290:**384
    assays
      calculations, **290:**399–400
      phosphomolybdate complex extrac-
        tion assay, **290:**398–399
      thin-layer chromatography,
        290:398–399
    chaperone/antichaperone activities,
      290:44–45
  cloning into expression vectors after am-
    plification, **290:**389–390
  domains, **290:**384

peptide-binding assays
  oligomer dissociation assay
    incubation conditions, **290:**407
    materials and solutions, **290:**406–407
    native polyacrylamide gel electropho-
      resis, **290:**406–408
    peptide induction, **290:**405–406
    reduced carboxymethylated lactalbu-
      min competition assay,
      290:408–409
  synthetic peptide preparation, **290:**404
purification of animal tissue protein
  anion-exchange chromatography,
    290:389
  ATP-affinity chromatography,
    290:388–389
  buffers and solutions, **290:**386–387
  microsome preparation
    beef liver, **290:**388
    dog pancreas, **290:**387
purification of domains
  ATPase
    expression of recombinant protein in
      Escherichia coli, **290:**396–397
    resolubilization from inclusion bod-
      ies, **290:**397
  enterokinase cleavage
    BiP purification, **290:**395
    cleavage reaction, **290:**395–396
    fragment purification, **290:**396
    materials and solutions, **290:**394
    site engineering between domains,
      290:394
purification of recombinant murine pro-
    tein in *Escherichia coli*
  anion-exchange chromatography,
    290:392
  ATP-affinity chromatography, **290:**392
  cell growth, **290:**392
  expression system, **290:**390–391
  histidine-tagged protein, **290:**393
  materials and solutions, **290:**391–392
substrate specificity
  functional conequences, **290:**384–385
  minimal substrate, **290:**384
phage display libraries, affinity bio-
    panning
  amplification of eluted phages,
    290:402–403
  applications, **290:**385–386

biotinylation of BiP, **290:**401–402
materials and solutions, **290:**400–401
panning reaction, **290:**402
positive phage identification by color-
    imetric assay, **290:**403
principle, **290:**400
single-stranded phage DNA isola-
    tion, **290:**403–404
BirA, *see* Biotin operon repressor
Birefringence, Gaussian beam optics,
    291:194–195
Bis(2,4-dinitrophenyl)oxalate, synthesis and
    educational chemiluminescence demon-
    strations, **305:**653–655
Bis(1-hydroxyheptyl)peroxide, synthesis,
    319:578–579, 581
Bis(1-hydroxynonyl)peroxide, synthesis,
    319:578–579, 581
Bis(*p*-nitrophenyl) phosphate, all-*trans*-reti-
    nyl ester hydrolase inhibition studies,
    316:391–392
Bistheonellide A, *see* Misakinolide A
Bis-2,4,6-trichlorophenoxalate, synthesis
    and educational chemiluminescence
    demonstrations, **305:**654–656
Black lipid membrane, *see* Planar lipid bi-
    layer
Bladder catheter biofilm
  bladder model for study
    apparatus, **310:**496–497
    applications, **310:**495–496
    assays of biofilm formation
      atomic absorption spectroscopy,
          310:501
      scanning electron microscopy,
          310:498–499
      transmission electron microscopy,
          310:499–500
      X-ray microanalysis, **310:**500–501
    growth media, **310:**497
    inoculation, **310:**497
  constant depth film fermentor
    antimicrobial evaluation,
        310:308–309
    atomic absorption spectroscopy,
        310:308
    culture, **310:**304–305
    steady state, **310:**309
  encrustment
    artificial urine models

dynamic encrustation model,
    310:563–565
modified Robbins device,
    310:562–563
static model, **310:**560–562
stimulated urine flow model,
    310:565
urine composition, **310:**560
atomic absorption spectroscopy,
    310:562
energy-dispersive X-ray analysis,
    310:561–562
infrared spectroscopy, **310:**561
mineral deposition, **310:**558
resistance of surfaces, **310:**558–559, 566
scanning electron microscopy, **310:**558
X-ray diffraction spectroscopy, **310:**561
infection risk, **310:**494–495, 558
BLAST, *see* Expressed sequence tag
Bleaching, *see* Antifading agents; Photo-
    bleaching; Photobleaching recovery,
    photoreceptors
Bleomycin–iron, cross-linking of chromatin
    complexes
  advantages, **304:**517
  cross-linking reaction conditions
    Ehrlich acites tumor cells, *in vivo*,
        304:520, 522
    isolated chromatin, **304:**519–520
  radical production, **304:**516–517
  two-dimensional gel electrophoresis analy-
      sis, **304:**520
BLM, *see* Black lipid membrane
Blood–brain barrier, oligonucleotide perme-
    ability, **314:**128–129
Blood phagocyte activation assays, *see*
    Phagocyte luminescence
Blot overlay assay
  F-actin
    advantages, **298:**32–33
    autoradiography, **298:**41
    denaturing polyacrylamide gel electro-
        phoresis of binding proteins,
        298:39–40
    electroblotting, **298:**40
    fixation of blots, **298:**40
    incubation of blot, **298:**40–41
    iodine-125 labeling of G-actin,
        298:36–39
    materials, **298:**35–36

membrane skeleton protein identifica-
    tion, **298:**33–34
nonradioactive detection, **298:**42
phosphorous-32 ATP complex visualiza-
    tion, **298:**41–42
polymerization of actin and yield,
    298:38–39
principle, **298:**33
paxillin
    autoradiography, **298:**85
    buffer preparation, **298:**84
    denaturing polyacrylamide gel electro-
        phoresis, **298:**84–85
    incubation conditions, **298:**85
    iodine-125 labeling of paxillin, **298:**85
solid-phase binding assays, **298:**113
vasodilator-stimulated phosphoprotein
    applications, **298:**111–112
    phosphorous-32 labeling of vasodilator-
        stimulated phosphoprotein,
        298:112–113
zyxin
    advantages, **298:**75
    autoradiography, **298:**75
    denaturing gel electrophoresis and blot-
        ting, **298:**75
    incubation conditions, **298:**75
    reagents and buffers, **298:**72–73
    zyxin radiolabeling
        iodine-125, 74
        phosphorous-32, **298:**73–75
Blue fluorescence protein
    absorption spectroscopy, **305:**166, 174
    bioluminescence activity, **305:**180
    fluorescence spectra, **305:**176
    green fluorescent protein variant,
        305:204, 209–210
    ligand exchange
        overview, **305:**170–172
        riboflavin replacement by Lum,
            305:176–178
    purification from *Vibrio fischeri*, **305:**172
    resonance energy transfer, **305:**165–166
    structure, **305:**165
BODIPY-labeled sphingolipids
    applications, **312:**294–295, 523, 525, 534
    commercial availability, **312:**523–524
    enzyme assays
        ceramidase, **312:**301
        β-galactosidase, **312:**300

β-glucosidase, **312:**300
sphingomyelinase
    lysosomal acid enzyme, **312:**299–300
    neutral magnesium-dependent en-
        zyme, **312:**300
sphingomyelin synthase assay on rep-
    lica colonies
    applications, **312:**530
    incubation on polyester disks,
        312:530–531
    lipid transfer from disk to thin-layer
        chromatography plate, **312:**531
    thin-layer chromatography, **312:**531,
        533
intracellular concentration measurement
    based on spectral shift
    bovine serum albumin conjugate prepa-
        ration, **312:**526
    endosomal transport, **312:**527
    fluorescence emission spectra of con-
        centration-dependence,
        312:525–526
    lysosomal transport, **312:**527
    prospects, **312:**534
    quantitative analysis
        early endosomes, **312:**528–529
        imaging instrumentation and analy-
            sis, **312:**527–528
        lysosomes, **312:**529–530
    uptake conditions, **312:**526–527
metabolic labeling of cells
    administration into cells and lyso-
        somes, **312:**301–302
    cell culture, **312:**302
    fluorescence analysis
        culture medium, **312:**304
        extracts, **312:**303
        intact cells, **312:**303
    termination of pulsing, **312:**302–303
synthesis
    acylation of lyso-sphingolipid, **312:**524
    ceramide dihexoside, **312:**298–299
    ceramide trihexoside, **312:**298–299
    ceramide, **312:**298
    dodecanoic acid conjugation for link-
        age, **312:**295–296
    glucocerebroside, **312:**298
    purification and storage, **312:**524–525
    sphingomyelin, **312:**332
Bohr effect

cyanomet intermediates of hemoglobin
    definition of Bohr effect, **295:**211
    deoxy/cyanomet model, **295:**211–212
    determination of Bohr effect,
        295:212–213
    pH effects, **295:**213–214
    molecular mechanism, **295:**210
    mutant hemoglobin studies, **295:**215–216,
        226–227
    oxygen ligation comparison, **295:**215
    pH measurement
        apparatus, **295:**219–220
        buffer capacity of oxyhemoglobin,
            295:220–221, 226
        small samples, **295:**221, 223–226
    preparation of hemoglobin for analysis,
        295:216–218
Bond energy/bond order vibrational analy-
    sis, transition-state structure
    ab initio optimized structures,
        308:394–395
    AMP deaminase, **308:**335
    BEBOVIB program
        advantages, **308:**372
        atomic mass lists, **308:**378
        force constants for each internal coordi-
            nate, **308:**377
        inputs, overview, **308:**373
        interaction force coefficients,
            308:377–378
        internal coordinates, **308:**375–377
        reaction coordinate, **308:**378–382
        structures of initial and final states,
            308:374
        temperature inputs, **308:**378
    computational power, **308:**356
    nucleoside hydrolase, **308:**337, 339–340
    reliability of technique, **308:**395–397
    software, **308:**371, 373
    structure interpolation
        Cartesian coordinates to transition-
            state structures to BEBOVIB for-
            mat using internal coordinates,
            308:383, 388–389
        internal coordinates, **308:**384–385
        nicotinamide adenine dinucleotide hy-
            drolysis application, **308:**385–388
        overview, **308:**368, 370–371, 382–383
        reaction space and interpolation philos-
            ophy, **308:**383–384

reference structures
    high energy intermediate, **308:**388
    hyperconjugation, **308:**390
    input in CTBI program, **308:**388–389
    parsing, **308:**390
    pruning to find rings, **308:**390–392
    reactant, **308:**387–388
    sources of structural information,
        308:387–388
    test structures, isotope effect calculations,
        308:368, 392–393
    transition-state imbalance, **308:**396–397
    unified models of reaction types, **308:**371,
        395
    Walden inversion motion, **308:**379–381
Bone marrow, gene therapy, *see* Gene ther-
    apy, *MDR1*
Boranophosphate oligonucleotides
    assays for quantification
        calibration of chemical assays,
            313:252–253
        circular dichroism, **313:**254
        mass spectrometry, **313:**254–255, 257
        molybdic acid reduction, **313:**252
        nuclear magnetic resonance,
            313:253–254
        phosphomolybdic acid reduction,
            313:252
        ultraviolet sectroscopy, **313:**254
    boron neutron capture therapy, **313:**228
    charge distribution, **313:**227–228, 257
    chemical synthesis
        by-products, **313:**237–238
        H-phosphonate aproach
            overview, **313:**230–231
            protecting groups, **313:**236–237
            solid-phase synthesis, **313:**232–233,
                235
            solution synthesis, **313:**231–232
        phosphoramidite approach, **313:**229–
            230, 236
        stereoselective synthesis, **313:**236
    enzymatic synthesis using boranophos-
        phate nucleoside triphosphate an-
        alogs
        advantages, **313:**246
        polymerase chain reaction, **313:**246–247
        primer extension reaction, **313:**246
    lipophilicity, **313:**228
    nucleoside triphosphate analog synthesis

applications, **313**:238–239

diastereomer separation with reversed-phase high-performance liquid chromatography, **313**:244

phosphorous-31 nuclear magnetic resonance, reaction monitoring, **313**:241, 243–244

solvents and reagents, **313**:239, 241

synthesis, **313**:241, 244

ribonuclease H cleavage of hybrid duplexes

assay

incubation conditions, **313**:247–248

product separation with high-performance liquid chromatography, **313**:248

endo/exo mechanisms in antisense effects, **313**:250–251

melting temperature determination, **313**:249

overview, **313**:229, 247

thermostability of hybrids, effect on hydrolysis rate, **313**:249–250

stability

nuclease stability, **313**:228–229

solution stability, **313**:251

structure, **313**:226–227, 257

Bovine pancreatic trypsin inhibitor, oxidation assay for protein disulfide isomerase, **290**:52

Bovine serum albumin, thiolation

2-iminothiolane reaction, **291**:157

thiol content, determination, **291**:157

BPTC, *see* 2-(4-Benzoylphenyl)thiazole-4-carboxylic acid

BPTI, *see* Bovine pancreatic trypsin inhibitor

Bradykinin, RNA ligand selection

binding reaction, **318**:233

N-bromoacetylbradykinin synthesis, **318**:230

guanosine 5′-monophosphorothioate

purification of RNA with thiopropyl-Sepharose, **318**:232–233

synthesis, **318**:230

transcription reaction, **318**:230–231, 233, 236–237

kinetic analysis of inhibition, **318**:234–237

polyacrylamide gel electrophoresis

[(β-acryloylamino)phenyl]mercuric chloride synthesis, **318**:231–232

guanosine 5′-monophosphorothioate–RNA separation, **318**:232–234

selection for catalysts, **318**:237

Branched RNA

comparative gel electrophoresis of global folding transitions

advantages, **317**:369

four-way junctions, **317**:371, 373–375

hammerhead ribozyme, **317**:375, 377

hybrid species in studies, **317**:373–374

retardation of bulged species, **317**:370–371

theory of gel electrophoresis, **317**:369–370

DNAzyme, mapping of branch points and other modifications, **317**:144–145

fluorescence resonance energy transfer of global folding transitions

anisotropy measurements, **317**:383–384

combination with other structural studies, **317**:392–393

comparative approach, **317**:388–389

fluorophores

labeling and purification of RNA, **317**:384–386, 388

selection of donor–acceptor pair, **317**:384

four-way junctions, **317**:378

instrumentation, **317**:379

normalized acceptor ratio method, **317**:381–383

principle, **317**:378

theory, **317**:379–381

transient electric birefringence analysis, **317**:392

types of branches, **317**:368–369

BRCA1, mouse versus human protein comparison, **314**:499–500

BRCA2

antisense knockdown in breast cancer cells

inhibition effects on mitoxantrone sensitivity, **314**:503

materials and reagents, **314**:501

oligonucleotide design, **314**:501–502

principle, **314**:500–501

ribonuclease protection assay of inhibition, **314**:502–503, 505–506

mouse versus human protein comparison,
        314:500
Breath hydrocarbons, lipid peroxidation
        markers
    flame ionization detection, **300:**457–458,
        462
    formation, **300:**457
    modulators, **300:**457, 460–461
    photoionization gas chromatography
        column selection, **300:**458–459
        confounding variables, **300:**460–461
        pentane measurement, **300:**459–460
        principle, **300:**458
        sensitivity, **300:**461–462
        smokers versus nonsmokers,
            300:461–462
Bromine, halogenation of pyrimidines,
        317:125–126, 160–161
Bromocresol purple, *see* Proton uptake, rho-
        dopsin activation
8-Bromo-cyclic AMP
    4,5-dimethoxy-2-nitrobenzyl caged com-
        pound
        calibration of flash-induced changes,
            291:427, 429–430
        concentration optimum determination,
            291:425
        design, **291:**417–418
        gated channel activation studies,
            291:426–427
        handling of solutions, **291:**425
        photochemical properties, **291:**421–422
        purity requirements, **291:**420–421, 430
        solubility, **291:**418, 421
        solvolysis, **291:**422–423
        synthesis, **291:**418–420
    phosphodiesterase resistance,
        291:416–417
8-Bromo-cyclic GMP
    4,5-dimethoxy-2-nitrobenzyl caged com-
        pound
        calibration of flash-induced changes,
            291:427, 429–430
        concentration optimum determination,
            291:425
        design, **291:**417–418
        gated channel activation studies,
            291:426–427
        handling of solutions, **291:**425
        photochemical properties, **291:**421–422

purity requirements, **291:**420–421, 430
        solubility, **291:**418, 421
        solvolysis, **291:**422–423
        synthesis, **291:**418–420
    phosphodiesterase resistance,
        291:416–417
1(*p*-Bromoacetamidobenzyl)-EDTA
    tethered iron(II) complex for ribonucleo-
        protein complex probing
        advantages, **318:**461–462
        buffers, **318:**463–464
        equipment and supplies, **318:**464
        footprinting S20 on 16S ribosomal
            RNA, **318:**468
        hydroxyl radical probing conditions,
            318:470
        iron(II) derivatization of protein,
            318:467
        linker–iron(II) complex preparation,
            318:466–467
        nuclease contamination prevention,
            318:462
        overview, **318:**461
        primer extension analysis
            extension reaction, **318:**472–473
            hybridization, **318:**472
            interpretation of data, **318:**473–475
            overview, **318:**471
            polyacrylamide gel electrophoresis
                analysis, **318:**473
            reagents and buffers, **318:**471–472
        purification of iron(II)–S20–16S ribo-
            somal RNA complex, **318:**468–469
        reagents, **318:**462–463
        reconstitution of 30S subunits con-
            taining iron(II)-derivatized pro-
            tein, **318:**469–470
        RNA isolation following probing,
            318:470–471
        S20 mutation for cysteine substitution
            accessibility assay, **318:**465–466
            site selection, **318:**464–465
        S20–16S ribosomal RNA complex for-
            mation, **318:**467–468
    tethered iron(II) complex for RNA struc-
        ture probing in nucleic acid com-
        plexes
        advantages, **318:**190
        buffers, **318:**176–177
        distance information, **318:**189–190

DNA oligonucleotide synthesis,
318:178

DNA–RNA duplex probing, **318:**181–
183, 189

enzymes and reagents, **318:**177–178,
189

radiolabeling of DNA, **318:**178–179

rationale, **318:**175–176

RNA–RNA duplex probing

anticodon stem–loop analogs,
318:186–188

environment surrounding 5′-end of
phenylalanine transfer RNA,
318:185

16S rRNA, **318:**183–184

tethering reaction, **318:**176, 180

transcription

efficiency of 5′-guanosine α-phospho-
rothioate incorporation, **318:**180

template preparation, **318:**179

in vitro reaction, **318:**179–180

4-Bromomethyl-3-nitrobenzoic acid succinimide ester

photochemical properties, **291:**159

reactivity, **291:**159

synthesis, **291:**156–159

5-Bromouridine, *see* 5-Halopyrimidines

Brønsted relation

linear-free energy relationship analysis of
hydrogen bonds, **308:**254–261, 269,
274–276

Marcus theory and proton transfer reactions, **308:**280–282, 291–292

Brown protein, *Drosophila melanogaster*

eye pigmentation role, **292:**213–215

gene cloning, **292:**214

hydropathy plot, **292:**215, 218

sequence alignment with other transport
proteins, **292:**215–217

site-directed mutagenesis, **292:**215,
218–219

White protein interaction sites,
292:221–223

BSA, *see* Bovine serum albumin

BuChE, *see* Butyrylcholinesterase

α-Bungarotoxin, acetylcholine receptor
assays

cell-surface binding assay, **293:**569–570

fluorescence microscopy, **293:**570–572

Butanoylhomoserine lactone

characterization, *see* Acylhomoserine lactone autoinducers

detection bioassay, **305:**293

Butyrylcholinesterase

caged ligands

carbamylcholine, **291:**266–267, 269,
273–276

choline, **291:**266–267, 269, 273–276

criteria for ideal compounds, **291:**273

inhibition analysis, **291:**272–274

kinetics of photolysis, **291:**273

noracetylcholine, **291:**266–267, 269

norbutyrylcholine, **291:**267, 269

photochemical properties, **291:**269–272

photolysis conditions, **291:**277–278

recovery of activity following photolysis, **291:**275–276, 278

physiological function, **291:**265

C

C14EO6, *see* Cloud point extraction, gangliosides

C62.5, *see* HtpG

Caenorhabditis elegans, *see also* Whole cell
patch clamp

advantages as research organism,
296:529–531

fire expression vectors, **296:**534–535

genome, **296:**531

immunofluorescence microscopy of vesicular transporters, **296:**542–544

laser ablation of cells, **296:**533–534

nervous system, **296:**530–531

plasma membrane neurotransmitter transporters, **296:**546–547

protein–protein interactions

suppressor screening, **296:**544–546

yeast two-hybrid system, **296:**544, 546

transformation, **296:**535–536

translational fusions, **296:**535

vesicular acetylcholine transporter

discovery, **296:**93–94

functional assays

behavioral assays, **296:**537–538

lethality rescue, **296:**538–539

mutation phenotypes, **296:**532

unc-17 gene, **296:**531–532

vesicular γ-aminobutyric acid transporter

functional assay, **296:**541
mutation phenotypes, **296:**533
unc-47 gene, **296:**532
vesicular glutamate transporter, functional assays, **296:**542
vesicular monoamine transporter
cat-1 gene, **296:**532–533
functional assays
food sensation assay, **296:**539–540
male mating efficiency, **296:**540–541
mutation phenotypes, **296:**532–533
CAF-1, *see* Chromatin assembly factor,
Caffeine
calcium release studies with caged inositol 1,4,5-triphosphate, **291:**393, 395
high-performance liquid chromatography analysis in tea with catechins
chromatography conditions, **299:**109, 111
comparison of green teas, **299:**112–113
instrumentation, **299:**108–109
quantitative analysis, **299:**111–112
sample extraction, **299:**109
standards, **299:**109
Caged compounds, *see also* Cyclic nucleotide-gated channel; Tubulin; *specific compounds and functional groups*
advantages in substrate release studies, **291:**1, 30, 64, 95–96
caged calcium, photolysis, **293:**376–377
criteria in design, **291:**1–2, 64, 120
signal transduction studies, *see* Calcium
sphingolipids
advantages in signal transduction studies, **312:**388, 399–400
applications, **312:**387–388
dihydrosphingosine and dihydrosphingosine 1-phosphate evoked calcium currents in neurons, **312:**391, 393–395, 397
flash photolysis, **312:**391
fumonisin B1 modulation of dihydrosphingosine responses, **312:**397–398, 400
historical perspective, **312:**388–389
sphingolipid receptor analysis, **312:**395
sphingomyelinase effects on evoked calcium currents in neurons, **312:**398–399

sphingosine evoked calcium currents in neurons, **312:**395, 397, 399
synthesis
analytical techniques, **312:**389–390
materials, **312:**389
2-*N*-(2-nitrobenzylation) of sphingosine derivatives, **312:**390
whole-cell patch-clamp recording, **312:**390–391
Calcein
compartmentation of ester-loaded fluorophore, **302:**349–350
membrane permeability studies, **302:**351–352
Calcineurin phosphatase, FKBP12 assay by inhibition with FK506 complex
incubation conditions, **290:**81–82
phosphate release assay, **290:**82–83
phosphopeptide substrate preparation, **290:**80–81
precautions, **290:**83
principle, **290:**79–80
Calcitonin fibril, X-ray fiber diffraction, **309:**528
Calcium, *see also* Cyclic nucleotide-gated channel; Sodium/calcium-potassium exchanger, photoreceptors
affinity for caged compounds, **291:**321, 324, 346–347
arterial endothelial cell, *in situ* calcium imaging with confocal microscopy
acetylcholine response, **307:**429, 431
data collection, **307:**429
fluo-3 loading, **307:**428
quantitative analysis, **307:**430
signal separation from smooth muscle cells, **307:**429
tissue preparation, **307:**428
ATPase, *see* Sarcoplasmic reticulum/endoplasmic reticulum calcium ATPase
caged, *see* Calcium, caged
calmodulin fluorescence assay, **294:**52
compartmentalization, *see* Inositol 1,4,5-triphosphate receptor
confocal microscopy imaging
line scanning and transient measurements, **302:**353–354
mitochondria measurements, **302:**356–358
nonratiometric imaging, **302:**352

ratio imaging with Indo-1, **302:**354, 356

simultaneous measurement with electrical potential, **302:**353–354

confocal microscopy imaging, custom system construction

  calcium transients, **307:**167–168

  cost, **307:**152–153

  dyes, **307:**164

  electronics, **307:**158, 161

  emission path, **307:**155, 157

  excitation path, **307:**155

  inflexibility of commercial systems, **307:**152–153

  morphological studies, **307:**168–169

  overview of design, **307:**153–155

  piezo-objective sytem, **307:**161–162

  resolution

    spatial, **307:**164–167

    temporal, **307:**167

  scan head, **307:**157–158

  software, **307:**162

  ultraviolet photolysis system, **307:**162–164

fluorescence assay with Fluo-3

  calibration, **294:**24–25

  detection wavelength, **294:**21

  dye loading, **294:**23–24

  microtiter plate reading, **294:**22–23

fluorescence assay with ratio fluorescence probes

  calibration of fluorescence signal, **294:**7–9, 24–25

  cell loading, **294:**4–5, 23–24

  hippocampal brain slice measurement, **294:**11–12

  indicator selection criteria, **294:**10–11

  instrumentation

    detectors, **294:**7

    filters, **294:**6

    light source, **294:**6

    microscope, **294:**5

    microtiter plate reader, **294:**21, 23

  presynapyic terminal measurements

    calibration of fluorescence signal, **294:**14–15

    data acquisition, **294:**13–14

    dye loading, **294:**12–13

    single terminal measurements in mossy fibers, **294:**15–19

  principle, **294:**4, 21–22

simultaneous patch-clamp analysis of plant cells

  caged calcium studies, **294:**437–440

  calibration of fluorescence signal, **294:**431–432

  cameras for imaging, **294:**433–434

  compartmentalization of dye, **294:**430

  confocal laser scanning microscopy, **294:**427–428, 430–432

  dye selection, **294:**426–427, 434–435

  Indo-1 loading, **294:**430–431

  instrumentation, **294:**426, 428, 430, 433–434

  light and heat responses of protoplasts, **294:**423–424

  photometry and imaging of ratio fluorescence dyes, **294:**432–435

  pipette resistance, **294:**428

  pipettes, **294:**422

  protoplast membrane exposure, **294:**411–419

  seal formation, **294:**422–424

fluorescent dyes, *see also* Calcium Green-1; Calcium Orange; Fluo-3; Fura Red; Indo-1

calibration

  calculations, **307:**460

  dual-wavelength methods, **307:**462–464

  importance, **307:**459

  single-wavelength methods, **307:**461–462

  in situ calibration and ionophores, **307:**460–461

  stock solutions, **307:**459

calmodulin–green fluorescent protein constructs

  advantages over aequorin systems, **307:**445

  calcium affinity, tuning, **307:**445–446

  comparison to chemical sensors, **307:**446

  confocal image acquisition requirements, **307:**468

  resonance energy transfer, **307:**445

  targeting to organelles, **307:**457–459

  transfection, **307:**450–451

  virus-mediated gene transfer, **307:**451

VP22-mediated protein transfer,
  307:451–452
loading of acetoxymethyl ester dyes
  dihydrorhod-2 loading of mitochon-
    dria, **307:**454–456
  Indo-1, loading of mitochondria and
    cytosolic quenching,
    307:453–454
  loading time, **307:**448
  principle, **307:**447
  rhod-2 loading of mitochondria,
    307:453
  solubility and dispersing agents,
    307:447–448
  temperature modulation,
    307:448–449
  multiple sensors in mitochondrial mea-
    surement, **307:**456–457
  overview of chemically synthesized sen-
    sors, **307:**442–444
fluorescent probe measurement
  apparatus for simultaneous electrophys-
    iological measurement,
    291:388–391
  response to calcium release by caged
    compounds
    cyclic ADP-ribose, **291:**410–413
    2,5-di(*tert*-butyl)hydroquinone caged
      compound, **291:**435–437, 439
    inositol 1,4,5-triphosphate, **291:**388–
      391, 395, 403
    nicotinic acid adenine dinucleotide
      phosphate, **291:**414–415
    plant hormones, **291:**478, 481–482
fractional ionic current determination for
  cyclic nucleotide-gated channel
  caged cyclic nucleotide photolysis
    8-bromo cyclic nucleotide features,
      315:803–805
    caging groups, **315:**802–805
    degradation following uncaging,
      315:803
    endogenous cyclic GMP, **315:**803
    solubility limitations, **315:**803–805
  calculations, **315:**800–802, 808, 810–811
  caveats
    assumptions, **315:**811–812
    cytoplasmic equilibrium with pipette
      solution, **315:**812–813
    Fura-2 controls, **315:**813–815

sampling space uniformity,
  315:816–817
unidirectionality of calcium flux,
  315:815–816
data acquisition, **315:**808
Fura-2 measurements of calcium,
  315:798, 800–802, 807–808, 810–
    811, 813–815
instrumentation, **315:**805, 807
overview, **315:**797–798
recording configuration, **315:**798–799
guanylate cyclase regulation, *see* Guanylate cyclase, photoreceptors
ion channel activity, drug screening with
  fluorescence assays
  advantages of assay, **294:**46–47
  α-amino-3-hydroxy-5-methylisoxazole-
    4-propioninc acid/kainate recep-
    tor, **294:**36
  cyclic nucleotide-gated channel,
    294:41
  limitations of assay, **294:**45–46
  N-methyl-D-aspartate receptor, **294:**32,
    34–35
  nicotinic acetylcholine receptor,
    294:36–37, 39, 41–45
  pharmacologic validation, **294:**26–27
  voltage-gated calcium channel, **294:**27,
    30–32
light adaptation role, **315:**708
mitochondrial flux, nitric oxide and peroxynitrite effects, **301:**385–386
multiphoton confocal microscopy,
  307:468–469
multipinhole confocal scanner for high-
  speed imaging
  charge-coupled device camera,
    307:434–435
  configuration of system, **307:**433–434
  fluo-4 imaging in mixed neuron and as-
    troglial cell culture, **307:**437,
    439–440
  principles, **307:**433
  temporal resolution, **307:**431, 433
  Z-axis scanning, **307:**435–437
nitric oxide synthase, dependence of isoforms, **301:**170–171, 505–506
oscillations and waves, rapid scan imaging, **307:**124, 420, 422, 425, 464–
    465, 468

polyphenol effects on flux in endothelial cells
aortic endothelial cell culture, **301:**529
fluorescence assay with Fura-2, **301:**530–532
sarcoplasmic reticulum release and fluorescence imaging, **307:**380
signaling, *see* Cyclic ADP-ribose; Inositol 1,4,5-triphosphate; Nicotinic acid adenine dinucleotide phosphate; Sarcoplasmic reticulum/endoplasmic reticulum calcium ATPase
tandem green fluorescent protein fusion proteins as probes, **302:**421–422
Calcium, caged
flash lamp photolysis studies with caged compounds, **291:**219–222
sodium, calcium-exchanger, DM-nitrophen caged calcium studies
electrophysiological measurements, **291:**302–306
photolysis kinetics, **291:**301
rate constant determinations, **291:**305–306
regulatory site studies, **291:**301–302, 304–305
photochemical conversion, quantification, **291:**339–340
two-photon photolysis
apparatus, **291:**361–365
applications, **291:**356–357, 380
calcium yield determination, **291:**358, 369–374
temporal behavior of calcium concentration distribution, **291:**374–379
uncaging action cross section determination, **291:**365–369
Calcium-activated photoprotein, *see also* Aequorin; Obelin
applications, **305:**225–226
definition, **305:**223–224
Calcium-ATPase, amyloid-β inhibition
assay, **309:**741
overview, **309:**739
Calcium-binding proteins, purification and identification in eye
conformational changes on calcium binding, **316:**112–114
flow dialysis analysis of calcium binding, **316:**110–112

hydrophobic interaction chromatography, **316:**105–106
protein–protein interactions
coimmunoprecipitation, **316:**117
cross-linking, **316:**117
gel overlay assay, **316:**115–117
identification of target proteins, **316:**117, 119–120
mass spectrometry identification of binding proteins, **316:**12
yeast two-hybrid system, **316:**119–120
reversed-phase high-performance liquid chromatography, **316:**107
sequence analysis and protein identification, **316:**107–109
uveal melanoma tissue extract preparation, **316:**105
Calcium/calmodulin-dependent protein kinase, inhibitors, **298:**548
Calcium channel, *see also* Calcium flux; Voltage-gated calcium channel
cut-open oocyte voltage clamp
calcium chelator removal of inactivation, **293:**317–318
current measurement after removing contaminating chloride currents, **293:**316–317
G protein inhibition, GTP perfusion analysis, **293:**330
speed of intracellular perfusion, **293:**328–329
whole-cell clamping comparison, **293:**326–328
types, **293:**371
Calcium-dependent guanylate cyclase activator protein
guanylate cyclase assay of activation
basal activity, **315:**734
calcium concentration manipulation and activation, **315:**731–732, 734, 737
comparison with other S100 proteins, **315:**739–741
radiolabeled GTP assay, **315:**731
rod outer segment preparation, **315:**731
immunoreactivity with anti-S100 antibodies, Western blot analysis, **315:**734, 737–739
purification from bovine extract

anion-exchange chromatography,
  315:732, 735
crude extract preparation, 315:732
gel electrophoresis
  analytical, 315:736–737
  preparative, 315:733, 737
gel filtration chromatography, 315:733,
  736
heat stability, 315:737
hydrophobic interaction chromatography, 315:732, 735–736
zinc affinity chromatography, 315:733,
  741
Calcium flux
action potential influx measurement
  action potential waveform voltage
    clamp
    applications, 293:353–354
    isolation of calcium currents, 293:354
    quality of voltage clamp,
      293:354–356
    signal-to-noise ratio, 293:364–366,
      371
    voltage template selection,
      293:356–358
  comparison of voltage clamp and fluo-
    rescence measurements,
    293:363–370
  fluorometric measurement
    calibration, 293:361–362
    detection systems, 293:362–363,
      368–369
    dye overload, 293:358–361
    dye types, 293:359
    quantification of fluorescence
      change, 293:358, 360–362
    signal-to-noise ratio, 293:366–368
  overview, 293:352
  simultaneous measurement of voltage
    clamp and fluorescence,
    293:369–370
bleaching adaptation role, 316:249–252
fluorescence indicators for photoreceptor
  studies, 316:146–147
photoreceptor measurements using spot
  confocal microscopy with Fluo-3
  advantages, 316:148–149
  argon ion lasers, 316:149, 151–154
  calibration, 316:159–162
  detectors, 316:154–156

dye loading, 316:157
instrumentation, 316:149
light-induced decline of calcium,
  316:157–159
photopigment cycle changes in calcium,
  316:162–163
photoreceptor preparation from tiger
  salamanders, 316:156–157
rationale, 316:147–148
phototransduction role, 316:133, 146
quantitative calcium introduction and
  membrane current measurement
  caged calcium photolysis, 293:376–377
  fluorescence imaging of subcellular cal-
    cium, 293:383
  fractional calcium currents,
    293:377–382
  Fura-2 quantification of calcium,
    293:372–373, 378–382
  iontophoresis, 293:374–375
  quantitative pressure injection, 293:375,
    378–379
  small cells, 293:382–383
  tail current measurement, 293:375–376
rod outer segment, cytoplasmic measurements
  circulating current relationship to free
    cyclic GMP, 316:137, 139
  current sources, 316:137, 139
  Indo measurements
    calcium dependence of cation ex-
      change rate, 316:145
    calibration, 316:141–142
    dextran-conjugated dye, 316:139–140
    light-evoked changes, 316:145–146
    optical setup for fluorescence mea-
      surements, 316:140–141
    snapshots of calcium, 316:142–145
  rod outer segment isolation, 316:134
  solution compositions, 316:134–135
  whole-cell patch-clamp recording,
    316:135–137
Calcium Green-1, loading of cells, 307:125
Calcium Orange, loading of cells, 307:125
Calcofluor, biofilm staining, 310:404, 406
Calexcitin
  microinjection in fibroblasts, 293:197–198,
    200
  patch clamp studies in fibroblasts,
    293:198–201

phosphorylation, **293:**195
potassium channel inhibition, **293:**195
purification by high-performance liquid
  chromatography, **293:**196–197
Calmodulin
calcium binding, **302:**87–88, 421–422
cyclic nucleotide-gated channel binding
calcium concentration buffering,
  315:784
deletion mutants of channels, electrophysiologic studies, **315:**824–825
dissociation constants of protein and
  peptide, **315:**786
effects on ligand binding, **315:**785
fluorescence binding assays with dansyl-
  calmodulin, **315:**823–824, 827
gel-overlay assay with heterologously-
  expressed channels, **315:**819–821
gel shift assay with channel peptides,
  315:821, 823
magnitude of response
channels versus channel peptides,
  315:825, 827
vision versus olfaction, **315:**818,
  827–828
olfactory versus rod channel inhibition,
  315:784
sites of interaction, **315:**783, 785–786,
  797, 818–819
state-dependent modification,
  315:783–784
surface plasma resonance spectroscopy
buffers and materials, **315:**787
comparison of binding constants
  with other techniques, **315:**827
controls, **315:**791
immobilization by amine coupling,
  315:789
immobilization by ligand thiol coupling, **315:**788
kinetic data interpretation, **315:**794,
  796
mass transfer limitations, **315:**786
overview of experimental design,
  315:788
principle, **315:**786–787
qualitative analysis, **315:**791–792
quantitative analysis, **315:**793–794
recording of titration series,
  315:789–791

Scatchard plot analysis, **315:**796–797
sensor chip geometry, **315:**791
functions, **302:**87–88
green fluorescent protein constructs, *see*
  Calmodulin–green fluorescent protein constructs
guanylate cyclase activation in non-mammals, **315:**730, 740
localization in yeast
fluorescence microscopy
cell preparation, **302:**95–96
imaging data acquisition, **302:**96
immunofluorescence microscopy in
  fixed cells, **302:**102
green fluorescent protein fusion
  protein
advantages of study, **302:**89–90
budding yeast constructs and expression, **302:**90–93
fission yeast constructs and expression, **302:**93–95
polarized growth sites, **302:**99–101
Spc110p association, **302:**88
spindle pole body, **302:**88–89, 97–99
temporal and spatial differentiation,
  302:101–102
phototransduction role, **316:**103–104
surface plasmon resonance
GRK1 binding kinetics analysis,
  316:30–31, 33
immobilization for studies,
  316:24–25
Calmodulin–green fluorescent protein constructs
advantages over aequorin systems in calcium imaging, **307:**445
calcium affinity, tuning, **307:**445–446
comparison to chemical calcium sensors,
  307:446
confocal image acquisition requirements,
  307:468
resonance energy transfer, **307:**445
targeting to organelles, **307:**457–459
transfection, **307:**450–451
virus-mediated gene transfer, **307:**451
VP22-mediated protein transfer,
  307:451–452
Calorimetry, *see* Differential scanning calorimetry; Microcalorimetry; Photoacoustic calorimetry; Titration calorimetry

Cameleons, *see* Calmodulin–green fluorescent protein constructs
CAMKII, *see* Calcium/calmodulin-dependent protein kinase
cAMP, *see* Cyclic AMP
Cancer, *see* Metastasis
Candida biofilm
  bacterial film comparison, **310:**645–646
  catheter disk culture
    antimicrobial susceptibility studies, **310:**654
    dry weight determination, **310:**647
    growth, **310:**647
    inoculation, **310:**646
    materials, **310:**646
    tetrazolium reduction assays of growth, **310:**648
    tritiated leucine incorporation, **310:**647–648
  cylindrical cellulose filter culture
    antimicrobial susceptibility studies, **310:**654, 656
    glucose-limited yeast preparation, **310:**651–652
    growth, **310:**652
    iron-limited yeast preparation, **310:**651
    overview, **310:**650–651
  medical devices, **310:**645
  perfused biofilm fermenter
    growth, **310:**650
    inoculation, **310:**649
    overview, **310:**648–649
  risk factors, **310:**644
  scanning electron microscopy, **310:**652–654
Capillary electrophoresis
  buffers, **294:**194
  fractured electrophoresis capillary fabrication, **294:**194–196
  patch-clamp detection
    buffer vial maintenance, **294:**200
    cell preparation, **294:**196–197
    current-to-voltage relationships, **294:**205
    electrophoretic migration rate calculation, **294:**204
    instrumentation, **294:**201–202
    neurotransmitter detection, principle, **294:**189–191, 193
    outside-out patch recording, **294:**205

patch clamping, **294:**199
    pipette solutions, **294:**200
    regeneration of capillary, **294:**200
    resolution in neurotransmitter separations, **294:**203
    sample volume calculation, **294:**202
    spectral analysis of currents, **294:**204
  principles
    electro-osmotic flow, **294:**192
    electrophoretic migration velocity, **294:**192
    electrophoretic mobility, **294:**192
    net migration velocity, **294:**192–193
    theoretical plate number, **294:**192
  quantitative analysis of oligonucleotides, **313:**209–210, 223
  two-electrode voltage-clamp detection
    cell preparation, **294:**197–199
    instrumentation, **294:**205
    neurotransmitter detection, principle, **294:**189–191, 193
    stability of detection system, **294:**206–207
  zone electrophoresis, resolution of protein folding intermediates, **309:**334
CAPP, *see* Ceramide-activated protein phosphatase
Capping free energy, helix–coil transition, **295:**3
Capsanthin, biosynthesis, **297:**257–258
Capsorubin, biosynthesis, **297:**257–258
Capsular polysaccharide, biofilms
  colorimetric assays
    acetylation, **310:**409–410
    neutral hexose, **310:**409
    pyruvilation, **310:**410
    uronic acid, **310:**409
  definition, **310:**146
  gel electrophoresis
    Eastern blot analysis, **310:**411–412
    running conditions, **310:**411
    staining, **310:**411
  hemagglutination inhibition assay of purity, **310:**408–409
  immunoelectron microscopy, **310:**406–407
  lectin staining, **310:**406
  light microscopy, **310:**404–406
  monosaccharide analysis, **310:**410
  probes, **310:**404–406
  purification, **310:**407–408

sulfate group detection by infrared spectroscopy, **310:**410
survival functions, **310:**403–404
Captive bubble contact angle, measurement of membrane hydrophobicity
advantages, **310:**479
appararatus, **310:**479–480
contact angle calculations, **310:**481
rationale, **310:**478
Carbamoyl phosphate synthase
catalytic activities and subunits, **308:**140–142
substrate channeling
kinetic studies, **308:**142
putative channel, **308:**141
rationale, **308:**140
Carbamylcholine
caged compound studies with crystallized cholinesterases, **291:**266–267, 269, 273–276
receptor, *see* Neurotransmitter receptor
Carbohydrate–carbohydrate interaction
cell adhesion systems, overview, **312:**447–448
Gb$_4$-dependent adhesion of human embryonal carcinoma 2102 cells
adherent cell measurements, **312:**457
adhesion conditions, **312:**454
caveats, **312:**457–458
overview, **312:**454
washing, **312:**454, 457
GM3-dependent adhesion of mouse B16 melanoma cells to endothelial cells
B16 cell adhesion to glycosphingolipid-coated plates
adhesion conditions, **312:**449
cell culture, **312:**449
glycosphingolipid coating of plates, **312:**448–449
caveats, **312:**457–458
cell adhesion conditions, **312:**451–453
endothelial cell monolayer preparation, **312:**451
GM3 liposome adhesion to glycosphingolipid-coated plates
adhesion conditions, **312:**451
liposome preparation, **312:**449, 451
inhibition by oligosaccharides and glycosphingolipid liposomes, **312:**454
overview, **312:**448

2β-Carbomethoxy-3β-(4-fluorophenyl)-tropane, dopamine transporter binding
equilibration kinetics, **296:**253
pH effects on sodium dependency, **296:**250–252
2β-Carbomethoxy-3β-(4-iodophenyl)-tropane, dopamine transporter binding
equilibration kinetics, **296:**253
photoaffinity labeling, **296:**220–221, 225, 227
Carbon dioxide
effects on peroxynitrite reactivity, overview, **301:**291, 354–355, 364–365
metabolism and transport, **301:**353–354
nitrosoperoxocarbonate adduct
decomposition pathways, **301:**362–363
glutathione trapping, **301:**359
half-life, **301:**364, 455
protein nitration, **301:**359–362, 366
peroxynitrite reaction
biological relevance, **301:**365–367
carbonate radical formation, **301:**363–364
mechanism, **301:**355
stopped-flow studies, **301:**351–352, 356–358
thermodynamic properties, **301:**358
solution preparation with defined carbon dioxide–bicarbonate concentrations, **301:**355–356
Carbon electrode
fast scan cyclic voltammetry measurement of dopamine uptake
brain region variability, **296:**704–707
evidence of uptake in measurements, **296:**701–702
inhibitors, assessment of action
bupriopion, **296:**703
cocaine, **296:**704
nomifensine, **296:**703–704
intact tissue measurements versus homogenates, **296:**691–693
Michaelis–Menten kinetics
correction for Nafion coating on carbon electrodes, **296:**697–701
inhibition analysis, **296:**703–704, 707
modeling of data, **296:**694–697
reverse transport measurement in *Planorbis* giant dopamine neuron amperometry, **296:**682

amphetamine treatment
  effect on quantal size, **296:**685
  reverse transport induction,
    296:687–688
cyclic voltammetry, **296:**682–683
data acquisition, **296:**680–682
disk electrode preparation, **296:**679
dissection, **296:**676–677
dopamine exocytosis, measurement,
  296:683–685
intracellular injection, **296:**680–681
resolution of dopamine release by reverse transport and exocytosis,
  296:688
ring electrode preparation, **296:**677–678
testing of electrodes, **296:**679–680
voltammetry, **296:**682, 687–688
Carbonic anhydrase
catalytic stages, **308:**286
free energy profile, **308:**13
His-64 in proton shuttling, **308:**286–287
isoelectric focusing of S-modified proteins, **301:**146–147, 150
Marcus parameters for proton transfer
  Brønsted plots, **308:**291–292
  His-64 proton transfer, **308:**291–293
  His-67 proton transfer, **308:**293
  human isozyme III as model system,
    308:287
  imidazole and lutidine buffers, proton
    transfer, **308:**293–294
  intrinsic kinetic barrier, **308:**295
  kinetic isotope effects, **308:**292–293
  mutagenesis and disruption of pK_a val-
    ues, **308:**288–289
  pH titration, **308:**290
  rate measurements, **308:**289–290
  work function dominance, **308:**290–291,
    295–297
native gel electrophoresis of aggregation
  intermediates, **309:**343
rate-limiting steps, **308:**285–286
Carbon monoxide, signaling, **300:**323
Carbonyl group, determination of oxidative
  stress in proteins, *see also* Protein carbonyl; Protein oxidation
2,4-dinitrophenylhydrazine assay
  immunohistochemistry, **309:**139–140
  rationale, **309:**138–139
  spectrophotometric assay, **309:**762–763

histofluorescence, **309:**764–765
immunochemistry, **309:**763–764
(2*S*,3*S*,4*R*)-2-(Carboxycyclopropyl)glycine,
  glutamate transporter inhibition,
  296:176, 180
α-Carboxy-*o*-nitrobenzyl carboxylates
γ-aminobutyric acid
  kinetics of photolysis, **291:**43
  synthesis, **291:**35–37
applications, **291:**31–32
suitability for neurotransmitter study
  with cell-flow, **291:**48–50
5,6-Carboxyfluorescein, vascular remodeling
  studies, **307:**259
5-(3-Carboxymethoxyphenyl)-2-(4,5-di-
  methylthiazolyl)-3-(4-sulfophe-
  nyl)tetrazolium, cell viability assay,
  309:719
Cardiac myocyte, cell volume measurements with confocal microscopy
background subtraction,
  307:282–283
calcein labeling, **307:**281
cell preparation, **307:**281
image acquisition, **307:**281
intensity segmenting, **307:**282–283, 285
motion artifacts, **307:**282
quantitative analysis of cell volume and
  surface, **307:**285, 287–288, 290–291
species variation, **307:**290–291
spherical aberration and accuracy of measurements, **307:**287–288, 290
three-dimensional visualization,
  307:283–285
thresholding, **307:**282–283
validation with conventional microscopy,
  307:291, 295
voltage clamp studies of volume change
  effects, **307:**293–294
Cardiovascular ion channel
cell culture
  cardiac cells, **293:**77–78
  coronary myocytes, **293:**78–79
  freezing, **293:**79
dissection and enzymatic isolation
  cardiac cells, **293:**76
  coronary myocytes, **293:**76–77
human tissue collection
  cardiac samples, **293:**73–75
  coronary tissue, **293:**75

immunofluorescence detection of proteins
indirect immunofluorescence, **293**:84–85
sections, **293**:85
single-cell immunofluorescence, **293**:85
recombinant channels
functional screening, **293**:86–87
selection of expression in transfected cells
antibiotic resistance, **293**:87
CD8 expression, **293**:87
green fluorescent protein expression, **293**:87
transcript detection
reverse transcriptase–polymerase chain reaction
RNA extracted from tissue, **293**:82–83
single-cell polymerase chain reaction, **293**:80–81, 84
RNA extraction, **293**:81–82
RNase protection assay, **293**:80, 83–84
Carnot cycle, comparison to biomolecular thermodynamics, **295**:331–332
Carotenoids, *see also* Cellular carotenoid-binding protein; Lycopene; Vitamin A; Zeaxanthin
biosynthesis
acyclic xanthophyll biosynthesis in *Rhodobacter*, **297**:258–260
capsanthin, **297**:257–258
capsorubin, **297**:257–258
cyclization reactions, **297**:252–255
early reactions, **297**:249
gene cloning in pathway
applications, **297**:245–246
color complementation technique, **297**:247–248
geranylgeranyl diphosphate synthase, **297**:248–249
lycophene β-cyclase, **297**:246–247, 252–254
phytoene desaturase, **297**:246–247, 251–252
phytoene synthase, **297**:247–248, 249, 251
Rhodobacter capsulatus, **297**:246
screening of complementary DNA libraries, **297**:248

transposon tagging technique, **297**:248
genetic manipulation of biosynthesis, **297**:262
inhibitors, **297**:262
lutein, **297**:256
lycophene ε-cyclase, **297**:254–255
neoxanthin, **297**:257
phytoene desaturation, **297**:251–252
phytoene synthesis, **297**:249, 251
regulation, **297**:260–262
secondary carotenoids, **297**:258
violaxanthin, **297**:256–257
zeaxanthin, **297**:256–257
energy transfer in singlet oxygen quenching, **319**:230
erythropoietic protoporphyria, β-carotene treatment, **319**:480–482, 484, 494
functions, **297**:245, **299**:408, 441–442
high-performance liquid chromatography, **299**:390, 392
high-performance liquid chromatography assay in skin
chromatography conditions, **319**:498–499
identification of peaks, **319**:499, 501
quantitative analysis, **319**:501–502
sample preparation, **319**:498
standards, **319**:499, 501
low-density lipoprotein oxidation
analysis by high-performance liquid chromatography, **301**:322–323
consumption and lipid peroxidation, **301**:331–332
effects of peroxynitrite, **301**:324, 327, 329, 331
lutein
distribution in retina, **299**:457
oxidative metabolites, **299**:460–461
photooxidation protection of retina, **299**:457–460
matrix-assisted laser desorption ionization–mass spectrometry
extraction, **299**:391–392
fragment ion mass analysis, **299**:394–395, 398–401, 406–408
green lettuce analysis, **299**:403, 406
instrumentation, **299**:392–393
molecular ion species and sensitivity, **299**:393–394

overview of ionization techniques,
  299:390–391
postsource decay, **299:**391, 393
standards and samples, **299:**391
tangerine juice concentrate analysis,
  299:401, 403
metabolism, **319:**495
photobleaching
  application to food systems, **299:**419
  degradation products, **299:**418–419
  flash photolysis, **299:**416–418
  mechanisms, **299:**409
  oxygen role, **299:**420
  radical efficiency assay, **299:**420
  steady-state photolysis
    quantum yield, **299:**409–410, 412
    sensitized photolysis, **299:**413–415
    unsensitized photolysis, **299:**410,
      412–413
reflection photometry assay of skin
  β-carotene supplementation, **319:**495
  data analysis, **319:**496
  distribution, **319:**497
  instrumentation, **319:**495–496
  serum levels, **319:**497
  validation, **319:**497–498
singlet oxygen scavenging rates in phospholipid membranes, **319:**104–106,
  108–109
structure, overview, **297:**244–245
Trolox equivalent antioxidant capacity
  assay
  decolorization assay, **299:**384, 389
  extraction and analysis of tomato,
    299:383–385, 387, 389
  materials, **299:**381
  preparation of preformed 2,2′-azino-
    bis(3-ethylbenzothiazoline-6-
    sulfonic acid) radical cation,
    299:381
  principle, **299:**380
  standards, preparation, **299:**381–383
types and structures, **301:**319
Casein kinase, inhibitors, **298:**548
Caspase
  caspase-3
    apoptosis role, **319:**332
    assay in irradiated HL-60 cells,
      319:340–341
    photodynamic therapy activation

Bap31 fragments, **319:**350–351
    inhibitor studies, **319:**349–350
CAT, *see* Cationic amino acid transporter;
  Chloramphenicol acetyltransferase
Catalase, singlet oxygen detection, *see also*
  Hydrogen peroxide
  activity assays
    electrochemical assay, **319:**116
    gel assay, **319:**116
  cell extract preparation, **319:**116
  hanging drop system for *in vitro* modifi-
    cation, **319:**117–118
  heme modification and electrophoretic
    mobility, **319:**114, 118
  Neurospora crassa
    conidia system, **319:**114
    heat shock, **319:**115
    paraquat induction, **319:**115
  purification for *in vitro* modification,
    319:117
  rationale, **319:**113
Catechins, *see also* Phenolic antioxidants
  absorption and excretion in humans,
    299:95
  antioxidant activity, **299:**207–208
  high-performance liquid chromatography
    chromatography conditions,
      299:103–105
    food sample preparation, **299:**100, 103
    retention times, **299:**105–106
    standards, **299:**100
    tea analysis with caffeine
      chromatography conditions, **299:**109,
        111, 203–204, 206
      comparison of green teas, **299:**112–
        113, 203
      extraction, **299:**204
      instrumentation, **299:**108–109
      precision, **299:**204, 206
      quantitative analysis,
        299:111–112
      sample extraction, **299:**109
      stability of samples, **299:**206
      standards, **299:**109, 204
    urine sample
      dietary manipulation, **299:**97, 100,
        106
      preparation, **299:**103
  inhibition of tyrosine nitration by peroxy-
    nitrite, **299:**215–216, 231–232

tea types and structures, **299:**107–108, 202–203

Cathepsin B, confocal microscopy of intracellular distribution, **307:**117

Catheter, *see* Bladder catheter biofilm

Cation-exchange chromatography
chaperonin 10, **290:**223, 225
GroES, **290:**126
heat shock protein 90, **290:**414
mammalian mitochondrial chaperonin 10, **290:**223, 226

Cationic amino acid transporter
expression patterns, **301:**87
ribonuclease protection assays, **301:**88–89
types, **301:**86–87
Western blot analysis, **301:**89–91

Cationic lipid:plasmid DNA
advantages in gene delivery, **292:**697–698
cystic fibrosis transmembrane regulator transfection into airway epithelial cells
cell culture, **292:**714
fluorescence assay of channel activity, **292:**714–716
neutral lipid enhancement of transfection, **292:**698
optimization of formulations
in vitro
cell culture, **292:**700
β-galactosidase reporter gene assay, **292:**700–701
liposome preparation, **292:**699–700
overview, **292:**698–699
ratios for optimal transfection efficiency, **292:**701, 704–705
structures of cationic lipids, **292:**701–702
in vivo
chloramphenicol acetyltransferase reporter gene assay, **292:**709–710
correlation to *in vitro* studies, **292:**712
intranasal instillation in mice, **292:**709–710
naked DNA control, **292:**710–711
ratios for optimal transfection efficiency, **292:**711–712
transfection efficiency, factors affecting
in vitro
cell lines, **292:**708

contact time, **292:**708
excipient selection, **292:**707–708
precipitation of complexes, **292:**705, 707
size of complex, **292:**707
vector promoters, **292:**708–709
in vivo
DNA purity, **292:**713
dose, **292:**713
toxicity, **292:**713

Caveolae, *see also* Detergent-insoluble membrane fraction
acid sphingomyelinase, biological significance, **311:**185–186
caveolae-related domains
cell distribution, **311:**184
formation of caveolae, **311:**185
detergent insolubility, **311:**185–186
membrane isolation
detergent insolubility method
advantages and disadvantages, **311:**188–190
buffers, **311:**186–187
detergent extraction, **311:**186
homogenization, **311:**187–188
metabolic labeling, **311:**187
sucrose density gradient centrifugation, **311:**188
sodium carbonate method, **311:**191–192
photoaffinity labeling of sphingolipid-binding proteins, **311:**585
signal transduction role, **311:**185
sphingomyelin assay in caveolin-enriched membranes
bacterial sphingomyelinase assay, **311:**192–193
thin-layer chromatography, **311:**193

C11-BODIPY
apoptosis studies in fibroblasts, **319:**607–608
cell labeling, **319:**606
confocal laser scanning microscopy and image processing, **319:**606–607
excimer formation, **319:**611–612
fluorescence spectra and properties, **319:**604, 606
multiwell screening of antioxidants
ascorbate, **319:**608
numerical analysis, **319:**610
pitfalls, **319:**611–612

tocopherol, **319:**608–609
photooxidation, **319:**612
quenching, **319:**611
reactive oxygen sensing, **319:**603–604
CCA, *see* Complementary chromatic adaptation
CCBP, *see* Cellular carotenoid-binding protein
CCD array, *see* Charge-coupled device array
CCK, *see* Cholecystokinin
C-cpn, *see* Cytosolic chaperonin
CD, *see* Circular dichroism
CD4, *see* Human immunodeficiency virus
CD8 expression, transfection screening, **293:**87
CD14, confocal microscopy imaging on neutrophils, **307:**385
Cdc2, *see* Cyclin-dependent kinase
CDFF, *see* Constant depth film fermentor
CD-GCAP, *see* Calcium-dependent guanylate cyclase activator protein
cDNA, *see* Complementary DNA
cDNA RDA, *see* Complementary DNA representational difference analysis
Cell adhesion molecule, *see also* Neural cell adhesion molecule; Intercellular adhesion molecule-1
    definition, **314:**213
    signal transduction, **314:**213
Cell–cell adhesion assay
    applications, **300:**396–397
    fluorescence assay
        anti-cell adhesion molecule antibody studies, **300:**401
        antioxidant regulation of tumor necrosis factor-α induced adhesion, **300:**400
        calcein labeling of cells, **300:**398
        cell culture, **300:**397–399
        microtiter plate reading, **300:**399
        oxidant-induced adherence, **300:**399–400
    overview of assays, **300:**397
Cell-column chromatography, endothelial barrier studies of nitric oxide
    advantages, **301:**4–5, 12
    apparatus, **301:**6–7
    cell culture, **301:**5–6
    continuous perfusion, **301:**8, 12–13

electrochemical detection of nitric oxide, **301:**10, 12
exposure of endothelial cells to nitric oxide and oxidants, **301:**8–10
Cell-flow, caged neurotransmitter characterization
    channel opening, rate and equilibrium constant determination, **291:**460
    instrumentation, **291:**46–47, 451–453
    quantification of photolysis, **291:**454
    suitability of compounds, determination, **291:**48–49
    theory, **291:**47, 447, 450–451
    validation, **291:**47–48
Cell-free translation system, *see also Escherichia coli* cell-free translation system; Rabbit reticulocyte lysate cell-free translation system
    applications, **290:**1–2
    binding affinity analysis for proteins
        data analysis, **290:**16–17
        equilibrium binding, **290:**15–16
        immunoadsorption trapping, **290:**16
    buffers and reagents, **290:**4–5
    commercially available systems, characteristics, **290:**3
    complexity, **290:**2
    immunoadsorption of polypeptides
        incubation conditions and product analysis, **290:**9
        support matrix, **290:**8–9
    kinetic analysis of protein folding, **290:**11–12, 14–15
    monoribosomal synthesis
        messenger RNA saturation, **290:**5
        ribonuclease treatment, **290:**6
    product analysis
        enzyme assay, **290:**25–26
        sodium dodecyl sulfate–polyacrylamide gel electrophoresis, **290:**24–25
    proteolysis of synthesized polypeptides, **290:**10–11
    quantification of products, **290:**4, 24–26
    ribosome–polypeptide complex
        dissociation with puromycin, **290:**7–8
        isolation, **290:**6–7
    sucrose density gradient centrifugation of products, **290:**5–6
Cell surface biotinylation, neurotransmitter transporters

γ-aminobutyric acid transporter, **296:**376, 378

applications, **296:**308, 315–318

biotinylation reaction, **296:**311–312

cell culture, **296:**310–311

cell lysis, **296:**312–313

norepinephrine transporter, **296:**315–318, 362, 364–366, 368

recovery of biotinylated proteins, **296:**313–314

structural modifications of biotin, **296:**309

Western blot analysis, **296:**314–315

Cellular carotenoid-binding protein

absorption spectra studies of binding, **299:**451–452, 455–456

binding assays, **299:**447–448, 452–454, 456

functions, **299:**456

gel electrophoresis, **299:**448, 453–454

iodine-125 labeling, **299:**449

ligand specificity, **299:**454–455

purification from ferret liver

animals and diet, **299:**443

anion-exchange chromatography, **299:**443–444, 449

apoprotein component release from complex, **299:**444

β-carotene affinity chromatography chromatography conditions, **299:**446–447, 450

column preparation, **299:**444–446

gel filtration, **299:**444, 449–450

homogenization, **299:**443

purification table, **299:**450–451

species comparison, **299:**442, 455

Cellular retinaldehyde-binding protein, preparation of recombinant apoprotein from *Escherichia coli*, **316:**332–333

Cell volume, confocal microscopy measurement

applications, **307:**312–313

cardiac myocyte measurements

background subtraction, **307:**282–283

calcein labeling, **307:**281

cell preparation, **307:**281

image acquisition, **307:**281

intensity segmenting, **307:**282–283, 285

motion artifacts, **307:**282

quantitative analysis of cell volume and surface, **307:**285, 287–288, 290–291

species variation, **307:**290–291

spherical aberration and accuracy of measurements, **307:**287–288, 290

three-dimensional visualization, **307:**283–285

thresholding, **307:**282–283

validation with conventional microscopy, **307:**291, 295

voltage clamp studies of volume change effects, **307:**293–294

error estimation

multicolor imaging, **307:**308

tissue compression, **307:**307

Z-axis distortion, **307:**307–308

limitations

data set size, **307:**308–310

image processing and display, **307:**313–315

time constraints, **307:**310–313

objective lens selection, **307:**297–299

optical sectioning, **307:**299–306

overview, **307:**274–276, 296–297

photobleaching minimization, **307:**278–279, 282, 295, 300

volume investigation

data acquisition and image stacks, **307:**277–279

overview, **307:**276–277

quantitative three-dimensional analysis, **307:**280

three-dimensional visualization of reconstructed object, **307:**279–280

two-dimensional segmentation, **307:**279

Cell volume regulation, *see* Chloride channel, swelling-activated

Centrifuge microscope

comparison to laser trap assays, **298:**413–414, 424

instrumentation, **298:**414–416

myosin motility assay

assay cell preparation, **298:**419

force-velocity relationships, **298:**423

latex bead coating with myosin, **298:**419

principle, **298:**413, 418

principle of force assays, **298:**413–414

sperm motility assay

assay cell preparation, **298:**421–422

axoneme surface drag force, **298:**426–427

demembranation of sperm, **298:**420

drag force of solution, **298:**422

force-velocity relationships and data
analysis, **298:**424–426

functional density of motor proteins,
298:423

goals, **298:**419

kinesin preparation, **298:**420

velocity measurement, **298:**420–421

theory, **298:**416–418

Centrosome

coverslips for centrifugation, **298:**233

functions, **298:**228

immunofluorescence and immunogold localization of proteins, **298:**234–235,
238

isolation of nucleation-competent
centrosomes

cell plating and treatments, **298:**230

cell washing and lysis, **298:**230

centrifugation on Ficoll, **298:**230–231

features of isolated centrosomes,
298:232

reagents and supplies, **298:**229

time requirements, **298:**229

yield, **298:**232

isolation of nuclei with attached
centrosomes

extraction and centrifugation, **298:**231

features of isolated centrosomes,
298:232–233

reagents and supplies, **298:**229, 231

time requirements, **298:**229

microtubule nucleation assay, **298:**235,
237

Western blot analysis of proteins,
298:237–238

Ceramidase, *see also* Acid ceramidase; Bile
salt-stimulated lipase

assays

fluorescent assay with BODIPY-sphin-
golipids, **312:**301

fluorescent assay with NBD-ceramide

incubation conditions, **311:**197–198

product separation and detection,
311:198

sample preparation, **311:**197

substrate delivery, **311:**197

plant enzyme, **311:**144–145

radiolabeled substrate assay

incubation detection, **311:**196

phase partitioning of product,
311:196–197

sample preparation, **311:**195–196

substrate preparation, **311:**196

thin-layer chromatography, **311:**197

in situ assays

fluorescent assay with NBD-ceramide, **311:**198–200

sphingosine mass measurement,
311:200

deficiency in Farber's disease, **311:**194–
195, 202

inhibitors, **311:**201

reaction catalyzed, **311:**194

regulators, **311:**200–201

substrate specificity

mammalian enzyme, **311:**200

plant enzyme, **311:**145

types in mammals, **311:**194–195

Ceramide

affinity chromatography, *see* Ceramide-
affinity chromatography

3-*O*-benzylceramides in sphingomyelin
synthesis, **311:**539–541, 544–546

chemical synthesis

D-*erythro*-C$_2$-ceramide synthesis,
311:527–528

D-*erythro*-C$_{16}$-ceramide synthesis,
311:528

column chromatography,
311:521–522

3-keto-C$_2$-ceramide synthesis,
311:528–530

(2*S*)-3-ketodihydro-C$_2$-ceramide synthesis, **311:**528–530

materials, **311:**519

1-*O* methyl derivative synthesis

2-*N-tert*-butyloxycarbonyl-1-*O-tert*-
butyldiphenylsilyl-2-*N*,3-*O*-
isopropylidene-D-*erythro*-
sphingosine, **311:**607–608

2-*N-tert*-butyloxycarbonyl-1-*O-tert*-
butyldiphenylsilyl-D-*erythro*-
sphingosine, **311:**607

2-*N-tert*-butyloxycarbonyl-2-*N*,3-*O*-
isopropylidene-D-*erythro*-
sphingosine, **311:**608

2-*N-tert*-butyloxycarbonyl-D-*erythro*-
sphingosine, **311:**607

column chromatography,
    311:603–604
materials, **311:**602–603
1-*O*-methyl-2-*N-tert*-butyloxy-
    carbonyl-2-*N*,3-*O*-isopro-
    pylidene-D-*erythro*-sphingo-
    sine, **311:**608
1-*O*-methyl-D-*erythro*-sphingosine,
    311:608–609
principle, **311:**604, 606
salt removal, **311:**603
D-*erythro*-sphingosine, **311:**606–607
structural characterization, **311:**604,
    610
thin-layer chromatography, **311:**603
overview, **311:**519, 523–525
purity analysis, **311:**522
thin-layer chromatography,
    311:520–522
fluorescence labeling, *see also* NBD-
    ceramide
BODIPY, **312:**298–299
Lissamine rhodamine, **312:**296–297
functions, **311:**117–118, 150, 156, 168–
    169, 207–208, 235, 285–286, **312:**16–
    17, 22–23, 32, 65, 407, 420–421, 429
glyconjugate synthesis, *see* Sphingosine
glycosylphosphatidylinositol anchor analy-
    sis in yeast
conconavalin A affinity chromatogra-
    phy, **312:**511–512
delipidation, **312:**509–511
lipid liberation
    acid hydrolysis, **312:**514
    nitrous acid, **312:**513–514
    phospholipase A$_2$, **312:**514
lipid structures, **312:**506–507
metabolic labeling, **312:**508, 510–511
octyl-Sepharose purification of anchor
    peptides, **312:**512–513
protease treatment, **312:**512
thin-layer chromatography,
    312:514–515
high-performance liquid chromatography
    mass analysis
advantages, **312:**17, 20
benzoylation, **312:**17–18, 21–22
chromatography conditions, **312:**18–20
column care, **312:**21–22
glycerol-containing samples, **312:**22

lipid extraction, **312:**17
standards, **312:**20–22
identification in pulse–chase studies of
    signaling, **311:**680–681
infrared spectroscopy, *see* Infrared spec-
    troscopy
mass spectrometry, **312:**23
measurement with diglyceride kinase
    data interpretation, **312:**30
    lipid extraction, **312:**24–25
    lipid solubilization with mixed micelles,
        312:25–26
    membrane preparation of enzyme,
        312:27–28
    reaction conditions, **312:**27–28
    standard curve preparation, **312:**26–27
    thin-layer chromatography, **312:**28–30
    validation, **312:**30–31
metabolism, **312:**304, 318–320, 407
overview of biosynthesis, **311:**22–23
photoaffinity labeling with 3-trifluoro-
    methyl-3-(*m*-[^{125}I]iodophenyl)-
    diazirine derivative
    competition analysis with purified pro-
        teins, **312:**437
    gel electrophoresis of labeled proteins,
        312:437
    labeling reaction, **312:**436–437
    overview, **312:**435–436
    radiolabeling of tin precursor, **312:**436
    uptake conditions, **312:**436–437
plant types, **311:**130–131
protein kinase C modulation, *see* GM3
protein kinase C translocation mutant
    Chinese hamster ovary cell analysis,
        312:314–316
radiolabeling
    amide labeling principles, **311:**503–504
    [3–^{3}H]ceramide synthesis, **311:**496
    N-[1–^{3}H]D-*erythro*-C$_2$-ceramide synthe-
        sis, **311:**505
    N-[1–^{14}C]D-*erythro*-C$_6$-ceramide syn-
        thesis, **311:**506–507
    N-[1–^{14}C]D-*erythro*-C$_{16}$-ceramide syn-
        thesis, **311:**507–508
    N-[9,10–^{3}H]D-*erythro*-C$_{16}$-ceramide
        synthesis, **311:**507–508
    enzymatic synthesis of radiolabeled
        compounds, **311:**658, 684–685
    extractions, **311:**487, 502

flash column chromatography, **311:**486,
    501
materials, **311:**500–501
purity analysis, **311:**487, 502
radioactivity determination and detec-
    tion, **311:**487, 502
thin-layer chromatography, **311:**485–
    487, 501
short-chain ceramides
    advantages, **312:**407–409
    analysis of quality and identity, **312:**410
    applications, **312:**409–410, 418
    cyclin-dependent kinase modulation,
        312:418
    delivery to cells, **312:**410–411
    diacylglycerol analog comparison,
        312:416–417
    effective concentration, **312:**411
    inhibitors of metabolizing enzymes,
        312:414–415, 419
    lipid extraction, **312:**414, 416
    lipid/bovine serum albumin complex
        preparation, **312:**411
    long-chain ceramide studies in verifica-
        tion of effects, **312:**419
    mechanism of action determination,
        312:417–418
    metabolism, **312:**413–414
    potency range, **312:**412–413
    recombinant enzymes of metabolism,
        studies of effects, **312:**419–420
    selectivity of target and response,
        312:417
    solubility, **312:**408
    specificity of action verification,
        312:416–417
    thin-layer chromatography, **312:**416
    uptake assay, **312:**411–412
solid-supported sphingosine synthesis,
    311:438–439
structure, **312:**64, 317, 320
tandem mass spectrometry, **312:**39–40
Ceramide-activated protein phosphatase
assays using myelin basic protein as sub-
    strate
    ceramide solubilization, **312:**425
    materials, **312:**242
    optimization for long-chain ceramide
        responsiveness, **312:**428
    protein phosphatase-1αc, **312:**425–426

protein phosphatase-1γc, **312:**426
    protein phosphatase 2A, **312:**426
    protein phosphatase 2Ac, **312:**426–427
    substrate labeling, **312:**424–425
    unit definition, **312:**423
ceramide activation
    manganese effects, **312:**427
    reducing agent effects, **312:**427–428
    salt effects, **312:**428
    stereospecificity, **312:**427
classes of phosphatases, **312:**421, 423
functions, **312:**423
short-chain ceramide studies, **312:**417–418
Ceramide-affinity chromatography
elution and binding specificity determina-
    tion, **312:**432, 434
extract preparation, **312:**433–434
loading and washing, **312:**434
matrix preparation, **312:**430–433
overview of ceramide-interacting pro-
    teins, **312:**429–430
regeneration and equilibration,
    312:434–435
Ceramide galactosyl-3′-sulfate, *see* Sulfatide
Ceramide galactosyltransferase
activity assays
    natural substrates, **311:**66–67
    short-chain ceramides as substrates
        incubation, **311:**68
        interference, **311:**69–70
        overview, **311:**67
        reagents, **311:**67–68
        reproducibility, **311:**69
        sample preparation, **311:**68
        thin-layer chromatography,
            311:68–69
biological significance, **311:**60–61
knockout mice, **311:**60
metabolic radiolabeling and gel electro-
    phoresis
    cell growth, **311:**71
    immunoprecipitation, **311:**72
    maturation analysis, **311:**72–73
    overview, **311:**70
    protease protection assay, **311:**71–72
    reagents, **311:**70–71
product detection
    characterization, **311:**63, 65
    extraction, **311:**62
    immunological detection, **311:**65–66

labeling, **311:**63
thin-layer chromatography, **311:**62–63
reaction catalyzed, **311:**60
subcellular localization, **311:**61, 70
Ceramide glucosyltransferase, *see* Glucosylceramide synthase
Ceramide glycanase
assay, **312:**198–199
incubation conditions, **311:**289
paper chromatography, **311:**289
substrate preparation, **311:**288–289
catalytic reaction, **312:**196
cation effects, **311:**292
detergent requirement, **311:**291
expression in embryogenesis and carcinogenesis, **311:**296
glycosphingolipid analysis
fluorophore-assisted carbohydrate electrophoresis, **312:**202, 205
oligosaccharide evaluation after hydrolysis, **312:**201
substrate specificity, **312:**200–201
hydrophobicity, **311:**293–295
inhibitors
alkyl and acyl chain length effects,
311:294–295
ceramide structural analogs, **311:**293–
294, 387
chemotherapeutic drugs, **311:**296–297
pH optima, **311:**291–292
properties of clam enzyme
detergent effects, **312:**199
heat stability, **312:**200
inhibitors, **312:**199
pH optimum, **312:**199
purification
clam, **312:**197–198
mammary tissue, **311:**289–290
reaction catalyzed, **311:**287
size, **311:**290–291
sources, **312:**197
stability, **311:**290
substrate specificity, **311:**292–293
tissue distribution, **311:**287–288
Ceramide kinase
assay
detergents, **311:**212
endogenous inhibitors, **311:**211–212
incubation conditions, **311:**210–211
oxidation prevention, **311:**212

salt effects, **311:**212
substrate preparation, **311:**210
thin-layer chromatography, **311:**211
calcium stimulation, **311:**209
extraction from membranes, **311:**212–213
isoforms, **311:**209–210
separation from diacylglycerol kinase and
sphingosine kinase, **311:**213–215
signal transduction role, **311:**208–209
storage, **311:**212
tissue distribution, **311:**209
Ceramide lactosyltransferase, *see* Lactosylceramide synthase
Ceramide-1-phosphate
chemical synthesis, **311:**519, 533–534
functions, **311:**234–235
identification in pulse–chase studies of
signaling, **311:**682
radiolabeling
[^{3}H]D-*erythro*-C$_{16}$-ceramide-1-phos-
phate synthesis, **311:**516
enzymatic synthesis, overview, **311:**513,
515, 658–659
phosphorous-32 compound,
311:237–238
Ceramide-1-phosphate phosphatase
assay
intact cell assay, **311:**241–243
substrate synthesis, **311:**237–238
in vitro assay, **311:**240–241
biological significance, **311:**234–235
requirements for catalysis, **311:**244
structure, **311:**236
substrate specificities, **311:**236–237
types in mammals, **311:**235–236
Ceramide phosphoethanolamine, biosynthesis, **311:**33
Ceramide phosphoethanolamine transferase
assay, **311:**38–39
biological significance, **311:**38
inhibitors, **311:**42
lipid and cation dependence, **311:**41–42
in situ activity assays, *see* Sphingomyelin
substrate specificity, **311:**41
Ceramide synthase
assays
endogenous substrate interference,
311:18
incubation conditions, **311:**16
optimization, **311:**17

product extraction, **311:**16–17
radiolabeled substrates, **311:**15–16
substrate addition, **311:**16
thin-layer chromatography, **311:**17
biological significance, **311:**15
inhibitors, **311:**20–21, 335, 346–347
kinetic properties, **311:**18–19
plant enzyme
  assay
    calculations, **311:**136
    incubation conditions,
      311:136–137
    optimization, **311:**137
    principle, **311:**135
    thin-layer chromatography,
      311:136–137
  biological significance, **311:**140
  fatty acid acyl donor, **311:**135
  substrate specificity, **311:**137–138
reaction catalyzed, **311:**15
stereoselectivity, **311:**19
subcellular localization, **311:**19–20
substrate specificity, **311:**15, 18–19
Cerebroside sulfotransferase, *see* Galactosylceramide : sulfotransferase
CFTR, *see* Cystic fibrosis transmembrane
  conductance regulator
CGalT, *see* Ceramide galactosyltransferase
cGMP, *see* Cyclic GMP
Channel density, determination in isolated
  vesicles, **294:**333–337
Channeling, *see* Substrate channeling
Chaperonin 10
  Chromatium vinosum protein
    amino acid analysis, **290:**150
    purification, **290:**148–149
    quaternary structure, **290:**149–150
    sequence analysis, **290:**150–151
  Escherichia coli protein, *see* GroES
  interaction with chaperonin 60, **290:**147–
    148, 151–152
  mammalian mitochondrial chaperonin 10,
    see Mammalian mitochondrial chap-
    eronin 10
  Rhodobacter sphaeroides protein
    electron microscopy, **290:**159
    gene cloning and expression in *Esche-
      richia coli*, **290:**155, 157
    physical properties, **290:**158
    purification, **290:**155–157

Rubisco refolding assay with chaperonin
  60, **290:**227–230
spinach chloroplast protein
  GroEL binding, **290:**220
  GroES homology, **290:**220
  oligomeric structure, **290:**228
  purification of recombinant protein
    from *Escherichia coli*
    anion-exchange chromatography,
      290:225–226
    cation-exchange chromatography,
      290:223, 225
    cell growth and optimization of ex-
      pression, **290:**222, 224
    extraction, **290:**224
    hydroxylapatite high-performance liq-
      uid chromatography, **290:**225
    materials, **290:**221
    plasmid constructs, **290:**223–224
    polyacrylamide gel electrophoresis,
      290:222–223
structure, **290:**147
Thermoanaerobium brockii protein
  function compared to *Escherichia coli*
    heat shock proteins, **290:**161–162
  oligomerization, **290:**168
  protein folding activity with chaper-
    onin 60, **290:**168–169
  purification
    anion-exchange chromatography,
      290:165
    cell growth and extraction, **290:**163
    GroEL affinity chromatography,
      290:166
    hydrophobic interaction chromatogra-
      phy, **290:**164–165
    piggy-back affinity chromatography
      with chaperonin 60, **290:**165–166
    storage, **290:**168
Thermus thermophilus protein
  assays with chaperonin 60 complex
    incubation conditions, **290:**173–174
    isopropylmalate dehydrogenase fold-
      ing assays, **290:**174–176
    lactate dehydrogenase heat denatur-
      ation protection assays,
      290:176–178
    substrate denaturation,
      290:172–173
  copurification with chaperonin 60

anion-exchange chromatography, **290:**171–172
cell growth and extraction, **290:**171
size-exclusion chromatography, **290:**172
storage, **290:**172
polyacrylamide gel electrophoresis and staining, **290:**170–171
thermostability, **290:**170
yeast protein
  biological function, **290:**195
  GroEL binding, **290:**194
  GroES homology, **290:**194
  purification of recombinant protein from *Escherichia coli*
    anion-exchange chromatography, **290:**200, 202
    cloning into expression vector, **290:**198–199
    extraction, **290:**199
    hydrophobic interaction chromatography, **290:**200–201
    nickel affinity chromatography of histidine-tagged protein, **290:**201–202
  structure, **290:**202
Chaperonin 60
  ATPase assay, **290:**152–153, 157–158
  Chromatium vinosum protein
    amino acid analysis, **290:**150
    comparison of levels with Rubisco, **290:**153–154
    purification, **290:**148–149
    quaternary structure, **290:**149–150
    sequence analysis, **290:**150–151
  Escherichia coli protein, *see* GroEL
  insect protein, *see* Symbionin
  interaction with chaperonin 10, **290:**147–148, 151–152
  mammalian cytosolic chaperonin, *see* Cytosolic chaperonin
  mammalian mitochondrial protein, *see* Mammalian mitochondrial chaperonin 60
  Rhodobacter sphaeroides protein
    comparison of levels with Rubisco, **290:**155, 160–161
    electron microscopy, **290:**159
    gene cloning and expression in *Escherichia coli*, **290:**155, 157

growth conditions and expression, **290:**159–160
homology with *Escherichia coli* GroEL, **290:**161
physical properties, **290:**158
purification, **290:**155–157
structure, **290:**147
Thermoanaerobium brockii protein
  ATPase activity, **290:**167
  function compared to *Escherichia coli* heat shock proteins, **290:**161–162
  homology with other chaperonin 60s, **290:**167
  oligomerization, **290:**166–167
  protein folding activity with chaperonin 10, **290:**168–169
  purification, **290:**163–164
  storage, **290:**168
Thermus thermophilus protein
  assays with chaperonin 10 complex
    incubation conditions, **290:**173–174
    isopropylmalate dehydrogenase folding assays, **290:**174–176
    lactate dehydrogenase heat denaturation protection assays, **290:**176–178
    substrate denaturation, **290:**172–173
  ATPase activity, **290:**170–171
  copurification with chaperonin 10
    anion-exchange chromatography, **290:**171–172
    cell growth and extraction, **290:**171
    size-exclusion chromatography, **290:**172
    storage, **290:**172
  monomer isolation, **290:**178–179
  polyacrylamide gel electrophoresis and staining, **290:**170–171
  rhodanese reactivation assay, **290:**179–180
  thermostability, **290:**170
  yeast protein, *see* Heat shock protein 60
Charge-coupled device array
  cooling, **305:**60, 83–84
  image intensifiers, **305:**53–54, 61, 83–84
  luminescence imaging, *see* Luminescence imaging
  noise, **305:**83
  overview of features, **305:**53

principle of detection, **305:**57, 59, 82–83
scan rate, **305:**83
standard versus back-illuminated chips,
    305:83
structure, **305:**57, 59, 82
three-phase system, **305:**59
variables, **305:**59–60
Charge coupling, modeling in steady-state
    biofilm systems, **310:**316–317
Charybdotoxin receptor, *see also* Scorpion
    venom peptides
calcium sensitivity, **294:**275, 287
comparison with μ-conotoxin GIIIA
    blocking of sodium channels,
    294:584–586
fusion of reconstituted receptors into
    lipid bilayers
    calcium regulation, **294:**284
    osmotic gradient enhancement, **294:**283
    patch-clamp analysis, **294:**283–284
    polarity, **294:**283
    tetraethylammonium blocking,
        294:284–285
kinetic basis of high-affinity binding,
    294:585–587
mechanism of blocking, **294:**582, 584–586
molecular dentistry concept of binding,
    294:581–582
purification from bovine tracheal smooth
    muscle
    anion-exchange chromatography,
        294:278–279
    cation-exchange chromatography,
        294:279–280
    charybdotoxin binding assay, **294:**275,
        277, 281–282
    hydroxylapatite chromatography,
        294:279
    membrane preparation, **294:**276
    solubilization, **294:**275–277
    sucrose density gradient centrifugation,
        294:279–282
    wheat germ agglutinin affinity chroma-
        tography, **294:**278
reconstitution into liposomes,
    294:282–283
reconstruction in planar bilayers,
    294:299–300
stoichiometry of binding, **294:**274
subunits

amino acid analysis, **294:**285–286
isolation, **294:**285
quaternary structure, **294:**281
T-tubule membrane content, **294:**292–293
CHDDE, *see* 1,3-Cyclohexadiene-1,4-dieth-
    anoate
Chemically initiated electron exchange lumi-
    nescence
development of theory, **305:**11–12
dioxetanone reactions, **305:**17
efficiency, **305:**13–16, 18
exothermicity of back-transferred elec-
    tron, **305:**12–13
generalization of luminescence mecha-
    nisms, **305:**4–5, 7, 18–19, 42
peroxyesters, **305:**16–17
phthaloyl peroxide reactions, **305:**17–18
pyrone endoperoxides, **305:**15–16
side reactions, **305:**15
triplet state products, **305:**14–15
Chemical modification interference, RNA–
    peptide interaction
overview, **318:**431–432
peptide affinity chromatography
    coupling peptide to solid support,
        318:432–434
    fractionation of modified RNAs,
        318:434
RNA modification, **318:**432
Chemiluminesce, *see also* Antioxidative ho-
    meostasis; Enhanced chemilumines-
    cence assay, tissue oxidation susceptibil-
    ity; Luminescence; Ultraweak photon
    emission
free radical measurement in lipoproteins,
    299:4–6
resonance energy transfer, applications,
    305:498
singlet oxygen, *see* Chemiluminescence,
    singlet oxygen
total peroxyl radical-trapping potential
    assay enhancement
    accuracy assessment, **299:**9
    contributions of individual compo-
        nents, **299:**10–14
    effects on results
        age, **299:**11–12
        antioxidant supplementation of diet,
            299:10
        disease states, **299:**12, 14

fasting, **299:**9–10
gender, **299:**11–12
metabolic rate, **299:**14
pH of assay buffer, **299:**13
smoking, **299:**10
low-density lipoprotein assay, **299:**8–9
plasma assay, **299:**7–8
principle, **299:**6–7
stoichiometric peroxyl radical-scaveng-
ing factors, **299:**10
total antioxidant reactivity, **299:**12–13
Chemiluminescence, singlet oxygen
dimol emission, **319:**13, 71, 75–77, 222–
223, 388–389
identification of singlet oxygen as emission source, **319:**66–67, 71
monomol emission, **319:**13, 71, 222–223,
388–389
near-infrared spectrometer
assignment of 1270-nm emission to singlet oxygen, **319:**66–67
calibration, **319:**63–64
detector, **319:**61
electronics, **319:**61–62
eosinophil peroxidase assays
kinetics, **319:**64
spectral analysis, **319:**64–65
injection port, **319:**61
interference filters, **319:**60–61
muon filter, **319:**62
schematic, **319:**60
sensitivity, **319:**62–63
ozone-induced singlet oxygen in plants
ascorbic acid reaction, **319:**512–513,
518–519
assay in leaves
carrier gas effects on emission intensity, **319:**517–518, 520
cell wall emission, **319:**516–517
gas volume per unit surface area effects, **319:**519–520
instrumentation, **319:**513–514
intensity of gas phase emission,
319:516
intercellular extract measurements,
319:515
kinetics, **319:**519
principle, **319:**513
spectral analysis, **319:**517
reactants in generation, **319:**514–515

photon generation, **319:**59
plant assays, **319:**78
prostaglandin endoperoxide H synthase,
assay of arachidonic acid oxidation
dimol emission, **319:**71, 75–77
filters, **319:**70
identification of singlet oxygen as emission source, **319:**71
inhibitors of chemiluminescence,
319:74–75
kinetics, **319:**72–73
oxygen requirement, **319:**73
peroxidase in singlet oxygen formation,
319:75–76
pH dependence, **319:**73
single photon-counting apparatus,
319:69–70
spectral analysis, **319:**73–74
quencher assay
biological quenchers, **319:**224–226
exogenous quenchers, **319:**225
generation of singlet oxygen, **319:**222,
224
instrumentation, **319:**223–224
principle, **319:**223
time-resolved measurements, *see* Time-
resolved near-infrared phosphores-
cence
Chemostat, biofilm culture, *see also* Constant depth film fermentor; Robbins device; Rototorque
continuous flow systems for oral biofilm
culture
aeration studies, **310:**294–296
antimicrobial agent evaluation
dental health care product additives,
310:293
pulsed dose concentration,
310:292–293
steady-state dose concentration,
310:292
conditioning films, **310:**289
mean generation time calculation,
310:283
multistage systems, **310:**286
pH effects, **310:**291–292
single stage reactor, **310:**282–283
steady state perturbation,
310:283, 286
substrata, **310:**283, 289

viable biomass balance calculation,
310:286
design, 310:254, 256
fluidized bed reactors, 310:298–299
Chevron analysis, circular dichroism kinetic
measurements of RNA folding,
317:405–406
Chicken retina
advantages as model system, 316:529–
530, 535–536
cone dominance, 316:527–529
development, 316:527–528
laminar microdissection
apparatus and instruments, 316:532
applications, 316:530–531
freeze-dried retinal section preparation, 316:531
section picking, 316:532, 534
oil droplets, 316:529
rd model, 316:534–535
rdd model, 316:535
structure, 316:528–529
Childhood-onset severe retinal dystrophy,
retinal pigment epithelium p65 mutations, 316:711–713
Chimeric norepinephrine transporter–
dopamine transporter
clones and plasmids, 296:467–468, 470
complementary DNA, size analysis,
296:468–471
expression analysis by transport and binding assays, 296:469–471
homology between transporters, 296:466
[125I]RTI-55 binding of chimeras with attenuated transport function, 296:470,
473–475
transformation of bacteria, 296:468, 470
transmembrane domain functional analysis, 296:479–480
Chimeric serotonin cross-species transporter
binding assays, 296:496
construction approaches
restriction enzyme-based chimeras,
296:485–486
restriction site-independent chimeras,
296:489–490, 492–493
restriction site introduction and complementary DNA cassette swapping,
296:486–489

electrophysiological characterization,
296:495
expression analysis, 296:496–497
imipramine binding, 296:481
ion dependence of transport, assays,
296:494
pharmacological profiles, 296:494–495
transmembrane domain functional
analysis
Drosophila–human transporter,
296:483
rat–human transporter, 296:483, 497
transport assays, 296:494
Chimeric serotonin transporter–norepinephrine transporter
construction approaches
restriction enzyme-based chimeras,
296:485–486
restriction site-independent chimeras,
296:489–490, 492–493
restriction site introduction and complementary DNA cassette swapping,
296:486–489
functionality, 296:479
Chinese Hamster ovary cell
advantages in recombinant protein expression, 306:20
amplification markers and enhancers,
306:20–21
dihydrofolate reductase-linked methotrexate mediated gene expression
cell culture, 306:22
cloning, 306:23–24
comparison with other expression systems, 306:36–41
mechanism, 306:21
methotrexate addition, 306:22–23
plasmid design, 306:21–22
primary selection, 306:23
transfection, 306:23
mutants, see Ceramide; Serine palmitoyltransferase
Chloramphenicol acetyltransferase
assay in Escherichia coli cell-free translation system, 290:25–26
reporter in mammalian cells compared
with firefly and bacterial luciferases
applications, 305:571–572
calibration curves, 305:559
expression levels, 305:560–563

materials, **305:**558–559
promoter for studies, **305:**558
rationale for fusion, **305:**558
sensitivity, **305:**559–560
stable transfection effects on assay re-
    producibility, **305:**564–567
thermal stability, **305:**563–564
transient transfected cell splitting, ef-
    fects on assay reproducibility,
    305:567, 569–571
streptococci fusion protein reporters
assay in biofilms, **310:**458–459
features, **310:**458
Chloride
confocal microscopy
data acquisition, **307:**474–477
dyes, overview, **307:**471–472
microscope design considerations,
    307:473–474, 480
photobleaching minimization, **307:**480
6-methoxy-*N*-ethylquinolinium iodide
    fluorescent dye
brain slice studies, **307:**480
calibration of intracellular fluorescence,
    307:477–479
chloride selectivity, **307:**479–480
confocal imaging, **307:**474–477
data analysis, **307:**477
kinetics of chloride response, **307:**481
loading of cells, **307:**473
reduction to membrane-permeable
    form, **307:**472–473
sample preparation, **307:**473
microelectrodes, **307:**471
ratiometric fluorescence indicator devel-
    opment, **307:**481
tracer measurement, **307:**471
Chloride channel, *see* Chloride channel,
    swelling-activated; Cystic fibrosis trans-
    membrane conductance regulator
Chloride channel, swelling-activated
P-glycoprotein regulation of cell volume
channel activation effects, **292:**365
channel activity measurement, **292:**361
channel sensitivity to hypoosmotic solu-
    tions, P-glycoprotein effects,
    292:367
electrophysiologic characterization,
    292:361, 363
evidence of interaction, **292:**360–361

mechanism of channel activity modula-
    tion, **292:**367–368
pharmacologic characterization,
    292:363, 365
phosphorylative modulation, **292:**369
physiologic significance, **292:**368–370
regulatory volume decrease, overview,
    292:359–360, 369
Chlorophyll
assay of light harvesting complex II abun-
    dance, **297:**230–231
biosynthesis
δ-aminolevulinic acid control
Arabidopsis mutants in feedback
    loop controlling synthesis,
    297:243–244
limitation of precursor for chloro-
    phyll biosynthesis, **297:**237,
    241–242
Arabidopsis thaliana as model system,
    297:238
overview, **297:**237
photo-oxidative damage during light-
    induced chloroplast formation
assay, **297:**240–242
mutant isolation, **297:**244
protochlorophyllide oxidoreductase me-
    diation
light-dependent isoforms, **297:**238
overview, **297:**237–238
POR A-deficient *Arabidopsis* analy-
    sis, **297:**239–240
xantha mutants, isolation and classifica-
    tion, **297:**242–243
fluorescence yield analysis of photosys-
    tem II
charge recombination rate with 3-
    (3′,4′-dichlorophenyl)-1,1′-di-
    methylurea, **297:**344–346
factors affecting yield, **297:**339–340
kinetic techniques, **297:**340–341
mutant characterization, **297:**341–346
secondary electron transfer rates,
    297:342–344
ligands in photosystem II reaction center,
    297:355–356
Chlorophyll protein, *see* CP47
Chloroplast
ATP synthase, *see* Chloroplast ATP syn-
    thase

differentiation from protoplastids, **297:**38

maize

  cloning of genes identified by mutations and targeted nuclear gene knockouts, **297:**56–57

  immunoprecipitation of radiolabeled proteins

    antibody binding reaction, **297:**52

    antigen elution and analytical electrophoresis, **297:**53

    membrane solubilization, **297:**52

    Staph A cell binding reaction, **297:**52–53

  Northern blot analysis of mutant phenotypes

    agarose gel electrophoresis, **297:**47

    buffers, **297:**47

    hybridization reaction, **297:**48

    RNA isolation, **297:**47

    transfer to membranes, **297:**48

  pulse-labeling of proteins in mutant phenotypes

    buffers, **297:**51–52

    in organello labeling, **297:**48–51

    plant growth conditions, **297:**49

    in vivo labeling, **297:**48–50

  ribosome–messenger RNA interaction assay for translation-defective mutants

    buffer preparation, **297:**55

    extraction, **297:**53–54

    Northern blot analysis, **297:**54

    puromycin control, **297:**54–55

    sucrose gradient centrifugation, **297:**54

  thylakoid membrane biogenesis, identification of gene mutations, **297:**40–41

  transcription run-on assays, **297:**56

  transposon mutagenesis and mutant propagation, **297:**41–42

  Western blot analysis of mutant phenotypes

    detection of bands, **297:**46

    extraction of proteins, **297:**43–44

    immunostaining, **297:**46

    overview, **297:**42–43

    plant growth conditions, **297:**43

    polyacrylamide gel electrophoresis, **297:**45

    quantification of extracts, **297:**44–45

    transfer to membranes, **297:**45

organism selection for biogenesis studies, **297:**38–39

reaction center mutagenesis by transformation in *Chlamydomonas reinhardtii*

  chimeric gene construction conferring antibiotic resistance, **297:**312–313, 315–316

  complications, **297:**311

  gene gun transformation, **297:**311, 317–318

  growth conditions, **297:**316–317

  growth media, **297:**316

  homologous recombination, **297:**311, 315

  recipient strains, **297:**316

  selection of transformants, **297:**312, 318–320

subfractionation for plant peptidylprolyl *cis–trans*-isomerase assay

  chloroplast stroma and thylakoids, **290:**85–86

  thylakoid grana and stroma membranes, **290:**86–88, 91

transformation, **297:**27, 39, 311, 317–318

Chloroplast ATP synthase

assays

  calcium ATPase activity, **297:**145–146

  magnesium ATPase activity, **297:**146

  proton permeability assay by fluorescence quenching, **297:**146

components, **297:**139

light activation, **297:**139–140

δ subunit, site-directed mutagenesis, **297:**148

ε subunit

  CF_1 $(-\ \varepsilon)$

    preparation, **297:**143

    reconstitution in solution with ε, **297:**144–145

    reconstitution in thylakoids, **297:**143–145

    storage and handling, **297:**143

  inhibition of ATPase , 140, 145, 147–148

  purification of overexpressed protein in *Escherichia coli*

    cell growth and harvesting, **297:**141

inclusion body purification,
 297:141–142
solubilization and refolding, **297:**142
storage and handling, **297:**143
subcloning of gene, **297:**141
site-directed mutagenesis, **297:**146–147
γ subunit reconstitution, **297:**148
3-Chlorotyrosine
atherosclerosis role, **301:**472
atherosclerotic lesion content, **300:**143
gas chromatography–mass spectrometry
amino acid analysis, **301:**473–474
arterial tissue analysis, **301:**481–483
derivatization with *N*-methyl-*N*-(*tert*-
butyldimethylsilyl)trifluoroacet-
amide and spectrometry, **301:**474–
475, 477
protein analysis, **301:**477, 480–481
high-performance liquid chromatography
assays
chlorination reactions
hypochlorous acid, **300:**115–116
myeloperoxidase, **300:**114–117
stimulated neutrophils, **300:**117
electrochemical detection, **300:**113,
119–120
fluorescence detection, **300:**113,
118–119
neutrophil isolation, **300:**115
protein hydrolysis, **300:**117
high-performance liquid chromatography
with electrochemical detection
chromatography conditions, **301:**156
electrochemical detection, **301:**156, 158
interpretation of results, **301:**158–160
protein hydrolysis, **301:**155–156
sample preparation, **301:**154
hypochlorous acid reaction in formation,
 300:112, 125, 140–141
isotope dilution negative-ion electron capture mass spectrometry
advantages in protein oxidation analysis, **300:**127–128
derivatization, **300:**133
gas chromatography, **300:**133–134
internal standard synthesis,
 300:128–129
mass spectrometry settings, **300:**134
protein and tissue hydrolysis,
 300:130–132

selected ion monitoring in quantification, **300:**134–136, 139
solid-phase extraction, **300:**132
tissue collection and preparation,
 300:129–130
protein oxidation marker, **300:**90
Chlorpromazine, induced phototoxicity for
sunscreen assay
applications, **319:**456
cell viability assay, **319:**45
chlorpromazine dose–response, **319:**455
Episkin preparation, **319:**454
formulas for testing, **319:**454–456
incubation, **319:**455
interleukin-1α assay, **319:**455
principle, **319:**453–454
ultraviolet A radiation source, **319:**454
CHO, *see* Chinese Hamster ovary cell
Cholecystokinin
antisense inhibition of PC1 and PC2 in
processing analysis
cell culture and transfection,
 314:107–108
controls, **314:**115
Northern blot analysis, **314:**108–109,
112
plasmid construction, **314:**107
procholecystokinin processing analysis,
 314:111–112, 115, 117
radioimmunoassay of cholecystokinin,
 314:110
rationale, **314:**106–107
Western blot analysis, **314:**109, 112,
115
forms, **314:**111
gel filtration of forms, **314:**110, 115
tissue distribution, **314:**110–111
tumor processing and expression, **314:**111
Cholera toxin
GM1 binding, **312:**459, 468, 535
retrograde intracellular glycolipid trafficking
overview, **312:**459, 466, 468–469
visualization of toxin-bound glycolipid,
 312:469–470
translocation assay of sphingolipids,
 312:574
Cholesterol
oxidation products, **319:**87, 188, 390–391
ozonization

high-performance liquid chromatography
electrochemical detection,
319:195–196
mass spectrometry detection, **319:**196
liposome cholesterol, **319:**194–195
products, **319:**188–189
singlet oxygen reporter
advantages, **319:**87
chemicals and reagents for assays,
319:88
high-performance liquid chromatography with electrochemical detection
of oxidation products
erythrocyte membranes,
319:94–97
operating conditions, **319:**93–94
photoxidized cells, **319:**98–100
iodometric assay, **319:**90
leukemia cell culture, **319:**89
lipid extraction, **319:**90
membrane preparation, **319:**88–89
photooxidation
cells, **319:**89–90
membranes, **319:**89
thin-layer chromatography of erythrocyte membrane oxidation products, **319:**91–93, 100
ultraviolet A irradiation products
high-performance liquid chromatography
chemiluminescence detection,
319:192–194
electrochemical detection, **319:**192
photoperoxidation reactions in cell-free
systems, **319:**191–192
skin types and carcinogenicity,
319:190–191
Choline, caged compound studies with crystallized cholinesterases, **291:**266–267,
269, 273–276
CHRAC, *see* Chromatin accessibility
complex
Chromatic aberration, confocal microscope
objective lenses, **307:**113–114, 299
Chromatin, *see also* Histone; Minichromosome; Nucleosome; Nucleosome core
particle; Solid-phase nucleus
assembly in *Xenopus* extracts
advantages of system, **304:**50–51

anion-exchange chromatography of
chromatin assembly components
chromatin assembly conditions,
304:58–60
chromatography conditions,
304:55–56
ATP requirement, **304:**53, 58
extract preparation, **304:**51–52
histones H2A/H2B, purification,
304:57–58
micrococcal nuclease analysis,
304:62–63
N1,N2–(H3,H4) complex
isolation, **304:**60–61
nucleosome assembly, **304:**61–62
nucleosome array formation
DNA concentration optimization,
304:55
DNA supercoiling assay, **304:**53,
62–63
time required for assembly, **304:**53
topoisomerase treatment of DNA,
304:53
cross-linking, *see* Cross-linking, chromatin complexes
electron microscopy, *see* Electron microscopy
immunoprecipitation
antibodies, **304:**84, 88
fixed chromatin
formaldehyde fixation, **304:**89–91
incubation conditions, **304:**91–92
unfixed chromatin
incubation conditions, **304:**87
protein A–agarose bead preparation,
304:85, 87
solution preparation, **304:**85
linker histone H5 incorporation into
model nucleosome arrays, **304:**34
mapping nonhistone chromatin-associated factors
applications, **304:**402–403, 405, 415,
417–418, 429–430
cell growth and lysis, **304:**402, 406–407
chromatin shearing, **304:**408, 422–423
confirmation of positive results,
304:427–429
cross-linking with formaldehyde,
304:401–402, 407, 415, 418
DNA purification, **304:**412, 424–426

extract collection, **304:**407–408

gel electrophoresis analysis of amplification products, **304:**414, 426–427

immunoprecipitation, **304:**402, 410–412, 423–424

materials and solutions, **304:**406, 410, 412–413

overview, **304:**401–402, 415

Polycomb mapping, **304:**405, 417

polymerase chain reaction, **304:**402, 405, 412–414, 426–427, 430

silent information regulator 3 mapping, **304:**402–403

spheroplast lysis, **304:**419–422

mapping with restriction endonucleases, **304:**383, 385, 392

native chromatin

definition, **304:**613

high-quality substrate preparation for structural studies, **304:**613–616

remodeling during nucleotide excision repair

DNA substrate preparation, **304:**334–335

Drosophila embryo extract preparation, **304:**335, 338

efficiency of repair, assay, **304:**337–338

extracts for study, **304:**333–334, 338, 350

micrococcal nuclease assay of nucleosome assembly

assembly reaction conditions and repair-independent suppression, **304:**341, 344

digestion reaction, **304:**344

gel electrophoresis and analysis, **304:**344–345

mechanism of nucleosome assembly from target site for repair, **304:**345–347, 351

principle, **304:**338, 341

Southern blot analysis, **304:**345

remodeling factors and functions, **304:**350

remodeling proteins, *see* Chromatin accessibility complex; Chromatin assembly factor 1; Histone acetyltransferase; Histone deacetylase; Nucleosome remodeling factor; SWI/SNF

replication-coupled assembly

cell culture, **304:**77–78

chromatin preparation with micrococcal nuclease digestion, **304:**80–81

dependence on factors, **304:**97, 99

DNA replication in isolated nuclei

labeling conditions, **304:**94–95

overview, **304:**92–93

reagents and solutions, **304:**93–94

histone acetylation in isolated nuclei, **304:**95–97

nuclei preparation from HeLa cells, **304:**79

overview, **304:**76–77

polyacrylamide gel electrophoresis

histone electrophoresis and transfer to membranes, **304:**84

nucleosomal DNA, **304:**82–83

radiolabeling

DNA, **304:**78

histone, **304:**78–79

liquid scintillation counting, **304:**82

metabolic inhibitors, **304:**79–80

structural overview, **304:**19, 50, 231–232, 377, 612, 696

structure probing with DNA methyltransferases

*Cvi*PI methyltransferase probing

cell maintenance and growth, **304:**445

characteristics of methyltransferase, **304:**443

specificity of methylation, **304:**445

Dam methyltransferase probing, **304:**431–433

interpretation of results, **304:**445–446

principle, **304:**431–432

resolution, **304:**447

sensitivity, **304:**446

*Sss*I methyltransferase probing

bisulfite genomic sequencing and positive display of methylcytosine, **304:**433–435, 437–439

cell maintenance and growth, **304:**435–436

deamination of DNA, **304:**436–437

DNA isolation, **304:**436

expression in yeast, **304:**435

gel electrophoresis analysis, **304:**441

polymerase chain reaction, **304:**438–439

validation of methodology,
304:441–442
transcription, see Nuclear run-on assay
Chromatin accessibility complex
ATPase assay, 304:752
discovery, 304:743
immobilized nucleosomes, interaction
assay
biotinylated DNA preparation, 304:753
DNase I footprinting, 304:754–756
magnetic bead coupling of nucleosomes, 304:754
nucleosome reconstitution, 304:753–754
overview, 304:753
nucleosome spacing assay, 304:750–752
overview of function, 304:742–743
restriction enzyme accessibility assay
gel electrophoresis, 304:749
incubation conditions, 304:748–750
nucleosome reconstitution
chromatin assembly, sarkosyl stripping, and purification,
304:746–747
dialysis conditions, 304:747–748
Drosophila embryo extract preparation, 304:744–746
Chromatin assembly factor 1, role in remodeling during nucleotide excision repair,
304:350
Chromatin remodeling protein 1, immunostaining and confocal microscopy,
307:77–78
Chromatosome, definition, 304:299
Chrome-gelatin, preparation for mounting
of amyloid immunohistochemistry samples, 309:17
Chromogranin B, see Human chromogranin
B–green fluorescent protein fusion
protein
Chromosome, structure
fluorescence in situ hybridization analysis,
see Fluorescence in situ hybridization
overview, 304:650
Chromosome spread
colchicine treatment of cells, 307:192
confocal microscopy
advantages, 307:190
prerequisites for chromosome analysis,
307:207
fixative

evaporation, 307:191–195
methanol:acetic acid, 307:193
histone removal, 307:191
mounting for light and electron microscopy
carbon films, spread preparation,
307:196
electron microscopy, 307:198–199
grid transfer, 307:196, 198
light microscopy, 307:198–199
oligonucleotide-primed in situ DNA synthesis, single copy gene localization
with confocal microscopy,
307:202–203
restriction enzyme–nick translation,
DNA detection with confocal microscopy
CpG island identification, 307:203
digestion, 307:200
immunostaining, 307:202
multiple fluorescence and reflectance
detection, 307:203–204, 206
nick translation, 307:200
overview, 307:199–200
slide preparation, 307:192–194
Chronic lymphocytic leukemia, see B-1 cell,
CD5+
Chronic myeloid leukemia, see c-ABL
Chronoamperometry
dopamine clearance measurement in
brain
anesthesia effects, 296:719
animal surgery
electrode placement verification,
296:715
mouse, 296:711–712
rat, 296:710–711
rhesus monkey, 296:712–713
brain slice analysis in vitro, 296:718
carbon electrode
calibration, 296:708
manufacture, 296:709–710
Nafion coating, 296:710
clearance effects on transporter activity, 296:716–717
data acquisition, 296:713–715
data analysis, 296:716
dopamine D_2 receptor antagonist effects, 296:718
dopamine injection, 296:709, 714

inhibition of non-dopamine transporters, **296**:709

micropipette preparation, **296**:710

potassium chloride injection, **296**:714–715

reproducibility, **296**:716

transporter inhibitor analysis, **296**:716, 718

dopamine clearance measurement in *Xenopus* oocyte, **296**:718–719

Chymotrypsin

mechanism, overview, **308**:207–208

organic solvent trapping of intermediates for crystallography

acyl enzyme intermediate, **308**:209–207

enzyme–product complex, **308**:211–212

first tetrahedral intermediate structure, **308**:208–209

motions occuring between intermediates, **308**:212–215

native enzyme, **308**:208

pre-second tetrahedral intermediate structure, **308**:211

CINC, *see* Cytokine-induced neutrophil chemoattractant

Circadian rhythm

cyanobacteria incorporation of *lux* genes for study

hit-and-run allele replacement, **305**:533–536

overview, **305**:516, 527–528

random insertion of *luxAB*, **305**:541–542

transposon mutagenesis of *Synechococcus* sp. strain PCC 9742 using *Escherichia coli* conjugation, **305**:536–537, 539, 541

vector, **305**:528–529, 531, 533

Gonyaulax polyedra bioluminescence

applications, overview, **305**:114–116, 119

constant light exposure experiments, **305**:113–114

controls, **305**:111, 113

overview of biological system, **305**:104, 260–261

period length determination, **305**:110–111

phase calculation, **305**:109

recording apparatus

automation, **305**:106–108

components and design, **305**:105–107

cycle of measurements, **305**:108

development, **305**:105

software, **305**:106–107, 109, 119

standards, **305**:108–109

relationship to other rhythms, **305**:116–118

synchronization of cycle, **305**:109

Circular dichroism

advantages and disadvantages in protein structure studies, **309**:608, 624

aggregation problems, **309**:631–632

baseline correction, **309**:629

boranophosphate oligonucleotides, **313**:254

cell selection, **309**:629–630

cystic fibrosis transmembrane regulator, nucleotide binding domain-1, **292**:681, 684, 694

DNA assemblies with tethered photooxidant, **319**:180

folding and stability, protein analysis, **309**:628

gramicidin A secondary structure, **294**:532–534

helix formation analysis, **295**:7–10, 12, 15, 28–29

instrument

calibration, **309**:628–629

oxygen purging, **309**:630

parameter setting for protein analysis, **309**:630

optical rotary dispersion comparison, **309**:623

partition coefficient determination, proteins in membranes, **295**:79, 82

peptide secondary structure on RNA binding, **318**:434–435

recoverin, calcium binding and conformational changes, **316**:113–114

rhodopsin peptides, initial analysis for nuclear magnetic resonance determination of structure, **315**:110

RNA, *see* Circular dichroism, RNA

sample preparation, **309**:629

secondary structure analysis

error sources, **309**:626–627

fibrils, **309**:312

peptides, **309**:627

representative spectra, **309:**624–625
spectra analysis, **309:**625–626
sphingolipids, *see* Circular dichroism,
     sphingolipids
tertiary structure analysis, **309:**627–628
theory, **309:**623–624
Circular dichroism, RNA
  folding transition monitoring
     chevron analysis for kinetic measure-
          ments, **317:**405–406
     magnesium-induced folding transition
          measurements
          dissociation constant determination,
               317:398–399
          Hill coefficient determination,
               317:398–399
          ribonuclease P, **317:**395–396, 398
          thermodynamic measurements,
               317:396
          titrations, **317:**396–398
          wavelength selection, **317:**395
     mutation effects on kinetics,
          317:407–408
     overview, **317:**393–394
     urea folding transitions
          data analysis, **317:**401, 403, 405
          titrations, **317:**400–401
  normalization of spectra, **317:**394–395
  spectrum acquisition, **317:**394–395
Circular dichroism, sphingolipids
  exciton chirality technique
     absolute configuration determination,
          312:219, 221
     chromophoric derivatization
          chromophore selection, **312:**223
          2,3-naphthalenedicarboxylic acid an-
               hydride preparation,
               312:223–225
          rationale, **312:**218–219
          two-step derivatization, **312:**225
     conformational flexibility and analysis,
          312:221
     dihydrosphingosine isomers,
          312:222–223
     overview, **312:**217–218
     sphingosine isomers, **312:**222–223, 227
     stereochemistry in biological function,
          312:217
Circular permutation analysis, RNA
  applications

folding of tRNA^{Phe}, **317:**326
R17 phage coat–RNA interactions,
     317:326–327
ribonuclease P
  pre-transfer RNA interactions,
     317:327
  quantitative analysis of cleavage,
     317:327–328, 330
circular permutation mixture
  active isomer separation, **317:**320
  generation, **317:**317, 320
  renaturation, **317:**320
cleaved RNA analysis, **317:**321–322
cloning of isomers, **317:**324–325
comparison with other modification meth-
     ods, **317:**313, 315
dephosphorylation coupling technique,
     317:322–323, 327
labeling of new ends with probes,
     317:324–325
principle, **317:**313, 315
quantitative analysis, **317:**327–328, 330
reverse transcription, determination of cir-
     cular permutation sites, **317:**321
RNA requirements, **317:**315
synthesis of circular RNA using T4 RNA
     ligase, **317:**315–317
5′-triphosphate removal from transcripts,
     317:325–326
*c*I repressor
differential scanning calorimetry
  assembly
     melting temperature relationship to
          assembly state, **295:**463–464
     stoichiometry for unbound repres-
          sors, **295:**462–463
  data collection and analysis, **295:**454
  DNA-binding analysis
     oligonucleotide synthesis,
          295:453–454
     right operator site-1 complexes,
          295:458–460, 464–465
  interdomain interaction free energies,
     295:460–461, 465–466
  linker arm transition, **295:**466–467
  unbound repressor denaturation
     isolated N-terminal domain, **295:**457
     mutant repressors , 457–458
     wild-type repressor, **295:**454–457, 462
domains, **295:**451–452

functional mutants, **295:**451–453
oligomerization, **295:**450–451
purification, **295:**453
Cis-regulatory DNA element analysis, *see*
Homologous recombination
Citrate synthase, *see also* Malate dehydrogenase–citrate synthase
reactivation after thermal inactivation, influence of chaperones
GroEL, **290:**332
heat shock protein 25, **290:**333
heat shock protein 70, **290:**333
heat shock protein 90, **290:**332–333
kinetics, **290:**331–332
reactivation reactions, **290:**334
renaturation after chemical denaturation,
influence of chaperones on aggregation and reactivation
denaturation reaction, **290:**338
GroEL, **290:**335, 337
heat shock protein 25,
290:336–337
heat shock protein 90,
290:335, 337
refolding pathway, **290:**334–336
renaturation conditions, **290:**338
thermal aggregation, influence of chaperones
chaperone preparation, **290:**328
citrate synthase preparation,
290:327–328
GroEL, **290:**327
heat shock protein 70, **290:**327
heat shock protein 90, **290:**327
incubation reaction, **290:**328
kinetics, **290:**326
light scattering measurement,
290:327–329
mechanism of aggregation suppression,
290:325–326
thermal denaturation and ligand stabilization, **290:**324–325
thermal inactivation, influence of chaperones
activity assay, **290:**330–331
GroEL, **290:**330
heat shock protein 25, **290:**330
heat shock protein 70, **290:**330
heat shock protein 90, **290:**330
incubation conditions, **290:**331

kinetics, **290:**329
CLIP-170, *see* Cytoplasmic linker protein-
170
Cloning vectors, gram-negative bacteria,
310:35, 37–38
Cloud point extraction, gangliosides
advantages over other extraction techniques, **312:**187–188
C14EO6 as detergent
availability, **312:**190
cloud point manipulation, **312:**190
handling, **312:**190
properties, **312:**189–190
cultured cell extraction, **312:**191, 193
desalting and detergent removal from
samples, **312:**194
principles, **312:**188–189
thin-layer chromatography analysis,
312:194, 196
tissue sample extraction, **312:**193
Cluster analysis, distribution analysis in confocal microscopy, **307:**220–221
CMCT, *see* 1-Cyclohexyl-3-(2-morpholino-
ethyl)carbodiimide metho-*p*-toluene sulfonate
CNB, *see* α-Carboxy-*o*-nitrobenzyl carboxylates
CNG channel, *see* Cyclic nucleotide-gated
channel
Cocaine, *see* Dopamine transporter
Coenzyme Q
functions, **299:**330–331
simultaneous determination with vitamin
E
animal maintenance, **299:**334–335
chromatography conditions, **299:**337–
338, 340, 344, 347
electrochemical detection, **299:**333–334,
342, 344–345
extraction, **299:**331–332
materials, **299:**334
plasma sample preparation, **299:**344
recovery and reproducibility, **299:**340–
341, 345
saponification, **299:**332–333
standards, **299:**336–337, 342–344,
347
storage of samples, **299:**332
Coenzyme Q10, topical antioxidant screening, **319:**469–471, 474–475

Colchicine
  cellular uptake and loss, **298:**263
  intracellular concentration determination
    nonradiolabeled compounds and high-
      performance liquid chromatogra-
      phy, **298:**273–274
    radiolabeled compounds
      monolayer culture directly in ster-
        ilized scintillation vials, **298:**273
      monolayer or suspension cultures,
        298:272–273
  microtubule binding
    binding site, **298:**251–252
    data analysis, **298:**263–264
    substoichiometric inhibition of polymer-
      ization, **298:**262–263
    suppression of dynamic instability,
      298:263
  photoaffinity labeling of P-glycoprotein
    with analogs, **292:**293–294, 300–301
  structure, **298:**256
  tubulin binding, **298:**261–264
Cold denaturation, artificial formulation,
  295:256–257
Color blindness, *see* Cone
Combinatorial mutagenesis, photosystem II
  genes in *Synechocystis* 6803, 301–302,
  303–305
Comet assay, *see* Single-cell gel electrophoresis
Commitment, *see* Rate-limiting step
Common Object Request Broker Architecture, ion channel database accession optimization, **294:**727, 730
Comparative sequence analysis, RNA
  alignment and ordering of sequences
    group I introns, **317:**495, 510
    importance, **317:**494–497
    phylogenetic ordering, **317:**495–497
  automation
    alignment, **317:**497–498
    covariation analysis, **317:**498–500
  base-pairing geometry inference from
    base variation patterns, **317:**500–502
  GAAA tetraloop, **317:**503–504
  GNRA loop, **317:**505, 510
  helix stacking motif recognition,
    317:502–503
  loop E motif, **317:**503–504
  modeling

    algorithms, **317:**508
    GYRA loops and receptors, **317:**507
    process as implemented in MANIP pro-
      gram, **317:**508–509
    Rev-binding site of human immunode-
      ficiency virus, **317:**506–507
    transfer RNA, Levitt's pair,
      317:505–506
  natural versus artificially selected mole-
    cules, **317:**493–494
  searching programs for motifs,
    317:504–505
  secondary structure analysis, **317:**491–492
  steps in tertiary structure analysis,
    317:492–493
  temporal considerations, **317:**509–510
Complementary chromatic adaptation
  biological function, **297:**280
  Calothrix, **297:**7601, 280–281
  Fremyella diplosiphon
    advantages and disadvantages as model
      system, **297:**280–281
    mutant generation
      cell growth, **297:**287–288
      electroporation, **297:**287–289
      overview of techniques, **297:**281–282
      selection and screening of trans-
        formants, **297:**289
    plasmid library construction
      amplification of library, **297:**286
      insert DNA preparation,
        297:282–284
      ligation and transformation,
        297:285–286
      plasmid map, **297:**282–283
      plasmid preparation, **297:**284–285
  phycobilisome structure alteration,
    297:279–280
Complementary DNA
  amplification from a small number of he-
    matopoietic cells
    adaptor ligation of complementary
      DNA, **303:**48–50
    complementary DNA synthesis
      first strand, **303:**47–48
      second strand, **303:**48
    materials and reagents, **303:**46
    messenger RNA preparation, **303:**47,
      55
    polymerase chain reaction

betaine and dimethyl sulfoxide improvement of efficiency, **303**:45–46, 50

differential display analysis of products, **303**:50–54

reaction conditions, **303**:50

direct selection, *see* Direct complementary DNA selection

expression cloning, library construction

functional screening, **296**:46–47

hybrid-depletion screening approach, **296**:49

library plating and DNA isolation from filters, **296**:42–44

ligation into plasmid vector, **296**:41

size selection by agarose gel electrophoresis, **296**:40–41

synthesis, **296**:36, 39

transfection of library in *Escherichia coli*, **296**:41–42

expression-based gene identification, **303**:84–85

full-length cloning with high efficiency

applications, **303**:19, 44

reverse transcriptase reaction

first-strand synthesis, **303**:29–31

trehalose stabilization of enzyme, **303**:19, 22

yield calculation, **303**:31

biotinylation

cap trapper in selection, **303**:19, 21

capture reaction, **303**:35

complementary DNA biotinylation reaction, **303**:33–34

complementary DNA release from beads, **303**:36

streptavidin-coated magnetic bed preparation, **303**:34–35

washing of beads, **303**:35

extraction and precipitation of complementary DNA, **303**:31–33, 38

λ cloning vector

cleaving of complementary DNA, **303**:41–42

ligation of complementary DNA, **303**:42–44

preparation, **303**:23–25

requirements, **303**:44

test insert insertion, **303**:25–26

materials, **303**:22–23

oligo(dG) tailing of first-strand complementary DNA, **303**:37–39

primer preparation, **303**:28–29

principle, **303**:19, 21–22

ribonuclease inhibition, **303**:26–27

RNA preparation, **303**:27–28

RNase I treatment, **303**:35

second-strand synthesis, **303**:39–40

spun column fractionation of complementary DNA, **303**:36–37

library construction for arraying

cloning, **303**:210

overview, **303**:208–209

plating, **303**:212–213

protein arrays, **303**:231

reverse transcription, **303**:209–210

RNA extraction, **303**:209

tissue collection, **303**:209

low stringency screening for neurotransmitter transporters

colony screening with radiolabeled oligonucleotides, **296**:54–56

libraries, **296**:52–53

materials and reagents, **296**:53–54

plaque screening by homology of libraries, **296**:56–59

screening of gene expression, *see* Complementary DNA representational difference analysis; DNA array; DNA microarray; Restriction endonucleolytic analysis of differentially expressed sequences; Sequencing by hybridization; Suppression subtractive hybridization; Targeted display

single cell analysis of messenger RNA

antisense RNA

amplification, **303**:4, 10–11

applications of amplified product, **303**:6, 11–12, 17

Northern blot analysis, **303**:10

radiolabeling, **303**:6, 11–12

complementary DNA synthesis, **303**:4, 6–7, 9–10

fixed tissue sample analysis with *in situ* transcription/antisense RNA amplification

amplification efficiency, **303**:16–17

applications and rationale, **303**:13–15

complementary DNA synthesis, **303**:16

fixation and embedding, **303:**15
immunohistochemical labeling,
   303:15–16
messenger RNA harvesting, **303:**7
simultaneous electrophysiological analysis, **303:**6–7
subcellular distribution analysis,
   303:12–13
transcript-aided drug design, **303:**18
Complementary DNA representational difference analysis
applications, **303:**325–326, 346–349
difference analysis
   adapters, change on a difference prod-
      uct, **303:**342–343
   driver generation, **303:**339–340
   first difference product generation,
      303:341–342
   polymerase chain reaction-coupled sub-
      tractive hybridization, **303:**327–
      330, 341
   second difference product generation,
      303:343
   tester generation, **303:**340–341
   third difference product generation,
      303:343
driver populations, combining, **303:**347
genomic representational difference analysis comparison, **303:**326–327
iterative technique to discover concealed
   genes, **303:**346–347
large numbers of cells as starting material
   complementary DNA template prepara-
      tion, **303:**338–339
   generation of representations, **303:**339
limitations, **303:**332
low numbers of cells as starting material
   complementary DNA preparation
      first-strand synthesis, **303:**334
      ligation to R linkers, **303:**335
      restriction enzyme digestion, **303:**335
      second-strand synthesis, **303:**334–335
   representative amplicon generation,
      303:335–336, 338
   RNA isolation
      genomic DNA removal, **303:**333–334
      messenger RNA purification, **303:**334
      total RNA preparation, **303:**333
multiple four-base cutters, **303:**349

principles, **303:**325–330
prokaryotes, **303:**349
screening of difference products,
   303:344–346
sensitivity, **303:**325, 331–332
starting material, **303:**330–331
stringency variation, **303:**348
subtraction types, **303:**330
time course analysis, **303:**347–348
troubleshooting, **303:**343–344
Complementary DNA selection, gene identification
bacterial artificial chromosome DNA
   biotinylation, **306:**84–85
   digestion, **306:**84
   preparation, **306:**84
binding, washing, and elution of complementary DNA fragments, **306:**85–87
hybridization
   first step, **306:**85
   second step, **306:**87
overview, **306:**81, 83
polymerase chain reaction
   complementary DNA fragments
      first step, **306:**86–87
      second step, **306:**87–88
   initial complementary DNA prepara-
      tion, **306:**83
repeat suppression, **306:**85
subcloning of products, **306:**88–89
Computational gene finding, *see also* Expressed sequence tag
annotation of genes, **303:**79–80, 82
annotation of sequences, **303:**159
consolidation with other techniques,
   303:146–147, 160–161
hypothesis status of predictions, **303:**77,
   80
Internet resources
   FgeneH, **303:**81–82
   Genie, **303:**81
   Genome Channel, **303:**82–83
   Genscan, **303:**81, 159–160
   GRAIL, **303:**81, 159
   MZEF, **303:**82
limitations in gene identification, **303:**146,
   159, 497
precautions, **303:**77–78
protein matches in validation, **303:**159
sequence database impact, **303:**78–79

splice site identification, **303**:205
transitive catastrophes, **303**:80–81
Conantokin
  N-methyl-D-aspartate receptor binding,
    294:612–613
  posttranslational processing, **294**:613
  structure, **294**:610, 612–613
  types, **294**:612
Concentration jump, *see* U-tube
Conductance, assays for enzymatic degrada-
    tion of biofilms
  advantages, **310**:353–354, 360–361
  biofilm culture for assay, **310**:354–357
  calibration curve, **310**:357–359, 361
  controls, **310**:359
  direct assay, **310**:356
  enumeration of cells, **310**:360–361
  indirect assay, **310**:355–356
Cone
  absorption of cone pigments
    peaks, **316**:631–632
    polymorphisms and variant human
      color vision, **316**:649–650
    recombinant pigments
      expression constructs, **316**:634
      human embryonic kidney cell trans-
        fection, **316**:635
      maximum wavelength determination,
        316:636–638
      mutation effects on maximum absorp-
        tion wavelength, **316**:649–650
      photobleaching difference spectra ac-
        quisition, **316**:635–636
      reconstitution, **316**:635
  chicken retina, cone dominance,
    316:527–529
  deuteranomaly
    green–red pigment hybrid character-
      ization
      long-range polymerase chain reac-
        tion amplification, **316**:663
      pulsed-field gel electrophoresis,
        316:663, 665
      sequencing and restriction enzyme
        analysis, **316**:665
      Southern blot analysis, **316**:665
    hybrid pigment genes, **316**:652–653,
      662–663
  electroretinograms, activation modeling,
    316:616–618

green and green–red hybrid pigment
  gene expression in arrays, polymer-
    ase chain reaction/single-strand con-
    formation polymorphism,
    316:668–670
green pigment
  deletion, **316**:652
  point mutation, **316**:654
opsin phosphorylation assay
  incubation conditions and gel autoradi-
    ography, **316**:551–552
  materials, **316**:551
  principle, **316**:551
  white light versus red light response,
    316:552
paired-flash electroretinography studies
  of photoreceptor response, subtraction of cone contribution, **316**:209–
  211, 218
photobleaching, *see* Photobleaching recovery, photoreceptors
recombination of pigments in red–green
  color vision defects, **316**:652
red–green hybrid gene, polymerase chain
  reaction amplification and detection
  in females, **316**:660, 662
red/green pigment transcripts, ratio determination in retina
  polymerase chain reaction/single-strand
    conformation polymorphism,
    316:667–668
  segregated expression,
    316:665, 667
red pigment Ser180Ala polymorphism
  color perception in males, slight varia-
    tions, **316**:653–654
  single-strand conformational polymor-
    phism analysis, **316**:659–660
sequence homology between pigments,
  316:651
spectral sensitivity determination
  chromatic adaptation and selective de-
    sensitization, **316**:628
  electroretinography, **316**:629–631,
    647–648
  fundus reflectometry, **316**:628
  microspectrophotometry, **316**:628–629,
    631, 648
  pigment absorption effects,
    316:631–632

psychosocial determination of M, L
and hybrid cones
comparison with *in vitro* determinations, **316:**648–649
flicker photometry, **316:**639–640
male dichromat studies, **316:**633, 639
maximum wavelength determination,
316:641
overview, **316:**638–639, 641
spectra calculation from corneal spectral sensitivities, **316:**641–644
psychosocial determination of S cones
flicker photometry, **316:**646
maximum wavelength determination,
316:646–647
overview, **316:**645–646
single-cell electrophysiology, **316:**629,
631, 648
variability sources, **316:**632–633
trichromacy, **316:**626, 628, 651
X-linked red–green photopigment genes,
determination of number and ratio
competitive polymerase chain reaction
amplification, **316:**658
data analysis, **316:**655–656
exons 2, **316:**4, and 5 analysis,
316:659
overview, **316:**655
primers, **316:**657–658
single-strand conformational polymorphism, **316:**655, 658–659
pulsed-field gel electrophoresis,
316:655
Southern blot analysis, **316:**655
Cone visual pigment
absorption spectroscopy, **315:**299
classification, **315:**312
comparison with rhodopsin, overview,
315:219–220, 293–294
meta intermediates
metarhodopsin-II decay rate and Glu-
122 role, **315:**294, 311–312
thermal behavior
absorption spectroscopy studies,
315:302–303, 305–308
cooling of samples, **315:**302–303
transducin activation assay, **315:**308–
309, 311
preparation from chicken
buffers, **315:**294, 296

concanavalin A affinity chromatography, **315:**296–297
ion-exchange chromatography,
315:296–298
transient transfection in HEK293S
cells, **315:**298–299
regeneration time
assay with absorption spectroscopy,
315:299–300, 302
comparison to rhodopsin, **315:**294, 302
Cone worse than rod disease, electroretinogram studies, **316:**625
Confocal laser scanning microscopy, *see
also* Cytotomography; Fluorescence *in
situ* hybridization; Green fluorescent
protein; Multiphoton excitation microscopy; Reflection confocal microscopy;
Two-photon excitation microscopy
acousto-optic deflectors and rapid scanning, **307:**400–402
advances in dyes and instrumentation,
310:143–144
advantages in biofilm studies, **310:**22, 101,
131–132, 170
alignment of system, **310:**135
androgen receptor–green fluorescent protein fusion protein, **302:**127–128
antisense oligonucleotide transfection
analysis, **313:**59–60, 65, 67, 346
archiving/printing of data, **310:**143
batch culture of biofilms, **310:**103–104
biofilm preparation
flat substratum samples, **310:**136
irregular substratum samples,
310:136–137
overview, **310:**135–136
upright versus inverted microscope
samples, **310:**136
bone cancer imaging, **307:**377–378
calcium imaging
line scanning and transient measurements, **302:**353–354
mitochondria measurements,
302:356–358
nonratiometric imaging, **302:**352
ratio imaging with Indo-1, **302:**354, 356
simultaneous measurement with electrical potential, **302:**353–354
cell preparations, overview, **302:**345–346,
307:119–121

cell volume measurement, *see* Cell volume, confocal microscopy measurement

chromosome spread, *see* Chromosome spread

coculture of cells, **307:**124–125, 134

compartmentation of ester-loaded fluorophores

  cold loading followed by warm incubation, **302:**350–351, 356–358

  warm versus cold loading, **302:**349–250

continuous flow systems, **310:**105

custom system construction

  calcium imaging

    dyes, **307:**164

    transients, **307:**167–168

  cost, **307:**152–153

  electronics, **307:**158, 161

  emission path, **307:**155, 157

  excitation path, **307:**155

  inflexibility of commercial systems, **307:**152–153

  morphological studies, **307:**168–169

  overview of design, **307:**153–155

  piezo-objective system, **307:**161–162

  resolution

    spatial, **307:**164–167

    temporal, **307:**167

  scan head, **307:**157–158

  software, **307:**162

  ultraviolet photolysis system, **307:**162–164

cytoskeleton imaging, *see* Cytoskeleton, confocal microscopy

data acquisition, **310:**139–140

data storage and processing, *see* Image data, confocal microscopy

deconvolution, **302:**83

digoxigenin-labeled phosphorothioate oligodeoxynucleotides, intracellular distribution, **313:**260–261, 265–266

distribution analysis

  cluster analysis, **307:**220–221

  estimator computation, **307:**219

  nearest neighbor distribution function, **307:**215, 217–220, 226–227

  overview, **307:**215, 217

  pair correlation function, **307:**218, 220, 227

  randomly distributed point objects, **307:**219–220

DNA detection, *see* Restriction enzyme–nick translation, DNA detection with confocal microscopy

dye loading, *see* Dye loading

dye selection, **307:**378–379

electrical potential, imaging with cationic fluorophores, **302:**346–348

electron microscopy of samples, *see* Electron microscopy

epifluorescence microscopy comparison, **307:**3, 370–373, 470

error estimation

  multicolor imaging, **307:**308

  tissue compression, **307:**307

  Z-axis distortion, **307:**307–308

eye imaging, *see* Eye

fluorescence recovery after photobleaching of potassium channels, **294:**54–56

G protein-coupled receptors and effectors fused with green fluorescent protein, **302:**162–163

green fluorescent protein imaging, *see* Green fluorescent protein

history of development, **307:**368–370

image analysis software, **310:**142

image processing and presentation, **310:**140–142

immunofluorescence localization in α-toxin permeabilized photoreceptors

  antibody preparation, **316:**272–273

  bovine retina sectioning and staining, **316:**273

  cell death analysis, **316:**277

  data collection, **316:**273, 275

  neurobiotin distribution following permeabilization, **316:**277–278

  toxin distribution, **316:**277

incoherent sources, **307:**17–18

instrumentation, **310:**133, 135

ion measurement, *see* Calcium; Chloride; pH

laser

  excitation, **307:**17, 375

  wavelengths, **307:**421–422

lectin staining

  applications, **310:**151–152

  controls, **310:**150–151

indirect sandwitch techniques,
    310:149–150
multiple lectin staining, **310**:149
single lectin staining, **310**:148
limitations in volume measurement
    data set size, **307**:308–310
    image processing and display,
        307:313–315
    time constraints, **307**:310–313
localization analysis
    calculation, **307**:214–215
    experimental determination, **307**:221
membrane permeability studies,
    302:351–352
membrane trafficking analysis with green
    fluorescent protein fusion proteins
    advantages over immunofluorescence
        microscopy, **302**:3
    combining with immunofluorescence,
        302:9–10
    data acquisition, **302**:6–7
    fluorescence recovery after photo-
        bleaching, **302**:7–9
    mounting, **302**:4, 6
metabolic studies, *see* Nicotinamide ade-
    nine dinucleotide(phosphate)
microcirculatory kinetics measurement,
    see Microcirculatory kinetics mea-
    surement
microtubule-associated protein 4–green
    fluorescent protein fusion protein
    cell preparation for microscopy,
        302:107–108
    image acquisition, **302**:108–109
    microtubule dynamics analysis
        dynamics parameters, calculation,
            302:113–116
        life history plots, **302**:113
        time series acquisition, **302**:111–112
        tracing of microtubules, **302**:112–113
monochromaticity of source, **307**:17
mounting of specimens, *see* Mounting,
    confocal microscopy
multicolor fluorescence experiments,
    302:346
multiphoton excitation, **310**:133, 135, 144
multipinhole confocal scanner for high-
    speed imaging
    charge-coupled device camera,
        307:434–435

configuration of system, **307**:433–434
fluo-4 imaging in mixed neuron and as-
    troglial cell culture, **307**:437,
    439–440
principles, **307**:433
temporal resolution, **307**:431, 433
Z-axis scanning, **307**:435–437
nuclear pore complex, *see* Nuclear pore
    complex
objective lenses, **310**:138–139, *see also* Ob-
    jective lens, confocal microscope
optical sectioning, **307**:4–6, 10–11, 299–
    306, 372–374
optics, overview, **307**:14–15, 20–21, 301,
    371–372, 425, 427, 441, 537–538
oxidative stress measurements,
    300:256–257
phagocytosis assay, *see* Phagocytosis, con-
    focal microscopy
photobleaching minimization,
    302:343–345
photoreceptor calcium flux measurements
    using spot confocal microscopy with
    Fluo-3
    advantages, **316**:148–149
    argon ion lasers, **316**:149, 151–154
    calibration, **316**:159–162
    detectors, **316**:154–156
    dye loading, **316**:157
    instrumentation, **316**:149
    light-induced decline of calcium,
        316:157–159
    photopigment cycle changes in calcium,
        316:162–163
    photoreceptor preparation from tiger
        salamanders, **316**:156–157
    rationale, **316**:147–148
pinhole size, **302**:341–343, **307**:12, 14, 372,
    374–375, 427
point source, **307**:3–4
point-spread function, *see* Point-spread
    function
principle, **302**:341–343, **310**:60, 101
rapid scan imaging, **307**:124, 134
ratio imaging of emission, principle,
    302:348–349
receptor imaging, *see* Receptor–ligand in-
    ternalization, confocal microscopy;
    specific receptors
refractive index mismatch, **307**:502–503

resolution
axial resolution, **307:**13–14, 52, 278,
303–304, 374–375, 378, 501–502
lateral resolution, **307:**13–14, 52, 278,
302–303, 501
limitations, **307:**135
temporal resolution, **307:**167, 301,
310–313
Robbins device biofilms, **310:**257–258
sampling rate, **310:**139
scanners
beam scanner, **307:**376
object scanner, **307:**376
pinhole type, **307:**398–399
rate of scanning, **307:**376–377, 397–399
slit type, **307:**399–400, 542–545
scanning, **302:**343
secretion assay, *see* Secretion
settings and filters for various probes,
307:121–122
simultaneous patch-clamp analysis of
plant cells, **294:**427–428, 430–432
speckle, **307:**17
staining options for biofilms, **310:**137–138
step size
control and time constraints,
307:311–313
matching optical section thickness,
307:305–306
teeth, *see* Dentin; Enamel
thickness measurement of tissues, *see*
also Cornea
image acquisition, **307:**234
objective lens design and control,
307:235
overview, **307:**230–231
reflected light confocal microscope de-
signs, **307:**231–234
thickness of sample, **307:**6, 427
two-color experiments, **307:**19, 378
types, O versus P, **307:**15
vascular casts, *see* Gelatin, vascular casts
with fluorochrome labeling
vascular remodeling studies, *see* Vascular
remodeling
virus studies, *see* Electron microscopy;
Vaccinia virus
Conformational change, *see also* Ligand-
induced conformational change; Osmotic stress

ATP-driven molecular discrimination, *see*
also ATP sulfurylase; DNA gyrase;
Nicotinate phosphoribosyltransferase
clock mechanism, **308:**34–35
conformational coupling, **308:**28–29,
50–52
conformer affinities for substrates and
products, **308:**31–34, 36
examples of enzymes, **308:**29, 49
ion transporter energetics, **308:**30–36
leaks, origin, **308:**52
necessary imperfection, **308:**49, 52, 70
optimization of ATP use, **308:**32–34
RecA, **308:**47–48
thermodynamic modeling, **308:**30–31,
35
free energy profiling, **308:**8–9, 11–12
tryptophan synthase, hydrophobic tunnel
substrate channeling, **308:**126–127
Congo red
absorbance spectra, amyloid binding
concentration dependence, **309:**289
fibril light scattering effects, **309:**289–
291, 294–296
shift on binding, **309:**289
amyloid-β fibril spectrophotometric assay
advantages, **309:**304–305
Alzheimer's disease applications,
309:285, 287
binding conditions, **309:**288
calculations, **309:**288–289, 292–298,
300–301
controls, **309:**288
difference spectra, **309:**291–292
dye concentration effects, **309:**297
fibril scattering correction, **309:**289–
291, 294–296
limiting concentration ranges of assay,
309:298
microtiter plate assay, **309:**298–299
modification for other amyloid fibrils,
309:303–304
nonturbid samples, **309:**301–303
stock solution preparation,
309:287–288
test compound interference,
309:299–300
binding to amyloid, **309:**4–5, 278,
285–286
biofilm staining, **310:**347, 404–405

counterstaining of immunohistochemistry
    samples, **309:**38–39
differentiation of amyloid forms,
    309:9
fibril isolate staining, **309:**41–42
flow cytometric characterization of amy-
    loid fibrils with thioflavin T and
    Congo red
    criteria for fibril detection, **309:**463
    data acquisition, **309:**462
    immunoglobulin light chain amy-
        loidosis, amyloid burden determi-
        nation, **309:**464–466
    instrument settings, **309:**461
    overview, **309:**460–461
    staining of sample, **309:**462–463
fluorescence detection
    overview, **309:**8–9, 278, 283–284
    spectral shift mechanism, **309:**278
history of use, **309:**3
insulin fibril binding, **309:**286
intensity of staining, **309:**7, 36
pitfalls, **309:**8
polarization microscopy, **309:**4–5, 7, 36–
    37, 68
solutions for staining, **309:**6, 36
specificity, **309:**7–8, 275, 285
staining reaction, **309:**6–7, 36
structure of dye, **309:**4
Sup35p NM fragment assembly assay,
    309:668–669
Conjugation
biofilm analysis with multicolor fluores-
    cence experiments, **310:**40–41
green fluorescent protein plasmid trans-
    fer assays, **310:**16, 30
transposon mutagenesis for conjugative
    plamid tagging, **310:**39–40
α-Conotoxin
IMI, **294:**611–612
MII, **294:**611
nicotinic acetylcholine receptor binding,
    294:608–612
sequences and binding specificities,
    294:608–610
αA-Conotoxin
nicotinic acetylcholine receptor binding,
    294:610–611
structures, **294:**610
δ-Conotoxin

inhibition of sodium channel inactivation,
    294:614, 617, 619–620
solubility, **294:**621
types and structures, **294:**619–620
κ-Conotoxin PVIIA, voltage-gated potas-
    sium channel binding, **294:**620–621
μ-Conotoxin, *see also* μ-Conotoxin GIIIA
arginine interactions in sodium channel
    binding, **294:**618–619
structures and sodium channel binding
    specificity, **294:**616
μO-Conotoxin
sodium channel binding, **294:**620
solubility, **294:**621
μ-Conotoxin GIIIA, sodium channel
    blocking
binding site, **294:**580, 584, 590, 602,
    604–605
comparison with charybdotoxin blocking
    of potassium channels, **294:**584–586
electrostatic compliance analysis, inter-
    residue distance measurement,
    294:597–598
kinetics of binding, **294:**585–586
mutagenesis effects on binding sensitivity
channel mutation
    alanine scanning mutagenesis,
        294:591
    chimera studies and block mutations,
        294:590–591
    conservative mutations, **294:**591
    cysteine scanning mutagenesis,
        294:591
    multiple primer techniques,
        294:592–593
    oligonucleotide-initiated polymeriza-
        tion of entire vector,
        294:592–594
    point mutations, **294:**579, 589–591
ligand mutation, **294:**583–584
mutant cycle analysis
    applications, **294:**596–597
    coupling coefficient calculation,
        294:595
    interaction energy, **294:**596
    principle, **294:**594–595
partial blocking toxins
    intrapore binding site determination,
        294:602
    mechanism of pore block, **294:**603–604

mutant types, **294:**600
selectivity and permeation factor analysis, **294:**602–603
voltage sensor movement and location analysis, **294:**601–602
peptide synthesis of analogs, **294:**587–589
specificity for skeletal muscle isoform, **294:**606
ω-Conotoxin, immunohistochemical analysis, **314:**302–303
ψ-Conotoxin
nicotinic acetylcholine receptor binding, **294:**611
structure, **294:**610
ω-Conotoxin
antibodies, **294:**693
types and structures, **294:**615, 618
voltage-gated calcium channel binding, **294:**614, 618
Constant depth film fermentor
adenylate measurements, **310:**306
applications, overview, **310:**265, 303–305
atmosphere regulation, **310:**271
cryosectioning of films, **310:**278–279, 305–306
design, **310:**265–267, 299–303
electron microscopy of films, **310:**277–278
indwelling bladder catheter biofilm
antimicrobial evaluation, **310:**308–309
atomic absorption spectroscopy, **310:**308
culture, **310:**304–305
steady state, **310:**309
inoculation, **310:**267–268, 271, 301–302
metal working biofilm
antimicrobial evaluation, **310:**309
culture, **310:**304
steady state, **310:**309
microelectrode measurements
oxygen gradient across biofilm, **310:**307
pH gradient across biofilm, **310:**307
oral biofilm studies
advantages, **310:**279
corrosion studies of dental materials, **310:**276–277
inhibitor analysis
light-activated antimicrobial agents, **310:**275–276
microcosm dental plaques, **310:**273–274

monospecies films, **310:**272
multispecies films, **310:**272–273
overview, **310:**271–272
pulsing of antimicrobial agents, **310:**275
sucrose supplementation effects, **310:**274
multispecies systems, **310:**268
nutrient source, **310:**268–270
steady-state oral community culture, **310:**303–304, 310
pan filling by biofilm, **310:**311–312
protein determination in films, **310:**305
river biofilm
culture, **310:**304
steady state, **310:**309
solute transport inside and outside of biofilm matrix, **310:**312
sterilization, **310:**301
substratum, **310:**267
temperature control, **310:**266, 270–271, 303
thickness of biofilms, regulation, **310:**267
vital staining of bacteria, **310:**277, 305
X-ray microanalysis, **310:**307–308
Conus venom peptides, *see also specific conotoxins*
complexity of mixtures, **294:**605–606
classification and nomenclature, **294:**606–608
solubility, **294:**621
storage, **294:**621–622
nonspecific adsorption, **294:**622–624
Copper
antimicrobial paint additive
efficacy, **310:**643–644
extracellular polysaccharide evaluation, **310:**640–643
modified Robbins device culture of biofilms, **310:**638–639
overview, **310:**637
substrata variation effects, **310:**641–643
viable cell counting, **310:**639–641
biofouling versus plastic, **310:**633, 636
Copper phenanthroline
cross-linking of chromatin complexes
advantages, **304:**517
cross-linking reaction conditions
Ehrlich acites tumor cells, *in vivo*, **304:**520, 522

isolated chromatin, **304:**519–520
DNA cleavage reaction, **304:**521
protein shadow hybridization assay,
  304:522
radical production, **304:**516–517
two-dimensional gel electrophoresis
  analysis, **304:**520
distance-dependence of cleavage,
  318:34–35
mechanism of RNA cleavage, **318:**33–34
RNA binding, **318:**34
RNA secondary structure and nucleotide
  distance determinations
  alkaline hydrolysis, **318:**37
  buffers, **318:**35–36
  data analysis
    distance between nucleotides,
      318:41–43
    secondary structure, **318:**39, 41
  end-labeling of RNA, **318:**36–37
  enzymes and reagents, **318:**35
  overview, **318:**35
  polyacrylamide gel electrophoresis,
    318:38–39
  principle, **318:**36
  probing reaction, **318:**38
  purification of labeled RNA, **318:**37
  ribonuclease T1 digestion, **318:**37
  specificity of RNA cleavage, **318:**33–35
Copper-regulated expression vectors, yeast
construction of vectors, **306:**152–153
copper
  metabolism in yeast, **306:**146
  transport system genes, **306:**147, 148
CUP1 promoter, **306:**146–147
induction/repression times, **306:**151
reporter gene assays, **306:**151
repression by transport gene promoters,
  306:148–149
restriction sites, **306:**149, 151
structure of vectors, **306:**149–150
Copper-responsive gene expression
Chlamydomonas reinhardtii
  advantages as model system,
    297:263–264
  copper-deficient media
    chelation of copper ions,
      297:267–268
    precautions in preparation,
      297:265–266

reagent preparation, **297:**266–267
solid medium preparation, **297:**267
copper proteins, **297:**264
culture conditions, **297:**268–269
nuclear run-on assays, **297:**270
RNA isolation and Northern blot analysis, **297:**270
copper measurement in samples
atomic absorption spectroscopy,
  297:277–278
inductively coupled plasma mass spectrometry, **297:**278
cytochrome c_6
  copper deficiency assays
    chemiluminescent heme staining,
      297:274–275
    Coomassie stain analysis in native
      gels, **297:**276
    peroxidase assay in gels, **297:**274
    rationale, **297:**264, 271
    reporter gene assays, **297:**276–277
    spectroscopic analysis of soluble ex-
      tracts, **297:**271–273
    Western blot analysis, **297:**273–274
  plastocyanin substitution, **297:**264
  promoter control of gene expression in
    transgenic organisms, **297:**279
plastocyanin assays
  Coomassie stain analysis in native gels,
    297:276
  spectroscopic analysis of soluble ex-
    tracts, **297:**271–273
  Western blot analysis, **297:**273–274
Scenedesmus obliquus
  culture conditions, **297:**269
  lysis, **297:**271
  RNA isolation, **297:**271
CORBA, *see* Common Object Request Broker Architecture
Cornea
  clinical confocal microscopy
    contact lens-associated changes,
      307:557–558
    data collection and analysis, **307:**551–
      553, 555, 557
    difficulty, **307:**536–537
    infection, **307:**558
    light source, **307:**550–551
    scanning slit microscope design,
      307:542–545, 563

characterization of mutants,
    297:336–337
control strain K3, construction,
    297:329–330
localized random mutagenesis,
    297:333–335
mutant verification, 297:333
partial deletion strains, construction for
    psbB, 297:330–331
site-directed mutagenesis, 297:331–332
strains and maintenance, 297:326–327
vector construction, 297:327–329
transmembrane topology, 297:321–322
CpG island, gene identification, 303:61, 85,
    145, 160
Cpn10, see Chaperonin 10
Cpn60, see Chaperonin 60
CPS, see Carbamoyl phosphate synthase
CRALBP, see Cellular retinaldehyde-
    binding protein
Creatine kinase
    amyloid-β inhibition and assay, 309:748,
        760
    isoenzyme characterization with firefly lu-
        ciferase ATP assay, 305:366–367
Cre recombinase
    recombinase-mediated cassette exchange,
        306:62–63
    selectable marker removal following ho-
        mologous recombination, 306:49–50,
        52
Cresol red, see Proton uptake, rhodopsin ac-
    tivation
CRET, see Chemiluminescence resonance
    energy transfer
Crk1, inhibition with P-ethoxy antisense oli-
    gonucleotides, 313:379, 382
Cross-linking, see also RNA–protein cross-
    linking; RNA–RNA cross-linking
    advantages and disadvantages in subunit
        interaction analysis, 293:15
    calcium-binding protein targets, 316:117
    chromatin complexes, see Cross-linking,
        chromatin complexes
    cyclic GMP phosphodiesterase Pγ to P$\alpha\beta$
        cross-linking reaction, 315:645
        4-(N-maleimido)benzophenone label-
            ing of Pγ mutants, 315:644
        sites of cross-linking, 315:645–646
    GroEL–GroES

buffer selection, 290:258
concentration of protein, 290:256
controls, 290:265–267
cross-linking reagent selection,
    290:256–258
glutaraldehyde, 290:256, 258–259, 262,
    265–268
heterooligomers, identification and
    quantification
    GroEL$_7$–GroES$_7$, 290:262–263, 265
    GroEL$_{14}$–GroES$_7$, 290:262
    GroEL$_{14}$–(GroES$_7$)$_2$, 290:262
homooligomers, identification and
    quantification, 290:260–262
importance for accurate determination
    of equilibrium state, 290:267–268
overview, 290:255
polyacrylamide gel electrophoresis,
    290:259–260
purity requirements, 290:255–256
temperature, 290:259
time of reaction, 290:259
GrpE, quaternary structure analysis,
    290:285–286
heterobifunctional photocleavable re-
    agents, see Cross-linking, heterobi-
    functional photocleavable reagents
quaternary structure determination, limi-
    tations, 290:277
S-modulin–rhodopsin kinase,
    316:19–20
sphingolipids to binding proteins,
    311:570–571
Cross-linking, chromatin complexes
    cyclobutane–pyrimidine dimer mapping
        in chromatin using photolyase
        DNA purification and digestion,
            304:449, 451, 460
        DNA repair, 304:447–448
        high-resolution techniques, 304:461
        indirect end-labeling, 304:449, 451,
            453–454, 457–461
        interpretaton of gels, 304:451, 453
        micrococcal nuclease digestion of chro-
            matin, 304:451, 453
        principle, 304:448–449
        ultraviolet irradiation and photoreacti-
            vation repair, 304:449, 454–455,
            459–460
    cysteine-substituted histones

binding to reconstituted nucleosomes, **304:**240

gene cloning, **304:**234

modification of cysteines with chemical probes
cross-linking reagent, **304:**236–237
reduction of histones, **304:**236

nucleosome dyad mapping with cysteine-substituted H4, 631

purification
cell lysis, **304:**235
chromatography, **304:**235
expression in *Escherichia coli*, **304:**234–235

rationale for construction, **304:**233

site-directed mutagenesis, **304:**233–234

dimethylsulfate cross-linking in replicating chromatin
cross-linking reaction, **304:**532
fixation of nuclei with formaldehyde, **304:**532
labeling newly replicating DNA with bromodeoxyuridine, **304:**531
overview, **304:**530–531
Western blot analysis, **304:**532–533

high mobility group proteins
HMG14/17 immunofractionation with cross-linking, **304:**153–154
HMG-I(Y) cross-linking to nucleosomes *in vivo*, **304:**182–183

mapping nonhistone chromatin-associated factors, **304:**401–402, 407, 415, 418

methylation of DNA and cross-linking to histones
application to non-histone systems, **304:**278
cross-linking reaction mechanism, **304:**518, 524
H2A contacts with DNA, **304:**275
H4 contacts with DNA, **304:**275
histidine linking to DNA, **304:**271
nuclei cross-linking studies, **304:**276–277
overview, **304:**264–265
proteolytic digestion of histones, **304:**271–272
two-dimensional gel electrophoresis analysis

DNA versus protein gels, **304:**265, 267
electrophoresis conditions, **304:**269–271
interpretation of bands, **304:**273, 275
linker histone binding on reconstituted nucleosomes, **304:**529
peptide resolution on gels, **304:**272–273
resolution in first dimension, **304:**267
short preliminary fractionation for separation time estimation, **304:**267–269

nucleosome reconstitution *in vitro* for cross-linking
dialysis conditions, **304:**239–240, 527
radiolabeling of DNA fragments, **304:**237–238
ratios of components, **304:**239

photoactive 5S probe synthesis and label-transfer cross-linking assay, **304:**529–530

protein image hybridization, *see* Protein image hybridization

radical-induced protein–DNA cross-linking
advantages, **304:**517
cross-linking reaction conditions
Ehrlich acites tumor cells, *in vivo*, **304:**520, 522
isolated chromatin, **304:**519–520
protein shadow hybridization assay with phenanthroline–copper, **304:**522
radical production
bleomycin–iron, **304:**516–517
phenanthroline–copper, **304:**516–517
two-dimensional gel electrophoresis analysis, **304:**520

RNA polymerase to DNA, **304:**352

site-directed photo-cross-linking
H2A cross-linking to DNA, **304:**245–246
linker histones and fragments
binding to reconstituted nucleosomes, **304:**528
preparation, **304:**527–528
nucleosome particles with random DNA sequences, **304:**242–244, 246, 526–527

nucleosomes reconstituted with defined
DNA sequences, **304:**241–242,
245–246
overview, **304:**240–241, 524–526
site of cross-linking, identification,
304:244–245, 250–251
4,5′,8-trimethylpsoralen cross-linking of
DNA, *see* 4,5′,8-Trimethylpsoralen
ultraviolet cross-linking of protein–DNA
complexes *in vivo*
applications, **304:**498
chromatin purification
cesium chloride gradient centrifugation, **304:**502–504
nuclei pelleting, **304:**502
protease inhibition, **304:**503
solution preparation, **304:**501–502
starting material, **304:**502
comparison with *in vitro* assays,
304:515
controls, **304:**511–512
density of protein bound to DNA,
304:513–514
Drosophila embryo irradiation,
304:500–501
efficiency of cross-linking, **304:**499,
510–511
immunoprecipitation of complexes,
304:504–507
limitations, **304:**517
overview, **304:**499–500, 618
resolution, **304:**497–498
restriction digestion of chromatin,
304:504–507
Southern blot analysis, **304:**507–510
specificity of binding, **304:**498–499,
511–512
Cross-linking, heterobifunctional photocleavable reagents
actin
caged G-actin complex preparation,
291:173–175
cross-linking with 4-bromomethyl-3-
nitrobenzoic acid succinimide ester, **291:**159–162
labeling with tetramethylrhodamine
iodoacetamide, **291:**157–158
labeling with thiol-reactive caged rhodamine reagent, **291:**170–171, 173
applications, **291:**155–156, 175

bovine serum albumin
thiolation with 2-iminothiolane,
291:157
thiol content, determination, **291:**157
4-bromomethyl-3-nitrobenzoic acid succinimide ester
photochemical properties,
291:159
reactivity, **291:**159
synthesis, **291:**156–159
criteria in design, **291:**162
1-(3,4-dimethoxy-6-nitrophenyl)-2,3-
epoxypropyl chloroformate
reactivity, **291:**168
synthesis, **291:**165
1-(3,4-dimethoxy-6-nitrophenyl)-2,3-
epoxypropyl hydroxide
absorption spectroscopy, **291:**167–168
reactivity with thiol groups, **291:**168
synthesis, **291:**165
N-1-([3,4-dimethoxy-6-nitrophenyl)-
2,3-epoxypropyl)oxy]carbonyl)4-
(*N*,*N*-dimethylamino)pyridinium,
3-oxo-2,5-diphenyl-2,3-dihydro-
thiophene 1,1-dioxide 4-hydroxide
caged G actin complex preparation,
291:173–175
solubility, **291:**169
synthesis, **291:**166
4-[1-(3,4-dimethoxy-6-nitrophenyl)-2,3-
epoxypropyl-1-oxycarbonyloxy]-3-
oxo-2,5-diphenyl-2,3-dihydrothio-
phene 1,1-dioxide
reactivity, **291:**168–169
rhodamine, preparation of thiol-reactive caged compound, **291:**169–170
synthesis, **291:**165–166
1-(3,4-dimethoxy-6-nitrophenyl)-2,3-
epoxypropylsuccinimidyl carbonate,
synthesis, **291:**166–167
1-hydroxyl-1-(3,4-dimethoxy-6-nitro-
phenyl)-2-propene, synthesis,
291:164
Crowding, *see* Molecular crowding
CRP, *see* CytR regulon
CRP1, *see* Chromatin remodeling protein 1
Cryoelectron microscopy, *see* Electron cryomicroscopy
Cryogenic photocalorimetry
efficiency factor, **315:**153

heat production measurement, **315:**149
instrumentation, **315:**148–149
partition functions, **315:**153
rhodopsin photobleaching energetics,
    315:156, 158
rhosopsin photoisomerization
    enthalpy changes, **315:**148, 151, 153
    quantum yields and photochemical pro-
        cesses, **315:**151
α-Crystallin
    chaperone activity assays
        disulfide bond-breaking and substrate
            unfolding
            insulin, **290:**380–382
            α-lactalbumin, **290:**380–383
        heat-denatured substrates
            buffer selection, **290:**377, 379
            substrate selection, **290:**379–380
            temperature response, **290:**376–377
        light scattering measurement, **290:**374,
            376
    genes, **290:**366
    purification from bovine lens
        homogenization of tissue, **290:**367
        lens age, considerations in preparation,
            290:369
        size-exclusion chromatography,
            290:366–367, 369
        storage, **290:**369
    purification of recombinant human pro-
        tein from *Escherichia coli*
        cell growth and lysis, **290:**371
        cloning of αB-crystallin, **290:**370
        column chromatography, **290:**372
        gel filtration elution profiles, **290:**374
        purity assessment, **290:**372, 374
        secondary structure of αA- and αB-crys-
            tallin, **290:**374
    tissue distribution, **290:**366
Crystallization, *see* Two-dimensional crystal-
    lization, membrane proteins
Crystallography, *see* Electron crystallogra-
    phy; X-ray crystallography
CS, *see* Citrate synthase
CSA, *see* Comparative sequence analysis
CsA, *see* Cyclosporin A
CTC, *see* 5-Cyano-2,3-ditoyltetrazolium
    chloride
CTP, *see* Cytidine-5′-triphosphate
Cupferron, nitric oxide release

electrolysis, **301:**235–237, 241
    measurements, **301:**237–239
Cupric sulfate, sphingolipid staining on
    thin-layer chromatography plates,
    312:78
Cut-open oocyte voltage clamp, *see Xeno-
    pus* oocyte
Cyanobacteria circadian rhythm, *see* Circa-
    dian rhythm, cyanobacteria
5-Cyano-2,3-ditoyltetrazolium chloride, re-
    spiratory activity assessment in biofilms
    disinfected biofilms, **310:**176–177
    drinking water biofilms,
        310:174–175
    enzyme-degraded biofilms, **310:**359
    industrial biofilms, **310:**175
    medical implant biofilms, **310:**548
Cyclic ADP-ribose
    caged 1-(2-nitrophenyl)ethyl compound
        calcium release analysis with fluores-
            cent probes, **291:**410–413
        photolysis conditions, **291:**409–410
        synthesis, **291:**405–407
        uncaging efficiency determination,
            291:411
    calcium release in sea urchin eggs,
        291:403–404
    synthesis, **291:**404
Cyclic AMP
    4-coumarinylmethy derivatives
        4-(7-acetoxycoumarinyl)methyladeno-
            sine derivative
            properties, **291:**51–52
            synthesis, **291:**51–54, 58
        biological activities, **291:**62–63
        4-(7-hydroxycoumarinyl)methyladeno-
            sine derivative
            properties, **291:**51–52
            synthesis, **291:**51–53, 56, 58
        4-(7-methoxycoumarinyl)methyladeno-
            sine derivative
            properties, **291:**51–52
            synthesis, **291:**50–54, 58
        photochemical and spectral properties,
            291:60–62
        4-(7-propionyloxycoumarinyl)-
            methyladenosine derivative
            properties, **291:**51–52
            synthesis, **291:**51–53, 55, 58
        purification, **291:**59

silver(I) oxide promoted condensation, **291:**51, 56–58

stability in dark, **291:**60

structural determination, **291:**59

desyl caged compound

  photolysis conditions, **291:**27

  properties and applications, **291:**4–6

  synthesis, **291:**19–20

hydrolysis-resistant analogs, *see* 8-Bromo-cyclic AMP

targets in signaling, **291:**415–416

Cyclic AMP-dependent protein kinase, *see* Protein kinase A

Cyclic GMP

accumulation in nitric oxide production assay, **301:**526–528

circulating current relationship to free cyclic GMP in isolated rod outer segments, **316:**137, 139

guanylate cyclase stimulation by nitric oxide, **301:**393, 395–396

heat of hydrolysis, *see* Microcalorimetry, time-resolved studies of phototransduction

hydrolysis-resistant analogs, *see* 8-Bromo-cyclic GMP

targets in signaling, **291:**415–416

Cyclic GMP phosphodiesterase

abundance in rod outer segment, **316:**253

assays for activity

  overview, **315:**610

  pH assay

    advantages, **315:**657

    incubation and flash conditions, **315:**615–616

    rod outer segment preparation, **315:**612–614

    solutions, **315:**615

  phosphate release colorimetric assay, **315:**611–612, 658–660

  radioassay with tritiated cyclic nucleotides, **315:**657–658

  regulatory effects on GTPase, *see* Transducin

  single rod outer segment assay

    advantages, **315:**751–752

    amphibian rod outer segment preparation, **315:**745–746

    calcium-dependent measurements, **315:**747–748, 750

    cone outer segment measurements, **315:**751

    inhibitor measurements, **315:**744

    overview, **315:**743–745

    pipette solutions, **315:**746

    washout minimization, **315:**751

calcium flux, bleaching adaptation role, **316:**249–252

calorimetric assay, *see* Microcalorimetry, time-resolved studies of phototransduction

cyclic GMP, noncatalytic binding

  dissociation kinetics

    cold chase experiment, **315:**667–668

    concentration jump experiments, **315:**667–668

  functions, **315:**599–600, 647, 671–672

  low-affinity binding assays

    centrifugal separation using silicone oil, **315:**669–671

    centrifugal ultrafiltration, **315:**669

    overview of equilibrium methods, **315:**668–669

  membrane filtration assays for high-affinity sites

    ammonium sulfate stabilization of bound ligand, **315:**665–667

    data analysis, **315:**662–663

    dissociation of bound ligand during separations, **315:**661–662, 665–667

    endogenous cyclic GMP destruction, **315:**663

    equilibrium state determination, **315:**661

    high-throughput screening assay, **315:**665

    incubation conditions, **315:**664

    inhibition of enzyme during assay, **315:**663

    materials, **315:**664

    nonspecific binding quantification, **315:**662, 664

    principles, **315:**660–661

    radioactivity quantification, **315:**665

    validation, **315:**665

  reciprocal relationship with Pγ binding, **315:**647

electrophysiology assay in photoreceptor

bleaching adaptation assays, **316:**243, 245, 249

calculations, **316:**242

rapid solution changes, **316:**239–241

families, **315:**597

genotyping of mouse retinitis pigmentosa models, **316:**523

$G\alpha_t/G\alpha_{i1}$ chimera mutant binding to Pγ subunit

activation assay of phosphodiesterase, **315:**551–552

fluorescence assay

dissociation constant calculation, **315:**550–551

incubation conditions, **315:**550

principle, **315:**549

probes, **315:**549

non-mutant binding, **315:**517–519

GTPase cycle in phototransduction, **315:**524–526, 742

inhibition by Pγ subunit

assays

competition assays, **315:**643–644

fluorescence assay, **315:**642–644

nonfluorescence assay as control for mutation and labeling effects, **315:**641–642

fluorescence labeling of mutant Pγ, **315:**640–641

functional assay of activity removal from rod outer segment membranes, **315:**616

interaction sites with P$\alpha\beta$, **315:**635–636, 645–646

mutant generation for assays

cysteine insertion sites, **315:**636–638

expression and isolation from *Escherichia coli*, **315:**640

polymerase chain reaction site-directed mutagenesis, **315:**639–640

vector design, **315:**637–639

photoreactive cross-linking to P$\alpha\beta$

cross-linking reaction, **315:**645

4-(*N*-maleimido)benzophenone labeling of Pγ mutants, **315:**644

sites of cross-linking, **315:**645–646

kinetic light scattering studies of rhodopsin/transducin interactions, phosphodiesterase signal

kinetic analysis, **315:**346–347

overview, **315:**345–346

phosphodiesterase preparation, **315:**335

membrane interactions, **315:**600–601

phototransduction role, **315:**597–598, 646

preparations from frog retina

determination of content in rod outer segment homogenates

Pγ subunit, **315:**652–654

phosphodiesterase, **315:**651–652

nonactivated phosphodiesterase, **315:**654–655

P$\alpha\beta$ dimers

catalytic dimer preparation with trypsin proteolysis, **315:**656–657

membrane-associated dimers, **315:**656

rod outer segment isolation

homogenization for assays, **315:**650–651

Percoll density gradient centrifugation, **315:**649–650

retina isolation, **315:**648–649

solutions, **315:**648

transducin-activated phosphodiesterase, **315:**655–656

purification from bovine retina

anion-exchange chromatography, **315:**603–604

cyclic GMP affinity chromatography, **315:**606–608

δ subunit, **315:**609–610

extraction, **315:**603

γ subunit, **315:**608–609

immunoaffinity chromatography, **315:**605–606

overview, **315:**602

partial purification from rods and cones, **315:**605

rod outer segment membrane preparation, **315:**602–603

solutions, **315:**602

storage, **315:**604

regulation in phototransduction, overview, **315:**646–647

rescue in mouse *rd* model of retinitis pigmentosa using adenovirus minichromosome

adenovirus minichromosome construction

bacterial expression, **316:**731
helper virus construction,
    316:731–732
purification on cesium chloride gradi-
    ents, **316:**738
transgene packaging, **316:**732, 734
agarose overlay and plaque formation,
    316:735
expression detection
    activity assay, **316:**741–742
    reverse transcriptase–polymerase
        chain reaction, **316:**740
    trypsin activation assay, **316:**742
    Western blot analysis,
        316:740–741
helper virus DNA preparation for
    transfection, **316:**739
lysate preparation following amplifica-
    tion, **316:**737–738
minichromosome/helper virus ratio de-
    termination, **316:**736–737
overview, **316:**725
plaque picking, **316:**735–736
safety, **316:**742–743
serial propagation, **316:**735, 737
transfections
    calcium phosphate, **316:**734–735
    lipofection, **316:**734
wild-type virus expression, **316:**739–740
RGS9 enhancement, **315:**527
S-modulin regulation of activation, **316:**3,
    14
substrate affinity in rods versus cones,
    315:599–600
subunits
    δ subunit functions, **315:**601–602
    γ subunit functions, **315:**601, 616,
        635–636
    overview, **315:**598
β subunit gene
    DNase I footprinting analysis
        binding reaction, **315:**634
        digestion reaction, **315:**634–635
        gel electrophoresis, **315:**635
        overview, **315:**633
        probe preparation, **315:**633–634
    gel mobility shift assay for transcrip-
        tion factor binding
        binding reaction, **315:**630, 633
        gel electrophoresis, **315:**633

nuclear extract preparation,
    315:627–628
overview, **315:**625
probe labeling, **315:**629–630
probe synthesis and purification,
    315:628–629
mutation in autosomal recessive retini-
    tis pigmentosa, **315:**617
reporter gene transfection assays for
    cis element analysis in 5′-flanking
    region
    cell lysate preparation, **315:**523
    β-galactosidase assay, **315:**624–
        625
    luciferase assay, **315:**623–624
    overview, **315:**618–619
    quantification of transcription levels,
        315:625
    transient transfection with calcium
        phosphate, **315:**621–623
    vector construction, **315:**619–621
γ subunit, surface plasmon resonance
    analysis of mutants, **316:**39
tissue distribution of type 6, **315:**598, 651
turnover numbers, **315:**598
zebrafish larva eye, assay
    incubation conditions and product sepa-
        ration, **316:**546–547
    light sensitivity determination,
        316:547
    materials, **316:**546
    principle, **316:**546
    time course of activation, **316:**547, 549
Cyclic nucleotide-gated channel
    antibodies, **294:**699, 703
    calmodulin binding
        calcium concentration buffering,
            315:784
        deletion mutants of channels, electro-
            physiologic studies, **315:**824–825
        dissociation constants of protein and
            peptide, **315:**786
        effects on ligand binding, **315:**785
        fluorescence binding assays with dansyl-
            calmodulin, **315:**823–824, 827
        gel shift assay with channel peptides,
            315:821, 823
        gel-overlay assay with heterologously-
            expressed channels, **315:**819–821
        magnitude of response

channels versus channel peptides,
    315:825, 827
vision versus olfaction, **315:**818,
    827–828
olfactory versus rod channel inhibition,
    315:784
sites of interaction, **315:**783, 785–786,
    797, 818–819
state-dependent modification,
    315:783–784
surface plasma resonance spectroscopy
    buffers and materials, **315:**787
    comparison of binding constants
        with other techniques, **315:**827
    controls, **315:**791
    immobilization by amine coupling,
        315:789
    immobilization by ligand thiol cou-
        pling, **315:**788
    kinetic data interpretation, **315:**794,
        796
    mass transfer limitations, **315:**786
    overview of experimental design,
        315:788
    principle, **315:**786–787
    qualitative analysis, **315:**791–792
    quantitative analysis, **315:**793–794
    recording of titration series,
        315:789–791
    Scatchard plot analysis, **315:**796–797
    sensor chip geometry, **315:**791
calmodulin regulation, **294:**249
cyclic GMP as state-specific modulator
    binding site identification, **315:**777–778
    mutant studies, **315:**777
    opening of channel, **315:**776–777
cysteinyl residues, reagents for state-
    specific modification, **315:**779–780
drug screening with fluorescence assay of
    calcium
    activity of various drugs, **294:**41
    advantages of assay, **294:**46–47
    limitations of assay, **294:**45–46
    pharmacologic validation, **294:**26–27
electrophysiologic data, relating to struc-
    tural models
    dose–response studies, **293:**634–636
    ligand analog data, **293:**637–640
    related channel data, **293:**640–642
energy minimization of molecular models

final minimization, **293:**630–631, 634
initial minimization
    adjustment of initial models,
        293:629–630, 633
    evaluation, **293:**628–629, 632–633
    template construction, **293:**627–628
secondary minimization, **293:**630,
    633–634
expression in *Escherichia coli*, **294:**521
expression in *Xenopus* oocyte for patch
    clamp studies
    artifacts, **315:**774–775
    heteromultimeric channels,
        315:775–776
    robust expression, **315:**773–774
fractional calcium ionic current determi-
    nation
    caged cyclic nucleotide photolysis
        8-bromo cyclic nucleotide features,
            315:803–805
        caging groups, **315:**802–805
        degradation following uncaging,
            315:803
        endogenous cyclic GMP, **315:**803
        solubility limitations, **315:**803–805
    calculations, **315:**800–802, 808, 810–811
    caveats
        assumptions, **315:**811–812
        cytoplasmic equilibrium with pipette
            solution, **315:**812–813
        Fura-2 controls, **315:**813–815
        sampling space uniformity,
            315:816–817
        unidirectionality of calcium flux,
            315:815–816
    data acquisition, **315:**808
    Fura-2 measurements of calcium,
        315:798, 800–802, 807–808, 810–
        811, 813–815
    instrumentation, **315:**805, 807
    overview, **315:**797–798
    recording configuration, **315:**798–799
functions, **294:**246–247
homology modeling
    comparison of olfactory and retina
        channels, **293:**640–642
    conserved residue placement, **293:**624
    deletions in alignment, **293:**623–624
    insertions in alignment, **293:**623–624,
        631

ligand docking, **293:**626, 637–640
limitations, **293:**642
loop regions, initial construction, **293:**624–626
pH setting, **293:**627
reference protein identification, **293:**620–623, 631
site-directed mutagenesis in model testing, **293:**642–643
water modeling, **293:**626–627, 631, 636
ion permeability, **315:**797
nickel binding
affinities, **315:**780
concentration response, **315:**781
controls for study, **315:**782
niflumic acid inclusion in pipette solution, **315:**782
partial agonist studies, **315:**781
state-specific binding, **315:**780–781
patch clamp recordings following covalent activation with 8-*p*-azidophenacylthio-cyclic GMP
conductance states, **315:**770–771
data acquisition, **315:**764–765, 768
functional heterogeneity in binding sites, **315:**767
ligand number assignment, **315:**768–770, 772
macroscopic currents, **315:**764–765, 767
principle, **315:**764
single-channel currents, **315:**768–771
photoaffinity labeling of rod channels
applications with other proteins, **315:**771–772
8-*p*-azidophenacylthio-cyclic GMP
phosphorous-32 labeling, **315:**759–760
purification, **315:**758–759
synthesis, **315:**756–758
cyanogen bromide cleavage of labeled subunits, **315:**761–762
gel electrophoresis
electroelution, **315:**761
peptide fragments, **315:**762
subunit separation, **315:**760–761
labels
labeling reaction, **315:**760
selection, **315:**756–757
partial purification of channels, **315:**760
principle, **315:**756

sequence analysis of labeled peptides, **315:**762
phototransduction role, **315:**755, 772–773, 785, 817–818, 831
purification of bovine rod channel
anion-exchange chromatography, **294:**254, 259
8-bromo-cyclic GMP affinity chromatography, **294:**259
calmodulin affinity chromatography, **294:**252–253, 255–256, 258
dye affinity chromatography, **294:**254, 259
gel electrophoresis in purity analysis, **294:**254–255, 257
immunoaffinity chromatography, **294:**253–254, 258–260
rod outer segment membranes
preparation, **294:**250–252
solubilization, **294:**252
solution preparation, **294:**249–250
purification from bovine rod outer segments with sodium/calcium-potassium exchanger
anion-exchange chromatography, **315:**836–837
calmodulin affinity chromatography, **315:**837–838
dye affinity chromatography, **315:**837
gel electrophoresis and Western blot analysis, **315:**841–843, 846–847
immunoaffinity methods
antibodies, **315:**838–839
immunoaffinity chromatography, **315:**840
immunoprecipitation, **315:**839–840
ricin-agarose chromatography, **315:**838
rod outer segment
isolation, **315:**833–835
solubilization, **315:**835–836
reconstitution
channel activity measurement, **294:**256–258
detergent dialysis, **294:**256
L-*cis*-diltiazem inhibition, **294:**257–258
reconstitution into liposomes
buffers, **315:**844
calcium flux measurements with Arsenazo III, **315:**844–846
incubation conditions, **315:**844

structure, **315:**755, 785, 831–833
subunits of sensory channels,
    294:247–249
tetracaine binding studies, **315:**782–783
Cyclic voltammetry, low molecular weight
    antioxidant quantification
accuracy and reproducibility, **300:**295–296
lipid-soluble compounds
    oxidative damage quantification,
        300:295
    oxidized compounds, **300:**294–295
    rat brain measurements, **300:**293–294
    tissue extraction, **300:**293
principle, **300:**286–288
water-soluble compounds
    biological fluids, **300:**288–289, 291
    tissues, **300:**291–293
Cyclin-dependent kinase
cdc2 phosphorylation of vimentin
    extraction of phosphoprotein, **298:**562
    phosphopeptide antibody preparation,
        298:566–567
    phosphorous-32 labeling
        in vitro, **298:**562–563
        in vivo, **298:**560–562
    phosphorylation site determination,
        298:563, 565–566
inhibitors, **298:**547–548
Cyclin T1
purification of human protein
    glutathione *S*-transferase–Tat affinity
        selection, **306:**359, 363
    TAR affinity chromatography, **306:**360,
        362–363
Tat interactions, **306:**352–353
Cyclobutane–pyrimidine dimer, mapping in
    chromatin using photolyase
DNA purification and digestion, **304:**449,
    451, 460
DNA repair, **304:**447–448
high-resolution techniques, **304:**461
indirect end-labeling, **304:**449, 451, 453–
    454, 457–461
interpretaton of gels, **304:**451, 453
micrococcal nuclease digestion of chroma-
    tin, **304:**451, 453
principle, **304:**448–449
ultraviolet irradiation and photoreactiva-
    tion repair, **304:**449, 454–455,
    459–460

Cyclodextrin, folding enhancement of inclu-
    sion body proteins, **309:**230–231
1,3-Cyclohexadiene-1,4-diethanoate
absorption, **319:**51, 228
disappearance pathways, **319:**52–54
nuclear magnetic resonance, **319:**52
peroxidation, **319:**54–55
physiochemical properties, **319:**52
synthesis of disodium salt, **319:**52, 57–58
trapped singlet oxygen measurement
    deuterium oxide assay, **319:**56
    equations, **319:**55–57
    instrumentation, **319:**57
    ordinary water assay, **319:**56–57
1-Cyclohexyl-3-(2-morpholinoethyl)carbo-
    diimide metho-*p*-toluene sulfonate
applications of RNA modification,
    318:9
primer extension analysis, **318:**16
RNA modification site, **318:**5
Cyclosporin A
inhibition of singlet oxygen-induced
    apoptosis, **319:**316, 321–322, 328–329
plant organelle peptidylprolyl *cis–trans*-
    isomerase binding assay
    chloroplast membrane preparation,
        290:91
    mitochondrial membrane preparation,
        290:91
    radioactive tagging,
        290:91–92
Cyclosporin synthetase, substrate channel-
    ing, **308:**143
Cysteine, caged protein conjugates
caging of *in vitro* translated protein,
    291:128–129
cyclic AMP-dependent protein kinase,
    291:129–130
engineering for site modification, **291:**118,
    134
heavy meromyosin
    functional analysis
        ATPase assay, **291:**110, 114
        in vitro motility assay, **291:**110,
            114–116
    photoactivation
        duration of irradiation, **291:**112–114
        transient kinetics analysis,
            291:113–116
    preparation

4,5-dimethoxy-2-nitrobenzyl bromide
    conjugation, **291:**111–112,
    114–115
    materials, **291:**97
    purification of protein, **291:**111–112
α-hemolysin R104C
    modification with 2-nitrobenzyl group,
        291:121, 128–129
    photolysis conditions, **291:**130–133
kemptide
    modification with 2-nitrobenzyl group,
        291:124–125
    photolysis conditions, **291:**130–132
mass spectrometry analysis, **291:**119, 124
2-nitrobenzyl reagents, efficiency in prepa-
    ration, **291:**120–121
Cysteine scanning mutagenesis, *see* Split re-
    ceptor, mapping tertiary interactions in
    rhodopsin; Substituted-cysteine accessi-
    bility method
Cystic fibrosis, *Pseudomonas aeruginosa* bio-
    film studies with Robbins device
    complement activation, **310:**211–212
    leukocyte response to biofilm,
        310:210–211
    pathogenesis of infection, **310:**210
Cystic fibrosis transmembrane conductance
    regulator
    antibody production, **294:**698, 703
        applications of antibodies, **292:**630, 652
        cross-reactivity characterization,
            292:648, 650–652
        fusion protein immunogen preparation,
            292:630–631, 633–635, 648
        immunization schedule, **292:**635–636
        monoclonal antibody
            hybridoma production, **292:**637
            hybridoma screening, **292:**637–639,
                648–649
            purification, **292:**639
        nucleotide binding domain-1, **292:**696
        polyclonal antibody, **292:**636
    antisense knockdown in renal tubular
        cells, **314:**380, 382
    ATPase activity, **294:**228, 241, 243–244
    ATP release
        assays
            bioluminescence assay, **292:**668–673
            cell culture, **292:**666–667
            patch-clamp analysis, **292:**673–674

radioactive trapping assay, **292:**667–
    668, 672–673
    functions, **292:**665, 674
    chloride channel
        assays, **294:**244–246
        function, **294:**227, 240–241
    copy number, **292:**616
    cystic fibrosis pathogenesis, **292:**652–653,
        675–676, 687, 717, 720–722
    developmental activity, **292:**720
    expression and purification in baculovi-
        rus–insect cell system
        gel filtration chromatography,
            294:233–235
        hydroxyapatatite chromatography,
            294:233, 235
        nickel affinity chromatography,
            294:236–237
        overview, **294:**228–230
        solubilization in detergents
            sodium dodecyl sulfate, **294:**232–233
            sodium pentadecafluorooctanoic
                acid, **294:**235–236
        transfer vector construction,
            294:230–231
    expression in *Saccharomyces cerevisiae*,
        292:626–628
    fluorescence assay of channel activity,
        292:714–716
    full-length complementary DNA assem-
        bly, **292:**617
    gene therapy, *see* Gene therapy, cystic fi-
        brosis transmembrane regulator
    immunolocalization, **292:**646–647
    immunoprecipitation, **292:**645–646
    inhibition by antibodies, **292:**646
    nucleotide binding domain-1
        ATP analog binding, fluorescence stud-
            ies, **292:**681, 683, 694
        circular dichroism spectroscopy of
            wild-type and ΔF508 mutant
            proteins, **292:**681, 684, 694
        glutathione *S*-transferase fusion protein
            expressed in baculovirus–insect
            cell system
            cell purification, **292:**692–693
            fluorescence-activated cell sorting in
                selection, **292:**692
            glutathione affinity chromatography,
                292:693

transfection, **292:**691–692
vector construction, **292:**691–692
verification of expression, **292:**693
glutathione *S*-transferase fusion protein
expressed in *Escherichia coli*
gel purification, **292:**690, 693–694
glutathione affinity chromatography,
292:690
inclusion body isolation and solubilization, **292:**690
overexpression system, **292:**689–690
storage, **292:**690
thrombin cleavage, **292:**690
vector construction, **292:**689
maltose-binding protein fusion protein
overexpression in *Escherichia coli*,
292:680–682
proteolytic cleavage with Factor X_a,
292:686
purification by amylose affinity chromatography, **292:**680–681
transformation, **292:**680
vector construction, **292:**678–680
mutation in cystic fibrosis, **292:**676–677,
688
patch-clamp studies
barrier models, **293:**174, 176
burst analysis, **293:**181–182
DRSCAN analysis of long records,
293:186–194
heterologous expression systems,
293:169–171
ion selectivity, **293:**172–174
macroscopic current recording from ex-
cised patches, **293:**176, 178
pipette preparation, **293:**171–172
pore size estimation, **293:**174–175
protein kinase A activation,
293:180–184
protein kinase C regulation,
293:182–184
protein phosphatase regulation,
293:185–186
rundown in membrane patches,
293:171
Src regulation, **293:**184–185
whole-cell currents, **293:**178–180
phosphorylation, **292:**343, **294:**241, 243
posttranslational processing, **292:**622,
641–642, 650

proteoliposome fusion assay with nystatin, **294:**241–242, 245–246
reconstitution
detergent removal, **294:**237–238
LiDS reconstitution, **294:**239
lipid selection, **294:**238–239
liposome preparation, **294:**240
sodium pentadecafluorooctanoic acid
reconstitution, **294:**239–240
reconstruction in planar bilayers, **294:**301,
303
renal endosome protein
endosomal acidification assay, **292:**655,
658–659
physiological function, **292:**661, 663
renal membrane isolation, **292:**653–654
solubilization and partial purification,
292:656, 659–661, 663
subcellular localization
immunohistochemistry, **292:**657–658,
661
marker enzyme analysis, **292:**655–
657, 661
Western blot analysis, **292:**655–657
sequence homology with other ATP-binding cassette transporters, **292:**109–
110, 162
stable expression in mammalian cells
advantages and disadvantages,
292:622–623
electrophysiologic characterization of
Chinese hamster ovary cells,
292:620, 622
histidine tagging for purification,
292:622
plasmids and selection, **292:**618, 620,
628
structural overview, **293:**169
substituted-cysteine accesibility method
analysis, **293:**124, 130, 135–136, 145
tissue distribution, **292:**653
topology, **292:**687–688
transgenic mice, **292:**628–629
transient expression in mammalian cells,
292:623
Western blot analysis, **292:**641–643, 645,
650
Xenopus oocyte expression after RNA
microinjection, **292:**623–625
Cytidine deaminase

molecular surface point similarity measures in binding energy analysis, **308**:405–407, 409

neural network analysis of binding energies, **308**:414–415, 424, 426

Cytidine-5'-triphosphate

isotopic labeling using glycolysis and pentose phosphate pathway enzymes

enzyme preparation and assays, **317**:23–28

monitoring of reaction mixtures, **317**:32

overview, **317**:19–23

preparation from uridine-5'-triphosphate, **317**:35–36

preparation of uridine-5'-triphosphate from glucose, **317**:29–32

purification from reaction pools, **317**:33–35

phosphorothioate-tagged nucleotide triphosphate analogs for nucleotide analog interference mapping, **317**:108

Cytidine-5'-triphosphate synthetase

assay, **317**:29

purification of recombinant enzyme from *Escherichia coli*, **317**:27–28

Cytochalasin, actin binding mechanisms, **298**:19, 24, 29

Cytochrome, *Rhodobacter*

allele exchange

gene transfer agent, **297**:90

suicide vectors, **297**:89–90

cytochrome *c*

oxidase staining, **297**:85

pathway dependence on growth conditions, **297**:81–82

types and properties, **297**:85–86

DNA transfer

conjugation, **297**:87–88

electroporation, **297**:88–89

fluorescence detection, **297**:85

genome sequencing, **297**:94

mutagenesis

chemical mutagenesis for random mutagenesis, **297**:91

combinatorial mutagenesis, **297**:94

mutator strains for localized mutagenesis, **297**:93

site-directed mutagenesis, **297**:93–94

transposon mutagenesis, **297**:91–92

peroxidase staining, **297**:83–84

prosthetic group and absorption maximum, **297**:82–83

Cytochrome *a*

fluorescence detection, **297**:85

peroxidase staining, **297**:83–84

prosthetic group and absorption maximum, **297**:82–83

Cytochrome *b*

fluorescence detection, **297**:85

nitric oxide inhibition, **301**:197

peroxidase staining, **297**:83–84

prosthetic group and absorption maximum, **297**:82–83

Cytochrome *c*

fluorescence detection, **297**:85

peroxidase staining, **297**:83–84

prosthetic group and absorption maximum, **297**:82–83

reduction assays of nitric oxide synthase isoforms

calculations, **301**:74

incubation conditions, **301**:73

specificity of reduction by isoforms, **301**:72, 74

Rhodobacter, see Cytochrome, *Rhodobacter*

Cytochrome c_6

copper deficiency assays

chemiluminescent heme staining, **297**:274–275

Coomassie stain analysis in native gels, **297**:276

peroxidase assay in gels, **297**:274

rationale, **297**:264, 271

reporter gene assays, **297**:276–277

spectroscopic analysis of soluble extracts, **297**:271–273

Western blot analysis, **297**:273–274

plastocyanin substitution, **297**:264

promoter control of gene expression in transgenic organisms, **297**:279

Cytochrome *d*, prosthetic group and absorption maximum, **297**:82–83

Cytochrome *o*

fluorescence detection, **297**:85

peroxidase staining, **297**:83–84

prosthetic group and absorption maximum, **297**:82–83

Cytochrome oxidase, nitric oxide inhibition, **301**:188, 194–196

Cytochrome P450cam, photoacoustic calorimetry of carbon monoxide dissociation, 295:325–327, 329–330

Cytokine-induced neutrophil chemoattractant, ozone induction in rat lung
  dexamethasone inhibition, 319:560, 562
  electrophoretic mobility shift assay, 319:557–558, 560
  neutrophilia role, 319:554–555
  Northern blot analysis, 319:556–557
  reverse transcriptase polymerase chain reaction, 319:556

Cytomegalovirus retinitis, antisense therapy, 313:31–32, 421, 314:168

Cytoplasmic linker protein-170
  functions, 298:197–198
  immunofluorescence localization, 298:203
  magnetic bead assay for microtubule–membrane interactions, 298:382
  microtubule-binding assay, 298:203, 206
  purification from placenta
    high-speed supernatant preparation, 298:199
    immunoaffinity chromatography, 298:198–200
    microtubule affinity purification and concentration, 298:200–202
  Western blot analysis, 298:202–203

Cytoskeleton, confocal microscopy
  actin detection with green fluorescent protein in multicolor experiments
    rhodamine-conjugated antibody detection, 307:617–618
    rhodamine-conjugated phalloidin detection, 307:616–617
  autofluorescence minimization, 307:184–185
  biopsy sample preparation, 307:179–180
  cell cultures
    monolayer cultures, 307:176–177
    multilayered cultures, 307:177
    three-dimensional cultures, 307:177–179
  fixation
    aldehydes, 307:181–184
    solvents, 307:181
  immunofluorescent labeling, 307:186–187
  mounting, 307:187
  optical sectioning limits, 307:188
  overview, 307:173

permeabilization, 307:180, 183–184
phalloidin, fluorescent labeling, 307:186
software-aided localization and colocalization of cytoskeletal elements, 307:188–189
storage of samples, 307:185
structure, function, and consequences for sample preparation, 307:173–176
triple labeling of actin, vimentin, and tubulin networks, 307:175–176

Cytosolic chaperonin
  cofactors, 290:231
  folding activity assay
    actin reaction conditions, 290:238
    efficiency of in vitro folding, 290:241
    principle, 290:234–235
    product analysis on nondenaturing polyacrylamide gels, 290:239–240
    sensitivity, 290:235, 240
    target proteins
      cloning, 290:236
      labeling, 290:236–238
  purification from rabbit reticulocyte lysate
    anion-exchange chromatography, 290:232–234
    ATP-affinity chromatography, 290:234
    lysate preparation, 290:232
    materials, 290:232
    size-exclusion chromatography, 290:234
  substrates, 290:230–231
  subunits, 290:230

Cytotomography
  data visualization, 307:115–117
  definition, 307:108
  filters, 307:112
  optical setup
    chromatic aberration, 307:113–114
    magnification calibration, 307:113, 118
    optical section number, 307:115
    point-spread function, 307:114–115
    scanner stability, 307:113
    spherical aberration, 307:113–114
  resolution optimization, 307:115
  specimen preparation
    cell surface staining, 307:112, 117
    fixation, 307:110
    living specimens, 307:110–112
    staining, 307:109–110

CytR regulon

CRP–CytR interactions
  CytR competition at *deoP2*, **295:**418,
    420–422
  footprint titration with DNase I
    extended footprint analysis and mul-
      tiple interactions, **295:**413–415
    free energy change evaluation,
      295:408–410
    heterologous cooperativity, thermody-
      namic analysis, **295:**405–408, 410,
      412–413, 423–424
  gel mobility shift assays of binding stoichi-
    ometry, **295:**415–418
  overview of regulation, **295:**403–405
  promoter sequences, **295:**404–405

D

D1
  genes, *see* psbA
  labeling *in vivo* with carbon-14 acetate
    and gel electrophoresis analysis,
    297:196–198
  mutant characterization by absorption
    spectroscopy, **297:**348–350
  site-directed mutagenesis, **297:**226–227
D2, genes, *see psbD*
DAB, *see* 3,3′-Diaminobenzidine tetrahydrochloride
DAN, *see* 2,3-Diaminonaphthalene
DanePy
  fluorescence measurements, **319:**83–84
  leaf studies of singlet oxygen production,
    319:83–85
  singlet oxygen reaction and fluorescence
    decay, **319:**82
  stability, **319:**82–83
DAPI, *see* 4′,6-Diamidino-2-phenylindole
DAT, *see* Dopamine transporter
Database, RNA structure, *see* RIBOWEB
Daunorubicin, photoaffinity labeling of P-
  glycoprotein with analogs, **292:**294, 301
DAZ, see Deleted in azospermia gene
DBMIB, *see* 2,5-Dibromo-3-methyl-6-isopro-
  pyl-*p*-benzoquinone
DCDHF, *see* 2,7-Dichlorodihydrofluo-
  rescein
DCMU, *see* 3-(3′,4′-Dichlorophenyl)-1,1′-
  dimethylurea

DD, *see* Differential display
N-Deacylase, translocation assay of sphingolipids, **312:**572–573
Deconvolution
  epifluorescence microscopy images and
    three-dimensional reconstruction,
    307:618–620, 622
  iterative multilevel thresholding and seg-
    mentation, **307:**271
Deconvolution fluorescence microscopy
  acquisition of data, **310:**182–183
  calibration
    two-dimensional axes, **310:**184
    Z-axis, **310:**187–188
  digitization of images, **310:**183
  epifluorescence microscope, **310:**179
  excitation sources
    continuous wave excitation, **310:**182
    flashlamp, **310:**180–181
    nitrogen laser-pumped dye laser,
      310:181–182
  flash control, **310:**183
  flow cell
    design of flow cell, **310:**190
    flow rate, **310:**191–192
    hydrodynamic property calculations,
      310:190–191
    operation and biofilm establishment,
      310:193–194
  multiple volume combination for
    multifluor imaging, **310:**189
  objective lenses, **310:**179
  overview, **310:**178–179
  rendered volume processing, **310:**188–189
  volume
    deconvolution, **310:**185
    postprocessing, **310:**186–187
    preprocessing, **310:**185
    sampling, **310:**184–185
    three-dimensional rendering, **310:**187
Degenerate polymerase chain reaction, *see*
  Polymerase chain reaction
Degradation, protein
  green fluorescent protein fusion protein
    guidelines
    confirmation of degradation, **302:**71–72
    degradation pathway identification,
      302:72
    fluorogenesis rate versus degradation
      rate, **302:**71

fusion site selection, **302:**72

variant selection of green fluorescent protein, **302:**73

HMG-CoA reductase–green fluorescent protein fusion protein

colony fluorescence analysis, **302:**68–70

constructs, **302:**61–62

cycloheximide effects, **302:**68

expression, **302:**62–63

flow cytometry, **302:**65–68

fluorescence microscopy imaging, **302:**63–65

HRD genes in degradation, **302:**60–61, 63, 67, 72

lovastatin effects on degradation, **302:**66–67

regulation of degradation, **302:**60

immunological assays, **302:**59

reporter fusion proteins, **302:**59–60

selectivity, **302:**58–59

Dehydration stress, *see* Freeze-drying, protein aggregation

Dehydroascorbate, half-life, **300:**78

Dehydroascorbic acid

half-life, **300:**78

high-performance liquid chromatography assays

coulometric electrochemical detection following reduction

hydrodynamic voltammetry plot, **299:**79, 82

instrumentation, **299:**67–68, 78–79

materials, **299:**68, 78

reduction reaction, **299:**75–76, 81

sample preparation, **299:**67, 72–73, 79–82

standards, **299:**74–75

derivatization, **299:**73–74

radiolabeled compound detection, **299:**75–76

recycling to ascorbic acid, *see* Ascorbic acid

stability in solution, **299:**65, 79–80

Dehydrosphingosine, chemical synthesis

column chromatography, **311:**521–522

D-*erythro*-4,5-dehydrosphingosine synthesis, **311:**525–526

materials, **311:**519

overview, **311:**519, 523–525

purity analysis, **311:**522

thin-layer chromatography, **311:**520–522

Delayed-type hypersensitivity, *see also* Skin, ultraviolet protection

effect of proteasome inhibitors, **300:**361–362

Deleted in azospermia gene, repeat identification, **303:**62

Denaturation, *see* Cold denaturation; Fluorescence titration

Dendrimer

antisense oligonucleotide delivery, **313:**352–354

modifications for receptor targeting, **313:**353–354

storage, **313:**354

structure, **313:**352–353

Dendrimer complexes, plasmids, **314:**36

Dendrotoxin

handling, **294:**650–651

potassium channel blocking

electrostatic interactions, **294:**656–658

kinetic analysis, **294:**653–655

mutation analysis of channel residue interactions, **294:**654–656

voltage dependence, **294:**659–670

Xenopus oocyte characterization, **294:**652–653, 657

recombinant protein expression systems for study, **294:**651

types and sequences, **294:**649–650, 661

Dental plaque, *see* Oral biofilm

Dentin, mineral loss measurement by confocal microscopy, **307:**495–496

DEPC, *see* Diethyl pyrocarbonate

Desyl group

carboxylate ester compound caging

γ-aminobutyric acid, **291:**10–11

applications, **291:**13–14, 16

L-glutamate

properties and applications, **291:**10–11, 13–14

synthesis, **291:**25–26, 37–38

photochemistry, **291:**11

phosphate compound caging

cyclic AMP

photolysis conditions, **291:**27

properties and applications, **291:**4–6

synthesis, **291:**19–20

desyl dihydrogen phosphate synthesis, **291:**17–19

limitations, **291:**7
photolysis, **291:**4, 6
photolysis of caged compounds,
    291:26–27
Detergent-insoluble membrane fraction
caveolae component, **312:**488
definition, **312:**488
glycosphingolipid-enriched microdomain
    separation from caveolae
    dot immunoblot analysis, **312:**490–491
    elution of beads, **312:**490
    GM3-enriched microdomain features,
        312:494
    immunobead separation, **312:**489–490
    lipid composition of subfractions from
        mouse melanoma B16 cells,
        312:491, 493–494
    Western blot analysis, **312:**490
preparation, **312:**488–489
Deuterium, solvent exchange in ribonucleo-
    tides, **317:**22–23
Dexamethasone, inhibition of ozone-in-
    duced transcripts
    cytokine-induced neutrophil chemoattrac-
        tant, **319:**560, 562
    nuclear factor-κB, **319:**560
DHFR, *see* Dihydrofolate reductase
DHLA, *see* Dihydrolipoic acid
DHPN, *see* N,N′-Di(2,3-dihydroxypropyl)-
    1,4-naphthalenedipropanamide
DHR, *see* Dihydrorhodamine 123
DHS-1-P, *see* Dihydrosphingosine-1-phos-
    phate phosphatase
Diacylglycerol kinase, sphingolipid identifi-
    cation, **311:**670
Dialysis-related amyloid fibril
circular dichroism analysis of secondary
    structure, **309:**312
electron microscopy, *in situ* observation
    data collection, **309:**504
    homology with connective tissue mi-
        crofibrils, **309:**501
    immunolabeling of β₂-microglobulin
        amyloid, **309:**507–508
    microscope maintenance, **309:**504
    photographic enlargement,
        309:505–506
    systemic microdissection, **309:**506
    ultrastructure and composition,
        309:498–499

kinetic analysis of formation with thio-
    flavin T fluorescence
    assumptions, **309:**311
    binding conditions, **309:**310–311
    first-order kinetic model of extension,
        309:313, 315–316
    initial rate measurement, **309:**316
    polymerization reaction
        composition, **309:**312–313
        initiation and termination, **309:**313
        reaction temperature, **309:**313
    wavelengths for detection, **309:**310
monomer purification, **309:**309–310
polarized light microscopy, **309:**312
protein concentration determination,
    309:310
purification, **309:**306–307
3,3′-Diaminobenzidine tetrahydrochloride,
    peroxidase substrate in amyloid immu-
    nohistochemistry, **309:**22–23
(S)-2,4-Diaminobutyric acid, γ-aminobu-
    tyric acid transporter inhibition,
    296:167–168
2,3-Diaminonaphthalene, S-nitrosothiol
    assay
    interferences, **301:**231–232
    linearity and detection limit, **301:**232
    principle, **301:**228–230
    proteins, **301:**204–205, 207, 211, 233–235
    reaction conditions, **301:**204–205, 207,
        211, 222–224, 228, 230–231
    sample preparation, **301:**232
Di-8-ANEPPS, loading of cells,
    307:131–132
N-(2-Diazo-3,3,3-trifluoropropionyl)-neur-
    aminyllactosylceramide, synthesis,
    311:593–594
(2-Diazo-3,3,3-trigluoropropionyl)[³H]lyso-
    GM2, synthesis, **311:**593
2′,7′-Dichlorofluorescin diacetate
cell loading and leakage, **300:**257–258,
    262
kinetics of oxidation
    data pooling for calculations,
        300:274–275
    dye leakage effects, **300:**269–273
    equations, **300:**264–269
    nonlinear least squares analysis,
        300:263–264
optical properties, **300:**258

oxidant sensitivity, **300**:259–261
structure and oxidation reaction, **300**:257, 259
2,5-Dibromo-3-methyl-6-isopropyl-*p*-benzo-
quinone, oxidation of plastoquinone
pool, **297**:225–226, 229, 231
2,7-Dichlorodihydrofluorescein
extinction coefficient, **301**:368
loading of cells, **301**:369–370
oxidation mechanisms, **301**:370–371
oxidizers, **301**:367–368, 370–371
peroxynitrite, assays of oxidation,
301:371–373
stock solution preparation, **301**:368–369
3-(3′,4′-Dichlorophenyl)-1,1′-dimethylurea
fluorescence analysis of charge recombination rate in photosystem II,
297:344–346
reduction of plastoquinone pool,
297:223–224
DIC microscopy, *see* Differential interference contrast microscopy
N,N′-Di(2,3-dihydroxypropyl)-1,4-naph-
thalenedipropanamide, *see* 1,4-Di-
methylnaphthalene, singlet oxygen carrier derivatives
Diethyldithiocarbamate, *see* Electron paramagnetic resonance
Diethyl pyrocarbonate
applications of RNA modification,
318:8–9
direct detection of RNA modification,
318:13
primer extension analysis, **318**:16
RNA modification site, **318**:5–6
water treatment for RNA studies,
318:119
Differential display, *see also* Messenger
RNA, differential display; Restriction
endonucleolytic analysis of differentially expressed sequences; RNA-
arbitrarily primed polymerase chain reaction; Solid-phase differential display
advantages and limitations, **303**:299
band analysis
cloning, **303**:255
direct sequencing, **303**:249–251
prioritization with reverse Northern
blots, **303**:257–258
strategies, overview, **303**:247–249

cancer research applications, **303**:234
cell and tissue selection, **303**:237
false positives, **303**:235–236
fluorescent differential display
cloning bands of interest, **303**:303, 308
detection system, **303**:300
gel electrophoresis, **303**:302–303, 307–308
overview of protocols, **303**:299–300
polymerase chain reaction, **303**:301–302, 306–307
reverse transcription, **303**:301, 306
RNA isolation, **303**:300, 305
selection of correct clones, **303**:305, 308
gel electrophoresis, **303**:246–247
hematopoietic cells, analysis of complementary DNA, **303**:50–54
high-throughput differential display,
303:257–258
polymerase chain reaction,
303:244–246
principle, **303**:235–236, 298–299
probe generation, **303**:234
reverse transcription reactions
incubation conditions, **303**:243
primer selection, **303**:240, 242
reproducibility, **303**:243–244
RNA isolation
amplification of limited samples,
303:240
commercial kits for preparation,
303:238
DNase I treatment, **303**:239
extraction, **303**:239
guanidine isothiocyanate cushion centrifugation, **303**:237–238
ribonuclease inhibition, **303**:240
sample preservation, **303**:239
verification of differential expression
Northern blot analysis, **303**:252–253
polymerase chain reaction
amplification of probes, **303**:252
quantitative analysis, **303**:255
reverse Northern dot blotting,
303:253–255
ribonuclease protection assay, **303**:255
Differential double-pulse amperometry, hydrogen peroxide measurement
amphetamine effects on brain release,
300:284

area of electrode, calculation,
    300:281–282
calibration of electrode, **300:**282–283
computer-controlled instrument,
    300:279–280
data acquisition, **300:**283
fabrication of platinum disk microelec-
    trode
    reference/counter electrode, **300:**279
    working electrode, **300:**277–278
implantation of microelectrode into rat
    brain, **300:**283
principle, **300:**276–277
shielded box, **300:**280–281
Differential interference contrast micros-
    copy, *see* Digitally-enhanced differen-
    tial interference contrast microscopy;
    Video-enhanced differential interfer-
    ence contrast microscopy
Differential scanning calorimetry
*c*I repressor
    assembly
        melting temperature relationship to
            assembly state, **295:**463–464
        stoichiometry for unbound repres-
            sors, **295:**462–463
    data collection and analysis, **295:**454
    DNA-binding analysis
        oligonucleotide synthesis,
            295:453–454
        right operator site-1 complexes,
            295:458–460, 464–465
    interdomain interaction free energies,
        295:460–461, 465–466
    linker arm transition, **295:**466–467
    unbound repressor denaturation
        isolated N-terminal domain, **295:**457
        mutant repressors , 457–458
        wild-type repressor, **295:**454–457, 462
    liquid–crystalline phase transition in
        model bilayers
        bilayer preparation, **295:**470
        data collection and analysis, **295:**472,
            474–475
        enthalpy change, **295:**474
        entropy change, **295:**474–475
        instrumentation, **295:**470–471
        mixed-chain-length phospholipid stud-
            ies, **295:**487, 490–493
        sample loading, **295:**471–472

van't Hoff enthalpy, **295:**474–475
Diffusion
    limited diffusion of nutrients and oxygen
        in biofilms, **310:**453–454
    modeling in steady-state biofilm systems,
        310:314–315, 318, 419
Diffuse reflectance, *see* Infrared spec-
    troscopy
Digitally-enhanced differential interference
    contrast microscopy
    alignment of microscope, **298:**321,
        323–324
    image acquisition and processing,
        298:327, 329
    simultaneous fluorescence imaging
        advantages, **298:**319
        schematic of apparatus, **298:**322
        yeast
            preparation for imaging, **298:**319–320
            data acquisition and analysis,
                298:329–331
Diglyceride kinase
    ceramide measurement
        data interpretation, **312:**30
        lipid extraction, **312:**24–25
        lipid solubilization with mixed micelles,
            312:25–26
        membrane preparation of enzyme,
            312:27–28
        reaction conditions, **312:**27–28
        standard curve preparation,
            312:26–27
        thin-layer chromatography, **312:**28–30
        validation, **312:**30–31
    substrate specificity, **312:**23
Digoxigenin labeling, *see* Phosphorothioate
    oligodeoxynucleotide
Dihydroceramide, radiolabeling
    [3–^{3}H]dihydroceramide synthesis,
        311:496–497
    [4,5–^{3}H]D-*erythro*-C$_{16}$-dihydroceramide
        synthesis
        D-*erythro*-C$_{16}$-ceramide as starting ma-
            terial, **311:**490–491
        principle, **311:**487–488, 519
    [4,5–^{3}H]D-*erythro*-sphinganine as start-
        ing material, **311:**491
    extractions, **311:**487
    flash column chromatography, **311:**486
    purity analysis, **311:**487

radioactivity determination and detection, **311**:487

thin-layer chromatography, **311**:485–487

Dihydroceramide desaturase

biological significance, **311**:22–23

ceramide formation assay

incubation conditions, **311**:26

microsome preparation from rat liver, **311**:24–25

principle, **311**:23–24

product extraction, **311**:27

substrate

radiolabeled substrate preparation, **311**:25–26

solubilization, **311**:26, 30

thin-layer chromatography

chromatography, **311**:27

sheet preparation, **311**:25

effector sites, **311**:29

localization and topology of complex, **311**:29–30

stereosecificity, **311**:30

water formation assay

incubation conditions, **311**:28

principle, **311**:27

radiolabeled substrate preparation and solubilization, **311**:28, 30

sample preparation, **311**:28

Dihydroethidium

cell loading, **300**:262

kinetics of oxidation

data pooling for calculations, **300**:274–275

dye leakage effects, **300**:269–273

equations, **300**:264–269

nonlinear least squares analysis, **300**:263–264

optical properties, **300**:258

oxidant sensitivity, **300**:259–261

structure and oxidation reaction, **300**:257, 259

Dihydrofolate reductase, *see also* Thymidylate synthase–dihydrofolate reductase

assay in *Escherichia coli* cell-free translation system, **290**:25

dihydrofolate pK_a determination at active site with difference Raman spectroscopy, **308**:193–195

electrospray ionization mass spectrometry of GroEL complex, **290**:299, 301, 307–310

methotrexate-linked expression systems, *see* Chinese Hamster ovary cell; Myeloma cell lines

Dihydrokainate, glutamate transporter inhibition, **296**:176, 180

Dihydrolipoic acid, high-performance liquid chromatography with electrochemical detection

cell culture, **299**:243

chromatography, **299**:241

coulometric detection, principle, **299**:240–241

current–voltage response curve, **299**:242

electrodes, **299**:240

extraction, **299**:243, 245–246

instrumentation, **299**:241

standards, **299**:242–243

Dihydromuscimol, γ-aminobutyric acid transporter inhibition, **296**:168

Dihydropyridine, photoaffinity labeling of P-glycoprotein with azidopine analog, **292**:294, 302, 304, 309–311, 437–438

Dihydrorhod-2, loading of mitochondria, **307**:454–456

Dihydrorhodamine 123

cell leakage, **300**:258–259

cell loading, **300**:262

extinction coefficient, **301**:368

Fe^{3+}-EDTA oxidation, **301**:439–441

hemoprotein oxidation, **301**:441

kinetics of oxidation

data pooling for calculations, **300**:274–275

dye leakage effects, **300**:269–273

equations, **300**:264–269

linear least squares analysis, **300**:263–264

loading of cells, **301**:369–370

optical properties, **300**:258

oxidant sensitivity, **300**:259–261

oxidation assay *in vivo*, **301**:430–433, 435

oxidation mechanisms, **301**:370–371

oxidizers, **301**:367–368, 370–371

peroxynitrite, assays of oxidation, **301**:303–304, 371–373

stock solution preparation, **301**:368–369

structure and oxidation reaction, **300:**257, 259

Dihydrosphingomyelin, *N*-palmitoyldihy-drosphingomyelin synthesis

3-*O*-benzylceramide preparation, **311:**539–541, 544–546

3-*O*-benzyldihydrosphingomyelin preparation and conversion, **311:**547

materials, **311:**543

overview, **311:**536–539, 541–543

phosphatidylating agents, **311:**541–542

thin-layer chromatography, **311:**543

Dihydrosphingosine, *see also* Sphinganine

caged compounds, *see* Caged sphingolipids

metabolism, **312:**304

stereochemistry

biological function, **312:**217

circular dichroism analysis, *see* Circular dichroism, sphingolipids

high-performance liquid chromatography analysis, **312:**227

Dihydrosphingosine 4–hydroxylase

biological significance, **311:**9–10

substrate specificity, **311:**14

yeast enzyme

assay

dihydroceramide as substrate, **311:**14

high-performance liquid chromatography, **311:**12, 14

incubation conditions, **311:**11–12

reagents, **311:**11

thin-layer chromatography, **311:**12

gene, **311:**10

microsome preparation, **311:**10–11

Dihydrosphingosine-1-phosphate lyase, *DPL1* cloning in yeast, **311:**329–330

Dihydrosphingosine-1-phosphate phosphatase

assays

dihydrosphingosine-1–^{32}P as substrate, **311:**228–229

D-*erythro*-dihydrosphingosine-1-phosphate as substrate, **311:**227–228

D-*erythro*-[4,5-^{3}H]dihydrosphingosine-1-phosphate as substrate, **311:**227–228

liver enzyme properties from rat, **311:**224

metabolic tritium labeling of precursors in yeast

dihydrosphingosine, **311:**230–231

palmitic acid, **311:**231

serine, **311:**231

reaction catalyzed, **311:**223

yeast enzymes

functions, **311:**224–225, 231–232

localization of green fluorescent protein fusion proteins, **311:**231–232

properties, **311:**224

YSR2

expression system, **311:**226

gene cloning, **311:**225–226, 329

heat shock response in knockouts, **311:**231

knockout constructs, **311:**229–230

purification of Flag-tagged protein, **311:**226–227

YSR3, gene cloning, **311:**329

α-Dihydrotetrabenazine, *see* Tetrabenazine

2,4′-Dihydroxyacetophenone, synthesis, **291:**24

Dimerizer-regulated gene expression

activation domain, **306:**266

AP1510

advantages over FK1012, **306:**264–265

modifications to prevent FKBP binding, **306:**265

AP1889

FKBP mutant dimerization, **306:**267, 279

storage and handling, **306:**272

DNA-binding domain, **306:**266

FK1012, mechanism of action, **306:**264

gene therapy application, **306:**266, 280

immunogenicity considerations, **306:**266

ligand-binding domain

configuration, **306:**266–267

modification, **306:**267

overview, **306:**263–264

plasmids for expression

target genes, **306:**271

transcription factors, **306:**269, 271

rapamycin as dimerizer, **306:**280–281

stable cell lines

decay of gene expression following dimerizer removal, **306:**278–279

dose response to AP1889, **306:**277

inhibitor response of gene expression,
  306:278
kinetics of expression, **306:**278
target plasmid integration, **306:**276–277
transcription factor fusion generation,
  306:275–276
target gene and promoter, **306:**268–269
transient transfection assays, **306:**272–274
4,5-Dimethoxy-2-nitrobenzyl bromide, *see*
  Cysteine, caged protein conjugates
4,5-Dimethoxy-2-nitrobenzyl carboxylate,
  applications, **291:**31
1-(3,4-Dimethoxy-6-nitrophenyl)-2,3-
  epoxypropyl chloroformate
  reactivity, **291:**168
  synthesis, **291:**165
1-(3,4-Dimethoxy-6-nitrophenyl)-2,3-
  epoxypropyl hydroxide
  absorption spectroscopy, **291:**167–168
  reactivity with thiol groups, **291:**168
  synthesis, **291:**165
N 1 ([3,4-Dimethoxy-6-nitrophcnyl)-2,3-
  epoxypropyl)oxy]carbonyl)4-(*N,N*-
  dimethylamino)pyridinium 3-oxo-2,5-
  diphenyl-2,3-dihydrothiophene 1,1-
  dioxide 4-hydroxide
  caged G-actin complex preparation,
    291:173–175
  solubility, **291:**169
  synthesis, **291:**166
4-[1-(3,4-Dimethoxy-6-nitrophenyl)-2,3-
  epoxypropyl-1-oxycarbonyloxy]-3-oxo-
  2,5-diphenyl-2,3-dihydrothiophene 1,1-
  dioxide
  reactivity, **291:**168–169
  rhodamine, preparation of thiol-reactive
    caged compound, **291:**169–170
  synthesis, **291:**165–166
4′,6-Diamidino-2-phenylindole, total cell
  counting within biofilms, **310:**171–172,
  359
1-(3,4-Dimethoxy-6-nitrophenyl)-2,3-
  epoxypropylsuccinimidyl carbonate,
  synthesis, **291:**166–167
1-(4,5-Dimethoxy-2-nitrophenyl)ethyl carboxylate
  A23187, synthesis, **291:**39–40
  applications, **291:**31
1,4-Dimethylnaphthalene, singlet oxygen
  carrier derivatives

bacteria killing
  cell growth, **319:**200
  N,N′-di(2,3-dihydroxypropyl)-1,4-
    naphthalenedipropanamide
    efficacy, **319:**203, 208
  exposure conditions, **319:**201
  overview, **319:**199
  survival assays, **319:**201–203
cell culture exposure, **319:**17–18
DNA reactions
  deoxyguanosine, **319:**14–15
  plasmid DNA, **319:**15–16
gene expression activation by 1,4-naphtha-
  lenedipropanoate, **319:**18–19
grafting sites for hydrophilic groups,
  319:6–7
history of development, **319:**5–6
hydrophilic functions insensitive to singlet oxygen, **319:**6
peroxidation
  N,N′-di(2,3-dihydroxypropyl)-1,4-
    naphthalenedipropanamide, **319:**12
  reactivity of compounds, **319:**9–10
  singlet oxygen sources
    molybdate-catalyzed disproportiona-
      tion of hydrogen peroxide, **319:**9
    photooxidation, **319:**9
  thermolysis rates, **319:**10–11
physiochemical properties, **319:**7–8, 20
singlet oxygen yield on thermolysis,
  319:198
stability, **319:**5
synthesis
  1-bromoethyl-4-chloromethyl-
    naphthalene, **319:**11
  N,N′-di(2,3-dihydroxypropyl)-1,4-
    naphthalenedipropanamide, **319:**7,
    11–12
  4-methyl-1-naphthalenepropanoate,
    319:6–7
  4-methyl-*N,N,N*-trimethyl-1-
    naphthaleneethanaminium, **319:**7
  1,4-naphthalenedimethanol, **319:**7
  1,4-naphthalenedipropanoate, **319:**7
virus inactivation
  cultivation of virus, **319:**200–201
  cytotoxicity assay, **319:**202–203
  N,N′-di(2,3-dihydroxypropyl)-1,4-
    naphthalenedipropanamide
    efficacy, **319:**204–207

enveloped virus inactivation,
    319:203–206
exposure conditions, **319**:201
nonenveloped virus inactivation,
    319:206
overview, **319**:16–17, 199
survival assays, **319**:201–202
Dimethyl suberimidate, chromatin fixation
    for transmission electron microscopy,
    304:224
Dimethyl sulfate
    antibiotic RNA aptamer modification
        hybridization, **318**:227–228
        modification reaction, **318**:226–229
        polyacrylamide gel electrophoresis anal-
            ysis, **318**:228
        primer extension, **318**:228
        renaturation, **318**:226
    applications of RNA modification,
        318:8–9
    cross-linking reaction mechanism,
        304:518, 524
    direct detection of RNA modification,
        318:12
    DNA–protein cross-linking in replicating
        chromatin
        cross-linking reaction, **304**:532
        fixation of nuclei with formaldehyde,
            304:532
        labeling newly replicating DNA with
            bromodeoxyuridine, **304**:531
        overview, **304**:530–531
        Western blot analysis, **304**:532–533
    footprinting
        genomic footprinting
            cell treatment with dimethyl sulfate,
                304:549, 552
            DNA isolation, **304**:552
            naked DNA controls, **304**:552–553
            human–mouse somatic cell hybrid foot-
                printing
                culture, **304**:574–575
                genomic DNA preparation,
                    304:575–576
                guanine–adenine-specific cleavage re-
                    action, **304**:576–577
                methylation reaction, **304**:575
                naked DNA methylation, **304**:576
        polymerase chain reaction analysis, *see*
            Polymerase chain reaction

transcription factor binding to nucleo-
    somes
    advantages, **304**:332
    binding conditions, **304**:329
    cleavage and separation of methyl-
        ated DNA, **304**:330
    methylation reaction, **304**:330
    principle, **304**:328
    safety, **304**:328
    solution preparation, **304**:329
mapping of RNA crystals for metal-
    binding site engineering,
    317:131–132
nucleosomal DNA methylation and
    cross-linking to histones
    application to non-histone systems,
        304:278
    H2A contacts with DNA, **304**:275
    H4 contacts with DNA, **304**:275
    histidine linking to DNA, **304**:271
    nuclei cross-linking studies,
        304:276–277
    overview, **304**:264–265
    proteolytic digestion of histones,
        304:271–272
    two-dimensional gel electrophoresis
        analysis
        DNA versus protein gels, **304**:265,
            267
        electrophoresis conditions,
            304:269–271
        interpretation of bands, **304**:273, 275
        linker histone binding on reconstitu-
            ted nucleosomes, **304**:529
        peptide resolution on gels,
            304:272–273
        resolution in first dimension, **304**:267
        short preliminary fractionation for
            separation time estimation,
            304:267–269
permeability of cell membranes, **304**:613
primer extension analysis, *in vitro*, **318**:16
protein image hybridization, *see* Protein
    image hybridization
RNA modification sites, **318**:5, 479–480
RNA structure probing *in vivo*
    advantages, **318**:480–481
    borohydride reduction and aniline
        cleavage at 7-methylguanosine,
        318:483–484

cell permeability, **318:**479–480
examples
  18S ribosomal RNA, **318:**489–490
  U2, 486, 489
  U3, 490, 492
primer extenson analysis
  extension reaction, **318:**485
  hybridization, **318:**484–485
  polyacrylamide gel electrophoresis,
   318:486
  primer labeling, **318:**484
  RNA extraction, **318:**482–483
  safety, **318:**481
  theory, **318:**480–481
  troubleshooting, **318:**492–493
  yeast culture treatment, **318:**481–482
uniformity of DNA reactions, **304:**614
Dimethylsulfoxide
carbon-centered radical generation in hydroxyl radical detection, **300:**203
water content and dye loading efficiency,
  307:24–25
3-(4,5-Dimethylthiazol-2-yl)-2,5-diphenyl-
  tetrazolium bromide
amyloid cytotoxicity assay
  amyloid-β treatment results, **309:**719,
   721–723
  5-(3-carboxymethoxyphenyl)-2-(4,5-
   dimethylthiazolyl)-3-(4-sulfo-
   phenyl)tetrazolium cell viability
   assay, **309:**719
  cell culture, **309:**717
  materials, **309:**717
  peptide preparation, **309:**718
  reduction assay, **309:**718–719, 761
  sensitivity of cell lines, **309:**722
amyloid inhibition of reduction, **309:**717,
  722–723
2,2′-Dinitrobenzhydryl carboxylate
applications, **291:**32
N-methyl-D-aspartate, synthesis,
  291:33–35
2,4-Dinitrophenylhydrazine, carbonyl assay
  in proteins
rationale, **309:**138–139
immunohistochemistry, **309:**139–140
spectrophotometric assay, **309:**762–763
Dinoflagellate bioluminescence
circadian rhythm, *see* Circadian rhythm,
  Gonyaulax polyedra bioluminescence

educational applications
advantages, **305:**664–666, 671–672
age-appropriate experiments, **305:**670
circadian rhythms, **305:**667–668
closedc ecosystems, **305:**666–667
commercial kits for study, **305:**679
division rates, **305:**667
immobilized metal affinity chromatography exercise, **305:**675–678
properties of protein for laboratory in-
  struction utilization, **305:**678–679
setup guidelines, **305:**669–670
toxicity experiments, **305:**668–669
World Wide Web sites, **305:**679–680
luciferase, *Gonyaulax polyedra*
assays, **305:**254, 262–264
expression of glutathione *S*-transferase
  fusion protein in *Escherichia coli*
  cleavage of protein, **305:**255–256
  glutathione affinity chromatography,
   305:255
  large-scale preparation, **305:**254
  optimization, **305:**256–258
  small-scale expression, **305:**253–254
  vector construction and transformation, **305:**252–253
gene structure, **305:**250–252
pH regulation, **305:**249–250, 260
reaction, **305:**258
luciferin-binding protein, *Gonyaulax
  polyedra*
assay, **305:**262–264
circadian regulation, **305:**260–261, 273
compartmentalization, **305:**259–260
gene cloning
  buffers, **305:**268, 270
  DNA purification, **305:**270
  inverse polymerase chain reaction,
   305:270–271
  restriction enzyme digestion,
   305:270–271
gene copy number estimation,
  305:271–272
immunolocalization, **305:**275–276
isoforms, **305:**261
luciferin stabilization, **305:**258
messenger RNA
  features, **305:**261, 272
  gene expression regulation,
   305:272–273

mobility shift assays of binding proteins, **305**:274–275

metabolic radiolabeling with sulfur-35 methionine, **305**:272

pH regulation, **305**:260

purification

ammonium sulfate precipitation, **305**:266

anion-exchange chromatography, **305**:266

buffers, **305**:264

cell harvesting and lysis, **305**:264, 266

fluorescence monitoring, **305**:268

gel filtration, **305**:266

hydroxyapatite chromatography, **305**:266

in situ hybridization, **305**:275–276

scintillon

compartmentalization of luminescence proteins, **305**:249, 259–260

purification, **305**:268

DiOC₆(3)

loading of cells, **307**:129–130

organelle staining, **307**:380

Dioxetane luminescence

applications of enzyme detection assays, **305**:411–412

emission characteristics, **305**:74

enzyme substrate incorporation and assay, **305**:381

gas phase decomposition studies, **305**:22

group substitution studies, **305**:20, 23

modeling of unique dioxetanes, **305**:25–26

molecular orbital calculations, **305**:20–21, 24

quantum yields, **305**:19

reporter gene assays

dual luciferase/β-galactosidase reporter gene assay, **305**:444–449

β-galactosidase, **305**:429–436

β-glucuronidase, **305**:437–440

instrumentation, **305**:449–450

overview, **305**:428–429

secreted placental alkaline phosphatase, **305**:440–444

simple versus electron-rich compounds, **305**:24

triplet products, **305**:22–23

Western blot chemiluminescent detection with substrates, **305**:417–427

9,10-Diphenylanthracene, singlet oxygen carrier, **319**:5

1,3-Diphenylisobenzofuran, singlet oxygen probe, **319**:22–24

Direct complementary DNA selection

applications, **303**:112–113, 139–142

artifacts, **303**:156

chromosome-specific library selection

advantages, **303**:121

cloning and characterization of selected library, **303**:138

efficiency monitoring, **303**:119

genomic template

artificial chromosome preparation, **303**:131

biotin labeling, **303**:115–116

concentration, **303**:129, 131

purification, **303**:115–116

hybridization selection on nylon paper disks

advantages, **303**:142–143

blocking/quenching reaction, **303**:133–135, 143

controlling parameters for selection, **303**:139

genomic DNA immobilization, **303**:134–135

human leukocyte antigen genes, **303**:140–141

hybridization of inserts to target genomic DNA, **303**:135

removal of nonspecific bound DNA, **303**:135–136

second cycle of selection, **303**:136

target-bound DNA elution and amplification, **303**:136

normalization of selections, **303**:118–121

polymerase chain reaction

selected complementary DNA, **303**:120

short-fragment random primed library, **303**:132–133

starting complementary DNA pools, **303**:117–118

principle, **303**:113–115, 128–129

screening for novel clones, **303**:138–139

short-fragment random primed library

limitations, **303:**142
preparation, **303:**132
Southern blot analysis, **303:**120–121,
137–138
cloning and analysis, **303:**155–156
complete coding regions, **303:**156
consolidation with other techniques,
303:146–147, 160–161
genomic DNA preparation, **303:**154
hybridizations, **303:**154–155
limitations in gene identification, **303:**146
single-stranded capture
capture reaction, **303:**124–125
efficiency, **303:**126
principle, **303:**121–122
repair reaction, **303:**125–126
single-stranded DNA preparation,
303:122–124
sources of complementary DNA,
303:153–154
Dissociation constant
derivation, **296:**271
determination, **296:**274–275
inhibition constant relationship,
296:271–272
interpretation, **296:**271–272
Disulfide bond formation
catalysis, *see* Protein disulfide isomerase
enzyme assays to detect defects in *Escherichia coli*
alkaline phosphatase, **290:**69–71
MalF–β-galactosidase fusion protein,
290:71
eukaryotes, **290:**47–48
gel separation of intermediates
native gel electrophoresis, **290:**68
sodium dodecyl sulfate gel electrophoresis, **290:**67–68
urea polyacrylamide gel electrophoresis, **290:**68–69 glutathione role,
290:27, 42–43
immunoprecipitation of OmpA,
290:65–67
prokaryotes, **290:**48, 59
redox equilibrium determination with glutathione
determination of reduced/oxidized glutathione ratio, **290:**73
equilibrium constant calculation,
290:72–74

tryptophan fluorescence measurement,
290:72–73
thiol trapping
2-aminoethylmethane thiosulfonate,
290:63
acid quenching, **290:**63–64
double trapping with iodoacetate and
iodoacetamide, **290:**61–62
N-ethylmaleimide, **290:**62–63
iodoacetamide, **290:**61
iodoacetate, **290:**60–61
metabolic labeling and trapping, **290:**65
Dithiolthione, *see* Anethole dithiolthione
DM-nitrophen
calcium uncaging efficiency by two-
photon photolysis, **291:**367–369, 379
cation chelation, **291:**240–241
flash photolysis and serotonin transport
in Retzius P cell synapse
flash photolysis setup, **296:**605–606
microinjection of DM-nitrophen,
296:604–605
photochemistry, **296:**603–604
pre- and postsynaptic current isolation
induced by flash photolysis,
296:602–603
two-electrode voltage clamp, **296:**601
Fourier transform infrared spectroscopy
difference spectra of photolysis,
291:241–242
photolysis reaction, **291:**241
sodium, calcium-exchanger, caged calcium studies
electrophysiological measurements,
291:302–306
photolysis kinetics, **291:**301
rate constant determinations,
291:305–306
regulatory site studies, **291:**301–302,
304–305
DMN, *see* 1,4-Dimethylnaphthalene
DMNB, *see* 4,5-Dimethoxy-2-nitrobenzyl
carboxylate
DMNPE, *see* 1-(4,5-Dimethoxy-2-nitrophe-
nyl)ethyl carboxylate
DMS, *see* Dimethyl sulfate
DMSO, *see* Dimethylsulfoxide
DNA
1,4-dimethylnaphthalene singlet oxygen
carrier derivative reactions

deoxyguanosine, **319:**14–15
plasmid DNA, **319:**15–16
gas chromatography–mass spectrometry,
    see Gas chromatography–mass spec-
    trometry
guanine, singlet oxygen oxidation chemis-
    try, **319:**149–151
HL-60 cell studies of ultraviolet A radia-
    tion, DNA analysis
    gel electrophoresis of fragments,
        319:333–334
    ladder detection, **319:**334–336
    8-oxodeoxyguanosine determination,
        319:333
8-hydroxydeoxyguanosine
    assay in DNA
        acidic hydrolysis, **319:**147–148
        cellular DNA, **319:**151–153
        enzymatic digestion, **319:**147
        extraction and isolation, **319:**145–147
        gas chromatography–mass spectrome-
            try, **319:**148–149, 152
        high-performance liquid chromatog-
            raphy, **319:**148, 152
        phosphorous-32 postlabeling,
            319:152–153
    singlet oxygen role in formation,
        319:144–145
isotopic labeling for kinetic isotope effect
    studies, **308:**316–318
long-range charge transport and oxida-
    tive damage
    assay using DNA assemblies with teth-
        ered photooxidant
        formamidopyrimidine-glycosylase di-
            gestion, **319:**183
        gel electrophoresis, **319:**183–184
        interduplex control reactions,
            319:184
        photolysis, **319:**182
        piperidine cleavage, **319:**183
        solution conditions, **319:**182
    construction of DNA assemblies with
        tethered photooxidant
        circular dichroism, **319:**180
        complementary oligonucleotide syn-
            thesis, **319:**180
        coupling of photooxidant, **319:**175,
            177–178
        design considerations, **319:**170–171

DNA synthesis, **319:**176–177
    enzymatic ligation of long oligonucle-
        otides, **319:**180–182
    ethidium intercalators, **319:**173, 175
    high-performance liquid chromatog-
        raphy, **319:**178–179
    intercalation site determination,
        319:171–173
    mass spectrometry, **319:**179
    rhodium intercalator as photooxi-
        dant, **319:**166–168, 171, 173
    ruthenium complexes, **319:**173
photooxidant types, **319:**168–171
principle, **319:**165–166
stacking dependence, **319:**168
thymine dimer, oxidative repair
    characteristics of repair, **319:**187
    cycloaddition reaction, **319:**184–185
    DNA construct and analysis,
        319:185, 187
    oxidative versus reductive repair,
        319:185
mitochondrial DNA, *see* Mitochondrial
    DNA
mutation spectra
    rationale for study, **319:**159, 164–165
    singlet-oxygen-induced DNA damage
        frequency of mutations, **319:**160
        protection, **319:**161
        transversion induction, **319:**161
    ultraviolet A-induced damage
        aprt gene in Chinese hamster ovary
            cells, **319:**163
        carcinogenicity, **319:**161–162
        frequency of mutations, **319:**160
        human embryonic kidney cells,
            319:162–164
        p53, **319:**163–164
        wavelength dependence, **319:**162
radical cation formation potential of
    bases, **319:**144
repair, *see* DNA repair
singlet oxygen damage, overview,
    319:154–155
transcriptional arrest as signal generator,
    319:257–258
ultraviolet damage
    action spectra
        alkaline elution assay, **319:**441–442,
            444–445

Chinese hamster ovary cell irradiation, **319:**439–440, 442
endonuclease analysis, **319:**438–439, 444
keratinocyte irradiation with sunlight, **319:**440, 442
overview, **319:**360–361, 437–438
wavelength-dependent mechanisms, **319:**442–444
cellular responses and repair, **319:**363, 437
cyclobutane pyrimidine dimers, **319:**360–361, 436
glutathione effects, **319:**361
8-hydroxyguanine formation and effects, **319:**361–362
overview of ultraviolet A damage, **319:**155–157
sequence specificity of ultraviolet A damage with photosensitizer
deuterium oxide effects, **319:**422–423
folic acid as photosensitizer, **319:**422–423
gel electrophoresis of DNA, **319:**420–422
hematoporphyrin as photosensitizer, **319:**422–423
irradiation conditions, **319:**419–420
menadione as photosensitizer, **319:**422–423
principle, **319:**418
radiolabeling of DNA, **319:**419
riboflavin as photosensitizer, **319:**421–423
sequencing of DNA, **319:**420–421
subcloning of *ras* fragments, **319:**418
type I versus type II photosensitization, sequence effects, **319:**423, 425–427
DNA array
analysis of libraries, overview, **303:**216–218
applications, **303:**206, 232–233
apoptotic cells, analysis of differential gene expression, **303:**389
arrayers
history of development, **303:**207
picker/spotter robot, **303:**212–213
arraying and gridding
colony membranes, **303:**214

high-density polymerase chain reaction membranes, **303:**215–216
complementary DNA library construction
cloning, **303:**210
overview, **303:**208–209
plating, **303:**212–213
reverse transcription, **303:**209–210
RNA extraction, **303:**209
tissue collection, **303:**209
complex probe analysis
arraying, **303:**228–229
evolution of technique, **303:**226–227
expression profiling, **303:**228
false positives and negatives, **303:**227–228
hybridization, **303:**230
imaging, **303:**230
oligonucleotide chip application, **303:**229
prerequisites for analysis, **303:**228
probe preparation, **303:**229–230
imaging
analysis of high-density grids, **303:**221–222
charge-coupled device camera, **303:**213–214
fluorescence detection with confocal microscopy, **303:**380–381
libraries, **303:**206, 208, 217, 223–224, 228, 381, 385–388
microarray, *see* DNA microarray
oligonucleotide fingerprinting
data capture and analysis, **303:**225–226
hybridization conditions, **303:**224–225
membrane generation, **303:**224
overlapping sequence identification, **303:**223–224
probes, **303:**222–223
throughput, **303:**224
reduced complexity probes in analysis
array selection, **303:**385–386
confirmation of differential expression, **303:**387–389
generation
overview of approaches, **303:**389–391
RNA-arbitrarily primed polymerase chain reaction, **303:**381, 383–384, 390

statistically primed polymerase chain
   reaction, **303**:383–384, 390
subtraction strategies, **303**:390–391
hybridization conditions, **303**:386
labeling
   fluorescent dyes, **303**:391
   radiolabeling of amplification prod-
      ucts, **303**:384–385
   posthybridization washes, **303**:386–387
   prewashing of filters for hybridization,
      303:386
   quenching of repetitive elements,
      303:391–392
single-copy probes, small-scale analysis
   databases, **303**:217
   high-stringency radioactive hybridiza-
      tion, **303**:218–219
   low-stringency radioactive hybridiza-
      tion, **303**:219
   nonradioactive hybridization,
      303:219–220
   stripping of membranes, **303**:220
   template radiolabeling, **303**:218
two hybrid screening, **303**:233
vector systems, **303**:211–212
DNA bending
HMG-D assay with ligase-mediated circularization, **304**:101, 103, 107, 123–124
HMG-I(Y) assays, **304**:177
DNA chip, *see* Microarray, DNA
DNA gyrase
ATP coupling
   ATPase activation, **308**:65–66
   efficiency, **308**:66–67
   regulation in function, **308**:67
   stoichiometry, **308**:66
DNA–ion binding
effect on hybridization, **295**:364–365
Poisson–Boltzmann analysis
   charge distribution effects on competi-
      tive binding, **295**:376
   competition between two counterion
      species
      cylindrical geometry regime,
         295:374–375
      planar geometry regime, **295**:372–373
   competitive binding isotherm analysis
      competitive binding of several coun-
         terion species, **295**:378
      two counterion species, **295**:377–378

counterion size effects on competitive
      binding, **295**:375–376
single counterion species near highly
      charged surface
   cylindrical geometry regime,
      295:371–372
   Debye screening length, **295**:369
   physical parameters for DNA in wa-
      ter, **295**:369–370
   planar geometry regime, **295**:370–371
   reducing potential, **295**:368
DnaJ
DnaK interaction, **290**:290–291
domains, **290**:286–287
quaternary structure, **290**:288
DnaK
DnaJ interaction, **290**:290–291
domains, **290**:288
GrpE interaction, **290**:289–290
quaternary structure, **290**:288–289
DNA methyltransferase, chromatin structure probing
*Cvi*PI methyltransferase probing
   cell maintenance and growth, **304**:445
   characteristics of methyltransferase,
      304:443
   specificity of methylation, **304**:445
Dam methyltransferase probing, **304**:431–
   433, 617
interpretation of results, **304**:445–446
principle, **304**:431–432, 617
resolution, **304**:447
sensitivity, **304**:446
*Sss*I methyltransferase probing
   bisulfite genomic sequencing and posi-
      tive display of methylcytosine,
      304:433–435, 437–439
   cell maintenance and growth,
      304:435–436
   deamination of DNA, **304**:436–437
   DNA isolation, **304**:436
   expression in yeast, **304**:435
   gel electrophoresis analysis, **304**:441
   polymerase chain reaction, **304**:438–439
   validation of methodology,
      304:441–442
DNA microarray
applications, **303**:179, 181, 204
arrayer, **303**:183–184, 188
controls in hybridization experiments

blocking controls, **303:**199
doped RNAs, **303:**198
normalization controls, **303:**199
RNA quality controls, **303:**198–199
genome-wide expression monitoring
   experimental questions, **303:**182
   throughput, **303:**204–205
   two-color hybridizations, **303:**181
   type I versus type II experiments,
      303:182–183
hybridization conditions, **303:**197, 203
imaging of slides and data reduction,
   303:200–201
overview, **303:**179–180
posthybridization washes, **303:**197–198
postprocessing
   blocking of free lysine groups, **303:**194
   denaturation, **303:**195–196
   rehydration of slides, **303:**193
   ultraviolet cross-linking, **303:**193
printing on slides
   etching of slides, **303:**192–193
   overview, **303:**188–189
   slide preparation, polylysine coating,
      303:187–188
   stability of prints, **303:**193
   tips
      cleaning, **303:**190, 192
      construction, **303:**189–190
      operation, **303:**189
      testing, **303:**190, 192
probe preparation
   labeling with fluorescent nucleotides
      dye selection, **303:**195–196
      Klenow labeling, **303:**197
      random-primed labeling,
         303:196–197
   RNA isolation, **303:**195
scanner, **303:**183–184
target DNA preparation
   choice of target, **303:**184–185
   polymerase chain reaction amplifica-
      tion, **303:**185
   precipitation and clean-up following
      amplification, **303:**185–186
   Sephacryl purification, **303:**186–187
troubleshooting
   high background, **303:**203–204
   locally weak signal, **303:**204
   postprocessing, **303:**201

prehybridized slides, **303:**202
spot shape problems, **303:**202–203
DNA oxidation, *see also* 8-Oxo-2'-deoxy-
   guanosine; 8-Oxoguanine
assays, overview, **300:**168–170, 175–177
effects of dietary antioxidants,
   300:182–184
hydroxyl radical reactions
   adenine reaction, product detection by
      high-performance liquid chroma-
      tography, **300:**199–201
   radical generation by photolysis of mer-
      capto *N*-oxides, **300:**195–197
   xanthine transient radicals, detection
      by electron paramagnetic reso-
      nance, **300:**197–199
nitroimidazole reduction derivatives, interaction with DNA, **300:**314–321
range of values for urinary biomarkers,
   300:168–169, 182
single-cell gel electrophoresis analysis of
   damage
   alkaline unwinding and electrophore-
      sis, **300:**189–190
   apoptotic DNA, **300:**193–194
   cell lysis, **300:**189
   cell types, **300:**185
   comet patterns, evaluation and quanti-
      fication, **300:**192–193
   DNA repair enzyme treatment,
      300:191
   inducers of DNA damage,
      300:185–187
   neutral assay, **300:**191–192
   slide preparation, **300:**187–189
   staining of slides, **300:**190–191
   types of repair products, **300:**168
DNA oxidation, *see* 8-Hydroxyguanosine
DNA photolyase, *see* Cyclobutane–pyrimidine dimer
DNA repair, *see also* Nucleotide excision
   repair
caged radioactive nucleotides in kinetic
   studies
   advantages and disadvantages, **291:**485
   apparatus design for studies,
      291:491–492
   dideoxy ATP
      analytical high-performance liquid
         chromatography, **291:**487–488

high specific activity α-^{32}P compound
    synthesis, **291:**486–487
  purification, **291:**487
  DNA damage and analysis, **291:**490,
    494
  electroporation of compounds,
    291:485–486, 492–493
  panic response, **291:**494–495
  photolysis conditions, **291:**488
  stability in mammalian cells,
    291:489–490
  thymidine triphosphate synthesis,
    291:489
  pathways, overview, **291:**483
  single cell versus multicellular organisms,
    291:484–485
  singlet-oxygen-induced DNA damage,
    319:157–158
  ultraviolet A-induced damage,
    319:158–159
DNA replication
  chromatin, replication-coupled assembly
    cell culture, **304:**77–78
    chromatin preparation with micrococ-
      cal nuclease digestion, **304:**80–81
    dependence on factors, **304:**97, 99
    DNA replication in isolated nuclei
      labeling conditions, **304:**94–95
      overview, **304:**92–93
      reagents and solutions, **304:**93–94
    histone acetylation in isolated nuclei,
      304:95–97
    nuclei preparation from HeLa cells,
      304:79
    overview, **304:**76–77
    polyacrylamide gel electrophoresis
      histone electrophoresis and transfer
        to membranes, **304:**84
      nucleosomal DNA, **304:**82–83
    radiolabeling
      DNA, **304:**78
      histone, **304:**78–79
      liquid scintillation counting,
        304:82
      metabolic inhibitors, **304:**79–80
  dimethylsulfate cross-linking in replicat-
    ing chromatin
    cross-linking reaction, **304:**532
    fixation of nuclei with formaldehyde,
      304:532

labeling newly replicating DNA with
    bromodeoxyuridine, **304:**531
  overview, **304:**530–531
  Western blot analysis, **304:**532–533
  solid-phase nucleus
    assays, **304:**70–72
    characteristics, **304:**64, 68, 70
    restriction digest analysis, **304:**70–71
DNase I footprinting
  advantages and limitations, **316:**600, 603
  chromatin structure probing, applications
    hypersensitive site mapping, **304:**622
    sensitive sites and chromatin conforma-
      tion, **304:**621–622
    transcription factor footprinting,
      304:622
  cyclic GMP phosphodiesterase β subunit
    gene analysis
    binding reaction, **315:**634
    digestion reaction, **315:**634–635
    gel electrophoresis, **315:**635
    overview, **315:**633
    probe preparation, **315:**633–634
  Drosophila embryo chromatin
    buffer preparation, **304:**477, 479
    digestion
      large-scale, **304:**471–473
      small-scale, **304:**473
    extent of digestion, **304:**468
    hsp26 gene mapping, **304:**479–480
    indirect end-labeling, **304:**466–468,
      475–477
    nuclei preparation
      buffers, **304:**464–465
      large-scale preparation, **304:**468–470
      small-scale preparation, **304:**470–471
    overview, **304:**465
  genomic footprinting of cell monolayers,
    304:554–556
  glucocorticoid receptor binding to nucleo-
    somes, analysis
    binding conditions, **304:**327
    digestion conditions and analysis,
      304:328
    principle, **304:**327
    solution preparation, **304:**327
  high-resolution mammalian nucleosome
    positioning, **304:**636–637
  HMG14/17, chromatin interaction analy-
    sis, **304:**134–135

HMG-I(Y) and directional substrate binding, **304:**169–172

immobilized nucleosomes, **304:**754–756

mammalian nucleosome positioning *in vitro*, **304:**628–629

periodicity of cleavage, **304:**378

permeabilization of cell membranes for nuclear entry, **304:**614

polymerase chain reaction analysis, *see* Polymerase chain reaction

principle, **316:**600, 603

retinal transcription factor characterization

binding reaction, **316:**608–609

digestion, **316:**609–610

gel electrophoresis and autoradiography, **316:**610

materials, **316:**605–606

radiolabeled DNA preparation with polymerase chain reaction, **316:**606–608

sequence ladder genration for both strands, **316:**608

rotational positioning analysis of DNA fragment on nucleosomes

DNA sequence selection, **304:**322–323

principle, **304:**322

temporal resolution, **317:**354

yeast chromatin mapping

agarose gel electrophoresis, **304:**369, 371

digestion conditions, **304:**369

inducible promoters, **304:**368

nuclei preparation

cell growth and lysis, **304:**366–367

centrifugation, **304:**367–368

solution preparation, **304:**366

principle, **304:**365

solution preparation, **304:**369

yeast nucleosome positioning assays

digestion conditions, **304:**389, 391–392

indirect end-labeling

digestion reaction, **304:**395

interpretation, **304:**381–382

principle, **304:**380

reagents, **304:**394–395

Southern blot analysis, **304:**395–396

nuclei isolation

cell growth and lysis, **304:**390–391

centrifugation, **304:**391

reagents, **304:**389

primer extension analysis

denaturing polyacrylamide gel electrophoresis, **304:**398–399

interpretation, **304:**385–387

multiple cycle primer extension for mapping low-copy sequences, **304:**398

one-cycle primer extension for mapping multicopy plasmids, **304:**397

principle, **304:**385

reagents, **304:**396–397

spheroplast preparation

cell growth and spheroplasting, **304:**392–393

centrifugation, **304:**393

digestion conditions, **304:**393–394

permeabilization, **304:**393

reagents, **304:**392

DNA supercoiling, psoralen assay

applications, **304:**639

calculations, **304:**641

cell culture, **304:**641–642

cross-linking reaction, **304:**642

DNA topological domain determination

calculations, **304:**647–650

principle, **304:**639–640

extraction of DNA, **304:**642–643

nicking of DNA, **304:**642

principle, **304:**640

single-strand breaks, measurement

eukaryotic cells, **304:**645

specific chromosomal loci, **304:**646

Southern blot analysis of cross-linked DNA, **304:**643–645

DNA topoisomerase, classification, **308:**63

reaction overview, **308:**63–65

subunits, **308:**65

DNA unwinding, temperature-invariant enthalpy, **295:**255–256

DNAzyme

advantages over ribozymes, **317:**140–141

RNA cleavage

analytical scale cleavage, **317:**143–144

crystallography applications, **317:**145

mapping branch points and other modifications, **317:**144–145

materials, **317:**142

preparative-scale cleavage, **317:**142–143

religation of products for modification,
    317:145
specificity, **317**:140–141
terminal products, **317**:141–142
structure, **317**:140–141
DNB, *see* 2,2′-Dinitrobenzhydryl carbox-
    ylate
DNPH, *see* 2,4-Dinitrophenylhydrazine
Domain deletion mutagenesis, photosystem
    II genes in *Synechocystis* 6803,
    297:301–303
Dominant negative suppression
    advantages and disadvantages, **293**:15
    intersubunit interaction, identification of
        regions, **293**:23–24
    principle, **293**:23
Dopamine D$_2$ receptor
    antogonist effects on dopamine clearance
        in brain, **296**:718
    substituted-cysteine accessibility method
        analysis, **296**:331, 335–337, 342–345
Dopamine receptor, antisense inhibition,
    313:29
Dopamine transporter, *see also* Vesicular
    monoamine transporter
    chimeric norepinephrine transporter–
        dopamine transporter
        clones and plasmids, **296**:467–468, 470
        complementary DNA, size analysis,
            296:468–471
        expression analysis by transport and
            binding assays, **296**:469–471
        homology between transporters,
            296:466
        [^{125}I]RTI-55 binding of chimeras with
            attenuated transport function,
            296:470, 473–475
        transformation of bacteria, **296**:468, 470
        transmembrane domain functional anal-
            ysis, **296**:479–480
    chronoamperometry, dopamine clearance
        measurement in brain
        anesthesia effects, **296**:719
        animal surgery
            electrode placement verification,
                296:715
            mouse, **296**:711–712
            rat, **296**:710–711
            rhesus monkey, **296**:712–713
        brain slice analysis *in vitro*, **296**:718

carbon electrode
    calibration, **296**:708
    manufacture, **296**:709–710
    Nafion coating, **296**:710
clearance effects on transporter activ-
    ity, **296**:716–717
data acquisition, **296**:713–715
data analysis, **296**:716
dopamine D$_2$ receptor antagonist ef-
    fects, **296**:718
dopamine injection, **296**:709, 714
inhibition of non-dopamine transport-
    ers, **296**:709
micropipette preparation, **296**:710
potassium chloride injection,
    296:714–715
reproducibility, **296**:716
transporter inhibitor analysis, **296**:716,
    718
cocaine addiction mechanism,
    296:456–458
cocaine antagonist screening
    assay buffer composition, **296**:250–252
    binding assay, **296**:257
    binding site, **296**:248–250
    data analysis, **296**:258
    logistics, **296**:255
    phenyltropane RTI analog screening,
        296:254
    striatal tissue preparation, **296**:256–257
    tube setup, **296**:256
    uptake assay, **296**:258
    uptake versus binding assays
        identical conditions and screening re-
            sults, **296**:253–254
        kinetic considerations, **296**:252–253
cocaine antagonists and disinhibitors
    approaches in identification,
        296:458–459
    classification, **296**:461
    nomenclature, **296**:458
    structure–function analysis, **296**:461–
        462, 464–465
    trihexyphenidyl and derivatives,
        296:461–462, 464
dopamine uptake in homogenized versus
    intact tissue, **296**:690–692
fast scan cyclic voltammetry measure-
    ment of dopamine uptake
    brain region variability, **296**:704–707

evidence of uptake in measurements, **296:**701–702

inhibitors, assessment of action

bupriopion, **296:**703

cocaine, **296:**704

nomifensine, **296:**703–704

intact tissue measurements versus homogenates, **296:**691–693

Michaelis–Menten kinetics

correction for Nafion coating on carbon electrodes, **296:**697–701

inhibition analysis, **296:**703–704, 707

modeling of data, **296:**694–697

GBR analog binding

assays

calculations, **296:**205

freezing effects on membrane preparations, **296:**206–207

incubation conditions, **296:**204

ligand purification, **296:**204–205

medium, ionic composition, **296:**207–211

membrane preparation, **296:**204–206

nonspecific binding, determination, **296:**211–213

temperature of incubation, **296:**211

binding specificity, **296:**213–216

GBR 12783, 203, 210–212, 217–218

GBR 12909, 213, 253–254

GBR 12935, 203, 212–218

sodium dependence, **296:**203, 217–218

genomic clones

characterization

confirmation of positive clones, **296:**502–504

exon-containing subclones, identification, **296:**507

exon/intron junction determination, **296:**508

intron sizing, **296:**507–508

native genomic DNA comparison, **296:**508–509

restriction site mapping, **296:**504–507

identification

library plating and screening, **296:**500

phage DNA isolation, **296:**501

polymerase chain reaction amplification of poorly represented species

Long and Accurate polymerase chain reaction, **296:**509–511

primer selection, **296:**512–513

glycosylation analysis, **296:**219, 227–229, 457

intracerebral microdialysis and dopamine detection

anesthesia, **296:**721–722

apparatus, **296:**726–727

contamination, **296:**723–724

guide cannula implantation, **296:**726

high-performance liquid chromatography with electrochemical detection

apparatus, **296:**728

mobile phase preparation, **296:**728

monoamine standard preparation, **296:**729

peak identification, **296:**730

running conditions, **296:**730

perfusion solution, **296:**722, 727

presentation of data, **296:**725–726

principle, **296:**720–721

probes

calibration, **296:**727

placement verification, **296:**730

recovery performance, **296:**723

types, **296:**722–723

sample collection, **296:**727–728

uptake blockers

administration routes, **296:**725

monoamine effect, evaluation, **296:**724

photoaffinity labeling

[^{125}I]DEEP, **296:**220–221, 225–227

immunoprecipitation, **296:**221–222

labeling conditions, **296:**220–221

proteolysis

gel purification and electroelution, **296:**223–225

in situ proteolysis, **296:**225–227

[^{125}I]RTI, **296:**220–221, 225, 227

phylogenetic analysis, **296:**428–436

reverse transport measurement in *Planorbis* giant dopamine neuron with carbon electrodes

amperometry, **296:**682

amphetamine treatment

effect on quantal size, **296:**685

reverse transport induction, **296:**687–688

cyclic voltammetry, **296:**682–683

data acquisition, **296:**680–682

disk electrode preparation, **296:**679
dissection, **296:**676–677
dopamine exocytosis, measurement,
    296:683–685
intracellular injection, **296:**680–681
resolution of dopamine release by re-
    verse transport and exocytosis,
    296:688
ring electrode preparation, **296:**677–678
testing of electrodes, **296:**679–680
voltammetry, **296:**682, 687–688
rotating disk electrode voltammetry
data acquisition, **296:**671–672
feedback mechanism studies,
    296:674–675
inhibitor analysis, **296:**674
Michaelis–Menten kinetics analysis,
    296:672
multisubstrate analyses, **296:**672–674
transfection of cells with human trans-
    porter, **296:**671
RTI-55 binding, equilibration kinetics,
    296:253
site-directed mutagenesis
glycosylation sites, **296:**459
transmembrane domains, **296:**460–461,
    465
substituted-cysteine accessibility method
analysis
binding assays, **296:**338–339, 341–342
cysteine modification with methane
    thiosulfonate reagents,
    296:339–340
kinetics of modification, **296:**340
native cysteine residues, identification
    and replacement, **296:**335–337
protection from modification,
    296:340–341
secondary structure determination,
    296:344–345
site-directed mutagenesis and func-
    tional effects, **296:**338, 343–344
stable transfection, **296:**338
transient transfection, **296:**338
transmembrane topology, **296:**345
uptake assays, **296:**339
transcription factor binding sites on gene,
    296:499
vaccinia virus–T7 RNA polymerase ex-
    pression system

Hela cell maintenance, **296:**438–439
infection and transfection, **296:**440–441
optimization of expression,
    296:441–442
overview, **296:**437–438
transport assay, **296:**442
virus preparation, **296:**439–440
WIN 35,248 binding
equilibration kinetics, **296:**253
pH effects on sodium dependency,
    296:250–252
Dot blot hybridization, human parvovirus
    B19 DNA detection with chemilumines-
    cent imaging, **305:**123–125, 128–129
Double-tagging assay, protein–protein inter-
    actions
advantages, **303:**445, 449
bait hybrid generation, **303:**441–442
filter preparation, **303:**443
fish fusion protein construction, **303:**443
limitations, **303:**449–450
p21, identification of binding domains
controls, **303:**447
cyclin binding, **303:**444–445, 447, 449
cyclin-dependent kinase binding,
    303:444–445, 447, 449
functions, **303:**444–445
plasmids, **303:**447
principle, **303:**441
screening, **303:**443–444
Doxorubicin, photoaffinity labeling of P-gly-
    coprotein with analogs, **292:**294, 301
DPL1, see Dihydrosphingosine-1-phosphate
    lyase
Drosophila melanogaster
chromatin accessibility, quantitative mea-
    surements with restriction enzymes
buffer preparation, **304:**495–496
digestion conditions, **304:**489–490, 493,
    495
DNA purification, **304:**493–495
hsp26 gene, **304:**490, 492
larva collection, **304:**492–493
nuclei isolation
salivary glands, **304:**494–495
whole larvae, **304:**493
overview, **304:**489–490
chromatin footprinting, *see* Dimethyl sul-
    fate; DNase I footprinting; Methidi-
    umpropyl-EDTA–Fe(II) footprint-

ing; Micrococcal nuclease; Permanganate footprinting; Ultraviolet footprinting

ectopic expression systems, *see Drosophila melanogaster*, ectopic expression systems

embryo production, large-scale, **304**:462

nuclei preparation from embryos
    buffers, **304**:464–465
    large-scale preparation, **304**:468–470
    small-scale preparation, **304**:470–471

P-element transformation, **304**:463–464

Drosophila melanogaster, ectopic expression systems
    applications, **306**:142
    constitutive and uniform expression promoters, **306**:130–132
    overview, **306**:129–130
    regulated expression
        combination of techniques, **306**:139–142
        enhancer piracy, **306**:132
        FLP-mediated recombination for single tissue expression, **306**:140–142
        GAL4 system for spatial regulation, **306**:137–139
        glass multimer reporter, **306**:132
        heat shock promoters
            conditional lethal mutant generation, **306**:134–135
            heat treatment, **306**:135, 137
            limitations, **306**:133–134
            restricted heat treatment of target cells, **306**:134
            vectors, **306**:135

Drosopterin
    Drosophila melanogaster eye color role, **292**:213
    transport, *see* Brown protein; White protein

DRSCAN
    ABFROUT module, **293**:193
    ANALYSIS module, **293**:191
    Disk Operating System interface, **293**:191
    EVNTANAL module, **293**:191–192
    file input/output module, **293**:193
    GRAFOUT module, **293**:194
    graphical interface module, **293**:193–194
    hardware and software requirements, **293**:188–189

HISTROUT module, **293**:192–193
language, **293**:189
long record analysis, **293**:186–188
output of single-channel current records for plotting, **293**:188
pCLAMP compatibility, **293**:186
text module interface module, **293**:189–191

DsbA, disulfide bond formation, **290**:48, 51
DSC, *see* Differential scanning calorimetry
DTH, *see* Delayed-type hypersensitivity
DTX, *see* Dendrotoxin

Dwell time
    simulation, **293**:733
    two-dimensional analysis, *see* Two-dimensional dwell-time analysis

Dye affinity chromatography
    GroEL, **290**:104, 108, 110–113, 118
    heat shock protein 104, **290**:435

Dye loading
    acetoxymethyl ester dyes
        cleavage, **307**:22–23
        loading time, **307**:448
        principle, **307**:447
        solubility and dispersing agents, **307**:447–448
        temperature modulation, **307**:448–449
    calcium dyes
        Calcium Green-1, **307**:125
        Calcium Orange, **307**:125
        dihydrorhod-2 loading of mitochondria, **307**:454–456
        Fluo-3, **307**:22–25, 126
        Fura Red, **307**:125–126
        Indo-1, loading of mitochondria and cytosolic quenching, **307**:453–454
        rhod-2 loading of mitochondria, **307**:453
    compartmentalization in cell, **307**:22
    di-8-ANEPPS, **307**:131–132
    dimethylsulfoxide, water content and dye loading efficiency, **307**:24–25
    DiOC$_6$(3), 129–130
    JC-1, **307**:132
    laser-induced rise of signals, **307**:22–24
    6-methoxy-*N*-ethylquinolinium iodide, **307**:473
    MitoFluor, **307**:130
    overview of techniques, **307**:446–447
    receptor labeling, **307**:132–133

SNARF-1, **307:**127–129
Sodium Green, **307:**127
Dynactin
  assays of activity, **298:**184
  function in dynein activity, **298:**171
  purification, large-scale from bovine
      brain
    anion-exchange chromatography,
        298:179–180
    cation-exchange chromatography,
        298:178
    equipment, **298:**174
    homogenization and high-speed super-
        natant production, **298:**177
    materials, **298:**174, 176–177
    overview, **298:**172–174
    scaling down of preparation, **298:**181
    storage, **298:**181
    sucrose gradient centrifugation,
        298:178–181
    yield, **298:**172
  purification by microtubule affinity
    overview, **298:**183–184
    yield, **298:**172
Dynamic bleaching
  conditions causing bleaching, **291:**184
  critical intensity estimation, **291:**184
  theory, **291:**182–183
Dynamin, green fluorescent protein fusion
    proteins in vesicle trafficking analysis,
    302:45
Dynein
  associated proteins, *see* Dynactin
  caged compounds in photolysis studies
    calcium chelators, **291:**321, 324
    nucleotides, **291:**321, 325
    purity requirements, **291:**321, 324–
        325
    table of properties, **291:**322–323
  cytoplasmic linker protein, *see* Cyto-
      plasmic linker protein-170
  inactivation in cytoplasm, **298:**386
  isolation from *Xenopus* metaphase and in-
      terphase cytosols, **298:**345–346
  mechanism of motion, **291:**308, 310
  photolysis applications, **291:**310
  purification, large-scale from bovine
      brain
    anion-exchange chromatography,
        298:179–180

cation-exchange chromatography,
    298:178
equipment, **298:**174
homogenization and high-speed super-
    natant production, **298:**177
materials, **298:**174, 176–177
overview, **298:**172–174
scaling down of preparation, **298:**181
storage, **298:**181
sucrose gradient centrifugation,
    298:178–181
yield, **298:**172
Western blotting, **298:**367–368, 372

E

E1, *see* Human papillomavirus
E2, *see* Human papillomavirus
EAA receptors, *see* Excitatory amino acid
    receptors
EAM, *see* Encapsidated adenovirus mini-
    chromosome
Early receptor current, rhodopsin
  advantages and limitations of study,
      315:290–292
  cones versus rods, **315:**277
  early receptor potential origins, **315:**270,
      292
  electrical structure–function approach to
      R_2 signal, **315:**286–287, 289
  fusion of cells, **315:**273–274, 278–279
  giant HEK293S recordings
    E134Q mutant, **315:**287, 289, 292
    transient transfection studies,
        315:289–290
    wild-type, **315:**284, 286
  immunocytochemistry, **315:**273, 276–277
  instrumentation
    flash apparatus, **315:**284
    microscope, **315:**281, 283–284
    patch-clamp, **315:**279, 281, 291
  materials for study, **315:**271–272
  polarity, **315:**270
  rationale for study, **315:**270–271, 292–293
  regeneration of pigment, **315:**274, 286–
      287, 291
  rod photoreceptor recordings, **315:**284, 286
  signal enhancement, **315:**277–279
  site-directed mutagenesis, **315:**272

transfection of HEK293S cells, **315:**272–273, 275–278, 290–291

whole-cell recording, **315:**274–275

Eastern blot analysis, exopolysaccharides of biofilms, **310:**411–412

EBNA1, *see oriP* plasmid

Ebselen

antioxidant activity, **299:**294

protection against peroxynitrite damage, **301:**301–302, 304–306, 309–311

Ectopic expression, *see Drosophila melanogaster*, ectopic expression systems

EDC, *see N*-Ethyl-1,3-[3-(diethylamino)propyl]carbodiimide

Edg receptors, *see* Sphingosine 1-phosphate

EDTAcyst(NPS), *see S*-(Nitrophenylsulfenyl)-cysteaminyl-EDTA

Educational demonstrations, luminescence, *see also* Acridine luminescence; Lucigenin; Luminol; Oxalate luminescence

acridan luminescence, **305:**651–652

bioluminescence demonstrations

appeal, **305:**660–661

dinoflagellates, *see* Dinoflagellate bioluminescence

firefly, **305:**666, 672

green fluorescent protein, *see* Green fluorescent protein

chemiluminescence demonstrations

comparison with other light-emitting reactions, **305:**658–659, 665

mechanism explanations, **305:**657–658

setup guidelines, **305:**633–634

electroluminescence, **305:**659

scientist involvement in public education

getting involved, **305:**662–664

need for scientists, **305:**661–662

tetrakis(dimethylamino)ethylene, **305:**646–647

triboluminescence, **305:**658–659

Effective charge

definition, **295:**494

evaluation, *see* Membrane-confined analytical electrophoresis

EF-Tu, *see* Elongation factor-Tu

EGF receptor, *see* Epidermal growth factor receptor

EIA, *see* Electroimmunoassay

Electrical potential, imaging with cationic fluorophores, **302:**346–348, 353–354

Electrochemical detection, *see* Cyclic voltammetry; Differential double-pulse amperometry; Electrochemical DNA biosensor; High-performance liquid chromatography

Electrochemical DNA biosensor

H-DNA detection, **300:**319

instrumentation, **300:**315–316

nitroimadizole interaction with DNA, mechanistic studies, **300:**316–321

Electrochemistry, *see* Carbon electrode; Chronoamperometry; Fast scan cyclic voltammetry; High-performance liquid chromatography with electrochemical detection; Rotating disk electrode voltammetry

Electroimmunoassay, serum amyloid P component, **309:**44–45

Electroluminescence, educational demonstrations, **305:**659

Electromagnetic field

bioelectric effect, biofilms

definition, **310:**657

overview, **310:**657–658

Pseudomonas aeruginosa biofilm

antibiotic treatment, **310:**661, 663

controls, **310:**664–665

culture, **310:**658–659

current dependence of effect, **310:**667–668

electric circuit, **310:**663–664

experimental chamber, **310:**660–661

oxygen intermediates in mechanism, **310:**666–667, 669–670

pH changes with electric current, **310:**665–666, 669

temperature changes with electric current, **310:**666, 669

bone fracture healing, **310:**657

Electron cryomicroscopy, chromatin and nucleosomes

advantages, **304:**191–192, 213

appearance of samples

chromatin, **304:**206–208

compaction, **304:**207–208

linker histone removal, **304:**208

nucleosomes, **304:**206

artifacts

cooling effects, **304:**10

interface effects, **304:**211–212

chromatin preparation
  buffers, **304**:202–203
  glutaraldehyde fixation, **304**:203
  micrococcal nuclease digestion,
    304:202–203
  overview, **304**:201–202
  starting concentrations of DNA,
    304:202
disadvantages, **304**:192, 213
film thickness and specimen distribution,
  304:212
frozen hydrated suspension preparation
  of chromatin
  apparatus, **304**:198–199
  blotting and film thickness,
    304:199–200
  cryogen removal, **304**:199
  overview, **304**:192–193
  plunger automation, **304**:199
  properties of specimen solution,
    304:201
  safety, **304**:201
  water evaporation, **304**:200–201
grids and support films, preparation
  bare grids, **304**:193
  carbon coating of film, **304**:195
  platinum coating of film, **304**:195
  film removal from slide, **304**:195
  grids with perforated film, overview,
    304:193–194
  hole formation, **304**:197–198
  reproducibility, **304**:197–198
  slide coating, **304**:194, 197
imaging conditions
  accelerating voltage, **304**:204
  contrast transfer function correction,
    304:204–205
  cryotransfer device, **304**:203
  initial survey, **304**:204
  temperature equilibration, **304**:203–204
  underfocusing, **304**:204–205
  recording of images, **304**:205–206
  temporal resolution, **304**:213
  three-dimensional images
    reconstruction, **304**:208–210
    tilt angles, **304**:205
Electron cryomicroscopy, structural analysis
  of membrane proteins
  advantages over other structure analysis
    techniques, **294**:135–137, 179

bacteriorhodopsin, **294**:179
cold stage insertion into microscope,
  294:158
cryotransfer of grid to cold stage,
  294:157–158
crystal quality and resolution, **294**:138
crystal transfer to grids, **294**:153
frozen specimen preparation,
  294:153–157
image appraisal, **294**:162–163, 165
image processing, **294**:165–167, 169, 171,
  173–175
low-dose cryomicroscopy
  exposure mode, **294**:160–161
  focus mode, **294**:159–160
  scanning and recording, **294**:161–162
  search mode, **294**:158–159
map interpretation
  3.5 Angström resolution or better,
    294:178
  5 to 10 Angström resolution,
    294:177–178
  10 to 15 Angström resolution, **294**:177
  15 to 30 Angström resolution, **294**:175
microscopes, **294**:138
overview, **294**:138, 140
sample quantity requirements, **294**:137
Electron cryomicroscopy, three-dimensional
  microscopy of ribosomes
  functional complexes, preparation
    antibiotic fixation, **317**:297–298
    A-site binding, PRE complex construc-
      tion, **317**:300–301
    buffers, **317**:299–300
    filter binding assay, **317**:301–302
    gel filtration, **317**:303
    Pi complex, **317**:300
    PRE and POST states, overview,
      317:297–299
    puromycin reaction, **317**:302
    spun column isolation, **317**:303–304
    stoichiometry determinations,
      317:304–305
    sucrose cushion centrifugation,
      317:302–303
    translocation and POST complex con-
      struction, **317**:301
  hurdles in obtaining higher resolution,
    317:291
  image processing

contrast transfer function estimation, **317:**285

micrograph selection and scanning, **317:**285

particle picking and optical density normalization, **317:**286–287

reconstruction of initial three-dimensional density map, **317:**287–288

reconstruction strategies, **317:**283–284

refinement, **317:**288–289

visualization and interpretation of three-dimensional density map, **317:**289–291

isolation and purification of 70S ribosomes

  bacteria growth, **317:**294

  crude 70S isolation, **317:**295

  materials and buffers, **317:**294

  overview and comparison of approaches, **317:**277–279

  reassociated ribosome, advantages and preparation, **317:**293, 296–297

  subunit isolation, **317:**295–296

  tight-coupled 70S isolation, **317:**295

low-dose electron microscopy, **317:**281

optical diffraction, **317:**282

overview, **317:**276–277

plunge-freezing, **317:**281

resolution, **317:**277, 291, 292–293

sample preparation, ligand complexes, **317:**279–281

specimen grid preparation, **317:**281

transfer RNA binding sites

  buffer composition, effects on binding localization, **317:**306–309

  charging state of transfer RNA, effects on binding localization, **317:**306–308

  E site role in protein synthesis, **317:**308–309

  localization, **317:**293, 305–306

  significance of additional sites, **317:**309

Electron crystallography

comparison with other structure elucidation methods, **315:**92–93

crystallization of rhodopsins

  induction by reconstitution

    added lipid reconstitution, **315:**99, 101

    natural lipid reconstitution, **315:**101, 106

    overview, **315:**99

  induction by selective extraction of membranes, **315:**96–97

  rod outer segment purification

    bovine, **315:**96

    frog, **315:**95

crystal requirements, **315:**92–93

galactosylceramide tubules

  crystallization trials, **312:**519

  doping, **312:**516

  materials, **312:**516, 518

  preparation of tubules, **312:**518–519

lipid layer crystallization principles, **312:**515–516

overview, **315:**93–94

resolution, **315:**106

rhodopsin three-dimensional structures, **315:**103–106

screening for two-dimensional crystals

  optical diffraction, **315:**102

  overview pictures, **315:**102

  sample preparation for electron microscopy, **315:**101–102

structures solved with technique, **315:**93

vitrification of specimens for cryomicroscopy, **315:**102

Electron microscopy, *see also* Electron cryomicroscopy; Immunocytochemistry; Transmission electron microscopy

amyloid-β fibrils

  immunoelectron microscopy

    isopeptide bond detection, **309:**182–183

    transglutaminase, **309:**182–183

  instrumentation and data collection, **309:**494

  limitations, **309:**512

  overview, **309:**180–182, 311, 497–498

  peptide preparation and fibril formation, **309:**491–492

  staining with uranyl acetate, **309:**494–496

amyloid fibrils, high-resolution *in situ*

  data collection, **309:**504

  immunolabeling and pretreatments, **309:**507–508

  microscope maintenance, **309:**504

  photographic enlargement, **309:**505–506

systemic microdissection, **309**:506
amyloid P component immunolabeling,
    309:508
amyloid ultrastructure comparison,
    309:311, 496–497
Archaeoglobus fulgidus biofilm,
    310:342–343
avidin–biotin–peroxidase complex for
    confocal imaging
  advantages and limitations,
    307:149–151
  confocal microscopy, **307**:145–146
  electron microscopy, **307**:147
  fixation, **307**:145
  incubation with complex, **307**:146
  overview, **307**:137, 145
  plastic embedding, **307**:146–147
  postfixation, **307**:146
  serotonin staining, **307**:147, 149
  tissue sectioning, **307**:145
biotin staining of neurons for confocal
    and electron microscopy
  back filling, **307**:138–139
  immunocytochemical staining, **307**:139
  intracellular labeling, **307**:138
caged fluorescein-labeled tubulin,
    291:352–353, 355–356
Candida biofilms, scanning electron mi-
    croscopy, **310**:652–654
catheter biofilm formation in bladder
    model
  scanning electron microscopy,
    310:498–499
  transmission electron microscopy,
    310:499–500
chaperonins
  data collection, **290**:248–249
  holey carbon film preparation,
    290:246–247
  principle, **290**:246
  specimen preparation, **290**:246–247
  time-resolved microscopy, **290**:247–248
chromosome spread, *see* Chromosome
    spread
correlative fluorescence and electron mi-
    croscopy of cytoskeleton
  cell culture and light microscopy,
    298:571–573
  criteria for success, **298**:570–571
  electron microscopy

actin depletion, **298**:574–575, 578
  critical point drying, **298**:585–588
  extraction, **298**:573–574
  fixation, **298**:580–581, 583
  immunostaining with colloidal gold-
    conjugated antibodies,
    298:583–585
  metal coating, **298**:588–589
  myosin subfragment 1 decoration of
    actin filaments, **298**:578–580
  replica preparation, **298**:589, 591–592
  spatial resolution, **298**:570
crystallography, *see* Electron crystallog-
    raphy
dialysis-related amyloid fibril, *in situ* ob-
    servation
  homology with connective tissue mi-
    crofibrils, **309**:501
  immunolabeling of β_2-microglobulin
    amyloid, **309**:507–508
  ultrastructure and composition,
    309:498–499
digoxigenin-labeled phosphorothioate oli-
    godeoxynucleotides, intracellular dis-
    tribution, **313**:261–262, 264, 267–268
environmental scanning electron micros-
    copy of opaque media biofilms,
    310:259
experimental murine AA amyloid, *in situ*
    observation
  cryofixation and freeze substitution of
    samples, **309**:499
  homology with connective tissue mi-
    crofibrils, **309**:501
  immunolabeling of amyloid A, **309**:507
  induction of amyloidosis, **309**:502–503
  isolated fibrils
    isolation, **309**:508–509
    sample preparation, **309**:509
    structure comparison with *in situ* fi-
      brils, **309**:502
  thin section preparation, **309**:503
  ultrastructure and composition,
    309:498–499
FtsZ
  carbon-coated copper grid preparation,
    298:301–302
  immunoelectron microscopy, **298**:296
  staining and visualization of polymers,
    298:302, 312

glycocalyx preservation in biofilms
  alcian blue
    glutaraldehyde–lysine fixation,
      310:196, 203
    paraformaldehyde–glutaraldehyde–
      lysine fixation, **310:**196, 198–199,
      203
    structure, **310:**195
  cell culture, **310:**197
  fixatives for staphylococcal glycocalyx,
    310:197–199
  importance, **310:**194–195
  lysine effects in fixation, **310:**199, 203
  ruthenium red
    glutaraldehyde–lysine fixation,
      310:196, 203
    paraformaldchyde–glutaraldehyde–
      lysine fixation, **310:**198, 203
    structure, **310:**195
grid preparation for fibrils
  cleaning, **309:**492
  coating
    carbon, **309:**493
    Triafol film, **309:**492–495
    ultrathin carbon, **309:**493
GroEL–GroES complex, **290:**250, 252–
  254, 260
image processing
  alignment and averaging,
    290:249–250
  correspondence analysis and image clas-
    sification, **290:**251
  film, **290:**249
  three-dimensional reconstruction
    tilted images, **290:**251–252
    untilted images containing randomly
      oriented views, **290:**252
immunoelectron microscopy of capsules,
  310:406–407
immunogold fluorescence staining for confocal and electron microscopy
  advantages and limitations,
    307:149–151
  confocal imaging, **307:**140, 142
  correlation of confocal and electron mi-
    croscopy images, **307:**144–145
  electron microscopy, **307:**144
  fixation, **307:**139
  incubation with immunogold-conju-
    gated streptavidin, **307:**140

  microtomy, confocal microscopy guid-
    ance, **307:**143–144
  overview, **307:**137–138
  permeabilization, **307:**140, 151
  plastic embedding, **307:**143
  postfixation for electron microscopy,
    307:142–143
  sectioning, **307:**139–140
  silver enhancement of gold particles,
    307:142
inclusion bodies, **309:**48
medical device implant biofilms
  scanning electron microscopy, **310:**551
  transmission electron microscopy,
    310:548–551
melanosomes, **298:**365
membrane morphology analysis in cell-
  free frog retinal system, **315:**86, 88
negative staining of chaperonins
  antibody labeling, **290:**244
  data collection, **290:**243–244
  principle, **290:**242–243
  resolution, **290:**242
  scanning transmission electron micros-
    copy and gold labeling, **290:**245
  specimen preparation, **290:**243
  stain types, **290:**243
  structural information, **290:**245
paired helical filaments, **309:**180–182
proteoglycans, immunolabeling, **309:**508
Pseudomonas aeruginosa adhesion to surfaces, **310:**554–555
quaternary structure determination, limitations, **290:**277
reflection confocal imaging of samples,
  307:136–137, 147
resolution compared to light microscopy,
  307:135–136
Rhodobacter sphaeroides chaperonins,
  290:159
Robbins device biofilms, **310:**257
scanning electron microscopy, three-
  dimensional imaging, **307:**497
serum amyloid P, immunoelectron microscopy, **309:**508, 508
Sup35p NM fragment assembly, **309:**671,
  673
thin section preparation of amyloids,
  309:503–504
transglutaminase, immunoelectron micros-

copy with paired helical filaments and tangles, **309**:182–183

γ tubulin ring complex, **298**:224–225

ultrastructural analysis of antisense effects in fetal brain culture

fixation and mounting, **314**:258–259

Reynold's lead citrate staining, **314**:260

uranyl acetate staining, **314**:259–260

urinary tract device encrustation, **310**:558

virus infection identification with confocal microscopy

advantages, **307**:573, 591

confocal microscopy

comparison with conventional histology, **307**:575

green fluorescent protein-tagged virus, **307**:580

immunofluorescence staining, **307**:580, 582

mounting, **307**:582

preembedment sectioning, **307**:578–579

propidium iodide staining, **307**:579–580

scanning, **307**:582–584

specimen selection and fixation, **307**:577–578

trimming of selected tissue specimens, **307**:584, 586

electron microscopy

data acquisition, **307**:590

overview, **307**:586, 588

processing and embedment, **307**:588–589

sectioning, **307**:589–590

principle of preembedment survey, **307**:573–574, 590–591

Electron nuclear double resonance, analysis of photosystem II mutants, **297**:353–356

Electron paramagnetic resonance

amyloid-β oxidation assays

advantages, **309**:754–755

spectral acquisition, **309**:757

spin labels and labeling, **309**:755, 757–758

spin trapping, **309**:758

stearic acid spin probes and lipid peroxidation evaluation, **309**:765–766

iron diethyldithiocarbamate, nitric oxide detection

cell-free assays, **301**:275–276

interaction with nonheme iron nitrosyl complexes and *S*-nitrosothiols, **301**:272–274

mononitrosyl iron complexes

formation and detection in macrophage culture, **301**:275, 276–278

formation and detection in mice, **301**:269, 274–275, 276–278

redox properties, **301**:272

signal, **301**:270–271

solubility, **301**:271–272

structure, **301**:270

synthesis in solution, **301**:274

potential applications, **301**:278–279

spectrometry, **301**:276

iron–sulfur cluster characterization in PsaC mutants, **297**:104–105, 111, 117, 121, 123

metal oxide oxidation product analysis, **319**:31, 35

nitric oxide fluorescent cheletropic traps, **301**:533, 535, 539

oxygen radical detection in photodynamic therapy, **319**:396–397

photodynamic therapy singlet oxygen detection, **319**:393

photosystem II mutants, **297**:351–353, 356

polyphenol response assay, nitric oxide detection, **301**:528–529

rhodopsin activation, conformation studies, **315**:179–180

TEMP, singlet oxygen detection in plants

buffers, **319**:78–79

hydroxylamine conversion prevention, **319**:79–80

leaf infiltration, **319**:81

photoinhibitory treatment, **319**:80–81

reaction conditions, **319**:79

spectra, **319**:79

trapping, **319**:78–79

Tepy, singlet oxygen detection in plants

leaf infiltration, **319**:81

trapping, **319**:78–79

topical antioxidant screening, **319**:478

xanthine transient radicals, detection, **300**:197–199

Electron spin echo envelope modulation,
    analysis of photosystem II mutants,
    297:353–354, 356
Electron spin resonance, *see* Spin-label elec-
    tron spin resonance
  hydroxyl radical scavenging assay
    ascorbic acid contribution, **299:**28, 32,
        34
    electron spin resonance measurements,
        299:29–30
    reaction conditions, **299:**30
    sample preparation, **299:**29
    standards, **299:**30
  nitric oxide radical detection,
        301:497–499
  reactive oxygen species measurement in
    tissues, **300:**256, 298
  superoxide anion generation by nitric ox-
    ide synthase
    advantages in detection, **301:**74, 170
    loop gap resonator measurements,
        301:172–173
    neuronal nitric oxide synthase assays,
        301:173–177
    spin trapping reagents, **301:**171–172
  superoxide anion radical scavenging
    assay
    ascorbic acid contribution, **299:**28, 32,
        34
    calibration, **299:**32
    electron spin resonance measurements,
        299:29–30
    reaction conditions, **299:**28–31
    sample preparation, **299:**29
Electron transfer chemiluminescence
  benzophenone reactions, **305:**8
  cyclic voltammetry, **305:**6
  9,10-diphenylanthracene, **305:**5
  pyrene reactions, **305:**7–8
  S-route, **305:**7
  theory, **305:**4–8
  N-p-toluenesulfonylcarbazole, **305:**6
Electrophoretic mobility, evaluation,
    295:494–495, 502–503
Electrophoretic mobility shift assay
  antisense oligodeoxynucleotide effects on
    ribonucleoprotein formation
    binding conditions, **313:**462–463
    extract preparation, **313:**461
    neuron culture, **313:**460–461

  overview, **313:**459–460
  polyacrylamide gel electrophoresis,
    313:463
  radiolabeling of probe, **313:**461–462
  reagents, **313:**460
calmodulin binding to cyclic nucleotide-
    gated channel peptides, **315:**821, 823
cyclic GMP phosphodiesterase, β subunit
    gene analysis
  binding reaction, **315:**630, 633
  gel electrophoresis, **315:**633
  nuclear extract preparation,
    315:627–628
  overview, **315:**625
  probe
    labeling, **315:**629–630
    synthesis and purification,
        315:628–629
excitatory amino acid receptor transcrip-
    tion factors
  acrylamide gel preparation,
    294:398–399
  electrophoresis, **294:**399, 403
  limited proteolysis of complexes in vali-
    dation, **294:**400
  nuclear extract preparation from
    mouse brain, **294:**397, 402–403
  probe preparation, **294:**397–398
  quantitative analysis, **294:**404
  receptor activation studies *in vivo*,
    294:401–402
  solution preparation, **294:**395–396
  supershift assay, **294:**400, 402
HMG14/17, nucleosome core particle in-
    teraction analysis
  binding conditions, **304:**137
  core particle preparation, **304:**136–137
  electrophoresis conditions, **304:**137
  overview, **304:**134–136
HMG-D
  binding reaction conditions, **304:**123
  cooperativity of DNA binding, analy-
    sis, **304:**113, 115–117
  data extraction and analysis,
    304:124–125
  DNA-binding affinities, **304:**107, 109,
    112
  DNA-binding specificity analysis,
    304:101
  electrophoresis conditions, **304:**119

kinetic analysis of DNA binding,
    304:112–113, 124
limitations, **304:**117–118
protein mutation effects on DNA bind-
    ing, **304:**112
HMG-I(Y)
nucleosome core particle complexes,
    304:175
supershift/ablation assays, **304:**183–185
nuclear factor-κB activation
aminopyropheophorbide, colon carci-
    noma cell photosensitization and
    gene activation, **319:**124–125, 127
binding reaction, **319:**596–598
cold competition assay, **319:**598–599
competion assay, **300:**366
electrophoresis, **300:**366
extract preparation, **319:**593–595
gel preparation, **319:**595–596
I-κB induction analysis, **300:**383–384
incubation conditions, **300:**365–366
materials, **319:**587
nuclear extract preparation, **300:**365
oligonucleotide labeling, **300:**365,
    319:589–591
oxidant dose response and kinetics,
    319:599
oxidant treatment of cells
    adherent cells, **319:**593
    suspended cells, **319:**592
ozone induction in rat lung, **319:**557–
    558, 560
principle, **300:**364–365
proteasome inhibitor analysis,
    300:351–352
sensitivity, **319:**586
stock solution preparation,
    319:587–589
supershift assay, **300:**367–368,
    319:598–599
troubleshooting
    background, **319:**601–602
    high activation variability, **319:**602
    protein content variability in ex-
        tracts, **319:**601
    quantitative analysis, **319:**602
    smeared lanes, **319:**599, 601
tumor necrosis factor activation,
    319:599
ultraviolet A activation, **319:**275–276

Western blot analysis, **319:**598
nucleosome core particle repositioning
    assay, **304:**17–19
principle, **316:**600
retinal transcription factor character-
    ization
binding reaction, **316:**602
gel electrophoresis and autoradiogra-
    phy, **316:**602
materials, **316:**600–601
mutant probes, determination of bind-
    ing site specificity, **316:**603
radiolabeling of probe, **316:**601–602
supershift assay for protein identifica-
    tion, **316:**603
transcription factor binding to nucleo-
    somes
affinity analysis, **304:**325–327
binding conditions, **304:**325
DNA sequence design, **304:**323–325
glucocorticoid receptor binding analy-
    sis, **304:**323–327
limitations, **304:**331–332
principle, **304:**323
triplex-forming oligonucleotide binding
    assays
electrophoresis conditions,
    313:82–83
incubations, **313:**80, 82
principle, **313:**79–80
restriction endonucleases, **313:**80–81
self-association structures, **313:**83
target concentration variation, **313:**83
ultraviolet A-induced transcription factor
    identification, **319:**288–290, 585–586
Electrophysiology, analysis of antisense
    knockdown of ion channels
controls, **314:**307
current delineation, **314:**307, 310
rationale, **314:**305–306
transfected cell identification,
    314:306–307
Electrophysiology, neurotransmitter trans-
    porters
applications, **296:**565
charge movement current
    characteristics, **296:**558–559
    curve fitting, **296:**563–565
    integration of transient current,
        296:563

isolation after induction by voltage and concentration jumps, **296**:559–560

time constant, **296**:560–561

concentration jump

principle, **296**:555–556

multiple-position valve system, **296**:556

separate valves and manifold system, **296**:556

current magnitude comparison to ion channels, **296**:552

cut-open oocyte technique, **296**:572

expression of transporters in *Xenopus* oocytes, **296**:552–554, 590

fluctuation models, **296**:583–585

glutamate transporter

D-aspartate transport and current, **296**:611–613

cell selection for analysis, **296**:618–619

cerebellar slices, preparation from rat, **296**:609–610

excitatory postsynaptic current in slices

comparison of climbing fiber decay with and without uptake inhibition, **296**:613–614, 616

double patch-clamping in decay measurement, **296**:614, 616

glial transport currents in microcultures

advantages of microcultures, **296**:633, 635, 645

cell preparation, **296**:637–639

characteristics of microcultures, **296**:632–633

dish preparation, **296**:636–637

extracellular solutions, **296**:640–641

glutamate transport during synaptic glutamate release, **296**:643–644

inward currents, **296**:643

media, **296**:636

patch clamp recording, **296**:639, 641–642

pipettes and solutions, **296**:639–640

glutamate release, sensing using glutamate-gated channels

blocking control, **296**:631

calibration of sensing cells, **296**:629

communication between releasing and sensing cells, **296**:629–630

selection of sensing cells, **296**:628–629

tonic release detection, **296**:631

intracellular solution, **296**:611

L-glutamate uptake and inhibition of other receptors, **296**:610–611

selective uptake blocking in Purkinje cells, **296**:613

stoichiometry, **296**:617

whole-cell patch clamping

electrode series resistance, **296**:620–621

forward uptake current recording, **296**:619–620

intracellular dialysis in analysis of countertransported substrates, **296**:620

reversed uptake current recording, **296**:621, 627

internal ion substitution, **296**:556–557

leakage current, **296**:551, 558, 579

ligand binding assays, **296**:554

macropatch clamping, **296**:555

norepinephrine transporter

four-state alternating access model, **296**:580–581

norepinephrine-induced currents

fluctuations, **296**:587, 589–590

inward currents, **296**:586–587

stable transfection in human embryonal kidney cells, **296**:578–579, 585–586

stoichiometry, **296**:579

transport rate, **296**:579, 582

two-state model, **296**:581–582

whole-cell voltage clamp, **296**:586

serotonin transporter

current to flux ratios, **296**:591–592

serotonin-induced inward currents, **296**:591

transient expression in *Xenopus* oocytes, **296**:590

transporter cycles per channel opening, **296**:592

simultaneous fluorescence microscopy in *Xenopus* oocytes, **296**:572–573, 576–578

small patch clamping, **296**:555

stoichiometry determination, **296**:565

transport current

estimation, **296**:579–580, 582–583

identification, **296**:557–558

transport-associated current, **296:**551, 557
turnover rate determination, **296:**565
two-electrode voltage clamp, **296:**554–555
two-microelectrode oocyte voltage clamp,
  296:572–573, 576
uptake measurements, **296:**554
Electroporation, caged substrates,
  291:283–285
Electroretinography
  applications, **316:**203–204, 222
  cone activation modeling, **316:**616–618
  cone worse than rod disease, **316:**625
  equal rod and cone disease, **316:**624
  GC-E knockout mouse recordings,
    316:560–564
  paired-flash studies of photoreceptor re-
    sponse
    amplitude–intensity relation, **316:**219,
      221–222
    blink reflex, **316:**209
    b-wave component, **316:**212
    clinical applications, **316:**222
    cone contribution, subtraction,
      316:209–211, 218
    data acquisition, **316:**207–208
    desensitization of rod-mediated probe
      response, **316:**214
    postreceptor negative components,
      316:212–214
    principle, **316:**204–205
    probe flash unit, **316:**208–209
    prospects, **316:**222–223
    time course of derived response,
      316:214, 216, 218–219
  peripherin/rds mutation studies,
    316:621–624
  photoresponses, **316:**612–614
  recording technique, overview, **316:**612
  rhodopsin mutation studies, **316:**618–621
  rod activation modeling, **316:**614–616
  rod worse than cone disease, **316:**624
  spectral sensitivity determination of
    cones, **316:**629–631, 647–648
  waves, **316:**202–203, 613
Electrospray ionization mass spectrometry,
  see also Mass spectrometry
  charge state distribution in protein con-
    formation analysis, **290:**305, 307–309,
    312–313
  GroEL complexes

binding contaminants, **290:**116–117
dihydrofolate reductase complex,
  290:299, 301, 307–310
homooligomers, **290:**311–312
nucleotide complexes, **290:**311–313
hydrogen exchange, measurement in proteins
  complexes of proteins, **290:**305
  conformational sensitivity, **290:**301
  GroEL–protein complexes,
    290:309–310
  nuclear magnetic resonance, compari-
    son of techniques, **290:**309–310
  peak width analysis, **290:**305, 307, 310
  principle, **290:**301–302
  sample preparation, **290:**302, 304–305
instrumentation, **290:**297–299
low-density lipoprotein cholesterol modification by molecular chlorine, characterization, **300:**105
operating conditions for proteins, **290:**299
Electrostatic compliance analysis, interresi-
  due distance measurement,
  294:597–598
Electrostatic interaction, energy for catalysis, **308:**21–22
ELF-97, alkaline phosphatase assay in phosphate-starved biofilms, **310:**175–176
ELISA, *see* Enzyme-linked immunosorbent
  assay
ELO2, cloning in yeast, **311:**330–331
ELO3, cloning in yeast, **311:**330–331
Elongation factor-3
  evolutionary relationship to ATP-binding
    cassette transporter, **292:**115
  yeast protein YEF3 functions, **292:**153
Elongation factor-Tu
  GTP hydrolysis dynamics, *see* $G_{i\alpha 1}$
  imperfect energetic coupling, *see* Kinetic
    proofreading
EM, *see* Electron microscopy
EMSA, *see* Electrophoretic mobility shift
  assay
Enamel
  mineral gain measurement by confocal
    microscopy
    autofluorescence measurement,
      307:494–495
    data acquisition and analysis,
      307:492–493

overview, **307:**491–492
remineralization regimen, **307:**492
validation, **307:**493–494
mineral loss measurement by confocal microscopy
data acquisition and analysis,
307:487–489
dyes, **307:**486, 489–490
fluoride effects, **307:**490
lesion formation, **307:**487
reflection microscopy, **307:**491
sample preparation, **307:**486–487
validation, **307:**488–491
transversal microradiography, **307:**485,
487–489, 492–494
Enamine luminescence, theory, **305:**35–36
Encapsidated adenovirus minichromosome
cyclic GMP phosphodiesterase rescue in
mouse *rd* model of retinitis pigmentosa
adenovirus minichromosome construction
bacterial expression, **316:**731
helper virus construction,
316:731–732
purification on cesium chloride gradients, **316:**738
transgene packaging, **316:**732, 734
agarose overlay and plaque formation,
316:735
expression detection
activity assay, **316:**741–742
reverse transcriptase–polymerase
chain reaction, **316:**740
trypsin activation assay, **316:**742
Western blot analysis, **316:**740–741
helper virus DNA preparation for
transfection, **316:**739
lysate preparation following amplification, **316:**737–738
minichromosome/helper virus ratio determination, **316:**736–737
overview, **316:**725
plaque picking, **316:**735–736
safety, **316:**742–743
serial propagation, **316:**735, 737
transfections
calcium phosphate, **316:**734–735
lipofection, **316:**734
wild-type virus expression, **316:**739–740

gene therapy applications, **316:**725
helper virus constructs, overview,
316:729–731
host immune response, **316:**726
plasmid constructs, overview,
316:728–729
replication
origin, **316:**725
requirements, **316:**726–728
Endocarditis
induced genes in *Streptococcus gordoni*
cloning, **310:**114, 116
selection, **310:**114
sequencing, **310:**114–116
pathogenesis, **310:**109–110
rabbit model, **310:**113–114
streptococcal gene expression assay with
in vitro expression technology
pAK36 vector characteristics, **310:**111
principle, **310:**110
reporter gene-fusion strain library construction, **310:**112–113
Endocytosis, green fluorescent protein
assays
Dictyostelium discoideum as model system, **302:**45–46, 49–50
dynamin fusion proteins, **302:**45
expression in phagosomes, **302:**44
fluid-phase tracers for pinocytosis measurement, **302:**47
fluorescence microscopy imaging,
302:46–47
labeled particles, **302:**43–44, 47
multicolor imaging, **302:**50
pH monitoring in endosomes with fluorescent dye, **302:**48
receptor fusion proteins, **302:**44–45
variants of protein, **302:**49–50
Endoglycoceramidase, sphingolipid identification, **311:**669
Endoglycoceramidase, translocation assay
of sphingolipids, **312:**572
Endoplasmic reticulum, aequorin targeting
for calcium measurements, **305:**481
Endoplasmic reticulum
fluorescent probes, **307:**121, 129–130, 380
network formation
chick embryo fibroblast cell-free
system
casein coating of slides, **298:**359

culture and harvesting, **298:**355–356
fluorescence microscopy assay,
  298:358
fractionation, **298:**356
homogenization, **298:**356
microtubule immobilization,
  298:358–359
network formation assay,
  298:356–357
solution preparation, **298:**354–355
variability between preparations,
  298:360
reconstitution *in vitro*, **298:**353
video-enhanced differential interference contrast microscopy, **298:**351,
  357
solute diffusion analysis with green fluorescent protein, **302:**257–258
ENDOR, *see* Electron nuclear double resonance
Endosome, magnetic bead assay for microtubule–membrane interactions
CLIP-170 identification, **298:**382
cytosol, preparation and manipulation,
  298:386–387
endosome preparation and manipulation,
  298:385–386
microtubule affinity matrix on magnetic
  beads, preparation, **298:**383–385
organelle binding to microtubule beads,
  298:387–390
principle, **298:**382–383
Endothelial cell, *see also* Polyphenols;
  Shear stress
immunoglobulin G internalization, observation with confocal microscopy,
  307:349–350
in situ calcium imaging with confocal microscopy
acetylcholine response,
  307:429, 431
data collection, **307:**429
dye loading, **307:**428
quantitative analysis, **307:**430
signal separation from smooth muscle
  cells, **307:**429
tissue preparation, **307:**428
Endothelin-1 receptor, confocal microscopy, **307:**132–133
Endothiapepsin

inhibitor design by rational mutagenesis,
  295:119–122
pepstatin, Gibbs energy of binding,
  295:114–116, 308–309
Energy-dispersive X-ray analysis, urinary
  tract device encrustation, **310:**561–562
Enhanced chemiluminescence assay, tissue
  oxidation susceptibility
applications, **300:**245–246, 252
calculations, **300:**247–249
incubation conditions, **300:**246–247
mitochondrial fraction susceptibility,
  300:251
mitochondrial susceptibility in various tissues, **300:**249, 251
sample preparation
mitochondrial fractions, **300:**246
tissues, **300:**246
tissue classification, **300:**249
Enkephalin, *see* Leucine-enkephalin
Enolase, energy for catalysis,
  308:21–22
Enthalpy change
Gibbs energy of binding
calculation, **295:**302–303, 314
contributions, **295:**101, 104–106,
  295–296
Kirchoff's law and temperature dependence of enthalpy change, **295:**91
liquid–crystalline phase transition in
  model bilayers, **295:**474
measured versus quantum mechanical calculation, **295:**230–232
temperature-invariant enthalpy
DNA unwinding, **295:**255–256
extrapolation using Planck–Benzinger
  thermal work function, **295:**234–
  236, 238–239, 241, 245
hydrogen bonding
carboxylic acid dimerization,
  295:264–269
water, **295:**261–264
pH effects on ribonuclease A stability,
  295:247–248
protein unfolding applications, **295:**251,
  253, 255–261
ribonuclease S′ system, peptide interactions, **295:**248–249, 251
self-associated protein systems,
  295:246–247

Entropy change
  Gibbs energy of binding
    calculation, **295:**303–305, 314
    contributions, **295:**101, 104, 106–107,
      295–296
  liquid–crystalline phase transition in
    model bilayers, **295:**474–475
ENU, *see* Ethylnitrosourea
Enzyme-linked immunosorbent assay
  advanced glycation end products
    hemoglobin assay
      clinical application, **309:**166
      hemolysate preparation from fresh
        blood, **309:**166–167
      hemolysate preparation from frozen
        and packed red blood cells,
        309:167
      incubations and washes, **309:**167–168
    incubations and washes, **309:**164
    low-density lipoprotein assay, **309:**168
    plate preparation, **309:**163–164
    serum assay
      clinical application, **309:**164–165
      proteinase K digestion of samples,
        309:165
  Arp2/3 complex, **298:**44–45
  chaperones in *Escherichia coli* cell-free
    translation system, **290:**21–22, 26
  cytokine induction by amyloid-β fibrils in
    human monocytes
    interleukin-1β, **309:**731–732
    materials, **309:**730
    principle, **309:**729–730
    solutions, **309:**730–731
  fumonisins, **311:**372–373
  galactosylceramide, soluble synthetic ana-
    logs, **311:**629–630
  gangliosides, **311:**91
  hybridoma clone screening for anti-glyco-
    sphingolipid monoclonal antibodies,
    312:164–165
  hybridoma screening, **309:**110–111
  8-oxo-2′-deoxyguanosine, **300:**169–170
  protein carbonyl assay
    albumin standards, **300:**107
    antibody incubation conditions, **300:**108
    blanks, **300:**109
    calibration, **300:**108–109
    clinical applications, **300:**110
    nucleic acid interference, **300:**111

    principle, **300:**106–107
    reagents, **300:**107
    sample preparation, **300:**108
    sensitivity and reproducibility, **300:**109
    validation, **300:**109–110
  receptor enzyme-linked immunosorbent
    assay of toxin–glycolipid binding
    antibody incubations and color develop-
      ment, **312:**466
    binding conditions, **312:**465
    blocking, **312:**465
    microtiter plate preparation,
      312:465–466
    overview, **312:**463–465
  rhodopsin fragments, **315:**65–66
  rhodopsin peptide ligand screening
    crude lysates, **315:**402–403
    overview, **315:**399, 402
    purified maltose-binding protein fusion
      proteins, **315:**403–404
  transthyretin amyloid, monoclonal anti-
    body probing of conformation
    cryptic epitope demonstration,
      309:600–601
    direct assay, **309:**595
    sandwich-type assay, **309:**596
Eosinophil peroxidase, singlet oxygen
  chemiluminescence assays
  kinetics, **319:**64
  spectral analysis, **319:**64–65
Epidermal growth factor receptor
  inhibitors, **319:**268
  radiation-induced signaling
    dephosphorylation assay in cell-free
      system
      membrane mix experiment, **319:**264
      membrane purification, **319:**263
      reaction conditions, **319:**263–264
    dephosphorylation assay *in vivo*,
      319:265
    dominant-negative mutants
      gel electrophoresis analysis, **319:**272
      materials, **319:**271–272
      principles of study, **319:**270–271
      transfection, **319:**272
    overview, **319:**258
    phosphorylation assay with Western
      blot analysis, **319:**261–262
Epifluorescence microscopy, *see also* Deconvolution fluorescence microscopy; Fluo-

rescence *in situ* hybridization; Immunofluorescence microscopy
antisense oligonucleotide subcellular distribution analysis, **313:**346, 376, 378
autofluorescence quenching, **310:**170–171
cryosectioning of biofilms, **310:**168–169
dye selection factors, **310:**167–168
enzyme-degraded biofilms, **310:**359
imaging, *see* Epifluorescence microscopy
imaging
medical device implant biofilms
cell counting with acridine orange,
310:545–548
viability assay with 5-cyano-2,3-ditoyl-
tetrazolium chloride, **310:**548
Pseudomonas aeruginosa biofilm, imaging
of green fluorescent protein,
310:50–51
Epifluorescence microscopy imaging
androgen receptor–green fluorescent protein fusion protein, **302:**128–131
Bacillus subtilis studies with green fluorescent protein
agarose pads on slides, **302:**144–145
applications, **302:**140–141, 150–153
cell growth conditions, **302:**142–143
double labeling, **302:**145–146
equipment, **302:**141–142, 146–147
fusion proteins, overview of types,
302:137–139
optimization, **302:**146–149
sample preparation, **302:**143
slide preparation, **302:**143–144
troubleshooting, **302:**149
green fluorescent protein-tagged sarcomeric proteins
cell culture on coverslips, **302:**183–184
data acquisition, **302:**184–185
HMG-CoA reductase–green fluorescent
protein fusion protein degradation
assay, **302:**63–65
microscope selection for green fluorescent protein, **302:**147
microspectrometry, **302:**152
plant gene transcription with green fluorescent protein, **302:**319–321, 325
Epilepsy, γ-aminobutyric acid-A receptor
knockdown studies, **314:**22–23
EPP, *see* Erythropoietic protoporphyria
EPR, *see* Electron paramagnetic resonance

EPS, *see* Exopolysaccharide
Epstein–Barr virus, *see* HEK 293–Epstein–
Barr virus expression system; *ori*P
plasmid
Equal rod and cone disease, electroretinogram studies, **316:**624
Equilibrium constant, derivation for enzymatic reactions, **308:**3
Equilibrium dialysis
partition coefficient determination, proteins in membranes
apparatus, **295:**74
calculations, **295:**75
high-performance liquid chromatography of peptides, **295:**74–75
quaternary structure determination, limitations, **290:**277
ER, *see* Endoplasmic reticulum; Estrogen
receptor
ERC, *see* Early receptor current, rhodopsin
ERG, *see* Electroretinography
Ergodic theorem, *see* Three-dimensional
structure, molecular modeling
ERK, *see* Extracellular signal regulated
kinase
Erythrocyte, chemiluminescence assay of
urine
advantages, **305:**410
clinical applications, **305:**402–403
comparison with automated imaging
assay, **305:**406
emission kinetics, **305:**406
incubation conditions and data collection,
305:405–406
instrumentation, **305:**405
interferences, **305:**406
reagents and solutions, **305:**403–405
Erythropoietic protoporphyria, *see also* Porphyrias
β-carotene treatment, **319:**480–482, 484,
494
cysteine treatment, **319:**482
gene defect, **319:**479
gene therapy
bone marrow targeting, **319:**482
fibroblasts, **319:**483
mouse model, **319:**483–484
peripheral blood stem cells,
319:483–484
laboratory testing, **319:**482

mouse model, antioxidant effects,
319:479–480
singlet oxygen role, **319:**479
Erythropoietin, receptor binding
overview of system, **295:**270, 293
surface plasma resonance biosensor analysis, *see also* Surface plasma resonance biosensor
analyte concentrations, **295:**276
capturing of ligand, **295:**274
data collection, **295:**277
equilibrium dissociation constant,
295:292
flow rate, **295:**275–276
mass transport coefficient,
295:291–292
model fitting, **295:**282–286
sample injection volume, **295:**277
Escherichia coli biofilm
cotton thread culture in rat subcutane-
ousa exudate
exudate preparation, **310:**580–581
inoculation, **310:**571–572
strain, **310:**581
lacZ fusion constructs for surface-acti-
vated gene studies
biofilm, gene expression monitoring,
310:72–73
controls, **310:**70–71
promoter screening, **310:**71–72
planktonic bacteria sensitivity assays
human polymorphonuclear leukocyte
preparation 586
hydrogen peroxide sensitivity assay,
310:588–589
phagocytosis assay, **310:**587–588
preparation, **310:**586
Escherichia coli cell-free translation system
chaperone-free system preparation
enzyme-linked immunosorbent assay of
chaperones, **290:**21–22, 26
incubation conditions, **290:**22–23
materials and reagents, **290:**19–20
size-exclusion chromatography, **290:**21
efficiency, **290:**3
incubation conditions, **290:**22
luciferase, β subunit folding analysis,
290:12, 14–15
ribosome, salt-washed preparation, **290:**21
S30 fraction

fractionation by centrifugation,
290:20–21
preparation, **290:**20
temperature optimum, **290:**3, 12
ESEEM, *see* Electron spin echo envelope
modulation
ESI–MS, *see* Electrospray ionization mass
spectrometry
ESR, *see* Electron spin resonance
EST, *see* Expressed sequence tag
Esterase, **305:**2-methyl-1-propenyl benzoate
chemiluminescence assay, **305:**383
Estrogen receptor, subcellular distribution,
302:74
Ethidium bromide, transmission electron microscopy support treatment, **304:**222
N-Ethyl-1,3-[3-(diethylamino)propyl]carbo-
diimide, cross-linking of photosystem I
subunits, **297:**133–134
N-Ethylmaleimide, thiol trapping,
290:62–63
Ethylnitrosourea
applications of RNA modification,
318:8–9
direct detection of RNA modification,
318:13
primer extension analysis, **318:**16
RNA modification site, **318:**6
Evanescent microscopy, *see* Total internal
reflection microscopy
Eve, ultraviolet cross-linking of protein–
DNA complexes *in vivo*, **304:**510–512
Excitatory amino acid receptors, *see also*
specific receptors
electrophoretic mobility shift assay of
transcription factors
acrylamide gel preparation,
294:398–399
electrophoresis, **294:**399, 403
limited proteolysis of complexes in validation, **294:**400
nuclear extract preparation from
mouse brain, **294:**397, 402–403
probe preparation, **294:**397–398
quantitative analysis, **294:**404
receptor activation studies *in vivo*,
294:401–402
solution preparation, **294:**395–396
supershift assay, **294:**400, 402
radioactive filter assays

ligand selection, **294:**388
reaction mixtures, **294:**388–389
synaptic membrane preparation from
   rat brain, **294:**386–387
Triton X-100 treatment of crude fractions, **294:**387–388
Western blot analysis
   blotting with semidry electrophoretic
      transfer, **294:**406, 409
   immunodetection, **294:**406–407, 410
   receptor activation studies *in vivo*,
      294:407–408
   sample preparation from mouse brain,
      294:405
   solution preparation, **294:**404–405
   troubleshooting, **294:**410
Exciton chirality, *see* Circular dichroism,
   sphingolipids
Exhaustive photon realignment microscopy,
   green fluorescent protein, **302:**152
Exocytosis, *see* Secretion
Exon
   exon-containing subclones, identification,
      296:507
   exon/intron junction determination in ge-
      nomic clones, **296:**508
Exon amplification
   advantages, **303:**157
   applications, **303:**99
   artifactual exon trapping products,
      303:88–89
   cloning of reverse transcription–poly-
      merase chain reaction products
      ligation-independent cloning with ura-
         cil DNA glycosylase, **303:**96–98
      overview of approaches, **303:**95–96
   consolidation with other techniques,
      303:146–147, 160–161
   DNA complexity and exon trapping prod-
      uct yield, **303:**92
   enhancement of effectiveness, **303:**89–90
   evolution of technique, **303:**86, 88–90
   large-insert trapping with cosmids
      consequences, **303:**101–102
      cosmid cloning and DNA isolation,
         303:103–104
      principle, **303:**102–103
      reverse transcription–polymerase chain
         reaction
         background reduction, **303:**108

electrophoretic analysis of products,
   303:109–110
primers, **303:**107–109
ribonuclease A treatmemt of products, **303:**108
RNA isolation, **303:**106–107
sequence analysis, **303:**109
transformation
   cell culture, **303:**104–105
   cell harvest, **303:**105
   electroporation, **303:**105–106
   vectors, **303:**100–103
limitations in gene identification, **303:**146
mapping of exon trapping products,
   303:98–99
principle, **303:**86, 90, 92, 100, 145
reverse transcription–polymerase chain
   reaction
   complementary DNA synthesis, **303:**97
   electrophoretic analysis of products,
      303:97
   primers, **303:**95–96, 158
   sequencing of products, **303:**158
sequencing of exon trapping products,
   303:98
vectors
   overview, **303:**156–158
   pSPL3 vector
      characteristics, **303:**92–93
      cloning, **303:**93–94
      electroporation into COS7 cells,
         303:94
      RNA isolation following transfec-
         tion, **303:**94–95
Exon trapping, *see* Exon amplification
Exonuclease III
   chromatin structure probing
      cell culture, **304:**586, 589
      digestion reaction, **304:**591–593
      entry site enzyme digestion,
         304:591–592
      mouse mammary tumor virus promoter
         analysis, **304:**585–586, 589, 597,
         599
      mung bean nuclease digestion, **304:**594
      nuclei isolation, **304:**589, 591
      principle, **304:**585–586, 623
      purification of digested DNA,
         304:593–594
      redigestion, *in vitro*, **304:**594–595

reiterative primer extension analysis
  amplification reaction, **304:**596–597
  end-labeling of oligonucleotides,
    304:596
  primer selection, **304:**595
  product analysis, **304:**597
  reagents, **304:**596
mammalian nucleosome positioning *in
  vitro*, **304:**630
protein image hybridization, DNA trimming, **304:**607–608, 611–612
translational nucleosome positioning
  assay
  digestion conditions and analysis,
    304:322
  principle, **304:**321–322
Exopolysaccharide, biofilms
  colorimetric assays
    acetylation, **310:**409–410
    neutral hexose, **310:**409
    pyruvilation, **310:**410
    uronic acid, **310:**409
  definition, **310:**146
  gel electrophoresis
    Eastern blot analysis, **310:**411–412
    running conditions, **310:**411
    staining, **310:**411
  hemagglutination inhibition assay of pu-
    rity, **310:**408–409
  immunoelectron microscopy, **310:**406–407
  lectin staining, **310:**406
  light microscopy, **310:**404–406
  monosaccharide analysis, **310:**410
  probes, **310:**404–406
  Pseudomonas aeruginosa biofilm analysis
    sample preparation, **310:**52
    total carbohydrate measurement,
      310:52–53
    uronic acids assay, **310:**53
  purification, **310:**407–408
  sulfate group detection by infrared spec-
    troscopy, **310:**410
  survival functions, **310:**403–404
Expectation–maximization algorithm, *see*
  Hidden Markov model
Experimental murine AA amyloid
  components of fibrils, **309:**707
  Congo red staining, **309:**710–711
  deposition quantification
    grading, **309:**714

immunocytochemistry, **309:**714–715
  standards, **309:**714
electron microscopy
  isolated fibrils
    isolation, **309:**508–509
    sample preparation, **309:**509
    structure comparison with *in situ* fi-
      brils, **309:**502
  in situ observation
    cryofixation and freeze substitution
      of samples, **309:**499
    data collection, **309:**504
    homology with connective tissue mi-
      crofibrils, **309:**501
    immunolabeling of amyloid A,
      309:507
    microscope maintenance, **309:**504
    photographic enlargement,
      309:505–506
    systemic microdissection, **309:**506
    thin section preparation, **309:**503
    ultrastructure and composition,
      309:498–499
human diseases of amyloid A amyloidosis, **309:**701–702
immunocytochemical detection,
  309:711–713
induction of amyloidosis
  amyloid enhancing factor, **309:**502–503,
    708–710
  casein, **309:**707–709
  overview, **309:**705, 707–708
inhibition of fibrillogenesis, **309:**715
isotypes of amyloid in different strains,
  309:702–705
organ deposition, **309:**702, 708
pathogenesis stages, **309:**705
radioiodinated serum amyloid P, detection of amyloid, **309:**713–714
utility with other amyloidogenic diseases,
  309:715–716
Expressed sequence tag
  applications, **303:**111–112
  ATP-binding cassette transporter gene
    identification
    contig assembly, **292:**123–124, 127
    full-length sequences, obtaining,
      292:125–126
    human genes deduced from yeast se-
      quences, **292:**160–162

pitfalls, **292:**126–127, 130
sequence searching with BLAST,
    292:122–123, 137–138
Xref database searching,
    292:124–125
computational gene finding, **303:**78–79
databases
    size of human database, **303:**127
    gene identification
        consolidation with other techniques,
            303:146–147, 160–161
        limitations, **303:**146, 151–152
        overview, **303:**144–145, 152–153
        radiation hybrid mapping data,
            303:152
    mouse database, **303:**153
disease gene identification, **303:**128
global approach in differential gene ex-
    pression, **303:**495
Human Genome Project progress,
    303:55–56, 77, 111–112
human libraries with bacterial contamina-
    tion, **292:**126
library characteristics, **303:**151
mapping, **303:**55–56, 85
Expression cloning, *see Xenopus laevis*
    oocyte
Expression systems, *see* Cell-free transla-
    tion system; Recombinant protein ex-
    pression systems
Extracellular signal regulated kinase
    assay by Western blot, **301:**520–521
    p21Ras regulation, **301:**514, 519–520
    shear stress effects, **301:**519–521
Eye
    cornea, *see* Cornea
    three-dimensional confocal microscopy of
        living eye
        correlative microscopy, **307:**562
        maximum intensity imaging,
            307:560–561
        volume rendering, **307:**560, 562
    three-dimensional reconstruction of con-
        focal data
        advanced reconstruction
            X, Y dimensions, **307:**508–511
            Z dimension, **307:**505–506, 508
        basic reconstruction, **307:**504–505
        optimization, **307:**501–504
        stereology incorporation, **307:**511–513

tissue preparation for confocal mi-
    croscopy
    dye incubation, **307:**498–499
    optic nerve head sections, connective
        tissue staining, **307:**499–500
    thickness limitations, **307:**498
    vimentin localization in trabecular net-
        work cells, **307:**500–501
in vivo confocal neuroimaging of retinal
    ganglia
    animal preparation, **307:**567–569
    applications, **307:**569–570
    equipment, **307:**565–567
    overview, **307:**563–565

F

Fabry disease, gene therapy, *see* Gene ther-
    apy, *MDR1*
FACE, *see* Fluorophore-assisted carbohy-
    drate electrophoresis
FACS, *see* Fluorescence-activated cell
    sorting
F-actin
    assembly assay, *see Listeria monocyto-
        genes* actin assembly assay
    blot overlay assay of binding proteins
        advantages, **298:**32–33
        autoradiography, **298:**41
        denaturing polyacrylamide gel electro-
            phoresis of binding proteins,
            298:39–40
        electroblotting, **298:**40
        fixation of blots, **298:**40
        incubation of blot, **298:**40–41
        iodine-125 labeling of G-actin,
            298:36–39
        materials, **298:**35–36
        membrane skeleton protein identifica-
            tion, **298:**33–34
        nonradioactive detection, **298:**42
        phosphorous-32 ATP complex visualiza-
            tion, **298:**41–42
        polymerization of actin and yield,
            298:38–39
        principle, **298:**33
    latrunculin-A disruption, **298:**20–21,
        23–25
    misakinolide A disruption, **298:**29–32

molluscan thick filament motility assay
  actin filament movements, **298:**428–429, 435
  buffer preparation, **298:**433
  calcium dependence, **298:**435
  flow cell setup, **298:**433
  isolation of thick filaments, **298:**430–432
  video-enhanced differential interference contrast microscopy, **298:**428, 433–435
myosin V binding
  affinity, **298:**5
  cosedimentation assays, **298:**17
myosin subfragment 1 decoration for electron microscopy, optical trapping of single filaments
  actin preparation, **298:**437
  bandwidth, **298:**454–455
  coverslip preparation, **298:**437–438
  detachment kinetics, **298:**458–459
  flow cell setup, **298:**445–446
  fluorescence labeling of actin, **298:**440
  instrumentation, **298:**441–443
  microsphere trapping by myosin affinity
    myosin labeling of beads, **298:**438–440
    promotion of attachment, **298:**447
  observed bead deflections
    decoupling diffusion and compliance measurement, **298:**449
    diffusion correlation and bead position, **298:**454
    modeling, **298:**448
    myosin concentration effects, **298:**449
    nonprocessive motors, **298:**451, 453
    poor myosin preparations, **298:**455
    processive motors, **298:**449–450
    signatures of protein attachment, **298:**451, 453–454
    thermal fluctuation, **298:**447
  simultaneous fluorescence and bright-field imaging, **298:**443–445
  statistical analysis, **298:**455, 457–459
  stiffness of all springs, **298:**458
rhodamine-phalloidin visualization, **298:**23–24
swinholide A, F-actin depletion *in vitro*, **298:**28–29, 31–32

FAK, *see* Focal adhesion kinase
Far-Eastern blotting, *see* High-performance thin-layer chromatography
Fast-atom bombardment–mass spectrometry, *see* Mass spectrometry
Fast protein liquid chromatography, recombinant 30S ribosomal subunit purification, **318:**452–454
Fast scan cyclic voltammetry
  advantages in neurotransmitter uptake measurements, **296:**692–694, 707
  dopamine uptake
    brain region variability, **296:**704–707
    evidence of uptake in measurements, **296:**701–702
    inhibitors, assessment of action
      bupriopion, **296:**703
      cocaine, **296:**704
      nomifensine, **296:**703–704
    intact tissue measurements versus homogenates, **296:**691–693
    Michaelis–Menten kinetics
      correction for Nafion coating on carbon electrodes, **296:**697–701
      inhibition analysis, **296:**703–704, 707
      modeling of data, **296:**694–697
    temporal resolution, **296:**693–694
Ferguson plot, quaternary structure determination, **290:**275–276
Ferric reducing/antioxidant power and ascorbic acid concentration assay
  calculations, **299:**25–27
  plasma assay, **299:**26–27
  principle, **299:**23–25
  standards, **299:**24–25
Ferric reducing/antioxidant power assay
  advantages in total antioxidant power determination, **299:**15–16
  automated assay, **299:**16, 18
  comparison with oxygen radical absorbance capacity assay, **299:**58–60
  manual assay, **299:**18–19
  plasma assay, **299:**22–23
  precision and sensitivity, **299:**21
  principle, **299:**16–17
  reaction characteristics of pure antioxidants, **299:**19–21
  reagent preparation, **299:**17
  sample preparation, **299:**18
  standards and controls, **299:**17–18

Ferrous oxidation in xylene orange, *see*
  Lipid hydroperoxide
Ferulic acid, *see also* Phenolic antioxidants
  antioxidant activity, **299:**96–97
  α_1-antiprotease protection against hypo-
      chlorous acid inactivation, **299:**209
  biosynthesis, **299:**208
  high-performance liquid chromatography
      chromatography conditions,
        299:103–105
      food sample preparation, **299:**100, 103
      retention times, **299:**105–106
      standards, **299:**100
      urine sample
        dietary manipulation, **299:**97, 100,
          106
        preparation, **299:**103
  peroxynitrite interactions, **299:**223, 228,
      234
Fetal brain, human culture antisense studies
  cell counting control, **314:**257–258
  cell treatment
    antisense administration, **314:**249–250
    uptake, **314:**250–252
  electron microscopy of ultrastructural ef-
      fects
    fixation and mounting, **314:**258–259
    Reynold's lead citrate staining, **314:**260
    uranyl acetate staining, **314:**259–260
  immunocytochemistry of targeted pro-
      teins
    antibody types, **314:**252–253
    double labeling, **314:**254–255
    fixation
      methanol, **314:**253–254
      overview, **314:**253
      paraformaldehyde, **314:**254
    immunoenzymatic detection, **314:**255
    pretreatment, **314:**254–255
  oligonucleotide design, **314:**248–249
  reverse transcription-polymerase chain re-
      action of messenger RNA,
      314:255–257
  specificity controls, **314:**249, 252
α-Fetoprotein, transcription in solid-phase
    nuclei
  chromatin assembly, **304:**66–68
  coupling with replication, **304:**72, 74–75
  gene constructs, **304:**65
  hepatoma factor effects, **304:**75

  methylation of gene, **304:**70
  repressed template transcription,
      304:74–75
FgeneH, computational gene finding,
    303:81–82
FIA, *see* Flow injection analysis
Fibroblast, *see* Patch clamp
Fibroblast growth factor-4
  antisense knockdown in limb culture
    lipofection, **314:**407–408, 410
    oligodeoxynucleotide design,
      314:407
    overview, **314:**406–407
    specificity, **314:**409
    targeted exposure of oligodeoxynucleo-
      tides, **314:**408–410
  apical ectodermal ridge expression in
      mouse limb bud, **314:**401
  limb culture for study
    advantages, **314:**402
    applications, **314:**405
    incubation conditions, **314:**403–404
    media, **314:**403–404
    plates, **314:**402–403
Fick's law, substrate flux modeling, **310:**419,
    454
Filamentous actin, *see* F-actin
Filter lift assay, protein–protein interac-
    tions
  advantages and limitations, **303:**440–441
  labeled protein probe preparation,
      303:438–439
  principle, **303:**438
  screening of library, **303:**439–440
Filter overlay assay, ion channel associated
    proteins, **294:**373–375
Fimbriae
  adhesin function, **310:**502–503
  oral biofilm, assays of initial formation
    hydroxylapatite beads assay,
      310:505–506
    nitrocellulose overlay assay,
      310:504–505
    polystyrene microtiter plate assay,
      310:506–507
    surface plasmon resonance
      advantages, **310:**508–509
      data acquisition, **310:**509–510
      dissociation constant determination,
        310:510–511

fimbrial binding affinity to host pro-
    teins, **310:**512–513
materials, **310:**509
principle, **310:**507–508
software for data analysis,
    310:511–512
purification and radiolabeling, **310:**503
Firefly luciferase
aequorin fusion protein
    plasma membrane, targeting, **305:**482,
        494
    subplasma membrane calcium and
        ATP measurements, **305:**494, 498
ATP assays
    biomass assays
        antibiotic sensitivity testing, **305:**361
        bacteriuria assays, **305:**360–361
        cytotoxicity and cell proliferation
            assays, **305:**361–362
        extraction of intracellular ATP,
            305:358–359
        hygiene monitoring, **305:**362–363
        neutralization of extractants,
            305:359–360
        overview, **305:**356–357
        sample collection and pretreatment,
            305:357
        sensitivity, **305:**360
        sterility testing, **305:**360
    calibration with ATP standard,
        305:354–356
    classification of reagents
        flash reagents, **305:**350, 369
        selection by application, **305:**369–370
        slow decay reagents, **305:**349–350
        stable light-emitting reagents,
            305:349, 369
    enzyme assays
        ATP/ADP/AMP production, simulta-
            neous measurement, **305:**364
        ATP/phosphocreatine production, si-
            multaneous measurement,
            305:364–365
        creatine kinase isoenzymes,
            305:366–367
        glycerol kinase quantification of glyc-
            erol, **305:**365–366
        overview, **305:**363–364
        oxidative phosphorylation assay,
            305:368–369

photophosphorylation assay,
    305:368–369
pyrophosphate quantification with
    ATP sulfurylase, **305:**367–368
urease qantification of urea, **305:**366
kinetics of luciferase reaction
    ATP response curve, **305:**353
    calculations, **305:**350–351
    peak intensity extrapolation,
        305:352–353
    total light emitted, **305:**352–353
principle, **305:**348–349
lantern collection, **305:**180
microinjection reporter in mammalian
    cells
    cell lines, **305:**572
    expression comparison between trans-
        fection and microinjection in fi-
        broblasts, **305:**573–575
    microinjection, **305:**573
    osteosarcoma cell versus fibroblast ex-
        pression levels, **305:**575–576
    vectors, **305:**572–573
optimization of reaction, **305:**347–348
peroxisomal targeting sequence, **305:**481
reaction steps, **305:**346–347
recombinant *Luciola mingrelica* enzyme
    expression in *Escherichia coli*
    cell growth, **305:**184–185
    host strain election, **305:**183
    lux vector system from *Vibrio fischeri*,
        305:181–183
    medium for growth, **305:**184
    purification
        ammonium sulfate fractionation,
            305:186
        anion-exchange chromatography,
            305:186, 188
        cell lysis, **305:**185
        dye affinity chromatography,
            305:186
reporter in mammalian cells compared
    with bacterial luciferase and chloram-
    phenicol acetyltransferase
    applications, **305:**571–572
    calibration curves, **305:**559
    expression levels, **305:**560–563
    materials, **305:**558–559
    promoter for studies, **305:**558
    sensitivity, **305:**559–560

stable transfection effects on assay re-
     producibility, **305:**564–567
thermal stability, **305:**563–564
transient transfected cell splitting, ef-
     fects on assay reproducibility,
     305:567, 569–571
FISH, *see* Fluorescence *in situ* hybridization
F₂-isoprostanes
   biosynthesis, **300:**3–4, 13
   gas chromatography/negative-ion chemi-
        cal ionization mass spectrometry
   derivatization, **300:**8–9, 15
   extraction and hydrolysis of phospholip-
        ids, **300:**6–7, 14–15
   instrument parameters, **300:**9, 15
   metabolites of F₂-isoprostanes, **300:**11
   plasma analysis, **300:**14–17
   purification
        aminopropyl chromatography, **300:**14
        gas chromatography, **300:**9, 15
        reversed-phase chromatography,
             300:7–8, 13–14
        thin-layer chromatography, **300:**7–9,
             13
   quantification, **300:**9–11, 15–16
   sample handling and storage, **300:**5–6
   sensitivity, **300:**7
   standards, **300:**7–8
   urine analysis, **300:**7, 11
   immunoassay, **300:**12
   lipid peroxidation and oxidant stress indi-
        cator, **300:**4–5, 13
Fixation, *see* Mounting
FK506
   FKBP12 assay by calcineurin phospha-
        tase inhibition with FK506 complex
   incubation conditions, **290:**81–82
   phosphate release assay, **290:**82–83
   phosphopeptide substrate preparation,
        290:80–81
   precautions, **290:**83
   principle, **290:**79–80
   plant organelle peptidylprolyl *cis–trans*-
        isomerase binding assays
   immunoblot assay, **290:**96–98
   radioactive assay, **290:**95–96
FK506 synthetase, substrate channeling,
        308:143–144
FK1012, mechanism of action, **306:**264
FKBP12

assays
   calcineurin phosphatase inhibition with
        FK506 complex
   incubation conditions, **290:**81–82
   phosphate release assay, **290:**82–83
   phosphopeptide substrate prepara-
        tion, **290:**80–81
   precautions, **290:**83
   principle, **290:**79–80
   overview, **290:**75–76
   peptidylprolyl isomerase, **290:**78
homology between human and yeast pro-
     teins, **290:**83
peptidylprolyl *cis–trans*-isomerase activ-
     ity, **290:**75
purification from *Saccharomyces cere-
     visiae*
   native protein, **290:**76–77
   recombinant protein, **290:**77–78
FKBP52
   assays, **290:**429
   biological functions, **290:**420–421
   domains, **290:**421
   heat shock protein 90 association,
        290:420
   immunosuppressant binding, **290:**420
   purification of recombinant rabbit pro-
        tein from *Escherichia coli*
   anion-exchange chromatography,
        290:423–424
   cell growth and lysis, **290:**422–423
   hydroxyapatite chromatography,
        290:423–424
Flagellar-mediated motility, assay, **310:**98
Flagellin, gene activation during agar
     growth, **310:**68–69
FLA-polysaccharides, *see* Fluorescently-
     labeled polysaccharides
FLASH, *see* Fluorescence arsenic helix
     binder technique
Flash lamp photolysis
   applications
        calcium flux in response to caged com-
             pounds, **291:**219–222
        chemotaxis, serine effects on bacteria,
             291:219
        microtubule dynamics, **291:**215
        muscle contraction, **291:**214–215, 320
        DM-nitrophen and serotonin transport in
             Retzius P cell synapse

flash photolysis setup, **296:**605–606
microinjection of DM-nitrophen,
   296:604–605
photochemistry, **296:**603–604
pre- and postsynaptic current isolation
   induced by flash photolysis,
   296:602–603
two-electrode voltage clamp,
   296:601
driving circuit, **291:**207–209
energy of pulse, measurement,
   291:213–214
lamp selection, **291:**204–205, 207
optics
   lenses, **291:**210–212
   light guides, **291:**213
   mirrors, **291:**209–210
   refractive optics versus ellipsoidal mir-
      rors, **291:**212–213
pulse duration, **291:**222
X-ray crystallography
   heat damage of crystals by flash,
      291:216–218
   measuring compounds released by pho-
      tolysis, **291:**219
   Ras GTPase studies in crystals,
      291:215–216
Flash photolysis, *see* Flash lamp photolysis;
   Laser photolysis
Flavonoids, *see also* Phenolic antioxidants;
   Polyphenols
absorption and excretion in humans,
   299:92–95
biological activities, overview, **301:**490–
   492, 504
dietary intake in Western countries,
   301:504
high-performance liquid chromatography
   chromatography conditions,
      299:103–105
   food sample preparation, **299:**100, 103
   retention times, **299:**105–106
   standards, **299:**100
   urine sample
      dietary manipulation, **299:**97, 100,
         106
      preparation, **299:**103
inducible nitric oxide synthase, effects on
   activity, **301:**509
nitric oxide radical scavenging

bioassay by aortic ring relaxation,
   301:492–494
competition assays, **301:**495–499
kinetic assays, **301:**499–503
physiological relevance, **301:**502–503
types and structures, **301:**490–491
Flicker photometry, *see* Cone
Flow cell, biofilm cultivation and analysis
advantages, **310:**3–4, 160
chemostat, *see* Chemostat, biofilm culture
delivery systems, **310:**161
cleaning and sterilization, **310:**161
culture selection, **310:**251–253
design and construction, **310:**4–5, 22, 24
disadvantages, **310:**160–161
fluorescence *in situ* hybridization, **310:**10,
   11
hydraulic diameter calculation,
   310:225
laminar flow chamber
   advantages, **310:**224–225
   cleaning of windows, **310:**227–228
   description, **310:**229–230
   destructive biofilm analysis,
      310:231–232
   fluorescent biomarker monitoring of
      films, **310:**226–229
   fouling studies, **310:**231
   inoculation, **310:**231
   laminar flow verification, **310:**226
   statistical analysis, **310:**232
   sterilization, **310:**231
mass transfer of nutrients, *see* Mass transfer-limited bioavailability, biofilm nutrients
medical device implant biofilm culture,
   310:540–541
metabolic monitoring with fluorescent
   dyes, **310:**18–20
microscopy, *see* Microscope flow cell
mixed culture film cultivation,
   310:253–254
once-flow-through continuous culture system, **310:**45–46
perfused biofilm fermenters, *see* Biofilm
   fermenter
Reynolds number characterization of
   flow conditions, **310:**225
Robbins device, *see* Robbins device
shear stress calculation, **310:**226

Flow cytometry, *see also* Fluorescence-acti-
    vated cell sorting
  amyloid fibril characterization with thio-
      flavin T and Congo red
    criteria for fibril detection, **309:**463
    data acquisition, **309:**462
    immunoglobulin light chain amyloid-
        osis, amyloid burden determina-
        tion, **309:**464–466
    instrument settings, **309:**461
    overview, **309:**460–461
    staining of sample, **309:**462–463
  antisense oligonucleotide transfection
      analysis, **313:**61, 67, 69
  apoptosis assays
    cell requirements, **319:**324–325
    forward light scattering, **319:**325, 329
    instrumentation, **319:**325
    mitochondria transmembrane potential
        analysis, **319:**326–328, 330
    side light-scattering,
        319:325, 329
  B-1 cell neoplasia, **314:**413, 421–422
  cell cycle analysis and bromodeoxyuri-
      dine uptake, **314:**437–439
  cell cycle analysis in glioma antisense
      model, **313:**367, 371–372
  cell sorting, **302:**297
  E-selectin expression, **314:**369–370
  fluorescent protein combination with
      dyes, **302:**197–198
  green fluorescent protein expression
      among individual bacterial cells in
      culture
    advantages and disadvantages,
        305:507–508, 512–513
    cell growth, **305:**508
    light scattering data collection,
        305:508–511
    plasmids and protein mutants,
        305:500–501, 512
    secondary fluorescence labels,
        305:511–512
  green fluorescent protein variants,
      302:189–190
  hematopoietic stem cell, fluorescence se-
      lection of acid sphingomyelinase ex-
      pression, **312:**333–335
  HL-60 cell studies of ultraviolet A radi-
      ation

    mitochondrial transmembrane po-
        tential
      DiOC$_6$, **319:**337–338
      JC-1, **319:**338–340
    peroxide detection, **319:**336
  keratinocytes, ultraviolet A-induced gene
      expression analysis, **319:**285
  laser scanning cytometry of androgen re-
      ceptor–green fluorescent protein fu-
      sion protein, **302:**128
  leukemia cells, methylphosphonodiester
      oligodeoxynucleotide uptake
    intracellular oligonucleotide concentra-
        tion determination, **313:**123–125
    two-color studies of uptake and viabil-
        ity, **313:**122–123
    viable cell assay,
        313:120
  mammalian cell transfection with green
      fluorescent protein vectors
    electroporation, **302:**190–192
    stable expression, **302:**192–193
  oligonucleotide uptake studies
    data acquisition, **313:**295–296, 345
    efflux analysis, **313:**291, 293–294, 296
    fluorescence advantages over radioac-
        tive labels, **313:**290
    internalization assay, **313:**295–296
    intracellular pH determination,
        313:296–297
    monensin treatment, backbone modifi-
        cation effects on uptake, **313:**291
    synthesis of 5′-fluorochrome labeled oli-
        gonucleotides, **313:**294–295
    washing of cells, **313:**290–291
  principle, **302:**296–297
  single-color analysis of fluorescent pro-
      teins, **302:**193–195
  specificity of fluorophores, **302:**297
  thiol assay
    advantages, **299:**247, 257
    cell preparation, **299:**248–249
    data collection, **299:**251–252
    differential assessment of thiols,
        299:249–251, 258
    monobromobimane modification of thi-
        ols, **299:**248–249, 251–253,
        257–258
    monochlorobimane detection of gluta-
        thione, **299:**248

three-color fluorescent protein detection, **302**:197

transcription efficiency assay with green fluorescence protein
applications, **302**:206
cytometer setup, **302**:200
data analysis, **302**:202–205
DNA constructs, **302**:201
overview, **302**:199–200
sensitivity, **302**:206
transfectants and controls in experimental design, **302**:201–202

transgenic plant green fluorescent protein expression
autofluorescence, **302**:310, 312
lasers for excitation, **302**:315
mutant green fluorescent protein applications, **302**:314–315
protoplast and subcellular organelle analysis
instrument settings, **302**:302–303
transfected protoplasts, **302**:305–306
transgenic plant nuclei, **302**:307–309
transgenic plant protoplasts, **302**:307
setting up
Coulter Elite, **302**:301–302
Cytomation MoFlo, **302**:302
sorting setup
Coulter Elite, **302**:303–304
Cytomation MoFlo, **302**:304–305

two-color fluorescent protein detection, **302**:195–197

Flow injection analysis, neurotransmitter transporter release and uptake, **296**:664–665

Flp recombinase
ectopic expression system in *Drosophila melanogaster*, **306**:140–142
recombinase-mediated cassette exchange, **306**:62–63
selectable marker removal following homologous recombination, **306**:49–50, 52

Fluid mosaic membrane model, limitations, **294**:208–209

Fluo-3, calcium measurements
acetoxymethyl ester cleavage, **307**:22–23
arterial endothelial cell, *in situ* calcium imaging with confocal microscopy
acetylcholine response, **307**:429, 431

data collection, **307**:429
dye loading, **307**:428
quantitative analysis, **307**:430
signal separation from smooth muscle cells, **307**:429
tissue preparation, **307**:428
calibration, **294**:24–25
compartmentalization in cell, **307**:22
detection wavelength, **294**:21
dye loading, **294**:23–24
laser-induced rise of signals, **307**:22–24
loading conditions, **307**:126, 428
microtiter plate reading, **294**:22–23
photoreceptor clacium flux measurements using spot confocal microscopy
advantages, **316**:148–149
argon ion lasers, **316**:149, 151–154
calibration, **316**:159–162
detectors, **316**:154–156
dye loading, **316**:157
instrumentation, **316**:149
light-induced decline of calcium, **316**:157–159
photopigment cycle changes in calcium, **316**:162–163
photoreceptor preparation from tiger salamanders, **316**:156–157
rationale, **316**:147–148

Fluorescein
caged compounds
caging group selection, **291**:67
practical considerations for use, **291**:67
protein labeling, **291**:72–74
synthesis, **291**:68
tubulin, *see* Tubulin
types and structures, **291**:64–66, 475
RNA labeling, **317**:421–422

Fluorescence-activated cell sorting, *see also* Flow cytometry
cystic fibrosis transmembrane regulator-expressing cells, **292**:692
hematopoietic stem cells, **292**:547–548, 550
MDR1-expressing cells, **292**:239, 242–243, 453, 469–471, 508–509, 531
gene therapy products, **292**:531, 533, 547–548, 550
transgenic mice, **292**:580–581, 583–584

Fluorescence anisotropy, GroEL–GroES complex, analysis of stoichiometry

anisotropy value, calculation, **290:**314–315

binding conditions and data collection, **290:**317

GroES labeling with succimidyl-1-pyrene butyrate, **290:**316–317

materials, **290:**315–316

Perrin–Weber equation, **290:**315

Fluorescence arsenic helix binder technique, green fluorescent protein imaging in living samples, **307:**111

Fluorescence lifetime

overview, **317:**412

pulse versus phase modulation methods, **317:**415

Fluorescence microscopy, *see also* Calcium; Confocal laser scanning microscopy; Deconvolution fluorescence microscopy; Epifluorescence microscopy; Fluorescence photoactivation imaging; Fluorescence *in situ* hybridization; Immunofluorescence microscopy; Intravital microscopy; *Listeria monocytogenes* actin assembly assay; *specific dyes*

centrosomes, immunofluorescence and immunogold localization of proteins, **298:**234–235, 238

correlative fluorescence and electron microscopy of cytoskeleton

cell culture and light microscopy, **298:**571–573

criteria for success, **298:**570–571

electron microscopy

actin depletion, **298:**574–575, 578

critical point drying, **298:**585–588

extraction, **298:**573–574

fixation, **298:**580–581, 583

immunostaining with colloidal gold-conjugated antibodies, **298:**583–585

metal coating, **298:**588–589

myosin subfragment 1 decoration of actin filaments, **298:**578–580

replica preparation, **298:**589, 591–592

spatial resolution, **298:**570

cytoplasmic linker protein-170 immunolocalization, **298:**203

endoplasmic reticulum network formation, **298:**356–357

focal adhesion kinase immunofluorescence, **298:**101–102

FtsZ, immunofluorescence localization, **298:**296, 302–305

interphase extracts, immunofluorescence microscopy, **298:**352

katanin immunolocalization, **298:**207

kinesin processive motility assay

axoneme preparation from sea urchin sperm and fluorescence labeling, **298:**166–167

bovine serum albumin preparation, **298:**168

data collection and analysis, **298:**169–171

incubation conditions, **298:**168–169

limitations, **298:**157–158

oxygen scavenger stocks, preparation, **298:**167

principle, **298:**156

sensitivity, **298:**156, 170

single fluorophore imaging, **298:**159–160

slide preparation, **298:**168

metaphase extracts, immunofluorescence microscopy, **298:**352

neurotransmitter transporters expressed in *Xenopus* oocytes

charge-coupled device detection, **296:**569, 571–572, 575

fluorescence labeling, **296:**573–574

ion measurements with fluorescent dyes

applications, **296:**575–576

loading of dye, **296:**576

simultaneous electrophysiological measurements, **296:**572–573, 576–578

optical measurements, **296:**567, 569–570, 574–575

photodiode measurements, **296:**570–571, 575

simultaneous digitally-enhanced differential interference contrast microscopy

advantages, **298:**319

schematic of apparatus, **298:**322

yeast

data acquisition and analysis, **298:**329–331

preparation for imaging, **298:**319–320

simultaneous fluorescence and bright-field imaging with optical traps, **298**:443–445
simultaneous fluorescence imaging of 5,7-dihydroxytryptamine uptake with voltage clamping in Retzius P cell synapse, **296**:598–601
vasodilator-stimulated phosphoprotein, immunofluorescence, **298**:110
Fluorescence photoactivation imaging
cameras, **291**:74
G-actin studies, **291**:104–105
green fluorescent protein, **291**:77–78
light sources, **291**:74–75
Fluorescence polarization
anisotropy overview, **317**:413, 415–416
instrumentation, **317**:415–416
Fluorescence quenching, *see* Lipid phase separation
Fluorescence recovery after photobleaching
applications, **294**:49
green fluorescent protein mutants, **302**:251–252, 261
green fluorescent protein–nuclear receptor fusion proteins, **302**:83–84
green fluorescent protein-tagged integral proteins, membrane trafficking, **302**:7–9
potassium channel–green fluorescent protein constructs
artifacts, corrections, and calibrations, **294**:56–58
channel mobility measurement, **294**:53–54, 58
confocal scanning laser microscopy measurements, **294**:54–56
solute diffusion analysis
complex phenomena, analysis, **302**:260
cytoplasm, **302**:254–255
diffusion coefficient, **302**:252
endoplasmic reticulum, **302**:257–258
membrane diffusion of fusion proteins, **302**:259–260
mitochondria, **302**:255–256, 258–259
principle, **302**:241
quantitative analysis in organelles, **302**:258–259
reversible versus irreversible photobleaching, analysis, **302**:253–254
Fluorescence resonance energy transfer

Bcl-2/Bax interaction analysis with green fluorescent protein mutants, **302**:422
branched RNA, global folding transition analysis
anisotropy measurements, **317**:383–384
combination with other structural studies, **317**:392–393
comparative approach, **317**:388–389
fluorophores
labeling and purification of RNA, **317**:384–386, 388
selection of donor–acceptor pair, **317**:384
four-way junctions, **317**:378
instrumentation, **317**:379
normalized acceptor ratio method, **317**:381–383
principle, **317**:378
theory, **317**:379–381
codon usage in mutant design, **302**:422–423
distance limitations, **317**:412, 430
efficiency of energy transfer, **302**:410–411, **317**:381, 430
green fluorescent protein with blue-shifted protein
calmodulin linker in calcium fluorescence assay, **294**:52
protease assays, **294**:52
protein–protein interactions in potassium channels, **294**:52–53
green fluorescent protein–nuclear receptor fusion proteins, **302**:84
normalized spectral overlap integral, **317**:380–381
nucleic acid labeling with fluorophores, **302**:409
principle, **294**:52, **302**:408–409
rate constant of energy transfer, **317**:380–381
RNA secondary structure monitoring, **317**:433–434
RNA tertiary structure monitoring
advantages and disadvantages, **317**:431, 433
applications, **317**:431
theory, **317**:430–431
selection of donor and acceptor, **302**:410–411
spectral overlap, **302**:414–415

tandem green fluorescent protein fusion
proteins
calcium probes, **302:**421–422
efficiency of various mutants
design of mutants, **302:**417–418, 420,
423
P43–Topaz, **302:**417
Sapphire–Topaz, **302:**414–417
table, **302:**413
W7–10cQ69K, **302:**417
linkers
length and dimerization, **302:**418–420
proteolysis, **302:**411, 413
protease assays, **302:**420–421
time-resolved analysis of conformational
isomers, hairpin ribozyme–substrate
complexes, **317:**434–435
Fluorescence *in situ* hybridization
antisense oligodeoxynucleotide effects on
ribonucleoprotein formation and
sorting
antisense oligodeoxynucleotide incubation, **313:**465
confocal microscopy, **313:**466
fixation, **313:**465
hybridization, **313:**466
immunocytochemistry, **313:**466
overview, **313:**464
probe labeling, **313:**465
reagents, **313:**464
applications, **304:**651
biofilm analysis, *see* Fluorescence *in situ*
hybridization, biofilm analysis
chromosome painting
DNA probe labeling
nick translation, **304:**654
overview, **304:**653
polymerase chain reaction labeling,
304:654
quantification of label, **304:**654–655
random priming, **304:**654
hybridization, **304:**655–656
image analysis, **304:**657–658
metaphase chromosome spread preparation, **304:**652–653
overview, **304:**651–652
signal detection, **304:**656–657
decondensed metaphase chromosomes
chromosome isolation, **304:**658–660
overview, **304:**658

protein removal, **304:**660–661
staining and visualization, **304:**661
gene mapping, **303:**56
homozygous transgenic mouse identification, **292:**578–579
nuclear matrix analysis, **304:**661–662
principle, **304:**650–651
RNA, *see* Fluorescence *in situ* hybridization, RNA
sensitivity and selectivity, **303:**3
yeast nuclear organization with immunofluorescence staining
antibody purification and specificity,
304:665–666
difficulty in analysis, **304:**663
fixation of cells, **304:**671
fluorophore selection, **304:**666, 672
green fluorescent protein applications,
304:663, 671
hybridization conditions,
304:669–670
immunofluorescence conditions,
304:667–669
nuclear integrity assays, **304:**672
overview of assays, **304:**663–664
probe preparation, **304:**666–667
troubleshooting, **304:**672
variations of technique, **304:**670–671
yeast strains and media, **304:**664–665
Fluorescence *in situ* hybridization, biofilm
analysis
confocal laser scanning microscopy,
310:91
embedding
agarose, **310:**24
polyacrylamide, **310:**24–26
polysaccharide-specific antibodies,
310:24–25
epifluorescence microscopy, **310:**90–91
melting temperature
factors affecting, **310:**87
formamide manipulation, **310:**87–88
multicolor experiments
β-galactosidase, **310:**39
green fluorescent protein, **310:**38–39,
77
peptide nucleic acids as probes, **310:**14
probe active count assay, **310:**90
ribosomal RNA probes
advantages, **310:**5–6, 80

background fluorescence reduction,
310:12–13
fixation
fixative selection, **310:**85–86
paraformaldehyde, **310:**8, 86–87
flow cell biofilm analysis
fixation, **310:**10, 25
hybridization, **310:**11, 26–28
fluorescence labeling
approaches, **310:**82–83
dye selection, **310:**7–8, 83
materials, **310:**84
oligonucleotide purification,
310:84–85
hybridization, **310:**9–10, 87–90
permeability enhancement of cells,
310:11
probe design, **310:**81–82
scraping of biofilms, **310:**13–14
steric hindrance of labeled probes,
310:11
weak signal enhancement, **310:**12
whole cell *in situ* hybridization, overview, **310:**7, 79–80
Fluorescence *in situ* hybridization, RNA
applications, **318:**493–494, 505–506
cultured cells
coverslip preparation, **318:**501–502
fixation, **318:**502
hybridization, **318:**502
washing, **318:**502
imaging and analysis, **318:**505
microscopes, **318:**504
oligonucleotide probes
advantages and disadvantages,
318:494–495
design, **318:**495
labeling with terminal transferase,
318:496
purification, **318:**495–496
synthesis, **318:**495
RNA probes
advantages and disadvantages,
318:496–497
design, **318:**497
fluorophore conjugation, **318:**498
purification, **318:**497–498
transcription, **318:**497
simultaneous immunofluorescence detection

primary antibody incubation and washing, **318:**503
secondary antibody incubation and
washing, **318:**503–504
yeast cells
coverslip preparation, **318:**498–499
fixation, **318:**499
hybridization, **318:**500–501
mounting medium preparation, **318:**501
probe preparation, **318:**500
spheroplasting, **318:**499–500
washing, **318:**501
Fluorescence spectroscopy
advantages, **317:**409–410
aromatic amino acid fluorescence properties, **309:**609, 614–615
data interpretation, general considerations, **317:**427–428
emission features, **317:**409, 411
fibril assays, *see* Congo red; Thioflavin T
instrumentation
commercial types, **317:**414
detector, **317:**413
kinetic accessories, **317:**416–417
light source, **317:**413
monochromator, **317:**413–414
optical accessories, **317:**415–416
spatial resolution instrumentation,
317:417–418
intrinsic protein fluorescence,
317:410–411
lifetime analysis, *see* Fluorescence
lifetime
ligand binding to RNA, **317:**435, 438
polarization, *see* Fluorescence polarization
prospects for RNA studies, **317:**438, 440
protein absorption studies
calibration of total internal reflection
fluorescence, **309:**426–427
emission spectra analysis, **309:**428
intrinsic protein fluorescence, total internal reflection fluorescence spectroscopy, **309:**424–426
kinetic analysis with total internal reflection fluorescence, **309:**427–428
overview, **309:**409–410
quenching studies, **309:**428–429
protein structure studies
advantages and disadvantages, **309:**608

aggregation problems, **309**:622
cuvettes, **309**:620
extrinsic fluorophores, **309**:615
fluorescence energy transfer,
  309:615–616
inner filter effect, **309**:621
instrumentation, **309**:620
photobleaching, **309**:621
polarization and anisotropy, principle,
  309:617–619
Raman peak of water, **309**:620–621
sample preparation, **309**:619
spectra interpretation for intrinsic fluorescence, **309**:616–617
temperature effects, **309**:621–622
quenching
  overview, **317**:411–412
  RNA secondary structure formation
    monitoring, **317**:428–430
reactive oxygen species detection with
  fluorescent probes, **309**:762
resonance energy transfer, *see* Fluorescence resonance energy transfer
RNA design for studies, **317**:427
RNA fluorescence derivatization
  2-aminopurine incorporation, **317**:422
  chemical synthesis with modified oligo-
    ribonucleotides, **317**:420–423
  cyanine derivatives, **317**:422
  deprotection and purification, **317**:426
  fluorescein tagging, **317**:421–422
  hexachlorofluorescein tagging, **317**:422
  postsynthetic labeling, **317**:423–424,
    426
  Wye base as intrinsic fluorophore,
    317:419
Stokes shift, **317**:411
Fluorescence titration
  automated protein denaturation studies
    chemical denaturation
      ammonium sulfate titrations,
        295:162–163
      apparent equilibrium constant calcu-
        lation, **295**:152–153
      background correction, **295**:155–156
      denaturant delivery, **295**:156–157,
        159
      free energy calculations, **295**:153
      reversibility testing, **295**:159–161
      sample preparation, **295**:154–155

schedule of titrant addition, **295**:151–
  152, 157–159
data processing, **295**:152
instrumentation, **295**:150–151
pH denaturation, **295**:166–167
thermal denaturation
  apparent equilibrium constant calcu-
    lation, **295**:153–154
  reversibility testing, **295**:165
  sample preparation, **295**:163–165
  van't Hoff enthalpy calculation,
    295:154
partition coefficient determination, proteins in membranes, **295**:78–82, 87
quenching with automated titrating fluo-
  rometer
  acrylamide quenching, **295**:169
  potassium iodide quenching,
    295:167–169
  Stern–Volmer plots, **295**:168–169
Fluorescent differential display, *see* Differential display
Fluorescent fatty acid, *see* C11-BODIPY
Fluorescently-labeled polysaccharides
  extracellular enzyme hydrolysis assay
    incubation, **310**:416
    molecular weight distribution analysis
      of products, **310**:416
    overview, **310**:413–414
    rate calcuations, **310**:417
  labeling density characterization,
    310:415
  storage and stability, **310**:415–416
  synthesis, **310**:414–415
Fluorescent nitric oxide cheletropic traps
  absorbance properties, **301**:534–535, 539
  electron paramagnetic resonance,
    301:533, 535, 539
  fluorescence properties, **301**:535–537, 539
  selectivity for nitric oxide, **301**:537–538
  sensitivity, **301**:533, 536–537
  stability in solution, **301**:535, 537
  synthesis, **301**:533
  trapping mechanism, **301**:532–534
Fluorescent probes, oxidative stress
  2',7'-dichlorofluorescin diacetate
    cell loading and leakage, **300**:257–258,
      262
    optical properties, **300**:258
    oxidant sensitivity, **300**:259–261

structure and oxidation reaction,
300:257, 259
dihydroethidium
cell loading, 300:262
optical properties, 300:258
oxidant sensitivity, 300:259–261
structure and oxidation reaction,
300:257, 259
dihydrorhodamine 123
cell leakage, 300:258–259
cell loading, 300:262
optical properties, 300:258
oxidant sensitivity, 300:259–261
structure and oxidation reaction,
300:257, 259
ideal properties, 300:257
kinetics of oxidation
data pooling for calculations,
300:274–275
dye leakage effects, 300:269–273
equations, 300:264–269
nonlinear least squares analysis,
300:263–264
5-Fluoroorotic acid, resistance test as yeast
reporter gene assay
cell culture, 303:420–421
principle, 303:420
Fluorophore-assisted carbohydrate electrophoresis
glycosphingolipid analysis with clam ceramide glycanase, 312:202, 205
oligosaccharides released from brain glycosphingolipids, 312:125–126
FNOCTs, see Fluorescent nitric oxide cheletropic traps
Focal adhesion kinase
activation by integrin binding, 298:89–90
antibodies
commercial availability, 298:94
types, 298:93, 96
domains, 298:91–92
functions, 298:90–91
immunofluorescence, 298:101–102
immunoprecipitation, 298:97
lysate preparation for assays, 298:95–96
paxillin phosphorylation assay, 298:99
phosphorylation
detection assay for autophosphorylation, 298:98–99
sites, 298:91–92

purification from baculovirus–insect cell
expression system
anion-exchange chromatography,
298:101
cell culture, 298:100–101
extraction, 298:101
gel filtration chromatography, 298:101
vector construction and expression,
298:100
Western blotting, 298:97–98
Focal adhesion kinase-related nonkinase
overexpression effects in chick embryo
fibroblasts, 298:90–91
tissue distribution, 298:98
Western blotting, 298:98
Folding, proteins, see also Fluorescence titration; Helix–coil transition; Partitioning, proteins in membranes
assays, see specific chaperones
chaperones, see specific chaperones
disulfide bond formation, see Disulfide
bond formation
extrapolation of temperature-invariant enthalpy using Planck–Benzinger thermal work function, 295:234–236,
238–239, 241, 245
Gibbs free energy change, 295:228–230
luciferase, β subunit folding analysis in
cell-free translation system, 290:12,
14–15
rate-limiting step in oligomer formation,
290:12
residue mutation analysis, 295:228–229
ribosome protein folding activity in Escherichia coli, 290:18
spontaneity, 295:227
steps, 290:11
temperature-invariant enthalpy of unfolding, 295:251, 253, 255–261
thermodynamic compensation,
295:245–246
Folding, RNA
circular dichroism of folding transitions,
see Circular dichroism, RNA
comparison with protein folding, 317:330
effectors, 317:331
footprinting analysis
definition of footprinting, 295:379–380
hydroxide radical cleavage and synchrotron X-ray footprinting

beamline set-up, **295:**393
dose–response of cleavage,
    295:382–383
factors affecting nucleic acid cleav-
    age, **295:**393–394
gel electrophoresis and analysis,
    295:400–401
kinetics of cleavage, **295:**382,
    386–388
magnesium-initiated RNA folding,
    295:399–400
manual mixing protocol, **295:**395–396
mechanism of nucleic acid cleavage,
    295:381–382, 386
millisecond exposure shutter,
    295:388–389
quenching of radical, **295:**387
radiolysis of water, **295:**382, 384–385
rapid mixing, device and protocol,
    295:389–392, 396–398
sample preparation, **295:**394–395
sources of radiation, **295:**388
Tetrahymena ribozyme folding analy-
    sis, **295:**402
quantitative footprinting, **295:**380–381
kinetic analysis, *see* Kinetic oligonucleo-
    tide hybridization; Synchrotron X-
    ray footprinting
ribozymes, **295:**379, 402
transitions in branched RNA, *see*
    Branched RNA
Folin–Ciocalteu reagent, total phenol
    analysis
advantages, **299:**177–178
chemistry of reaction, **299:**160–161
cinchonine precipitations, **299:**175–176
comparison with other techniques,
    299:169–171
flavonoid–nonflavonoid separations,
    299:170–173
flow automation, **299:**156–158
Folin–Denis reagent comparison,
    299:154–155
incubation conditions and color develop-
    ment, **299:**155–156
interference
    ascorbic acid, **299:**166–167
    nucleic acids, **299:**164
    oxidation, **299:**164
    proteins, **299:**164–165

sugars, **299:**165–166
sulfite, **299:**167–169
sulfur dioxide, **299:**167–168
molar absorptivity of specific phenols,
    299:161–163
phloroglucinol addition, **299:**173
preparation of reagent, **299:**155
sample preparation, **299:**159
standards and blanks, **299:**158–159
tannin separations, **299:**174–175
values for various wines, **299:**176–177
volume of reaction, **299:**156
Footprinting, *see also* Dimethyl sulfate;
    DNase I footprinting; Hydroxyl radical
    footprinting; Methidiumpropyl-EDTA–
    Fe(II) footprinting; Micrococcal
    nuclease; Permanganate footprinting;
    Synchrotron X-ray footprinting; Ultravi-
    olet footprinting
BiRA binding to biotin operator, thermo-
    dynamic analysis with DNase I,
    295:441–450
CytR–CRP cooperativity analysis with
    DNase I footprinting
    extended footprint analysis and multi-
        ple interactions, **295:**413–415
    free energy change evaluation,
        295:408–410
definition of footprinting, **295:**379–380
hydroxide radical cleavage and synchro-
    tron X-ray footprinting
    beamline set-up, **295:**393
    dose–response of cleavage,
        295:382–383
    factors affecting nucleic acid cleavage,
        295:393–394
    gel electrophoresis and analysis,
        295:400–401
    kinetics of cleavage, **295:**382, 386–388
    magnesium-initiated RNA folding,
        295:399–400
    manual mixing protocol, **295:**395–396
    mechanism of nucleic acid cleavage,
        295:381–382, 386
    millisecond exposure shutter,
        295:388–389
    quenching of radical, **295:**387
    radiolysis of water, **295:**382, 384–385
    rapid mixing, device and protocol,
        295:389–392, 396–398

sample preparation, **295:**394–395
sources of radiation, **295:**388
Tetrahymena ribozyme folding analysis, **295:**402
quantitative footprinting, **295:**380–381
Force assay, *see* Motility assay
Force clamp, *see* Optical trap
Formaldehyde
chromatin fixation for transmission electron microscopy, **304:**223–224
DNA–protein cross-linking, **304:**618
mapping nonhistone chromatin-associated factors, **304:**401–402, 407, 415, 418
Formamide, ladder analysis of modified RNA, **318:**14
N-Formyl peptide receptor, confocal microscopy imaging on neutrophils, **307:**385
Forskolin, photoaffinity labeling of P-glycoprotein with analogs, **292:**309
Forward genetics
MscL studies, **294:**479–480, 524
overview of microbial ion channel studies, **294:**522–524
plate phenotypes of microbes, **294:**521–522
c-Fos
antisense studies of learning and memory function, **314:**276, 290
pharmacokinetics of antisense inhibition, **314:**263
Fouling, *see* Biofouling, membranes
Fourier transform infrared spectroscopy
amide band assignments, **294:**79–83
amide hydrogen–deuterium exchange, **294:**83–84
angular orientation of secondary structural elements to membrane, **294:**87–91
attenuated total reflection, **294:**77, 90
caged GTP photolysis
band assignment by isotopic labeling, **291:**237–240
difference spectra, **291:**231
dithiothreitol effects on spectra, **291:**233–234
intermediates in decay, **291:**234–235
magnesium effects on spectra, **291:**232
photolysis reaction, **291:**231
synthesis of compounds, **291:**225

caged phosphate photolysis
band assignment by isotopic labeling, **291:**227–229
difference spectra, **291:**226–229
magnesium effects on spectra, **291:**229–231
difference spectroscopy of rhodopsin mutants
active state characterization
aspartate/glutamate mutants, **315:**184
dark state, **315:**183
H211N mutant, **315:**184–185
lipid vesicle versus detergent spectra, **315:**183–184
metarhodopsin-II, **315:**181, 183
attenuated total reflection sampling principle, **315:**180, 192–193
proton exchange studies, **315:**193–195
band assignment, **315:**195
molecular changes leading to activation, **315:**185–186
sample preparation, **315:**180–181
Schiff base environment studies
Glu-113, mutation and role in stabilization/activation, **315:**186–187, 189–190
Gly-90 mutants, **315:**190–192
low-temperature studies, **315:**191–192
water molecule interactions, **315:**188–189
DM-nitrophen photolysis
cation chelation, **291:**240–241
difference spectra, **291:**241–242
photolysis reaction, **291:**241
Fourier deconvolution, band narrowing, and band fitting, **294:**78–79
hydrogen-bonding network of water molecules in opsin mutants, **315:**129
infrared dichroism, **294:**84–87
instrumentation, **294:**76–77
instrumentation for photolysis studies, **291:**223–224, 244
ion channel applications, **294:**59–60, 75–76, 91
protein deposits
aggregation kinetics, **309:**569
attenuated total reflectance spectroscopy

data acquisition, **309:**565–566
powdered proteins, **309:**567
principle, **309:**564–565
sensitivity, **309:**565
thin film spectroscopy, **309:**566–567
diffuse reflectance spectroscopy
data collection, **309:**567–568
signal components, **309:**567
isotope-edited spectra, **309:**570
microscopy, **309:**569–570
prion protein, isotopically labeled peptides
intermolecular interactions,
309:588–589
secondary structure analysis,
309:585–588
sample preparation
mull preparation, **309:**562
overview, **309:**559
potassium bromide pellets,
309:562–564
spectra analysis
deconvolution, **309:**571–572
factor analysis, **309:**574, 576
overview, **309:**570–571
peak assignment, **309:**572–574
secondary structure, **309:**572–574
transmission mode for thin films
data acquisition, **309:**568–569
protein gels, **309:**568
water absorption minimization and suppression, **309:**560–562, 570
Ras studies with caged GTP
data collection, **291:**224–225
difference spectra, **291:**243–245
photosystem II mutants, **297:**357–358
rhodopsin activation studies, overview,
315:195–196
sample cells, **294:**77
sensitivity, **294:**59
stable-isotope labeling of opsins in baculovirus–insect cell recombinant expression system, **315:**25–26
time-resolved measurement
data collection techniques, **291:**232–233
temporal resolution, **291:**223
14-3-3, kinase, *see* Sphingosine-dependent
kinase 1
FOX, *see* Ferrous oxidation in xylene
orange

FPLC, *see* Fast protein liquid chromatography
Fractionation factor, *see* Low-barrier hydrogen bond
Frameshifting assay, RNA–protein interactions
applications, **318:**366–367, 374
cell culture and gene transfer,
318:370–371
efficiency of frameshifting, calculation,
318:373
extract preparation, **318:**372
iron in cells and frameshifting efficiency,
318:371–372
plasmids, **318:**370
principles, **318:**363, 365–370
programmed frameshifting, **318:**363, 365
reporter assays
β-galactosidase, **318:**372–373
luciferase, **318:**372
Western blot, **318:**373
reporter genes, **318:**365–366, 369–370
RNA structure analysis, **318:**366–367
FRAP, *see* Fluorescence recovery after photobleaching
FRAP assay, *see* Ferric reducing/antioxi-
dant power assay
FRASC assay, *see* Ferric reducing/antioxi-
dant power and ascorbic acid concentration assay
Free energy
activation, *see* Activation energy
hydrogen bonds, linear-free energy relationships
Brønsted coefficients and slopes,
308:254–261, 269, 274–276
intramolecular hydrogen bond
strength, **308:**261–263, 266–268
substituted phenols, **308:**254
profiles
activation energy, **308:**4–5
carbonic anhydrase, **308:**13
conformational change profiling,
308:8–9, 11–12
enzymatic bimolecular reactions,
308:7–9
enzymatic unimolecular reaction,
308:5–6
nonenzymatic unimolecular reaction,
308:4

orotidylate decarboxylase, **308:**11
partition ratios of intermediates, **308:**9
product concentration effects, **308:**6, 11
proline racemase, **308:**11–13
pyruvate phosphate dikinase, construc-
    tion of profile
calculations, **308:**172–174
dissociation constant determinations,
    308:164–165
equilibrium constant determinations,
    308:167–168
kinetic model in simplification,
    308:163
physiological conditions, modeling,
    308:174, 176
rate constant determinations,
    308:160, 162, 164–166
time courses for single turnover
    reactions/simulations,
    308:168–172
rate-limiting step, **308:**17, 19
software for construction, **308:**24,
    26–27
structure–function study applications,
    308:163
substrate concentration effects, **308:**6–7
Freeze-drying, protein aggregation
formulation development process for ther-
    apeutic proteins, **309:**248–249
glass transition temperature for protein
    formulations, **309:**249–250
mechanisms, **309:**246–247
prevention and stabilizers, **309:**247–249
sugar stabilizer crystallization avoidance,
    309:250–251
Freezing stress, protein aggregation
mechanism, **309:**245–246
prevention, **309:**245
therapeutic proteins, **309:**244–245
Frequency doubling, theory, **291:**196–198,
    201
Fresh frozen plasma, *see* Virus inactivation
FRET, *see* Fluorescence resonance energy
    transfer
FRNK, *see* Focal adhesion kinase-related
    nonkinase
FSCV, *see* Fast scan cyclic voltammetry
FTIR, *see* Fourier transform infrared spec-
    troscopy
FtsZ

electron microscopy
carbon-coated copper grid preparation,
    298:301–302
immunoelectron microscopy, **298:**296
staining and visualization of polymers,
    298:302, 312
GTPase assays, **298:**299–300, 308–309
immunofluorescence localization, **298:**296,
    302–305
polymerization
assembly conditions, **298:**309, 312
DEAE-dextran method, **298:**300–301
nucleotide dependence of conforma-
    tion, **298:**313
protofilament sheets, **298:**312–313
quantitative assay, **298:**301
slightly acidic conditions, **298:**300–301
tubes, **298:**313
purification
ammonium sulfate fractionation,
    298:298–299, 306–307
anion-exchange chromatography,
    298:299
cation-exchange chromatography,
    298:299
cell lysis, **298:**298, 306
overexpression in *Escherichia coli*,
    298:297, 305–306
protein concentration assay, **298:**299,
    308
sequence homology with tubulin,
    298:296–297, 305
Ftz, ultraviolet cross-linking of protein–
    DNA complexes *in vivo*, **304:**511–512
Fuc-GM1
nomenclature, **311:**639
radiolabeled compound synthesis
materials, **311:**640, 642
overview, **311:**639–641
stability, **311:**655–656
tritium labeling at position 3 of sphin-
    gosine, **311:**647–649
Fumonisins
B_1
ceramide synthase inhibition, **311:**20–
    21, 131, 335, 346–347
isolation and purification, **311:**365–369
mass spectrometry, **311:**363–364
B_2, isolation and purification,
    311:365–369

B_3, isolation and purification, **311**:366–367, 369

B_4, isolation and purification, **311**:366, 369

colored pigment removal from preparations, **311**:369

detection, **311**:363

disease associations, **311**:361–363

enzyme-linked immunosorbent assay, **311**:372–373

Fusarium synthesis
  cultures
    corn, **311**:364–365
    liquid, **311**:365
    rice, **311**:365
  species and type distribution, **311**:361
  high-performance liquid chromatography
    derivatization, **311**:370
    evaporative light-scattering detection, **311**:371
    reversed-phase chromatography, **311**:370–371
  structures, **311**:361–363
  thin-layer chromatography, **311**:371–372

Functional genomics, overview, **314**:148–149

Functional tagging
  equations in analysis, **293**:28
  potassium channel assembly, **293**:31
  principle, **293**:28–29
  subunit stoichiometry determination, **293**:27

Fura Red, loading of cells, **307**:125–126

Fura-2, *see also* Calcium flux
  calibration, **294**:7–9, 24–25
  cell loading, **294**:4–5, 21, 23–24
  instrumentation
    detectors, **294**:7
    filters, **294**:6, 22–23
    light source, **294**:6
    microscope, **294**:5
    microtiter plate reader, **294**:21, 23
  presynaptic terminals
    population measurements
      calibration of fluorescence signal, **294**:14–15
      data acquisition, **294**:13–14
      dye loading, **294**:12–13
    single terminal measurements in mossy fibers

advantages, **294**:15–16
calibration of fluorescence signal, **294**:19
data acquisition, **294**:17–18
dye loading, **294**:17
principle, **294**:4, 21–22
simultaneous patch-clamp analysis of plant cells, **294**:433–435

G

$G\alpha_0$, antisense knockdown studies, **313**:151–152

GAAA tetraloop, modeling with comparative sequence analysis, **317**:503–504

GABA, *see* γ-Aminobutyric acid

GABA$_A$, *see* γ-Aminobutyric acid receptor type A

GABA$_B$, *see* γ-Aminobutyric acid receptor type B

GABA transporter, *see* γ-Aminobutyric acid transporter; Vesicular γ-aminobutyric acid transporter

G-actin
  blot overlay assay of binding proteins, **298**:41
  iodine-125 Bolton–Hunter labeling, **298**:36–39
  latrunculin-A binding
    cytoskeleton effects, **298**:18–19
    mechanism, **298**:19, 29
    mutagenesis analysis of binding sites on actin, **298**:19–20
  6-nitroveratryloxycarbonyl chloride caged conjugate
    fluorescence photoactivation microscopy, **291**:104–105
    functional analysis of caged and decaged proteins, **291**:105–107
    labeling ratio, **291**:101–102
    photoactivation conditions, **291**:104
    polymerization competence, **291**:102–103
    preparation, **291**:101
    tetramethylrhodamine labeling of caged protein, **291**:103
  preparation for *Listeria monocytogenes* actin assembly assay, **298**:118–119

GALA peptide, transport across cell membranes, **313:**355

Galactose oxidase, translocation assay of sphingolipids, **312:**573

β-Galactosidase

chemiluminescence assay substrates

imidopyrazine substrate, **305:**384

indole substrates, **305:**383–384

luminol substrates, **305:**381

NADH assay, **305:**387

conjugated secondary antibody chemiluminescence detection in Western blot

dual detection, **305:**424–425

single detection, **305:**424

cyanobacteria, *in vivo* assays of reporter, **305:**526–527

flagellin gene induction reporting, **310:**68–69

flow cell biofilm analysis, **310:**32–35

fluorescent sphingolipid assay, **312:**300

lasR expression reporting, **310:**125, 127

lateral flagellar gene activation reporting, **310:**64–65

loading of substrate dyes into bacteria with hypotonic shock, **310:**33–36

multicolor experiments with fluorescence *in situ* hybridization of biofilms, **310:**39, 77

photoaffinity labeling, **311:**580

quorum sensing autoinducer-fusion protein detection, **310:**120–122

reporter gene assay with chemiluminescence

cell extract preparation

tissue cultures, **305:**432

yeast, **305:**432–433

controls, **305:**436

dioxetane substrates, **305:**429–431

dual reporter gene assays with luciferase

cell extract preparation, **305:**447

controls, **305:**449

detection, **305:**447–448

overview, **305:**428–429, 444–446

reagents, **305:**446–447

heat inactivation of endogenous enzyme, **305:**436

instrumentation, **305:**449–450

microplate luminometer detection

direct lysis of microplate cultures, **305:**435–436

Galacton-Plus, **305:**435

Galacton-*Star*, **305:**434–435

overview, **305:**428–429

reagents, **305:**431–432

sensitivity, **305:**450

tube luminometer detection

Galacton-Plus, **305:**433–434

Galacton-*Star*, **305:**433

S fimbrial adhesin gene activation reporting, **310:**68

solid surface genetic responses, fusion protein studies

Escherichia coli

controls, **310:**70–71

promoter screening, **310:**71–72

biofilm, gene expression monitoring, **310:**72–73

marine bacteria, **310:**70

Pseudomonas aeruginosa, alginate gene expression, **310:**74–76

sphingolipid identification, **311:**667–668

streptococci fusion protein reporters

assay in biofilms, **310:**459

features, **310:**458

fluorescence detection, **310:**460

substrate dyes

fluorescein di-β-D-galactopyranoside assay, **310:**59–61

methylumbelliferyl-β-D-galactoside, **310:**61, 75

o-nitrophenyl-β-D-galactoside chromogenic assay, **310:**59, 65, 72–74

overview, **310:**32–33

β-Galactosylceramidase, deficiency in Krabbe's disease, **311:**297

Galactosylceramide

[^{14}C]galactosylceramide, enzymatic synthesis and purification, **311:**685–687

human immunodeficiency virus studies with microdomains

cholesterol–galactosylceramide reconstitution, **312:**499

gp120

binding site mapping, **312:**500

interaction kinetics, **312:**498–499

interactions, **311:**627

overview, **312:**495–496

surface pressure measurements,
    312:496–497
soluble synthetic analogs
    anti-human immunodeficiency virus-1
        activity
        clinical applications, 311:638
        mucosal epithelial cell infection,
            311:634, 636
        peripheral blood mononuclear cell in-
            fection, 311:636
        screening throughut, 311:636–637
        structure–activity relationships,
            311:636–637
    cytotoxicity assays, 311:636
    design strategy, 311:628–629
    enzyme-linked immunosorbent assay,
        311:629–630
    phase behavior, 311:629, 637
sulfation, see Sulfatide
synthesis, see Ceramide galactosyltrans-
    ferase
tubules
    doping, 312:516
    electron crystallization trials, 312:519
    materials, 312:516, 518
    preparation of tubules, 312:518–519
viral envelope glycoprotein binding
    assays
    direct assay, 311:630–631
    inhibition assay, 311:631–633
viral membrane fusion promotion,
    312:495
Galactosylceramide:sulfotransferase
    activators, 311:104–105
    assay, 311:103–104
    kinetic parameters, 311:104
    reaction catalyzed, 311:103
    sequence analysis, 311:104
Galanin
    biosynthesis, 314:186
    feeding behavior control
        antisense effects in colchicine-induced
            hyperphagia model
            antisense oligodeoxynucleotide de-
                sign, 314:195–196
            overview of model, 314:186–187, 196
            specificity of effects, 314:196
        overview of control, 314:186, 195
    luteinizing hormone-releasing hormone
        regulation

antisense effects in ovariectomized rats
    intracerebroventricular infusion,
        314:191
    luteinizing hormone pulsatility ef-
        fects, 314:191–192
    oligonucleotide design, 314:191
    overview, 314:186
    receptors, 314:186
Galbonolide A, see Rustmicin
GalNAc-T, see N-Acetylgalactosaminyltran-
    sferase
GalT-2, see Lactosylceramide synthase
Gamete lytic enzyme, preparation and cell
    treatment for nuclear transformation,
    297:34–35
Ganglioside, see Glycosphingolipid; specific
    gangliosides
Ganglioside glycosyltransferases
    N-acetylgalactosaminyltransferases
        assay
            incubation conditions, 311:86–87
            materials, 311:85
            product identification and structural
                analysis, 311:89–91
            product separations, 311:88, 93–94
            quantitative analysis, 311:88–89
            stock solutions, 311:86
        knockout mice, 311:82
        reaction catalyzed, 311:83
    galactosyltransferases
        assay
            incubation conditions, 311:86–87
            materials, 311:85
            product identification and structural
                analysis, 311:89–91
            product separations, 311:88, 93–94
            quantitative analysis, 311:88–89
            stock solutions, 311:86
        reaction catalyzed, 311:83
    sialyltransferases
        assay
            incubation conditions, 311:87–88
            materials, 311:85
            product identification and structural
                analysis, 311:89–91
            product separations, 311:88, 93–94
            quantitative analysis, 311:88–89
            stock solutions, 311:87
        reaction catalyzed, 311:85
    topology, 311:82

Ganglioside sialidases
  assays
    fluorescent versus radioactive assays,
      312:346
    4-methylumbelliferyl-*N*-acetyl-neur-
      aminic acid as substrate, **312:**352–
      353
    sialic acid determination with periodic/
      thiobarbituric acid, **312:**351–352
    sialic acid radiolabeled substrates
      assay, **312:**358
      sialyltransferase labeling, **312:**357
      tritium labeling by oxidation/reduc-
        tion, **312:**357–358
      tritium-labeled *N*-acetylneuramine lacti-
        tol as substrate, **312:**353–354
      tritium-labeled sphingosines as sub-
        strates
        assay, **312:**356
        C-3 labeling, **312:**354–355
        double-bond labeling, **312:**355–356
  mammalian enzymes
    properties, **312:**340–346
    subcellular localization, **312:**340
  nomenclature, **312:**339
  preparation for activity assays
    cultured cell lysates, **312:**347
    tissue
      cytosolic sialidases, **312:**347–348
      lysosomal sialidases, **312:**348–349
      myelin-associated sialidase,
        312:350–351
      nuclear sialidases, **312:**351
      plasma membrane-bound-sialidases,
        312:349–350
Gap junction, two-dimensional crystallization
  electron cryomicroscopy, image pro-
    cessing, **294:**165–167, 169, 171,
    173–175
  membrane isolation from enriched recom-
    binant preparations, **294:**139–142
  in situ crystallization,
    294:144, 147
Gas chromatography, *see* Gas chromatography–mass spectrometry; Isotope dilution negative-ion electron capture mass
  spectrometry; Negative-ion chemical
  ionization mass spectrometry; Photoionization gas chromatography

Gas chromatography headspace assay, volatile lipid oxidation products
  diet effects on lipid peroxidation,
    299:198–200
  instrumentation, **299:**197
  overview, **299:**196–197
  phenolic antioxidant effects, *in vitro*
    assays, **299:**200
  sample preparation, **299:**197–198
  wine phenolics, evaluation, **299:**199–201
Gas chromatography–mass spectrometry
  DNA analysis
    acid hydrolysis, **319:**411
    applications, **319:**401–403, 416–417
    blood analysis
      antioxidant addition during extrac-
        tion, **319:**408–409
      blood preparation, **319:**404
      DNA extraction, **319:**404–405,
        410–411
      materials, **319:**403–404
      oxygen exclusion during extraction,
        319:408
      ribonuclease digestion, **319:**405,
        407–408
      storage of blood samples, **319:**409–
        410, 416
    derivatization, **319:**411–412, 415
    8-hydroxyguanine, **319:**148–149, 152,
      402
    quantitative analysis, **319:**415–416
    running conditions, **319:**415
  metal oxide oxidation products, **319:**31,
    33–34
5-nitro-γ-tocopherol, **301:**487, 489
phenolic antioxidants in wine
  costs, **299:**151
  extraction and characterization,
    299:142–143
  gas chromatography conditions,
    299:141
  instrumentation, **299:**139, 141–142
  linearity, **299:**148
  mass spectrometry conditions, **299:**142
  materials, **299:**139
  peak identification, **299:**143–145
  precision, **299:**149–150
  recovery, **299:**148–149
  resolution, **299:**145, 147–148
  sample preparation, **299:**139

sensitivity, **299**:149–150
standards, **299**:139–140
α-tocopherol
  calibration curve preparation, **299**:311,
    317–318
  extraction and derivatization,
    299:311–312
  instrumentation, **299**:312
  oxidation product analysis from rat
    liver, **299**:314–317
  selected ion monitoring, **299**:313–314
tyrosine derivatives
  amino acid analysis, **301**:473–474
  arterial tissue analysis, **301**:481–483
  derivatization with *N*-methyl-*N*-(*tert*-
    butyldimethylsilyl)trifluoroacet-
    amide and spectrometry, **301**:474–
    475, 477
  protein analysis, **301**:477, 480–481
vitamin E ozonation analysis,
  319:550–551
GAT-1, *see also* Vesicular γ-aminobutyric
  acid transporter
  epitope tagging in transmembrane topol-
    ogy analysis
    carbonate extraction, **296**:301–302
    expression in *Xenopus* oocytes,
      296:299–300
    glycosylation analysis, **296**:302
    immunoprecipitation of fusion pro-
      teins, **296**:300–301
    plasmid construction, **296**:296–298
    principle, **296**:294–295
    protease protection assay, **296**:294–296,
      300, 302–305, 307
    RNA transcription, **296**:298
  fusion proteins in transmembrane topol-
    ogy analysis, **296**:293–294
  hydropathy analysis, **296**:295–296
  membrane assembly, **296**:296, 304–305
  phylogenetic analysis, **296**:428–436
  stoichiometry of transport, **296**:551
Gating current
  charge per channel, **293**:334
  data analysis, *see* Hidden Markov model;
    Two-dimensional dwell-time analysis;
    Voltage clamp
  instrumentation for recording
    amplifiers, **293**:341–342
    clamping configurations, **293**:341

computer control, **293**:343
filters, **293**:342
schematic, **293**:340
system performance, **293**:343–344
waveform generation, **293**:342–343
modeling, **293**:332–333
patch clamp measurement, **293**:338–340
separation from other currents in detection
  capacitive current, **293**:335–338
  ionic current, **293**:335–336
Shaker-type potassium channel, gating
  current measurement
  analog compensation of linear compo-
    nents, **293**:345–346
  elementary gating event, **293**:348–350
  expression systems, **293**:344–345
  frequency domain recording, **293**:350,
    352
  inactivation
    charge mobilization, **293**:347–348
    mutant, **293**:345–347
  leak correction, **293**:405
  signal-to-noise ratio in detection,
    293:334–335
  voltage sensors, **293**:331–332
Gaussian beam optics
  beam radius, **291**:187
  birefringence, **291**:194–195
  focusing, **291**:187
  Rayleigh range, **291**:189
  reflection, **291**:192–193
  refraction, **291**:194
  Self's equations, **291**:188–190
  waist position, **291**:188, 190
  waist radius magnification, **291**:189
  wave polarization, **291**:190–191
Gb$_3$
  human immunodeficiency virus interac-
    tions, **312**:503–505
  verotoxin binding in platelets, **312**:463
Gb$_4$
  dependent adhesion of human embryonal
    carcinoma 2102 cells
    adherent cell measurements, **312**:457
    adhesion conditions, **312**:454
    caveats, **312**:457–458
    overview, **312**:454
    washing, **312**:454, 457
  verotoxin binding, **312**:459

GBR 12783, dopamine transporter binding, **296:**203, 210–212, 217–218
GBR 12909, dopamine transporter binding, **296:**213, 253–254
GBR 12935, dopamine transporter binding, **296:**203, 212–218
GCAP1, *see* Guanylate cyclase-activating protein 1
GCAP2, *see* Guanylate cyclase-activating protein 2
GC–MS, *see* Gas chromatography–mass spectrometry
Gcn5
  human enzyme
    affinity competition assay, **304:**709–711
    assay of histone acetyltransferase, **304:**703–705, 707
    expression and purification of recombinant proteins
      glutathione *S*-transferase fusion proteins, **304:**701–702
      histidine-tagged enzyme, **304:**699–700
    expression quantification by sulfur-35 labeling and immunoprecipitation, **304:**712–713
    immunodepletion analysis, **304:**709
    phosphorylative regulation
      DNA–PK phosphorylation *in vitro*, **304:**703, 706–707
      immunodepletion of DNA–PK, **304:**709
      overview, **304:**698
      phosphatase effects on activity, **304:**705, 707
      phosphate-32 labeling *in vivo*, **304:**711, 713–714
      Western blot analysis, **304:**706
    transfection of COS cells, **304:**711–712
  recruitment in yeast, **304:**697–698
GCS, *see* Glucosylceramide synthase
GD1a
  nomenclature, **311:**639
  radiolabeled compound synthesis
    materials, **311:**640, 642
    overview, **311:**639–641
    stability, **311:**655–656
    tritium labeling at position 3 of sphingosine, **311:**647–649
GD1b

nomenclature, **311:**639
radiolabeled compound synthesis
  materials, **311:**640, 642
  overview, **311:**639–641
  stability, **311:**655–656
  tritium labeling at position 6 of terminal galactose and *N*-acetylgalactosamine residues, **311:**642–645
GD3 synthase, photoaffinity labeling, **311:**577, 593
GDRDA, *see* Genetically directed representational difference analysis
Gelatin, vascular casts with fluorochrome labeling
  advantages, **307:**98–99
  animal perfusion, **307:**105
  antifading agents, **307:**99–100
  confocal microscopy, **307:**102–103
  fetus perfusion, **307:**105–106
  fluorescence labeling of gelatin, **307:**104–105
  immunostaining comparison and application, **307:**98, 107
  mounting in polyester resin, **307:**99–102, 107
  organ perfusion, **307:**106
  replica technique comparison, **307:**97
Gel filtration
  advantages and disadvantages in subunit interaction analysis, **293:**15
  cell-free translation system
    chaperones, **290:**21
    products, **290:**6–7
  chaperonin 10, **290:**172
  chaperonin 60, **290:**172
  α-crystallin, **290:**366–367, 369
  cytosolic chaperonin, **290:**234
  dissociation constant determination, *see* Gel filtration, subunit dissociation constant determination
  GLYT2, **290:**8–9
  GroEL–GroES complex, **290:**128–129
  GroEL, **290:**104, 108, 110
  GrpE, quaternary structure analysis, **290:**286
  heat shock protein 60, **290:**198
  heat shock protein 90, **290:**415, 417
  Hop/p60, **290:**428
  mammalian mitochondrial chaperonin 10, **290:**226–227

mammalian mitochondrial chaperonin 60, **290:**209, 211

p23, **290:**426

protein aggregation, *see* Gel filtration, protein aggregation

quaternary structure determination, overview, **290:**275–277

SecB, **290:**449–450

symbionin, **290:**184

Gel filtration, protein aggregation

advantages for study, **309:**318

principle, **309:**318–319

small zone chromatography

Bence Jones proteins, **309:**329–330, 332

column preparation, **309:**323–324

elution behavior, **309:**322

large zone comparison, **309:**319–320

V_L proteins

concentration dependence of elution profiles, **309:**320

dimerization constant determination, **309:**328

dimerization simulation, **309:**325–326

experimental elution profile generation, **309:**324–325

Gel filtration, subunit dissociation constant determination

hemoglobins

bovine hemoglobin, **306:**127

hemoglobin A, **306:**123–124

hemoglobin F, **306:**124–125

hemoglobin F_1, **306:**125

hemoglobin S, **306:**123–124

preparation, **306:**114–115

recombinant hemoglobins with substitutions in $\alpha_1\beta_2$ interface, **306:**126

recombinant sickle hemoglobins, **306:**125–126

small zone gel filtration on Superose-12

analysis time, **306:**118

calculations, **306:**121–122

care of column, **306:**118–119

elution conditions, **306:**117

elution profiles and curve analysis, **306:**117, 119–122

overview, **306:**114

peak concentration analysis, **306:**117–118

sample dilution and peak broadening, **306:**127–129

standards and calibration, **306:**115–117

systematic errors, **306:**122

Gel overlay, calcium-binding protein target identification, **316:**115–117

Gene gun, transformation of *Chlamydomonas reinhardtii* chloroplasts, **297:**311, 317–318

Gene identification, *see also* Computational gene finding; Direct complementary DNA selection; Exon amplification; Expressed sequence tag

combination of techniques, **303:**146–147, 160–161

contigs as starting materials, **303:**149–151

preliminary considerations in technique selection

assessing success, **303:**149

genomic region

size, **303:**147

type of chromosome band, **303:**147–148

verification, **303:**147

Gene mapping, *see* Fluorescent *in situ* hybridization; Optical mapping

GENESIS simulation program, **294:**707, 709–710, 713, 726–727, 730

Gene switching

advantages, **313:**178

delivery of oligonucleotides, **313:**178–179, 185–187

dual luciferase assay, **313:**189

HeLa cell system, **313:**188–189

leader sequence targeting, **313:**179, 186

materials and reagents, **313:**186–187

modified oligonucleotides for study, **313:**175–176, 182–185

optimization, **313:**185

overview, **313:**174–175, 177–178

plasmid design, **313:**179–180

positive test system, **313:**187–188

reporter genes, **313:**180–181, 188

requirements for culture system, **313:**174–175, 178–179

steric blockers, **313:**181–182

Gene therapy, *see also* Adeno-associated virus; Encapsidated adenovirus minichromosome; Ribozyme

cystic fibrosis transmembrane regulator, *see* Gene therapy, cystic fibrosis transmembrane regulator

dimerizer-regulated gene expression,
    306:266, 280
erythropoietic protoporphyria
    bone marrow targeting, **319:**482
    fibroblasts, **319:**483
    mouse model, **319:**483–484
    peripheral blood stem cells,
        319:483–484
gene switch, ideal criteria, **306:**282
green fluorescent protein as marker
    adeno-associated viral vectors, **302:**437
    adenoviral vectors, **302:**437
    detection and cell sorting, **302:**425, 427
    herpes simplex virus vector
        amplicon vector concentration,
            302:434
        cells and viruses, **302:**433
        cytotoxicity of vector, **302:**435–437
        gene transfer *in vitro*, **302:**434–435
        pHE vector construction, **302:**433
        titration, **302:**434
        virus production, **302:**433–434
    lentiviral vectors, **302:**437
    red-shifted mutant, **302:**424–425
    retroviral vectors, *see also* Retrovirus,
        green fluorescent protein ex-
        pression
        efficiency, **302:**427
        human peripheral blood lymphocyte
            transduction efficiency, **302:**430
        lymphocyte suicide gene therapy in
            relapsed leukemia, **302:**427–428
        rat lymphocyte transduction effi-
            ciency, **302:**428–430
herpes simplex virus vectors
    amplicon-based vectors, **302:**431–433
    genome, **302:**430–431
    helper genes, **302:**431
human immunodeficiency virus treatment
        with ribozymes
    clinical trials, **306:**224
    preclinical studies, **306:**222–224
    retroviral vectors for gene delivery,
        306:223–224
    strains and ribozyme cocktail treat-
        ment, **306:**222–223
    targets, **306:**222–223
intracellular trafficking of RNA tran-
        scripts, **306:**248, 251–252
MDR1, *see* Gene therapy, *MDR1*

promoters in effector RNA expression
    overview, **306:**247–248
    RNA polymerase I promoters,
        306:248–249
    RNA polymerase II promoters
        messenger RNA, **306:**249–251
        U1, 251–252
    RNA polymerase III promoters
        overview of types, **306:**252–253
        type 2 promoters, **306:**253, 255, 257
        type 3 promoters, **306:**257
retroviral vectors, translational control
        with iron response element
    clinical application, **306:**241
    design of vector, **306:**227, 229, 240
    particle generation and cell transduc-
        tions, **306:**230–232
    plasmid and vector constructions,
        306:229–230
    rationale, **306:**226
    regulated expression analysis, **306:**232,
        234, 240–241
    stability of vector, **306:**234, 236
    stable producer cell analysis,
        306:236–237
RNA applications
    antisense RNAs, **306:**244
    decoys and aptamers, **306:**244–245
    overview, **306:**207, 242–243, 259–260
    ribozyme applications and develop-
        ment, **306:**207–208, 225, 245–247
stability of recombinant RNA transcripts,
    306:248, 250–252, 255
viral vector optimization, **302:**424
viral vector selection for effector RNA
    expression, **306:**257–259
Gene therapy, cystic fibrosis transmem-
        brane regulator
adeno-associated virus vectors
    genomic features of virus, **292:**725–726
    insertion locus, **292:**725, 728
    latency, **292:**725
    proliferating cell transfection efficiency,
        292:728
    size limitations, **292:**727
    transfer efficiencies in various cells,
        292:727–728
    in vitro transfer, **292:**728–729
    in vivo transfer
        clinical trials, **292:**731–732

macaque, **292:**729, 731
rabbit, **292:**729
adenovirus vectors, **292:**723–724
airway epithelium factors
  electrolyte transport, pumps and chan-
    nels, **292:**719–720
  histology, **292:**718–719
  pathogen-binding theory and transfec-
    tion efficiency requirements,
    292:720–722
cationic lipid:plasmid DNA formulations
    in gene delivery
  advantages, **292:**697–698
  neutral lipid enhancement of transfec-
    tion, **292:**698
  optimization of formulations, *in vitro*
    cell culture, **292:**700
    β-galactosidase reporter gene assay,
      292:700–701
    liposome preparation, **292:**699–700
    overview, **292:**698–699
    ratios for optimal transfection effi-
      ciency, **292:**701, 704–705
    structures of cationic lipids,
      292:701–702
  optimization of formulations, *in vivo*
    chloramphenicol acetyltransferase re-
      porter gene assay, **292:**709–710
    correlation to *in vitro* studies,
      292:712
    intranasal instillation in mice,
      292:709–710
    naked DNA control, **292:**710–711
    ratios for optimal transfection effi-
      ciency, **292:**711–712
  transfection, airway epithelial cells
    cell culture, **292:**714
    fluorescence assay of channel activ-
      ity, **292:**714–716
  transfection efficiency, factors affecting
    in vitro
    cell lines, **292:**708
    contact time, **292:**708
    excipient selection, **292:**707–708
    precipitation of complexes, **292:**705,
      707
    size of complex, **292:**707
    vector promoters, **292:**708–709
  transfection efficiency, factors affecting
    in vivo

DNA purity, **292:**713
dose, **292:**713
toxicity, **292:**713
Moloney murine leukemia virus vector,
    292:722–723
Gene therapy, *MDR1*
  adeno-associated virus vectors
    integration locus, **292:**539
    integration of *rep* negative vectors,
      292:544–545
    inverted terminal repeat sequences,
      292:539, 544
    packaging size and site, **292:**539–540,
      545
    recombinant virion production,
      292:540–543
    titration of vector, **292:**543
    transduction and selection of express-
      ing cells, **292:**543, 545
  application for cancer chemotherapy pa-
    tients, **292:**524, 526, 538, 547, 557,
    572
  bicistronic vectors
    cell culture, **292:**530
    construction
      bicistronic vectors, **292:**528–530
      subcloning vectors, **292:**526–528
    fluorescence activated cell sorting anal-
      ysis of expression, **292:**531, 533
    α-galactosidase bicistronic vector
      assay of α-galactosidase, **292:**532
      Fabry disease treatment rationale,
        292:525–526
      protein expression analysis,
        292:533–535
    herpes simplex virus thymidine kinase
      bicistronic vector
      ganciclovir selection, **292:**524–525
      protein expression analysis,
        292:535–536
      transduction and transplantation of
        bone marrow cells, **292:**532,
        536–537
    internal ribosome entry site utilization,
      292:525–526
    NADPH oxidase gp91 bicistronic vec-
      tor for X-linked chronic granulo-
      matous disease therapy, **292:**537
    overview of vectors, **292:**525–526, 537
    retrovirus transduction, **292:**531, 533

transfection with calcium phosphate,
  292:531, 533
vincristine sensitivity assay, **292:**530,
  534–535
bone marrow targeting, **292:**524–525
dilution effect, **292:**523
retroviral vectors
  limitations, **292:**538
  transfer of human P-glycoprotein to hu-
    man T lymphocytes
    applications, **292:**557–558
    cell culture, **292:**560
    characterization of transduced cells,
      292:570–571
    packaging cells, **292:**558, 560–561
    safety, **292:**571–572
    screening of producer clones,
      292:565–567
    transduction efficiency, **292:**558–559,
      571
    transduction of packaging cells,
      292:564–565
    transduction of T cells, **292:**567,
      569–571
    transfection of producer cells,
      292:561, 563
  transfer of human P-glycoprotein to
    murine hematopoietic stem cells
    cell culture, **292:**255–256, 550–551
    colony assays of multidrug-resistant
      stem cells, **292:**257–258
    fluorescence-activated cell sorting,
      292:547–548, 550
    gene transfer, **292:**256, 550–552
    immature cell phenotype, preserva-
      tion in culture, **292:**552, 554
    immunomagnetic cell fractionation,
      292:256, 548, 550
    protein expression assays in trans-
      duced bone marrow cells,
      292:257, 554–555
    protein expression assays in trans-
      planted transduced bone mar-
      row cells, **292:**555–556
Genetically directed representational difference analysis
  controls, **306:**76
  driver pools, **306:**75
  oligonucleotides and enzymes, **306:**75–77
  overview, **306:**74–75

Genie, computational gene finding, **303:**81
Genome Channel, computational gene find-
  ing, **303:**82–83
Genomic sequence analysis, *see* Computa-
  tional gene finding
Genscan, computational gene finding,
  303:81, 159–160
Geranylgeranyl diphosphate synthase, gene
  cloning and carotenoid biosynthesis,
  297:248–249
GFP, *see* Green fluorescent protein
G$_{i\alpha1}$
  ADP-ribosylation, transition-state imbal-
    ance, **308:**396
  GTP hydrolysis
    associative versus disociative mecha-
      nism, **308:**72–73
    magnesium ions in catalysis, **308:**91
    microscopic reversibility, **308:**91–92
    overview, **308:**71–72
    stages, **308:**90
    transition-state, **308:**72
    X-ray crystallography with ligands
      crystallization conditions, **308:**74–75
      data collection, **308:**74, 76
      data sets, **308:**73–74
      domains, **308:**76
      GDP complex, **308:**86–88
      GDṖP$_i$ ternary complex trapping,
        308:85–86, 91
      ground state enzyme–substrate com-
        plex, **308:**78, 80–81, 90
      hydrogen bonding interactions,
        308:76–78
      intermediates, forcing within crystals,
        308:88–90
      regulators of G-protein signaling,
        mechanism of action, **308:**92
      transition-state as modeled by
        GDṖAlF$_4^-$·Mg^{2+}, **308:**81–85,
        90–91
  kinetic isotope effect analysis of transi-
    tion-state, **308:**322
  protein–protein interactions in signal
    transduction, **308:**71
Giant dopamine neuron, *see* Neuron
Giant membrane patch clamp
  applications and limitations, **293:**267
  bath solution switching, **293:**271–274
  capacitance measurement, **293:**276–277

half-cell giant patch, **293:**270–271
noise minimization, **293:**222–223
phospholipid exchange, **293:**277–280
pipette
    fabrication, **293:**268–269
    solution switching, **293:**274, 276
seal formation, **293:**269–270
Giauque function, derivation, **295:**232
Gibberellin, caged
calcium response measurement with fluo-
        rescent probes, **291:**478, 481–482
injection into plant cells, **291:**477–478
photolysis conditions, **291:**478–480
structure, **291:**475
synthesis, **291:**475–476
Gibbs energy of binding
calculation, **295:**104, 295
comparison of experimental and pre-
        dicted values, **295:**308–312, 314–315
contributions
    Coulombic contributions, **295:**108
    enthalpy change, **295:**101, 104–106,
        295–296
    entropy change, **295:**101, 104, 106–107,
        295–296
    heat capacity change, **295:**101, 104,
        107–108, 295–296
    protonation contributions, **295:**108–109
    translational entropy, **295:**109–110
DNase I footprinting analysis
    BiRA binding to biotin operator,
        295:441–450
    CytR–CRP cooperativity analysis
        extended footprint analysis and mul-
            tiple interactions, **295:**413–415
        free energy change evaluation,
            295:408–410
energy minimization of peptide ligand
        conformations, **295:**118–119
folding relationship to binding,
        295:297–298
ligand design, **295:**101
parameter calculations in energy pre-
        diction
    enthalpy change, **295:**302–303, 314
    entropy change, **295:**303–305, 314
    heat capacity change, **295:**301, 313–314
    reparameterization, **295:**312–314
prediction for human immunodeficiency
        virus-1 protease ligands

conformational change of protease
        upon ligand binding, **295:**118
hydrophobic effect in binding,
        295:112–113
inhibitor design by rational mutagene-
        sis, overview for aspartic prote-
        ases, **295:**119–122
pepstatin, **295:**114–116
residue stability constants for protease,
        295:117–118
resistance to inhibitors, molecular
        mechanism, **295:**125–126
structural mapping of binding energet-
        ics, **295:**115–118, 124–125
substrate binding, modeling
    chromogenic substrate assays and
        thermodynamic analysis,
        295:122–123
    natural substrate binding complex
        and thermodynamic analysis,
        295:123–125
values for various ligands, **295:**112–114
proton linkage effects in determination,
        295:305–308
solvent accessibility, calculation for pro-
        teins, **295:**110–111, 297–301
structural parameterization, **295:**104–105
Gibbs–Helmholtz equation, temperature de-
        pendence of equilibrium constant,
        295:89–90
Ginkgo biloba
flavonoid antioxidants in extracts, absorp-
        tion and excretion in humans, **299:**94
inducible nitric oxide synthase, effects on
        activity, **301:**510, 512–513
medicinal uses, **301:**505
GIRK, *see* G-protein-coupled inward recti-
        fier potassium channel
Glass bead, nuclear transformation of *Chla-
        mydomonas reinhardtii*
cell growth, **297:**33–34
gamete lytic enzyme, preparation and cell
        treatment, **297:**34–35
materials, **297:**33
GLE, *see* Gamete lytic enzyme
Glibenclamide
radioligand synthesis
    azidoiodoglibenclamide, **294:**449–451
    iodoglibenclamide, **294:**446–448
SUR1 binding assays

particulate receptors, **294:**457
solubilized receptors, **294:**457–458
Glioma, *see* Microtubule-associated protein
1A, antisense inhibition in glioma
model
α-Globin, locus control region
binding proteins
dimethylsulfate footprinting and ligation-mediated polymerase chain reaction analysis, **304:**580, 582
novel protein identification,
304:582–584
overview, **304:**573
DNase I hypersensitivity, **304:**572
guanine–adenine ligation-mediated polymerase chain reaction and dimethyl
sulfate footprinting, **304:**578, 580–583
β-Globin
antisense correction of thalassemic mutants
HeLa cell system
cell culture, **313:**510
lipofection, **313:**510–511
overview of experimental system,
313:508–509
phosphorothioate oligoribonucleotide, **313:**508, 510
principle, **313:**506–507, 509–510
reverse transcription-polymerase
chain reaction analysis of RNA,
313:511–512
RNA isolation, **313:**511 ⁻
Western blot analysis, **313:**512–514
luciferase reporter gene assay
cell lines, **313:**520
luminescence detection, **313:**521
oligonucleotide treatment,
313:520–521
plasmids, **313:**520
principle, **313:**518–520
protein assay, **313:**521
U7 small nuclear RNA system
cell culture, **313:**517
overview, **313:**514–516
reverse transcription-polymerase
chain reaction analysis of RNA,
313:518
stable expression cell line selection,
313:517–518
transient expression, **313:**515

U7.324 construct, **313:**516–517
gene and histone acetylation analysis in
chick embryo, **304:**542, 547
IVS2-654 mutant in thalassemia,
313:507–508
locus control region
binding proteins
dimethylsulfate footprinting and ligation-mediated polymerase chain
reaction analysis, **304:**580–582
novel protein identification,
304:582–583
overview, **304:**573
DNase I hypersensitivity, **304:**572
guanine–adenine ligation-mediated
polymerase chain reaction and dimethyl sulfate footprinting,
304:578, 580–584
Globotetoraosylceramide, enzymatic synthesis and purification of carbon-14 compound, **311:**685–687
Glucocerebrosidase
affinity labeling, **311:**572–573
translocation assay of sphingolipids,
312:572
Glucocorticoid receptor
DNA-binding to nucleosomes, *see* Glucocorticoid receptor, DNA-binding to
nucleosomes
green fluorescent protein fusion protein
dexamethasone effects, **302:**75
DNA recognition specificity, **302:**79
expression, **302:**82
fusion sites and functionality,
302:75–77
ligand affinity, **302:**79, 83
transactivation potential, **302:**79, 83
Western blot analysis, **302:**79
subcellular distribution, **302:**73–74
Glucocorticoid receptor, DNA-binding to
nucleosomes
dimethyl sulfate methylation protection
assay
advantages, **304:**332
binding conditions, **304:**329
cleavage and separation of methylated
DNA, **304:**330
methylationreaction, **304:**330
principle, **304:**328
safety, **304:**328

solution preparation, **304:**329
DNase I footprinting
  binding conditions, **304:**327
  digestion conditions and analysis,
    304:328
  principle, **304:**327
  solution preparation, **304:**327
electrophoretic mobility shift assay
  affinity analysis, **304:**325–327
  binding conditions, **304:**325
  DNA sequence design, **304:**323–325
  limitations, **304:**331–332
  principle, **304:**323
  site of binding, **304:**321
Glucose, isotopic precursors for enzymatic
  ribonucleotide synthesis, **317:**19–23
Glucose 6-phosphate, caged compound studies *in vivo*
  electroporation of caged substrate into
    sea urchin eggs, **291:**283–285
  glucose 6-phosphate dehydrogenase
    assay, *in vivo*, **291:**285–286
    fertilization of sea urchin eggs, effect
      on activity, **291:**286–288
    inertness testing of caged compound,
      291:282–283
  photolysis conditions, **291:**285
  synthesis of caged compound
    2-nitroacetophenone hydrazone prepa-
      ration and reaction with glucose 6-
      phosphate, **291:**281
    radiolabeling, **291:**280
Glucose 6-phosphate dehydrogenase, chemiluminescence assay, **305:**385, 387
Glucose transporter, amyloid-β inhibition
  assay, **309:**741, 743
  mechanism, **309:**740
β-Glucosidase, assay with fluorescent sphingolipids, **312:**300
Glucosylceramidase
  assay of plant enzyme, **311:**144–145
  intestinal enzyme
    assay, **311:**284
    biological significance, **311:**284–285
    pH optimum, **311:**283
    purification, **311:**283–284
  substrate specificity of plant enzyme,
    311:145
Glucosylceramide, tandem mass spectrometry, **312:**40–42

Glucosylceramide synthase
  assays
    fluorescence assays
      incubation conditions, **311:**52, 306
      modifications, **311:**52–53
      NBD-ceramide–bovine serum albu-
        min complex preparation,
        311:51–52
      NBD substrate preparation, **311:**306
      thin-layer chromatography, **311:**52,
        306
    overview, **311:**51, 305
    plant enzyme, **311:**140–141
    radiolabeled substrates
      incubation conditions, **311:**45, 305
      principle, **311:**43–44
      product extraction, **311:**45–46
      reagents, **311:**44
      sample preparation, **311:**45, 305
      substrate preparation, **311:**44–45
      thin-layer chromatography, **311:**306
  biological significance, **311:**42–43, 49–50
  cloning of complementary DNA
    deficient cell line preparation, **311:**308
    libraries, **311:**308
    overview, **311:**307–308
    panning enrichment of positive clones,
      311:309–311
    plasmids and vectors, **311:**308–309
    sequencing, **311:**313
    SIB selection, **311:**311, 313
    transfections, **311:**309, 311
  divalent cation effects, **311:**47
  gene
    isolation, **311:**317
    structure analysis, **311:**317–318
  Golgi fraction preparation, **311:**54–56
  induction by ceramide, **311:**48
  inhibitors, *see also* 1-Phenyl-2-decanoyl-
    amino-3-morpholino-1-propanol
    clinical applications, **311:**43, 49–50, 373,
      384, 387
    enzyme induction, **311:**48–49
  liposomal composition effects on activity,
    311:46
  mutant GM-95 cells, **312:**306
  nucleotide effects, **311:**46–47
  pH optimum, **311:**47
  purification techniques, **311:**58–59
  reaction catalyzed, **311:**42, 50

recombinant enzyme expression
  Escherichia coli, **311:**313–314
  mammalian cells
    cell line development, **311:**314–315
    lipid analysis, **311:**315, 317
  recombinant human protein expression in
    bacteria, **311:**56, 59
  reconstitution, **311:**58
  solubilization in detergents, **311:**57–58
  substrate specificity
    mammalian enzyme, **311:**46
    plant enzyme, **311:**141–142
  tissue distribution, **311:**53–54
  Western blot analysis, **311:**53
Glucosyloctanoylsphingosine, synthesis,
  311:44–45
α-Glucosyl rutin, topical antioxidant screening, **319:**469–471, 474, 476–477
Glucosylthioceramide
  chemical synthesis
    3-*O*-benzoyl-1-*O-tert*-butyldiphenylsilyl-
      2-dichloroacetamido-4-octadecen-
      1,3-diol, **311:**613
    3-*O*-benzoyl-2-dichloroacetamido-1-
      iodo-4-octadecen-3-ol, **311:**614–615
    3-*O*-benzoyl-2-dichloroacetamido-1-*O*-
      methanesulfonyl-4-octadecen-1,3-
      diol, **311:**614
    3-*O*-benzoyl-2-dichloroacetamido-4-
      octadecen-1,3-diol, **311:**613–614
    1-*O-tert*-butyldiphenylsilyl-2-
      dichloroacetamido-4-octadecen-
      1,3-diol, **311:**613
    column chromatography, **311:**603–604
    2-dichloroacetamido-4-octadecen-1,3-
      diol, **311:**610, 613
    fluorescent compound, **311:**616–617
    glucosylthiosphingosine, **311:**616
    materials, **311:**602–603
    principle, **311:**610
    protected glucosylthiosphingosine syn-
      thesis and characterization,
      311:615–616
    radioactive compound, **311:**617–618
    salt removal, **311:**603
    structural characterization, **311:**604
    thin-layer chromatography, **311:**603
  stability, **311:**610
Glucosyltransferase, mutans streptococci
  assay, **310:**516–517

mutagenesis in *Streptococcus mutans*
  insertional inactivation of *gtfB* and
    gtfC, **310:**521–522
  rationale, **310:**519
  reconstruction of mutants, **310:**522–523
  transformation, **310:**520
plaque formation role, **310:**513–514, 516
purification of cell-associated enzyme
  anion-exchange chromatography,
    310:518
  cell growth, **310:**518
  extraction, **310:**518
  hydroxylapatite chromatography,
    310:519
  types, **310:**517–518
β-Glucuronidase, reporter gene assay with
  chemiluminescence
  cell extract preparation,
    305:438
  controls, **305:**440
  dual reporter gene assays with luciferase
    cell extract preparation, **305:**447
    controls, **305:**449
    detection, **305:**448–449
    overview, **305:**428–429, 444–446
    reagents, **305:**446–447
  instrumentation, **305:**449–450
  microplate luminometer detection,
    305:439
  overview, **305:**428–429, 437
  reagents, **305:**437–438
  sensitivity, **305:**450
  tube luminometer detection, **305:**438–439
γ-helix, structure compared to α-helix,
  295:26
GLUT4, green fluorescent protein fusion
  proteins in vesicle trafficking analysis,
  302:44–45
Glutamate, conformational analysis
  conformer interconversions, **296:**177–178
  conformer library creation by low energy
    conformation calculations,
    296:178–179
  overlaying of analogs, **296:**179–181
L-Glutamate
  desyl caged compound
    properties and applications, **291:**10–11,
      13–14
    synthesis, **291:**25–26, 37–38
  p-hydroxyphenacyl caged compound

properties and applications, **291:**10–11, 14, 16

synthesis, **291:**24–25

o-nitromandelyloxycarbonyl caged compound

electrophysiology studies of glutamate receptor effects, **291:**440–443

photochemistry, **291:**440

synthesis, **291:**434–435

receptor, *see* Glutamate receptor; Neurotransmitter receptor

Glutamate dehydrogenase, isoergonic cooperativity

assumptions of model, **295:**338–339

basic element of theory, **295:**337–338

experimental support

calorimetry data, **295:**334–337, 340–343

enzyme kinetics, **295:**341–342

X-ray crystallography, **295:**340

experimental system overview, **295:**332–333

linkage between chemical events and cleft-closing in catalytic cycle

isoergonic cooperativity model, **295:**347–348

Lumry scheme, **295:**344–347

overview of conformational changes, **295:**343–344

model expansion, **295:**338–340

q dependence on L_T for two-subunit case, **295:**339

Glutamate receptor, *see also* Excitatory amino acid receptors

ab initio modeling of structure, **293:**602–604

binding energy transduction to ion channel opening, **293:**616–619

classification, **293:**590–591

GluR1 phosphorylation

extraction of labeled receptors, **294:**355–356

functional analysis by patch-clamp whole-cell recording

basal phosphorylation, **294:**368, 370

intracellular patch perfusion solutions, **294:**367–368

site-directed mutants, **294:**370

stimulation of phosphorylation, **294:**366–368

fusion proteins

phosphorylation *in vitro*, **294:**362–363

rationale for study, **294:**361–362

immunoprecipitation of labeled receptors, **294:**354, 356–357

investigation in native tissues, **294:**353–354

phosphoaminoacid analysis, **294:**360–361

phosphopeptide mapping

overview, **294:**357–358

proteolytic digestion, **294:**358

two-dimensional thin-layer chromatography, **294:**359–360

radiolabeling in cell culture, **294:**354–355

site-directed mutagenesis of phosphorylation sites, **294:**363–364, 370

site-specific antibodies, generation and application, **294:**364–366

homology modeling of structure, **293:**600–602

ligand docking in molecular modeling

hot spot determination, **293:**607–610, 613–615

refinement of binding sites by alignment

S1, **293:**610–611

S2, **293:**611

S3, **293:**611

S4, **293:**612

S5, **293:**612

S6, **293:**612

radioactive filter assays

ligand selection, **294:**388

reaction mixtures, **294:**388–389

synaptic membrane preparation from rat brain, **294:**386–387

Triton X-100 treatment of crude fractions, **294:**387–388

transmembrane topology, molecular modeling, **293:**591–593

types, **294:**385

Glutamate transporter, *see also* Vesicular glutamate transport system

amyloid-β inhibition

assay, **309:**741, 743

overview, **309:**740–741

antisense oligodeoxynucleotide inhibition of subtypes

advantages and disadvantages in pro-
    tein function analysis, **296:**515
control sequences, **296:**518
dosing and delivery, **296:**522–523
glutamate uptake assay in analysis,
    296:527–529
inhibition efficiency, Western blot anal-
    ysis, **296:**523–526
mechanisms of inhibition, **296:**516
modifications of oligonucleotides
    cell entry enhancement, **296:**520
    stabilization against nucleases,
        296:516, 518, 520
preparation of oligonucleotides for
    chronic intraventricular fusion,
    296:520–522
sequence selection, **296:**517–518
assay of heterogeneity in crude synapto-
    somes
calculations, **296:**195–196
differentiation of pure blockers from
    substrate inhibitors, **296:**200–201
incubation conditions, **296:**194–195
multiple subtypes in synaptosomal
    preparation, **296:**201–202
pharmacological analyses, **296:**197–199
pipetting guide, **296:**192, 194
potential indirect effects, **296:**199–200
reagents, **296:**191–192
sodium dependence, **296:**196–197
synaptosome preparation, **296:**192–194
electrophysiology
cell selection for analysis, **296:**618–619
cerebellar slices, preparation from rat,
    296:609–610
D-aspartate transport and current,
    296:611–613
excitatory postsynaptic current in slices
    comparison of climbing fiber decay
        with and without uptake inhibi-
        tion, **296:**613–614, 616
    double patch-clamping in decay mea-
        surement, **296:**614, 616
glial transport currents in microcul-
    tures
    advantages of microcultures, **296:**633,
        635, 645
    cell preparation, **296:**637–639
    characteristics of microcultures,
        296:632–633

dish preparation, **296:**636–637
extracellular solutions, **296:**640–641
glutamate transport during synaptic
    glutamate release, **296:**643–644
inward currents, **296:**643
media, **296:**636
patch clamp recording, **296:**639,
    641–642
pipettes and solutions, **296:**639–640
glutamate release, sensing using gluta-
    mate-gated channels
blocking control, **296:**631
calibration of sensing cells,
    296:629
communication between releasing
    and sensing cells, **296:**629–630
selection of sensing cells,
    296:628–629
tonic release detection, **296:**631
intracellular solution, **296:**611
L-glutamate uptake and inhibition of
    other receptors, **296:**610–611
selective uptake blocking in Purkinje
    cells, **296:**613
stoichiometry of transport, **296:**617
whole-cell patch clamping
electrode series resistance,
    296:620–621
forward uptake current recording,
    296:619–620
intracellular dialysis in analysis of
    countertransported substrates,
    296:620
reversed uptake current recording,
    296:621, 627
functional assays in *Caenorhabditis ele-*
    gans, **296:**542
inhibition by conformationally-con-
    strained compounds
cis-1-aminocyclobutane 1,3-dicarbox-
    ylate, **296:**176, 180, 187–188
(2*S*,3*S*,4*R*)-2-(carboxycyclopropyl)-
    glycine, **296:**176, 180
design and synthesis of new com-
    pounds, **296:**181–184
dihydrokainate, **296:**176, 180
iterative strategy for discovery,
    296:175–176
meso-2,4-methano-2,4-pyrrolidine,
    296:182–183

L-*anti-endo*-3,4-methano-pyrrolidine dicarboxylate, **296:**182–183
L-*trans*-2,4-pyrrolidine, **296:**176, 180,
187–188
overlaying of structures with glutamate
conformers, **296:**179–181
synaptosomal exchange assay of transport, **296:**186–189
synaptosomal transport assay of inhibition, **296:**185–186
inhibitor analysis, **296:**155, 185–186,
197–201
intracellular pH changes measured with
2′,7′-bis(carboxyethyl)carboxy-
fluorescein during glutamate
transport
acidification and glutamate transport,
296:623
loading of cells, **296:**622
ratio fluorescence, **296:**622–623
ion coupling stoichiometry determination
electrophysiological measurements,
296:246–247
flux ratio determination, **296:**239–240
ion-sensitive microelectrodes for extracellular measurements during glutamate
transport
anion measurements, **296:**626–627
fabrication, **296:**625–626
pH measurements, **296:**626
potassium measurements, **296:**626
nomenclature of transporter subtypes,
296:515
Glutamine synthetase, amyloid-β inhibition
and assay, **309:**748, 752, 759–760
Glutaraldehyde, chromatin fixation
cryoelectron microscopy, **304:**203
transmission electron microscopy,
304:223–224
Glutaraldehyde
cross-linking of GroEL–GroES, **290:**256,
258–259, 262, 265–268
cross-linking of photosystem I subunits,
297:133
Glutathione
depletion in disease, **299:**258
disulfide bond formation in proteins,
290:27, 42–43
DNA protection against ultraviolet radiation, **319:**361

flow cytometry assay
advantages, **299:**247, 257
cell preparation, **299:**248–249
data collection, **299:**251–252
differential assessment of thiols,
299:249–251, 258
monobromobimane modification of thiols, **299:**248–249, 251–253,
257–258
monochlorobimane detection of glutathione, **299:**248
high-performance liquid chromatography
analysis of ozone depletion in skin
equipment, **319:**541
operation, **319:**541
reagents, **319:**540
sample preparation, **319:**542
standards, **319:**541–542
high-performance liquid chromatography
assay with electrochemical detection
cell culture, **299:**243
chromatography, **299:**241
coulometric detection, principle,
299:240–241
current–voltage response curve,
299:242
electrodes, **299:**240
extraction, **299:**243, 245–246
instrumentation, **299:**241
overview, **299:**81–82
standards, **299:**242–243
high-performance liquid chromatography
of N-(1-pyrenyl)maleimide derivatives
derivatization reaction, **299:**259,
261–264
glutathione disulfide quantitation,
299:259–260, 262
materials, **299:**260
metabolic modulation studies,
299:265–266
sample preparation, **299:**264
standard curves, **299:**261–262
lipoic acid effect on cellular levels,
299:253–254
nitric oxide donor effects on intracellular
concentration, **301:**416–417
nitrosative stress resistance, **300:**392–393
oxidative stress inhibition, **319:**260–261
recycling, **299:**247

redox equilibrium determination in disulfide bond formation
determination of reduced/oxidized glutathione ratio, **290:**73
equilibrium constant calculation, **290:**72–74
tryptophan fluorescence measurement, **290:**72–73
7-trifluoromethyl-4-chloro-*N*-methylquinolium colorimetric assay
absorbance properties, **299:**277, 279–281
incubation conditions, **299:**284–286
instrumentation, **299:**283
interferences, **299:**281–283
reaction mechanism, **299:**280, 286
reagents, **299:**283
reproducibility, **299:**281
sample preparation, **299:**283–284
sensitivity, **299:**281, 286
Glutathione disulfide/glutathione ratio
apoptosis correlation, **299:**275–276
glutathione determination
blood sample preparation, **299:**271
glutathione *S*-transferase assay, **299:**271–273
protein precipitation, **299:**268, 271
tissue sample preparation, **299:**271–272
glutathione disulfide determination
accuracy requirements, **299:**276
blood sample preparation, **299:**268–269
N-ethylmaleimide trapping of glutathione, **299:**267–270
high-performance liquid chromatography, **299:**268, 270–271
tissue sample preparation, **299:**269
mitochondrial DNA oxidation, correlation, **299:**274–275
oxidative stress indicator, **299:**267, 274
validation of assay, **299:**273–274
Glutathione peroxidase, protection against peroxynitrite damage, **301:**301–302, 304–306, 309–311
Glycerol gradient centrifugation, separation of spliceosomal small nuclear ribonucleoproteins, **318:**392–393
Glycerol kinase, quantification of glycerol with firefly luciferase ATP assay, **305:**365–366

Glycerol 3-phosphate, *see also* Uptake glycerol phosphate system
metabolism in bacteria, **292:**40
uptake systems in bacteria, **292:**40–41
Glycine receptor, *see* Neurotransmitter receptor
Glycine receptor
antibodies, **294:**685, 701
baculovirus–insect cell expression system, **294:**273
purification from mammalian spinal cord
aminostrychnine affinity chromatography, **294:**262–263
membrane preparation, **294:**261–262
solubilization, **294:**262
strychnine binding, **294:**260
subunits in spinal cord, **294:**260
transient expression systems
human embryonic kidney cells
calcium phosphate transfection, **294:**270, 272–273
cell culture and harvesting, **294:**271–272
plasmid preparation, **294:**270–271
vectors, **294:**270
Xenopus oocytes
applications of expression system, **294:**263–264
injection capillary preparation, **294:**268
injection of cytoplasma, **294:**269
injection of nucleus, **294:**269–270
messenger RNA isolation from neural tissue, **294:**264
preparation of oocytes, **294:**266–267
transcription of complementary DNA *in vitro*, **294:**264–266
vector linearization, **294:**265
Glycine transporter
assay of transport in reconstituted liposomes, **296:**7
glycosylation of GLYT2, structural analysis
carbohydrate moiety, **296:**13, 15–16
functional analysis, **296:**16–17
glycan detection, **296:**14–15
glycosidase treatment, **296:**13–16
Western blot analysis, **296:**14
hydrodynamic characterization of GLYT2

detergent complexes, preparation,
    296:8
gel filtration chromatography, **296:**8–9
molecular weight calculation,
    296:10–12
sucrose density gradient centrifugation,
    296:9–10
phylogenetic analysis, **296:**428–436
purification of GLYT2 from pig brain
    stem
    crude synaptic plasma membrane prep-
        aration, **296:**4–5
    hydrophobic interaction chromatogra-
        phy, **296:**6
    hydroxylapatite chromatography,
        296:6–7
    solubilization of membranes, **296:**5
    sucrose density gradient fractionation,
        296:7
    wheat germ agglutinin affinity chroma-
        tography, **296:**6
    yield, **296:**7–8
types GLYT1 and GLYT2
    homology, **296:**426–427
    inhibitor specificity, **296:**3
    tissue distribution, **296:**3
Glyco-*S*-nitrosothiols, *see S*-Nitroso-*N*-ace-
    tylpenicillamine
Glycogen phosphorylase, photolysis studies
    in crystals, **291:**252–253
Glycogen phosphorylase *b*, S-nitrosylation
    alkylation with iodoacetate, **301:**146–147
    isoelectric focusing, **301:**145–150
Glycosphingolipid, *see also specific glyco-
    sphingolipids*
    biosynthesis pathways, **311:**82–84,
        312:339–340
    cell adhesion assays, *see* Carbohydrate–
        carbohydrate interaction; Lectin
    ceramide glycanase analysis, *see* Cer-
        amide glycanase
    classification, **311:**59
    cloud point extraction of gangliosides, *see*
        Cloud point extraction, gangliosides
    concentration of gangliosides in biologic
        samples, **312:**135
    discovery, **312:**45
    enriched microdomain separation from
        caveolae, *see* Detergent-insoluble
        membrane fraction

extraction, **312:**48–49, 54
Far-Eastern blotting
    advantages, **312:**149
    apparatus, **312:**149
    band marking on plates, **312:**147–148
    blotting, **312:**148
    detection on blots, **312:**148–149
    glycosphingolipid purification
        blotting, **312:**151–152
        extraction, **312:**152
        materials, **312:**151
    mass spectrometry coupling,
        312:152–154
    materials, **312:**147
    overlays on blots, **312:**154–156
    overview, **312:**145–147
    sialidase assay on blots,
        312:156–157
functions, **311:**60, **312:**47–48, 339
high-performance liquid chromatography
    acute myelogenous leukemia ganglio-
        side analysis, **312:**143
    aminopropyl chromatography of gangli-
        osides, **312:**142
    anion-exchange chromatography
        brain gangliosides from mouse,
            312:55–57
        column care, **312:**55
        GM3 gangliosides, **312:**60, 62–63
        granulocyte monosialogangliosides
            from humans, **312:**57–60
        materials, **312:**53–54
        overview, **312:**49
        preparative chromatography, **312:**53,
            62–63
    ceramide subspecies preparation for bi-
        ologic assays, **312:**143–145
    native compound separation systems,
        312:51–53
    precolumn derivatization, **312:**50–51
    reversed-phase chromatography of gan-
        gliosides, **312:**142–143
high-performance thin-layer chromatogra-
    phy for ganglioside analysis
    anion-exchange chromatography,
        312:117–118
    autoradiography, **312:**139–140
    desalting, **312:**118
    detection, **312:**122–123, 125–126
    extraction, **312:**117–118

mass spectroscopy coupling, **312:**116, 132–134
overlay technique
analysis of *in situ* reaction products, **312:**128–131
antibody overlay, **312:**127
glycolipid overlay, **312:**131–132
overview, **312:**115–116, 126
quantitative analysis, **312:**127–128
plates, **312:**118–119
solvent systems
one-dimensional, **312:**119
two-dimensional, **312:**119, 122
high-performance thin-layer chromatography immunostaining of glycosphingolipids
antibody incubation and detection, **312:**159
overview, **312:**157–158
reagents, **312:**158–159
immunolocalization of gangliosides, *see also* GM1
antibodies
commercial sources, **312:**180
factors affecting binding in membranes, **312:**180–183
fixatives, **312:**181, 183
hippocampal neuron studes, **312:**184–186
plasma membrane association, **312:**115, 179
postfixation immunolabeling, **312:**184, 186
prefixation with formaldehyde, **312:**183–184
primary antibody incubation, **312:**184
metabolic radiolabeling of cellular gangliosides
applications, **312:**140–141
cell growth and labeling, **312:**139
high-performance thin-layer chromatography autoradiography, **312:**139–140
marker preparation, **312:**140
monoclonal antibodies, *see* Monoclonal antibody, anti-glycosphingolipid antibodies
nomenclature, **312:**46–47
purification of gangliosides by three-step method

cerebrospinal fluid analysis in cancer, **312:**137–138
gel filtration, **312:**136–137
organic phase partition, **312:**136
overview, **312:**135
total lipid extraction, **312:**135–136
reactions on silica plates for analysis
fucosylation, **312:**130–131
β-galactosidase treatment, **312:**129
glycosphingolipid glucosyltransferase, **312:**129–130
neuraminidase treatment, **312:**128–129
sphingosine oxidation for glycoconjugate synthesis, *see* Sphingosine
structures
brain gangliosides, **312:**47, 56, 64–65, 116
determination by nuclear magnetic resonance, *see* Nuclear magnetic resonance, gangliosides
elucidation techniques, **312:**49–50
granulocyte monosialogangliosides, **312:**47, 58
overview, **311:**59, **312:**45–46, 64–65
supramolecular organization, **312:**47
thin-layer chromatography, **312:**49, 54, 194, 196
toxin binding, *see* Cholera toxin; Verotoxin
Glycosylation
dopamine transporter analysis, **296:**219, 227–229
GAT-1, **296:**302
GLYT2, structural analysis
carbohydrate moiety, **296:**13, 15–16
functional analysis, **296:**16–17
glycan detection, **296:**14–15
glycosidase treatment, **296:**13–16
Western blot analysis, **296:**14
ion channel studies in heterologous expression systems, **293:**481–483
norepinephrine transporter
site-directed mutagenesis of N-glycosylation sites, **296:**360, 362
trafficking role, **296:**362, 364–366, 368
Western blot analysis, **296:**355–357
peptide antigens, **292:**752–753
P-glycoprotein
assay for defective mutants, **292:**486–487

N-linked glycosylation detection, **292**:274–275, 277, 283–284
sulfonylurea receptor assay, **292**:741
Glycosylphosphatidylinositol anchor
abundance in yeast, **312**:506
ceramide analysis in yeast
conconavalin A affinity chromatography, **312**:511–512
delipidation, **312**:509–511
lipid liberation
acid hydrolysis, **312**:514
nitrous acid, **312**:513–514
phospholipase A$_2$, **312**:514
lipid structures, **312**:506–507
metabolic labeling, **312**:508, 510–511
octyl-Sepharose purification of anchor peptides, **312**:512–513
protease treatment, **312**:512
thin-layer chromatography, **312**:514–515
lipid remodeling, **312**:507–508
GLYT1, *see* Glycine transporter
GLYT2, *see* Glycine transporter
GM1
biotinylation
biotinylation reaction, **312**:543–544
chromatography, **312**:538
deacylation, **312**:540, 542
desalting, **312**:537–538
enzymatic degradation, **312**:544–545
GM1 preparation, **312**:538–539
mass spectrometry, **312**:538, 543–544
materials, **312**:536–537
nuclear magnetic resonance, **312**:538, 544
principle, **312**:540
radioactivity quantification, **312**:537
solvents, **312**:537
thin-layer chromatography, **312**:537, 542–543
cholera toxin binding, **312**:459, 468, 535
delayed ion extraction matrix-assisted desorption ionization time-of-flight mass spectrometry, **312**:93, 96
fatty acid radiolabeling of deacylated ganglioside, **312**:539–540, 542
immunoelectron microscopy localization of biotinylated ganglioside
LR Gold embedded samples
cell culture, **312**:553

fixation and postfixation, **312**:553–554
principle, **312**:553
progressive lowering of temperature embedding, **312**:554
sectioning and immunolabeling, **312**:554–555
overview, **312**:547–548
preembedding labeling
cell culture, **312**:549
Epon embedding, **312**:551–552
fixation, **312**:549
immunolabeling, **312**:551
principle, **312**:548–549
silver enhancement, **312**:552
sodium borohydride treatment of cells, **312**:551
transport studies, applications and limitations, **312**:560–561
ultrathin cryosections
cell culture and uptake, **312**:557, 559
cryoprotectant infiltration, **312**:559
cryosectioning, **312**:560
fixation, **312**:559
immunolabeling, **312**:560
methylcellulose solution preparation, **312**:559–560
principle, **312**:555, 557
nomenclature, **311**:639
photoaffinity labels, **311**:577–578, 583–586, 592
radiolabeled compound synthesis
double labeling, **311**:653–654
fatty acid radiolabeling, **311**:649–652
[^{14}C]GM1a, enzymatic synthesis and purification, **311**:685–687
materials, **311**:640, 642
overview, **311**:639–641
sialic acid acetyl group radiolabeling, **311**:652–653
stability, **311**:655–656
tritium labeling at position 3 of sphingosine, **311**:647–649, 658
tritium labeling at position 6 of terminal galactose and *N*-acetylgalactosamine residues, **311**:642–645
uptake of biotinylated ganglioside in fibroblasts
cell culture, **312**:540
incubation conditions, **312**:547

principle, **312:**545–547
GM2
  nomenclature, **311:**639
  photoaffinity labels, **311:**593–594
  radiolabeled compound synthesis
    double labeling, **311:**653–654
    fatty acid radiolabeling, **311:**653–654
    materials, **311:**640, 642
    overview, **311:**639–641
    sialic acid acetyl group radiolabeling, **311:**652–653
    stability, **311:**655–656
    tritium labeling at position 3 of sphingosine, **311:**647–649
    tritium labeling at position 6 of terminal galactose and *N*-acetylgalactosamine residues, **311:**642–645
GM2 activator protein
  affinity labeling, **311:**571
  assay of GM2 degradation
    calculations, **311:**262
    incubation conditions, **311:**261
    materials, **311:**261–262
    principle, **311:**260–261
    product separation, **311:**262
  fibroblast assay
    GM1 loading studies in GM2 activator-deficient cells, **311:**272
    principle, **311:**271
    uptake of recombinant protein, **311:**273
  hexosaminidase requirement, **311:**257
  mechanism of action, **311:**255, 257
  mutations in disease, **311:**257–258
  photoaffinity labeling, **311:**586
GM2 synthase, photoaffinity labeling, **311:**577
GM3
  analog synthesis with 2-(tetradecyl)hexadecyl residue in place of ceramide, **311:**566, 56
  analysis in serine palmitoyltransferase Chinese hamster ovary cell mutants, **312:**314–316
  anion-exchange high-performance liquid chromatography, **312:**60, 62–63
  CD4 interactions with microdomains, **312:**500–50
  delayed ion extraction matrix-assisted desorption ionization time-of-flight mass spectrometry, **312:**89–90, 93

dependent adhesion of mouse B16 melanoma cells to endothelial cells
    B16 cell adhesion to glycosphingolipid-coated plates
      adhesion conditions, **312:**449
      cell culture, **312:**449
      glycosphingolipid coating of plates, **312:**448–449
    caveats, **312:**457–458
    cell adhesion conditions, **312:**451–453
    endothelial cell monolayer preparation, **312:**451
    GM3 liposome adhesion to glycosphingolipid-coated plates
      adhesion conditions, **312:**451
      liposome preparation, **312:**449, 451
    inhibition by oligosaccharides and glycosphingolipid liposomes, **312:**454
    overview, **312:**448
  enriched microdomain features, **312:**494
  functions, **311:**564
  human immunodeficiency virus-1 binding, **311:**637–638
  immunosuppression, **311:**564, 566
  nomenclature, **311:**639
  radiolabeled compound synthesis
    double labeling, **311:**653–654
    fatty acid radiolabeling, **311:**653–654
    materials, **311:**640, 642
    overview, **311:**639–641
    sialic acid acetyl group radiolabeling, **311:**652–653
    stability, **311:**655–656
    tritium labeling at position 3 of sphingosine, **311:**647–649
GM3 synthase, photoaffinity labeling, **311:**587–588
GM4
  analog synthesis with 2-(tetradecyl)hexadecyl residue in place of ceramide, **311:**566, 568
  immunosuppression, **311:**564, 566
GMPS, Guanosine 5′-monophosphorothioate
GNRA loop, modeling with comparative sequence analysis, **317:**505, 510
Golgi apparatus, fluorescent probes, **307:**121
gp120, *see* Human immunodeficiency virus

GPI anchor, *see* Glycosylphosphatidylinositol anchor

G-protein, *see also* $G_{i\alpha1}$

antisense experimental design for receptor studies, **314:**92–93

antisense RNA analysis of neuronal $G\alpha$ signal transduction

advantages, **314:**136–137

cell culture, **314:**144–145

cloning strategy, **314:**143–144

controls, **314:**141, 147

design of probes

length, **314:**137–138

secondary structures, **314:**138

targeting sites, **314:**138

$G\alpha_q$ expression analysis, **314:**141–142, 146–147

immunostaining, **314:**146–147

intranuclear injection, **314:**139, 142, 145–146

materials and reagents, **314:**142–143

mechanism of action, **314:**137

plasmid purification, **314:**144

RNA probe advantages over DNA, **314:**147–148

specificity assessment, **314:**139, 141

conformational changes with receptor activation, **315:**178–179, 364, 490

coupled receptors, *see also* Opioid receptor

$G\alpha$

carboxyl-terminal interactions with receptors, **315:**388, 490

chimera applications, **315:**502–503

subunit structure, **315:**363–364, 502

kinetic isotope effect analysis of transition-state, **308:**320–322

regulators of G-protein signaling, mechanism of action, **308:**92

signal transduction, **314:**136

subunit substitutions between species, **314:**91

superfamily of coupled receptors, **315:**12, 91, 251–252

timing of filament assembly reactions, **308:**103

transducin activation of rhodopsin, *see* Transducin

transitory nature of receptor interactions, **315:**389

types, **314:**136

G-protein-coupled inward rectifier potassium channel

nonsense codon suppression and unnatural amino acid incorporation in *Xenopus* oocytes, **293:**523

recombinant protein expression with adenovirus, **293:**496–497, 499

unidirectional flux measurement of *Xenopus* oocyte channels

efflux measurement, **293:**561–564

influx measurement

nonchannel flux correction, **293:**560

potassium-42 counting and specific activity, **293:**559–560

reflux, **293:**560

tail current error, **293:**560–561

voltage clamp measurement, **293:**557–558

G protein-coupled receptor

desensitization

agonist dependence, **302:**156

arrestin function and regulation, **302:**154–156, 165

kinase role, **302:**154–156, 165

mechanism overview, **302:**154

sequestration, **302:**156–157

epitope tagging studies of sequestration, **302:**157

functions, overview, **302:**153–154

green fluorescent protein fusion proteins

β_2-adrenergic receptor, **302:**158–159, 161, 163–165

β-arrestins, **302:**159–160, 162, 165–167, 169

confocal laser scanning microscopy, **302:**162–163

design, **302:**158–160

functional characterization, **302:**160–162, 170

GRK2, 159, 161–162, 165–166

sensitivity limitations, **302:**169–170

subcellular distribution and desensitization

arrestin , 165–167, 169

kinase, **302:**165–166

receptor, **302:**163–165

immunofluorescence analysis of transfectants, **302:**157–158

resensitization, **302:**156–157

signaling impairment by amyloid-β
GTPase assay, **309:**744–746
muscarinic receptor, **309:**743–744
thrombin receptor, **309:**744
structure, **302:**154–155
G-quartet
formation with antisense oligonucleotides, **314:**78–79, 314–315, 557–558
oligonucleotide purification, **313:**223
protein interactions, **313:**20
GR, *see* Glucocorticoid receptor
GRAIL, computational gene finding,
303:81, 159
Gramicidin
applications in research, **294:**525–526
conformational analysis of analogs
circular dichroism spectroscopy,
294:532–534
molecular dynamics simulations,
294:539–542
nuclear magnetic resonance
solid-state studies, **294:**536–539
solution studies, **294:**535–536
single-channel measurements
electrodes, **294:**542
functional approach to structure identification by dimerization,
294:545–547, 549–550
membrane capacitance minimization,
294:542–543
purity requirements, **294:**543–544
reconstitution, **294:**543
size-exclusion high-performance liquid
chromatography, **294:**534–535
molecular force transducer in membranes
changes in membrane deformation energy, **294:**220–222
channel lifetime measurement,
294:222–223
heterodimer formation approach,
294:223
membrane conductance measurements,
294:220–222
monomer–dimer equilibrium, **294:**217–
218, 220
phenomological spring constant determination, **294:**221, 223
molecular modeling, **293:**694–695, 721
peptide synthesis of analogs
acylation, **294:**531

cleavage of peptide from resin,
294:530–531
N-Fmoc-amino acid synthesis,
294:527–528
N-formylamino acid preparation,
294:526–527
purification of product, **294:**531–532
resin selection, **294:**528
solid-phase synthesis, **294:**528–530
sequence of gramicidin A, **294:**525
Grb2, inhibition with P-ethoxy antisense oligonucleotides
Philadelphia chromosome-positive leuke-
mic cell inhibition, **313:**382–383
protein expression analysis, **313:**379, 382
Green fluorescent protein
assays
absorption, **305:**213–214
fluorescence, **305:**214–215
Bacillus subtilis studies
applications, **302:**140–141, 150–153
fluorescence microscopy imaging
agarose pads on slides, **302:**144–145
cell growth conditions, **302:**142–143
double labeling, **302:**145–146
equipment, **302:**141–142, 146–147
optimization, **302:**146–149
sample preparation, **302:**143
slide preparation, **302:**143–144
troubleshooting, **302:**149
folding and fluorogenesis time, **302:**136
fusion proteins, overview of types,
302:137–139
immunofluorescence microscopy,
302:149–150
baculovirus expression, *see* Baculovirus–
insect cell system, green fluorescent
protein expression
biofilm formation reporter assays
biofilm-specific regulatory element isolation, **310:**15–16
overview, **310:**14–15, 29–30
random mutation of protein, **310:**16–17
chitinase fusion proteins, **310:**76
chromophore
amino acids, **302:**265, 285, 373, 445
environment, **305:**203, 206, 208
formation, **302:**379–380, 444–445,
307:608
photoisomerization, **305:**206

structure, **302:**383–384
structure, **305:**197–198
commercial vectors for fusion protein
    construction, **294:**49–50
cyanobacteria, *in vivo* assays of reporter,
    305:526–527
degradation assay, *see* HMG-CoA reduc-
    tase–green fluorescent protein fusion
    protein
dimerization, **302:**418–420
educational demonstrations
    college courses, overview and history
        of applications, **305:**673–675
    grade school, **305:**671
endocytosis assay, *see* Endocytosis
environmental monitoring with reporter
    gene induction, **310:**41–42
expression in fungi
    confocal microscopy
        image acquisition, **307:**613
        instrumentation, **307:**612–613
        propidium iodide multicolor staining,
            307:615–616
        rhodamine-conjugated antibody, ac-
            tin staining in multicolor detec-
            tion, **307:**617–618
        rhodamine-conjugated phalloidin, ac-
            tin staining in multicolor detec-
            tion, **307:**616–617
    epifluorescence microscopy and digital
        deconvolution, **307:**618–620, 622
    fixation, **307:**612
    mutants and detection wavelengths,
        307:611–612
    quantitative analysis of *Aureobasidium*
        pullulans on leaves, **307:**620–621
    species of fungi, **307:**609
    transcriptional control, **307:**610
    transformation, **307:**610
    vectors, **307:**611
fluorescence intensity, effects
    formaldehyde fixation, **302:**148
    sodium azide, **302:**148
    ultraviolet light pre-exposure, **302:**148
fluorescence photoactivation imaging,
    291:77–78
fluorescence resonance energy transfer
    with blue-shifted proteins
    calmodulin linker in calcium fluores-
        cence assay, **294:**52

protease assays, **294:**52
protein–protein interactions in potas-
    sium channels, **294:**52–53
folding, **305:**201–202, 212
functions, **305:**673
gene therapy marker, *see* Gene therapy
G protein-coupled receptor fusion pro-
    teins, *see* G protein-coupled receptor
history of study, **305:**196–198, 212
host–pathogen interaction studies, *see*
    Host–pathogen interaction
imaging in living samples, **307:**111
linkers in fusion protein design, **302:**78,
    160, 173–174, 411, 413, 418–420
maturation
    chromophore, **302:**379–380, **310:**32
    temperature effects, **302:**373–374
metastasis marker, *see* Metastasis
multicolor experiments with fluorescence
    in situ hybridization of biofilms,
    310:38–39, 77
mutants and properties
    Bcl-2/Bax interaction analysis with mu-
        tants, **302:**422
    blue variant, **302:**332
    classification, **302:**385, 388–392
    cyan variant, **302:**332, 391
    destabilized enhanced variant
        applications, **302:**438–439, 444
        d2EGFP properties, **302:**287, 322
        degradation domain inclusion,
            302:322, 439
        NF-κB activation marker,
            32–38
        transfection and fluorescence analy-
            sis, **302:**440–441
        Western blot analysis of degradation,
            302:441, 443–444
    detection enhancement, **310:**15–16,
        30–31
    enhanced variant, **302:**39–40, 207,
        331–332
    expression and purification of histi-
        dine-tagged proteins from *Esche-*
        richia coli, **302:**446–447
    fluorescence resonance energy transfer
        with tandem green fluorescent pro-
        tein fusion proteins
        calcium probes, **302:**421–422

design of mutants, **302:**417–418, 420, 422–423

efficiency of various mutants, table, **302:**413

linker length and dimerization, **302:**418–420

linker proteolysis, **302:**411, 413

P43–Topaz pair, **302:**417

protease assays, **302:**420–421

Sapphire–Topaz pair, **302:**414–417

W7–10cQ69K pair, **302:**417

photoactivated mutant, **302:**392

S65A, **302:**448

S65T, **302:**21, 39–40, 49, 79–81, 158, 190, 314, 331, 445–449

S147P/S65T

fluorescence properties, **302:**375–376, 378

maturation properties, **302:**376–378

T203I, **302:**314

tables, **302:**286, 331, 385–387, 412, 447

Tyr-66

anionic phenolic group, **302:**389–390

aromatic residue, **302:**203, 390–391

neutral and anionic phenolic group, **302:**389

neutral phenolic group, **302:**390

nonaromatic residue substitution, **302:**392

W7 variant, **302:**81

Y66F, **302:**392, 448

Y66H, **302:**391–392, 448

Y66W, **302:**391

yellow variant, **302:**81, 190, 332

nuclear receptor fusion proteins, *see specific receptors*

on/off switching behavior, **305:**208–209

pH sensing

intracellular targeting studies, **302:**264–265

mutant sensitivity to pH changes, **302:**263–264

wild-type protein spectra studies, **302:**381–382, 393–394

photoactivation, **305:**209

photobleaching, *see* Photobleaching, green fluorescent protein

plasmid transfer assay, **310:**16, 30

potassium channel fusion protein expression, **294:**51–52

Pseudomonas aeruginosa biofilm imaging with confocal laser scanning microscopy

advantages, **310:**47–48

electroporation, **310:**48–50

epifluorescence microscopy imaging, **310:**50–51

filters, **310:**51

plasmids, **310:**48

quorum sensing mutants, **310:**51–52

purification of recombinant protein from *Escherichia coli*

ammonium sulfate precipitation, **305:**220

anion-exchange chromatography, **305:**222

cell growth, **305:**216–217

cell lysis, **305:**217–218

concentrating of protein, **305:**221–222

DNA precipitation with protamine sulfate, **305:**218–220

ease of purification, **305:**223

expression vector, **305:**216

gel filtration, **305:**220–222

overview, **305:**212–213

reporter gene expression among individual bacterial cells in culture

cell suspension fluorescence measurements, **305:**501, 503–504

flow cytometry analysis

advantages and disadvantages, **305:**507–508, 512–513

cell growth, **305:**508

light scattering data collection, **305:**508–511

secondary fluorescence labels, **305:**511–512

fluorescence microscopy imaging, **305:**504–506

overview, **305:**499–500

plasmids and protein mutants, **305:**500–501, 512

retroviral expression, *see* Retrovirus, green fluorescent protein expression

ribosome synthesis reporter assay, **310:**42

rotational correlation time, **302:**252, 256

site-directed mutagenesis and spectral variants

blue fluorescent protein, **305:**204

overview, **305:**200–201, 206, 208

pH titration applications, **305:**210–211
resonance energy transfer applications, **305:**209–210
yellow fluorescent protein, **305:**204–206
size, **302:**265–266
spectral variants, **294:**49–51
spectrum
    concentration effects, **302:**393
    ionic strength effects, **302:**393
    pH effects, **302:**263–264, 381–382, 393–394
    properties, **305:**198–199, 215
    temperature effects, **302:**393
    wild-type protein, **302:**285, 381–382, 444
stability, **302:**379, 438–439, **305:**212, **310:**15, 32, 62
structure, **302:**21, 78, 160, 382–384, 438, 446, **305:**201–203
toluene degradation gene promotor fusions, **310:**76–77
transcription reporter assays, *see* Transcription, green fluorescent protein as reporter
transfection marker, *see* Transfection
transfection screening for ion channel expression, **293:**87, 201, 210, 216, 270, 495
transgene expression, *see* Transgenic mouse; Transgenic plant
transposon mutagenesis of marine bacteria, **310:**76
virus tagging, **307:**580
wavelengths for excitation and emission, **310:**61 62
Green fluorescent protein–annexin V fusion protein, *see* Annexin V–green fluorescent protein fusion protein
Green fluorescent protein–calmodulin fusion protein, *see* Calmodulin
Green fluorescent protein–HMG-CoA reductase fusion protein, *see* HMG-CoA reductase–green fluorescent protein fusion protein
Green fluorescent protein–human chromogranin B fusion protein, *see* Human chromogranin B–green fluorescent protein fusion protein
Green fluorescent protein–inositol 1,4,5-tri-

phosphate receptor fusion protein, *see* Inositol 1,4,5-triphosphate receptor
Green fluorescent protein–microtubule-associated protein 4 fusion protein, *see* Microtubule-associated protein 4
Green fluorescent protein–protein A fusion protein, *see* Protein A–green fluorescent protein fusion protein
Green fluorescent protein–ratTGN38 fusion protein, *see* RatTGN38–green fluorescent protein fusion protein
Green fluorescent protein–sarcomeric protein fusion proteins, *see* Myocyte
Griess reagent, *see* Nitric oxide synthase; *S*-Nitrosoglutathione
GRK1
    adenosine nucleotide effects on protein–protein interactions, **316:**34, 36
    autophosphorylation, **316:**36
    binding kinetics analysis with recoverin or calmodulin using surface plasmon resonance, **316:**30–31, 33
    fragment binding to recoverin, **316:**39–40
    protein–protein interactions, **316:**34
GroEL
    application in native tryptophan fluorescence folding studies, **290:**101
    ATP-dependence of chaperone activity, **290:**218, 293
    ATPase assay, **290:**117
    Chromatium vinosum protein, *see* Chaperonin 60
    citrate synthase, effects on refolding
        reactivation after thermal inactivation, **290:**332
        renaturation after chemical denaturation, **290:**335, 337
        thermal aggregation, **290:**327
        thermal inactivation, **290:**330
    contaminating tryptophan fluorescence emission and purity, **290:**115–116, 136–137, 139–140
    electrospray ionization mass spectrometry of binding contaminants, **290:**116–117
    electrospray ionization mass spectrometry of complexes
        binding contaminants, **290:**116–117
        dihydrofolate reductase complex, **290:**299, 301, 307–310

homooligomers, **290:**311–312
nucleotide complexes, **290:**311–313
folding efficiency in rapidly growing cells, **290:**18
gene, **290:**119
GroES complex, *see* GroEL–GroES complex
mechanism of action, **290:**100–101, 120
mixed-ring GroEL complexes
   design, **290:**145–146
   purification, **290:**146
nucleotide requirement for refolding activity, **290:**130–131
purification of low fluorescence background protein
   ammonium sulfate fractionation, **290:**104
   anion-exchange chromatography, **290:**104, 107–110, 118
   cell lysis, **290:**104
   dye affinity chromatography, **290:**104, 108, 110–113, 118
   ion-exchange chromatography in methanol, **290:**137–139
   overexpression system and cell growth, **290:**101–102, 104
   size-exclusion chromatography, **290:**104, 108, 110
single-ring version SR1
   design, **290:**141–142
   expression, **290:**142–143
   GroES binding, **290:**144–145
   purification, **290:**143–144
sodium dodecyl sulfate–polyacrylamide gel electrophoresis and staining of contaminants, **290:**113, 115, 136–137, 140–141
structure, **290:**121
GroEL–GroES complex
analytical ultracentrifugation analysis, **290:**292–293
cross-linking
   buffer selection, **290:**258
   concentration of protein, **290:**256
   controls, **290:**265–267
   cross-linking reagent selection, **290:**256–258
   glutaraldehyde, **290:**256, 258–259, 262, 265–268

heterooligomers, identification and quantification
   GroEL$_7$–GroES$_7$, **290:**262–263, 265
   GroEL$_{14}$–GroES$_7$, **290:**262
   GroEL$_{14}$–(GroES$_7$)$_2$, **290:**262
homooligomers, identification and quantification, **290:**260–262
importance for accurate determination of equilibrium state, **290:**267–268
overview, **290:**255
polyacrylamide gel electrophoresis, **290:**259–260
purity requirements, **290:**255–256
temperature, **290:**259
time of reaction, **290:**259
dimerization effects
   ATPase, **290:**129–130, 292
   folding activity, **290:**131–133
electron microscopy, **290:**250, 252–254, 260
fluorescence anisotropy analysis of stoichiometry
   anisotropy value, calculation, **290:**314–315
   binding conditions and data collection, **290:**317
   GroES labeling with succimidyl-1-pyrene butyrate, **290:**316–317
   materials, **290:**315–316
   Perrin–Weber equation, **290:**315
formation with adenosine nucleotides, size-exclusion chromatography, **290:**128–129
interacting residues, **290:**313–314
X-ray crystallography, **290:**254
GroES
Chromatium vinosum protein, *see* Chaperonin 10
functions, **290:**120–121, 253
gene, **290:**119
GroEL complex, *see* GroEL–GroES complex
oligomerization, assays, **290:**133–134
photolabeling with bisANS
   binding conditions, **290:**320
   conformational change analysis in GroEL–GroES complex, **290:**322–323
   fluorescent properties of fluorophore, **290:**318–319

hydrophobic binding site identification, **290**:318, 321–322

photolabeling conditions, **290**:320–321

purification of recombinant protein

ammonium sulfate fractionation, **290**:124–125

anion-exchange chromatography, **290**:125–126

cation-exchange chromatography, **290**:126

cell growth, **290**:123–124

heat treatment, **290**:127–128

nickel affinity chromatography, **290**:124, 127

vectors and strains for overexpression, **290**:122–123

structure, **290**:121

Ground state destabilization, enzyme catalysis, **308**:21

Group I intron ribozyme, *see* Ribozyme

Growth hormone, gene therapy, **314**:39

GrpE

DnaK interaction, **290**:289–291

quaternary structure analysis

analytical ultracentrifugation, **290**:285–286, 295

cross-linking, **290**:285–286

size-exclusion chromatography, **290**:286

GSH, *see* Glutathione

GSL, *see* Glycosphingolipid

GT1aα

myelin-associated glycoprotein binding, **311**:560–561

synthesis, **311**:561–562, 564

GTP

hydrolysis by G-proteins, *see* $G_{i\alpha1}$

imperfect energetic coupling, *see* ATP sulfurylase; Kinetic proofreading

isotopic labeling using glycolysis and pentose phosphate pathway enzymes

enzyme preparation and assays, **317**:23–28

monitoring of reaction mixtures, **317**:32

overview, **317**:19–23

preparation from glucose, **317**:29–32

purification from reaction pools, **317**:33–35

phosphorothioate-tagged nucleotide triphosphate analogs for nucleotide an-

alog interference mapping, **317**:106–108

tubulin polymerization role, *see* Tubulin polymerization

GTP, caged

Fourier transform infrared spectroscopy of photolysis

band assignment by isotopic labeling, **291**:237–240

difference spectra, **291**:231

dithiothreitol effects on spectra, **291**:233–234

intermediates in decay, **291**:234–235

magnesium effects on spectra, **291**:232

photolysis reaction, **291**:231

Ras studies, **291**:224–225, 243–245

synthesis of compounds, **291**:225

synthesis of compounds, **291**:225

Guanidinium hydrochloride

folding enhancement of inclusion body proteins, **309**:229

inclusion body solubilization, **309**:218–219

Guanine–adenine ligation-mediated polymerase chain reaction, *see* Polymerase chain reaction

Guanosine 5′-monophosphorothioate

bradykinin ligand screening

purification of RNA with thiopropyl-Sepharose, **318**:232–233

transcription reaction, **318**:230–231, 233, 236–237

incorporation efficiency, *in vitro* transcription, **318**:180

synthesis, **318**:230

Guanylate cyclase, photoreceptors

activating proteins, *see also* Calcium-dependent guanylate cyclase activator protein; Guanylate cyclase activating proteins

calmodulin activation in non-mammals, **315**:730, 740

overview of types, **315**:674, 676, 690–691, 709

S100A1, 739–741

S100β, **315**:739–742

assays

avoidance of rod outer segment light activation, **315**:691–692, 710–711, 718

cyclic GMP radioimmunoassay, **315:**711
(S_p)-GTPαS as substrate
advantages, **315:**692–693, 698–699, 711
anion-exchange chromatography for product separation, **315:**696
calculations, **315:**697–698
incubation conditions, **315:**697
kinetic, **315:**693
materials and solutions, **315:**694–695
principle, **315:**693–694
rod outer segment preparation and washing, **315:**695–696
unreacted nucleotide adsorption on neutral alumina, **315:**696–697
kinase activity
autophosphorylation assay, **315:**685–687
kinetics, **315:**689
myelin basic protein as substrate, **315:**685, 687–688
pyrophosphate production assay
GCAP1 purification for assay, **315:**720
intact membrane assays and interfering enzymes, **315:**721–723, 729
overview, **315:**711, 719, 729
principle of coupled reaction assay, **315:**720
rod outer segment membranes, purification and stripping, **315:**719–720
spectrophotometer requirements and modification, **315:**720–721, 729
stripped membrane assay reconstituted with GCAP1, 725, 727–729
washed membrane assay, **315:**723–725
radiolabeled GTP as substrate, **315:**679, 731
single rod outer segment assay
advantages, **315:**751–752
amphibian rod outer segment preparation, **315:**745–746
calcium-dependent measurements, **315:**748, 750
cone outer segment measurements, **315:**751
overview, **315:**743, 745
pipette solutions, **315:**746

washout minimization, **315:**751
ATP regulation, **315:**676, 684–685
calcium regulation, **315:**690, 698, 708–709, 717, 730
comparison with soluble cyclases, **315:**689–690, 718
divalent cation dependence, **315:**698
domains, **315:**674, 683–684
expression of human protein in transfected HEK293 cells
cell culture, **315:**712
membrane preparations, **315:**713
transfection, **315:**712–713
vector, **315:**712
inhibitors, **315:**698
isoforms, **315:**673–674, 708
mutation in disease, **315:**718
nomenclature, **315:**708
phosphorylation
autophosphorylation
evidence, **315:**685
preparation of autophosphorylated protein, **315:**685–687
sites, **315:**688–689
stoichiometry, **315:**688
kinases, **315:**676, 684
sites, **315:**676
purification of GC-E from bovine retina
GTP affinity chromatography
binding and elution, **315:**682–683
buffers, **315:**681
manganese dependence of binding, **315:**681
kinetic parameters, **315:**683
overview, **315:**677–678
rod outer segment membranes
preparation, **315:**679–680
solubilization, **315:**680–681
yield, **315:**683–684
Guanylate cyclase-activating protein, *see also* Guanylate cyclase-activating protein 1; Guanylate cyclase-activating protein 2
assays, **315:**699, 701, 716
calcium binding, **315:**690, 698, 708–709, 717
GCAP1 reconstitution with guanylate cyclase in stripped membranes assay, **315:**725, 727–729
purification for assay, **315:**720

localization in retina, **315:**691, 709–710
overview of types, **315:**674, 676, 690–691, 709
purification from bovine retina
    assay, **315:**699, 701
    GCAP1 purification from rod outer segments
        anion-exchange chromatography, **315:**703
        extract preparation, **315:**703
        hydroxylapatite chromatography, **315:**703
        overview, **315:**702–703
        reversed-phase high-performance liquid chromatography, **315:**704
    immunoaffinity chromatography
        antibodies, **315:**702, 709
        GCAP1, 705–707
        GCAP2, 706–707
        immunoadsorbent preparation, **315:**704–705
        recombinant protein purification, **315:**707
        retinal extract preparation, **315:**705
    overview, **315:**701–702
purification of recombinant proteins from *Escherichia coli*
    gel filtration chromatography, **315:**715
    inclusion body solubilization, **315:**715
    induction, **315:**715
    myristoylation, **315:**716–717
    strains of bacteria, **315:**714
    transformation, **315:**714
    vectors, **315:**714
S100, *see* Calcium-dependent guanylate cyclase activator protein
Guanylate cyclase-activating protein 1, calcium effects, **316:**103
Guanylate cyclase-activating protein 2
    calcium dependency of guanylate cyclase activity modulation, **316:**123
    homology with other guanylate cyclase-activating proteins, **316:**132
    myristoylation, **316:**123
    structure determination by nuclear magnetic resonance
        data acquisition and analysis, **316:**125–126
        isotopic labeling and purification of recombinant protein, **316:**124–125

sample preparation, **316:**125
three-dimensional structures
    recoverin homology, **316:**131–132
    unmyristoylated protein with calcium, **316:**130–131
Guanylyl cyclase
    electrophysiology assay in photoreceptors
        bleaching adaptation assays, **316:**245, 247, 249
        calculations, **316:**242
        rapid solution changes, **316:**239–241
    GC-E knockout mouse
        electroretinogram recordings, **316:**560–564
        generation, **316:**559
        histological analysis, **316:**560–561
        single-cell electrophysiology, **316:**560–561, 563–564
    gene loci, **316:**558
    mutation in disease, **316:**558–559
    types, **316:**558
    zebrafish larva eye, assay
        calcium dependence, **316:**550–551
        incubation conditions and product separation, **316:**550
        materials, **316:**549–550
        principle, **316:**549
Guvacine, γ-aminobutyric acid transporter inhibition, **296:**168–173
Gx
    nomenclature, **311:**639
    radiolabeled compound synthesis, overview, **311:**639–641
GYRA loop, modeling with comparative sequence analysis, **317:**507

H

HACCP, *see* Hazard Analysis Critical Control Point
Hairpin ribozyme
    cleavage assay
        product analysis, **306:**215, 217
        in vitro conditions, **306:**213, 215
    gene expression inhibition *in vivo*
        endogenous versus exogenous delivery, **306:**218–219
        expression detection of ribozyme or substrate

Northern blot analysis, **306:**221
reverse transcription polymerase
chain reaction, **306:**220–221
ribonuclease protection assay,
306:220–221
expression vector design and construction, **306:**219
phenotypic change analysis, **306:**221,
225
requirements, **306:**218
transfection and stable expression,
306:219–220
gene therapy
applications and ribozyme development, **306:**207–208, 225
human immunodeficiency virus
clinical trials, **306:**224
limitations of current treatments,
306:221–222
preclinical studies, **306:**222–224
retroviral vectors for gene delivery,
306:223–224
strains and ribozyme cocktail treatment, **306:**222–223
targets, **306:**222–223
recycling, **306:**208
RNA preparation for ribozyme and substrate
preparative polyacrylamide gel electrophoresis, **306:**212–213
transcription, *in vitro*,
306:212
structure, **306:**208
template DNA construction for ribozyme
and substrate, **306:**210, 212
time-resolved fluorescence resonance energy transfer analysis of conformational isomers of substrate complexes, **317:**434–435
Haldane relationship, constraints on enzymatic reactions, **308:**3, 15
5-Halopyrimidines
mechanism of protein cross-linking with
ultraviolet irradiation, **318:**89–90
MS2 bacteriophage coat protein, cross-
linking to substituted RNAs
controls, **318:**97
dissociation constant determination,
318:94, 103
equipment, **318:**100

irradiation conditions, **318:**94, 96,
101–104
materials, **318:**99
polyacrylamide gel electrophoresis of
cross-linked products, **318:**94
protein sequencing, **318:**98–99
substituted RNA preparation by *in
vitro* transcription, **318:**100–101
photoreactive amino acids, **318:**89–90, 95,
97
structure of nucleoprotein cross-links,
318:89–90
substitution site selection for cross-
linking studies, **318:**92–94
ultraviolet absorption and light sources
for cross-linking, **318:**90–92
Hammerhead ribozyme, *see also* Ribozyme
comparative gel electrophoresis of global
folding transitions, **317:**375, 377
disulfide cross-linking
alkaline hydrolysis analysis,
318:172–173
2′-amino nucleotide modification with
isothiocyanates, **318:**167, 169
catalytic activity of cross-linked ribozymes, **318:**173–174
cross-linking reaction, **318:**169
polyacrylamide gel electrophoresis analysis, **318:**169–173
radiolabeling of RNA, **318:**168
rationale, **318:**166–167
RNA synthesis, **318:**168
engineering of metal-binding sites
overview, **317:**130
religation of DNAzyme products for
modification, **317:**145
religation of ribonuclease H products,
317:137–139
phosphoramidite-modified pyrimidine
base studies, **317:**53–54
quantitative structure–activity relationship, three-dimensional modeling
base modifications for position 7 substitution, **317:**460–463, 466–468
limitations of crystallography,
317:458–459
logical–combinatorial approach,
317:461, 463
mathematical model generation,
317:463–464

physical meaning of model,
    317:468–470
statistical significance and predictivity,
    317:464–466
site-specific RNA cleavage, 317:133–134
structure, 318:165–166
X-ray crystallography of small ribozymes
    chemical synthesis
        automated synthesis, 317:151–152
        desalting, 317:153
        overview, 317:151
        reversed-phase high-performance liq-
            uid chromatography,
            317:152–153
    crystallization
        cations, 317:155–158, 182
        precipitants, 317:154–155, 182
        screening, 317:154, 182–183
    crystal packing promotion, 317:150–
        151, 181–182
    diffraction data acquisition
        cryoprotection, 317:159–160, 184
        crystal requirements, 317:158
        radiation sensitivity, 317:158–159
        synthetic mother liquors, 317:159
    heavy atom derivatization, 317:160–
        162, 183–184
    initial electron density maps, 317:164–
        165, 184–185
    intermolecular contacts in crystal lat-
        tice, 317:181–182
    modified ribozymes, 317:458–459
    Patterson methods for heavy atom posi-
        tions, 317:162–164
    refinement, 317:165–168, 185–186
    time-resolved crystallography
        catalysis initiation in crystals,
            317:187–188
        difference Fourier analysis,
            317:193–194
        extent of cleavage, analysis in crys-
            tals, 317:196–198
        kinetic assays for cleavage in crys-
            tals, 317:194–196
        Laue diffraction observation of tran-
            sient intermediates, 317:192–193
        overview, 317:186–187
        termination of catalysis in crystals,
            317:196
    trapping of reaction intermediates

cryocrystallographic trapping,
    317:191–192, 468–469
kinetic bottleneck modifications,
    317:188–191
X-ray crystallography versus fluorescence
    resonance energy transfer structures,
    318:165–166, 174
Hard particle model, see Molecular
    crowding
HAT, see Histone acetyltransferase
Hat1p, recombinant enzyme production,
    304:717–718
HAT tag, see Transposon mutagenesis
Hazard Analysis Critical Control Point,
    firefly luciferase ATP assay applica-
    tion, 305:362
HBV, see Hepatitis B virus
hcf, see High chlorophyll fluorescence
hCgB, see Human chromogranin B
HDAC, see Histone deacetylase
Heat shock protein 18.1
    assays
        luciferase refolding assay,
            290:363–365
        substrate binding
            incubation conditions, 290:359–360
            nondenaturing polyacrylamide gel
                electrophoresis, 290:361–363
            size-exclusion high-performance liq-
                uid chromatography,
                290:360–361
    oligomeric structure determination by
        nondenaturing pore-exclusion poly-
        acrylamide gel electrophoresis,
        290:352–355
    purification of Pisum sativum recombi-
        nant protein in Escherichia coli
        ammonium sulfate fractionation,
            290:357
        anion-exchange chromatography,
            290:357–358
        cell harvest and lysis, 290:356–357
        expression and initial analysis,
            290:351–352, 356
        hydroxylapatite chromatography,
            290:358
        reagents and materials, 290:355–356
        solubilization, 290:352
        sucrose gradient fractionation, 290:357
        vectors, 290:351

temperature-induced conformational
change, **290:**359
Heat shock protein 25
citrate synthase, effects on refolding
reactivation after thermal inactivation,
290:333
renaturation after chemical denaturation, **290:**336–337
thermal aggregation, **290:**349
thermal inactivation, **290:**330
expression in stress, **290:**340–341
purification of recombinant protein from
Escherichia coli
column chromatography, **290:**347–348
extraction, **290:**347
overexpression system, **290:**346–347
Heat shock protein 26
assay with citrate synthase, **290:**349
expression in stress, **290:**340–341
purification of recombinant protein from
Saccharomyces cerevisiae
column chromatography, **290:**345–346
extraction, **290:**344
overexpression system, **290:**343–344
Heat shock protein 27
assay with citrate synthase, **290:**349
expression in stress, **290:**340–341
purification of recombinant protein from
Escherichia coli
column chromatography, **290:**347–348
extraction, **290:**347
overexpression system, **290:**346–347
Heat shock protein 47, antisense knockdown in kidney, **314:**387
Heat shock protein 60
biological function, **290:**195
homology with GroEL, **290:**194
purification of recombinant yeast protein
from *Pichia pastoris*
anion-exchange chromatography,
290:197–198
cytosol extraction, **290:**196–197
expression system generation, **290:**196
size-exclusion chromatography, **290:**198
structure, **290:**202
Heat shock protein 70
citrate synthase, effects on refolding
reactivation after thermal inactivation,
290:333
thermal aggregation, **290:**327

thermal inactivation, **290:**330
in situ hybridization
administration in rat brain,
314:239–240
frozen sections
preparation, **314:**240
processing, **314:**241–242, 247
hybridization mix, **314:**241
localization
fluorescent probe detection,
314:243–244
radiolabeled probes, **314:**243
oligonucleotide design, **314:**239
radiolabeling of probe, **314:**240–241
stability analysis, **314:**245
subcellular localization, **314:**246–247
vibratome sections
preparation, **314:**240
processing, **314:**242–243, 247
Heat shock protein 90
biological functions, **290:**409–410, 412–
413, 415
citrate synthase, effects on refolding
reactivation after thermal inactivation,
290:332–333
renaturation after chemical denaturation, **290:**335, 337
thermal aggregation, **290:**327, 418
thermal inactivation, **290:**330, 418
Escherichia coli protein, *see* HtpG
essentiality for survival, **290:**409
expression levels, **290:**410, 412, 415–416
kinase association, **290:**419
partner proteins, *see* FKBP52; Hop/p60;
p23
purification of recombinant human protein in baculovirus system
anion-exchange chromatography,
290:417
cell growth and lysis, **290:**416
hydroxyapatite chromatography,
290:417
size-exclusion chromatography, **290:**417
purification of recombinant yeast protein
in yeast
anion-exchange chromatography,
290:414
cation-exchange chromatography,
290:414
cell growth and lysis, **290:**413–414

hydroxyapatite chromatography,
    290:414–415
size-exclusion chromatography, **290:**415
steroid hormone receptor association,
    290:418–419
Heat shock protein 100, recombinant pro-
    tein expression in yeast, **290:**437–438
Heat shock protein 104
    assays
        ATPase, **290:**442–443
        oligomerization, **290:**443–444
    ATPase activity, **290:**431
    biological functions, **290:**430
    concentrating and storage of purified
        preparations, **290:**441–442
    domains, **290:**431
    physical properties, **290:**430–431
    purification of recombinant yeast protein
        from *Escherichia coli*
        anion-exchange chromatography,
            290:440–441
        buffer and solution preparation,
            290:439–440
        cell growth and lysis, **290:**439–440
        nickel affinity chromatography, **290:**440
        plasmid and strain construction,
            290:438–439
    purification of recombinant yeast protein
        from yeast
        anion-exchange chromatography,
            290:435–436
        cell growth, **290:**433–434
        cell lysis, **290:**434–435
        dye affinity chromatography, **290:**435
        hydroxyapatite chromatography,
            290:435–436
        mutant proteins, **290:**438
        plasmid and strain construction,
            290:431–432
        scaling of preparations, **290:**437
        solutions and equipment, **290:**432–433
        yield, **290:**436–437
Heavy atom derivatives, RNA
    cocrystallization with heavy atoms,
        317:160
    crystal analysis for engineering
        dimethylsulfate mapping, **317:**131–132
        rationale, **317:**130–131
    engineering of RNA, advantages,
        317:125, 183–184

halogenation of pyrimidines with bro-
    mine or iodine, **317:**125–126,
    160–161
large RNA engineering, **317:**128–130
mercuration of pyrimidine C-5 position,
    317:127
phasing requirements for large RNAs,
    317:173–175
phosphorothioate incorporation for mer-
    cury binding, **317:**126, 161–162, 184
site selection for metal binding engi-
    neering, **317:**128
soaking of crystals, **317:**124–125, 160
Heavy meromyosin, caged protein con-
    jugate
    functional analysis
        ATPase assay, **291:**110, 114
        in vitro motility assay, **291:**110,
            114–116
    photoactivation
        duration of irradiation, **291:**112–114
        transient kinetics analysis, **291:**113–116
    preparation
        4,5-dimethoxy-2-nitrobenzyl bromide
            conjugation, **291:**111–112, 114–115
        materials, **291:**97
        purification of protein, **291:**111–112
HEK 293–Epstein–Barr virus expression
    system
    cell culture, **306:**31
    comparison with other expression sys-
        tems, **306:**36–41
    immortalization of cells, **306:**29–30
    plasmid design, **306:**30–31
    selection for stable episomes, **306:**31
    transfection, **306:**31
Helix–coil transition
    C-cap propensity, **295:**5, 27, 31
    complex sequences, predicting helix pro-
        pensity
        model systems, **295:**33–40
        parameters in modeling, **295:**25–26
        reference systems, **295:**23–24
        side-chain interactions, **295:**24–25
    continuity, **295:**1–2
    control peptides in analysis, **295:**17–18
    cooperativity, **295:**1–2
    free energy
        capping, **295:**3, 27
        nucleation, **295:**5–6, 30–31

propagation, **295:**5, 27, 30–32
side-chain interactions, **295:**27–28
helix formation, quantitation
circular dichroism, **295:**7–10, 12, 15, 28–29
nuclear magnetic resonance, amide proton exchange, **295:**7, 10–12, 15, 29
Lifson–Roig model
glutamine–aspartate hydrogen bond interactions, analysis, **295:**19–21
homopolymer approximation, **295:**20–23
implementation, **295:**4–5
modifications, **295:**3, 25
overview, **295:**2–3
Zimm–Bragg model comparison, **295:**30
N-cap propensity, **295:**5–7, 17, 27, 31
partition function calculation, **295:**4–5
peptide design in analysis
chromophore incorporation, **295:**13–14
helix content optimization, **295:**14–15
ideal peptide criteria, **295:**12
repeating sequences, **295:**13, 34
solubility, **295:**14
stabilizing interactions in peptides, **295:**15–17
unnatural residues in analysis
helix propensity predictions for natural amino acids, **295:**33–35, 37–39, 41
N-4-hydroxybutyl-L-glutamine, **295:**32
N^5-3-hydroxypropyl-L-glutamine, **295:**32
Hematopoietic cell
complementary DNA amplification from a small number of cells
adaptor ligation of complementary DNA, **303:**48–50
complementary DNA synthesis
first strand, **303:**47–48
second strand, **303:**48
full-length complementary DNA cloning with high efficiency, second-strand synthesis, **303:**39–40
materials and reagents, **303:**46
messenger RNA preparation, **303:**47, 55
polymerase chain reaction

betaine and dimethyl sulfoxide improvement of efficiency, **303:**45–46, 50
differential display analysis of products, **303:**50–54
reaction conditions, **303:**50
solid-phase differential display in gene expression analysis, **303:**496
Hematopoietic stem cell, fluorescence selection of acid sphingomyelinase expression
acid sphingomyelinase assays, **312:**333–334
applications, **312:**330–331, 337–338
BODIPY-sphingomyelin
preparation, **312:**332
pulse–chase labeling, **312:**332–333
cell preparation, **312:**331
c-Kit immunolabeling, **312:**332, 335
colony-forming units–spleen assay, **312:**334, 336
enzyme deficiency in Niemann-Pick disease, **312:**331, 337–338
fluorescence-activated cell sorting, **312:**333–335
lineage depletion, **312:**331–332
retrovirus
amphotropic versus ecotropic viruses, **312:**336
transduction, **312:**333
Heme oxygenase
assays
bilirubin spectrophotometric assay, **300:**323–324
high-performance liquid chromatography assay
applications, **300:**335–336
assembly of *in vitro* heme oxygenase reactions, **300:**329–330
calibration and calculations, **300:**331–333
cell culture, **300:**325–326
chromatography system preparation, **300:**331
microsomal protein extract preparation, **300:**326–328
principle, **300:**325
sodium arsenite treatment, **300:**326
troubleshooting, **300:**333–335
radioisotope assays, **300:**324

functions, **300:**322–323

immunochemical assay, **309:**148–150

isozymes, **300:**322

messenger RNA analysis of induction, **300:**335

Hemoglobin, *see also* S-Nitrosohemoglobin

advanced glycation end products, enzyme-linked immunosorbent assay

clinical application, **309:**166

hemolysate preparation from fresh blood, **309:**166–167

hemolysate preparation from frozen and packed red blood cells, **309:**167

incubations and washes, **309:**167–168

Bohr effect

cyanomet intermediates

definition of Bohr effect, **295:**211

deoxy/cyanomet model, **295:**211–212

determination of Bohr effect, **295:**212–213

pH effects, **295:**213–214

molecular mechanism, **295:**210

mutant hemoglobin studies, **295:**215–216, 226–227

oxygen ligation comparison, **295:**215

pH measurement

apparatus, **295:**219–220

buffer capacity of oxyhemoglobin, **295:**220–221, 226

small samples, **295:**221, 223–226

preparation of hemoglobin for analysis, **295:**216–218

carbon monoxide binding, **295:**209–210

Co^{2+}/Fe^{2+} hybrids, oxygen binding analysis

affinity, **295:**192

binding isotherm construction, **295:**194–199

concentration-dependent isotherms, fitting, **295:**200–201, 203–204

cooperative free energy, **295:**205–208

fitting function for asymmetric mixed metal hybrid, **295:**199–200

ligation states, **295:**193–194

dihydrorhodamine oxidation, **301:**441

hydrogen peroxide interactions, **301:**437–438

isoelectric focusing of species, **295:**218–219

ligation microstates and stoichiometric reaction sequence of oxygen binding, **295:**191, 208–209

nitric oxide effects on oxidation, **301:**437–444

nitric oxide radical scavenging, **301:**497–498

spectrophotometric determination of oxidation status, **301:**441–443

spectroscopic studies of oxygen binding

degeneracy of Adair constants, **295:**191–192

overview, **295:**190–191

subunit dissociation constant determination

bovine hemoglobin, **306:**127

hemoglobin A, **306:**123–124

hemoglobin F, **306:**124–125

hemoglobin F_1, **306:**125

hemoglobin S, **306:**123–124

preparation of hemoglobins, **306:**114–115

recombinant hemoglobins with substitutions in $\alpha_1\beta_2$ interface, **306:**126

recombinant sickle hemoglobins, **306:**125–126

small zone gel filtration on Superose-12

analysis time, **306:**118

calculations, **306:**121–122

care of column, **306:**118–119

elution conditions, **306:**117

elution profiles and curve analysis, **306:**117, 119–122

overview, **306:**114

peak concentration analysis, **306:**117–118

sample dilution and peak broadening, **306:**127–129

standards and calibration, **306:**115–117

systematic errors, **306:**122

Hemolysin B

hemolysin A transport

assays

gel electrophoresis of secreted hemolysin A, **292:**57–59

hemolytic zone assay, **292:**55–57

secreted hemolysin A measurement

in liquid culture by absorption
    spectroscopy, **292:**57
energetics, **292:**68
overview, **292:**51–52, 55
signal sequence
    genetic complementation analysis of
        hemolysin B interaction, **292:**53–
        55, 60–61, 63–66
    modeling of hemolysin B interaction,
        292:65–66
    mutations resulting in transport defi-
        ciency, **292:**60–61, 64–65
    structure, **292:**53
phylogenetic analysis, **292:**52–53
α-Hemolysin R104C
    modification with 2-nitrobenzyl group,
        291:121, 128–129
    photolysis of caged protein, **291:**130–133
Hepatitis B virus
    clinical features of infection, **314:**524
    duck model
        antisense inhibition
            hepatocyte isolation and culture,
                314:528–529
            liver targeting of oligodeoxynucleo-
                tides, **314:**535–536
            mechanisms of action, **314:**532–533
            optimal oligodeoxynucleotide crite-
                ria, **314:**533–534
            phosphorothioate oligodeoxynucleo-
                tide synthesis, **314:**526–527
            rationale, **314:**526
            transient effects, **314:**534–535
            viral target selection, **314:**527–528
            in vivo studies of viral replication in-
                hibition, **314:**529–531
        drug testing, **314:**525–526
        overview, **314:**525
    glycopeptide YEE(ahGalNAc)₃-modified
        antisense oligodeoxynucleotide tar-
        geting to asialoglycoprotein receptor
        cell culture, **313:**308
        controls, **313:**320
        DNA analysis, **313:**308–309, 319
        dose–response, **313:**319
        stability of neoglycoconjugate, **313:**309
        surface antigen assay, **313:**308, 319–320
        target sequences, **313:**307–308
        toxicity analysis, **313:**209, 320–321
    life cycle, **314:**525

prevalence of infection, **314:**524
structure, **314:**525
therapeutic strategies, **314:**524–525
Herpes simplex virus
    bacterial artificial chromosome muta-
        genesis
        DNA purification, **306:**341
        electrocompetent cell preparation,
            306:341
        electroporation, **306:**341
        gene replacement
            day-to-day protocol, **306:**347
            screening and characterization,
                306:348
            vector construction, **306:**344–347
        infectious viral DNA preparation,
            306:349–350
        media and growth conditions,
            306:340
        overview, **306:**337, 339–340
        plasmid
            construction, **306:**342–344
            transfection into mammalian cells,
                306:348
        rescue and growth of virus,
            306:348–349
        strains of bacteria, **306:**340
        vhs gene deletion, **306:**350, 352
    gene therapy vector
        amplicon-based vectors, **302:**431–433
        genome, **302:**430–431
        green fluorescent protein as marker
            amplicon vector concentration,
                302:434
            cells and viruses, **302:**433
            cytotoxicity of vector, **302:**435–437
            gene transfer *in vitro*, **302:**434–435
            pHE vector construction, **302:**433
            titration, **302:**434
            virus production, **302:**433–434
        helper genes, **302:**431
    mutant generation, overview, **306:**337
Hexachlorofluorescein, RNA labeling,
    317:422
Hexanoic acid, preparation of *N*-hydroxy-
    succinimide ester, **311:**534–535
Hexanoic ceramide, carbon-14 radiolabeling
    for sphingomyelin synthase assay,
    311:35
Hexanoylhomoserine lactone

characterization, *see* Acylhomoserine lactone autoinducers

detection bioassay, **305:**293

N-Hexanoyl-D-*erythro*-sphinganine, tritiated substrate preparation, **311:**28

β-Hexosaminidase B, photoaffinity labeling, **311:**579–580

HGP, *see* Human Genome Project progress, **303:**55–56

HIC, *see* Hydrophobic interaction chromatography

Hidden Markov model

comparison to other noise reduction methods, **293:**420–421, 429, 432

computational costs, **293:**436–437

expectation–maximization algorithm computation, **293:**425–427

E step, **293:**424–425

M step, **293:**425

overview, **293:**423–424

likelihood function estimation, **293:**427–428, 436

model evaluation applications, **293:**435–436

modifications of signal model, **293:**429, 432, 434–435

signal model construction

assumptions, **293:**422–423

discrete time, **293:**421

finite-state, **293:**421–422

first-order process, **293:**422

parameter requirements, **293:**422

states, estimation of number, **293:**428–429

High chlorophyll fluorescence, screening for mutants, **297:**40, 193, 195–196

High mobility group proteins

DNA binding

motif, sequence conservation, **304:**100

specificity, **304:**100–101

HMG14/17

DNase I footprinting of nucleosome interactions, **304:**134–135

electrophoretic mobility shift assay of nucleosome core particle interactions

binding conditions, **304:**137

core particle preparation, **304:**136–137

electrophoresis conditions, **304:**137

overview, **304:**134–136

functions, **304:**133–134, 142–143

hydroxyl radical footprinting analysis of chromatin interactions

chromatin particle purification, **304:**139

footprinting conditions, **304:**139

gel electrophoresis, **304:**139–140

nucleosome complex preparation, **304:**139

overview, **304:**134, 137–138

nucleosome core particle binding, **304:**133

organization in cellular chromatin

clustering and quantification, **304:**150–151

gel electrophoresis and analysis, **304:**154–155

immunfractionation, **304:**150–151

immunofractionation with cross-linking, **304:**153–154

oligonucleosome preparation, **304:**151

reconstitution into chromatin

assembly reaction, **304:**146

functions in *in vivo* versus *in vitro*-assembled chromatin, **304:**143

micrococcal nuclease digestion studies, **304:**147–148, 150

sedimentation velocity centrifugation analysis, **304:**146–148

topoisomer analysis of assembly effects, **304:**144, 146

Xenopus egg extract preparation, **304:**143–144, 146

thermal denaturation assay of nucleosome interactions, **304:**134

two-dimensional gel analysis of nucleosome core particle interactions

antibody preparation and immunofractionation, **304:**141–142

electrophoresis conditions, **304:**142

overview, **304:**134, 140–141

HMG-D

DNA bending assay with ligase-mediated circularization, **304:**101, 103, 107, 123–124

DNA preparation for binding studies

extraction, **304:**122

oligonucleotide synthesis, **304:**122–123

restriction digestion, **304**:122
electrophoretic mobility shift assay
    binding reaction conditions, **304**:123
    cooperativity of DNA binding, analysis, **304**:113, 115–117
    data extraction and analysis, **304**:124–125
    DNA-binding affinities, **304**:107, 109, 112
    DNA-binding specificity analysis, **304**:101
    electrophoresis conditions, **304**:119
    kinetic analysis of DNA binding, **304**:112–113, 124
    limitations, **304**:117–118
    protein mutation effects on DNA binding, **304**:112
immunoprecipitation, amplification, and sequencing of bound DNA, **304**:121
purification of recombinant protein from *Escherichia coli*
    chromatography, **304**:120
    expression system and cell growth, **304**:119
    isoforms from oxidation of intercalating methionine, **304**:118–120
    lysis and ammonium sulfate fractionation, **304**:120
    mass spectrometry, **304**:120–121
    quantification, **304**:121
structure, **304**:101
Western blot analysis of chromatin interactions
    blocking and probing of membranes, **304**:133
    Drosophila embryo collection, **304**:129
    gel electrophoresis, **304**:131
    membrane transfer, **304**:131–132
    micrococcal nuclease digestion of chromatin, **304**:130–131
    nuclei isolation, **304**:130
    principle, **304**:126–127
    rationale, **304**:125–126, 129
HMG-I(Y)
    binding site mutagenesis and antisense transfection assays, **304**:185–188
    cross-linking to nucleosomes *in vivo*, **304**:182–183

DNA bending assays, **304**:177
DNA-binding specificity, **304**:156, 169–171
electrophoretic mobility shift assay
    nucleosome core particle complexes, **304**:175
    supershift/ablation assays, **304**:183–185
fluorescence competition binding assays, **304**:172–173
fluorescence *in situ* immunolocalization, **304**:181–182
footprinting
    DNase I footprinting and directional substrate binding, **304**:169–172
    hydroxyl radical footprinting, **304**:172, 179
    reconstituted nucleosomes with defined sequence DNAs, **304**:176
four-way junction binding assays, **304**:179
functions, **304**:155–156, 179–180
nucleosome-binding assays, **304**:173, 175
purification
    acid extraction, **304**:157–158
    cation-exchange chromatography, **304**:166
    cell starting materials, **304**:158–159
    gel electrophoresis, **304**:164–165
    lysis, **304**:159
    polybuffer exchanger purification, **304**:165
    quantification by absorbance, **304**:163–164
    recombinant protein from *Escherichia coli*, **304**:161–162, 166–167
    reversed-phase chromatography, **304**:163, 166–167
structure, **304**:156–157
supercoiling assay with two-dimensional gel electrophoresis, **304**:177, 179
types, **304**:155
Western blot analysis
    antibody production, **304**:167–169
    incubations and washings, **304**:169
overview of types and function, **304**:100, 133, 155

High-performance electrophoresis chromatography, β-tubulin, **298:**241–242, 246
High-performance liquid chromatography,
see also High-performance liquid chromatography with electrochemical detection; Reversed-phase high-performance
liquid chromatography
adenine reaction with hydroxyl radical,
product detection, **300:**199–201
advanced glycation end products, **309:**158
ω-agatoxin-IVB, **294:**105–106
amyloid-β peptide purity analysis,
309:192–193, 486
antioxidant depletion with ozone in murine skin, analysis
advantages, **319:**546
ascorbate and urate analysis
equipment, **319:**539
operation, **319:**540
reagents, **319:**539
sample preparation, **319:**540
standards, **319:**540
extraction, overview, **319:**538
glutathione analysis
equipment, **319:**541
operation, **319:**541
reagents, **319:**540
sample preparation, **319:**542
standards, **319:**541–542
overview, **319:**537
ozone exposure, **319:**537
peak identification and quantification,
319:544–545
tape stripping, **319:**537–538
troubleshooting, **319:**545
vitamin E analysis
equipment, **319:**543
operation, **319:**543
reagents, **319:**542
sample preparation, **319:**543–544
standards, **319:**543
boranophosphate nucleoside triphosphate
analog diastereomer separation,
313:244
carotenoids, **299:**390, 392
ceramide mass analysis
advantages, **312:**17, 20
benzoylation, **312:**17–18, 21–22
chromatography conditions, **312:**18–20
column care, **312:**21–22

glycerol-containing samples, **312:**22
lipid extraction, **312:**17
standards, **312:**20–22
3-chlorotyrosine assays
chlorination reactions
hypochlorous acid, **300:**115–116
myeloperoxidase, **300:**114–117
stimulated neutrophils, **300:**117
electrochemical detection, **300:**113,
119–120
fluorescence detection, **300:**113,
118–119
neutrophil isolation, **300:**115
protein hydrolysis, **300:**117
cholesteryl linoleate hydroperoxide assay,
see Lipid hydroperoxide
conformational analysis of gramicidin analogs with size-exclusion chromatography, **294:**534–535
dehydroascorbic acid assays
coulometric electrochemical detection
following reduction
hydrodynamic voltammetry plot,
299:79, 82
instrumentation, **299:**67–68, 78–79
materials, **299:**68, 78
reduction reaction, **299:**75–76, 81
sample preparation, **299:**67, 72–73,
79–82
standards, **299:**74–75
derivatization, **299:**73–74
radiolabeled compound detection,
299:75–76
dihydroxybenzoate product measurement
in hydroxyl radical assay, **300:**253,
299, 301–302, 307–308
DNA assemblies with tethered photooxidant, **319:**178–179
enzyme assay
all-*trans* retinol dehydrogenase,
316:365, 368, 370–371
11-*cis*-retinol dehydrogenase, **316:**365–
366, 368, 370–371
glutathione
electrochemical detection, **299:**81–82
N-(1-pyrenyl)maleimide derivatives
derivatization reaction, **299:**259,
261–264
glutathione disulfide quantitation,
299:259–260, 262

materials, **299**:260
metabolic modulation studies,
    299:265–266
sample preparation, **299**:264
standard curves, **299**:261–262
glycosphingolipids
acute myelogenous leukemia ganglio-
    side analysis, **312**:143
aminopropyl chromatography of gangli-
    osides, **312**:142
anion-exchange chromatography
    brain gangliosides from mouse,
        312:55–57
    column care, **312**:55
    GM3 gangliosides, **312**:60, 62–63
    granulocyte monosialogangliosides
        from humans, **312**:57–60
    materials, **312**:53–54
    overview, **312**:49
    preparative chromatography, **312**:53,
        62–63
ceramide subspecies preparation for bi-
    ologic assays, **312**:143–145
native compound separation systems,
    312:51–53
precolumn derivatization, **312**:50–51
reversed-phase chromatography of gan-
    gliosides, **312**:142–143
heme oxygenase assay, *see* Heme oxy-
    genase
8-hydroxyguanosine separation and detec-
    tion, **319**:148, 152
lipid hydroperoxide, *see* Lipid hydroper-
    oxide
methylphosphonodiester oligodeoxy-
    nucleotides
analytical chromatography
    anion-exchange chromatography,
        313:103–104
    column regeneration, **313**:104
    reversed-phase chromatography,
        313:103
purification by anion-exchange chroma-
    tography, **313**:104–105
minor groove binder–oligodeoxynucleo-
    tide conjugate, **313**:173
nitrate and nitrite determinations
biological systems
    chromatography, **301**:265–268
    sample preparation, **301**:263–265

detection methods, **301**:259, 262,
    266–269
ion-exchange chromatography,
    301:259–262, 265–268
reversed-phase chromatography,
    301:259
simple chemical systems
    chromatography, **301**:260–262
    radiolysis for free-radical generation,
        301:260
nitrated hydroxycinnamic acids, **299**:214,
    219, 222–223, 228
nitrated lipids, **301**:460, 464–465, 467–468
phenolic antioxidants
catechin analysis in tea with caffeine
    chromatography conditions, **299**:109,
        111, 203–204, 206
    comparison of green teas, **299**:112–
        113, 203
    extraction, **299**:204
    instrumentation, **299**:108–109
    precision, **299**:204, 206
    quantitative analysis, **299**:111–112
    sample extraction, **299**:109
    stability of samples, **299**:206
    standards, **299**:109, 204
chromatography conditions,
    299:103–105
food sample preparation, **299**:100, 103
retention times, **299**:105–106
standards, **299**:100
urine sample
    dietary manipulation, **299**:97, 100,
        106
    preparation, **299**:103
wine analysis
    absorbance characterization of
        peaks, **299**:119–120, 132–133
    anthocyanins, **299**:119–120, 135–136
    catechins, **299**:119
    difficulty of separations, **299**:113–114
    flavonols, **299**:121
    instrumentation, **299**:124, 131
    linearity, **299**:125, 127–129, 132
    low pH gradients, **299**:114, 116–117,
        124–125, 131
    pH shift gradient, **299**:114–116
    polystyrene reversed-phase chroma-
        tography, **299**:117
    precision, **299**:129, 132

procyanidins, **299:**137
recovery, **299:**129, 132
resolution, **299:**125, 132
sample preparation, **299:**123
sensitivity, **299:**122–123, 130, 132
standards, **299:**118–119, 123–124,
    130–131, 133, 135
phospholipid hydroperoxide glutathione
    peroxidase assay, *see* Phospholipid
    hydroperoxide glutathione perox-
    idase
phospholipid peroxidation assay, *see* Phos-
    pholipid peroxidation
phosphorothioate oligodeoxynucleotides
  anion-exchange chromatography
    advantages, **313:**226
    analytical chromatography, **313:**209
    G-quartets, **313:**223
    principle, **313:**205
    trityl-off purification of gram quanti-
        ties, **313:**218–220
    trityl-on purification, **313:**216–217
  chemicals and buffers, **313:**206
  columns and media, **313:**207–208
  displacement chromatography, **313:**220,
      222
  gel permeation chromatography,
      313:205
  hydrophobic interaction chromatogra-
      phy, **313:**205, 217–218
  instrumentation, **313:**207
  postpurification processing,
      313:223–224
  quantitative analysis
    capillary gel electrophoresis,
        313:209–210, 223
    ultraviolet spectroscopy, **313:**208
  reversed-phase chromatography
    advantages, **313:**225
    detritylation on-column, **313:**212–213
    gram scale purification, **313:**213, 215
    ion-pairing agents, **313:**205–206, 225
    principle, **313:**204–205
    trityl-on purification of 20-mer,
        313:210, 212
  synthesis, **313:**206
retinoic acid isomer separation
  chromatography conditions and instru-
      mentation, **299:**434–437, 441
  data analysis, **299:**437–438

extraction, **299:**433–435, 441
resolution of peaks, **299:**438–440
sensitivity and reproducibility,
    299:440–441
standards, **299:**431, 433
retinoid analysis in visual cycle, *see* Reti-
    noids, high-performance liquid chro-
    matography analysis
scorpion venom peptide inhibitors 628
squalene peroxide analysis, **319:**616–617,
    619–621
stereochemical analysis of sphingolipids,
    312:227
tandem mass spectrometry coupling for
    sphingolipid analysis, **312:**35–36, 45
tannins, size separation
  degree of polymerization, analysis,
      299:182, 184
  gradient, **299:**182
  peak identification, **299:**180–181
  sample preparation, **299:**180–181
tocopherol oxidation products,
    301:486–487
topical antioxidant analysis in tape-
    stripped samples, **319:**618
vitamin E
  diol column chromatography, **299:**320,
      323–325
  fluorescence detection, **299:**325
  materials, **299:**321
  milk replacer diet effects in piglet tis-
      sues, **299:**328–329
  oils and diets, extraction, **299:**322
  separation of components, **299:**323–324
  simultaneous determination with coen-
      zyme Q
    animal maintenance, **299:**334–335
    chromatography conditions, **299:**337–
        338, 340, 344, 347
    electrochemical detection, **299:**333–
        334, 342, 344–345
    extraction, **299:**331–332
    materials, **299:**334
    plasma sample preparation, **299:**344
    recovery and reproducibility,
        299:340–341, 345
    saponification, **299:**332–333
    standards, **299:**336–337, 342–344, 347
    storage of samples, **299:**332
  standards, **299:**322–323

tissue pulverization and extraction,
299:321–322, 329
triacylglycerol influence on quantitation, **299:**325–329
High-performance liquid chromatography
with electrochemical detection
Anethole dithiolthione
cell culture, **299:**302
chromatography conditions,
299:302–303
extraction, **299:**301–306
instrumentation, **299:**301
ascorbic acid assay
amperometric detection, **299:**66
chromatography conditions, **299:**68,
70–72
coulometric detection, **299:**66–67, 72
hydrodynamic voltammetry plot,
299:79, 82
instrumentation, **299:**67–68, 78–79
materials, **299:**68, 78
passivation, **299:**72
recycling assay, **299:**86–87
sample preparation, **299:**67, 72–73,
79–82
standards, **299:**72
cholesterol oxidation products
electrochemical detection
erythrocyte membranes, **319:**94–97
operating conditions, **319:**93–94
photoxidized cells, **319:**98–100
ozonization products
electrochemical detection,
319:195–196
mass spectrometry detection, **319:**196
ultraviolet A irradiation products
chemilumiscence detection,
319:192–194
electrochemical detection, **319:**192
DNA oxidation products
analysis of immunoaffinity column samples, **300:**164–165
artifacts, **300:**181
calculations, **300:**163
cells, DNA extraction and hydrolysis,
300:178
chromatography, **300:**162, 179–180
DNA extraction
chaotropic sodium iodide, **300:**158–
159, 165

phenol extraction, **300:**159–160, 165
chloroform/isoamyl alcohol extraction, **300:**177
DNA hydrolysis, **300:**160–161
instrumentation, **300:**161–162, 179
sensitivity, **300:**162–163
tissue homogenization, **300:**157–158,
177
urine analysis by column switching,
300:170–174
monoamine analysis in microdiasylates
apparatus, **296:**728
mobile phase preparation, **296:**728
monoamine standard preparation,
296:729
peak identification, **296:**730
running conditions, **296:**730
nitrotyrosine and derivatives
acetylation and extraction of amino
acids, **301:**32–33
acid hydrolysis, **301:**155–156
deacetylation and dithionite reduction,
301:29–30, 33–34
electrochemical detection, **301:**156, 158
enzymatic hydrolysis, **301:**32, 155
goals, **301:**29
interpretation of results, **301:**158–160
mobile phase preparation and gradient
elution, **301:**34–36, 156
prevention of *ex vivo* nitration,
301:39–40
sample preparation, **301:**31, 154–155
selectivity and sensitivity, **301:**38–40,
153–154
standards, **301:**36–38
thiol and disulfide analysis with electrochemical detection
cell culture, **299:**243
chromatography, **299:**241
coulometric detection, principle,
299:240–241
current–voltage response curve,
299:242
electrodes, **299:**240
extraction, **299:**243, 245–246
instrumentation, **299:**241
standards, **299:**242–243
High-performance thin-layer chromatography
Far-Eastern blotting

advantages, **312:**149

apparatus, **312:**149

band marking on plates, **312:**147–148

blotting, **312:**148

detection on blots, **312:**148–149

glycosphingolipid purification

  blotting, **312:**151–152

  extraction, **312:**152

  materials, **312:**151

  mass spectrometry coupling,
    312:152–154

materials, **312:**147

overlays on blots, **312:**154–156

overview, **312:**145–147

sialidase assay on blots,
  312:156–157

ganglioside analysis

  anion-exchange chromatography,
    312:117–118

  autoradiography, **312:**139–140

  desalting, **312:**118

  detection, **312:**122–123, 125–126

  extraction, **312:**117–118

  mass spectroscopy coupling, **312:**116,
    132–134

  overlay technique

    analysis of *in situ* reaction products,
      312:128–131

    antibody overlay, **312:**127

    glycolipid overlay, **312:**131–132

    overview, **312:**115–116, 126

    quantitative analysis, **312:**127–128

  plates, **312:**118–119

  solvent systems

    one-dimensional, **312:**119

    two-dimensional, **312:**119, 122

immunostaining of glycosphingolipids

  antibody incubation and detection,
    312:159

  overview, **312:**157–158

  reagents, **312:**158–159

two-dimensional technique, **312:**115

Hill coefficient, determination from isothermal titration calorimetry, **295:**363

Hill plot, cooperativity of high mobility
  group protein binding to DNA,
  304:113, 115, 117

Hine equation, rate enhancement calculation, **308:**257, 274–276

Hippocampal brain slice

calcium measurement with fluorecent
  probes, **294:**11–12

recombinant adenovirus infection,
  293:499–500, 502–503

Hisactophilin

acidic lipid binding, **302:**53

functions, **302:**57

green fluorescent protein fusion protein
  constructs, **302:**53, 55

  pH-dependent translocation, **302:**55

  quantitative fluorescence imaging,
    302:56–57

histidine role in pH sensing, **302:**51

isoforms, **302:**52–53

myristoylation, **302:**57

structure, **302:**52–53

translocation and pH sensitivity, **302:**51

Histidine oxidation, proteins, *see* 2-Oxohis-
  tidine

Histidine transporter, *Salmonella typhi-
  murium*

ATP-binding protein

  membrane topology, **292:**8–9

  sequence homology with other pro-
    teins, **292:**3, 7–8

  three-dimensional structure, **292:**9–10

domain organization, **292:**5–6

ligand transport

  assay systems

    ATP addition to inside-out vesicles,
      292:16

    reconstitution into proteoliposomes,
      292:16–17

    ligand-binding protein mutation stud-
      ies, **292:**13

membrane channel protein, **292:**6–7

subunits, **292:**5, 32

Histone, *see also* Chromatin; Nucleosome;
  Nucleosome core particle

acetylation mapping, *see* Histone acetyltransferase; Histone deacetylase

cross-linking, *see* Cross-linking, chromatin complexes

cysteine-substituted histones

  binding to reconstituted nucleosomes,
    304:240

  gene cloning, **304:**234

  modification of cysteines with chemical
    probes

    cleavage reagent, **304:**236–237

cross-linking reagent, **304**:236–237
reduction of histones, **304**:236
nucleosome dyad mapping with cyste-
    ine-substituted H4, 631
purification
    cell lysis, **304**:235
    chromatography, **304**:235
    expression in *Escherichia coli*,
        304:234–235
rationale for construction, **304**:233
site-directed hydroxyl radical mapping
    5S nucleosome studies, **304**:249–250
    agarose gel electrophoresis, **304**:247,
        249
    binding reaction, **304**:246–247
    cleavage reaction, **304**:246
    sequence analysis of cleaved DNA,
        304:250–251
site-directed mutagenesis, **304**:233–234
functions in chromatin, **304**:50, 298
H2A/H2B, purification from *Xenopus* ex-
    tracts, **304**:57–58
hydroxyl radical footprinting, *see* Hy-
    droxyl radical footprinting
linker histones
    H5 incorporation into model nucleo-
        some arrays, **304**:34
    removal, **304**:208
    structure, **304**:232
N1,N2–(H3,H4) complex from *Xenopus*
    extracts
    isolation, **304**:60–61
    nucleosome assembly, **304**:61–62
nucleosome core particle preparation
    from recombinant histones
    advantages, **304**:3
    dialysis conditions for reconstitution,
        304:11, 13
    DNA fragment preparation, **304**:11
    histone purification
        anion-exchange chromatography,
            304:8–9
        expression in *Escherichia coli*,
            304:5–6
        gel filtration, **304**:7–8
        inclusion body preparation, **304**:6–7
        yield, **304**:5
    octamer refolding, **304**:9–10
    overview, **304**:3–4
    purification

anion-exchange chromatography,
    304:13–16
modified complexes, **304**:14
preparative gel electrophoresis,
    304:13–14, 16–17
octamer components, **304**:19
polyacrylamide gel electrophoresis and
    transfer to membranes, **304**:84
radiolabeling, **304**:78–79
reconstitution of nucleosome arrays from
    purified components
    dialysis conditions for reconstitution,
        304:24, 26
    DNA template purification, **304**:22–23
    histone octamer purification, **304**:23–24
    histone saturation of reconstituted
        arrays
        enrichment of saturated arrays,
            304:32–34
        importance of determination, **304**:26
        overview of assays, **304**:26–27
        quantitative agarose gel electrophore-
            sis assay, **304**:27, 30
        restriction enzyme digestion assay,
            304:30–31
        sedimentation velocity ultracentrifu-
            gation assay, **304**:31–32
        states of saturation, **304**:20
    instrumentation, **304**:22
    materials and reagents, **304**:21–22
Histone acetyltransferase, *see also* Gcn5;
    Hat1p
    acetylation mapping of histones
        antibody production against chemically
            acetylated H4
            affinity purification, **304**:536–537
            enzyme-linked immunosorbent assay
                of effectiveness, **304**:537
            immunization, **304**:535
        chicken embryo erythrocyte analysis
            β-globin locus analysis, **304**:542, 547
            housekeeping gene acetylation,
                304:546–547
            hybridization analysis, **304**:541–542
            locus-wide mapping, **304**:542–545
            probe, **304**:540–541
            structural changes in acetylated
                nucleosomes, **304**:545–546
        chromatin preparation for immunose-
            lection

microccocal nuclease digestion,
  304:537–538
mononucleosomes, **304:**538
salt-soluble chromatin, **304:**537–538
directed acetylation at active genes,
  304:547
immunoselection
  DNA analysis, **304:**539–540
  immunoprecipitation, **304:**538
  principle, **304:**533–535
  protein analysis, **304:**539
isolated nuclei, **304:**95–97
N-terminal microsequencing
  chromatin substrate preparation,
    304:693
  high-performance liquid chromatog-
    raphy to resolve core histones,
    304:694–695
  histone substrate preparation,
    304:693
  interpretation of results, **304:**695
  overview, **304:**692–693
  sequencing, **304:**695
acetyl-CoA labeling for assays,
  304:681–682
activity gel assay
  acetylation reaction, **304:**690
  controls, **304:**691
  electrophoresis, **304:**690
  gel preparation, **304:**689–690
  interpretation of results, **304:**691–692
  overview, **304:**689
  staining and fluorography, **304:**691
classification and types, **304:**675–677
function, overview, **304:**675, 696–697
liquid assay for free histone or chromatin
    substrates
  filter spotting and washing, **304:**684
  incubation conditions, **304:**683–684
  overview, **304:**682
  pH and ionic strength effects, **304:**684,
    686–687
  product analysis, **304:**688
  radiolabel quantification, **304:**684
  reagents, **304:**682–683
  structure of substrates in assay
    chromatin, **304:**687–688
    histone, **304:**687
liquid assay for peptide substrates,
  304:688–689

regulation, **304:**697–698
substrates
  chromatin preparation
    micrococcal nuclease digestion of
      chicken erythrocyte nuclei,
      304:679–680
    overview, **304:**679
    sucrose gradient centrifugation,
      304:680–681
  free histone preparation, **304:**678
  requirements for assay, **304:**677–678
  synthetic peptide preparation, **304:**679
  types, **304:**675–677
Histone deacetylase
  acetylated substrate preparation for assay
    histone acetyltransferase modification
      of purified histones
      acetylation reaction, **304:**718
      overview, **304:**716–717
      recombinant Hat1p production,
        304:717–718
    in vivo, **304:**716
  assay conditions, **304:**720
  phosphorylative regulation, **304:**698
  purification of Rpd3-like complex from
      Xenopus extracts
    advantages of *Xenopus* system,
      304:720–721
    egg extract production, **304:**721
    ion-exchange chromatography,
      304:724–725
    oocyte extract production,
      304:721, 724
    storage, **304:**725
    sucrose gradient density centrifugation,
      304:725
    yield, **304:**725
  recruitment, **304:**697
  structural classification, **304:**715–716
Histone H1 kinase, assay of cell cycle status
    in *Xenopus* extracts, **298:**346–348
HIV, *see* Human immunodeficiency virus
HLA, *see* Human leukocyte antigen
HMG-CoA reductase–green fluorescent
    protein fusion protein
  constructs, **302:**61–62
  degradation assay
    colony fluorescence analysis, **302:**68–70
    cycloheximide effects, **302:**68
    flow cytometry, **302:**65–68

fluorescence microscopy imaging,
  302:63–65
HRD genes in degradation, **302:**60–61,
  63, 67, 72
lovastatin effects on degradation,
  302:66–67
regulation of degradation, **302:**60
expression, **302:**62–63
HMG proteins, *see* High mobility group
  proteins
HMM, *see* Heavy meromyosin; Hidden
  Markov model
Hoechst 33342, vascular remodeling studies,
  307:253, 255–256, 262–263
HOG1, photodynamic therapy activation,
  319:356–357
Homologous recombination, cis-regulatory
  DNA element analysis
  advantages over other techniques
    overview, **306:**42–43, 66
    transgenic mouse analysis, **306:**43–45
    transient expression assays, **306:**43
  selectable markers
    effect on gene regulation following ho-
      mologous recombination,
      306:47–49
    removal with recombinases
      Cre recombinase, **306:**49–50, 52
      deletion/insertion reaction, **306:**50,
        52
      efficiency of deletion, **306:**52
      Flp recombinase, **306:**49–50, 52
      inversion reaction, **306:**50
      oocytes, **306:**54
      stable integration of recombinases,
        306:53
      target sites, **306:**50, 52
      transgenic mice, **306:**53–55
    types, **306:**52–53
  systems for multiple modifications at a
    single locus
    double replacement, **306:**58, 60
    overview, **306:**58–59
    plug and socket strategy, **306:**60, 62
    recombinase-mediated cassette ex-
      change, **306:**62–66
  systems for mutational analysis of ele-
    ments
    primary cell lines, **306:**45–46
    somatic cell hybrids, **306:**47

transformed cell lines, **306:**45–46
targeted sites, generating without selectable markers, **306:**55–58
Homology modeling, *see* Three-dimensional
  structure, molecular modeling
Hop/p60
  purification of recombinant human pro-
    tein from *Escherichia coli*
    anion-exchange chromatography,
      290:427
    cell growth and lysis, **290:**426–427
    hydroxyapatite chromatography,
      290:428
    polyacrylamide gel electrophoresis,
      290:429
    size-exclusion chromatography,
      290:428
  structure, **290:**422
Horner reaction, seunthesis of 11-*cis*-retinal
  analogs, **315:**234–235
Horseradish peroxidase
  chemiluminescence assay
    acridan carboxylates as substrates,
      305:381, 383
    chemiluminescence enhancers,
      305:379–380
    luminescence-enhancing reagent sys-
      tem, **305:**389–390
  in situ hybridization chemiluminescence
    detection, **305:**579, 583–584
Host–pathogen interaction, green fluorescent protein studies
  bacterial localization by green fluorescent
    protein expression
    plasmid construction, **302:**293
    in vitro, **302:**291–292
    in vivo, **302:**292
  detection
    flow cytometry, **302:**290
    fluorescence microscopy, **302:**288
    fluorometric detection, **302:**290
  epithelial cell infection, **302:**291
  expression optimization of mutants,
    302:287–288
  gene expression determination
    extracellular genes, **302:**293–294
    intracellular genes, **302:**294–295
  host fusion protein localization, **302:**295
  macrophage infection, **302:**290–291
  plasmid construction, **302:**288

HPEC, *see* High-performance electrophoresis chromatography

HPLC, *see* High-performance liquid chromatography

HPLC–EC, *see* High-performance liquid chromatography with electrochemical detection

HPTLC, *see* High-performance thin-layer chromatography

HPV, *see* Human papillomavirus

HR, *see* Homologous recombination

H-ras, isoelectric focusing of S-nitrosylated protein, **301:**150–151

HSC, *see* Hematopoietic stem cell

Hsc70, quaternary structure, **290:**289

HSP18.1, *see* Heat shock protein 18.1

Hsp25, *see* Heat shock protein 25

Hsp26, *see* Heat shock protein 26

Hsp27, *see* Heat shock protein 27

Hsp60, *see* Heat shock protein 60

Hsp70, *see* Heat shock protein 70

Hsp90, *see* Heat shock protein 90

Hsp100, *see* Heat shock protein 100

Hsp104, *see* Heat shock protein 104

HSV, *see* Herpes simplex virus

Hst-1, *see* Fibroblast growth factor-4

5-HT$_{2A}$, *see* Serotonin 2A receptor

HtpG
   biological function, **290:**410
   expression levels, **290:**410
   purification of recombinant protein from *Escherichia coli*
      ammonium sulfate precipitation, **290:**411
      anion-exchange chromatography, **290:**411–412
      cell growth and lysis, **290:**411
      hydroxyapatite chromatography, **290:**412
      overexpression system, **290:**410–411

Human chromogranin B–green fluorescent protein fusion protein
   constitutive secretion assay
      cell culture and expression, **302:**14
      cell lines, **302:**12, 14
      confocal microscopy, **302:**15–16
      constructs, **302:**12
      culture chambers, **302:**15
      fluorescence video microscopy, **302:**16
      fluorometry in a cuvette, **302:**17

image processing, **302:**16–17
principle, **302:**12–13
secretion block and release, **302:**15
sodium butyrate treatment, **302:**14–15
tracking vesicles, **302:**17

regulated secretion assay
   cell culture, **302:**19
   confocal microscopy, **302:**19
   rationale, **302:**17–19

Human embryonic kidney cell, *see* HEK 293–Epstein–Barr virus expression system

Human Genome Project
   annotation of genes, **303:**79–80, 82
   progress, **303:**55–56, 77, 111–112, 127, 158
   protein function elucidation, overview of approaches, **303:**497

Human immunodeficiency virus, *see also* Rev; TAR RNA
   antisense disruption of TAR structure, **313:**10–11
   fusion complex
      CD4 interactions, **312:**503–505
      gp120 coreceptors, **312:**504–505
      modeling, **312:**503–506
      reconstitution in glycosphingolipid microdomains, **312:**502–503
   galactosylceramide microdomain studies
      cholesterol–galactosylceramide reconstitution, **312:**499
      gp120
         binding site mapping, **312:**500
         interaction kinetics, **312:**498–499
         overview, **312:**495–496
         surface pressure measurements, **312:**496–497
   Gb$_3$ interactions, **312:**503–505
   gene therapy with ribozymes
      clinical trials, **306:**224
      preclinical studies, **306:**222–224
      retroviral vectors for gene delivery, **306:**223–224
      strains and ribozyme cocktail treatment, **306:**222–223
      targets, **306:**222–223
   gene therapy with RNA decoys, **306:**244
   GM3 binding, **311:**637–638
   GM3 microdomains, CD4 interactions, **312:**500–502

gp120 interactions with galactosylceramide, **311:**627
limitations of current treatments,
306:221–222
membrane fusion, overview, **312:**495
neoglycoconjugate binding assays with
gp120
neoglycoprotein conjugates,
312:486–487
neohydrocarbon conjugates,
312:485–486
receptors, **311:**633–634, 637–638
soluble synthetic galactosylceramide analogs, anti-viral activity
clinical applications, **311:**638
mucosal epithelial cell infection,
311:634, 636
peripheral blood mononuclear cell infection, **311:**636
screening throughut, **311:**636–637
structure–activity relationships,
311:636–637
Tat
associated kinase complex, **306:**352–
353, 359–360, 364
cyclin T1
purification by glutathione *S*-trans-
ferase–Tat affinity selection,
306:359, 363
purification by TAR affinity chromatography, **306:**360, 362–363
Tat interactions, **306:**352–353
TAR-dependent transactivation, *in
vitro* assay
glutathione *S*-transferase–Tat fusion
protein preparation, **306:**356
HeLa cell nucear extract preparation, **306:**354–356
incubation conditions, **306:**357–358
polyacrylamide gel electrophoresis,
306:358
solution preparation, **306:**356
TAR competition assays, **306:**358
Tat transport-enhancing fragment
delivery of antisense oligonucleotides
assay, **313:**357–358
rationale, **313:**355, 357
synthesis of conjugate, **313:**358
sequence, **313:**355
Human immunodeficiency virus-1 protease

conformational change upon ligand binding, **295:**118
Gibbs energy of binding
hydrophobic effect in binding,
295:112–113
pepstatin, **295:**114–116
structural mapping of binding energetics, **295:**115–118, 124–125
values for various ligands, **295:**112–114
inhibitors
design by rational mutagenesis, overview for aspartic proteases,
295:119–122
resistance, molecular mechanism,
295:125–126
residue stability constants, **295:**117–118
substrate binding, modeling
chromogenic substrate assays and thermodynamic analysis, **295:**122–123
natural substrate binding complex and
thermodynamic analysis,
295:123–125
Human leukocyte antigen, direct complementary DNA selection, **303:**140–141
Human papillomavirus
diseases, **306:**328
E1, structure and function, **306:**328–330
E2
ATPase assay, **306:**335
DNA helicase assay, **306:**335–336
expression systems
baculovirus–insect cell expression system, **306:**334–335
mammalian cell lines, **306:**332–334
plasmid and cell lines, **306:**330–331
transient transfection and replication
assays, **306:**331–332
FLAG epitope tagging, **306:**331
purification from Sf21 cells, **306:**335
structure and function, **306:**328–330
Western blot analysis, **306:**332
genome, **306:**328
replication factors, **306:**328–329
Huntington's disease, *see* Polyglutamine-
containing protein aggregates
Hydrazine, direct detection of RNA modification, **318:**13
Hydrazide luminescence
efficiency, **305:**29
mechanism, **305:**31–32

structure requirements, **305:**29–30
Hydride transfer reaction
    quantum mechanical tunneling complica-
        tions in kinetic isotope effects,
        308:324–325
    Swain–Schadd relationship,
        308:323–324
Hydrogen bond, *see also* Low-barrier hydro-
    gen bond
    acid–base catalysis, **308:**24
    Badger–Bauer relationships in Raman
        spectroscopy, **308:**183–185
    energy for catalysis, **308:**22, 246, 248,
        254–257, 261, 304–305
    fluoride bonding with ethanol, energetics,
        308:247
    linear-free energy relationships
        Brønsted coefficients and slopes,
            308:254–261, 269, 274–276
        substituted phenols, **308:**254
    strength, **308:**22, 219, 221
    transition-state binding, **308:**304–305
Hydrogen bonding, temperature-invariant
    enthalpy
    carboxylic acid dimerization, **295:**264–269
    water, **295:**261–264
Hydrogen exchange, amyloidogenic protein
    analysis with electrospray mass spec-
    trometry
    advantages over nuclear magnetic reso-
        nance techniques, **309:**634–635
    data acquisition, **309:**638
    data analysis, **309:**638–639
    exchange reaction, **309:**637–638
    insulin
        exchange properties, **309:**643, 645
        preparation of bovine samples, **309:**637
    localization of exchange site with peptic
        digestion, **309:**635
    lysozyme
        exchange properties of wild-type and
            variant proteins, **309:**639
    mass distribution of conformations,
        309:634–635
    preparation of samples, **309:**637
    principle, **309:**633–634
    prospects of technique, **309:**645–646
    sample introduction for mass spectrome-
        try, **309:**635–636
    transthyretin

exchange properties of wild-type and
    variant proteins, **309:**639, 641
    monomer analysis, **309:**641, 643
    preparation of samples, **309:**637
Hydrogen exchange, electrospray ionization
    mass spectrometry measurement in pro-
    teins
    complexes of proteins, **290:**305
    conformational sensitivity, **290:**301
    GroEL–protein complexes, **290:**309–310
    nuclear magnetic resonance, comparison
        of techniques, **290:**309–310
    peak width analysis, **290:**305, 307, 310
    principle, **290:**301–302
    sample preparation, **290:**302, 304–305
Hydrogen peroxide
    differential double-pulse amperometry
        measurement
        amphetamine effects on brain release,
            300:284
        area of electrode, calculation,
            300:281–282
        calibration of electrode, **300:**282–283
        computer-controlled instrument,
            300:279–280
        data acquisition, **300:**283
        fabrication of platinum disk microelec-
            trode
            reference/counter electrode, **300:**279
            working electrode, **300:**277–278
        implantation of microelectrode into rat
            brain, **300:**283
        principle, **300:**276–277
        shielded box, **300:**280–281
    Escherichia coli sensitivity assay,
        310:588–589
    generation, **301:**439
    hydroxyl radical formation, **300:**275, 297
    nitric oxide donor enhancement of toxic-
        ity, **301:**416
    nitric oxide protection against toxicity,
        301:414–421, 427
    Pseudomonas aeruginosa biofilm resis-
        tance
        catalase
            native gel assay, **310:**603–605
            spectrophotometric assay, **310:**602–
                603, 605
            types, **310:**600, 604
        culture, **310:**600–601

planktonic cell resistance compared to
  biofilm, **310:**601–602, 604
superoxide dismutase
  expression in biofilms, **310:**605–606
  native gel assay, **310:**603–604
  spectrophotometric assay,
    310:602–603
  types, **310:**604
sources in cells, **300:**275–276
stimulated chemiluminescence, measurement of oxidative damage
  apparatus, **300:**451
  blood ultraviolet irradiation effects,
    300:451
  correlation with antioxidative parame-
    ters, **300:**452–453
  smoking effects, **300:**453
Hydropathy profile, molecular modeling
  of transmembrane topology,
  293:591–593
Hydrophobic interaction chromatography,
  chaperonin 10, **290:**164–165, 200–201
Hydrostatic pressure, folding enhancement
  of inclusion body proteins, **309:**232
Hydroxamic acid, nitroxyl synthesis by oxidation in presence of amines,
  301:218–219
Hydroxide radical, *see* Synchrotron X-ray
  footprinting
Hydroxyapatite chromatography
  chaperonin 10, **290:**225
  FKBP52, **290:**423–424
  heat shock protein 18.1, **290:**358
  heat shock protein 90, **290:**414–415, 417
  heat shock protein 104, **290:**435–436
  Hop/p60, **290:**428
  HtpG, **290:**412
  mammalian mitochondrial chaperonin 10,
    290:226
N-Hydroxybenzenesulfonamide, *see* Piloty's
  acid
4-Hydroxybenzoic acid, hydroxyl radical
  assay
  advantages and limitations, **300:**307
  brain tissue measurements, **300:**308, 310
  dihydroxybenzoate product measurement
    by high-performance liquid chroma-
    tography, **300:**307–308
  trapping efficiency, **300:**307
3-Hydroxybutanoylhomoserine lactone

characterization, *see* Acylhomoserine lactone autoinducers
detection bioassay, **305:**293
Hydroxycinnamic acids, *see also* Phenolic
  antioxidants; *specific compounds*
antioxidant activity, **299:**96–97
biosynthesis, **299:**208
high-performance liquid chromatography
  chromatography conditions,
    299:103–105
  food sample preparation, **299:**100,
    103
  retention times, **299:**105–106
  standards, **299:**100
  urine sample
    dietary manipulation, **299:**97, 100,
      106
    preparation, **299:**103
inhibition of tyrosine nitration by peroxynitrite, **299:**216, 219, 232–233
peroxynitrite interactions
  m-coumaric acid, **299:**223, 234
  o-coumaric acid, **299:**222–223,
    233–234
  p-coumaric acid, **299:**219–222, 233
  ferulic acid, **299:**223, 228, 234
  high-performance liquid chromatogra-
    phy analysis of products, **299:**214,
    219, 222–223, 228
  mass spectrometry analysis of products,
    299:214–215, 221–222, 228
  reaction conditions, **299:**214
  spectral characterization of products,
    299:214, 219–221, 223, 228
2-Hydroxyethylamidoethyl thionophosphate
  antioxidant activity mechanism,
    299:294–295
  peroxyl radical reactivity, **299:**299
  reactivity
    hydrogen peroxide decomposition,
      299:296
    lipid hydroperoxide reduction,
      299:297–300
    sodium hypochlorite, **299:**296–297
  synthesis, **299:**295
2-Hydroxyethylmethane thiosulfonate, synthesis, **293:**125–126
8-Hydroxyguanine, *see* DNA
8-Hydroxyguanosine, immunochemical
  assay, **309:**146–147

1-Hydroxyl-1-(3,4-dimethoxy-6-nitrophe-
    nyl)-2-propene, synthesis, **291:**164
Hydroxyl radical
    clinical importance of oxidative damage,
        300:297–298
    electron spin resonance assay of scaveng-
        ing activity
        ascorbic acid contribution, **299:**28, 32,
            34
        electron spin resonance measurements,
            299:29–30
        reaction conditions, **299:**30
        sample preparation, **299:**29
        standards, **299:**30
    fluorescence assay
        calculations, **300:**205–208
        cellular determinations, **300:**211–212
        dimethylsulfoxide in carbon-centered
            radical generation, **300:**203
        high-performance liquid chromatogra-
            phy, **300:**208, 210
        interferences
            bioreduction of probe, **300:**213–214
            superoxide reaction, **300:**214–216
        methyl radical trapping reagent
            product of methyl radical reaction,
                characterization, **300:**209–210
            structure and application,
                300:203–204
            synthesis and characterization,
                300:208–209
        principle, **300:**202–203
        superoxide driven Fenton reaction, hy-
            droxyl radical detection,
            300:210–211
    formation in cells, **299:**3
    generation, *in vitro*
        photolysis of mercapto *N*-oxides,
            300:195–197
        radiolysis of water, **300:**195
    guanidino compound effects on brain rad-
        ical generation, **300:**254–255
    half-life, **300:**297
    4-hydroxybenzoic acid assay
        advantages and limitations, **300:**307
        brain tissue measurements, **300:**308,
            310
        dihydroxybenzoate product measure-
            ment by high-performance liquid
            chromatography, **300:**307–308

trapping efficiency, **300:**307
markers of damage, **300:**310, 312–313
phenylalanine modification, **301:**471–472
purine reactions
    adenine reaction, product detection by
        high-performance liquid chroma-
        tography, **300:**199–201
    radical generation by photolysis of mer-
        capto *N*-oxides, **300:**195–197
    xanthine transient radicals, detection
        by electron paramagnetic reso-
        nance, **300:**197–199
salicylic acid assay
    dihydroxybenzoate product measure-
        ment by high-performance liquid
        chromatography, **300:**253, 299,
        301–302
    electrochemical detection of other me-
        tabolites, **300:**304
    interferences, **300:**303
    microdialysis perfusion application,
        300:302–303, 305
    oxidative injury assessment in brain tis-
        sue, **300:**305–307
    rat preparation for intact brain mea-
        surements, **300:**253
    reactivity of salicylic acid, **300:**299–302
    sensitivity, **300:**302
sources in cells, **300:**275, 297–298
spin trapping for electron spin resonance,
    300:298
Hydroxyl radical footprinting, *see also*
    Iron(II)–EDTA
generation of radicals, **318:**6
HMG14/17, chromatin interaction
    analysis
    chromatin particle purification, **304:**139
    footprinting conditions, **304:**139
    gel electrophoresis, **304:**139–140
    nucleosome complex preparation,
        304:139
    overview, **304:**134, 137–138
HMG-I(Y), 172, 179
radical reactivity with DNA, **304:**252
ribozyme structure probing, **317:**392
RNA modification site, **318:**6
site-directed hydroxyl radical mapping,
    histones
    cysteine-substituted histones
        5S nucleosome studies, **304:**249–250

agarose gel electrophoresis, **304:**247, 249
binding reaction, **304:**246–247
cleavage reaction, **304:**246
modification of cysteines with cleavage reagent, **304:**236–237
sequence analysis of cleaved DNA, **304:**250–251
histone octamers
H4 cysteine modification with cleavage reagent, **304:**253, 256–257
iron ion loading, **304:**258–259
S-(nitrophenylsulfenyl)-cysteaminyl-EDTA synthesis, **304:**253, 255
nucleosome assembly and gel electrophoresis analysis, **304:**257–258
rate of reaction, **304:**259
reaction conditions, **304:**259–260
sequence analysis of cleaved DNA and mapping, **304:**260–263
translational position determination, **304:**252
X-ray footprinting, *see also* Synchrotron X-ray footprinting
4-Hydroxynonenal, *see* Lipid peroxidation
p-Hydroxyphenacyl group
carboxylate ester compound caging
γ-aminobutyric acid, **291:**10–11
applications, **291:**13–14, 16
L-glutamate
properties and applications, **291:**10–11, 14, 16
synthesis, **291:**24–25
peptides, **291:**11–13
photochemistry, **291:**11
phosphate compound caging
ADP coupling in synthesis, **291:**22–24
ATP
photolysis conditions, **291:**28–29
properties and applications, **291:**8
stability testing, **291:**29
p-hydroxyphenacyl phosphate
photolysis, **291:**28
synthesis, **291:**20–22
photolysis reactions, **291:**8–10
photolysis of caged compounds, **291:**26–27
5-Hydroxysphinganine, *see* Phytosphingosine
5-Hydroxy-1-tetralone, ganglioside detection on high-performance thin-layer chromatography plates, **312:**125
Hypoxanthine–guanine phosphoribosyltransferase, low-barrier hydrogen bonds in catalysis, **308:**304–305

I

IAA, *see* Iodoacetate
IAM, *see* Iodoacetamide
IAPP, *see* Islet amyloid polypeptide
IbpB, *see* Inclusion body protein B
ICAM-1, *see* Intercellular cell adhesion molecule-1
ICON, *see* In vivo confocal neuroimaging
ICP-MS, *see* Inductively coupled plasma mass spectrometry
IEF, *see* Isoelectric focusing
IGORPRO, *see* Markov model, ion channel kinetics
I-κB, *see* Nuclear factor-κB
IL-10, *see* Interleukin-10
Image data, confocal microscopy
animation, **307:**116–117
Cavalieri principle in object volume estimation, **307:**309–310
compression formats, **307:**40–43, 54
cytoskeletal elements, localization and colocalization with software, **307:**188–189
filtering and preprocessing of images, **307:**51–53
hardware selection for processing, **307:**49–50, 54–55
intensity levels, **307:**29, 377, 497
Kalman filter averaging, **307:**613
limitations in processing and display, **307:**313–315
morphometry in three dimensions
line length, **307:**54
surface area, **307:**53–54
volume, **307:**53, 116
rendering, *see also* Cytotomography
ion-selective dyes and volume rendering, **307:**123–124
simple projections, **307:**45, 47
surface extraction, **307:**44
three-dimensional rendering, overview of stereo pairs, **307:**43–44

weighted projection and alpha blending, **307:**47, 49
software selection for processing, **307:**49–
51, 55, 116
storage of data
CD-ROM, **307:**37–39
costs, **307:**33–34
digital video disk, **307:**38
floppy diskettes and Zip disks, **307:**34–
35, 39–40
hard disks, **307:**35–36
magnetic tape, **307:**30–33, 39
magneto-optical media, **307:**36–37
permanence, **307:**30
random access memory, **307:**29
space requirements, **307:**30
speed of connection with computer,
307:34
write once read many drives, **307:**37
three-dimensional reconstruction
advanced reconstruction
X, Y dimensions, **307:**508–511
Z dimension, **307:**505–506, 508
basic reconstruction, **307:**504–505
optimization, **307:**501–504
stereology incorporation, **307:**511–513
volume investigation
data acquisition and image stacks,
307:277–279
overview, **307:**276–277
quantitative three-dimensional analysis,
307:280
three-dimensional visualization of reconstructed object, **307:**279–280
two-dimensional segmentation, **307:**279
Imidazole, *see* Ribonuclease A mimics
Imine luminescence, theory, **305:**35–36
Immediate pigment darkening, *see* Skin, ultraviolet protection
Immunoadsorption, polypeptides in cell-
free translation systems
binding affinity analysis, **290:**16
incubation conditions and product analysis, **290:**9
support matrix, **290:**8–9
Immunoaffinity chromatography
cytoplasmic linker protein-170, 198–200
kinectin, **298:**188–191, 196–197
paxillin, **298:**79
septin complexes, **298:**280–283, 540–541

γ tubulin ring complex, **298:**219, 222
vasodilator-stimulated phosphoprotein,
298:108–109
Immunoassay, luminescence
bioluminescent protein reporters,
305:337–338
commercial kits, **305:**336–337
electrochemiluminescent immunoassay,
305:339
enzyme reaction products, **305:**390
enzyme substrates, **305:**335–336, 339, 342
luminescent oxygen channeling immunoassay, **305:**337
popularity, **305:**334–335
protein kinase activity assay, **305:**410–416
Immunoblotting, *see* Western blot
Immunochemical assay
advanced glycation end products
applications, **309:**157–158
epitopes, **309:**156–157
nervous tissue samples, **309:**144–145,
149
slide preparation and staining, **309:**169
advanced lipoxidation end products,
309:144–146, 149–150
alkaline phosphatase, amyloid detection
blocking of endogenous enzyme,
309:19
primary versus secondary detection systems, **309:**21–22
substrates, **309:**23–24
amyloids
antibodies
absorption, **309:**24–25
dilution, **309:**19–20
primary antibody incubation, **309:**20,
38
purification, **309:**20
types, **309:**14–15, 37
antigenic epitope retrieval, **309:**18
controls, **309:**24, 39
coupling enzymes
blocking of endogenous enzymes,
309:19, 37–38
types and substrates, **309:**22–24
detection systems, **309:**20–22
fixation, **309:**15–16
mounting, **309:**16–17, 24
sectioning, **309:**16
wash buffers, **309:**17–18, 38

carbonyl groups in proteins, **309:**139–140, 763–764

Congo red counterstaining of samples, **309:**38–39

4-hydroxynonenal–protein conjugates, **309:**738

oxidative stress markers

advanced glycation end products, **309:**144–145, 149

advanced lipoxidation end products, **309:**144–146, 149–150

Alzheimer's disease samples, **309:**149–150

antigen retrieval on tissue sections, **309:**143

complications and artifacts, **309:**151–152

heme oxygenase-1, **309:**148–150

8-hydroxyguanosine, **309:**146–147

imaging of optical density, **309:**150–151

iron regulatory proteins, **309:**147–149

peroxidase-coupled secondary antibody, **309:**142–143

tyrosine nitration assay, **309:**143–144

peroxidase in amyloid immunohistochemistry detection

blocking of endogenous enzyme, **309:**19, 37–38

primary versus secondary detection systems, **309:**21–22

substrates, **309:**22–23

transglutaminase, **309:**176, 178–179

Immunocytochemistry, neurotransmitter transporter localization, *see also* Immunofluorescence microscopy

γ-aminobutyric acid transporter expressed in polarized epithelial cells, **296:**376–378

antibody

affinity purification, **296:**398–400, 413–416

immunization, **296:**393–394, 412–413

immunoglobulin G isolation from affinity-purified antibodies, **296:**400

specificity testing

immunoblotting, **296:**394–395, 397–398, 417–420

immunoprecipitation, **296:**417

antigen

coupling to carrier, **296:**390–392, 407–408

glutathione *S*-transferase fusion proteins as antigens

advantages, **296:**408–409

plasmid construction, **296:**409–410

purification of fusion proteins, **296:**412

recombinant protein expression in *Escherichia coli*, **296:**410–412

solubilization of inclusion bodies and affinity column preparation, **296:**416–417

selection for antigenicity, **296:**390

sources, **296:**389–390

fixation

function, **296:**402

paraformaldehyde, depolymerization, **296:**402–403

perfusion fixation, **296:**403, 420–421

freeze-substitution postembedding electron microscopy, **296:**405–406

labeling

incubations and washes, **296:**403–404

preembedding versus postembedding labeling, **296:**401–402

neurotransmitter transporters expressed in neurons, **296:**385–387

preembedding electron microscopy, **296:**404–405

sectioning, **296:**400–401, 421

staining

electron microscopy, **296:**400

light microscopy, **296:**421–422

Immunoelectron microscopy, *see* Electron microscopy; GM1

Immunofluorescence microscopy

androgen receptor–green fluorescent protein fusion protein

comparison to intrinsic fluorescence localization, **302:**126–127

materials, **302:**125–126

permeabilization conditions, **302:**126–127

background fluorescence minimization, **293:**540–541

Bacillus subtilis, expressed green fluorescent protein, **302:**149–150

biofilm characterization

advantages and disadvantages, **310:**17, 18

fixation, antibody incubations, and washings, **310:**18

calmodulin–green fluorescent protein fusion protein, **302:**102

cardiovascular ion channel detection

indirect immunofluorescence, **293:**84–85

sections, **293:**85

single-cell immunofluorescence, **293:**85

cloned ion channel distribution in neuronal tissue, **293:**155–156

colocalization of ion channel associated proteins, **294:**372, 380–381

confocal microscopy, **293:**547–548

epitope tagging of proteins and staining, **293:**538–540

frozen sections, **293:**544–547

G protein-coupled receptor transfectants, **302:**157–158

monoclonal versus polyclonal antibodies, **293:**537–538

myocyte staining, **302:**185

neurotransmitter transporter localization

antibody preparation, **296:**60–62

fixation, **296:**62–63

sectioning, **296:**63

staining, **296:**63–64

vesicular transporters in *Caenorhabditis elegans*, **296:**542–544

prosthetic joint infection detection, **310:**575–576

RatTGN38–green fluorescent protein fusion protein, **302:**9–10

topology staining, **293:**542–544

Xenopus oocyte preparation, **293:**539

Immunoglobulin G, *see* Receptor–ligand internalization, confocal microscopy

Immunoglobulin light chain amyloidosis, *see also* Bence Jones proteins

agitation-stimulated fibrillogenesis

aggregation conditions, **309:**207

cuvette conditions, **309:**215–216

kinetic analysis, **309:**207–209

orbital shaker conditions, **309:**214–215

peptide synthesis, **309:**206–207

protein quantification, **309:**209

recombinant V_L proteins, **309:**212–214

thioflavin T monitoring, **309:**209–212, 216

association constants for variable domains, **309:**323

flow cytometric characterization of amyloid burden, **309:**464–466

pathologies, **309:**206

protein structure, **309:**205

small zone chromatography

column preparation, **309:**323–324

concentration dependence of elution profiles, **309:**320

dimerization constant determination, **309:**328

dimerization simulation, **309:**325–326

elution behavior, **309:**322

experimental elution profile generation, **309:**324–325

large zone comparison, **309:**319–320

X-ray fiber diffraction, **309:**526

Immunogold, *see* Electron microscopy

Immunohistochemistry, *see also* Subcellular localization

chemiluminescence detection

alkaline phosphatase detection, **305:**578–579, 588

antibody dilutions, **305:**587

applications, **305:**589

controls, **305:**587

gastric mucosa antigens, **305:**587–589

horseradish peroxidase detection, **305:**579

imaging, **305:**579–580, 590

quantitative analysis, **305:**580–581, 589

sensitivity, **305:**578

specimen preparation, **305:**586–587

luciferin-binding protein from *Gonyaulax polyedra*, **305:**275–276

principle, **305:**577–578

Immunomagnetic cell fractionation, hematopoietic stem cells, **292:**256, 548, 550

Immunophilins, *see also* Peptidylprolyl *cis–trans*-isomerase; *specific proteins*

drug binding, **290:**74, 84

types, **290:**75

Immunoprecipitation

acetylation mapping, **304:**538

advantages and disadvantages in subunit interaction analysis, **293:**14

a-factor, **292:**201–202

avidity assay for subunit interactions,
  293:24
chromatin
  antibodies, **304:**84, 88
  fixed chromatin
    formaldehyde fixation, **304:**89–91
    incubation conditions, **304:**91–92
  unfixed chromatin
    incubation conditions, **304:**87
    protein A–agarose bead preparation,
      304:85, 87
    solution preparation, **304:**85
coimmunoprecipitation of ion channel associated proteins
  full-length proteins expressed in heter-
    ologous cells, **294:**377–380
  native tissue analysis, **294:**381–384
  principle, **294:**372
cysteine-modified neurotransmitter transporters, **296:**322, 324
cystic fibrosis transmembrane regulator,
  292:645–646
cytoskeletal proteins with peptide antibodies, **298:**534
cytosolic versus membrane proteins,
  293:548–549
focal adhesion kinase, **298:**97
GAT-1, **296:**300–301
HMG-D and bound DNA, **304:**121
I-κB–NF-κB complexes,
  300:379–381
incubation conditions, **293:**551–552
intersubunit interaction, identification of
  regions, **293:**20–22
mapping nonhistone chromatin-associated factors, **304:**402, 410–412,
  423–424
MDR1, 297–298
membrane isolation by sucrose gradient
  centrifugation, **293:**551
norepinephrine transporter, **296:**354,
  357–360
OmpA in disulfide bond formation studies, **290:**65–67
oocyte homogenates, **293:**22, 548–550
polyacrylamide gel electrophoresis in
  analysis, **293:**47, 552
potassium channel subunits from lysed
  cells
  antibody incubation, **293:**45–46

fixed *Staphylococcus aureus* as matrix,
  293:43, 45
gel electrophoresis, **293:**47
imaging of gels
  autoradiography, **293:**47–48
  fluorography, **293:**48
  phosphorimaging, **293:**48–49
  precoating, **293:**44–45
  protein A agarose as matrix, **293:**43–44
  protein G agarose as matrix, **293:**44
radiolabeled chloroplast proteins in
  maize membrane
  antibody binding reaction, **297:**52
  antigen elution and analytical electro-
    phoresis, **297:**53
  solubilization, **297:**52
  Staph A cell binding reaction,
    297:52–53
Ste6p, **292:**205–206
sulfonylurea receptor, **292:**741
ultraviolet cross-linked DNA–protein
  complexes, **304:**504–507
vasodilator-stimulated phosphoprotein,
  298:110–111
in vitro translation system products,
  293:21–22
Immunostaining, ABCR localization in eye,
  315:883–884
IMTS, *see* Iterative multilevel thresholding
  and segmentation
Inclusion body
  aggregation during refolding
    features of aggregates, **309:**223–224
    kinetic competition between folding
      and aggregation, **309:**224, 235–236
    prevention strategies
      antibodies, **309:**231–232
      chaperones and foldases,
        309:233–234
      fusion protein constructs,
        309:233
      hydrostatic pressure, **309:**232
      low molecular weight folding en-
        hancers, **309:**227–231
      pulse renaturation, **309:**224–226
      site-directed mutagenesis, **309:**232
      solid support matrices, **309:**232–233
  cell lysis for isolation, **309:**54–55, 57,
    217–218
  centrifugation

low-speed centrifugation, **309:**51, 55, 57, 217–218

sucrose density gradient centrifugation, **309:**55–58

chaperone gene mutations in formation, **309:**53

cross-flow filtration, **309:**56

detergent extraction, **309:**56–57

electron microscopy, **309:**48

growth conditions in optimization, **309:**54

homogeneity of protein composition, **309:**49, 51, 55

periplasmic versus cytoplasmic bodies, **309:**53–54

promoters for formation, **309:**52–53

quality, factors affecting, **309:**49

renaturation

disulfide bond formation, **309:**220–223

folding of nondisulfide-bonded proteins, **309:**223

impurity effects, **309:**49, 51–52

optimization of folding conditions, **309:**234–235

overview, **309:**48–49

transfer of protein into refolding buffer, **309:**219–220

solubilization, **309:**218–219

Inclusion body protein B

functions, **290:**339

purification of recombinant protein from *Escherichia coli*

column chromatography, **290:**343

extraction, **290:**342

overexpression system, **290:**341–342

Indigo, singlet oxygen quenching, **319:**239 240

Indirect end-labeling

cyclobutane–pyrimidine dimer, mapping in chromatin using photolyase, **304:**449, 451, 453–454, 457–461

DNase I footprinting of *Drosophila* embryo chromatin, **304:**466–468, 475–477

low-resolution mammalian nucleosome positioning

micrococcal nuclease digestion, **304:**633–634

nuclei isolation, **304:**632–633, 635

optimization, **304:**635–636

overview, **304:**631–632

Southern blot analysis, **304:**634–635

methidiumpropyl-EDTA–Fe(II) footprinting of *Drosophila* embryo chromatin, **304:**466–468, 475–477

micrococcal nuclease

chromatin structural probing, **304:**619–620

Drosophila embryo chromatin analysis, **304:**466–468, 475–477

4,5′,8-trimethylpsoralen, nucleosome positioning, **304:**636

yeast nucleosome positioning assays

digestion reaction, **304:**395

interpretation, **304:**381–382

principle, **304:**380

reagents, **304:**394–395

Southern blot analysis, **304:**395–396

Indo-1, *see also* Calcium

compartmentation of ester-loaded fluorophore, **302:**350

ratio confocal microscopy imaging, **302:**354, 356

Indo-dextran, calcium flux measurements in rod outer segments

calcium dependence of cation exchange rate, **316:**145

calibration, **316:**141–142

dextran-conjugated dye, **316:**139–140

light-evoked changes, **316:**145–146

optical setup for fluorescence measurements, **316:**140–141

ratiometric detection, advantages, **307:**24

snapshots of calcium, **316:**142–145

Indolicidin, partition coefficient determination, **295:**75

Inducible nitric oxide synthase, *see* Nitric oxide synthase

Inductively coupled plasma mass spectrometry, copper determination, **297:**278

Indwelling urethral catheter, *see* Bladder catheter biofilm

Influenza A virus, *see* M2

Influenza virus hemagglutinin, transport across cell membranes, **313:**354

Infrared spectroscopy, *see also* Fourier transform infrared spectroscopy

urinary tract device encrustation, **310:**561

advantages in lipid studies, **312:**228, 246–247

ceramide analysis in stratum corneum
  ceramide types, **312**:228–229
  horizontal attenuated total reflection,
    312:231–232
  sample preparation, **312**:231–232
  spectral regions sensitive to structure
    and organization
    CD$_2$ rocking modes, **312**:241–243
    methylene scissoring modes,
      312:236–238
    methylene stretching modes,
      312:233–236
    methylene wagging modes,
      312:238–241
    overview, **312**:232–234
    polar region vibrations, **312**:243–244
  transmission spectroscopy, **312**:231–232
Connes advantage, **312**:230
deuterium substitutions,
  312:228
Fellgett advantage, **312**:231
imaging with array detector, **312**:245–246
infrared reflection–absorption spectroscopy of monomolecular films,
  312:244–245
spectrometer components and principles,
  312:229–230
bands in secondary structure analysis,
  309:559–560
protein adsorption studies, **309**:410–411
protein aggregate structure characterization, **309**:254–255
protein structure studies, advantages and
  disadvantages, **309**:608
wave numbers, **309**:559
Inosine, isotopic labeling for kinetic isotope
  effect studies, **308**:308–309
Inosine-containing messenger RNA
  adenosine deaminase substrate identifi-
    cation
    arbitrarily primed polymerase chain re-
      action, **318**:64–65
    characteristics of substrates, **318**:66–67
    complementary DNA synthesis, first
      strand, **318**:62–63
    confirmation of candidates
      confirmation of A to G changes,
        318:66
      principle, **318**:65–66
      screening out false positives, **318**:66

differential display, statistical analysis,
  318:67
polyacrylamide gel electrophoresis of
  amplification products, **318**:65
polyadenylation reaction, **318**:62
principle, **318**:61–62
types of substrates, **318**:48–49
detection of inosine at candidate editing
  sites
  complementary DNA synthesis, first
    strand, **318**:58–59
  ligation of anchor, **318**:56, 58
  polymerase chain reaction
    amplification of cleavage products,
      318:59
    cloning and sequencing of products,
      318:59, 61
  principle, **318**:56
ribonuclease T1, specific cleavage of inosine 3′-phosphodiester bond
  cellular RNA purification, **318**:50–51
  control RNA synthesis, **318**:49–50
  digestion reaction, **318**:54–55
  glyoxal reaction, **318**:54
  optimization, **318**:53–54
  oxidation of 3′-hydroxyl groups,
    318:51–52
  postcleavage processing
    cyclic phosphate removal, **318**:55
    glyoxal adduct removal, **318**:55–56
  principle, **318**:52–53
Inosine–uridine nucleoside hydrolase
  difference Raman spectroscopy, electro-
    static interactions and inhibitor de-
    sign, **308**:200–201
  neural network analysis of binding en-
    ergies
    accuracy, **308**:419, 424
    binding sites, **308**:416
    orientation of ligand binding, **308**:417
    substrate specificity, **308**:416–417
    training of network, **308**:418
    transition-state structures and affinities,
      308:418–419, 422–424
  transition-state structure, **308**:199
Inositol 1,4,5-triphosphate
  caged compound
    applications, **291**:381–382
    calcium fluorescent probe measure-
      ment with electrophysiological

measurement, **291**:388–391, 395, 403
kinetics of calcium release, **291**:395–397
light sources for photolysis, **291**:386–388
linearity and calibration of photore-lease, **291**:384–386
photochemistry, **291**:382–383
spatial control and heterogeneity of cal-cium release sites, **291**:397–402
Xenopus oocyte studies
  caffeine studies, **291**:393, 395
  dose–response relationship of cal-cium release, **291**:391–392
  InsP_4 agonist studies, **291**:393
  microinjection, **291**:383–384
mediation of calcium release, **291**:380–381, 404
Inositol 1,4,5-triphosphate 3-kinase, green fluorescent protein vector analysis of neuron transfectants, **302**:223–224
Inositol 1,4,5-triphosphate receptor
green fluorescence protein fusion protein
  expression levels, **302**:230–231
  plasmid, **302**:227
  subcellular localization, **302**:226, 232–233
  Western blot analysis, **302**:230
smooth endoplasmic reticulum associa-tion, **302**:226
types, **302**:226
Inositolphosphoryl ceramide synthase
assays
  fluorescent assay with NBD-ceramide, **311**:125, 143
  plant enzyme, **311**:142–143
  radiolabeled phosphatidylinositol
    incubation conditions, **311**:124, 142–143
    product extraction and quantifica-tion, **311**:124–125
    substrate preparation, **311**:123–124
gene in yeast, **311**:123
inhibitors, **311**:29, 335, 348
ion dependence, **311**:129
pH optimum, **311**:129
purification from yeast
  anion-exchange chromatography, **311**:126–129

cell growth, **311**:123
detergent extract preparation, **311**:126
extract preparation, **311**:125–126
hydrophobic affinity chromatography, **311**:127–128
microsome preparation, **311**:126
reaction catalyzed, **311**:123
substrate specificity of plant enzyme, **311**:143–144
synthetic and analytical uses, **311**:130
Insulin
α-crystallin chaperone assay by disulfide bond-breaking, **290**:380–382
antisense RNA expression in plasmids, **314**:33
reduction assay for protein disulfide isomerase, **290**:33, 52
Insulin fibril
hydrogen exchange measured with elec-trospray mass spectrometry
  exchange properties, **309**:643, 645
  preparation of bovine samples, **309**:637
X-ray fiber diffraction, **309**:528
Insulin-like growth factor I receptor, anti-sense RNA expression in plasmids, **314**:33
Insulin receptor, *see* Receptor–ligand inter-nalization, confocal microscopy
Intercellular adhesion molecule-1
adhesion assay, *see* Cell–cell adhesion assay
antisense targeting
  lipid encapsulation effects on efficacy, **313**:340
  renal tubular cells, **314**:382
  target selection, **313**:15, 32
assays of nuclear factor-κB activation, **300**:354–356, 373–374
dual-radiolabeled monoclonal antibody technique for *in vivo* measurement
  advantages, **301**:15–16
  backflush procedure, **301**:18
  calculations, **301**:19–20
  expression kinetics, **301**:21
  exsanguination, **301**:18
  injection of antibodies into mice, **301**:16–18
  radioiodination of antibodies, **301**:16
  tissue preparation and analysis, **301**:18–19

expression on endothelial cells, **301:**15
ligands, **301:**14
liver ischemia/reperfusion damage role, **300:**480
Interfacial stress, protein aggregation mechanism, **309:**251
surfactants in prevention, **309:**251–254
Interferon-γ, aggregation pathway, **309:**244, 255
Interleukin-1 receptor antagonist, aggregation in solution and stabilizers, **309:**242
Interleukin-1β, enzyme-linked immunosorbent assay of induction by amyloid-β fibrils in human monocytes, **309:**729–732
Interleukin-2 receptor, triplex-forming oligonucleotide inhibition, **313:**93
Interleukin-10
  antisense knockdown
    B cell lymphoma therapy, **314:**417–418
    B-1 cell neoplasia in mice
      clinical implications and costs, **314:**426–429
      flow cytometry analysis, **314:**421–422
      histologic analysis, **314:**422–424
      infusion pump delivery, **314:**420–421
      lipofection with cochleates, **314:**418–421, 423–425
      oligonucleotide design, **314:**419
      survival effects, **314:**422, 425–426
      toxicity, **314:**428
  B-1 cell growth factor, **314:**411, 415
  immunosuppression, **314:**417
  sequence homology with Epstein–Barr virus proteins, **314:**417–418
  T cell synthesis, **314:**416–417
Internet resources, ion channels
  accession optimization by object specification and indexing, **294:**726–728, 730–731
  Biology Workbench project, federation of databases, **294:**706–707
  Common Object Request Broker Architecture, **294:**727, 730
  compacting channel-specific information in federated cell electrical stimulation databases, **294:**705, 707, 709–713, 715
  developmental gene expression data, **294:**715, 718

GENESIS simulation program, **294:**707, 709–710, 713, 726–727, 730
genome information strategies, **294:**704–707
intuitive access and declarative queries, supporting data structures, **294:**723–726
Ion Channel Network site, **294:**720
peer review, **294:**706
prediction of signal-transducing molecular assemblies containing channel effectors, **294:**718–719
search engines, **294:**729
systematic gene family nomenclature and interoperability of data sets, **294:**719–720, 722–723, 730
TRANSFAC database, **294:**718
universally addressable bibliographic data collection, **294:**728–729
Interphase extract
applications, **298:**331–332, 338–339
immunofluorescence microscopy, **298:**352
motility assay with video-enhanced differential interference contrast microscopy
  coverslips, **298:**348
  image acquisition and processing, **298:**349
  instrumentation, **298:**439
  organelle movement in *Xenopus* extracts
    direction of movement, **298:**351
    features, **298:**349–350
    quantitation of movement, **298:**350–351
    standard assay, **298:**350
preparation from HeLa cells
  culture and harvest of cells, **298:**333–334
  cytochalasin B washing, **298:**336
  homogenization, **298:**336
  immunodepletion of proteins, **298:**337–338
  NuMa protein assembly, **298:**336–337
  reagents, **298:**333
preparation from *Xenopus* eggs
  advantages of *Xenopus* system, **298:**340, 352–353
  calcium treatment, **298:**340–341, 343

centrifugation and crushing, **298**:342–343

cytosol fraction preparation, **298**:343–345

cytostatic factor induction of metaphase arrest, **298**:340–342

egg production, **298**:341–342

histone H1 kinase assay of cell cycle status, **298**:346–348

isolation of soluble microtubule motors, **298**:345–346

materials, **298**:341

organelle fraction preparation, **298**:343–345

Intracellular calcium, *see* Calcium flux

Intracerebral microdialysis

anesthesia, **296**:721–722

apparatus, **296**:726–727

applications, overview, **296**:719–720

contamination, **296**:723–724

guide cannula implantation, **296**:726

high-performance liquid chromatography with electrochemical detection

apparatus, **296**:728

mobile phase preparation, **296**:728

monoamine standard preparation, **296**:729

peak identification, **296**:730

running conditions, **296**:730

perfusion solution, **296**:722, 727

presentation of data, **296**:725–726

principle, **296**:720–721

probes

calibration, **296**:727

placement verification, **296**:730

recovery performance, **296**:723

types, **296**:722–723

sample collection, **296**:727–728

uptake blockers

administration routes, **296**:725

monoamine effect, evaluation, **296**:724

Intravital microscopy

dorsal skin fold chamber model

history, **300**:463–464

microscopic observations, **300**:469–470

reactive oxygen species and microcirculatory dysfunction, **300**:470, 472

surgical implantation

chambers, **300**:464–466

indwelling catheters, **300**:467–469

liver surface, fluorescence microscopy

history, **300**:472–473

ischemia/reperfusion studies of reactive oxygen species, **300**:478–481

microscopic observations, **300**:476–478

surgical technique, **300**:473–474, 476

microvascular damage from reactive oxygen species, **300**:462–463

nitric oxide antioxidant and oxidant studies

dihydrorhodamine 123 oxidation assay

in vivo, **301**:430–433, 435

microscopy, **301**:430–431

surgery, **301**:430

Intron

exon/intron junction determination in genomic clones, **296**:508

sizing in genomic clones, **296**:507–508

Invertase, *see* Signal sequence trap

Invertin, chemiluminescence assay, **305**:388

Iodide channel

iodine uptake assays

inward potassium gradient assay, **294**:306–307

outward anionic gradient assay of selectivity, **294**:314–315

outward iodide gradient assay

membrane preparations, **294**:307–308

reconstituted proteoliposomes, **294**:312–313

reconstitution into proteoliposomes

iodide uptake

activation energy, **294**:318

imposed iodide gradient dependence, **294**:314

kinetic properties, **294**:316–318

lipid dependency, **294**:313

protein dependency, **294**:313

selectivity and specificity, **294**:314–316

liposome preparation, **294**:310–311

plasma membrane vesicle preparation from thyroid, **294**:305–306

proteoliposome reconstitution, **294**:311–312

solubilization, **294**:308–310

thyroid function, **294**:304

Iodination, radiolabeling of proteins

amyloid-β, **309**:361–363, 470–471

chloramine T labeling, **309**:413

iodine monochloride labeling, **309:**412–413

isotope features, **309:**411–412

serum amyloid P radiolabeling, detection of amyloid, **309:**713–714

Iodine

halogenation of pyrimidines, **317:**125–126, 160–161

sphingolipid staining on thin-layer chromatography plates, **312:**78

Iodixanol gradient, *see* Adeno-associated virus

Iodoacetamide, thiol trapping, **290:**61–62

Iodoacetate, thiol trapping, **290:**60–62

5-Iodocytidine, *see* 5-Halopyrimidines

Iodoglibenclamide, radioligand synthesis, **294:**446–448

Iodoglyburide

photoaffinity labeling of sulfonylurea receptor

membrane preparations, **292:**737, 739

specificity, **292:**738

whole cells, **292:**737–738

synthesis of radiolabeled iodine-125 compound, **292:**733–734

4-Iodophenol, chemiluminescence enhancement of enzyme assays, **305:**389

5-Iodouridine, *see* 5-Halopyrimidines

Ion channel associated proteins

coimmunoprecipitation

full-length proteins expressed in heterologous cells, **294:**377–380

native tissue analysis, **294:**381–384

principle, **294:**372

comparison to ion channel subunits, **294:**371

filter overlay assay, **294:**373–375

genetic approaches in identification, **294:**372

immunofluorescence colocalization, **294:**372, 380–381

pull-down assay, **294:**373, 375–377

yeast two-hybrid screening in identification, **294:**371, 373

Ion channel permeability, *see also* Pore size

multi-ion pores, **293:**388

one-ion pores, **293:**387–388

pore occupancy determination, **293:**389–390

pore-blocking studies, **293:**388–389

selectivity

relative ion permeability, **293:**386–387

reversed potential calculation, **293:**385–386

single-ion Nernst potential, **293:**384–385

Ion dependence, neurotransmitter transporters

chemical potential, **296:**237

coupling stoichiometry

determination

electrophysiological measurements, **296:**245–247

flux ratios, **296:**238–240

potassium antiport, **296:**244, 259

static head assay, **296:**244–245

steady-state gradients in vesicles, **296:**240–244

valinomycin application for electrogenic transporters, **296:**243

glutamate excitotoxity relevance, **296:**234

kinetics of ion dependence, **296:**235–236

thermodynamics, **296:**236–238

electrical potential, **296:**237

sodium dependence

GBR analog binding to dopamine transporter, **296:**203, 217–218

vesicular glutamate transport system assay, **296:**196–197

sodium, potassium ATPase generation of gradients, **296:**233

Ion-exchange chromatography, *see* Anion-exchange chromatography; Cation-exchange chromatography

Ionizing radiation, oxidative stress, **319:**260

Ion pump, *see also* Calcium; Sodium, calcium-exchanger; Sodium, potassium-ATPase

electrophysiological measurements with caged substrates, **291:**289–290

overview of caged substrate studies, **291:**290, 305–306

Iontophoresis, calcium introduction and membrane current measurement, **293:**374–375

IP$_3$ receptor, *see* Inositol 1,4,5-triphosphate receptor

IPMDH, *see* Isopropylmalate dehydrogenase

IR, *see* Infrared spectroscopy
IRE, *see* Iron response element
Iron, role in singlet oxygen effects
  calcein assay, **319:**295
  chelator studies, **319:**294
  iron regulatory protein measurements,
    319:295
  lipid peroxidation, **319:**294, 296
Iron diethyldithiocarbamate, *see* Electron
  paramagnetic resonance
Iron(II)–EDTA
  applications of RNA modification,
    318:8–10
  hydroxyl radicals in RNA cleavage,
    318:175–176
  RNA modification site, **318:**6–7
  tethered complex for ribonucleoprotein
    complex probing
    advantages, **318:**461–462
    buffers, **318:**463–464
    equipment and supplies, **318:**464
    footprinting S20 on 16S ribosomal
      RNA, **318:**468
    hydroxyl radical probing conditions,
      318:470
    iron(II) derivatization of protein,
      318:467
    linker–iron(II) complex preparation,
      318:466–467
    nuclease contamination prevention,
      318:462
    overview, **318:**461
    primer extension analysis
      extension reaction, **318:**472–473
      hybridization, **318:**472
      interpretation of data, **318:**473–475
      overview, **318:**471
      polyacrylamide gel electrophoresis
        analysis, **318:**473
      reagents and buffers, **318:**471–472
    purification of iron(II)–S20–16S ribo-
      somal RNA complex, **318:**468–469
    reagents, **318:**462–463
    reconstitution of 30S subunits con-
      taining iron(II)-derivatized pro-
      tein, **318:**469–470
    RNA isolation following probing,
      318:470–471
    S20 mutation for cysteine substitution
      accessibility assay, **318:**465–466

    site selection, **318:**464–465
    S20–16S ribosomal RNA complex for-
      mation, **318:**467–468
  tethered complex for RNA structure
    probing in nucleic acid complexes
    advantages, **318:**190
    buffers, **318:**176–177
    distance information, **318:**189–190
    DNA oligonucleotide synthesis,
      318:178
    DNA–RNA duplex probing, **318:**181–
      183, 189
    enzymes and reagents, **318:**177–178,
      189
    radiolabeling of DNA,
      318:178–179
    rationale, **318:**175–176
    RNA–RNA duplex probing
      anticodon stem–loop analogs,
        318:186–188
      environment surrounding 5′-end of
        phenylalanine transfer RNA,
        318:185
      16S rRNA, **318:**183–184
    tethering reaction, **318:**176, 180
    transcription
      efficiency of 5′-guanosine α-phospho-
        rothioate incorporation, **318:**180
      template preparation, **318:**179
      in vitro reaction, **318:**179–180
Iron(II)/iron(III), oxidative stress measurement
  chelation and binding, **309:**142
  cytochemical detection, **309:**141
  oxidation, *in situ*, **309:**141–142
  rationale, **309:**140–141
Iron regulatory proteins, immunochemical
  assay, **309:**147–149
Iron response element
  binding proteins and translational con-
    trol, **306:**226–227
  retroviral vector modification for transla-
    tional control
    clinical application, **306:**241
    design of vector, **306:**227, 229, 240
    particle generation and cell transduc-
      tions, **306:**230–232
    plasmid and vector constructions,
      306:229–230
    rationale, **306:**226

regulated expression analysis, **306:**232, 234, 240–241

stability of vector, **306:**234, 236

stable producer cell analysis, **306:**236–237

stem sequence importance to function, **306:**237–238, 240

Iron–sulfur cluster, *see* PsaC

Ischemia–reperfusion injury, nitric oxide role, **301:**422, 425

ISH, *see in Situ* hybridization

Islet amyloid polypeptide

diabetes type II pathogenesis role, **309:**152

inflammatory response in human monocytes, **309:**724–725

Isocitrate dehydrogenase, kinetic bottleneck trapping of intermediates in crystal studies, **317:**189

Isoelectric focusing, *see also* Two-dimensional gel electrophoresis

phosphorylated rhodopsins, **315:**71–72, 76

S-nitrosylated proteins

carbonic anhydrase III, **301:**146–147, 150

glycogen phosphorylase *b*, **301:**145–150

H-ras, **301:**150–141

Isoergonic cooperativity

definition, **295:**337

glutamate dehydrogenase

assumptions of model, **295:**338–339

basic element of theory, **295:**337–338

experimental support

calorimetry data, **295:**334–337, 340–343

enzyme kinetics, **295:**341–342

X-ray crystallography, **295:**340

experimental system overview, **295:**332–333

linkage between chemical events and cleft-closing in catalytic cycle

isoergonic cooperativity model, **295:**347–348

Lumry scheme, **295:**344–347

overview of conformational changes, **295:**343–344

model expansion, **295:**338–340

q dependence on L_T for two-subunit case, **295:**339

isothermal calorimetric titration using Wiseman isotherm, **295:**333–334

scope of phenomena, **295:**348–349

Isomerhydrolase

assay

enzyme preparation, **316:**326

high-performance liquid chromatography of products, **316:**327

incubation conditions, **316:**326–327

materials, **316:**325

principle, **316:**325

tracer concentrations of substrate, **316:**327–328

11-*cis*-retinol

generation, **316:**325

product inhibition, **316:**328–330

simultaneous detection with lecithin : retinol acyltransferase and retinyl ester hydrolase activities

all-*trans*-retinol preparation, **316:**333

cellular retinaldehyde-binding protein preparation and effects, **316:**332–333, 340–341

denaturing gel electrophoresis, **316:**339

high-performance liquid chromatography

calculations, **316:**339

extraction of retinoids, **316:**333–334

retinyl ester hydrolysis, **316:**337

separation of retinoids, **316:**334–335, 340

incubation conditions, **316:**333

kinetic effects of phosphates, ATP, and alcohols, **316:**341–343

materials, **316:**332

stability of proteins, **316:**339–340

ultraviolet treatment of retinal pigment epithelium microsomes, **316:**332, 340

visual cycle role, **316:**325, 331

Isopeptidase

assay, **309:**173

digestion of paired helical filaments and tangles, **309:**184–185

mechanism of cleavage in assay system, **309:**183–184

Isopropylmalate dehydrogenase, protein folding assays, **290:**174–176

Isothermal titration calorimetry, *see* Titration calorimetry

Isothromboxanes, lipid peroxidation and oxidant stress indicator, **300:**12
Isotope dilution negative-ion electron capture mass spectrometry, tyrosine oxidation products in proteins
  advantages in protein oxidation analysis, **300:**127–128
  derivatization, **300:**133
  gas chromatography, **300:**133–134
  internal standard synthesis, **300:**128–129
  mass spectrometry settings, **300:**134
  protein and tissue hydrolysis, **300:**130–132
  selected ion monitoring in quantification, **300:**134–138
  solid-phase extraction, **300:**132
  tissue collection and preparation, **300:**129–130
Isotope effect, *see* Kinetic isotope effect
ISP-I, *see* Myriocin
IST, *see in Situ* transcription
Iterative multilevel thresholding and segmentation, deconvolution of confocal images, **307:**271

J

Jasmonic acid, caged
  calcium response measurement with fluorescent probes, **291:**478, 481–482
  injection into plant cells, **291:**477–478
  photolysis conditions, **291:**478–480
  structure, **291:**475
  synthesis, **291:**475–476
JC 1, loading of cells, **307:**132
JNK, *see* c-Jun N-terminal kinase
JNK, *see* c-Jun N-terminal kinase
JNK, *see* Jun N-terminal kinase
Joint prosthesis biofilm, *see* Prosthetic joint biofilm
JPEG, compression of confocal images, **307:**40–43, 54
c-Jun
  antisense RNA expression in plasmids, **314:**33
  antisense studies of learning and memory function, **314:**276
Jun-B, antisense studies of learning and memory function, **314:**276

c-Jun N-terminal kinase
  activation, **314:**363
  antisense knockdown, JNK1 and JNK2 in cancer cells
    inhibition duration assay, **314:**353–354
    lipofection
      efficiency assay, **314:**346–348
      protein analysis, **314:**346
      transfection conditions, **314:**344–345
    phenotypic assays of cells
      anchorage-independent growth, **314:**357, 359–360
      proliferation, **314:**356–357
      transformation, **314:**356
      xenograft formation, **314:**360–362
    prospects, **314:**362
  antisense knockdown, reoxygenation-induced apoptosis effects in kidney cells
    cell culture, **314:**366
    fluorescence microscopy assay of apoptosis, **314:**370–371
    JNK isoform specificity, **314:**372, 374
    knockdown efficacy, **314:**371–372
    oligonucleotide treatment, **314:**366–367
    rationale for study, **314:**371
  antisense knockdown, tumor necrosis factor-α-induced E-selectin expression effects in endothelial cells
    cell culture, **314:**366
    flow cytometry analysis, **314:**369–370
    JNK isoform specificity, **314:**377–378
    knockdown efficacy, **314:**375, 377
    oligonucleotide treatment, **314:**366–367
    rationale for study, **314:**374–375
  assays
    activity assays
      isoform-specific assay, **314:**368–369
      total activity assay in cell lysates, **314:**355–356
    cell extract preparation, **300:**341–342
    decreased electrophoretic mobility assay of phosphorylation, **300:**345
    immune-complex kinase assay, **300:**342
    in-gel kinase assay, **300:**344–345
    materials, **300:**340
    tumor necrosis factor activation, characterization, **300:**342–344
  controls in antisense studies, **314:**364–365

half-lives of isoforms, determination,
    314:351, 353
isoforms, **314:**363–364, 372
knockout mouse findings, **314:**343
Northern blot analysis of antisense knock-
    down, **314:**367–368
oligonucleotide design
    chemical modifications, **314:**366
    cross-inhibition verification for isoform
        inhibition, **314:**349, 351
    gene-walk screening, **314:**348–349
    synthesis, **314:**365–366
    target selection, **314:**365–366
pathway function overview, **314:**342–344,
    363
purification of glutathione *S*-transferase
    fusion protein, **300:**340–342
redox sensitivity, **300:**339
shear stress effects, **301:**514–515
Western blot, **301:**520–521, **314:**368

K

Kainate receptor, *see also* Excitatory amino
    acid receptors; Neurotransmitter re-
    ceptor
domain labeling with radioligands,
    294:394
drug screening with fluorescence assay of
    calcium
    activity of various drugs, **294:**36
    advantages of assay, **294:**46–47
    limitations of assay, **294:**45–46
    pharmacologic validation, **294:**26–27
radioactive filter assays
    ligand selection, **294:**388
    reaction mixtures, **294:**388–389
    synaptic membrane preparation from
        rat brain, **294:**386–387
    Triton X-100 treatment of crude frac-
        tions, **294:**387–388
stable cell line expression of recombinant
    channels, **294:**25–26
subunits, **294:**386
Kallikreins, confocal microscopy imaging
blood vessel and atheromatous plaque tis-
    sue kallikrein, **307:**391
brain tissue kallikrein
    antibodies, **307:**382–383

astrocytoma tissue kallikrein, **307:**390
distribution and cellular localization,
    307:388–389
histology, **307:**382
tissue collection and processing,
    307:381–382
Helicobacter pylori-associated gastric ul-
    cer disease, **307:**391–392
kidney allograft tissue kallikrein,
    307:393–394
neutrophils, **307:**386–387
Kat1, heterologous expression in yeast,
    293:94, 98–99, 101–103
Katanin
ATP dependence, **298:**206–207
immunolocalization, **298:**207
isolation from sea urchin eggs
    anion-exchange chromatography,
        298:209, 216–218
    buffer preparation, **298:**208
    cation-exchange chromatography,
        298:209, 214, 217–218
    collection of eggs, **298:**208, 217
    high-speed supernatant preparation,
        298:214–215
    hydroxyapatite chromatography,
        298:209, 215–216
microtubule shearing
    assay
        flow cell preparation and incubation,
            298:209–210, 217
        limiting dilution in quantitation,
            298:212
        materials, **298:**207–208
        monitoring of reaction, **298:**210–212
        photodamage of microtubules,
            298:217
        principle, **298:**209
        solution assays, **298:**212–213
    mechanism, **298:**206–207
K_D, *see* Dissociation constant
Kemptide
modification with 2-nitrobenzyl group,
    291:124–125
photolysis of caged peptide, **291:**130–132
thiophosphorylation and modification,
    291:125–126
Keratin intermediate filament
metabolic phosphorous-32 labeling,
    298:568–569

phosphorylation and disassembly,
  298:567–568
Keratinocyte
  DNA damage from irradiation with sun-
    light, **319:**440, 442
  mitochondrial DNA deletion, **319:**376
  ultraviolet A-induced gene expression
    AP-2, **319:**280, 288, 290
    cell culture, **319:**281
    deuterium oxide studies, **319:**284
    differential reverse transcriptase poly-
      merase chain reaction, messenger
      RNA expression analysis,
      319:285–286
    electrophoretic mobility shift assay for
      transcription factor identification,
      319:288–290
    fluorescence-activated cell sorting, pro-
      tein expression analysis, **319:**285
    irradiation, **319:**281–282
    materials, **319:**284–285
    mimicing with singlet oxygen genera-
      tors, **319:**284
    reporter gene assays, **319:**286–288
    singlet oxygen quencher effects,
      319:282–283
Ketanserin, vesicular monoamine transporter binding
  distinguishing from 5-HT2$_A$ receptor bind-
    ing, **296:**76
  photoaffinity labeling with 7-azido[8-
    ^{125}I]iodoketanserin
    purification of labeled protein, **296:**79–
      81, 83
    quantitative analysis, **296:**77–79
    reaction conditions, **296:**77, 79
    synthesis, **296:**77
Kethoxal
  applications of RNA modification,
    318:9
  primer extension analysis, **318:**16
  RNA modification site, **318:**5
3-Ketosphinganine reductase, assay, **311:**135
3-Ketosphinganine synthase, *see* Serine palmitoyltransferase
Ketosteroid isomerase, hydrogen bonds and
  energy for catalysis, **308:**22–23
Khafrefungin
  bioassay for sphingolipid inhibition in
    Candida albicans, **311:**337–339

inositol phosphoceramide synthase inhibition, **311:**335, 348
preparation, **311:**344
structure, **311:**337
KIE, *see* Kinetic isotope effect
Kinectin
  cosedimentation with kinesin,
    298:191–192
  functions, **298:**185–186
  identification of soluble kinesin–
    membrane protein complex,
    298:190–191
  monoclonal antibody preparation,
    298:189–190
  purification from chick embryo brain
    biotinylation of vesicle surface pro-
      teins, **298:**188, 197
    collection of brains,
      298:187
    immunoaffinity chromatography,
      298:188–191, 196–197
    kinesin affinity chromatography,
      298:186–187
    microsomal fraction preparation,
      298:187–188
    solution preparation, **298:**187
  vesicle motility assay
    antibody inhibition, **298:**195–196
    cell culture, **298:**192–193
    data analysis, **298:**195
    flow chamber incubation, **298:**194–195
    vesicle preparation, **298:**193–194
    video-enhanced differential interfer-
      ence contrast microscopy,
      298:194–195
Kinesin
  associated proteins, *see* Kinectin
  ATP-binding core, **291:**308
  caged compounds in photolysis studies
    calcium chelators, **291:**321, 324
    nucleotides, **291:**321, 325
    purity requirements, **291:**321, 324–325
    table of properties, **291:**322–323
  fluorescence labeling
    chemical labeling compared to green
      fluorescent protein fusion con-
      structs, **298:**158
    cysteine labeling with Cy3
      K560Cys construct, rationale and
        preparation, **298:**160–162

labeling reaction, **298:**162–163
green fluorescent protein fusion
protein
expression by *in vitro* translation,
298:165–166
kinesin-GFP expression and purification, **298:**163–164
Ncd-GFP preparation, **298:**165
spectral properties of mutants,
298:160
Unc104-GFP preparation,
298:164–165
immunoaffinity purification of XKCM1,
539–540
isolation from *Xenopus* metaphase and interphase cytosols, **298:**345–346
kinesin-II
functions, **298:**145–146
purification from sea urchin eggs
anion-exchange chromatography,
298:143–144
buffers and solution, preparation,
298:139
gamete collection, **298:**138–139
gel filtration chromatography,
298:143
homogenization, **298:**140
microtubule/microtubule-associated
protein purification, **298:**140–141
overview, **298:**138
sucrose density gradient centrifugation, **298:**144
ultracentrifugation, **298:**141, 143
sequence analysis, **298:**144–145
subunits, **298:**144–145
tracking rate, **298:**145
KLP61F
functions, **298:**153
purification from *Drosophila melanogaster* embryos
cytosol production, **298:**146, 149
embryo collection, **298:**148–149
gel filtration chromatography,
298:151–152
maintenance of flies, **298:**147–148
microtubule/microtubule-associated
protein purification, **298:**149, 151
sucrose density gradient centrifugation, **298:**152
subunit structure, **298:**153

mechanism of motion, **291:**307–310
motility assay
apparatus for combined photolysis studies, **291:**336–337
bead assay for force measurement,
291:338
optical trap bead assay of motility,
298:482–489
pan-kinesin antibody production,
298:136–138
photolysis applications, **291:**310
preparation from bovine brain,
291:335
processive motility assays
fluorescence microscopy assay
axoneme preparation from sea urchin sperm and fluorescence labeling, **298:**166–167
bovine serum albumin preparation,
298:168
data collection and analysis,
298:169–171
incubation conditions, **298:**168–169
limitations, **298:**157–158
oxygen scavenger stocks, preparation, **298:**167
principle, **298:**156
sensitivity, **298:**156, 170
single fluorophore imaging,
298:159–160
slide preparation, **298:**168
microsphere assay, **298:**156–157
microtubule gliding assay, **298:**155–156
processivity of kinesins, **298:**154–155
translocation inhibition by caged ATP,
291:346
Western blotting, **298:**367–368, 372
Kinetic isotope effect
Badger's rule, relating bond order and
bond strength, **308:**383–384
contributions to isotope effects,
308:359–361
excited-state energy, calculation, **308:**362
Heisenberg uncertainty principle application to vibrational frequencies,
308:359
k_{cat}/K_M and irreversible step, **308:**365
Marcus theory correlation of proton
transfer reactions, **308:**284–285,
292–293

moment of inertia, calculation, **308**:361–
    362, 382
Pauling bond order, relating bond length
    and bond order, **308**:363
rate-limiting step determination, **308**:15–
    16, 19
ricin, irreversible step and isotope effect,
    308:367–368
transition-state structure analysis
    S-adenosylmethionine synthetase,
        308:343
    AMP deaminase
        catalytic mechanism,
            308:333, 335
        inhibitor design, **308**:348–349, 394
        isotope effects and transition-state
            modeling, **308**:335–336
        molecular electrostatic potential sur-
            face of transition-state,
            308:347–348
    binding isotope effects, **308**:330–332
    bond energy/bond order vibrational
        analysis of transition-state
        structure
        ab initio optimized structures,
            308:394–395
        AMP deaminase, **308**:335
        BEBOVIB inputs, **308**:373–382
        computational power, **308**:356
        nucleoside hydrolase, **308**:337,
            339–340
        reliability of technique, **308**:395–397
        software, **308**:371, 373
        test structures, isotope effect calcula-
            tions, **308**:368, 392–393
        transition-state imbalance,
            308:396–397
        unified models of reaction types,
            308:371, 395
        Walden inversion motion, .
            308:379–381
    commitment to catalyis
        chemical versus nonchemical steps,
            308:365–366
        forward commitment, **308**:325–327
        large commitment factors, overcom-
            ing with transient kinetic analy-
            sis, **308**:328–332
        minimization for isotope effect stud-
            ies, **308**:365

reverse commitment, **308**:327–328,
    366
competitive radiolabeled technique,
    308:306–307, 364–365
computational approaches to structure
    elucidation, overview, **308**:333, 368
correction for isotopic depletion,
    308:322–323
distance sensitivity of substituted iso-
    tope centers, **308**:345–346
DNA and RNA processing enzymes,
    308:320
error sources, **308**:393–394
generalizations in interpretation,
    308:305–306
hydride transfer
    commitment factor measurement,
        308:325
    quantum mechanical tunneling com-
        plications, **308**:324–325
    Swain–Schadd relationship,
        308:323–324
labeling of substrates
    nicotinamide adenine dinucleotide,
        308:313, 316, 356–357
    nucleosides and nucleotides,
        308:308–310, 312
    oligonucleotides, **308**:316–317
    purification, **308**:317–318
magnitude of effects, **308**:306
mass spectrometry accuracy, **308**:307
molecular electrostatic potential sur-
    face of substrates and transition-
    state, **308**:346–348, 355
nucleoside hydrolase
    inhibitor design, **308**:354–355, 394
    isotope effect measurement, experi-
        mental setup, **308**:318, 320
    molecular electrostatic potential
        surface of transition-state,
        308:346
    transition-state modeling, **308**:336–
        340, 342–343
out of plane C—H bending, **308**:344
overview of transition-state analysis
    and inhibitor design, **308**:307
primary carbon-14 isotope effect,
    308:344
protein-modifying enzymes,
    308:320–322

purine nucleoside phosphorylase
  commitment factors, **308:**341
  functions and inhibition rationale, **308:**340
  inhibitor design, **308:**349–350, 353–354, 394
  isotope effects for arsenolysis, **308:**341
  transition-state modeling, **308:**341–342
quantum mechanical calculations, **308:**373
reaction coordinate motion at transition-state, **308:**344, 358, 378–382
α secondary isotope effect, **308:**344
β secondary isotope effect, **308:**345
structure interpolation
  Cartesian coordinates to transition-state structures to BEBOVIB format using internal coordinates, **308:**383
  internal coordinates, **308:**384–385
  nicotinamide adenine dinucleotide hydrolysis application, **308:**385–388, 395–396
  overview, **308:**368, 370–371, 382–383
  reaction space and interpolation philosophy, **308:**383–384
  reference structures, **308:**387–392
vibrational energy relationship to frequency, **308:**359
vibrational modeling with balls and strings, **308:**362–363
zero point energy, calculation, **308:**361
Kinetic light scattering
advantages in membrane protein–protein interaction studies, **315:**330
arrestin–rhodopsin complex studies
  arrestin preparation, **315:**335
  modeling of binding reaction, **315:**342
instrumentation, **315:**333–334
light-induced binding of soluble proteins to rhodopsin disk membranes
  bimolecular analysis, **315:**338–339
  centrifugation assay comparison, **315:**336, 338, 343
  data acquisition, **315:**336
  overview, **315:**335–336
rhodopsin kinase–rhodopsin complex studies

bimolecular rate constant, **315:**341–342
dissociation constant determination, **315:**339, 341
rhodopsin kinase preparation, **315:**335
rhodopsin-containing membrane preparation, **315:**334–335
theory
  difference scattering curve, **315:**331–332
  intensity of scattering, **315:**332
  light scattering by vesicles, **315:**330–331
  relative mass change versus relative intensity change, **315:**332
  triggered scattering change, **315:**331–332
transducin association/dissociation with rhodopsin
  dissociation signal
    kinetic analysis, **315:**346–347
    overview, **315:**344–345
  light-independent binding, **315:**343
  modeling of binding reaction, **315:**343–344
  nucleotide affinity, **315:**342–343
  origin of signal, **315:**329–330
  phosphodiesterase signal
    kinetic analysis, **315:**346–347
    overview, **315:**345–346
    phosphodiesterase preparation, **315:**335
  rates and dissociation constant determination, **315:**344
  sample preparation, **315:**334–335
Kinetic oligonucleotide hybridization, RNA folding analysis
comparison with other folding analysis methods, **317:**352–353
DNA probe
  optimization, **317:**342, 344
  preparation, **317:**340–341
equipment, **317:**340
gel electrophoresis, **317:**341–342
initiation of folding with effectors, **317:**331–332, 334
kinetic constraints, **317:**335–336
kinetic folding assay
  incubation conditions and quenching, **317:**347–348, 350
  principle, **317:**346–347
oligonucleotide screening, **317:**336–337

overview, **317:**331–334

rapid quench reaction

constraints on temporal resolution, **317:**335–336

optimization, **317:**342, 344–346

principle, **317:**333–335

reaction conditions, **317:**341

reagents, **317:**337–338, 340

ribonuclease P, **317:**331

RNA preparation, **317:**340

spatial resolution, **317:**334–335

temporal resolution, **317:**331, 334, 352

Tetrahymena group I ribozyme, **317:**331, 337–338, 340–342, 346, 350–352

thermodynamic analysis, **317:**350–351

in vitro selection, **317:**351–352

Kinetic proofreading

imperfect energetic coupling, **308:**68–69

overview, **308:**67–69

stoichiometry of polypeptide elongation reaction, **308:**69

Kininogen, confocal microscopy imaging on neutrophils, **307:**386–388

Kinin receptors, confocal microscopy imaging

acute renal transplant rejection, B2 receptor expression, **307:**394

astrocytoma receptors, **307:**390

blood vessel and atheromatous plaque receptors, **307:**391

gastric mucosa receptors, **307:**392–393

neuron receptors, **307:**389–390

neutrophil B2 receptors, **307:**387

Kirchoff's law, temperature dependence of enthalpy change, **295:**91

KLP61F, *see* Kinesins

KLS, *see* Kinetic light scattering

K_m, *see* Michaelis constant, carrier-mediated transport

L

Lac repressor, *see also* Peptides-on-plasmids library screening

density of binding on DNA, **304:**513–514

α-Lactalbumin

α-crystallin chaperone assay by disulfide bond-breaking, **290:**380–383

reduced carboxymethylated lactalbumin

competition assay of BiP ligands, **290:**408–409

β-Lactamase, induction and quantification in *Pseudomonas aeruginosa* biofilms, **310:**213–216

Lactate dehydrogenase

active site structure, **308:**187–188

difference Raman spectroscopy

stereospecific hydride transfer, **308:**188–189

substrate binding energetics, **308:**188

transition-state stabilization, **308:**189–191

heat denaturation protection assays, **290:**176–178

Lactosylceramide synthase

affinity labeling, **311:**572

assay

high-performance thin-layer chromatography, **311:**76

incubation conditions, **311:**76

principle, **311:**75

product characterization, **311:**76–77

reagents, **311:**75

sample preparation, **311:**75–76

biological significance

atherosclerosis, overview, **311:**73–75

cell adhesion, **311:**81

cell proliferation, **311:**78–81

signal transduction, **311:**78

detergent dependence, **311:**73, 77

divalent metal dependence, **311:**73, 77

inhibitors, **311:**74–75, 77

isoelectric point, **311:**77

kinetic parameters, **311:**77

reaction catalyzed, **311:**73

size, **311:**77

stability and storage, **311:**77–78

substrate specificity, **311:**77

Lactosylthioceramide

chemical synthesis

column chromatography, **311:**603–604

materials, **311:**602–603

O-(2,3-di-*O*-benzoyl-β-D-galactopyranosyl)-(1→1)-3-*O*-benzoyl-2-dichloroacetamido-4-octadecen-1,3-diol, **311:**622–623

O-(2,3-di-*O*-benzoyl-4,6-*O*-*p*-methoxybenzylidine-β-D-galactopyranosyl)-(1→1)-3-*O*-ben-

zoyl-2-dichloroacetamido-4-octa-
decen-1,3-diol, **311:**622

O-(β-D-galactopyranosyl)-(1→1)-2-
dichloroacetamido-4-octadecen-
1,3-diol, **311:**618, 620

S-(β-D-galactopyranosyl)-(1→4)-O-(4-
thio-β-D-glucopyranosyl)-(1→1)-2-
amino-4-octadecen-1,3-diol,
311:625

S-(β-D-galactopyranosyl)-(1→4)-O-(4-
thio-β-D-glucopyranosyl)-(1→1)-2-
(8-N-(7-nitrobenz-1, **311:**3-diazol-2-
oxa-4-yl)amino)octanamido-4-
octadecen-1,3-diol, **311:**625

S-(β-D-galactopyranosyl)-(1→4)-O-(4-
thio-β-D-glucopyranosyl)-(1→1)-2-
[1-^{14}C]-octanamido-4-octadecen-
1,3-diol, **311:**626

O-(4,6-O-p-methoxybenzylidine-β-D-
galactopyranosyl)-(1→1)-2-
dichloroacetamido-4-octadecen-
1,3-diol, **311:**621–622

principle, **311:**618

salt removal, **311:**603

structural characterization, **311:**604

O-(2,3,4,6-tetra-O-acetyl-β-D-galacto-
pyranosyl)-(1→4)-O-(2,3,6-tri-O-
benzoyl-4-thio-β-D-galac-
topyranosyl)-(1→1)-3-O-ben-
zoyl-2-dichloroacetamido-4-octa-
decen-1,3-diol, **311:**624–625

thin-layer chromatography, **311:**603

O-(2,3,6-tri-O-benzoyl-β-D-galacto-
pyranosyl)-(1→1)-3-O-benzoyl-
2-dichloroacetamido-4-octa-
decen-1,3-diol, **311:**623

O-(2,3,6-tri-O-benzoyl-4-O-trifluoro-
methanesulfonyl-β-D-galacto-
pyranosyl)-(1→1)-3-O-benzoyl-
2-dichloroacetamido-4-octa-
decen-1,3-diol, **311:**623

stability, **311:**610, 618

λ repressor, density of binding on DNA,
304:513–514

Laminar flow chamber, *see* Flow cell

Laminin, sample mounting for confocal microscopy, **307:**85, 88–89, 96

Lamm equation, density gradient sedimentation, **290:**272, 274

Laser

confocal microscopy
excitation, **307:**17, 375
wavelengths, **307:**421–422

multiphoton excitation microscopy
pulse picker to minimize damage,
307:532, 534–536
sources, **307:**469, 522–523, 526
power reduction in photobleaching minimization, **307:**73, 77–78

Laser-induced optoacoustic spectroscopy
energy storage studies
bacteriorhodopsin, **315:**147–148
rhodopsin, **315:**148, 154
instrumentation, **315:**145
principles, **315:**143–145
reference compounds, **315:**145–146
rhodopsin photobleaching energetics,
315:156, 158
temperature dependence of signal,
315:146

Laser photolysis
caged carboxylates, **291:**40
characteristics of laser types, **291:**203, 319
comparison to flash lamp systems,
291:203–204
components of basic laser, **291:**185–186
dynamic bleaching
conditions causing bleaching, **291:**184
critical intensity estimation, **291:**184
theory, **291:**182–183
frequency doubling, theory, **291:**196–198,
201

Gaussian beam optics
beam radius, **291:**187
birefringence, **291:**194–195
focusing, **291:**187
Rayleigh range, **291:**189
reflection, **291:**192–193
refraction, **291:**194
Self's equations, **291:**188–190
waist position, **291:**188, 190
waist radius magnification, **291:**189
wave polarization, **291:**190–191

group velocity dispersion, **291:**199–200

mode locking, theory, **291:**198–199

Q-switching, production, **291:**195

safety
fail-safe protection, **291:**186
goggles, **291:**183
voltage supply, **291:**186

single-photon photolysis
    development, **291:**175–176
    laser systems, **291:**176, 201–202
    spatial resolution, **291:**359
  stimulated emission, theory, **291:**179–182
  two-photon photolysis, *see* Two-photon
    photolysis
Laser scanning confocal microscopy, *see*
    Confocal laser scanning microscopy
Laser trap, *see* Optical trap
Latrunculin-A
  actin monomer binding
    cytoskeleton effects, **298:**18–19
    mechanism, **298:**19, 29
    mutagenesis analysis of binding sites
      on actin, **298:**19–20
  mammalian cell cytoskeleton analysis
    addition to cells, **298:**25
    comparison to cytochalasin D,
      298:24–25
  Saccharomyces cerevisiae cytoskeleton
    analysis
    addition to cells, **298:**21
    comparison with other yeast species,
      298:24
    disassembly of F-actin structures,
      298:20–21
    halo assays of drug sensitivity,
      298:21–22
    reversibility of drug action, **298:**20–21
    rhodamine-phalloidin visualization of
      F-actin disruption, **298:**23–24
Laue diffraction, *see* X-ray crystallography,
    photolysis studies; X-ray crystallogra-
    phy, RNA
LBHB, *see* Low-barrier hydrogen bond
Lcb1
  gene cloning in yeast, **311:**328–329
  serine palmitoyltransferase association,
    311:3, 5
Lcb2
  gene cloning in yeast, **311:**328–329
  serine palmitoyltransferase association,
    311:3, 5, 9
Lcb4, *see* Sphingosine kinase
Lcb5, *see* Sphingosine kinase
LCR, *see* Locus control region
LDH, *see* Lactate dehydrogenase
LDL, *see* Low-density lipoprotein
Lead(II) acetate

boundary mapping of antibiotic RNA ap-
    tamers, **318:**223, 225–226
  direct detection of RNA modification,
    318:12
Leadzyme, X-ray crystallography of small ri-
    bozymes
  chemical synthesis
    automated synthesis, **317:**151–152
    desalting, **317:**153
    overview, **317:**151
    reversed-phase high-performance liq-
      uid chromatography, **317:**152–153
  crystallization
    cations, **317:**155–158
    precipitants, **317:**154–155
    screening, **317:**154
  crystal packing promotion, **317:**150–151
  diffraction data acquisition
    cryoprotection, **317:**159–160
    crystal requirements, **317:**158
    radiation sensitivity, **317:**158–159
    synthetic mother liquors, **317:**159
  heavy atom derivatization, **317:**160–162
  initial electron density maps, **317:**164–165
  Patterson methods for heavy atom posi-
    tions, **317:**162–164
  refinement, **317:**165–168
Leakage current, *see* Electrophysiology,
    neurotransmitter transporters
Leber's congenital amaurosis, retinal pig-
    ment epithelium p65 mutations,
    316:711–714
Lecithin : retinol acyltransferase
  catalytic reaction, **316:**401
  complementary DNA
    characterization, **316:**406–407
    library screening, **316:**405
  expression in human embryonic kidney
    cells
    subcloning into expression vector,
      316:407
    transfection, **316:**407–408
  inhibitors, sulfhydryl-directed,
    316:401–402
  messenger RNA isolation, **316:**402–403
  Northern blot analysis of tissue distribu-
    tion, **316:**408–410
  retinal pigment epithelial cell preparation
    for complementary DNA library
    preparation, **316:**402

reverse transcriptase–polymerase chain
reaction amplification of RNA,
316:403–405
sequence analysis
nucleic acid, **316:**412–413
protein, **316:**410–412
simultaneous detection with retinyl ester
hydrolase and retinol isomerase activities
all-*trans*-retinol preparation, **316:**333
cellular retinaldehyde-binding protein
preparation and effects, **316:**332–
333, 340–341
denaturing gel electrophoresis, **316:**339
high-performance liquid chromatography
calculations, **316:**339
extraction of retinoids, **316:**333–334
retinyl ester hydrolysis, **316:**337
separation of retinoids, **316:**334–335,
340
incubation conditions, **316:**333
kinetic effects of phosphates, ATP, and
alcohols, **316:**341–343
materials, **316:**332
stability of proteins, **316:**339–340
ultraviolet treatment of retinal pigment
epithelium microsomes, **316:**332,
340
visual cycle role, **316:**325, 331, 401
Western blot analysis, **316:**410
Lectin
high-throughput assay for immobilized
glycosphingolipid binding to
transfected cells
adhesion
conditions, **312:**442–443
quantification, **312:**443, 445
glycosphingolipid coating of micro-
wells, **312:**440–441
precautions, **312:**445–446
transfection of cells, **312:**439–440
translocation assay of sphingolipids,
312:574
valency of carbohydrate binding, **312:**438
Lectin affinity chromatography
protein disulfide isomerase, **290:**53
SUR1, 455
Lectin staining, biofilms
advantages over immunostaining, **310:**147

confocal microscopy
applications, **310:**151–152
controls, **310:**150–151
indirect sandwitch techniques,
310:149–150
multiple lectin staining, **310:**149
single lectin staining, **310:**148
extracellular polymeric substances,
310:406
lectin stock solution preparation,
310:147–148
safety precautions, **310:**147
specificity of carbohydrate interactions,
310:145
Leech neuron, *see* Retzius P cell synapse
Legionella pneumophila biofilm
antimicrobial susceptibility assays
chlorine dioxide treatments,
310:632–633
coupon sampling, **310:**631
inoculation, **310:**630
propidium iodide imaging with fluorescence microscopy, **310:**631–632
test rigs, **310:**630, 635
viability culture, **310:**632, 636
copper versus plastic biofouling, **310:**633,
636
Legionnaire's disease role, **310:**629
Leucine-enkephalin, photocleavable biotin
labeling
avidin affinity purification, **291:**144–145,
148–149
conjugation, **291:**144
fluorescamine assay, **291:**145
photocleavage kinetics, **291:**146–148
Leukemia inhibitory factor, recombinant human protein expression, **306:**36–41
Leukocyte, chemiluminescence assay of
urine
advantages, **305:**410
cell prelysing, **305:**409
clinical applications, **305:**402, 406–407
comparison with automated imaging
assay, **305:**409–410
incubation conditions and data collection,
305:408
instrumentation, **305:**408
interferences, **305:**409
reagents and solutions, **305:**407–408
specificity, **305:**408–409

urine osmolality effects, **305:**409

Leukocyte adhesion, nitric oxide effects, **301:**422–423, 428–429

Leukocyte rolling, adhesion molecule mediation, **300:**395–396

LIF, *see* Leukemia inhibitory factor

Lifetime fluorescence imaging, multiphoton excitation microscopy, **307:**528, 532

Lifson–Roig model, *see* Helix–coil transition

Ligand-induced conformational change
BirA, **295:**439–440
glutamate dehydrogenase, **295:**343–344
isoergonic cooperativity model, **295:**347–348
Lumry scheme, **295:**344–347

Ligation-mediated polymerase chain reaction, *see* Polymerase chain reaction

Light chain amyloidosis, *see* Immunoglobulin light chain amyloidosis

Light harvesting complex I, *see* B875

Light harvesting complex II, *see also* B800-850
chlorophyll per cell assay of abundance, **297:**230–231
herbicide effects on expression, **297:**231
signal transduction of photoacclimation, **297:**232–234

Light-regulated gene expression, *see* Complementary chromatic adaptation; Phytochrome; *psbA*

Light scattering, *see also* Kinetic light scattering; Quasielastic light-scattering spectroscopy
fibril correction in Congo red assay, **309:**289–291, 294–296
static light scattering applications, **309:**431–432
turbidity measurements of fibrillogenesis, **309:**431, 480–481, 486–488

Limb culture, *see* Fibroblast growth factor-4

LIOAS, *see* Laser-induced optoacoustic spectroscopy

Lipid bilayer, *see also* Liquid–crystalline phase transition
molecular dynamics simulated annealing with restrained molecular dynamics, **293:**682–684

Lipid hydroperoxide, *see also* Low-density lipoprotein

chemiluminescence detection
calibration curves, **300:**38–40
data collection, **300:**41–42
kinetic modeling of enmission, **300:**37–38
luminol, mechanism of photon emission, **300:**35–37
materials, **300:**40
sample preparation, **300:**40–41
single-photon counting instrumentation and sensitivity, **300:**34, 42
specificity, **300:**34–35, 42

cholesteryl linoleate hydroperoxide assay by reversed-phase high-performance liquid chromatography
calibration, **300:**48–49
chromatography conditions, **300:**47–48
detection, **300:**49–50
extraction, **300:**46
internal standard preparation, **300:**44–45, 50
limitations of assay, **300:**50
lipoprotein isolation and oxidation *in vitro*, **300:**46
peak identification, **300:**48–49

ferrous oxidation in xylene orange, plasma assay
applications, **300:**61
confounding factors, **300:**61–62
incubation conditions, **300:**60
principle, **300:**58–59
reagents, **300:**59–60
triphenylphosphate treatment, **300:**59–60

high-performance liquid chromatography with mercury cathode electrochemical detection
advantages over other detection techniques, **300:**24
extraction of samples, **300:**25
instrumentation, **300:**25–26
oxidatively stressed cells
photooxidation, **300:**26–27
profiles of lipid hydroperoxides, **300:**27, 29
phospholipid hydroperoxide glutathione peroxidase assay
advantages over coupled assay, **300:**32
decay profile, **300:**30–32

lipid hydroperoxide substrate preparation, **300:**30
reaction mixture, **300:**30
reagents, **300:**24–25
iodometric assay
overview, **300:**24, 26
skin oxidation status
aging effects, **300:**435–436
electrochemical detection, **300:**435, 437
γ-irradiation effects, **300:**436
principle, **300:**434
plasma values, range, **300:**58
toxicity mechanisms, **300:**26, 33–34
Lipid nitration
atherosclerosis role, **301:**456–457
bioactivity, **301:**465
high-performance liquid chromatography, **301:**460, 464–465, 467–468
mass spectrometry, **301:**463, 468–469
nitric oxide reaction with lipid peroxyl radicals
anaerobic nitric oxide solution preparation, **301:**466
fatty acid preparation, **301:**466
initiation of peroxidation, **301:**457
kinetic analysis, **301:**458
oxygraphy, **301:**458, 466–467
nitrous acid, nitrosation of lipid hydroperoxide, **301:**463–464, 469–470
unsaturated lipid reactions
nitrogen dioxide, **301:**462–463, 469
nitronium ion, **301:**462–463, 469
peroxynitrite
extraction and chromatographic analysis, **301:**460, 467–468
incubation conditions, **301:**459–460
synthesis and infusion of peroxynitrite, **301:**460–461, 467
Lipid peroxidation, *see also* Lipid hydroperoxide; Low-density lipoprotein; Phospholipid peroxidation
breath hydrocarbons as markers
flame ionization detection, **300:**457–458, 462
formation, **300:**457
modulators, **300:**457, 460–461
photoionization gas chromatography
column selection, **300:**458–459
confounding variables, **300:**460–461

pentane measurement, **300:**459–460
principle, **300:**458
sensitivity, **300:**461–462
smokers versus nonsmokers, **300:**461–462
indicators, *see* Alkanal; F$_2$-isoprostanes; Malondialdehyde
Lipid peroxidation, *see also* Low-density lipoprotein
amidothionophosphate reduction, **299:**297–300
amyloid-β induction
free fatty acid analysis, **309:**766–767
glutathione protection, **309:**737
4-hydroxynonenal–protein conjugates
glucose transporter inhibition, **309:**740
immunohistochemistry, **309:**738
immunoprecipitation, **309:**737–738, 766
induction, **309:**734–735
Western blot analysis, **309:**738
malondialdehyde assay, **309:**737
overview, **309:**734–735
stearic acid spin probes in evaluation, **309:**765–766
assays
gas chromatography headspace assay of volatile oxidation products
diet effects on lipid peroxidation, **299:**198–200
instrumentation, **299:**197
overview, **299:**196–197
phenolic antioxidant effects, *in vitro* assays, **299:**200
sample preparation, **299:**197–198
wine phenolics, evaluation, **299:**199–201
overview, **299:**191–192
thiobarbituric acid assay, **299:**191–193
Criegee ozonation, **319:**572
4-hydroxy-2-nonenal adduct detection
applications, **319:**563–564
cytotoxicity, **319:**562–564
enzyme-linked immunosorbent assay, **319:**569
immunohistochemical analysis, **319:**569–570
polyclonal antibody preparation

amino acid–HNE adduct antibody
    characterization, **319:**565
fluorescent lysine–HNE antibody
    preparation, **319:**566–567
principle, **319:**564–565
reductively-stabilized amino acid–
    HNE adduct antibody character-
    ization, **319:**565–566
protein adducts, **319:**563, 567
reduction of protein adducts
    sodium borohydride, **319:**567–568
    sodium cyanoborohydride, **319:**568
Western blot analysis, **319:**568–569
initiation and propagation, **299:**190–191
iron role, **319:**294, 296
polyunsaturated fatty acid chemistry and
    antioxidant action, **299:**190–191
retinol interactions with α-tocopherol in
    liposome oxidation prevention
chemicals and equipment, **299:**422
oxidation conditions
    liposomes, **299:**422–423
    retinal membrane oxidation,
        299:423–424
regeneration, **299:**421–422, 428, 430
retinal membrane protection, **299:**425,
    427–429
soybean phosphatidylcholine liposome
    protection, **299:**424–425, 429
vitamin analysis with high-performance
    liquid chromatography, **299:**424
β scission pathways leading to aldehydes,
    319:573
signal transducers
classes, **319:**570–571
formation, **319:**571–574
synthesis
    bis(1-hydroxyheptyl)peroxide,
        319:578–579, 581
    bis(1-hydroxynonyl)peroxide,
        319:578–579, 581
    glycerophospholipid hydroxyhydro-
        peroxides, **319:**574–575
    1-hydroxy-1-hydroperoxyalkanes,
        319:575–576
    materials, **319:**576
    1-palmitoyl-2-(1-hydroxy,1-
        hydroper-oxynonanoyl)-sn-
        glycero-3-phosphocholine,
        319:578

1-palmitoyl-2-[3-(8-octanoyl)-5-(1-
    octyl)-1,2,4-trioxolane]-sn-
    glycero-3-phosphocholine,
    319:577
1-palmitoyl-2-(9-oxononanoyl)-sn-
    glycero-3-phosphocholine,
    319:577–578
phospholipid aldehydes,
    319:574–575
X-ray crystallography, **319:**581–582
skin equivalent assay
Episkin preparation, **319:**457
K assay, **319:**458
sunscreen evaluation, **319:**458–459
ultraviolet A source, **319:**457–458
skin surface product analysis, *see* Skin, ox-
    idation product measurement
Lipid phase separation
biological significance,
    312:272–273
composition effects, **312:**274
fluorescence quenching assay
advantages, **312:**274
data analysis, **312:**287–288
fluorescence measurements,
    312:286–287
lipid composition effects
    assay, **312:**282–283
    fully miscible lipid mixtures,
        312:276–277
    partially immiscible lipid mixtures,
        312:277–279
principle, **312:**273
sample preparation
    lipid dispersion at ambient tempera-
        ture, **312:**285
    lipid dispersion at elevated tempera-
        ture, **312:**285–286
    lipid mixing, **312:**284–285
sensitivity, limiting factors, **312:**288–290
temperature
    effects on fluorescence, **312:**279–281
    variance in assay, **312:**283–284
ternary lipid mixture analysis,
    312:290
ordered versus fluid phases, **312:**274
relationship between lipid structure, com-
    position, and domain formation,
    312:281–282
temperature effects, **312:**274, 276

Lipid phosphate phosphatase, *see* Ceramide-1-phosphate phosphatase; Sphingosine-1-phosphate phosphatase
Lipid tubules, *see* Electron crystallography
Lipofection, *see also* Transfection, antisense oligonucleotides
  myocytes, **302:**180, 182
  neurons, **302:**221–222
Lipoic acid
  antioxidant activity, **299:**240
  cellular reduction to dihydrolipoic acid, **299:**246
  high-performance liquid chromatography with electrochemical detection
    cell culture, **299:**243
    chromatography, **299:**241
    coulometric detection, principle, **299:**240–241
    current–voltage response curve, **299:**242
    electrodes, **299:**240
    extraction, **299:**243, 245–246
    instrumentation, **299:**241
    standards, **299:**242–243
  protein modification, **299:**239
  Tetrahymena thermophila assay
    basal medium for assay, **299:**287–288
    growth culture, **299:**287
    maintenance culture, **299:**287
    principle, **299:**287
    sample preparation
      blood, **299:**292
      cerebrospinal fluid, **299:**292
      liver, **299:**293
      overview, **299:**290–291
      urine, **299:**292
    standards, **299:**288–289
Liposome gene delivery, *see* Cationic lipid:plasmid DNA
Lipoxamycin, serine palmitoyltransferase inhibition, **311:**335, 350
Liquid–crystalline phase transition
  differential scanning calorimetry
    bilayer preparation, **295:**470
    data collection and analysis, **295:**472, 474–475
    enthalpy change, **295:**474
    entropy change, **295:**474–475
    instrumentation, **295:**470–471
    mixed-chain-length phospholipid studies, **295:**487, 490–493
    sample loading, **295:**471–472
    van't Hoff enthalpy, **295:**474–475
  mixed-chain-length phospholipids, modeling of thermodynamic behavior, **295:**487–489
  phospholipid hydrocarbon chain conformation, **295:**468
  physiological relevance, **295:**468–469
  Raman spectroscopy
    C–C stretching mode region, **295:**480
    C–H stretching mode region, **295:**478–480
    data collection and analysis, **295:**477–478
    instrumentation, **295:**475–477
    mixed-chain-length phospholipid studies, **295:**490, 492–493
    sample preparation, **295:**475
  X-ray diffraction
    bilayer partial thickness estimation, **295:**486
    data collection, **295:**483–484
    instrumentation, **295:**482–483
    low-angle scattering analysis, **295:**484
    mixed-chain-length phospholipid studies, **295:**490–491
    sample preparation, **295:**480–482
    wide-angle scattering analysis, **295:**484–487
Lissamine rhodamine-labeled sphingolipids
  applications, **312:**294–295
  enzyme assays
    β-galactosidase, **312:**300
    β-glucosidase, **312:**300
    ceramidase, **312:**301
    sphingomyelinase
      lysosomal acid enzyme, **312:**299–300
      neutral magnesium-dependent enzyme, **312:**300
  metabolic labeling of cells
    administration into cells and lysosomes, **312:**301–302
    cell culture, **312:**302
    fluorescence analysis
      culture medium, **312:**304
      extracts, **312:**303
      intact cells, **312:**303
      termination of pulsing, **312:**302–303

synthesis
  ceramide dihexoside, **312:**297
  ceramide trihexoside, **312:**297
  ceramide, **312:**296
  dodecanoic acid conjugation for link-
    age, **312:**295–296
  galactocerebroside, **312:**296
  glucocerebroside, **312:**296
  sphingomyelin, **312:**297
  sulfatide, **312:**297–298
Listeria monocytogenes actin assembly
  assay
  actin preparation and labeling,
    298:118–119
  alternative bacteria, **298:**116–117
  applications, **298:**115–116
  Arp2/3 assay
    fluorescence detection, **298:**61
    incubation conditions, **298:**61
    killed fluorescence-labeled bacteria
      preparation, **298:**59–60
  host cell cytoplasm/protein preparation,
    298:119
  motility assays, **298:**119–122
  preparation of bacteria,
    298:117–118
  principle, **298:**114–115
  quantitative analysis, **298:**120–122
  simultaneous phase contrast/fluorescence
    microscopy measurements,
    298:120–121
LIVE/DEAD *Bac*Light Viability kit, membrane integrity assessment in biofilms,
  310:173–174, 277
Liver, oxygen-free radical formation in isolated and perfused rat liver with chemiluminescent imaging, **305:**123–126,
  130–131
Loading, *see* Dye loading
Locus control region, *see* α-Globin; β-
  Globin
Long-term potentiation, hippocampal slice
  studies with recombinant adenovirus infection, **293:**499–500, 502
Loop E motif, modeling with comparative
  sequence analysis, **317:**503–504
Low-barrier hydrogen bond
  characteristics, **308:**22, 219, 221
  energy for catalysis, **308:**22–23, 246, 248,
    254–257, 261

geometrical effects on energetics,
  308:270–271
hypoxanthine–guanine phosphoribosyltransferase, **308:**304–305
mutation analysis of enzymes,
  308:273–274
proton nuclear magnetic resonance
  bond lengths, **308:**224–226
  criteria for identification,
    308:222–223
  data acquisition, **308:**225–229
  deshielding of resonance, **308:**224
  fractionation factor
    bond length determination,
      308:234–235
    deuterium exchange reaction,
      308:232–233
    factors less than unity, **308:**233
    measurement, data acquisition and
      calculations, **308:**235–237
    quantum mechanical theory,
      308:233–234
  free energy of formation
    mutation studies of enzymes,
      308:243–244
    pH titration in measurement,
      308:244–245
  overview of enzymes, **308:**221–222,
    230, 245
  proton exchange rates
    limiting mechanisms, **308:**239
    measurement, data acquisition and
      calculations, **308:**240–243
    mechanism representation in pro-
      teins, **308:**238–239
    protection factor, **308:**237–239
  resonance assignment
    J coupling, **308:**229
    mutagenesis of enzyme, **308:**229, 231
    nuclear Overhauser effects, **308:**232
    temperature measurement and calibra-
      tion, **308:**227–228
    water suppression, **308:**228–229
serine protease mechanism, **308:**23–24,
  244
staphylococcal nuclease, **308:**269
strength, **308:**22, 219
triosephosphate isomerase
  active site interactions with transition-
    state, **308:**272–273

catalytic problem description,
  308:248–250
entropic barrier, lowering by enzyme,
  308:272
strengthening of hydrogen bonds ac-
  companying charge rearrangement
  basis for effect, **308**:251–254
  catalytic contribution, quantitative es-
    timation, **308**:254–257, 261,
    266–268
  dielectric environment lowering by
    enzyme, **308**:261, 264–266, 274
  Hine equation and rate enhance-
    ment calculation, **308**:257,
    274–276
  linear-free energy analysis of intra-
    molecular hydrogen bond
    strength, **308**:261–263, 266–268
  multiple hydrogen bond analysis,
    308:269–270
  overview, **308**:250–251
Low-density lipoprotein
  advanced glycation end product assay
    with enzyme-linked immunosorbent
    assay, **309**:168
  antioxidant depletion after copper oxida-
    tion, **301**:321
  chemiluminescence, free radical measure-
    ment, **299**:4–6
  cholesteryl ester hydroperoxide content
    determination, **300**:19–20
  cholesteryl linoleate hydroperoxide assay
    by reversed-phase high-performance
    liquid chromatography
    calibration, **300**:48–49
    chromatography conditions,
      300:47–48
    detection, **300**:49–50
    extraction, **300**:46
    internal standard preparation, **300**:44–
      45, 50
    limitations of assay, **300**:50
    lipoprotein isolation and oxidation *in
      vitro*, **300**:46
    peak identification, **300**:48–49
  dietary effects on fatty acid composition
    and oxidation, **299**:194–195
  lipid hydroperoxide-free preparation
    advantages and comparison with other
      methods, **300**:20–23

discontinuous density gradient centrifugation, **300**:18
metal ion removal, **300**:18–19
size-exclusion chromatography,
  300:18–19
myeloperoxidase modifications
p-hydroxyphenylacetaldehyde–lysine
  modification
  chloride-free sodium hypochlorite
    preparation, **300**:92
  chlorine-demand free conditions,
    300:91–92
  extraction of protein hydrozylates,
    300:98
  gas chromatography–mass spectrome-
    try, **300**:98–101
  mechanism of aldehyde generation,
    300:90
  mechanism, **300**:90
  myeloperoxidase purification,
    300:92–93
  protein hydrolysis, **300**:98
  sensitivity, **300**:101
  specificity, **300**:101
  stability, **300**:101
  synthesis of lysine adduct, **300**:95–97
  tissue collection, storage, and pro-
    cessing, **300**:97–98
low-density lipoprotein isolation
  intima, **300**:93–95
  plasma, **300**:95
molecular chlorine oxidation of cholesterol
  characterization by electrospray ion-
    ization mass spectrometry,
    300:105
  gas chromatography–mass spectrome-
    try of phospholipid liposomes,
    300:102–103
  incubation conditions, **300**:103–105
nitric oxide protection against oxidation
  atherosclerosis implications,
    301:423–424
  copper-dependent oxidation,
    301:446–449
  delivery of nitric oxide, **301**:444–446
  kinetic modeling of oxidation, **301**:450–
    451, 453
  peroxyl radical-mediated oxidation,
    301:449–450

thiobarbituric acid-reactive substances, **301**:447–448

α-tocopherol depletion, **301**:448–449

oxidation assays

gas chromatography headspace assay of volatile oxidation products

diet effects on lipid peroxidation, **299**:198–200

instrumentation, **299**:197

overview, **299**:196–197

phenolic antioxidant effects, *in vitro* assays, **299**:200

sample preparation, **299**:197–198

wine phenolics, evaluation, **299**:199–201

overview, **299**:193–196

oxidation in atherosclerosis, **299**:35, 192, 362, **300**:43–44, 88–89, 456

oxidation in isolation from plasma, **300**:17–18

peroxynitrite oxidation

carotenoid and tocopherols

analysis by high-performance liquid chromatography, **301**:322–323

consumption and lipid peroxidation, **301**:331–332

effects of peroxynitrite, **301**:324, 327, 329, 331

conjugated diene assay, **301**:323, 327

isolation of lipoprotein, **301**:322

protein modifications, **301**:321, 332

thiobarbituric acid-reactive substances, **301**:323–324, 327, 329

plasma oxidizability assays

Alzheimer's disease patients, **299**:46

coronary heart disease patients, **299**:46, 48

hyperlipidemic patients, **299**:46, 48

overview, **299**:36–37

photometric detection of conjugated dienes

comparison with other indices of lipid peroxidation, **299**:41, 43–44, 47

data acquisition, **299**:38–39

isolated low-density lipoprotein assay, **299**:44–45

kinetics, **299**:39–40, 48

oxidants, **299**:37–40

reproducibility, **299**:40–41

sample preparation, **299**:38, 40, 49

relationship to antioxidant and fatty acid composition, **299**:45, 47–48

Watanabe heritable hyperlipidemic rabbits, **299**:46–47

protein oxidation, roles

hydroxyl radical, **300**:143

metal ions, **300**:143

myeloperoxidase, **300**:142–143

reactive nitrogen species, **300**:144

susceptibility to oxidation, **299**:35–36, 191, 194

α-tocopherol, prooxidant activity

antioxidant versus prooxidant activity, **299**:362–364, 373, 375

assay

high-performance liquid chromatography, **299**:368–370

reagents, **299**:364–365

dose response of lipid oxidizability, **299**:372–373

inverse deuterium kinetic isotope effect, **299**:371–372

lipoprotein preparation

deuterium-labeled low-density lipoprotein, **299**:367–368

native and tocopherol-enriched low-density lipoprotein, **299**:365–366

tocopherol-depleted low-density lipoprotein, **299**:366–367

tocopherol-replenished low-density lipoprotein, **299**:367

mechanism, **299**:362–364

phase transfer activity effects, **299**:370–371

total peroxyl radical-trapping potential assay, **299**:8–9

ubiquinol-10 loss in oxidation, **300**:20–22

vitamin E loss in oxidation, **300**:21–22

LPP, *see* Lipid phosphate phosphatase

LRAT, *see* Lecithin:retinol acyltransferase

LSM1

caged tyrosine peptide

design of inactive caged peptide, **291**:88–90

kinetics of photolysis, **291**:87–88

myosin light-chain kinase inhibition, **291**:93–94

spectral and photochemical properties, **291**:85–87

synthesis, **291:**80–85
structure and function, **291:**79–80
LTP, *see* Long-term potentiation
Luciferase, *see also* Firefly luciferase
  activation by caged ATP, **291:**246, 248,
    250
  advantages as reporter, **310:**152
  bacteria, *see* Bacterial bioluminescence;
    lux control system, *Vibrio*
  bioelectronic imaging and pattern recogni-
    tion applications, **291:**246–247, 250
  dinoflagellate, *see* Dinoflagellate biolumi-
    nescence
  dual reporter gene assays with chemilumi-
    nescence
    cell extract preparation, **305:**447
    controls, **305:**449
    detection
      luciferase/β-galactosidase,
        305:447–448
      luciferase/β-glucuronidase,
        305:448–449
    instrumentation, **305:**449–450
    overview, **305:**428–429, 444–446
    reagents, **305:**446–447
    sensitivity, **305:**450
  firefly, *see* Firefly luciferase
  immobilization in agarose gels,
    291:249–250
  inhibition by caged acetate pH effects,
    291:246, 248–249
  lateral flagellar gene activation reporting,
    310:64–65
  oxygen role in luminescence reaction,
    305:4
  photon-counting microscopy
    absolute photon flux determination,
      310:160
    background light levels, **310:**156
    correlative microscopy,
      310:159–160
    dark box, **310:**154–155
    data acquisition, **310:**156–159
    hardware, **310:**152–154
    light leak testing, **310:**155–156
    quantitative analysis, **310:**159
    sensitivity, **310:**153
    software, **310:**154
    Vibrio fischeri cells, **310:**155–159
  reaction catalyzed, **291:**245

refolding assay with heat shock protein
  18.1, **290:**363–365
reporter gene assays in transgenic *Arabidopsis thaliana*
  in vitro assay, **297:**189–190
  in vivo imaging, **297:**190–191
substrate, *see* Luciferin
β subunit folding analysis in cell-free
  translation system, **290:**12, 14–15
Vibrio fisheri genes as reporters,
  310:63–64
Luciferin
  emission characteristics, **305:**74
  firefly luciferin
    adenylate derivatives, synthesis,
      305:190
    analogs
      activity with firefly luciferase,
        305:192
      2-(4-benzoylphenyl)thiazole-4-
        carboxylic acid, **305:**195
      2-cyano-6-hydroxynaphthalene inter-
        mediate, synthesis, **305:**191–193
      2-cyano-6-hydroxyquinoline interme-
        diate, synthesis, **305:**193
      D-naphthylluciferin synthesis,
        305:191, 193
      D-quinolylluciferin synthesis, **305:**191,
        194
      structures, **305:**190–191
    binding site structure, **305:**195
    caged substrate, **305:**190
    chemical synthesis, **305:**189–190
    structure, **305:**188–189
  luminescence mechanisms
    categorization by mechanisms,
      305:46–47
    firefly
      color variation, **305:**38–40
      efficiency, **305:**36–37
      oxygen reactions, **305:**37–38
      triplet products, **305:**37
    imidazolopyrazine luciferins, **305:**40–42
  oxygen sensitivity, **305:**258
Luciferin-binding protein, *Gonyaulax
  polyedra*
  assay, **305:**262–264
  circadian regulation, **305:**260–261, 273
  compartmentalization, **305:**259–260
  gene cloning

buffers, **305:**268, 270
DNA purification, **305:**270
inverse polymerase chain reaction,
    305:270–271
restriction enzyme digestion,
    305:270–271
gene copy number estimation,
    305:271–272
immunolocalization, **305:**275–276
isoforms, **305:**261
luciferin stabilization, **305:**258
messenger RNA
    features, **305:**261, 272
    gene expression regulation,
        305:272–273
    mobility shift assays of binding pro-
        teins, **305:**274–275
metabolic radiolabeling with sulfur-35
    methionine, **305:**272
pH regulation, **305:**260
purification
    ammonium sulfate precipitation,
        305:266
    anion-exchange chromatography,
        305:266
    buffers, **305:**264
    cell harvesting and lysis, **305:**264, 266
    fluorescence monitoring, **305:**268
    gel filtration, **305:**266
    hydroxyapatite chromatography,
        305:266
in situ hybridization, **305:**275–276
Lucigenin
    chemiluminescence theory, *see* Acridine
        luminescence
    educational demonstrations
        features of luminescence, **305:**643–644
        mechanism of luminescence,
            305:644–645
        sign demonstration
            apparatus, **305:**645–646
            materials, **305:**645
    enhancement of chemiluminescence,
        305:344, 388
    phagocyte assay, *see* Phagocyte lumines-
        cence
Lumazine proteins
    absorption spectroscopy, **305:**166, 174
    bioluminescence activity, **305:**180
    fluorescence spectra, **305:**176

ligand binding site,
    305:179–180
ligand exchange, **305:**170–172
recombinant *Photobacterium* protein ex-
    pression in *Escherichia coli*
    cell growth and extraction
        complete medium, **305:**168
        defined media, **305:**168–169
    growth media, **305:**167
    purification of labeled riboflavin-bound
        protein
        anion-exchange chromatography,
            305:169–170
        size-exclusion chromatography,
            305:170
    selection of colonies, **305:**167
    stability, **305:**173
resonance energy transfer, **305:**165–166
structure, **305:**165
Luminescence
    applications, *see also specific applications*
        benefits of assays, **305:**334
        biomedical research, **305:**340–345
        blots, **305:**341–343
        enzyme activity assays
            direct versus coupled assays,
                305:369–370
            table of available assays by enzyme
                class, **305:**372–375
        imaging, **305:**345
        laboratory medicine, **305:**334–340
        overview, **305:**334–335, 370
        reporter genes, **305:**343–344
        therapeutics, **305:**345
    blot detection, *see* Dot blot hybridization;
        Southern blot; Western blot
    dinoflagellates, *see* Dinoflagellate biolumi-
        nescence
    educational demonstrations, *see also* Acri-
        dine luminescence; Lucigenin; Lumi-
        nol; Oxalate luminescence
        acridan luminescence, **305:**651–652
        bioluminescence demonstrations
            appeal, **305:**660–661
            dinoflagellates, *see* Dinoflagellate bio-
                luminescence
            firefly, **305:**666, 672
            green fluorescent protein, *see* Green
                fluorescent protein
        chemiluminescence demonstrations

comparison with other light-emitting reactions, **305:**658–659, 665
mechanism explanations, **305:**657–658
setup guidelines, **305:**633–634
electroluminescence, **305:**659
scientist involvement in public education
getting involved, **305:**662–664
need for scientists, **305:**661–662
tetrakis(dimethylamino)ethylene, **305:**646–647
triboluminescence, **305:**658–659
excited state generation mechanisms, *see* Acridine luminescence; Bacterial bioluminescence; Chemically initiated electron exchange luminescence; Dioxetane luminescence; Electron transfer chemiluminescence; Enamine luminescence; Hydrazide luminescence; Imine luminescence; Luciferin; Oxalate luminescence; Peroxide chemiluminescence
imaging, *see* Luminescence imaging
nomenclature reform, **305:**47–48
phagocytes, *see* Phagocyte luminescence
units and standards, *see also* Luminometer
calibration standards, **305:**49–50, 90–96
photon units and measurement losses, **305:**48–49, 68
relative light units, **305:**49, 68, 87
Luminescence imaging
applications
aequorin for calcium measurements, **305:**486, 495–498
dot blot hybridization for human parvovirus B19 DNA detection, **305:**123–125, 128–129
immunohistochemistry, **305:**579–580, 590
overview, **305:**120–121, 131–132
oxygen-free radical formation in rat liver, **305:**123–126, 130–131
reporter gene analysis, **305:**87
in situ hybridization, **305:**579–580, 590
commercial systems, **305:**98
cyanobacteria, luciferase luminescence, **305:**524–525
detector, *see* Charge-coupled device array

instrumentation, **305:**121–122
macroscopic imaging system, **305:**84
microtiter plate imaging
acetylcholinesterase inhibitor evaluation, **305:**123–124, 126, 128
advantages, **305:**121
antioxidant activity evaluation, **305:**123–124, 126
software, **305:**85–87
Luminol
absolute calibration of luminometers
aqueous solution, **305:**90–91
dimethyl sulfoxide solution, **305:**91–92, 90–91
chemiluminescence detection of lipid hydroperoxide
calibration curves, **300:**38–40
data collection, **300:**41–42
kinetic modeling of enmission, **300:**37–38
materials, **300:**40
photon emission mechanism, **300:**35–37
sample preparation, **300:**40–41
single-photon counting instrumentation and sensitivity, **300:**34, 42
specificity, **300:**34–35, 42
educational demonstrations
ammonia fountain, **305:**642–643
clock reaction
demonstration, **305:**640–641
mechanism, **305:**641–642
overview, **305:**640
safety, **305:**640
variations, **305:**642
dissolving in base with bleach addition, **305:**638
dissolving in dipolar aprotic solvent with addition of strong base
luminescence generation, **305:**636–637
materials, **305:**635–636
mechanism, **305:**637–638
safety, **305:**636
synthesis, **305:**635
transition metal-induced luminescence
demonstration, **305:**639
materials, **305:**638–639
safety, **305:**639
three-color experiment, **305:**639
emission characteristics, **305:**74, 90

enhancement of chemiluminescence,
  305:344, 379–380, 389
enzyme substrate incorporation and
  assay, **305**:380–381
4-iodophenol enhancement of peroxi-
  dase-catalyzed oxidation, **305**:389
phagocyte assay, *see* Phagocyte luminescence
phospholipid peroxidation detection by
  high-performance liquid chromatography with postcolumn chemiluminescence detection
  cell culture, **300**:54
  chemiluminescence detection,
    300:52–53
  extraction, **300**:54–55
  instrumentation, **300**:52
  phosphate content assay, **300**:55–56
  reagents and materials, **300**:53–54
  standardization, **300**:56–57
quantum yield, **305**:89
Luminometer
automatic tube luminometers, **305**:78
calibration
  absolute calibration
    definition, **305**:70, 87–88
    detector geometry challenges,
      305:88–89
    lamp calibration to blackbody,
      305:88
    luminol standards, **305**:89–92
  day-to-day calibration
    encapsulated radioactive phosphor
      calibrators, **305**:93–94
    radioactive liquid calibrators,
      305:92–93
    uranyl salt calibrators, **305**:94–95
  internal calibration, **305**:71
  light calibrators, **305**:70–71
  standardization of standards, **305**:49–
    50, 95–96
  system performance check, **305**:70
detection, *see* Avalanche photodiode;
  Charge-coupled device array; Photomultiplier tube
dynamic range and sensitivity, **305**:69–70
glow versus flash detection, **305**:73–74, 97
imaging, *see* Luminescence imaging
manufacturers, **305**:98–103
measuring chamber, **305**:65–66

microplate luminometers, *see also* Luminescence imaging
  cross-talk, **305**:79–81, 97
  injection of reagents, **305**:81, 97–98
  overview, **305**:62, 67, 78–79, 97
  multifunctional instruments, **305**:81–82
  practical hints, **305**:73
  reagent injector, **305**:66–67
scintillation counters as luminometers,
  305:449–450
selection of instrument
  ATP measurements, **305**:75–76
  DNA probe techniques, **305**:77
  enzyme and metabolite assays, **305**:76
  factors in instrument selection, **305**:73
  immunoassays, **305**:76–77
  phagocytosis assay, **305**:76
  reporter gene assays, **305**:75, 449
  sensitivity determination, **305**:71–72
single sample tube luminometers
  overview, **305**:77
  portable instruments, **305**:77–78
LumP, *see* Lumazine proteins
Lumry–Eyring model, protein aggregation,
  309:242–244
Lumry theory, application to enzymes,
  295:344–347
Lutein, *see* Carotenoids
Luteinizing hormone-releasing hormone,
  control of release, *see* Galanin; Neuropeptide Y
lux control system, *Vibrio*, *see also* Bacteria
  bioluminescence
autoinducer activity assays with *Esche-*
  richia coli sensor strains
  autoinducer specificity, **305**:286
  lux gene inhibition assay, **305**:285–286
  lux gene stimulation assay,
    305:284–285
  principles, **305**:283–284
  reporter plasmid design, **305**:279–280,
    282
  test sample preparation, **305**:284
autoinducers, *see* Acylhomoserine lactone autoinducers
cyanobacteria incorporation of *lux* genes
  aldehyde supply
    endogenous, **305**:520
    exogenous, **305**:518, 520
  biosensing application, **305**:516–517

circadian rhythm studies
  hit-and-run allele replacement,
    305:533–536
  overview, **305:**516, 527–528
  random insertion of *luxAB*,
    305:541–542
  transposon mutagenesis of *Synecho-
    coccus* sp. strain PCC 9742 using
    Escherichia coli conjugation,
    305:536–537, 539, 541
  vector, **305:**528–529, 531, 533
constructs, **305:**518–519
luciferase assays *in vivo*
  colonies, **305:**523–524
  comparison with other reporters,
    305:526–527
  heterocyst considerations,
    305:525–526
  imaging of single cells, **305:**524–525
  suspended cells, **305:**521, 523
  reporters of gene expression during dif-
    ferentiation, **305:**514–516
genes and proteins, **305:**513
Lux protein screening, **305:**287
quorum sensing response, **305:**279–281
recombinant protein expression in *Escherichia coli*
  advantages, **305:**315, 329
  cell lysis, **305:**328–329
  culture conditions, **305:**315–316, 320–
    321, 326–328
  firefly luciferase, **305:**181–183
  gene topology and expression, **305:**320
  promoter characteristics, **305:**316–318
  vector
    characteristics, **305:**318–320
    construction, **305:**321–325
    polymerase chain reaction amplifica-
      tion of gene, **305:**322, 324–325
Vibrio harveyi system comparison with
  Vibrio fischeri, **305:**287
Lycopene, *see also* Carotenoids
  photolysis, **299:**419–420
  simultaneous determination with vitamin
    E, vitamin A, and xanthophylls with
    reversed-phase high-performance liq-
    uid chromatography
    calibration, **299:**352
    extraction from plasma and serum,
      299:352–354, 356

instrumentation and chromatography
  conditions, **299:**350–352
peak identification, **299:**357–360
quality assurance, **299:**353, 356–357,
  361–362
reagents and solutions, **299:**349
stability of samples, **299:**360–361
standards, **299:**350
Lycophene β-cyclase, gene cloning and carotenoid biosynthesis, **297:**246–247,
  252–254
Lycophene ε-cyclase, carotenoid biosynthesis, **297:**254–255
Lysenin, resistant cell variant selection,
  312:308–310
Lysine, caged protein conjugates
  G-actin
    fluorescence photoactivation micros-
      copy, **291:**104–105
    functional analysis of caged and de-
      caged proteins, **291:**105–107
    labeling ratio, **291:**101–102
    photoactivation conditions, **291:**104
    polymerization competence,
      291:102–103
    preparation of 6-nitroveratryloxycar-
      bonyl chloride conjugate, **291:**101
    tetramethylrhodamine labeling of
      caged protein, **291:**103
  inactivation of proteins, **291:**98–99
  photolytic mechanism, **291:**98
  preparation of 6-nitroveratryloxycarbonyl
    chloride conjugates
    4-dimethylaminopyridine salt prepara-
      tion, **291:**100
    materials, **291:**97
    quantification of conjugation,
      291:99–101
  profilin
    functional analysis of caged protein,
      291:108–110
    preparation of 6-nitroveratryloxycar-
      bonyl chloride conjugate, **291:**108
    recombinant protein purification,
      291:107–108
Lysosphingomyelin
  chemical synthesis
    [^{14}C-choline]lysosphingomyelin synthe-
      sis, **311:**512–513
    lysosphingomyelin, **311:**533

enzymatic synthesis with sphingolipid cer-
    amide *N*-deacylase, **311:**302–303
functions, **311:**297–298
Lysozyme
    hydrogen exchange measured with elec-
        trospray mass spectrometry
        exchange properties of wild-type and
            variant proteins, **309:**639
        preparation of samples, **309:**637
    organic solvent transfer of crystals,
        308:205
    phage T4 enzyme, temperature-invariant
        enthalpy in unfolding, **295:**258–261
    refolding assay for protein disulfide iso-
        merase, **290:**51–52

M

M2
    amantadine inhibition, **294:**491
    expression in mouse erythroleukemia
        cells
        gene cloning and amplification, **294:**494
        plasmid preparation, **294:**495
        rationale, **294:**493
        time course of expression, **294:**493–494
        transfection, **294:**495
        Western blot analysis, **294:**495–496
        whole cell patch-clamp analysis
            background permeability,
                294:497–498
            cell-to-cell variation, **294:**498–499
            combined measurements with fluo-
                rescent pH indicators,
                294:505–506
            current/voltage relation, **294:**499–501
            ion selectivity, estimation from re-
                verse potentials, **294:**501,
                503–505
            overview, **294:**496
            perfusion, **294:**497
            single-channel parameter estimation,
                294:505
            solution composition, **294:**496–497
    function in influenza A virus infection,
        294:490–491
    molecular modeling, **293:**664–666, 668
    permeability characteristics, **294:**492–493,
        501, 503–505

structure, **294:**491
MAD, *see* Multiwavelength anomalous dif-
    fraction
MAG, *see* Myelin-associated glycoprotein
Magnesium
    affinity for caged ATP, **291:**321, 324,
        346–347
    effects on caged GTP infrared spectra,
        291:229–231
Magnetic bead, assay for microtubule–
    membrane interactions
    CLIP-170 identification, **298:**382
    cytosol, preparation and manipulation,
        298:386–387
    endosome preparation and manipulation,
        298:385–386
    microtubule affinity matrix on magnetic
        beads, preparation, **298:**383–385
    organelle binding to microtubule beads,
        298:387–390
    principle, **298:**382–383
Magnification, calibration for confocal mi-
    croscope, **307:**113, 118
Major histocompatibility peptide transport
    system, *see* Transporters associated
    with antigen processing
Malachite green, dye-binding assay for
    ATP hydrolysis, **292:**38–39, 519–520
Malate dehydrogenase–citrate synthase, sub-
    strate channeling, **308:**142–143
MALDI-MS, *see* Matrix-assisted laser
    desorption/ionization mass spec-
    trometry
4-(*N*-Maleimido)benzophenone, *see* Cross-
    linking
m-Maleimidobenzoyl-*N*-hydroxysuccinimide
    ester, antigen coupling to carrier,
    296:391–392
MalF–β-galactosidase fusion protein, disul-
    fide bond formation defect in *Esche-
    richia coli*, detection, **290:**71
Malondialdehyde
    gas chromatography–negative-ion chemi-
        cal ionization mass spectrometry
        assay
        derivatization, **300:**72–73
        gas chromatography, **300:**75–76
        internal standard synthesis, **300:**72
        malonaldehyde standard preparation,
            300:72

sample preparation
  cultured cells, **300:**71
  organs, **300:**71–72
  plasma, **300:**71
  sperm, **300:**71
  standard curve generation and calculations, **300:**74–75
lipid peroxidation marker, **300:**70
thiobarbituric acid assay, **300:**70–71, 78
Maltose transporter, *Escherichia coli*
ATP-binding protein
  ATPase activity and assays, **292:**9, 17–18, 27–29, 37–39
  membrane topology, **292:**8–9
  sequence homology with other proteins, **292:**3, 7–8
domain organization, **292:**5–6
inner membrane transport complex purification
  bacteria growth, **292:**22
  high-performance liquid chromatography, **292:**24
  solubilization, **292:**23–24
  vesicle preparation, **292:**23
ligand transport
  assay systems
    accumulation in intact cells, **292:**13–14
    ATP addition to inside-out vesicles, **292:**16
    binding protein addition to cells from external medium, **292:**14
    binding protein addition to right-side-out vesicles, **292:**15
    substrate uptake in proteoliposomes, **292:**16–17, 27–29
  mechanism, **292:**18–20
maltose-binding protein
  binding modes, **292:**10–11, 21
  functional exchange with UgpC, **292:**49–50
  mutational analysis, **292:**11–13, 21
  purification, **292:**24–26
membrane channel proteins, **292:**6
mutants with uncoupled ATPase activity
  ATP hydrolysis assays, **292:**37–39
  selection for restored transport in *malE* strain, **292:**33–34
  transport property characterization, **292:**34, 36–37

reconstitution into proteoliposomes, **292:**16–17, 21, 26
subunits, **292:**5, 21, 31–32
Mammalian cytosolic chaperonin, *see* Cytosolic chaperonin
Mammalian mitochondrial chaperonin 10
  biological functions, **290:**220–221
  GroEL binding, **290:**220, 229–230
  GroES homology, **290:**220
  oligomeric structure, **290:**227–228
  protein folding assay with chaperonin 60, **290:** 214
  purification of recombinant mouse protein from *Escherichia coli*
    cation-exchange chromatography, **290:**223, 226
    cell growth and optimization of expression, **290:**222, 224
    extraction, **290:**224
    hydroxylapatite high-performance liquid chromatography, **290:**226
    materials, **290:**221
    plasmid constructs, **290:**223–224
    polyacrylamide gel electrophoresis, **290:**222–223
    size-exclusion chromatography, **290:**226–227
  Rubisco refolding assay with chaperonin 60, **290:** 227–230
Mammalian mitochondrial chaperonin 60
  GroES binding, **290:**204
  oligomeric structure, **290:**214–215, 217
  protein folding assay with chaperonin 10, **290:** 214
  purification of recombinant protein from *Escherichia coli*
    anion-exchange chromatography, **290:**206–208
    cell growth and lysis, **290:**205–206
    dissociation of oligomer, **290:**204–207
    expression system, **290:**205
    materials, **290:**205
    polyacrylamide gel electrophoresis, **290:**206
    reconstitution of oligomers
      ATP dependence, **290:**209, 211
      reaction conditions, **290:**211
      size-exclusion chromatography, **290:**211
      storage, **290:**212

size-exclusion chromatography, **290:**209
Rubisco binding, **290:**204, 214
stability of *in vitro*-reconstituted particles,
    290:212, 214, 217
MAP 1A, *see* Microtubule-associated protein 1A
MAP4, *see* Microtubule-associated protein 4
MAPK, *see* Mitogen-activated protein kinase
MARCKS, *see* Myristoylated alanine-rich C kinase substrate
Marcus rate theory, *see* Proton transfer
Markov model, ion channel kinetics
    comparison to experimental data,
        293:734–735
    comparison to non-Markov models,
        293:724–725
    data requirements for model development
        channel class homogeneity,
            293:729–730
        voltage clamp fidelity, **293:**730–731
    digitization of time courses, **293:**396
    dwell time simulation, **293:**733
    heuristic approach in kinetic model development, **293:**725–728
    hidden Markov model, *see* Hidden Markov model
    number of states, estimation, **293:**437–438, 731–732
    probability function
        analytical solution, **293:**396–397
        equilibrium distribution, **293:**397
        numerical solution, **293:**396
    scopes of kinetic models, **293:**729
    simulation software
        commercial programs, **293:**736
        custom package for implementation in
            IGORPRO
            advantages, **293:**736–737
            model creation, **293:**738–739
            routines description, **293:**739–740
            running simulation, **293:**739
            source code, **293:**737, 741–759
            theory of operation, **293:**737
    single-channel time series simulation,
        293:733–734
    statistical approach in kinetic model development, **293:**725–726, 728

time homogeneity assumption, **293:**724–725, 737
vector of state probabilities, **293:**395
voltage and concentration jumps to isolate specific rate constants,
    293:732–733
Mass spectrometry, *see also* Electrospray ionization mass spectrometry; Gas chromatography–mass spectrometry; Isotope dilution negative-ion electron capture mass spectrometry; Matrix-assisted laser desorption ionization–mass spectrometry; Negative-ion chemical ionization mass spectrometry
    acylhomoserine lactone autoinducer structure elucidation, **305:**296–297, 312
    advanced glycation end products,
        309:158, 170
    amyloid-β peptide purity analysis,
        309:192, 472
    boranophosphate oligonucleotides,
        313:254–255, 257
    calcium-binding proteins
        binding protein identification, **316:**120
        identification of calcium-binding proteins, **316:**108–109
    ceramide, **312:**23
    cholesterol ozonation product detection,
        319:196
    delayed ion extraction matrix-assisted desorption ionization time-of-flight mass spectrometry of sphingolipids
        gala-series glycosphingolipids, **312:**88, 91
        ganglio-series glycosphingolipids,
            312:91, 93, 96, 99
        globo-series glycosphingolipids, **312:**91
        GM1, 93, 96
        GM3, 89–90, 93
        identification of individual species,
            312:81
        lysoglobotriaosylceramide, **312:**86
        lysolactosylceramide, **312:**84
        lysosphingolipid
            mass spectrometry, **312:**87–88
            preparation, **312:**81, 85, 87
        lysosulfatide, **312:**83
        multiple sphingolipid extraction and analysis from single samples
            clinical applications, **312:**99

Fabry disease necrotic tissues, **312:**100–101

human platelets, **312:**100

principle, **312:**80–81

psychosine, **312:**82

sphingomyelin, **312:**92, 99

DNA, long-range charge transport and oxidative damage, **319:**179

electron impact ionization of sphingolipids, **312:**33

electrospray ionization of sphingolipids, **312:**35–36

Far-Eastern blotting coupling, **312:**152–154

fumonisin B$_1$, **311:**363–364

ganglioside structure analysis with fast atom bombardment, **311:**89–91

high-performance thin-layer chromatography coupling, **312:**116, 132–134

hydrogen exchange of amyloidogenic proteins measured with electrospray mass spectrometry

advantages over nuclear magnetic resonance techniques, **309:**634–635

data acquisition, **309:**638

data analysis, **309:**638–639

exchange reaction, **309:**637–638

insulin

exchange properties, **309:**643, 645

preparation of bovine samples, **309:**637

localization of exchange site with peptic digestion, **309:**635

lysozyme

exchange properties of wild-type and variant proteins, **309:**639

preparation of samples, **309:**637

mass distribution of conformations, **309:**634–635

principle, **309:**633–634

prospects of technique, **309:**645–646

sample introduction for mass spectrometry, **309:**635–636

transthyretin

exchange properties of wild-type and variant proteins, **309:**639, 641

monomer analysis, **309:**641, 643

preparation of samples, **309:**637

microsequencing analysis of U4/U6U5 proteins, **318:**395

minimally modified antisense oligonucleotides, **313:**272

myriocin, **311:**358–359

nitrated lipids after liquid chromatography separation, **301:**463, 468–469

2-nitrobenzyl caged protein analysis, **291:**119, 124

nomenclature of complex sphingolipid product ions, **312:**34

peptide mapping by electrospray ionization–mass spectrometry, **301:**106

peptide mass fingerprinting of retinal proteins separated by two-dimensional gel electrophoresis

applications, **316:**494–495, 510–511

in-gel digestion

protease selection, **316:**502–503

trypsin digestion, **316:**503–504

matrix-assisted laser desorption ionization–time of flight analysis

criteria for identification using MS-Fit, **316:**508–509

database searching, **316:**505–506

equipment, **316:**505

large fragment fingerprinting, **316:**509–510

principle, **316:**504–505

spectra acquisition, **316:**505

prospects, **316:**510–511

phospholipids, fast-atom bombardment–mass spectrometry, **300:**64–67, 69

α_2-Macroglobulin receptor/low-density lipoprotein receptor-related protein, photodynamic therapy response, **319:**356

phosphopeptides

data analysis, **292:**350–353

electrospray ionization, **292:**348, 355

sensitivity, **292:**348

sequencing, **292:**347

phosphoramidite-modified oligoribonucleotides, **317:**52

photoaffinity-labeled peptides, **311:**596

prenylation analysis

gas chromatography–mass spectrometry analysis of cleaved prenyl groups, **316:**444–445

high-performance liquid chromatography–electrospray ionization mass spectrometry of peptides, **316:**448–449, 451

prion protein, characterization of purity and isotope label, **309:**581–584

rhodopsin phosphopeptide analysis, **316:**485, 487–488, 492

RNA–protein cross-link site analysis, *see* Matrix-assisted laser desorption/ionization mass spectrometry

soft ionization techniques for sphingolipids, **312:**33–35

sphingolipids, tandem mass spectrometry
advantages over single scans, **312:**36, 43–44
ceramides, **312:**39–40
constant neutral loss scan, **312:**37–38
free sphingoid bases, **312:**38–39, 42
glucosylceramides, **312:**40–42
high-performance liquid chromatography coupling, **312:**35–36, 45
multiple reaction monitoring, **312:**38, 44
precursor ion scan, **312:**36–37
product ion scan, **312:**36
sphingomyelin, **312:**42

sphingosine 1-phosphate, **312:**10–11

Mass tagging, subunit stoichiometry determination, **293:**26–27

Mass transfer-limited bioavailability, biofilm nutrients
biofilm coupons
counting of bacteria, **310:**402
glass surface, **310:**397–398
hydrophobic organic compound aging, **310:**399
iron oxide deposition, **310:**398
microscopy, **310:**401–402
natural organic matter deposition, **310:**398–399
preparation, **310:**398
storage, **310:**399
dimensionless parameters in substrate flux studies, **310:**417–418
equations, **310:**394
interphase mass transfer modeling
bioenhancement, **310:**425–426
dimensionless parameters, **310:**422–425
external mass transport resistance, **310:**421
flux equation, **310:**421
pollutant bioavailability analysis, **310:**421–422

intraphase mass transfer modeling with dimensionless parameters, **310:**419–421
overview, **310:**394–395
reactor system for study
design, **310:**395–397
fluid residence time, **310:**400
inoculation, **310:**399–400
nutrient feed, **310:**401
operation, **310:**400–401

Mast cell
inflammatory mediator production, **301:**22–23
ruthenium red assay of activation
advantages and limitations, **301:**26–27
principle, **301:**23–24
quantitative analysis, **301:**24, 26

Matrix-assisted laser desorption ionization–mass spectrometry
carotenoids
extraction, **299:**391–392
fragment ion mass analysis, **299:**394–395, 398–401, 406–408
green lettuce analysis, **299:**403, 406
instrumentation, **299:**392–393
molecular ion species and sensitivity, **299:**393–394
overview of ionization techniques, **299:**390–391
postsource decay, **299:**391, 393
standards and samples, **299:**391
tangerine juice concentrate analysis, **299:**401, 403

macromolecular characterization, **318:**438–439

RNA–protein cross-link site determination
amino terminal sequence analysis with Edman degradation, **318:**445
ammonium hydroxide-hydrolyzed complex, mass analysis, **318:**443–444
data acquisition, **318:**442
overview, **318:**439–441
phosphodiesterase-hydrolyzed complex, mass analysis, **318:**442, 444–445
prospects, **318:**445–446
reversed-phase high-performance liquid chromatography of complexes, **318:**441–442
sample preparation, **318:**442

total complex, mass analysis,
    318:442–443
sequencing applications, **318:**439
Maximum velocity, carrier-mediated
    transport
derivation, **296:**270
interpretation, **296:**272–273
turnover number determination, **296:**273
Maxi-potassium channel, *see* Charybdotoxin
    receptor
MBS, *see m*-Maleimidobenzoyl-*N*-hydroxy-
    succinimide ester
McGhee–von Hippel binding isotherm,
    anticooperative binding analysis,
    295:365–366
MD, *see* Molecular dynamics
MDM2, inhibition and p53 induction,
    314:167
MDR, *see also* MDR1; *MDR1*; *mdr1*, mu-
    rine; *mdr2*, murine; *mdr3*, murine; P-
    glycoprotein; Multidrug-resistant cells
associated protein, *see* Multidrug-resis-
    tance associated protein
lipid substrates, **312:**564
subfamily in *Saccharomyces cerevisiae*,
    292:155–158
MDR1, *see also* P-glycoprotein
assays of expression and drug transport
colony assay for rhodamine-123 efflux,
    292:246–247
fluorescence-activated cell sorting,
    292:239, 242–243, 453, 469–471,
    508–509, 531
indirect immunofluorescence and flow
    cytometry, **292:**233–234
radiolabeled drug accumulation,
    292:471–472, 500–503
ATPase
assays, **292:**438–441, 467–469, 489–490,
    499–500, 507
baculovirus–insect cell protein charac-
    terization, **292:**438–441
characterization of expressed protein
    from rodent cultured cells
activity dependence on ATPase ex-
    pression level, **292:**511–512
contaminating ATPase inhibition,
    292:510–511, 513
crude membrane preparation,
    292:505–506

inhibitor analysis, **292:**509–511
inorganic phosphate colorimetric
    assay, **292:**507
phosphorylation-deficient mutants,
    292:512–513
kinetic parameters, **292:**397–398
purified or reconstituted proteins,
    292:499–500
site-directed mutant characterization,
    292:490, 492, 505
dihydrofolate reductase fusion protein
mammalian cell transfection library of
    mutants, **292:**479–480
methotrexate resistance in selection,
    292:475, 477
polymerase chain reaction mutagenesis,
    292:478–479
vector construction, **292:**476–477
drug specificity, **292:**226–227, 492–493
expression in baculovirus–insect cell
    system
advantages, **292:**428–429
ATPase characterization, **292:**438–441
cell infection and culture, **292:**434–435
histidine-tagged protein
expression system, **292:**319–320
insect cell membrane preparation,
    292:320–321
metal affinity chromatography,
    292:321–322
membrane isolation, **292:**436–437
recombinant baculovirus construction
amplification of virus, **292:**433–434
calcium phosphate tranfection,
    292:432
DNA preparation, **292:**431–432
purification of virus, **292:**433
site-directed mutants, **292:**483–484
vector construction for expression,
    292:429–431
expression in *Escherichia coli*,
    292:380–381
expression in vaccinia virus–T7 RNA
    polymerase system
advantages, **292:**442–443, 455, 457
characterization of expressed protein,
    292:453, 463–473
disadvantages, **292:**454–455, 473
efficiency, **292:**443–444, 455–457
infection of cells, **292:**452–453, 460–462

membrane preparation, **292**:462–463
plasmid construction, **292**:444, 446,
  459–460
principle, **292**:443
recombinant virus screening,
  292:449–450
safety issues, **292**:445–446, 453
transfection with lipofectin, **292**:447–
  448, 461
virus amplification and purification,
  292:450–451
virus generation, **292**:444, 446–447
virus isolation by plaque purification,
  292:444, 448–450
virus titer determination,
  292:451–452
gene, *see MDR1*
heterologous expression systems, overview, **292**:473
monoclonal antibodies
  applications
    detection of drug-resistant cells,
      292:264
    fluorescence-activated cell sorting of
      expressing cells, **292**:469–470
    gene cloning, **292**:263
    modulation of transport,
      292:263–264
    protein purification, **292**:263
    selective killing of cells, **292**:264–265
  generation, **292**:260–263
mutation and drug transport specificity,
  292:226–227
phosphorylation
  ATPase activity of mutants,
    292:512–513
  confirmation of site phosphorylation by
    site-directed mutagenesis,
    292:341–342
  peptide mapping
    amino acid sequence of Lys-C pep-
      tide, **292**:334–335
    proteolysis following reduction and
      carboxamidomethylation,
      292:332–333
    reversed-phase high-performance liq-
      uid chromatography,
      292:333–335
    starting material requirements,
      292:328–330

tryptic phosphopeptides, purification
  and sequences, **292**:336, 338–340
protein kinase C phosphorylation reaction, **292**:329–331, 341
purification of phosphorylated protein
  by gel electrophoresis,
  292:331–332
photoaffinity labeling in drug-binding site
  identification
  amino acid sequence analysis, **292**:312
  assumptions, **292**:290, 308–309
  chimeric protein studies, **292**:316–317
  denaturing polyacrylamide gel electro-
    phoresis, **292**:297–298, 310, 324
  immunological mapping
    high-resolution mapping,
      292:314–315
    low-resolution mapping, **292**:313–314
    principle, **292**:312–313
  immunoprecipitiation, **292**:297–298
  inhibition with unlabeled drugs,
    292:298–300
  intact cell labeling, **292**:466
  membrane labeling, **292**:467
  membrane preparations, **292**:311
  peptide mapping, **292**:305–306, 312
  probe selection, **292**:308–309
  site-directed mutagenesis, **292**:304–305,
    315–316
  synthesis of analogs and labeling
    ATP, **292**:464–465
    colchicine, **292**:293–294, 300–301
    daunorubicin, **292**:294, 301
    dihydropyridine, **292**:294, 302, 304,
      309–311, 437–438, 465–467
    doxorubicin, **292**:294, 301
    forskolin, **292**:309
    labeling reactions, **292**:296–297, 310,
      322, 324, 465–467
    phenothiazines, **292**:302–303
    prazosin, **292**:305–306, 309, 313–314,
      322, 324–325, 327, 437–438,
      465–467
    verapamil, **292**:294–296, 302
    vinblastine, **292**:291, 293, 299–300,
      304
  vanadate trapping, effects on labeling,
    292:325–328
purification from multidrug-resistant KB-
  V1 cells

anion-exchange chromatography,
    292:495–496, 498, 503
characterization of purified protein,
    292:499–504
orientation determination in proteolipo-
    somes, **292:**499
plasma membrane vesicle isolation,
    292:494–495
reconstitution, **292:**496–497, 499
solubilization, **292:**495
storage, **292:**503
Western blot analysis, **292:**497–499
wheat germ agglutinin affinity chroma-
    tography, **292:**495–496, 498, 503
site-directed mutagenesis
    applicability of MDR1 to mutagenesis
        studies, **292:**480–481
    assay for glycosylation-defective mu-
        tants, **292:**486–487
    baculovirus–insect cell system expres-
        sion, **292:**483–484
    overview of mutagenesis, **292:**481–482
    stable expression in mammalian cells,
        292:482–483, 505
    transient expression in human embry-
        onic kidney cells
        advantages, **292:**484
        ATPase assay of reconstituted pro-
            tein, **292:**489–490, 492
        purification of histidine-tagged mu-
            tants, **292:**487–489
        transfection, **292:**484–485
    Western blot analysis, **292:**463–464,
        507–508
MDR1
    gene therapy, *see* Gene therapy, *MDR1*
    genetic suppressor elements
        definition, **292:**225
        isolation from retroviral expression li-
            braries
            functional testing in selection,
                292:239, 242–243, 246–247
            library construction, **292:**239–242
            overview, **292:**227–228, 238–239
            recovery of complementary DNA,
                292:243–246
    mutant isolation
        drug selection and recovery of comple-
            mentary DNA inserts,
            292:234–237

infection of recipient cells, **292:**233
mutagenized plasmid preparation,
    292:230–231
overview, **292:**229–230
sequencing of clones, **292:**237–238
transfection of packaging cells, **292:**232
protein, *see* MDR1
RNA mapping for antisense sequence se-
    lection, **314:**183
transduction into bone marrow cells,
    292:255–257
transgenic mice
    actin promoter and stability of expres-
        sion, **292:**573–575
    Eμ transgenic mice production, over-
        view, **292:**572, 575–576
    expression assays
        immunofluorescence and fluores-
            cence-activated cell sorting,
            292:580–581, 583–584
        Northern blot analysis of RNA,
            292:580
        ribonuclease protection assay,
            292:581–582
        Western blot analysis, **292:**584
    functional assays
        drug resistance measurements in
            white blood cells, **292:**589–593
        drug sources, preparation, and doses,
            292:586, 588–589
        rhodamine-123 efflux, **292:**584–586,
            593
    homozygous transgenic mouse identifi-
        cation, **292:**578–580
    protection against toxic substrates,
        292:546, 572, 589–593
    reversal agent testing, **292:**588, 593
    screening of founder mice, **292:**577–578
    strain selection, **292:**576
    tissue expression and targeting,
        292:574, 582–583
mdr1, murine
    expression in *Escherichia coli*
        advantages, **292:**370–371
        bioenergetics analysis, **292:**379–382
        drug susceptibility testing, **292:**376–377
        functional assays
            leaky mutant generation for outer
                membrane permeability,
                292:375–376

photolabeling, **292:**373–374
transport assay with tetraphenylar-
sonium, **292:**374–375
limitations, **292:**371–372
plasmids, **292:**372
reversal of drug resistance,
292:377–378
strains and growth conditions,
292:372–373
topology studies, **292:**378–379
expression in yeast, *see* Secretory vesicle,
Saccharomyces cerevisiae
genes, **292:**290
knockout mice, **292:**573
linker region of protein
expression and purification of recombinant Mdr1a, **292:**353
mass spectrometry of phosphopeptides
data analysis, **292:**350–353
electrospray ionization, **292:**348, 355
sensitivity, **292:**348
sequence analysis, **292:**347
peptide mapping, **292:**353–355
phosphorylation *in vitro* with kinases,
292:353
sequences, **292:**343–344
site-directed mutagenesis of phosphorylation sites, **292:**355, 357–358
phosphorylation of protein
cyanogen bromide digestion,
292:344–347
labeling *in vitro* with kinases, **292:**344–
346, 350
peptide mapping, **292:**348–349
phosphorous-32, metabolic labeling,
292:345–346
mdr2, murine
expression in yeast, *see* Secretory vesicle,
Saccharomyces cerevisiae
phosphatidylcholine translocation assay
in yeast secretory vesicles
fluorescent phosphatidylcholine,
292:391–392
principle, **292:**391–393
reaction conditions, **292:**395–396
secretory vesicle preparation with
outer leaflet label, **292:**393, 395
mdr3, murine
ATPase
assay, **292:**402–403

contaminating ATPase inhibition,
292:411–412
kinetic parameters, **292:**410–411
pH optimum, **292:**406–407
valinomycin stimulation, **292:**409
vanadate inhibition, **292:**410–411
verapamil stimulation, **292:**408–409
vinblastine stimulation, **292:**409
expression in yeast
Pichia pastoris
advantages, **292:**413
alcohol oxidase promoter in overexpression, **292:**399, 403, 412–413
contaminating ATPase inhibition,
292:411–412
efficiency of expression system,
292:412–413
induction, **292:**404
membrane preparation, **292:**400–401,
406–407
plasmid construction, **292:**399, 403
screening of clones, **292:**403–404
transformation, **292:**399–400
Saccharomyces cerevisiae, see also Secretory vesicle, *Saccharomyces cerevisiae*
chimeric amino-terminal nucleotide-
binding domain protein construction, **292:**415–421
membrane isolation, **292:**425–426
site-directed mutagenesis,
292:417–418
functional assay by Chinese hamster
ovary cell transfection, **292:**421–423
functional replacement of *STE6*,
292:134–135, 157, 159, 194, 211, 382,
398, 421, 423–425
photoaffinity labeling of protein with io-
doarylazidoprazosin, **292:**401–402,
404–406
solubilization and reconstitution of protein in *Escherichia coli* lipids,
292:402, 407–408, 412
Mechanosensitive channels
Escherichia coli, see MscL
gene identification in *Caenorhabditis elegans*, **294:**459
overview, **294:**458–459
pressure clamp
construction tips, **294:**487–489

electronic control, **294:**485, 487
mechanical arrangement, **294:**483–485
principle, **294:**482–483
system volume minimization,
  294:484–485
transducer/mixer chamber, **294:**485
valves, **294:**483–484
Saccharomyces cerevisiae channels, patch-
  clamp analysis, **294:**511–514
Medical device implant, *see also* Bladder
  catheter biofilm; *Candida* biofilm; Prosthetic joint biofilm
biocompatibility, **310:**534–535
biofilm culture
  continuous culture flow cell,
    310:540–541
  quantitative recovery of cells,
    310:542–543
  semibatch culture, **310:**539
biofilm elimination concentration evaluation, **310:**543–544
complexity of biofilms, **310:**577–578
electron microscopy of biofilms
  scanning electron microscopy, **310:**551
  transmission electron microscopy,
    310:548–551
epifluorescence microscopy of biofilms
  cell counting with acridine orange,
    310:545–548
  viability assay with 5-cyano-2,3-ditoyl-
    tetrazolium chloride, **310:**548
infection
  sources, **310:**535–536
  treatment, **310:**536–537
market, **310:**535
perioperative sampling for bacteria recovery
  endoscopic brush technique,
    310:538–539
  Maki roll technique, **310:**537–538
Medical Internal Radiation Dose, *see* Oligonucleotide radiotherapy
MEL, *see* Murine erythroleukemic cells
Melanophore, *see* Melanopsin
Melanopsin
function, **316:**279
gene cloning, **316:**279–282
G protein coupling, **316:**294
tissue distribution in amphibians, **316:**291
Xenopus melanophore

advantages and prospects for system,
  316:308–309
isolation from embryos
  dissociation of tissue, **316:**300
  Ficoll gradient purification,
    316:300–301
  maintenance of cultures, **316:**301–302
  materials, **316:**299–300
  primary culture, **316:**300
microtiter plate reader absorbance
  assay of responses
  data acquisition, **316:**308
  illumination artifacts, **316:**306
  materials, **316:**305–306
  sample preparation, **316:**307–308
phenotype switching in development,
  316:293–294
response to light and scoring,
  316:291–293
transgenic embryo preparation for melanopsin expression
  materials, **316:**296–297
  nuclear transplantation, **316:**298–299
  overview, **316:**294–296
  restriction-enzyme-mediated integra-
    tion, **316:**298
  sperm and egg preparation,
    316:297–298
  transgene incorporation analysis,
    316:302
video microscopy assay of light response
  image acquisition, **316:**303–304
  materials, **316:**302–303
  transgenic melanophore responses,
    316:304–305
Melanosome
electron microscopy characterization,
  298:365
function, **298:**361, 389
motility assay in cell-free *Xenopus*
  system
  axoneme preparation, **298:**368–369
  cyclic AMP effects, **298:**362
  substrate preparation, **298:**369
  video-enhanced differential interfer-
    ence contrast microscopy,
    298:369–371
motor proteins and Western blotting,
  298:362, 367–368, 372

protein composition analysis by denaturing gel electrophoresis, **298:**366–367
purification from cultured melanophores
cell culture, **298:**362–363
cell harvesting, **298:**363
density gradient centrifugation, **298:**364, 372
homogenization, **298:**363–364
yield, **298:**364–365
vesicle transport reactivation in lysed teleost melanophores
digitonin lysis and reactivation, **298:**390, 393–396
melanophore preparation from scales
epidermis removal, **298:**391–393
sources, **298:**390–391
Nonidet P-40 lysis and reactivation, **298:**390, 396–397
quantitative measurements of pigment granule transport, **298:**399
troubleshooting reactivation, **298:**398–399
Melittin
aggregation, **295:**81, 86
partitioning–folding coupling in membranes
helix formation, **295:**63–64
thermodynamics, **295:**64–65
Membrane bilayer, *see also* Liquid–crystalline phase transition; Planar lipid bilayer
carriers versus channels as conformation reporter proteins, **294:**216–217
deformation energy, **294:**214–216
fluid mosaic membrane model, limitations, **294:**208–209
gramicidin as molecular force transducer
changes in membrane deformation energy, **294:**220–222
channel lifetime measurement, **294:**222–223
heterodimer formation approach, **294:**223
membrane conductance measurements, **294:**220–222
monomer–dimer equilibrium, **294:**217–218, 220
phenomological spring constant determination, **294:**221, 223

perturbation and channel function, **294:**211–212, 214
protein conformational changes and bilayer perturbations, **294:**210–211, 223–224
Membrane biofouling, *see* Biofouling, membranes
Membrane-confined analytical electrophoresis
absorbance profiles, acquisition, **295:**500–501
apparatus
accuracy and precision of absorbance readings, **295:**497–498
electrophoresis cell, **295:**498–499
overview of design, **295:**496
buffer flow, **295:**499
bulk fluid flow, error source and detection, **295:**514–515
concentration gradient instability, **295:**515
diffusion coefficient determination
absence of electric field, **295:**507–508
coupled flow effects, **295:**511, 513
nonzero electric field, **295:**508–511
effective charge, evaluation, **295:**503, 505–506
electric field, establishing and measuring, **295:**499–500
electrophoretic mobility, evaluation, **295:**494–495, 502–503
information available from experiments, **295:**494–496
simulation, **295:**516–518
theory, **295:**501–502
Membrane fusion, *see* Rod outer segment
Membrane potential
assessment in disinfected biofilms, **310:**176–177
fluorescent probes, **307:**131–132
Membrane–protein interaction, *see* Partitioning, proteins in membranes
Membrane trafficking, rhodopsin
carboxy-terminal domain role, **315:**79, 88
cell-free frog retinal system for post-trans-Golgi network trafficking assay
applications, **315:**77–79, 88
electron microscopy analysis of membrane morphology, **315:**86, 88
polyacrylamide gel electrophoresis

analysis of ATP-dependent transfer, **315:**86

post-Golgi carrier membrane formation, **315:**81, 83–85

pulse-labeling of proteins in cultured retina, **315:**79–80

rod outer segment removal and enrichment, **315:**80–81

sucrose density gradient fractionation of postnuclear supernatant, **315:**85–86

Western blot analysis, **315:**86

fractionation studies, **315:**77

opsin fragments, **315:**69

MEP, *see* Molecular electrostatic potential

MEQ, *see* 6-Methoxy-*N*-ethylquinolinium iodide

Mercapto *N*-oxides, photolysis in hydroxyl radical generation, **300:**195–197

Mercury

mercuration of pyrimidine C-5 position, **317:**127

RNA phosphorothioate incorporation for mercury binding, **317:**126, 161–162, 184

Meromyosin, *see* Heavy meromyosin

Messenger RNA

abundance, estimation in neurons, **303:**3

differential display, *see* Messenger RNA, differential display

differential expression analysis, *see* Complementary DNA representational difference analysis; Differential display; DNA array; DNA microarray; Restriction endonucleolytic analysis of differentially expressed sequences; RNA-arbitrarily primed polymerase chain reaction; Sequencing by hybridization; Subtractive hybridization; Targeted display

inosine bases, *see* Inosine-containing messenger RNA

protein fusion for selection, *see* RNA–protein fusion

reverse transcription–polymerase chain reaction, *see* Polymerase chain reaction

single cell analysis

antisense RNA

amplification, **303:**4, 10–11

applications of amplified product, **303:**6, 11–12, 17

Northern blot analysis, **303:**10

radiolabeling, **303:**6, 11–12

complementary DNA synthesis, **303:**4, 6–7, 9–10

fixed tissue sample analysis with *in situ* transcription/antisense RNA amplification

amplification efficiency, **303:**16–17

applications and rationale, **303:**13–15

complementary DNA synthesis, **303:**16

fixation and embedding, **303:**15

immunohistochemical labeling, **303:**15–16

messenger RNA harvesting, **303:**7

simultaneous electrophysiological analysis, **303:**6–7

subcellular distribution analysis, **303:**12–13

transcript-aided drug design, **303:**18

splicing, *see* RNA splicing

Messenger RNA, differential display

complementary DNA

amplification, **300:**406

cloning, **300:**408–410

electrophoretic separation, **300:**406–407

excision from gel and amplification, **300:**408

identification of fragments, **300:**407

synthesis, **300:**405

extraction of RNA, **300:**404–405

overview of antioxidant regulation assays, **300:**402–404

principle, **300:**402

Metachromic leukodystrophy

arylsulfatase A deficiency, **311:**95

diagnosis, **311:**95–96, 100–101

Metal working biofilm, constant depth film fermentor

antimicrobial evaluation, **310:**309

culture, **310:**304

steady state, **310:**309

Metalloprotease transporter, *Erwinia chrysanthemi*

assays by protease detection

cup–plate assay, **292:**71

gelatin hydrolysis following gel electrophoresis, **292:**71

skim milk agar plate assay, **292:**70
cell compartment markers in secretion
    specificity determination, **292:**71–72
comparison to HasA secretion system,
    292:75
domain organization, **292:**67–68
function, **292:**67
gene organization, **292:**68–69
metalloprotease signal sequence, **292:**72
metalloprotease specificity, **292:**68, 70–72
proteins
    antibody studies, **292:**73–74
    expression in *Escherichia coli*, **292:**73–
        74, 76
    topology, **292:**75
PrtD
    ATPase
        activity in various preparations,
            292:79–81
        assay, **292:**79
        inhibitors, **292:**81
        reconstition, **292:**81
    overexpression system, **292:**76
    purification
        cytoplasmic membrane vesicles,
            292:76–77
        phosphocellulose chromatography,
            292:78–79
        solubilization, **292:**77–78
    site-directed mutants of Walker A do-
        main conserved lysine, **292:**76
Metallothionein, *S*-nitrosylation, **301:**202,
    210–211
Metaphase extract
    applications, **298:**331–332, 338–339
    immunofluorescence microscopy, **298:**352
    motility assay with video-enhanced differ-
        ential interference contrast mi-
        croscopy
        coverslips, **298:**348
        image acquisition and processing,
            298:349
        instrumentation, **298:**439
        organelle movement in *Xenopus* ex-
            tracts
            direction of movement, **298:**351
            features, **298:**349–350
            quantitation of movement,
                298:350–351
            standard assay, **298:**350

preparation from HeLa cells
    culture and harvest of cells,
        298:333–334
    cytochalasin B washing,
        298:334–335
    homogenization, **298:**335
    immunodepletion of proteins,
        298:337–338
    mitotic aster formation, **298:**335–336
    reagents, **298:**333
preparation from *Xenopus* eggs
    advantages of *Xenopus* system,
        298:340, 352–353
    centrifugation and crushing,
        298:342–343
    cytosol fraction preparation,
        298:343–345
    cytostatic factor induction of meta-
        phase arrest, **298:**340–342
    egg production, **298:**341–342
    histone H1 kinase assay of cell cycle
        status, **298:**346–348
    isolation of soluble microtubule mo-
        tors, **298:**345–346
    materials, **298:**341
    organelle fraction preparation,
        298:343–345
Metastasis
    β-galactosidase as micrometastasis
        marker, **302:**21, 30
    green fluorescent protein marker
        advantages over cell tagging with dye,
            302:30–31
        cell culture, transfection, and selection,
            302:23, 29
        doubling time of stable clones and iso-
            lation, **302:**23–24, 26
        host tissue specificity for successful col-
            onization, **302:**28–29
        intravenous injection and metastasis,
            302:25, 27–28, 30
        overview of assays, **302:**20–22
        patterns of tumor metastasis, **302:**27
        sectioning and fluorescence micros-
            copy, **302:**25
        stability of protein expression, **302:**25–
            26, 30–31
        subcutaneous tumor growth, **302:**24
        surgical orthotopic implantation in
            nude mice

CHO-K1 Clone-38 tumor fragments, **302**:24, 27, 29

human lung cancer tumor fragments, **302**:24, 27, 30

tumor progression in histoculture, **302**:25–26, 28, 31

vector constructs, **302**:23

Methanesulfohydroxyamic acid, synthesis, **301**:214, 216–217

meso-2,4-Methano-2,4-pyrrolidine, glutamate transporter inhibition, **296**:182–183

L-*anti-endo*-3,4-Methano-pyrrolidine dicarboxylate, glutamate transporter inhibition, **296**:182–183

Methidiumpropyl-EDTA–Fe(II) footprinting

cleavage

reaction mechanism, **304**:666, 620

sequence specificity, **304**:462–463, 466, 620

Drosophila embryo chromatin

buffer preparation, **304**:477, 479

digestion reaction, **304**:473–475

extent of digestion, **304**:468

hsp26 gene mapping, **304**:479–480

indirect end-labeling, **304**:466–468, 475–477

hormone-induced hypersensitive sites, **304**:620–621

Methionine sulfoxide reductase

assays, **300**:243–244

functions, **300**:239–241

purification of recombinant proteins

bovine enzyme, **300**:242–243

yeast enzyme, **300**:241–242

thioredoxin oxidation, **300**:240–241

6-Methoxy-*N*-ethylquinolinium iodide

brain slice studies, **307**:480

calibration of intracellular fluorescence, **307**:477–479

chloride selectivity, **307**:479–480

confocal imaging, **307**:474–477

kinetics of chloride response, **307**:481

loading of cells, **307**:473

reduction to membrane-permeable form, **307**:472–473

sample preparation for confocal microscopy, **307**:473

2-Methoxy-5-nitrophenyl carboxylate

β-alanine

kinetics of photolysis, **291**:40–41

quantum yield determination, **291**:43, 45

synthesis, **291**:38–39

applications, **291**:32

Methylamine tungstate, chromatin staining for transmission electron microscopy, **304**:229

N-Methyl-D-aspartate, 2,2′-dinitrobenzhydryl carboxylate synthesis, **291**:33–35

N-Methyl-D-aspartate receptor, *see also* Excitatory amino acid receptors

α-actinin-2 association, pull-down assay, **294**:375–377

antibodies, **294**:687, 701–702

conantokin binding, **294**:612–613

domain labeling with radioligands

glycine recognition domain, **294**:391–392

ion channel domain, **294**:392–393

N-methyl-D-aspartate recognition domain, **294**:389–390

polyamine domain, **294**:393–394

drug screening with fluorescence assay of calcium

activity of various drugs, **294**:32, 34–35

advantages of assay, **294**:46–47

limitations of assay, **294**:45–46

pharmacologic validation, **294**:26–27

PSD-95 association

filter overlay assay, **294**:373–375

immunofluorescence colocalization, **294**:380–381

radioactive filter assays

ligand selection, **294**:388

reaction mixtures, **294**:388–389

synaptic membrane preparation from rat brain, **294**:386–387

Triton X-100 treatment of crude fractions, **294**:387–388

stable cell line expression of recombinant channels, **294**:25–26

subunits

interactions studied by yeast two-hybrid system, **293**:105–109

types, **294**:385

Methylation, proteins

electrospray ionization mass spectrometry, analysis of peptide methyl esters

high-performance liquid chromatography–mass spectrometry, **316:**448–449, 451
preparation of peptides, **316:**445–447
rhodopsin kinase trypsinization, **316:**447–448
examples of vision proteins, **316:**452
functions, **316:**452–453
transducin, *see* Transducin
N-Methyl-*N*-(*tert*-butyldimethylsilyl)trifluoroacetamide, derivatization of tyrosine derivatives for gas chromatography–mass spectrometry, **301:**474–475, 477
6-Methylcytidine phosphoramidite, synthesis, **317:**49–51, 62–65
4-Methylumbelliferyl-*N*-acetyl-neuraminic acid, ganglioside sialidase assay, **312:**352–353
Methylene blue
    fresh frozen plasma treatment for virus inactivation
        absorption, **319:**208
        photoproducts, **319:**212
        principle, **319:**208
        production, **319:**212–216
        tolerability, **319:**208, 212
        virus sensitivity, **319:**208–209, 216
    RNA–protein cross-linking
        advantages and disadvantages, **318:**32
        binding reaction, **318:**25–26
        cross-linking reaction, **318:**26
        dye titration, **318:**26–27
        gel electrophoresis, **318:**26
        immunoprecipitation, **318:**30
        labeled nucleotide selection, **318:**29
        light source and illumination duration, **318:**28
        materials and reagents, **318:**25
        mechanism and lack of base specificity, **318:**30–31
        overview, **318:**23–24
        purified complexes, cross-linking, **318:**29–30
        ribonuclease selection, **318:**29
        side reaction suppression, **318:**27–28
        thionine dye comparison, **318:**30
Methylumbelliferyl substrates
    methylumbelliferyl-β-D-galactoside hydrolysis by β-galactosidase, **310:**61, 75

monosaccharide dimer hydrolysis by biofilms, **310:**413
4-Methyl-1-naphthalenepropanoate, *see* 1,4-Dimethylnaphthalene, singlet oxygen carrier derivatives
3-(4′-Methyl-1′-naphthyl)propionic acid
    bacteria killing activity, **319:**220–222
    dioxygenation of lipids in organic solvents, **319:**218–220
    human umbilical vein endothelial cell toxicity, **319:**220
    squalene susceptibility to oxidation, **319:**219–220
    thermolysis, stoichiometry and kinetics, **319:**217–218
Methylphosphonodiester oligodeoxynucleotide
    control design in antisense studies, **313:**130–131
    efficacy of antisense inhibition compared to other modifications, **313:**129–130
    high-performance liquid chromatography
        analytical chromatography
            anion-exchange chromatography, **313:**103–104
            column regeneration, **313:**104
            reversed-phase chromatography, **313:**103
        purification by anion-exchange chromatography, **313:**104–105
    leukemia cell uptake and effects
        electroporation of oligonucleotides, **313:**121–122
        flow cytometry analysis
            intracellular oligonucleotide concentration determination, **313:**123–125
            two-color studies of uptake and viability, **313:**122–123
            viable cells, **313:**120
        fluorescence microscopy of subcellular localization, **313:**125–126
        half-life analysis of oligonucleotide stability, **313:**128–129
        streptolysin O permeabilization, **313:**120–121
        Western blot analysis, **313:**126–128
    rationale for antisense application, **313:**97
    ribonuclease H cleavage of complexes, assays

Nonidet P-40-lysed cell assay
  advantages, **313**:118
  incubations and lysis, **313**:118
  Northern blot analysis, **313**:119–120
reverse ligation-mediated polymerase
    chain reaction of cleavage reaction
    products
  amplifications, **313**:134
  ligation reaction, **313**:133
  overview, **313**:131
  reverse transcription, **313**:133–134
  RNA linker synthesis, **313**:131–133
  sequencing, **313**:134–135
synthetic RNA assay with *Escherichia
    coli* ribonuclease H
  gel electrophoresis, **313**:108–109
  incubation conditions, **313**:108
in vitro-transcribed RNA assay with
    cell extract ribonuclease H
  degradation analysis of chimeric oli-
      gonucleotides, **313**:117–118
  extract preparation, **313**:113–114
  formaldehyde gel electrophoresis
      and Northern blotting, **313**:116
  point mutation selectivity of chi-
      meric oligonucleotides, **313**:113
  ribonuclease class identification,
      313:116–117
  RNA purification, **313**:115–116
in vitro-transcribed RNA assay with
    Escherichia coli ribonuclease H
  coupling transcription with cleavage
      reaction, **313**:112–113
  formaldehyde gel electrophoresis
      and Northern blotting,
      313:110–112
  incubation conditions, **313**:109–110
  open loop region identification in
      RNA, **313**:112
  transcription, **313**:109
synthesis
  deprotection, **313**:99–100
  detritylation, **313**:101–102
  fluorescein-labeled oligodeoxynucleo-
      tide, **313**:106–108
  phosphodiester central region,
      313:98–99
  phosphorothiodiester purification,
      313:99
  reversed-phase purification

    trityl-off, **313**:102–103
    trityl-on, **313**:100–101
4-Methyl-*N,N,N*-trimethyl-1-naphthalene-
    ethanaminium, *see* 1,4-Dimethyl-
    naphthalene, singlet oxygen carrier
    derivatives
6-Methyluridine phosphoramidite, synthe-
    sis, **317**:49–51, 62–64
Metronidazole
  electrochemical reduction, **300**:314–315
  interaction with DNA, **300**:316–321
  structure, **300**:314
MFC, *see* Microscope flow cell
Michaelis constant, carrier-mediated
    transport
  derivation, **296**:270
  interpretation, **296**:271–272
Michaelis–Menten kinetics
  fast scan cyclic voltammetry measure-
      ment of dopamine uptake
    correction for Nafion coating on car-
        bon electrodes, **296**:697–701
    inhibition analysis, **296**:703–704, 707
    modeling of data, **296**:694–697
  modifications for transporters,
      296:662–664
  radioactive neurotransmitter uptake,
      296:689–690
  rotating disk electrode voltammetry, data
      analysis, **296**:655, 658, 672
Microarray, DNA
  applications, **306**:3–4, 18
  density, **306**:3
  fluorescence-based transcription assay
    array design, **306**:10
    complementary DNA
      amplification by *in vitro* transcrip-
          tion, **306**:13–14
      direct labeling for hybridization to
          oligonucleotide microarrays,
          306:11–13
      preparation from RNA, **306**:11–12
      single-round reverse transcription,
          306:14
    cross-hybridization, **306**:18
    data analysis, **306**:16–18
    detection, **306**:5, 15–16
    hybridization and washing, **306**:14–15
    messenger RNA
      direct labeling for hybridization to

oligonucleotide microarrays,
306:10–11
preparation from cells, **306:**9
normalization from scan-to-scan,
306:17–18
optimization, **306:**18
polymerase chain reaction amplification of clones, **306:**7–9
slide processing, **306:**8
spotting, **306:**4–7
two-color analysis, **306:**9–10
nomenclature, **306:**4–5
oligonucleotide arrays
comparison to DNA arrays, **306:**4
hybridization conditions, **306:**6
sources of oligonucleotides, **306:**5–6
Microcalorimetry
adhesion analysis of biofilms
Bacillus subtilis river biofilms,
310:366–368
overview, **310:**366
Thiomonas biofilms, **310:**368–369
advantages in biofilm analysis, **310:**361–
362, 374
ampoule units, **310:**364–365
biocide efficacy testing, **310:**371–374
combined calorimetric units, **310:**365–366
corrosion activity quantification for biofilms, **310:**369–371
instrumentation, **310:**362, 364
principles, **310:**362–364
time-resolved studies of phototransduction
brass blocks in instrumentation,
316:256
cyclic GMP hydrolysis enthalpy,
316:253–254
cyclic GMP phosphodiesterase quantification, **316:**265–266, 269
electronics, **316:**260–261
GTP analog studies, **316:**266–268
impulse response, time resolution, and
calibration, **316:**261–262, 264
polystyrene block in Faraday cage,
316:257
polyvinylidene difluoride film, electric
dipole and principle of operation,
316:257–260
sample chamber, **316:**254–256
sample preparation, **316:**265–266

sensitivity, **316:**268–269
Microcirculatory kinetics
confocal microscopy
alveolar septal microvasculature, cell
movement, **307:**419
confocal scanning laser Doppler flow-
metry, **307:**419
dyes, **307:**421–422
overview, **307:**396–397
pulmonary acinar microcirculation
erythrocyte behavior, **307:**408–409
leukocyte behavior and adhesion
molecule function, **307:**408,
411–415
microvessel responses to physiological stimuli, **307:**415–418
resolution of structures, **307:**409–410
shear rates, **307:**410–411
systemic microcirculation comparison, **307:**411–412
real-time confocal laser scanning microscopes, **307:**397–404
scan speeds, **307:**397
single-line imaging approach, **307:**419
tandem scanning reflected light microscope design, **307:**404–407
epifluorescence microscopy limitations,
307:395–396, 407–408
measurement techniques, overview,
307:394–395
Micrococcal nuclease
DNA cleavage characteristics, **304:**378,
619–620
Drosophila embryo chromatin analysis
buffer preparation, **304:**477, 479
digestion
large-scale, **304:**471–473
small-scale, **304:**473
extent of digestion, **304:**468
indirect end-labeling, **304:**466–468,
475–477
nuclei preparation
buffers, **304:**464–465
large-scale preparation, **304:**468–470
small-scale preparation, **304:**470–471
overview, **304:**465
indirect end-labeling probing of chromatin structure, **304:**619–620
mammalian nucleosome positioning
high-resolution, **304:**638

in vitro, **304:**629–630
nucleosome assembly assay
  assembly reaction conditions and re-
      pair-independent suppression,
      304:341, 344
  digestion reaction, **304:**344
  gel electrophoresis and analysis,
      304:344–345
  mechanism of nucleosome assembly
      from target site for repair,
      304:345–347, 351
  principle, **304:**338, 341
  Southern blot analysis, **304:**345
nucleosome boundary mapping at high
    resolution, **304:**625–626
permeabilization of cell membranes for
    nuclear entry, **304:**614
sequence specificity, **304:**371, 380, 382
yeast mononucleosome analysis
  agarose gel electrophoresis, **304:**374
  digestion conditions, **304:**371, 373
  nuclei preparation
    cell growth and lysis, **304:**366–367
    centrifugation, **304:**367–368
    solution preparation, **304:**366
  principle, **304:**365, 371
  solution preparation, **304:**371
yeast nucleosome positioning assays
  digestion conditions, **304:**389, 391–392
  indirect end-labeling
    digestion reaction, **304:**395
    interpretation, **304:**381–382
    principle, **304:**380
    reagents, **304:**394–395
    Southern blot analysis, **304:**395–396
  nuclei isolation
    cell growth and lysis, **304:**390–391
    centrifugation, **304:**391
    reagents, **304:**389
  primer extension analysis
    denaturing polyacrylamide gel elec-
        trophoresis, **304:**398–399
    interpretation, **304:**385–387
    multiple cycle primer extension for
        mapping low-copy sequences,
        304:398
    one-cycle primer extension for map-
        ping multicopy plasmids,
        304:397
    principle, **304:**385

reagents, **304:**396–397
spheroplast preparation
  cell growth and spheroplasting,
      304:392–393
  centrifugation, **304:**393
  digestion conditions, **304:**393–394
  permeabilization, **304:**393
  reagents, **304:**392
Microdialysis, *see* Intracerebral microdi-
    alysis
β_2-Microglobulin amyloid, *see* Dialysis-re-
    lated amyloid fibril
Microinjection, *see also* Transgenic mouse;
    Xenopus oocyte
  antisense oligonucleotides, **313:**153–154
  fibroblasts, **293:**197–198, 200
  firefly luciferase reporter in mammalian
      cells
    cell lines, **305:**572
    expression comparison between trans-
        fection and microinjection in fi-
        broblasts, **305:**573–575
    microinjection, **305:**573
    osteosarcoma cell versus fibroblast ex-
        pression levels, **305:**575–576
    vectors, **305:**572–573
  neuron injection with neurotransmitter
      transporter plasmid DNA
    cell culture, **296:**382–384
    DNA preparation, **296:**383
    immunofluorescence microscopy of ex-
        pressed protein, **296:**385–387
    loading pipettes and microinjection,
        296:384–385
    micropipette production, **296:**382
  Xenopus oocytes, **293:**283–284, 523,
      535–536
Microplate luminometer
  cross-talk, **305:**79–81
  injection of reagents, **305:**81
  overview, **305:**62, 67, 78–79
Microscope flow cell
  advantages, **310:**329
  applications, **310:**190
  cleaning and sterilization, **310:**161, 326
  comparison to membrane separation sys-
      tems, **310:**192–193
  construction
    assembly, **310:**324–325
    biofilm chamber, **310:**325

design considerations, **310:**162, 164
materials, commercial suppliers and description, **310:**166, 329, 331–332
mold pouring, **310:**322, 324
tower assemblies, **310:**325–326
disadvantages, **310:**160–161
experimental setup, **310:**164–165
fluorescence microscopy
confocal laser scanning microscopy, **310:**105
deconvolution fluorescence microscopy
design of flow cell, **310:**190
flow rate, **310:**191–192
hydrodynamic property claculations, **310:**190–191
operation and biofilm establishment, **310:**193–194
selection of technique, **310:**22
inoculation, **310:**165
light transmission features, **310:**162
liquid delivery systems, **310:**161
objective lenses and working distance, **310:**162, 164
oral *Streptococcus*
confocal microscopy
growth analysis, **310:**328–329
immunofluorescent staining, **310:**327–328
culture, **310:**326–327
parallel plate flow chamber
advantages and disadvantages, **310:**534
biofilm eradication studies, **310:**530
design, **310:**527–528
initial microbial adhesion
experiment design, **310:**528–529
plasma coating effect on intial staphylococcal adhesion, **310:**530–531
Pseudomonas aeruginosa detachment from contact lenses, **310:**532, 534
Staphylococcus epidermidis adhesion and surface growth, **310:**531–532
surface growth studies, **310:**529–530
Microsomal membranes, cell-free translation of P-glycoprotein, **292:**272–273, 281
Microsome, preparation
plants, **311:**147–148
rat liver, **311:**24–25
yeast, **311:**10–11, 126
Microtubule, *see also* Tubulin

antimitotic drugs, *see* Colchicine; Nocodazole; Podophyllotoxin; Taxol; Vinblastine
assembly dynamics
dynamic instability, **298:**254
kinetic parameters, **298:**254–256
treadmilling (flux), 253–254
colchicine-binding site, **298:**251–252
differential interference contrast microscopy, **302:**103
dynamic instability, **302:**103
mass determination, **298:**274–275
microtubule-based transport in eggs and early embryos
Drosophila melanogaster, **298:**133–136
sea urchin, **298:**134–135
motility assay
apparatus for combined photolysis studies, **291:**336–337
gliding assay, **291:**337
motor proteins, *see* Dynein; Kinesins
nucleation assays, **298:**225, 235, 237
organelle translocation, *see* Organelle motility
phagosome binding, *see* Phagosome
preparation from tubulin, **291:**335
severing protein, *see* Katanin
taxol photoaffinity labeling
direct photolabeling, **298:**239
^{3}H-3′-(*p*-azidobenzamido)taxol
labeling reaction, **298:**241, 243–245
site of photincorporation, **298:**245–246, 250–251
structure, **298:**240
^{3}H-2-(*m*-azidobenzoyl)taxol
digestion of β-tubulin, **298:**242–243, 246, 248
labeling reaction, **298:**241, 243–245
peptide analysis, **298:**242
site of photincorporation, **298:**246, 248–251
structure, **298:**240
high-performance electrophoresis chromatography of labeled β-tubulin, **298:**241–242, 246
microtubule assembly assay, **298:**240
β-tubulin isolation and formic acid digestion, **298:**241, 245–246
vinblastine-binding site, **298:**251–252

Microtubule-associated protein 1A, anti-
    sense inhibition in glioma model
  cell preparation, **313:**363–364
  colony-forming assay, **313:**365, 367
  flow cytometry analysis of cell cycle,
    313:367, 371–372
  immunohistochemical staining, **313:**364–
    365, 371
  lipofection with miniosmotic pump,
    313:367, 370
  rationale, **313:**362–363, 371
  target sequence, **313:**363
  uptake assay, **313:**364
Microtubule-associated protein 4
  green fluorescent protein fusion protein
    cell preparation for microscopy,
      302:107–108
    confocal laser scanning microscopy,
      302:108–109
    constructs, **302:**104–106
    dual labeling studies, **302:**116–117
    dynamics
      dynamics parameters, calculation,
        302:113–116
      life history plots, **302:**113
      time series acquisition, **302:**111–112
      tracing of microtubules, **302:**112–113
    microtubule arrays, morphological phe-
      notypes of constructs, **302:**109–110
    microtubule binding, **302:**104
    microtubule organization analysis with
      binding domain construct, **302:**117,
      119–120
    transfection, **302:**106–107
  microtubule assembly promotion, **302:**104
Microwave treatment, antigen recovery in
    amyloid immunohistochemistry sam-
    ples, **309:**18
Mineralocorticoid receptor
  green fluorescent protein fusion protein,
    302:77
  subcellular distribution, **302:**74, 77
Minichromosome
  intactness assay, **304:**49
  isolation from yeast
    endogenous plasmid elimination,
      304:39–41
    growth of cells, **304:**42
    nuclei
      lysis, **304:**45–48

preparation from spheroplasts,
      304:43–45
    overview of protocols, **304:**35–36
    plasmid design, **304:**36–38
    spheroplast preparation, **304:**42–43
    transformation by electroporation,
      304:41–42
    yeast strain selection, **304:**38–39
    yield, **304:**49
  quantitative analysis, **304:**49
Minimoidin, fatty acid elongation pathway
    inhibition, **311:**335
Minor groove binder–oligodeoxynucleotide
    conjugate
  high-performance liquid chromatography,
    313:173
  overview of conjugation chemistry,
    313:164–165
  stabilization of DNA–RNA hybrids,
    313:165
  structure, **313:**165
  synthesis
    acylation of 5′-alkylamine-modified oli-
      gonucleotides on solid supports
      overview, **313:**165
      reagents, **313:**173
      steps, **313:**170–171, 163
    anhydrous conditions
      equipment, **313:**171–172
      overview, **313:**165
      reagents, **313:**172
      steps, **313:**167–170, 172–173
Misakinolide A
  actin binding
    assays, **298:**30
    mechanism, **298:**26–27
  actin dimer formation assay, **298:**30–31
  addition to cells, **298:**31
  applications, **298:**29–32
  effects on actin organization in living
    cells, **298:**27–28
  solubility, **298:**30
  structure, **298:**26
Mitochondria
  ATP-binding cassette proteins, *see* Mito-
    chondria, ATP-binding cassette pro-
    teins
  calcium dyes
    loading
      dihydrorhod-2, **307:**454–456

rhod-2, **307:**453
multiple sensors in measuring,
    307:456–457
fluorescent probes, **307:**121, 130
isolation, **301:**382
isolation and subfractionation for plant
    peptidylprolyl *cis–trans*-isomerase
    assay, **290:**88–91
nitric oxide and peroxynitrite effects
    calcium flux, **301:**385–386
    incubation conditions, **301:**382–383
    intactness, **301:**387
    membrane potential, **301:**383–384
    oxygen consumption, **301:**386–387
    pyridine nucleotide content,
        301:388–389
nitric oxide synthase
    citrulline assay, **301:**391–392
    membrane potential regulation,
        301:392
    oxygen consumption regulation,
        301:392
    oxyhemoglobin assay of activation,
        301:389–391
respiration
    measurement, **301:**189–194
    nitric oxide inhibition, **301:**188,
        194–198
    regulation of oxygen uptake,
        301:197–198
solute diffusion analysis with green fluo-
    rescent protein, **302:**255–256,
    258–259
Mitochondria, ATP-binding cassette pro-
    teins
Atm1p
    epitope tagging, **292:**780–782
    functions in yeast, **292:**785–786
    immunofluorescence
        antibody treatment, **292:**783
        fixing cells, **292:**782
        treatment of slides, **292:**782
    submitochondrial localization,
        292:783–785
    gene disruption using polymerase chain
        reaction fragments, **292:**779–780,
        785–786
    polymerase chain reaction search for
        genes, **292:**776–779
Mitochondrial DNA

genome, **319:**367
large-scale deletions in photoaging
    fibroblast cultures
        culture and medium, **319:**374
        deuterium oxide effects, **319:**375
        irradiation conditions, **319:**374–375
        singlet oxygen generator effects,
            319:375
        singlet oxygen quencher effects,
            319:375
        ultraviolet A dose response of viabil-
            ity, **319:**373–374
        keratinocytes, **319:**376
    polymerase chain reaction detection
        advantages, **319:**367
        DNA extraction, **319:**368
        nested polymerase chain reaction,
            319:369–370
        primers, **319:**370
        principle, **319:**368
        quantitative analysis, **319:**371
        short-cycle polymerase chain reac-
            tion, **319:**368–369
        tissue distribution of deletions,
            319:372–373
    mutation frequency, **319:**367
Mitochondrial transmembrane potential
apoptosis role, **319:**332
flow cytometry assays, **319:**326–328, 330,
    336–340
HL-60 cell studies of ultraviolet A radi-
    ation
        DiOC$_6$, **319:**337–338
        JC-1, **319:**338–340
    T cell apoptosis changes, **319:**307
    topical antioxidant screening, **319:**467–
        468, 473
MitoFluor, loading of cells, **307:**130
Mitogen-activated protein kinase
    AP-1 activation, **319:**586
    inhibitors, **298:**547–548, **319:**269
    Jun N-terminal kinase pathway, *see* c-Jun
        N-terminal kinase
    pathways, overview, **314:**342–343, 363
    photodynamic therapy stimulation,
        319:358
    redox sensitivity, **300:**339
    singlet oxygen induction
        generation of singlet oxygen
            5-aminolevulinate, **319:**137

cellular accumulation of sources, **319:**137–138

chemical generation, **319:**135–136

rose bengal, **319:**136–137

photodynamic therapy effects, **319:**141–142

specificity of kinase activation in skin fibroblasts, **319:**138–141, 143

ultraviolet A induction

cell culture, **319:**132

deuterium oxide effects, **319:**133

immunoprecipitation and assay of kinases, **319:**134–135

irradiation conditions, **319:**132

overview, **319:**131, 262

singlet oxygen quencher effects, **319:**132–138

specificity of kinase activation, **319:**135, 142

Western blot analysis of phosphorylated kinases, **319:**133–134

Mitoxantrone, BRCA2 knockdown effects on BxPC-3 cell sensitivity, **314:**503

MK-801, neurotransmitter receptor inhibition

dissociation constant determinations, **291:**464–465

effects on opening and closing rates, **291:**463–464

MLCK, *see* Myosin light chain kinase

MMTV, *see* Mouse mammary tumor virus

MNase, *see* Micrococcal nuclease

MNEA, *see* 4-Methyl-*N,N,N*-trimethyl-1-naphthaleneethanaminium

MNP, *see* 2-Methoxy-5-nitrophenyl carboxylates; 4-Methyl-1-naphthalenepropanoate

Mode locking, theory, **291:**198–199

S-Modulin

cyclic GMP phosphodiesterase, regulation of activation, **316:**3, 14

discovery, **316:**3

electrophysiological detection in truncated rod outer segment

internal perfusion and current measurements, **316:**6–9

rod outer segment

isolation, **316:**4

truncation, **316:**5–6

suction electrode preparation, **316:**4–5

localization in retina, **316:**14

physical properties, **316:**13

purification

anion-exchange chromatography, **316:**11–12

frog retina soluble protein preparation, **316:**9–10

hydrophobic affinity chromatography, **316:**10–11

recombinant protein from *Escherichia coli*, **316:**13

rhodopsin phosphorylation regulation analysis

phosphorous-32 radiolabeling and gel electrophoresis, **316:**18–19

rhodopsin kinase cross-linking, **316:**19–20

rod outer segment preparation from frog, **316:**16–18

Molecular crowding

controls in osmotic stress measurement of proteins, **295:**51, 53

differential hydration versus differential volume exclusion, calculation of cosolute effects on reaction equilibria, **295:**147–149

effects of background solutes

reaction equilibria, **295:**128–130, 145–149

reaction rates, **295:**130–132

hard particle models

calculation of volume exclusion effects on reaction equilibria, **295:**145–147

closed equations

derived from theories of hard particle fluids, **295:**138–140

hard-sphere model, **295:**137–138

experimental justification, **295:**141–142, 144–145

overview, **295:**132–133

two-body interaction coefficient calculation

hard particle model, **295:**133–134

hard-sphere random coil, **295:**136

interacting hard spheres, **295:**134

Molecular dynamics, *see* Three-dimensional structure, molecular modeling; Topology, transmembrane

Molecular electrostatic potential, surface of

substrates and transition-state,
308:346–348, 355
Molecular surface point similarity measures, binding energy analysis
accuracy, **308:**409
adenosine deaminase, **308:**404–405, 407,
409
AMP nucleosidase, **308:**405, 407,
409
cytidine deaminase, **308:**405–407, 409
indications for use, **308:**425
overview, **308:**401–404
quantum mechanical descriptions of molecules, **308:**400–401
similarity as a function of α, **308:**402–403
Monoamine transporter, *see* Vesicular
monoamine transporter
Monoclonal antibody, anti-glycosphingolipid antibodies
applications, **312:**160
characterization
carbohydrate determinant antibodies,
312:179
ganglio-series glycolipid antibodies,
312:166
globo-series glycolipid antibodies,
312:166, 178
lacto-series glycolipid antibodies,
312:178
neolacto-series glycolipid antibodies,
312:178
sulfated determinant antibodies,
312:179
table of antibodies and properties,
312:167–172
generation
glycolipid synthetic antigens,
312:161–162
high-performance thin-layer chromatography immunostaining of glycolipid antigens, **312:**165
hybridoma clone screening by enzyme-
linked immunosorbent assay,
312:164–165
immunization with pure carbohydrate
antigens, **312:**161
immunization with whole cells or tissues, **312:**160–161
mouse strains for immunization,
312:162–163

Salmonella minnesota culture and immunization, **312:**163–164
immunolocalization of gangliosides
antibodies
commercial sources, **312:**180
factors affecting binding in membranes, **312:**180–183
fixatives, **312:**181, 183
hippocampal neuron studes,
312:184–186
plasma membrane association, **312:**115,
179
postfixation immunolabeling, **312:**184,
186
prefixation with formaldehyde,
312:183–184
primary antibody incubation, **312:**184
surface plasmon resonance analysis of
binding, **312:**212
translocation assay of sphingolipids,
312:574
Monomeric actin, *see* G-actin
Monte Carlo simulation
simultaneous data fitting, **293:**416–417
three-dimensional structure, molecular
modeling
boundary conditions, **293:**713–715
moving of molecules, **293:**709–710
overview, **293:**699
sampling of phase space, **293:**710–711
3-Morpholinosydnonimine
effects on oxidative stress, **301:**416–417
peroxynitrite generation, **301:**400–401,
523
tocopherol reactions, **301:**485–486
Mössbauer spectroscopy, analysis of photosystem II mutants, **297:**359
Motility assay, *see* Centrifuge microscope;
Myosin; Optical trap; Organelle motility; Traction force assay
Mounting, confocal microscopy
antifading agents, *see* Antifading agents
cytoskeleton samples, *see* Cytoskeleton,
confocal microscopy
epoxy resin embedding, **307:**83–84
extracellular matrix preservation for
thick sections and whole mounts
fixation, **307:**87–90, 103
immunostaining, **307:**95–96, 104
laminin, **307:**85, 88–89, 96

pretreatment
  detergents, **307:**91–92, 94–95
  proteinase treatment, **307:**92–93
  purposes, **307:**91
sectioning, **307:**90, 104
tenascin, **307:**85, 87
vascular casts with fluorochrome-labeled gelatin
  advantages, **307:**98–99
  animal perfusion, **307:**105
  antifading agents, **307:**99–100
  confocal microscopy, **307:**102–103
  fetus perfusion, **307:**105–106
  fluorescence labeling of gelatin, **307:**104–105
  immunostaining comparison and application, **307:**98, 107
  mounting in polyester resin, **307:**99–102, 107
  organ perfusion, **307:**106
  replica technique comparison, **307:**97
fixation, **307:**57–59, 80, 110
immunofluorescence samples, **307:**56–59, 65–67
thick sections, **307:**81–82
thin sections, **307:**80–81
whole mount specimens, **307:**81–82, 84
Mouse mammary tumor virus, promoter analysis by exonuclease III probing, **304:**585–586, 589, 597, 599
MPE–Fe(II), *see* Methidiumpropyl-EDTA–Fe(II) footprinting
MR, *see* Mineralocorticoid receptor
mRNA, *see* Messenger RNA, differential display
MRP, *see* Multidrug-resistance associated protein
MS, *see* Mass spectrometry
MscL
  antibody generation and applications, **294:**474–475
  cross-linking of subunits, **294:**475–477
  enrichment of channel activity, **294:**469–470
  forward genetics studies, **294:**479–480, 524
  gene cloning and heterologous expression in yeast or cell-free reticulocyte lysate systems, **294:**471–473

homolog cloning from other bacteria, **294:**473–474
membrane topology analysis using alkaline phosphatase as reporter gene, **294:**477–478
patch clamping of bacteria
  filament generation from *Escherichia coli*
    magnesium formation, **294:**462
    mutants, **294:**462
    snaking approach, **294:**460–461
    ultraviolet formation, **294:**462
  protoplast generation, **294:**462–463
  reconstituted bacterial membranes, **294:**466–468
  reconstituted solubilized membrane proteins, **294:**468–469
  seal formation, **294:**463–466
  spheroplast formation, **294:**461–462
purification of recombinant channels, **294:**470–471
reverse genetics studies, **294:**479
solute efflux measurement following osmotic downshock, **294:**480–482
structure, **294:**459
msDNA, *see* Multicopy single-stranded DNA
MsrA, *see* Methionine sulfoxide reductase
Mt-cpn10, *see* Mammalian mitochondrial chaperonin 10
Mt-cpn60, *see* Mammalian mitochondrial chaperonin 60
MTBSTFA, *see* N-Methyl-N-(*tert*-butyldimethylsilyl)trifluoroacetamide
MTS, *see* 5-(3-Carboxymethoxyphenyl)-2-(4,5-dimethylthiazolyl)-3-(4-sulfophenyl)tetrazolium
MTSEH, *see* 2-Hydroxyethylmethane thiosulfonate
MTT, *see* 3-(4,5-Dimethylthiazol-2-yl)-2,5-diphenyltetrazolium bromide
Müller cell opsin, *see* Retinal pigment epithelium retinal G-protein-coupled receptor
Multicopy single-stranded DNA
  antisense DNA synthesis
    gel electrophoresis analysis of expression, **313:**483
    lpp expression inhibition, **313:**485
    principle, **313:**480, 482

reverse transcriptase extension assay, **313**:483–485
vector construction, **313**:482
bacterial functions, **313**:477
structure, **313**:477
synthesis from retrons, **313**:477, 480
Multidrug-resistance associated protein
discovery, **292**:595
functional assays of transfected HeLa
cells
ATP depletion assay, **292**:606–607
chemosensitivity testing, **292**:604–605
drug accumulation and efflux assays,
292:605–606
functional homology with P-glycoprotein,
292:594–595
gene cloning
differential display, **292**:596–597
differential hybridization, **292**:596–600
full-length complementary DNA isolation, **292**:601–602
messenger RNA probes and screening,
292:598–600
Northern blot, **292**:603–604
positive clone analysis, **292**:600
subtractive hybridization, **292**:596
Western blot analysis, **292**:603–604
gene transfection
calcium phosphate transfection,
292:603
vector construction, **292**:602–603
inhibitors, **292**:614–616
isoforms, **292**:607–608
kinetic properties of MRP1 transport,
292:612–614
leukotriene C$_4$, assay of ATP-dependent
transport, **292**:608, 611–612
multidrug-resistance associated protein/
Cystic fibrosis transmembrane regulator subfamily, ATP-binding cassette
transporters in *Saccharomyces cerevisiae*, **292**:154–155
physiological function, **292**:607
plasma membrane vesicle preparation,
292:609–611
sequence homology with other ATP-binding cassette transporters, **292**:109–
110, 162, 595
substrate specificity, **292**:608–609,
612–614

tissue distribution, **292**:607
Multidrug-resistant cells
colony assays of stem cells, **292**:257–258
MDR1 transduction into bone marrow
cells, **292**:255–257
selection
clones
cross-resistance screening,
292:253–254
drug concentration and handling,
292:251–252
expansion, **292**:252
freezing, **292**:253
maintenance, **292**:254
media, **292**:251
overview, **292**:248–249
plating, **292**:250–251
mass populations and multifactorial
multidrug-resistant cells,
292:249–250
starting cell lines, **292**:250
transfection and transduction of multidrug resistance genes, **292**:254–255
Multidrug-resistance transporter, *see* MDR
Multiphoton excitation microscopy, *see also*
Two-photon excitation microscopy
advantages and limitations, **307**:15–16,
18, 469, 513, 516
calcium imaging, **307**:468–469
detectors, **307**:525–527
instrumentation overview,
307:524–528
laser sources, **307**:469, 522–523, 526
pulse compensation, **307**:524
video rate data acquisition,
307:529–530
Multiple antigenic peptide system, antigen
coupling to carrier, **296**:390–392
Multiwavelength anomalous diffraction,
phase determination for large RNAs,
317:173–175, 179–180
Murine erythroleukemic cells
MEL-E9, heterologous protein expression
cell culture, **306**:27
characteristics of cells, **306**:26–27
comparison with other expression systems, **306**:36–41
primary selection and cloning, **306**:27
transfection, **306**:27

recombinant protein production, overview, **306:**26
Murine senile amyloid fibril, *see also* Senescence-accelerated mouse
circular dichroism analysis of secondary
structure, **309:**312
deposition sites, **309:**680
electron microscopy, **309:**311
isoforms in different strains, **309:**681–683
kinetic analysis of formation with thioflavin T fluorescence
assumptions, **309:**311
binding conditions, **309:**310–311
first-order kinetic model of extension,
309:313, 315–316
initial rate measurement, **309:**316
polymerization reaction
composition, **309:**312–313
initiation and termination, **309:**313
reaction temperature, **309:**313
wavelengths for detection, **309:**310
polarized light microscopy, **309:**312
protein concentration determination,
309:310
purification
apolipoprotein A-II monomer purification, **309:**308–309, 684–685
liver fibrils in R1.P1-*Apoa2^c* mice,
309:683–684
liver fibrils in SAMP1 mice,
309:306–307
mouse strain selection for purification,
309:681, 683
proapolipoprotein A-II, **309:**685–686
Muscarinic receptor, signaling impairment
by amyloid-β, **309:**743–744
Muscimol, γ-aminobutyric acid transporter
inhibition, **296:**168
Muscle fiber, *see also* Actin; Heavy meromyosin; Myosin
apparatus for photolysis experiments
light sources, **291:**315, 319–320
setups, **291:**313, 315
biophysical signals recorded in conjunc-
tion with photolysis, **291:**315–317
caged ATP
actomyosin studies
ATPase assays, **291:**332–333
ATP binding, **291:**330–332
caged calcium studies, **291:**332

cross-bridge detachment, **291:**330,
334, 344–346
relaxation of tension, **291:**330, 334
muscle shortening velocity, inhibition
before photolysis, **291:**343–346
caged compounds in photolysis studies
calcium chelators, **291:**321, 324
nucleotides, **291:**321, 325
purity requirements, **291:**321, 324–325
table of properties, **291:**322–323
motility assays, **291:**334–335
photochemical conversion of caged compounds, quantification
ATP, **291:**338–341
calcium, **291:**339–340
phosphate, **291:**339
photolysis
applications, **291:**214–215, 253–254, 310
uniformity in sample chamber,
291:341–342
preparation for photolysis experiments
detergent treatment, **291:**311
glycerol extraction, **291:**311
mounting, **291:**312–313
solutions for photolysis experiments
buffer, **291:**327
cations, **291:**326
changes following photolysis,
291:328–329
computer analysis, **291:**325–326
free radical scavengers, **291:**327–328
ionic strength, **291:**325–327
nucleotide depletion, **291:**328
Muscle, *see* Myocyte
Mutant cycle analysis
applications, **294:**596–597
coupling coefficient calculation, **294:**595
interaction energy, **294:**596
principle, **294:**594–595
c-Myb, antisense inhibition, **313:**28, 30
c-Myc
antisense inhibition in cell lines
Burkitt lymphoma cells, **314:**546–547
HL-60 cells, **314:**545–546, 569
peripheral blood lymphocytes,
314:545–546
rat glioma cells
cell culture, **314:**125
cell growth studies, **314:**125
comparison of probes, **314:**127

oligodeoxynucleotide sequences,
    314:124–125
Western blot analysis of expression,
    314:125–127
antisense inhibition in Eμ-*myc* transgenic
mice
aptamer effects, **314:**557–558
controls, **314:**572–573
efficacy of knockdown, **314:**548
immunostimulatory effects, **314:**558–
    560, 567–568, 573, 576–578
lymphoma cell transplant model
    clinical effects, **314:**564, 566–567,
        569–571, 573–574
    duration of therapy, **314:**571–572,
        578–579
    intraperitoneal administration,
        314:552
    micropump administration, **314:**552
    overview, **314:**548–551
    prolonged release pellet administra-
        tion, **314:**551
    subcutaneous administration,
        314:551–552
mechanisms of action, **314:**575–579
Northern blot analysis of expression in
    spleen and tumor cells, **314:**562,
    564
overview of model, **314:**547–548
pharmacokinetic modeling,
    314:553–554
statistical analysis of antisense effects,
    314:544–545, 562–563
toxicology of phosphorothioate-modi-
    fied oligodeoxynucleotides,
    314:555–557, 573
tumor onset, **314:**548
Western blot analysis of expression in
    spleen and tumor cells, **314:**560–
    561, 563–564, 571–572
antisense inhibition in lymphoma cells
cationic lipid delivery
    efficacy, **313:**395–396
    incubation conditions, **313:**391
    Lipofectamine versus Lipofectin,
        313:391, 395–396
    optimization, **313:**391–392, 394
    toxicity, **313:**392, 394
cell culture, **313:**390
controls, **313:**396

overview, **313:**389–390
sequence targeting, **313:**390–391
antisense oligodeoxynucleotide design
modifications for stability, **314:**539–541
sequence selection, **314:**124–125,
    538–539
synthesis and purification of phosphoro-
    thioate-modified oligodeoxynucleo-
    tides, **314:**541–543
target sites on messenger RNA,
    314:538–539
carcinogenesis role, **314:**538
clinical applications of antisense inhibi-
    tion, **314:**579–580
gene translocations in Burkitt's
    lymphoma, **313:**388–389
intraparenchymal delivery of antisense
    phosphorothioate oligonucleotides
    autoradiography, **314:**132–132
    infusion, **314:**131–132
    nucleic acid extraction and analysis
        from brain and cerebrospinal fluid,
        314:133–134
    oligonucleotide preparation, **314:**131
    rat preparation, **314:**131
    volume of distribution and brain tissue
        content of oligonucleotide,
        314:134–135
Max binding, **314:**537–538
transcriptional activation, **314:**538
translocation screening, **314:**579
Myelin-associated glycoprotein, ligands,
    311:560–561
Myeloma cell lines
dihydrofolate reductase-mediated gene
    expression in Sp2/0 cells
    cell culture, **306:**25
    cloning, **306:**25
    comparison with other expression sys-
        tems, **306:**36–41
    mechanism, **306:**24–25
    methotrexate addition, **306:**25
    primary selection, **306:**25
    transfection, **306:**25
recombinant protein production, over-
    view, **306:**24
Myeloperoxidase, *see also* Phagocyte lumi-
    nescence
catalysis of protein nitration, **301:**152–
    153, 289–290

3-chlorotyrosine modification of proteins,
    see 3-Chlorotyrosine
p-hydroxyphenylacetaldehyde generation
    from free tyrosine, assay of lysine
    modification
    chloride-free sodium hypochlorite prep-
        aration, **300**:92
    chlorine-demand free conditions,
        300:91–92
    extraction of protein hydrozylates,
        300:98
    gas chromatography–mass spec-
        trometry
        calibration, **300**:99
        derivatization of amino acids,
            300:98–99
        instrumentation, **300**:99
        lysine adduct detection, **300**:99–101
    isolation
        intima, **300**:93–95
        plasma, **300**:95
    lysine modification
        mechanism, **300**:90
        stability, **300**:101
        synthesis of lysine adduct, **300**:95–97
    mechanism of aldehyde generation,
        300:90
    protein hydrolysis, **300**:98
    sensitivity, **300**:101
    specificity, **300**:101
    tissue collection, storage, and pro-
        cessing, **300**:97–98
low-density lipoprotein, molecular chlo-
    rine oxidation of cholesterol
    characterization by electrospray ioniza-
        tion mass spectrometry, **300**:105
    gas chromatography–mass spectrome-
        try of phospholipid liposomes,
        300:102–103
    incubation conditions, **300**:103–105
    molecular chlorine generation,
        300:101–102
    oxidative damage role, **300**:89, 125,
        139–141
    purification, **300**:92–93, 114–115
    reaction catalyzed, **300**:89, 112
Myocyte
    fluorescence dye tagging of sarcomeric
        proteins and microinjection,
        302:171–172

green fluorescent protein-tagged sarco-
    meric proteins
    actin, **302**:173–174
    α-actinin, **302**:173
    applications in myofibril formation
        studies, **302**:172
    site selection for linkage, **302**:173–175
    titin fragments, **302**:173–174
    vectors, **302**:175
immunofluorescence staining, **302**:185
isolation and culture of chick embryonic
    cells
    cardiomyocytes, **302**:175–178
    egg incubation, **302**:176
    embryo extract preparation, **302**:176,
        181
    expression levels in transfected cells,
        302:185–186
    fluorescence imaging
        cell culture on coverslips,
            302:183–184
        data acquisition, **302**:184–185
        photobleaching, **302**:186
        skeletal muscle cells, **302**:176, 178–180
    transfection
        calcium phosphate, **302**:182–183
        efficiency, **302**:178
        lipofection, **302**:180, 182
        microinjection, **302**:183
Myoglobin
    dihydrorhodamine oxidation, **301**:441
    histidine pK_a values, **295**:184
    hydrogen peroxide interactions,
        301:437–438
    nitric oxide effects on oxidation,
        301:437–444
    photoacoustic calorimetry of carbon mon-
        oxide dissociation, **295**:321–325
    spectrophotometric determination of oxi-
        dation status, **301**:441–443
Myosin, *see also* Muscle fiber; Myosin V
    ATP-binding core, **291**:308
    caged compounds in photolysis studies
        calcium chelators, **291**:321, 324
        nucleotides, **291**:321, 325
        purity requirements, **291**:321, 324–325
        table of properties, **291**:322–323
    centrifuge microscope motility assay
        assay cell preparation, **298**:419
        force-velocity relationships, **298**:423

latex bead coating with myosin,
298:419
principle, 298:413, 418
mechanism of motion, 291:307–309
molluscan thick filament assay
actin filament movements, 298:428–
429, 435
buffer preparation, 298:433
calcium dependence, 298:435
flow cell setup, 298:433
isolation of thick filaments,
298:430–432
video-enhanced differential interference contrast microscopy, 298:428,
433–435
optical trapping of actin filaments, see Optical trap
photolysis applications, 291:310
subfragment 1 decoration of actin filaments for electron microscopy,
298:578–580
Myosin V
assays
cosedimentation assays with F-actin,
298:17
sliding filament motility assay, 298:17
steady-state ATPase, 298:16
calpain cleavage, 298:18
F-actin binding, 298:5
functions, 298:3
genes, 298:3
purification from chick brain
buffer preparation, 298:9–10
centrifugation and sodium chloride precipitation, 298:11
denaturing polyacrylamide gel electro
phoresis, 298:13–14, 16
dissection and homogenization,
298:10–11
gel filtration chromatography,
298:12–13
ion-exchange chromatography,
298:13–14
overview, 298:5–7, 9
scaling up, 298:15–16
starting tissue, 298:9
storage, 298:15
time considerations, 298:15
Triton X-100 extraction, 298:11–12
ultracentrifugation, 298:12

yield, 298:14–15
subunit composition, 298:4–5
tissue distribution, 298:3–4
Myosin light chain kinase, inhibitors,
298:548
Myriocin
nomenclature, 311:349
preparation from Isaria sinclairii
chromatographic analysis, 311:358–359
fermentation, 311:354–355
overview, 311:350, 352
partial purification, 311:357–358
starter cultures, 311:352–354
stock cultures, 311:352
serine palmitoyltransferase inhibition
applications, 311:359–361
bioassay, 311:355, 357
overview, 311:335, 349
Myristoylated alanine-rich C kinase substrate, green fluorescent protein fusion
protein, 302:58
MZEF, computational gene finding, 303:82

N

NA, see Numerical aperture
NAADP, see Nicotinic acid adenine dinucleotide phosphate
NAC, see N-Acetylcysteine
NAChR, see Nicotinic acetylcholine receptor
NAD(P)H, see Nicotinamide adenine dinu-
cleotide(phosphate)
NADPH oxidase, see Phagocyte luminescence
NAIM, see Nucleotide analog interference
mapping
1,4-Naphthalenedimethanol, see 1,4-Dimeth-
ylnaphthalene, singlet oxygen carrier
derivatives
1,4-Naphthalenedipropanoate, see 1,4-Di-
methylnaphthalene, singlet oxygen carrier derivatives
D-Naphthylluciferin, synthesis, 305:191, 193
NAPRTase, see Nicotinate phosphoribosyltransferase
NB protein pore, molecular dynamics simulated annealing with restrained molecular dynamics, 293:676–679

NBC, *see* Sodium/bicarbonate cotransporter
NBD-ceramide
ceramidase assay
incubation conditions, **311**:197–198
product separation and detection,
311:198
sample preparation, **311**:197
substrate delivery, **311**:197
chemical synthesis of 1-*O* methyl derivative
column chromatography, **311**:603–604
materials, **311**:602–603
1-*O*-methyl-D-*erythro*-sphingosine as
starting material, **311**:608–609
principle, **311**:604, 606
salt removal, **311**:603
structural characterization, **311**:604
thin-layer chromatography, **311**:603
glucosylceramide synthase assays
incubation conditions, **311**:52, 306
modifications, **311**:52–53
NBD-ceramide–bovine serum albumin
complex preparation, **311**:51–52
substrate preparation, **311**:306
thin-layer chromatography, **311**:52, 306
inositolphosphoryl ceramide synthase
assay, **311**:125, 143
sphingomyelin synthase assay
incubation conditions, **311**:37
optimization, **311**:38
quantification, **311**:37–38
in situ synthesis, **311**:39–40
substrate, **311**:37
NCAM, *see* Neural cell adhesion molecule
Ncd, *see* Kinesins
NCP, *see* Nucleosome core particle
NCX, *see* Sodium–calcium exchanger
NDMOL, *see* 1,4-Naphthalenedimethanol
NDP kinase, *see* Nucleoside diphosphate
kinase
NDP, *see* 1,4-Naphthalenedipropanoate
Nearest neighbor distribution function, distribution analysis in confocal microscopy, **307**:215, 217–220, 226–227
Negative-ion chemical ionization mass spec-
trometry
alkanal assay, **300**:76–78
F$_2$-isoprostane assay
derivatization, **300**:8–9, 15

extraction and hydrolysis of phospholipids, **300**:6–7, 14–15
instrument parameters, **300**:9, 15
metabolites of F$_2$-isoprostanes, **300**:11
plasma analysis, **300**:14–17
purification
aminopropyl chromatography, **300**:14
gas chromatography, **300**:9, 15
reversed-phase chromatography,
300:7–8, 13–14
thin-layer chromatography, **300**:7–9,
13
quantification, **300**:9–11, 15–16
sample handling and storage, **300**:5–6
sensitivity, **300**:7
standards, **300**:7–8
urine analysis, **300**:7, 11
malondialdehyde assay
derivatization, **300**:72–73
gas chromatography, **300**:75–76
internal standard synthesis, **300**:72
malonaldehyde standard preparation,
300:72
sample preparation
cultured cells, **300**:71
organs, **300**:71–72
plasma, **300**:71
sperm, **300**:71
standard curve generation and calculations, **300**:74–75
NEM, *see* N-Ethylmaleimide
Neoxanthin, biosynthesis, **297**:257
NEPO, *see* 3-(4'-Methyl-1'-naphthyl)propi-
onic acid
NER, *see* Nucleotide excision repair
Nernst equation, electrical potential imaging with cationic fluorophores,
302:346
Nernst potential, single-ion, **293**:384
Nerve growth factor
antisense knockdown
controls, **314**:165–166
oligonucleotide design, **314**:161
phosphorothioate oligonucleotide toxicity in COS cells, **314**:161–162,
165–166
Western blot analysis of expression,
314:162–163, 165
chicken protein expression in COS cells,
314:160–161

neuronal differentiation role,
    314:149–150
NET, *see* Norepinephrine transporter
Network formation, *see* Endoplasmic reticu-
    lum, network formation
Neural cell adhesion molecule
  antisense studies in chick learning model
    behavioral effects, **314:**218–219, 221
    chick model overview, **314:**216
    intracerebral injection, **314:**216–217,
        222
    oligodeoxynucleotide design and purifi-
        cation, **314:**217–218
    specificity of effects, **314:**221–223
  isoforms and functions, **314:**214
  receptor binding, **314:**214
Neural network, binding energy calculation
  advantages, **308:**410, 425–426
  architecture of network, **308:**412–414
  back propagation, **308:**414
  cytidine deaminase analysis, **308:**414–415,
      424, 426
  hidden layer magnitudes and inhibitors
      design, **308:**423–424
  inosine–uridine nucleoside hydrolase
      analysis
    accuracy, **308:**419, 424
    binding sites, **308:**416
    orientation of ligand binding,
        308:417
    substrate specificity, **308:**416–417
    training of network, **308:**418
    transition-state structures and affinities,
        308:418–419, 422–424
  inputs and mapping, **308:**411–412
  learning, **308:**410, 414
  output layer, **308:**413
  overview of biological applications,
      308:410–411
  pattern classification, **308:**425
  weights and biases, **308:**413–414
Neuraminidase
  photoaffinity labeling of human lyso-
      somal enzyme, **311:**580–581
  translocation assay of sphingolipids,
      312:572
Neurofibrillary tangle, *see also* Alzheimer's
    disease
  electron microscopy, ultrastructure analy-
      sis, **309:**180–182

immunoelectron microscopy of transglu-
    taminase, **309:**182–183
isolation, **309:**179–180
isopeptidase digestion, **309:**184–185
purification, **309:**180
Neuron, *see also* Whole cell patch clamp
  avidin–biotin–peroxidase complex for
      confocal imaging
    advantages and limitations,
        307:149–151
    confocal microscopy, **307:**145–146
    electron microscopy, **307:**147
    fixation, **307:**145
    incubation with complex, **307:**146
    overview, **307:**137, 145
    plastic embedding, **307:**146–147
    postfixation, **307:**146
    serotonin staining, **307:**147, 149
    tissue sectioning, **307:**145
  biotin staining of neurons for confocal
      and electron microscopy
    back filling, **307:**138–139
    immunocytochemical staining, **307:**139
    intracellular labeling, **307:**138
  calcium phosphate transfection, **302:**221
  cerebellar granule neurons, primary
      culture
    dissection, **302:**215
    polylysine coating of culture surfaces,
        302:214–215
    supplements, **302:**216
    trypsinization/trituration, **302:**215–216,
        218
  confocal neuroimaging of retinal ganglia,
      in vivo
    animal preparation, **307:**567–569
    applications, **307:**569–570
    equipment, **307:**565–567
    overview, **307:**563–565
  fiber-optic confocal imaging,
      307:564
  fluo-4 imaging in mixed neuron and astro-
      glial cell culture, **307:**437, 439–440
  green fluorescent protein vector analysis
      of transfectant
    advantages, **302:**222–225
    constructs, **302:**218–221
    glutamate receptor subunits, **302:**224
    inositol triphosphate 3-kinase,
        302:223–224

immunogold fluorescence staining for con-
    focal and electron microscopy
  advantages and limitations,
    307:149–151
  confocal imaging, **307:**140, 142
  correlation of confocal and electron mi-
    croscopy images, **307:**144–145
  electron microscopy, **307:**144
  fixation, **307:**139
  incubation with immunogold-conju-
    gated streptavidin, **307:**140
  microtomy, confocal microscopy guid-
    ance, **307:**143–144
  overview, **307:**137–138
  permeabilization, **307:**140, 151
  plastic embedding, **307:**143
  postfixation for electron microscopy,
    307:142–143
  sectioning, **307:**139–140
  silver enhancement of gold particles,
    307:142
kinin receptor distribution, confocal mi-
    croscopy imaging, **307:**389–390
leech neuron, *see* Retzius P cell synapse
lipofection, **302:**221–222
messenger RNA abundance, estimation,
    303:3
reverse transport measurement in *Pla-
    norbis* giant dopamine neuron with
    carbon electrodes
  amperometry, **296:**682
  amphetamine treatment
    effect on quantal size, **296:**685
    reverse transport induction,
        296:687–688
  cyclic voltammetry, **296:**682–683
  data acquisition, **296:**680–682
  disk electrode preparation, **296:**679
  dissection, **296:**676–677
  dopamine exocytosis, measurement,
        296:683–685
  intracellular injection, **296:**680–681
  resolution of dopamine release by re-
        verse transport and exocytosis,
        296:688
  ring electrode preparation, **296:**677–678
  testing of electrodes, **296:**679–680
  voltammetry, **296:**682, 687–688
single cell analysis of messenger RNA
  antisense RNA

  amplification, **303:**4, 10–11
  applications of amplified product,
        303:6, 11–12, 17
    Northern blot analysis, **303:**10
    radiolabeling, **303:**6, 11–12
  complementary DNA synthesis, **303:**4,
        6–7, 9–10
  fixed tissue sample analysis with *in situ*
    transcription/antisense RNA ampli-
    fication
    amplification efficiency, **303:**16–17
    applications and rationale, **303:**13–15
    complementary DNA synthesis,
        303:16
    fixation and embedding, **303:**15
    immunohistochemical labeling,
        303:15–16
  messenger RNA harvesting, **303:**7
  simultaneous electrophysiological analy-
    sis, **303:**6–7
  subcellular distribution analysis,
        303:12–13
  transcript-aided drug design, **303:**18
stable versus transient transfection,
    302:213
transient expression of neurotransmitter
    transporters
  microinjection of plasmid DNA
    cell culture, **296:**382–384
    DNA preparation, **296:**383
    immunofluorescence microscopy of
        expressed protein, **296:**385–387
    loading pipettes and microinjection,
        296:384–385
    micropipette production, **296:**382
    viral transfection, **296:**381
Neuronal nitric oxide synthase, *see* Nitric
    oxide synthase
Neuropathic pain, *see* Peripheral nerve in-
    jury; Peripheral nerve 3 protein
Neuropeptide Y
  antisense inhibition of receptor, **313:**29
  biosynthesis, **314:**184
  feeding behavior control
    antisense oligodeoxynucleotide effects
        colchicine-induced hyperphagia
            model, **314:**186–187
        neuropeptide antisense, **314:**192–195
        receptor antisense, **314:**197–200
    overview, **314:**184–185, 192

luteinizing hormone-releasing hormone
  regulation
  antisense effects in ovariectomized rats
    blockade of luteinizing hormone
      surge, **314:**188–189
    dose, **314:**188
    inhibition of neuropeptide Y induc-
      tion by progesterone or nalox-
      one, **314:**189–191
    luteinizing hormone pulsatility ef-
      fects, **314:**191–192
    oligonucleotide design, **314:**188
    overview, **314:**185, 187
  receptors, **314:**185–186, 197–198
Neurotransmitter receptor
  caged compounds
    apparatus for photolysis, **291:**453
    α-carboxy-2-nitrophenyl neurotransmit-
      ters, photolytic properties, **291:**453,
      455–458
    cell-flow quantification of photolysis,
      291:454
    channel opening, rate and equilibrium
      constant determination, **291:**457,
      459–460
    inhibitor studies
      dissociation constant determinations,
        291:464–465
      effects on opening and closing rates,
        291:460–461, 463–464
    2-methoxy-5-nitrophenyl neurotransmit-
      ters, photolytic properties, **291:**453,
      457, 459
    microsecond electrophysiology mea-
      surements, **291:**468, 470
    neuronal circuit function analysis,
      291:467–468
    Xenopus oocyte studies, **291:**470–472
  cell-flow studies
    apparatus, **291:**451–453
    channel opening, rate and equilibrium
      constant determination, **291:**460
    principle, **291:**447, 450–451
  classification, **291:**444–445
  malfunctions in disease, **291:**443–444
  minimum reaction scheme for mediated
    reactions
    characteristics, **291:**445–446
    equations for analysis, **291:**448–449
  single cell excitatory and inhibitory neuro-

    transmitter signal integration,
    291:465–467
Neutral sphingomyelinase, *see also* Sphingomyelinase
  activation, **311:**168
  apoptosis role, **311:**156–157, 168
  assays
    cell lysate preparation, **311:**169–170
    controls, **311:**171
    incubation conditions, **311:**166
    precipitation assay for high throughput
      automated assays, **311:**184
      background reduction,
        311:182–183
      enzyme preparation from rat brain,
        311:177–179
      incubation conditions, **311:**179–180
      kinetic parameters, **311:**182
      manual assays, **311:**183
      principle, **311:**176–177
      product characterization,
        311:180–181
      radioactivity quantification, **311:**180
      reagents, **311:**179
      validation, **311:**181–182
    principle, **311:**165, 169
    radioactivity quantification, **311:**166–
      167, 170
    reagents, **311:**165–166
    robotic assay for inhibitor screening
      controls, **311:**174
      incubation conditions, **311:**173
      multiwell plates, **311:**171–172
      performance, **311:**174–176
      robot, **311:**172
      upper phase aspiration, **311:**173–174
      vaidation, **311:**175
    sample preparation, **311:**165
    substrate preparation, **311:**170
    substrate/detergent-mixed micelle prep-
      aration, **311:**166
  gene, **311:**164
  lipid cofactors, **311:**160–162
  pH optimum, **311:**159–160
  purification from rat brain
    chromatography, **311:**158–159
    detergent extract preparation, **311:**158
    overview, **311:**157
    reagents, **311:**157–158
  stability, **311:**159

Neutrophil
  confocal microscopy analysis
    immunocytochemistry
      antibodies, **307:**382–383
      CD14 receptor, **307:**385
      cell collection and processing,
        307:381
      fixed cells, **307:**383
      N-formyl peptide receptor, **307:**385
      image analysis, **307:**384–385
      kallikreins, **307:**386–387
      kinin B2 receptor, **307:**387
      kininogen and kinin, **307:**386–388
      nonfixed cells, **307:**383
    phagocytosis of *Staphylococcus aureus*
      controls, **307:**327
      data analysis and interpretation,
        307:327–328
      modifications for other cell types,
        307:325–326
      neutrophil isolation, **307:**325–326
      phagocytosis conditions, **307:**326–327
      stopping of phagocytosis, **307:**327
  restriction endonucleolytic analysis of dif-
    ferentially expressed sequences
    activator effects
      gel display, **303:**294
      incubation with neutrophils,
        303:277–278
    advantages of neutrophil studies,
      303:273
    cloning of amplification products,
      303:290
    complementary DNA synthesis,
      303:275, 279–280, 291
    database comparisons, **303:**297
    fly adaptor
      ligation with digested complemen-
        tary DNA, **303:**275
      preparation, **303:**273–274
    gel analysis
      autoradiography, **303:**276, 285
      band recovery from dried gels,
        303:285–286
      data documentation, **303:**290
      materials, **303:**276
      polymerase chain reaction products,
        303:283–284
      sequencing, **303:**284–285, 289–290
    neutrophil isolation, **303:**274, 276–277

    oligonucleotides, **303:**273
    polymerase chain reaction
      fly adaptor-ligated complementary
        DNA, **303:**282–283, 291–293
      gel-eluted complementary DNA,
        303:286–287, 293
      materials, **303:**275–276
      primer FA-1, end labeling, **303:**275
      product purification, **303:**283
      sequencing of products, **303:**289–290,
        293
      unlabeled primers and complemen-
        tary DNA amplification,
        303:287–289
    reagents, **303:**274–276
    redundant band sources, **303:**294, 296
    repetitive sequence bands, **303:**297
    reproducibility, **303:**294
    restriction enzyme digestion, **303:**275,
      280–281
    RNA isolation, **303:**278–279, 291
New fuchsin–naphthol phosphate, alkaline
  phosphatase substrate in amyloid im-
  munohistochemistry, **309:**23–24
NF-1, *see* Nuclear factor 1
NF-κB, *see* Nuclear factor-κb
NFT, *see* Neurofibrillary tangle
NGF, *see* Nerve growth factor
Nickel affinity chromatography, histidine-
  tagged proteins
  chaperonin 10, **290:**201–202
  GroES, **290:**124, 127
  heat shock protein 104, **290:**440
Nicotinamide adenine dinucleotide(phosphate)
  caged 1-(2-nitrophenyl)ethyl compound
    calcium release analysis with fluores-
      cent probes, **291:**414–415
    high-performance liquid chromatogra-
      phy, **291:**407–409
    photolyis conditions, **291:**414
    structural characterization, **291:**409
    synthesis, **291:**407
  calcium release in sea urchin eggs,
    291:403–404
  isotopic labeling for kinetic isotope effect
    studies, **308:**313, 316, 318, 356–357
  metabolic fluorescence imaging, over-
    view, **307:**351–353
  photobleaching, **307:**352–353

two-photon excitation microscopy
  principle, **307**:353–354
  excitation wavelength, **307**:353–354
  localization of excitation, **307**:354
  resolution, **307**:354
  instrumentation for confocal micros-
    copy, **307**:355–357
  linearity of detection, **307**:358–359
  photobleaching correction, **307**:359
  viability controls, **307**:358
  data analysis and quantification,
    307:359
  pancreatic β cells, glucose response
    analysis
    immunostaining of enzymes and hor-
      mones, **307**:362–365
    islet cell studies, **307**:360–362
    principle, **307**:360
  muscle metabolism imaging,
    307:365–368
  cornea imaging, **307**:517, 519
  skin metabolism imaging
    correlative microscopy, **307**:515
    data acquisition, **307**:531–532
    multiphoton process, experimental
      characterization, **307**:515
    validation, **307**:519
    wavelengths, **307**:517
  ultraviolet confocal fluorescence micros-
    copy, **307**:519–522
Nicotinamide guanine dinucleotide, isotopic
  labeling for kinetic isotope effect studies, **308**:313
Nicotinate phosphoribosyltransferase
  active site, **308**:40–41
  assay, **308**:39–40
  ATP dependence and independence,
    308:37–39, 43
  ATP-driven molecular discrimination
    energetics, **308**:44–46
    physiological significance, **308**:46–47
    RecA comparison, **308**:47–48
  cloning and overproduction of recombi-
    nant *Salmonella* enzyme, **308**:39
  conformational changes, **308**:44
  ligand affinities, **308**:41
  pH effects, **308**:40–41
  rapid quench studies, **308**:42
  reaction cycle, **308**:36–37, 40–42
  side reactions, **308**:43

stoichiometry of reaction, **308**:43–44
Nicotinic acetylcholine receptor, *see also*
  Neurotransmitter receptor
  antibodies
    autoimmunity, **294**:701
    muscle receptors
      mouse monoclonal antibodies,
        294:681, 700–701
      rabbit polyclonal antibodies, **294**:682,
        701
      rat monoclonal antibodies, **294**:679–
        680, 700
    neuronal receptors
      monoclonal antibodies, **294**:683
      rabbit polyclonal antibodies raised
        against extracellular domain pep-
        tides, **294**:684
  binding isotherms, analysis
    competitive inhibition data analysis,
      294:130–131
    curve fitting, **294**:129–130
    subtraction of nonspecific binding,
      294:128–129
  binding sites and heterogeneity,
    294:117–118
  conotoxin binding
    α-conotoxin, **294**:608–612
    αA-conotoxin, **294**:610–611
    ψ-conotoxin, **294**:611
  desensitization, **294**:117–118
  drug screening with fluorescence assay of
    calcium
    activity of various drugs, **294**:36–37, 39,
      41–45
    advantages of assay, **294**:46–47
    limitations of assay, **294**:45–46
    pharmacologic validation,
      294:26–27
  fluorescence binding assays
    advantages and disadvantages,
      294:118–119
    ligand selection
      acetylcholine antagonists, **294**:126
      noncompetitive site ligands, **294**:119,
        126–127
    membrane preparation, **294**:119
    noncompetitive binding assay, **294**:127
  heterologous expression in *Chara
    corallina*, **294**:521
  radioligand binding assays

acetylcholine assays for allosteric interaction analysis, **294:**121
α-bungarotoxin assays
cells, **294:**124–125
membranes, **294:**122–124
ligand selection, **294:**119
membrane preparation, **294:**119
microcentrifuge assay, **294:**120–121
noncompetitive agonists, effects on
binding to agonist site, **294:**125
noncompetitive binding analysis with
phencyclidine or ethidium,
294:121–122
sensitivity, **294:**118
stable cell line expression of recombinant
channels, **294:**25–26
thermodynamic cycle analysis
allosteric effects on second ligand binding, **294:**132–133
double-mutant analysis, **294:**133–135
principle, **294:**132
Niemann-Pick disease, *see* Hematopoietic
stem cell, fluorescence selection of acid
sphingomyelinase expression
NinaA, functions, **315:**456
Nipecotic acid, γ-aminobutyric acid transporter inhibition, **296:**168–173
Nitrate
contamination in peroxynitrite preparations
calibration curves, **301:**184
effect of preparation age, **301:**182–183
effect of synthesis method,
301:185–187
ultraviolet second-derivative spectroscopy, **301:**179–180, 184–185
high-performance liquid chromatography
biological systems
chromatography, **301:**265–268
sample preparation, **301:**263–265
detection methods, **301:**259, 262,
266–269
ion-exchange chromatography,
301:259–262, 265–268
reversed-phase chromatography,
301:259
simple chemical systems
chromatography, **301:**260–262
radiolysis for free-radical generation,
301:260

nitric oxide, assays of whole-body synthesis
applications in clinical studies, **301:**49–
50, 92
L-[^{15}N$_2$]arginine assay
calculations, **301:**96–97
mass spectrometry, **301:**95–96, 98
tracer infusion studies, **301:**96
urine nitrate measurements, **301:**93–
95, 97
dietary effects, **301:**50, 52–53, 93, 98
low-nitrate diets
liquid diets, **301:**55
solid standardized diet, **301:**53–54
solid variable diet, **301:**54–55
microtiter plate Griess reagent assay
analysis of data, **301:**59–60
blood collection, **301:**57
cerebrospinal fluid collection, **301:**58
nitrite conversion and measurement,
301:58–59
urine collection, **301:**55–57
Nitric oxide synthase
calcium dependence of isoforms,
301:170–171, 505–506
carbon monoxide difference spectroscopy
following NADPH reduction, isoform differences, **301:**77
cytochrome *c* reduction assays of isoforms
calculations, **301:**74
incubation conditions, **301:**73
specificity of reduction by isoforms,
301:72, 74
cytoprotective action of isoforms, **301:**436
domains, **301:**70, 114, 171
inducible enzyme
antisense knockdown in kidney,
314:387–389
assay of activity, **301:**507–508
effects on activity
flavonoids, **301:**509
Gingko biloba extract, **301:**510, 512
pine bark extract, **301:**510, 512–513
expression in disease, **301:**49, 60–61,
506
modulation in therapy, rationale,
301:506–507
preparation from macrophages, **301:**507
isoforms in renal function, **314:**389–390

long-term potentiation role, **293**:500, 502
mitochondrial enzyme
   citrulline assay, **301**:391–392
   membrane potential regulation,
     301:392
   oxygen consumption regulation,
     301:392
   oxyhemoglobin assay of activation,
     301:389–391
neuronal enzyme, antisense knockdown
   kidney
     activity assay, **314**:395
     animal care, **314**:390
     mean arterial pressure effects,
       314:396–398, 400
     7-nitroindazole inhibition, compari-
       son with antisense knockdown
       effects, **314**:390, 395–396,
       398–400
     oligonucleotide design, **314**:395
     rationale, **314**:390
     renal medullary interstitial infusion,
       314:391–394
     Western blot analysis, **314**:394,
       397–398
   opioid tolerance studies, **314**:60
neuronal enzyme, peptides-on-plasmids library screening of PDZ-binding proteins
   affinity panning
     efficiency monitoring, **303**:461
     lysis of library, **303**:457
     principle, **303**:455
     reagents, **303**:455
     rounds, **303**:455–458
   amplification of interacting plasmids
     DNA precipitation, **303**:458–459
     electroporation, **303**:459–460
     overview, **303**:458
     reagents, **303**:458
     rounds, **303**:460–461
   binding clone identification
     enzyme-linked immunosorbent assay,
       303:463, 465–466
     lysis of LacI clones, **303**:464–465
     overview, **303**:463
   complexity of library, **303**:453
   enrichment determination, **303**:461–463
   immobilization of panning proteins,
     303:454

   LacI fusion proteins, **303**:452
   principle, **303**:452
   purification of panning proteins,
     303:453–454
   sequence analysis and interacting pro-
     tein search, **303**:466–467
   prosthetic groups, **301**:70–71, 114,
     171
recombinant expression of isoforms in
   Escherichia coli, **301**:71–72, 114
scintillation proximity assay of neuronal
   heme domain ligands
   equilibrium binding assays,
     301:117–118
   kinetics of binding, **301**:118, 121–122,
     124–125
   nitroarginine, **301**:117–122
   principle, **301**:115–116
   reagents, **301**:116
   recombinant domain expression and pu-
     rification, **301**:116
   tetrahydrobiopterin, **301**:117–118,
     121–124
superoxide anion generation
   adrenochrome assay, **301**:75–76
   electron acceptor-stimulated superox-
     ide generation, **301**:175–177
   electron spin resonance detection
     advantages, **301**:74, 170
     loop gap resonator measurements,
       301:172–173
     neuronal nitric oxide synthase
       assays, **301**:173–177
     spin trapping reagents, **301**:171–172
   specificity of production by isoforms,
     301:75–76
Nitric oxide
   amperometric determination of saturated
     solution concentration
   advantages, **301**:160–161
   chronoamperometric curve evaluation,
     301:164–166
   data acquisition, **301**:163
   electrode and sensor fabrication,
     301:162, 166–167
   interference from other species,
     301:166–168
   solution preparation, **301**:162
   temperature effects, **301**:168
   theory, **301**:161

validation with Griess reagent,
301:163–164, 168
antioxidant versus oxidant actions,
301:413–433, 435–436
antioxidative gene induction, 300:393
apoptosis protection
exposure of cells, 301:394–395
guanylate cyclase stimulation, 301:393,
395–396
L-arginine transport, regulation of synthesis, 301:78–79, 91
cell-column chromatography in endothelial barrier studies
advantages, 301:4–5, 12
apparatus, 301:6–7
cell culture, 301:5–6
continuous perfusion, 301:8, 12–13
electrochemical detection of nitric oxide, 301:10, 12
exposure of endothelial cells to nitric
oxide and oxidants, 301:8–10
cell injury, see Nitrosative stress
delivery techniques, overview,
301:444–446
difficulty of study, 301:3–4
diffusion chambers, 301:4–5
donors from peroxynitrite reactions
detection of nitric oxide release
electrochemical detection,
301:294–296
vasorelaxation bioassay, 301:296–297
synthesis, 301:293–294, 298
electrochemical detection, 301:64–66, 204,
208–209, 294–296, 499–500
electron paramagnetic resonance detection with iron diethyldithiocarbamate
cell-free assays, 301:275–276
interaction with nonheme iron nitrosyl
complexes and S-nitrosothiols,
301:272–274
mononitrosyl iron complexes
formation and detection in macrophage culture, 301:275, 276–278
formation and detection in mice,
301:269, 274–275, 276–278
redox properties, 301:272
signal, 301:270–271
solubility, 301:271–272
structure, 301:270

synthesis in solution, 301:274
potential applications, 301:278–279
spectrometry, 301:276
fluorescent cheletropic spin traps
absorbance properties, 301:534–535,
539
electron paramagnetic resonance,
301:533, 535, 539
fluorescence properties, 301:535–537,
539
selectivity for nitric oxide, 301:537–538
sensitivity, 301:533, 536–537
stability in solution, 301:535, 537
synthesis, 301:533
trapping mechanism, 301:532–534
hemoglobin modification, see S-Nitrosohemoglobin
I-κB induction, assays
electrophoretic mobility shift assay of
nuclear factor-κB, 300:383–384
immunoprecipitation of I-κB–NF-κB
complexes, 300:379–381
luciferase as reporter gene,
300:387–388
Northern blot analysis, 300:375,
381–383
nuclear run-on assay, 300:385–386
transient tranfection assay, 300:376,
386–388
Western blot analysis, 300:375–379
interaction with oxidants, 300:389–391
lipid peroxyl radical reactions
anaerobic nitric oxide solution preparation, 301:466
fatty acid preparation, 301:466
initiation of peroxidation, 301:457
kinetic analysis, 301:458
oxygraphy, 301:458, 466–467
low-density lipoprotein, protection
against oxidation
atherosclerosis implications,
301:423–424
copper-dependent oxidation,
301:446–449
delivery of nitric oxide, 301:444–446
kinetic modeling of oxidation, 301:450–
451, 453
peroxyl radical-mediated oxidation,
301:449–450
mitochondria effects

calcium flux, **301:**385–386
incubation conditions, **301:**382–383
intactness, **301:**387
membrane potential, **301:**383–384
oxygen consumption, **301:**386–387
pyridine nucleotide content,
    301:388–389
nitrate assays of whole-body synthesis
  applications in clinical studies, **301:**49–
    50, 92
  L-[^{15}N$_2$]arginine assay
    calculations, **301:**96–97
    mass spectrometry, **301:**95–96, 98
    tracer infusion studies, **301:**96
    urine nitrate measurements, **301:**93–
      95, 97
  dietary effects, **301:**50, 52–53, 93, 98
  low-nitrate diets
    liquid diets, **301:**55
    solid standardized diet, **301:**53–54
    solid variable diet, **301:**54–55
  microtiter plate Griess reagent assay
    analysis of data, **301:**59–60
    blood collection, **301:**57
    cerebrospinal fluid collection, **301:**58
    nitrite conversion and measurement,
      301:58–59
    urine collection, **301:**55–57
N-nitrosamine formation, **301:**41–42
peroxynitrite formation, *see* Peroxynitrite
photodynamic therapy apoptosis role,
    319:353
platelets
  assay of nitric oxide
    platelet preparation, **301:**63
    aggregation measurement, **301:**63–64
    electrochemical detection, **301:**64–66
  nitric oxide effects, **301:**61–63
polyphenol effects on endothelial cell re-
    lease, **301:**522, 525–529
radical scavenging of flavonoids, *see* Fla-
    vonoids
respiration inhibition, **301:**195–198
sheared cell release, **301:**519
signal transduction, **300:**389
solution preparation, **301:**162, 381–382
spermine adduct in generation,
    301:439
zinc homeostasis effects, *see* Zinc
Nitrite

contamination in peroxynitrite prepara-
    tions
  calibration curves, **301:**184
  effect of preparation age, **301:**182–183
  effect of synthesis method, **301:**185–187
  ultraviolet second-derivative spectros-
    copy, **301:**179–180, 184–185
Griess reaction, principle, **301:**497–498
high-performance liquid chromatography
  biological systems
    chromatography, **301:**265–268
    sample preparation, **301:**263–265
  detection methods, **301:**259, 262,
    266–269
  ion-exchange chromatography,
    301:259–262, 265–268
  reversed-phase chromatography,
    301:259
  simple chemical systems
    chromatography, **301:**260–262
    radiolysis for free-radical generation,
      301:260
Nitroarginine, scintillation proximity assay
    of binding to neuronal nitric oxide syn-
    thase heme domain, **301:**117–122
7-Nitrobenz-2-oxa-1,3-diazole-ceramide, *see*
    NBD-ceramide
2-Nitrobenzyl group
  applications in caged compounds,
    291:3–4, 64, 118–119
  limitations in photochemical applications,
    291:3–4
  mass spectrometry analysis of modified
    proteins, **291:**119, 124
  photolysis reaction, **291:**2–3, 226
  rate of photolysis, **291:**122
  scavenging free radicals after photolysis,
    291:134, 233, 328
Nitrogen dioxide, unsaturated lipid reac-
    tions, **301:**462–463, 469
3-Nitro-4-hydroxyphenylacetic acid, high-
    performance liquid chromatography
    with electrochemical detection
  chromatography conditions, **301:**156
  electrochemical detection, **301:**156, 158
  interpretation of results, **301:**158–160
  sample preparation, **301:**154
o-Nitromandelyloxycarbonyl group
  2,5-di(*tert*-butyl)hydroquinone caged com-
    pound

calcium studies with fluorescent indicators, **291:**435–437, 439
cell loading, **291:**435
synthesis, **291:**432
glutamate caged compound
electrophysiology studies of glutamate
receptor effects, **291:**440–443
photochemistry, **291:**440
synthesis, **291:**434–435
imidazole
reactivity, **291:**431–432
synthesis, **291:**431–434
Nitronium ion, unsaturated lipid reactions,
301:462–463, 469
S-(Nitrophenylsulfenyl)-cysteaminyl-EDTA
histone modification, **304:**253, 256–257
site-directed hydroxyl radical mapping,
histones
iron ion loading, **304:**258–259
nucleosome assembly and gel electrophoresis analysis, **304:**257–258
rate of reaction, **304:**259
reaction conditions, **304:**259–260
sequence analysis of cleaved DNA and
mapping, **304:**260–263
synthesis
cysteine blocking, **304:**253
linkage to EDTA, **304:**255
mass spectrometry characterization,
304:255
2-nitrophenylsulfenyl exchange,
304:255
2-Nitrophenethyl group
limitations in photochemical applications,
291:3–4
photolysis reaction, **291:**2
Nitrosative stress
assessment, **300:**394–395
manifestations of nitrosylation,
300:391–392
oxidant interactions with nitrosants,
300:389–391
resistance mechanisms, **300:**392–394
S-Nitroso-*N*-acetylpenicillamine
effects on oxidative stress, **301:**416–417
sugar–SNAP
kinetics of nitric oxide releae,
301:248–249
rationale for therapy, **301:**242–243
stability of compounds, **301:**247–248

synthesis
3-acetamido-4,4-dimethylthietan-2-
one, **301:**244
fructose–2-SNAP, **301:**245–246
galactose–2-SNAP, **301:**246
glucose–1-SNAP, **301:**244
glucose–2-SNAP, **301:**245
glucose–6-SNAP, **301:**245
overview, **301:**243
synthesis and addition to cell cultures,
301:127–128, 132, 446
N-Nitrosoamino acids, *see also* Nitrosoproline test
gas chromatography and detection,
301:43, 45–46
indices of endogenous nitrosation,
301:43–44
kinetics of formation, **301:**43
types in urine, **301:**43
N-Nitrosoaniline, *see* Peptidyl *N*-nitrosoani-
lines
S-Nitrosocysteine, synthesis and addition to
cell cultures, **301:**127–128, 130, 133, 446
p-Nitrosodimethylaniline, bleaching, **319:**23,
25
S-Nitrosoglutathione
addition to cell cultures, **301:**127–128
assays
colorimetric analysis with neutral
Griess reagent, **301:**203–207, 209
fluorometric analysis with 2,3-diamino-
naphthalene
interferences, **301:**231–232
linearity and detection limit,
301:232
principle, **301:**228–230
reaction conditions, **301:**204–205,
207, 211, 222–224, 228, 230–231
sample preparation, **301:**232
reagents, **301:**203
copper cleavage, **301:**221
copper,zinc-superoxide dismutase, effects
on decomposition
assay, **301:**222–224
glutathione effects, **301:**224–226
physiological implications, **301:**226–227
hexose–monophosphate pathway regulation, **301:**221
nitric oxide delivery in lipid peroxidation
studies, **301:**445–446

protection of cells against nitric oxide tox-
    icity, **301:**201–202
synthesis, **301:**127, 203, 222
S-Nitrosohemoglobin
  formation
    conformation effect on formation rate,
      301:103–104, 111
    nitric oxide scavenging, **301:**99
    S-nitrosothiol group transfer,
      301:99–103
    structural basis for modification,
      301:105–106
  implications for oxygen delivery in nor-
    mal and disease states, **301:**113–114
  local blood flow regulation, oxygen de-
    pendence, **301:**109–113
  measurement with nitrosylhemoglobin, in-
    fluence of oxygen tension, **301:**107,
    109, 111
  peptide mapping by electrospray ioniza-
    tion–mass spectrometry, **301:**106
  synthesis
    S-nitrosodeoxyhemoglobin, **301:**101
    S-nitrosomethemoglobin, **301:**101–102
    S-nitrosooxyhemoglobin, **301:**101
  vasoactivity in isolated vascular rings, oxy-
    gen dependence, **301:**109
N-Nitroso-N-oxybenzenamines
  nitric oxide release
    electrolysis, **301:**235–237, 241
    measurements, **301:**237–239
  synthesis, **301:**240–241
Nitrosoperoxocarbonate adduct
  decomposition pathways,
    301:362–363
  glutathione trapping, **301:**359
  half-life, **301:**364, 455
  protein nitration, **301:**359–362, 366
Nitrosoproline test
  gas chromatography and detection,
    301:43, 45–46
  infection studies of nitrosation in ani-
    mals, **301:**46–49
  kinetics of proline nitrosation, **301:**42
  loading methods, **301:**44
  rationale, **301:**42–43
  urine collection, **301:**44–45
S-Nitrosylated protein, *see also* Glycogen
    phosphorylase *b*; S-Nitrosohemoglobin
  assays

colorimetric analysis with neutral
    Griess reagent, **301:**203–207, 209
fluorometric analysis, **301:**204–205, 207,
    211, 233–235
reagents, **301:**203
N-methyl-N-nitrosoaniline inhibition of
    cysteine proteases, **301:**249–250
nitrosating agents, overview, **301:**249–250
Nitrotyrosine
  atherosclerotic lesion content, **300:**144
  catalysts of protein nitration, **301:**152–153
  detection and quantification, overview of
    methods, **301:**28–29
  formation in proteins from peroxynitrite,
    300:125–126, 142
  gas chromatography–mass spectrometry
    amino acid analysis, **301:**473–474
    arterial tissue analysis, **301:**481–483
    derivatization with N-methyl-N-(*tert*-bu-
      tyldimethylsilyl)trifluoroacetamide
      and spectrometry, **301:**474–475,
      477
    protein analysis, **301:**477, 480–481
  high-performance liquid chromatography
    with electrochemical detection
    acetylation and extraction of amino
      acids, **301:**32–33
    acid hydrolysis, **301:**155–156
    deacetylation and dithionite reduction,
      301:29–30, 33–34
    electrochemical detection, **301:**156, 158
    enzymatic hydrolysis, **301:**32, 155
    goals, **301:**29
    interpretation of results, **301:**158–160
    mobile phase preparation and gradient
      elution, **301:**34–36, 156
    prevention of *ex vivo* nitration,
      301:39–40
    sample preparation, **301:**31,
      154–155
    selectivity and sensitivity, **301:**38–40
    sensitivity, **301:**153–154
    standards, **301:**36–38
  immunochemical assay, **309:**143–144
  immunohistochemical analysis
    antibody preparation, **301:**375–376
    blocking with free nitrotyrosine,
      301:376–377, 380
    controls for binding specificity,
      301:379–380

cultured cell immunofluorescence,
  301:378–379
frozen tissue sample immunofluorescence, **301**:379
limitations, **301**:28–29
paraffin-embedded tissues, **301**:377–378
positive controls, **301**:380
immunoprecipitation of modified proteins
antibody preparation, **301**:138
applications, **301**:141–145
controls, **301**:139–140
elution from beads, **301**:137, 139
extraction, **301**:136–138
incubation conditions, **301**:137
microsequencing of proteins,
  301:140–141
preclearing of samples, **301**:137–138
Western blot analysis, **301**:137, 139
immunoreactivity in disease, **301**:135–136,
  144–145, 374–375, 454, 472–473, 484
isotope dilution negative-ion electron capture mass spectrometry
advantages in protein oxidation analysis, **300**:127–128
derivatization, **300**:133
gas chromatography, **300**:133–134
internal standard synthesis,
  300:128–129
mass spectrometry settings, **300**:134
protein and tissue hydrolysis,
  300:130–132
selected ion monitoring in quantification, **300**:134–136, 138–139
solid-phase extraction, **300**:132
tissue collection and preparation,
  300:129–130
manganase superoxide dismutase nitration in renal allograft rejection,
  301:136, 141, 144
mechanisms of formation, **301**:27–28, 135,
  142–144, 289–291, 301, 307–311,
  334–335, 373–374, 456, 472–473
nitrosoperoxocarbonate adduct reaction
  in formation, **301**:359–362
oxidative stress marker, **301**:151–152
peroxynitrite
scavenger assay by nitration inhibition
high-performance liquid chromatography determination, **301**:337
incubation conditions, **301**:336–337

scavenger identification, **301**:337–338
treatment of cells and product analysis,
  301:142–144, 307–309
Nitrous acid, nitrosation of lipid hydroperoxide, **301**:463–464, 469–470
6-Nitroveratryloxycarbonyl chloride conjugate, *see* Lysine, caged protein conjugates
Nitroxyl
cellular sources, **301**:280
electronic states, **301**:282
intermediate in nitric oxide synthesis,
  301:212
reactivity, **301**:212, 281–282, 454–455
stability in solution, **301**:212, 279
catalase-mediated oxidation of cyan-
  amide, **301**:213
nucleophilic substitution of acylnitroso
  compounds
hydroxamic acid oxidation in presence of amines, **301**:218–219
overview, **301**:217–218
thermal decomposition of *N*-acyl-3,6-
  dihydro-1,2-oxazines,
  301:219–220
releasing compounds
Angeli's salt, **301**:214–216, 282–285,
  287
methanesulfohydroxyamic acid,
  301:214, 216–217
overview, **301**:212–215
Piloty's acid, **301**:214–215, 217,
  286–287
synthesis
NLS, *see* Nuclear localization sequence
NMDA, *see* *N*-Methyl-D-aspartate
NMDA receptor, *see* *N*-Methyl-D-aspartate
  receptor
Nmoc, *see* *o*-Nitromandelyloxycarbonyl
  group
NMR, *see* Nuclear magnetic resonance
NNDF, *see* Nearest neighbor distribution
  function
nNOS, *see* Neuronal nitric oxide synthase
NO, *see* Nitric oxide
Nocodazole
intracellular concentration determination
nonradiolabeled compounds and high-
  performance liquid chromatography, **298**:273–274

radiolabeled compounds
   monolayer culture directly in steril-
      ized scintillation vials, **298:**273
   monolayer or suspension cultures,
      298:272–273
microtubule binding
   cell migration study, **298:**267
   depolymerization effects, **298:**265–266
   suppression of dynamic instability,
      298:266–267
   structure, **298:**256
NONOates
   addition to cell cultures, **301:**128–129
   decay rates, **301:**394
   effects on oxidative stress, **301:**414, 418
   nitric oxide delivery in lipid peroxidation
      studies, **301:**446
Nonstationary noise analysis, application to
   nonideal voltage clamp data,
   293:417–419
2-Nonyl-4-hydroxyquinoline *N* oxide, oxidation of plastoquinone pool, **297:**225
Noracetylcholine, caged compound studies
   with crystallized cholinesterases,
   291:266–267, 269
Norbutyrylcholine, caged compound studies
   with crystallized cholinesterases,
   291:267, 269
Norepinephrine transporter, *see also* Vesicular monoamine transporter
   baculovirus–insect cell expression system
      baculovirus production
         amplification, **296:**449
         overview, **296:**445–446
         plaque purification, **296:**447–448
         recombinant virus construction,
            296:447
         titration, **296:**448–449
      cell line maintenance, **296:**446
      infection, **296:**449–450
      inhibitor binding assays, **296:**453–455
      membrane preparation, **296:**450
      principle, **296:**444–445
      uptake assays, **296:**452–453
      Western blot analysis of neurotransmit-
         ter transporter expression,
         296:450–452
   binding assays
      PC12 cell plasma membrane vesicles,
         296:277–278

      placenta brush border membrane vesi-
         cles, **296:**285–286
      principle, **296:**274–275
      transfected cells, **296:**351, 353–354
   carrier-mediated transport, characteris-
      tics, **296:**267
   cell surface biotinylation, **296:**315–318,
      362, 364–366, 368
   chimeric norepinephrine transporter–
      dopamine transporter
      clones and plasmids, **296:**467–468, 470
      complementary DNA, size analysis,
         296:468–471
      expression analysis by transport and
         binding assays, **296:**469–471
      homology between transporters,
         296:466
      [^{125}I]RTI-55 binding of chimeras with
         attenuated transport function,
         296:470, 473–475
      transformation of bacteria, **296:**468, 470
      transmembrane domain functional anal-
         ysis, **296:**479–480
   data analysis of transport and binding
      assays, **296:**275, 277
   electrophysiology
      four-state alternating access model,
         296:580–581
      norepinephrine-induced currents
         fluctuations, **296:**587, 589–590
         inward currents, **296:**586–587
      stable transfection in human em-
         bryonal kidney cells, **296:**578–579,
         585–586
      stoichiometry of transport, **296:**579
      transport rate, **296:**579, 582
      two-state model, **296:**581–582
      whole-cell voltage clamp, **296:**586
   glycosylation of human protein
      site-directed mutagenesis of N-glycosyl-
         ation sites, **296:**360, 362
      trafficking role, **296:**362, 364–366, 368
      Western blot analysis, **296:**355–357
   immunoprecipitation and pulse-chase la-
      beling, **296:**354, 357–360
   intracerebral microdialysis and norepi-
      nephrine detection
      anesthesia, **296:**721–722
      apparatus, **296:**726–727
      contamination, **296:**723–724

guide cannula implantation, **296:**726
high-performance liquid chromatogra-
    phy with electrochemical detection
    apparatus, **296:**728
    mobile phase preparation, **296:**728
    monoamine standard preparation,
        296:729
    peak identification, **296:**730
    running conditions, **296:**730
perfusion solution, **296:**722, 727
presentation of data, **296:**725–726
principle, **296:**720–721
probes
    calibration, **296:**727
    placement verification, **296:**730
    recovery performance, **296:**723
    types, **296:**722–723
sample collection, **296:**727–728
uptake blockers
    administration routes, **296:**725
    monoamine effect, evaluation,
        296:724
intravesicular water space determination,
    296:266
kinetic theory
    binding, **296:**270–271
    interpretation of kinetic constants,
        296:271–273
    transport, **296:**269–270
phylogenetic analysis, **296:**428–436
placenta brush border membrane vesicle
    preparation
    advantages as starting tissue, **296:**278–
        280, 289–290
    centrifugation, **296:**282–283
    dissection, **296:**281
    federal regulations, **296:**278
    flow chart, **296:**282
    gross structure, **296:**280–281
    homogenization, **296:**283
    marker enzymes, **296:**284
    storage, **296:**284
    tissue collection, **296:**280
plasma membrane vesicle isolation from
    PC12 cells
    applications of isolated vesicles,
        296:261
    cell culture, **296:**261–262
    homogenization, **296:**262–264
    marker enzymes, **296:**265–266

sucrose gradient centrifugation,
    296:263–265
ultracentrifugation, **296:**263
protein kinase C, regulation of biosynthe-
    sis and surface expression, **296:**368,
    370
rotating disk electrode voltammetry, up-
    take assay
l-amphetamine competition with dopa-
    mine, **296:**658–659
comparison to radioactive catechola-
    mine kinetic analysis, **296:**649–650
electrode preparation, **296:**652–654
incubation conditions, **296:**653
instrumentation, **296:**650–652
Michaelis–Menten kinetics analysis,
    296:655, 658
oxidative current monitoring, **296:**654
transfection of cells with human trans-
    porter, **296:**650, 652
two-substrate model, **296:**658–659
site-directed mutagenesis of conserved
    aspartate in transmembrane domain
    1, **296:**477–479
temperature dependence of uptake,
    296:266–267
transient transfection and cell culture,
    296:350
translocation mechanism, overview,
    296:267–268
uptake assay
    PC12 cell plasma membrane vesicles,
        296:277
    placenta brush border membrane vesi-
        cles, **296:**284–285
    principle, **296:**273
    transfected cells, **296:**351
Western blot analysis, **296:**354–357
Northern blot analysis
    antisense knockdown of transcripts
    c-Jun N-terminal kinase, **314:**367–368
    c-Myc, **314:**562, 564
    proprotein convertases, **314:**108–109,
        112
    Raf-1, **314:**447
    transforming growth factor-β,
        314:495–496
    antisense oligonucleotide transfection
        analysis, **313:**60–63
    antisense RNA

analysis from single cells, **303:**10
expression, **313:**471–472
Arabidopsis thaliana, ozone-sensitive mutants, **319:**533–534
chloroplast mutant phenotypes
agarose gel electrophoresis, **297:**47
buffers, **297:**47
hybridization reaction, **297:**48
RNA isolation, **297:**47
transfer to membranes, **297:**48
cytokine-induced neutrophil chemoattractant induction by ozone, **319:**556–557
differential display, verification of differential expression
Northern blot analysis, **303:**252–253
reverse Northern dot blotting,
303:253–25
flagellin gene activation during agar
growth, **310:**69
I-κB induction by nitric oxide, **300:**375,
381–383
lecithin:retinol acyltransferase tissue distribution, **316:**408–410
light-activated gene transcription in
Arabidopsis thaliana
hybridization, **297:**185
overview, **297:**184–185
reagents, **297:**175–186
MDR1 RNA expression in transgenic
mice, **292:**580
multidrug-resistance associated protein
messenger RNA expression,
292:603–604
psbA transcripts, **297:**201, 203
retinitis pigmentosa GTPase regulator,
316:703
ribonuclease H assays, *see* Ribonuclease
H
ribozymes or substrates, **306:**221
short-chain dehydrogenase/reductase,
316:379–380
UAS$_G$–*SNR6* transcript quantification,
303:419 420
NOS, *see* Nitric oxide synthase
NPC, *see* Nuclear pore complex
NPEGTA, calcium uncaging efficiency by
two-photon photolysis, **291:**367–369
NPM, *see* *N*-(1-Pyrenyl)maleimide
NPRO test, *see* Nitrosoproline test
NPY, *see* Neuropeptide Y

NQNO, *see* 2-Nonyl-4-hydroxyquinoline *N*
oxide
NTT4
genes, **296:**427
phylogenetic analysis, **296:**428–436
Nuclear factor 1, DNA-binding site on
nucleosomes, **304:**321
Nuclear factor-κB
activation mechanism, **300:**346, 363–364,
375
activators, overview, **319:**120–121, 273
aminopyropheophorbide, colon carcinoma cell photosensitization and
gene activation
antioxidant effects, **319:**128
deuterium oxide effects, **319:**127, 129
electrophoretic mobility shift assay,
319:124–125, 127
irradiation conditions, **319:**124
Anethole dithiolthione effects,
299:300–301
classification, **300:**374–375
cytokine gene regulation, **319:**122
domains, **319:**119
electrophoretic mobility shift assay
binding reaction, **319:**596–598
cold competition assay, **319:**598–599
extract preparation, **319:**593–595
gel preparation, **319:**595–596
materials, **319:**587
oligonucleotide labeling, **319:**589–591
oxidant dose response and kinetics,
319:599
oxidant treatment of cells
adherent cells, **319:**593
suspended cells, **319:**592
sensitivity, **319:**586
stock solution preparation,
319:587–589
supershift assay, **319:**598–599
troubleshooting
background, **319:**601–602
high activation variability, **319:**602
protein content variability in extracts, **319:**601
quantitative analysis, **319:**602
smeared lanes, **319:**599, 601
tumor necrosis factor activation,
319:599
Western blot analysis, **319:**598

electrophoretic mobility shift assay of activation
competition assay, **300:**366
electrophoresis, **300:**366
I-κB induction analysis, **300:**383–384
incubation conditions, **300:**365–366
nuclear extract preparation, **300:**365
oligonucleotide labeling, **300:**365
principle, **300:**364–365
proteasome inhibitor analysis, **300:**351–352
supershift assay, **300:**367–368
green fluorescent protein reporter assay
advantages, **302:**32–33
destabilized protein for rapid turnover, **302:**33, 37
high-throughput drug screening, **302:**38
principle, **302:**33
tumor necrosis factor-α induction
dose response, **302:**35–36
time course, **302:**34–35
inflammation role, **300:**345–346
inhibitors
degradation, **319:**120, 274, 277–278, 585
kinase assay, **319:**278–279
phosphorylation, **319:**273–274, 277–279, 585
types, **319:**119
inhibitory protein I-κB, *see also* I-κB
activation regulation, **300:**346, 364, 375
induction by nitric oxide
electrophoretic mobility shift assay of nuclear factor-κB, **300:**383–384
immunoprecipitation of I-κB–NF-κB complexes, **300:**379–381
luciferase as reporter gene, **300:**387–388
Northern blot analysis, **300:**375, 381–383
nuclear run-on assay, **300:**385–386
transient tranfection assay, **300:**376, 386–388
Western blot analysis, **300:**375–379
phosphorylation and degradation, **300:**348, 363
proteolysis by ubiquitin–proteasome pathway, **300:**346–348
intercellular adhesion molecule-1 assays of activation, **300:**354–356, 373–374

nuclear translocation of p65 in activation, assay
immunofluorescence, **300:**369
Western blot analysis, **300:**369
ozone induction in rat lung
dexamethasone inhibition, **319:**560
electrophoretic mobility shift assay, **319:**557–558, 560
inflammation role, **319:**553–554
photodynamic therapy response, **319:**355
posttranslational processing, **319:**119–120, 585
proteasome inhibitors, inhibition of nuclear factor-κB activation
effect on bacterial cell wall-induced arthritis
leukocyte isolation, **300:**358–359
polyarthritis induction in rats, **300:**357–358
prevention of arthritis by inhibitors, **300:**360–361
residual 20S proteasome activity, assay, **300:**359–360
effect on delayed-type hypersensitivity, **300:**361–362
electrophoretic mobility shift assay, **300:**351–352
fluorescence immunoassay of cell adhesion molecule expression, **300:**354–356
immunoassay of interleukin-2 expression, **300:**356–357
Western blot analysis of I-κB degradation, **300:**353–354
subcellular localization, **319:**128
target genes, **300:**372–373, **319:**273
transactivation assays with reporter genes, **300:**369–372
ultraviolet A activation
electrophoretic mobility shift assay, **319:**275–276
extract preparations
lysates for luciferase assay, **319:**275
nuclear extracts, **319:**274–275
whole cell extracts, **319:**275
inhibitor analysis
glutathione *S*-transferase fusion, **319:**279
kinase assay, **319:**278–279
Western blot analysis of degradation

and phosphorylation,
    319:277–278
luciferase reporter gene assay,
    319:276–277
nuclear translocation detection by indi-
    rect immunofluorescence, **319:**276
Nuclear localization sequence, transport
    mechanism, **307:**208–209
Nuclear magnetic resonance
    acylhomoserine lactone autoinducer
        spectra
        carbon-13 spectroscopy, **305:**312
        proton spectroscopy, **305:**311–312
        7,8-*cis*-tetradecenoylhomoserine lac-
            tone, structure elucidation,
            305:297, 299
    amyloid-β peptide purity analysis,
        309:193–196
    boranophosphate oligonucleotides
        nucleotide synthesis monitoring with
            phosphorous-31 spectroscopy,
            313:241, 243–244
        quantitative analysis, **313:**253–254
    gangliosides, *see* Nuclear magnetic reso-
        nance, gangliosides
    gramicidin A conformational analysis
        solid-state studies, **294:**536–539
        solution studies, **294:**535–536
    guanylate cyclase-activating protein 2
        structure determination
        data acquisition and analysis,
            316:125–126
        isotopic labeling and purification of re-
            combinant protein, **316:**124–125
        sample preparation, **316:**125
        three-dimensional structures
            recoverin homology, **316:**131–132
            unmyristoylated protein with cal-
                cium, **316:**130–131
    helix formation and amide proton ex-
        change, **295:**7, 10–12, 15, 29
    low-barrier hydrogen bond studies by pro-
        ton resonance
        bond lengths, **308:**224–226
        criteria for identification, **308:**222–223
        data acquisition, **308:**225–229
        deshielding of resonance, **308:**224
        fractionation factor
            bond length determination,
                308:234–235

deuterium exchange reaction,
    308:232–233
factors less than unity, **308:**233
measurement, data acquisition and
    calculations, **308:**235–237
quantum mechanical theory,
    308:233–234
free energy of formation
    mutation studies of enzymes,
        308:243–244
    pH titration in measurement,
        308:244–245
overview of enzymes, **308:**221–222,
    230, 245
proton exchange rates
    limiting mechanisms, **308:**239
    measurement, data acquisition and
        calculations, **308:**240–243
    mechanism representation in pro-
        teins, **308:**238–239
    protection factor, **308:**237–239
resonance assignment
    J coupling, **308:**229
    mutagenesis of enzyme, **308:**229,
        231
    nuclear Overhauser effects, **308:**232
temperature measurement and calibra-
    tion, **308:**227–228
water suppression, **308:**228–229
opaque biofilms
    diffuse filtering technique, **310:**263
    growth monitoring, **310:**263
    overview, **310:**259
    resolution of imaging, **310:**262
    water signal, **310:**260, 262–263
peptide ion channel ligands, structure de-
    termination
    buffers, **294:**111
    dynamic analysis with relaxation stud-
        ies, **294:**112–116
    heteronuclear-based determinations,
        294:97–98, 111–112
    homonuclear proton-based determina-
        tions, **294:**95–97, 109–110
    isotope labeling, **294:**99–100, 116–117
    relaxation times, senstivity to correla-
        tion time, **294:**98
    rigidity in solution, **294:**93–95
    suitability for nuclear magnetic reso-
        nance, **294:**92–93

protein structure studies, advantages and disadvantages, **309**:608
recoverin structure determination
data acquisition and analysis, **316**:125–126
isotopic labeling and purification of recombinant protein, **316**:124–125
sample preparation, **316**:125
three-dimensional structures
calcium-induced conformational changes, **316**:128–129
myristoylated protein with and without calcium, **316**:127
unmyristoylated protein, **316**:126–127
rhodopsin, structure determination of domains
applications, **315**:115
assembly of structure, **315**:107–108
cytoplasmic surface structure elucidation, **315**:112–113
data acquisition, **315**:110
initial analysis, **315**:109–110
long-range constraints, **315**:114–115
peak assignment, **315**:110–111
peptides
selection for analysis, **315**:108–109
structures, **315**:111–112
proline kinks, **315**:114
secondary structure analysis, **315**:110, 114
short-range interactions, **315**:113–114
validity of whole protein structure determination, **315**:107–108
RNA, *see* Nuclear magnetic resonance, RNA
solid state, *see* Nuclear magnetic resonance, solid state
stable-isotope labeling of opsins in baculovirus–insect cell recombinant expression system, **315**:25–26
tryptophan synthase, solid-state studies of substrate channeling
advantages over solution-state studies, **308**:128
aminoacrylate intermediate, **308**:128–132
serine binding site, **308**:132
Nuclear magnetic resonance, gangliosides
aggregate features, **312**:249–250, 265
biotinylated GM1, 538, 544

data acquisition modes for structural analysis
COSY, **312**:255, 262
heteronuclear single quantum coherence, **312**:261
HOHAHA, **312**:252–253, 264
NOESY, **312**:268
one-dimensional proton resonance, **312**:255
ROESY, **312**:253, 265, 267–268, 270
TOCSY, **312**:261–262
primary structure
ceramide moiety, **312**:264
linkage position and sugar sequence, **312**:261–262, 264
monosaccharide composition, **312**:255, 260–261
sugar configuration, **312**:255, 260–261
proton chemical shifts
labile protons, **312**:258
nonlabile protons in water, **312**:259
nonlabile protons, **312**:256–257
solvent effects, **312**:253–254
temperature effects, **312**:254
sample preparation
desalting, **312**:251
dimethylsulfoxide dissolution, **312**:251
ganglioside/dodecylphosphocholine mixed micelles, **312**:251
lipid-modified ganglioside micelles, **312**:251–252
purity requirements, **312**:249–251
secondary structure
conformation, **312**:265–270
dynamics, **312**:270–272
model system, **312**:264–265
structural information available, **312**:247–248
Nuclear magnetic resonance, RNA
alignment of molecules for dipolar coupling interactions
advantages of dipolar coupling studies, **317**:220–221
calculations, **317**:233
liquid crystalline phases, **317**:214–215, 221–222
magnetic field strength dependence, **317**:221
paramagnetic-induced alignment, **317**:221

Pf1 filamentous phage
  alignment assay for phage prepara-
    tions, **317:**227
  buffers, **317:**230, 232
  cell growth and harvesting, **317:**225
  comparison with other phages,
    317:223
  concentration optimization,
    317:228–229
  flexibility and advantages of ap-
    proach, **317:**222, 224–225,
    239–240
  homogeneity assay of phage prepara-
    tion, **317:**226–227
  IRE-I RNA ^{1}H–^{15}N coupling, dis-
    tance and angle information,
    317:233–237, 239
  line broadening, **317:**235
  mechanism of nucleic acid align-
    ment, **317:**222–223
  potassium bromide density gradient
    centrifugation of phage, **317:**226
  precipitation of phage, **317:**225
  safety in preparation, **317:**229
  sample preparation, **317:**227–228
  stability of samples and recycling of
    components, **317:**232–233
  storage of phage, **317:**226
  surface area of phage, **317:**223–224
  temperature dependence, **317:**222,
    224–225, 230
  viscosity effects, **317:**235
  water alignment, **317:**230
  proton–proton interactions in dipolar
    coupling spectroscopy, **317:**237,
    239
aminoglycoside–ribosomal RNA com-
    plexes
  affinity parameters, determination,
    317:250–251
  aminoglycoside substituent effects,
    317:259–260
  biological prediction testing, **317:**258
  buffers for data collection, **317:**244
  chemical exchange, slow, intermediate,
    and fast, **317:**251–253
  chemical modification of RNA,
    317:242–243
  design of RNA, **317:**241
  drug design application, **317:**260
  labeling and synthesis of RNA,
    317:244–245
  mapping of antibiotic-binding site on
    RNA chemical shift, **317:**253–254
  mechanism of antibiotic action, **317:**240
  model comparison with ribosome,
    317:243
  mutant oligonucleotides, **317:**243–244,
    258–259
  nuclear Overhauser enhancement spec-
    troscopy, **317:**254–255
  prospects, **317:**261
  resonance assignments
    antibiotic, **317:**246–248
    exchangeable protons, **317:**245–246
    RNA protons, **317:**245
  structure
    calculations, **317:**255–256
    comparison with chemical modifica-
      tion data, **317:**256–257
    comparison with mutational data,
      317:257–258
  titration of antibiotic into RNA,
    317:249–250
deuteration for structure studies, ratio-
    nale, **317:**18–19
protein–RNA complexes
  chemical shift perturbation analysis,
    mapping interaction surfaces
    data acquisition, **317:**200–201
    heteronuclear single quantum coher-
      ence, **317:**200
    limitations, **317:**201, 203
    principle, **317:**199
    RNA-dependent protein–protein in-
      terfaces, **317:**201
  deuteration of proteins for structure
    and dynamics analysis,
    317:206–208
  difficulty of study, **317:**198–199
  long-range intermolecular distance de-
    termination with paramagnetic
    spin labels
    heteronuclear single quantum coher-
      ence, **317:**210–211
    labeling of protein or RNA,
      317:209–210
    principle, **317:**208–209
    sensitivity, **317:**212
    structure elucidations, **317:**212–213

TROSY, **317**:211
nuclear Overhauser enhancement spectroscopy for structure determination
carbon-filtered spectra, **317**:203–204
limitations, **317**:208
nitrogen-edited experiments,
317:204–206
principle, **317**:203
prospects, **317**:218–220
residual dipolar couplings
alignment of molecules, **317**:214–215
calculations, **317**:217
heteronuclear single quantum coherence, **317**:215
principle, **317**:213–214
refinement of structures, **317**:215,
217–219
TAR of human immunodeficiency virus-
2, structure determination with isotopically-labeled RNA
heteronuclear single quantum coherence, **317**:37–38
nuclear Overhauser enhancement spectroscopy, **317**:37
synthesis from isotopically-labeled ribonucleotides, **317**:36
Nuclear magnetic resonance, solid state
advantages in protein aggregate studies,
309:536–537
amyloid-β fragments, **309**:557
chemical shielding, **309**:538
chemical shifts and extracting structural
parameters
backbone angles, **309**:544–546
secondary structure, **309**:542–544
dihedral angle determination experiments, **309**:546–547
dipolar couplings, **309**:540–541
distance determination experiments
dipolar recoupling with a windowless
sequence, **309**:553–554
dipolar recovery at the magic angle,
309:553
radio frequency-driven recoupling,
309:554–555
rotational echo double resonance,
309:551, 553
rotational resonance tickling,
309:550–551

rotational resonance, **309**:548–550
simple excitation for the dephasing of
rotational-echo amplitudes,
309:553
transferred-echo double resonance,
309:553
heteronuclear decoupling, **309**:541
homonuclear decoupling, **309**:541–542
instrumentation and field strength,
309:537, 555
isotopic enrichment of samples,
309:555–556
large protein systems, **309**:558–559
line shapes and magic angle spinning,
309:538–540
prion protein, **309**:557–558
prion protein, isotopically labeled peptides, **309**:589, 591
Nuclear pore complex
confocal microscopy
accuracy of localization
calculation, **307**:214–215
experimental determination, **307**:221
distribution analysis
cluster analysis, **307**:220–221
estimator computation, **307**:219
nearest neighbor distribution function, **307**:215, 217–220, 226–227
overview, **307**:215, 217
pair correlation function, **307**:218,
220, 227
randomly distributed point objects,
307:219–220
fluorescent labeling, **307**:221–222
localization
accuracy, **307**:226
data processing, **307**:223, 225–226
distribution analysis, **307**:226–227,
229
representation, **307**:226
optimization of imaging conditions,
307:222–223
point spread function approximation,
307:211–212
signal-to-noise ratio, **307**:212, 215
functions
nuclear morphogenesis, **307**:209–210
translocation of proteins, **307**:207–209
lateral mobility, **307**:210–211, 229
structure, **307**:207–208

Nuclear run-on assay
  comparison to other transcription assays,
    304:351–352
  Drosophila assays
    buffers and reagents, **304:**354
    hybridization of labeled run-on RNA
      to specific sequences, **304:**357–358
    nuclei preparation
      adult flies, **304:**355–356
      cell cultures, **304:**354–355
      lysis of cells, **304:**354
      overview, **304:**353–354
    run-on reaction, **304:**356–357
  elongation rate and interpretation of re-
    sults, **304:**361
  kinetic analysis of polymerase associa-
    tions, **304:**362
  mammalian cell assays
    buffers and reagents, **304:**358–359
    nuclei preparation, **304:**359
    run-on reaction, **304:**359–360
  principle, **304:**352–353
  quantitative comparison of different
    genes, **304:**361–362
  yeast assays
    buffers and reagents, **304:**360
    permeabilized cell preparation,
      304:360
    run-on reaction, **304:**360
Nuclear transformation, *Chlamydomonas
  reinhardtii*
  applications, **297:**27–28
  frequency, **297:**27, 31
  integration site, **297:**27
  photosynthesis applications, **297:**37–38
  plasmid preparation, **297:**32
  promoters, **297:**30
  reporters, **297:**29–30
  selectable markers, **297:**27–29, 36–37
  strains, **297:**31–32
  transformation techniques
    glass bead transformation
      cell growth, **297:**33–34
      gamete lytic enzyme, preparation
        and cell treatment, **297:**34–35
      materials, **297:**33
      overview, **297:**31
      particle bombardment, **297:**35–36
Nucleation, protein aggregates, *see* Aggregation kinetics

Nucleolin, translational repression for RNA
  binding screening, **318:**329
Nucleoside diphosphate kinase
  assay and kinetic parameters for bovine
    retina enzyme, **316:**92–93, 99
  expression of bovine retina enzyme in
    Escherichia coli
    cell growth, **316:**95–96
    cloning of isoforms, **316:**93–94
    physical properties, **316:**96
    purification, **316:**96
    vectors, **316:**94–95
  glycosylation analysis, bovine retina en-
    zyme, **316:**90, 92, 98
  homology between species, **316:**98–99
  isoform comparisons of bovine retina en-
    zyme, **316:**96–98
  phototransduction role, **316:**87
  purification from bovine retina,
    316:88–89
  quaternary structure, bovine retina en-
    zyme, **316:**89–90
Nucleoside hydrolase, *see also* Inosine–
  uridine nucleoside hydrolase
  inhibitor design, **308:**354–355, 394
  isotope effect measurement, experimental
    setup, **308:**318, 320
  molecular electrostatic potential surface
    of transition-state, **308:**346
  transition-state modeling, **308:**336–340,
    342–343
Nucleoside 5′-*O*-(1-thiotriphosphate), synthesis, **317:**82, 96–98
Nucleosome, *see also* Chromatin; Histone;
  Nucleosome core particle
  array formation from *Xenopus* extracts
    DNA concentration optimization,
      304:55
    DNA supercoiling assay, **304:**53, 62–63
  array reconstitution from purified compo-
    nents
    dialysis conditions for reconstitution,
      304:24, 26
    DNA template purification, **304:**22–23
    histone octamer purification, **304:**23–24
    histone saturation of reconstituted
      arrays
      enrichment of saturated arrays,
        304:32–34
      importance of determination, **304:**26

overview of assays, **304:**26–27
quantitative agarose gel electrophoresis assay, **304:**27, 30
restriction enzyme digestion assay,
304:30–31
sedimentation velocity ultracentrifugation assay, **304:**31–32
states of saturation, **304:**20
instrumentation, **304:**22
materials and reagents, **304:**21–22
DNA accessibility, restricting factors,
304:313
DNA pitch, **304:**251–252
electron microscopy, *see* Electron microscopy
footprinting, *see* Dimethyl sulfate; DNase
I footprinting; Hydroxyl radical footprinting; Methidiumpropyl-EDTA–
Fe(II) footprinting; Micrococcal
nuclease; Permanganate footprinting;
Restriction enzyme accessibility mapping; Ultraviolet footprinting
hydroxyl radical footprinting, *see* Hydroxyl radical footprinting
linker DNA in array formation, **304:**19,
298
linker histone H5 incorporation into
model nucleosome arrays, **304:**34
oligonucleosome preparation from rat
liver nuclei
gel filtration chromatography, **304:**317
liver homogenization, **304:**314–315
nuclei isolation and micrococcal
nuclease digestion, **304:**315–317
solution preparation, **304:**314
polyacrylamide gel electrophoresis, *see
also* Electrophoretic mobility shift
assay
advantages, **304:**312
native nucleosomes
cage effect in stabilization of complexes, **304:**300
casting of gels, **304:**300
comparison to standard DNA gel
electrophoresis, **304:**299–300
effectiveness, **304:**304
electrophoresis conditions, **304:**300
glycerol inclusion, **304:**302–303
histone acetylation detection,
304:303–304

optimization, **304:**301
particle mass separations, **304:**303
staining, **304:**301
nucleosome mobility assays
interpretation of bands, **304:**311
principle, **304:**309, 312
two-dimensional gel electrophoresis,
304:309–311
positioned nucleosomes
preparative gel electrophoresis,
304:307–308
principle of separation, **304:**304–307
positioning
exonuclease III assay
digestion conditions and analysis,
304:322
principle, **304:**321–322
high-resolution mammalian mapping
DNaseI footprinting, **304:**636–637
double-stranded micrococcal
nuclease footprinting, **304:**638
ligation-mediated polymerase chain
reaction, **304:**637–638
importance of studies, **304:**626
low-resolution mammalian mapping
indirect end-labeling, overview,
304:631–632
micrococcal nuclease digestion,
304:633–634
nuclei isolation, **304:**632–633, 635
optimization, **304:**635–636
Southern blot analysis, **304:**634–635
mammalian mapping, *in vitro*
core particle end detection with exonuclease III, **304:**630
core particle end generation with micrococcal nuclease, **304:**629–630
nucleosome dyad mapping with cysteine-substituted H4, 631
reconstitution of nucleosomes,
304:627
rotational positions and dyads
mapped with DNase I,
304:628–629
rotational versus translational, **304:**377
yeast assay
cleavage reagents in assays, **304:**378,
380, 387, 389
digestion reactions, **304:**391–392
indirect end-labeling for low-resolu-

tion mapping, **304**:380–382,
    394–396
nuclei isolation, **304**:389–391
positioning mechanisms, **304**:377–378
primer extension analysis for high-
    resolution mapping, **304**:385–
    387, 396–399
restriction endonuclease mapping,
    304:383, 385
spheroplast preparation and diges-
    tion, **304**:392–394
reconstitution by histone exchange
DNA sequence selection for transcrip-
    tion factor interaction assays,
    304:320–321
glycerol gradient centrifugation,
    304:318–320
incubation conditions, **304**:318
solution preparation, **304**:318
structure, overview, **304**:231–232, 298,
    313, 377
transcription factor binding, *see* Electro-
    phoretic mobility shift assay; Di-
    methyl sulfate, footprinting; DNase I
    footprinting
Nucleosome core particle
definition, **304**:299
gel shift assay and repositioning of parti-
    cles, **304**:17–19
high mobility group protein binding, *see*
    High mobility group proteins
preparation from recombinant histones
advantages, overview, **304**:3–4
dialysis conditions for reconstitution,
    304:11, 13
DNA fragment preparation, **304**:11
histone purification
    anion-exchange chromatography,
        304:8–9
    expression in *Escherichia coli*,
        304:5–6
    gel filtration, **304**:7–8
    inclusion body preparation,
        304:6–7
    yield, **304**:5
octamer refolding, **304**:9–10
purification
    anion-exchange chromatography,
        304:13–16
    modified complexes, **304**:14

preparative gel electrophoresis,
    304:13–14, 16–17
Nucleosome remodeling factor
overview of function, **304**:742, 757–758
purification
    cellulose phosphate chromatography,
        304:764
    Drosophila extract preparation,
        304:759–762
    glycerol gradient centrifugation,
        304:764–765
    hydroxylapatite chromatography,
        304:763–764
    ion-exchange chromatography,
        304:762–763
    overview, **304**:758–759
    yield, **304**:759
subunits, **304**:758
Nucleotide analog interference mapping,
    see also Phosphorothioate interference
    analysis
nucleotides and applications,
    317:80–81
phosphorothioate-tagged nucleotide tri-
    phosphate analogs
    adenosine, **317**:104–106
    cytidine, **317**:108
    guanosine, **317**:106–108
    uridine, **317**:108–109
principle, **317**:79, 81, 92–93
Tetrahymena group I intron analysis
materials and reagents, **317**:93, 95
nucleotide incorporation into RNA
    transcripts with mutant T7 RNA
    polymerase
normalization of incorporation effi-
    ciency, **317**:99–100
nucleotides and reactions, **317**:98–99
phosphorothioate-tagged nucleotide tri-
    phosphate synthesis, **317**:96–98
quantitative analysis, **317**:102, 104
RNA selection assay, **317**:100–102
transfer RNA aminoacylation analysis
aminoacylation reaction, **317**:87–88
critical phosphate and 2′-hydroxyl posi-
    tion identification, **317**:90–91
iodine cleavage and cleavage pattern
    analysis, **317**:89–90
materials, **317**:83–84
overview, **317**:82–83

separation of charged and uncharged RNAs, **317**:88–89

transcript

dephosphorylation, **317**:86–87

nucleotide incorporation into RNA transcripts with mutant T7 RNA polymerase, **317**:85–86

phosphorous-32 labeling, **317**:87

Nucleotide excision repair

chromatin remodeling during repair

DNA substrate preparation, **304**:334–335

Drosophila embryo extract preparation, **304**:335, 338

efficiency of repair, assay, **304**:337–338

extracts for study, **304**:333–334, 338, 350

micrococcal nuclease assay of nucleosome assembly

assembly reaction conditions and repair-independent suppression, **304**:341, 344

digestion reaction, **304**:344

gel electrophoresis and analysis, **304**:344–345

mechanism of nucleosome assembly from target site for repair, **304**:345–347, 351

principle, **304**:338, 341

Southern blot analysis, **304**:345

remodeling factors and functions, **304**:350

lesion targets, **304**:333, 447–448

Nucleus, *see also* Solid-phase nucleus

aequorin targeting for calcium measurements, **305**:482

fluorescent probes, **307**:121

Numerical aperture

pinhole size in confocal microscopy, **307**:12

point-spread function dependence, **307**:8–9

resolution of microscopes, **307**:19, 298

NURF, *see* Nucleosome remodeling factor

Nystatin

cystic fibrosis transmembrane conductance regulator assay by proteoliposome fusion, **294**:241–242, 245–246

nystatin/ergosterol-induced vesicle fusion with planar bilayers

channel density calculation, **294**:333–337

enhancing vesicle delivery to bilayers, **294**:329–330

limitations, **294**:337–338

lipid-to-protein ratio, **294**:332–333

native fusigenic vesicle preparation, **294**:330–334

osmotic gradient, **294**:328, 338

overview, **294**:320–321

stirring, **294**:329, 338

vesicle lipids, **294**:321–322

vesicle size, **294**:322–325

O

Obelin

advantages as calcium reporter, **305**:226

calcium indicator compared with aequorin

calcium concentration–effect curves, **305**:243–245

fractional rate of discharge, expression of results, **305**:240–241

magnesium effects, **305**:245–248

overview, **305**:224–225, 248–249

rapid kinetics measured by stopped-flow, **305**:245–248

chromophore formation, **305**:225

complementary DNA cloning from *Obelia longissima*

bioluminescence expression cloning, **305**:228–230

library construction, **305**:226–228

sequence analysis, **305**:230–231

expression and purification of recombinant *Obelia longissima* apoobelin in *Escherichia coli*

anion-exchange chromatography, **305**:234–235, 237

cell growth and harvesting, **305**:232–233

cell lysis, **305**:233–234

denaturing gel electrophoresis analysis, **305**:237–238

expression vector, **305**:231–232

inclusion body solubilization, **305**:233–234

yield, **305**:238

ligand charging, **305:**239
photocyte localization, **305:**224
stability, **305:**248
structure, **305:**224–225
Objective lens, confocal microscope
    aberrations, *see* Chromatic aberration;
        Spherical aberration
    calcium imaging, **307:**427
    design and control in tissue thickness
        measurement, **307:**235
    numerical aperture, *see* Numerical ap-
        erture
    selection factors, **307:**297–299
Octanoylsphingosine, synthesis, **311:**44
Octanoyl-D-*erythro*-sphinganine, carbon-14
    labeled substrate preparation,
    311:25–26
Oligonucleotide, *see also* Antisense oligonu-
    cleotide; Boranophosphate oligonucleo-
    tides; Methylphosphonodiester oligo-
    deoxynucleotide; Minor groove
    binder–oligodeoxynucleotide conju-
    gate; P-ethoxy oligonucleotide; Phos-
    phoramidate oligonucleotide; Phospho-
    rothioate oligodeoxynucleotide; RNA
    oligonucleotide
    isotopic labeling for kinetic isotope effect
        studies, **308:**316–318
    photocleavable biotin labeling
        affinity purification, **291:**145, 150–151
        high-performance liquid chromatogra-
            phy, **291:**149–150
        phosphorylation at 5′-end, **291:**153
        photocleavage, **291:**145–146, 150–151
        polyacrylamide gel electrophoresis anal-
            ysis, **291:**151–152
        synthesis, **291:**145, 149, 153
Oligonucleotide-primed *in situ* DNA synthe-
    sis, single copy gene localization with
    confocal microscopy, **307:**202–203
Oligonucleotide radiotherapy
    applications, **314:**519, 521, 523–524
    dose modeling
        Medical Internal Radiation Dose
            scheme, **314:**510–511, 513
        nuclear *S* factor calculations,
            314:514–517
        organ dose estimation, **314:**511–513,
            522
        overview, **314:**509–510

phosphorothioate oligodeoxynucleotide
        pharmacokinetics, **314:**521–522
    protein association effects, **314:**520
    relative advantage factor, **314:**515, 517
    relative efficacy calculation, **314:**515,
        517
    tumor dose estimation, **314:**512, 514
    dual labeling, **314:**507, 522–523
    necrotic cell imaging, **314:**521
    oligodeoxyribonucleotide design,
        314:507–509
    radionuclides and labeling, **314:**506–509,
        514–515, 519–523
    radioresponsiveness, **314:**518
    radiosensitivity, **314:**518
    rationale, **314:**506–508, 518–519
    uptake of oligonucleotides, **314:**518, 520
OmpA, immunoprecipitation in disulfide
    bond formation studies, **290:**65–67
OP-Cu, *see* Copper phenanthroline
Opioid receptor
    antisense oligodeoxynucleotide targeting
        administration in mice and rats,
            314:6–7, 98–99
        antisense mapping
            DOR-1 δ receptor, **314:**57–58, 95
            MOR-1 μ receptor, **314:**58, 60
        applications
            analgesia pathway delineation,
                314:99–100
            correlation of cloned and pharmaco-
                logically defined receptors,
                314:18
            G protein assignment to receptors,
                314:18–19, 99–100
        assay system sensitivity, **314:**54–55
        controls, **314:**56, 97–98
        dose of oligonucleotide, **314:**55
        fluorescein-labeled oligodeoxynucleo-
            tide distribution in central nervous
            system, **314:**8–9
        frequency of antisense treatment,
            314:55
        Gα subunit targeting
            dose, **314:**96
            sequences, **314:**6
            specificity validation, **314:**101
            in vivo knockdown, **314:**98–99
        immunodetection of targeted proteins
            antibody injection, **314:**9

antibody purification, **314:**10–11
antibody radioiodination, **314:**11
autoradiography and analysis of expression, **314:**11–13
knockout mouse comparison, **314:**57
media for cell culture studies, **314:**94
messenger RNA abundance effects on
knockout, **314:**53–54
nonspecific effects, **314:**56–57
oligonucleotide stability in central nervous system, **314:**53
optimization of conditions
in vitro studies, **314:**93–97
in vivo studies, **314:**100–101
periventricular cellular structure after
subchronic intracerebroventricular treatment in mice, **314:**7
primer design, **314:**56
synthesis of oligodeoxynucleotides,
314:4–5
targeting sequences for μ and δ opioid
receptors, **314:**5
uptake of oligodeoxynucleotides,
314:97
Western blot analysis
Gα subunits, **314:**15, 17, 101
opioid receptors, **314:**13–15
cloning, **314:**3
G protein coupling, **314:**3–4
neuronal nitric oxide synthase role in opioid tolerance, **314:**60
turnover rates with G proteins, **314:**55,
96–97
Opsin, *see also specific opsins*
basal activity, **315:**232, 238
bovine opsin expression in *Pichia pastoris*
absorption spectroscopy characterization, **315:**8
advantages of system, **315:**10
chromophore formation, **315:**8, 11
expression vector construction,
315:4–5, 10
G-protein activation, **315:**9
glycosylation analysis, **315:**7, 10
large-scale culture, **315:**5–6
media, **315:**4
purification
buffers, **315:**6–7
immunoaffinity chromatography,
315:8

subcellular localization, **315:**7
transformation by electroporation,
315:5
yield, **315:**5, 11
constitutively active opsin mutants
absorbance spectroscopy, **315:**216–217
all-*trans*-retinal activation
basal activity correlation with *trans*-
retinal activation, **315:**258–260
immunoaffinity chromatography,
315:254
membrane preparation, **315:**253–254
modeling of activation, **315:**258–260
retinal analog studies of activation,
315:264, 266
Schiff base formation in mutants,
315:260–263
transducin activation, **315:**255–256
transient transfection in COS-1 cells,
315:253
definition, **315:**208
Glu113Gln mutant, **315:**218
immunoaffinity chromatography,
315:211–212, 216
Lys296Gly mutant, **315:**218
materials for functional assays,
315:208–209
membrane preparation, **315:**210–211
mutagenesis of bovine gene, **315:**209
reconstitution into lipid vesicles,
315:212
rhodopsin kinase assay, **315:**217–218
Schiff base complex preparation,
315:212
transducin activation assay
filter binding assay, **315:**213–214,
255–256
intrinsic tryptophan fluorescence
assay, **315:**256
pH dependence, **315:**208, 214–215
transducin purification, **315:**213
transient expression in COS-1 cells,
315:209–210
Western blot analysis of expression,
315:212
generation from rhodopsin, **315:**219, 236
phylogenetic relationships of retinal and
extraretinal proteins, **316:**279–280
promoter analysis in transiently
transfected *Xenopus* embryos, **316:**57

fluorescence labeling of actin, **298**:440
instrumentation, **298**:441–443
microsphere trapping by myosin af-
    finity
    myosin labeling of beads,
        298:438–440
    promotion of attachment, **298**:447
observed bead deflections
    decoupling diffusion and compliance
        measurement, **298**:449
    diffusion correlation and bead posi-
        tion, **298**:454
    modeling, **298**:448
    myosin concentration effects, **298**:449
    nonprocessive motors, **298**:451, 453
    poor myosin preparations, **298**:455
    processive motors, **298**:449–450
    signatures of protein attachment,
        298:451, 453–454
    thermal fluctuation, **298**:447
    simultaneous fluorescence and bright-
        field imaging, **298**:443–445
    statistical analysis, **298**:455, 457–459
    stiffness of all springs, **298**:458
stiffness of traps, **298**:461–462, 483–484
time-sharing instrument
    acousto-optical deflector, **298**:474–479
    closed-loop feedback arrangements,
        298:479–481
    multiple traps, **298**:477–479
    optical trap, **298**:474–476
    overview, **298**:473–474
    position detector, **298**:476–477
Optic nerve, *see* Eye
ORAC, *see* Oxygen radical absorbance ca-
    pacity assay
Oral biofilm, *see also Streptococcus* biofilm;
    Saliva
    acid adaptation of streptococci biofilms
        acid tolerance response in plaque,
            310:446–447
        alkali production assay, **310**:453
        buffering capacity assays, **310**:449–450
        killing resistance assay, **310**:447–448
        pH change across biofilms, assessment,
            310:452–453
        pH measurement within biofilms,
            310:451–452
        proton movement into biofilms, assay,
            310:450–451

proton-translocating ATPase assay,
            310:448–449
anaerobes in mouth, **310**:281, 442
atmosphere
    oxygenation and biofilm growth,
        310:294–296
    selection for growth, **310**:271, 290–291
bacterial genera in oral cavity, **310**:279–
    280, 441
chemostat continuous flow systems
    aeration studies, **310**:294–296
    antimicrobial agent evaluation
        dental health care product additives,
            310:293
        pulsed dose concentration,
            310:292–293
        steady-state dose concentration,
            310:292
    conditioning films, **310**:289
    mean generation time calculation,
        310:283
    multistage systems, **310**:286
    pH effects, **310**:291–292
    single stage reactor, **310**:282–283
    steady state perturbation, **310**:283, 286
    substrata, **310**:283, 289
    viable biomass balance calculation,
        310:286
constant depth film fermentor
    advantages, **310**:279
    applications, **310**:265
    atmosphere regulation, **310**:271
    corrosion studies of dental materials,
        310:276–277
    cryosectioning of films, **310**:278–279
    design, **310**:265–267
    electron microscopy of films,
        310:277–278
    inhibitor analysis
        light-activated antimicrobial agents,
            310:275–276
        microcosm dental plaques,
            310:273–274
        monospecies films, **310**:272
        multispecies films, **310**:272–273
        overview, **310**:271–272
        pulsing of antimicrobial agents,
            310:275
        sucrose supplementation effects,
            310:274

inoculation, **310:**267–268, 271
multispecies systems, **310:**268
nutrient source, **310:**268–270
steady-state oral community culture,
    310:303–304, 310
substratum, **310:**267
temperature control, **310:**266, 270–271
thickness of biofilms, regulation,
    310:267
vital staining of bacteria, **310:**277
controlled environment model
    accumulation time, **310:**221–223
    advantages and limitations,
        310:222–224
    aims, **310:**216
    design, **310:**216–218
    dilution rate, **310:**221
    nutrient control and media composi-
        tion, **310:**218–220
    planktonic environment, **310:**218
    shear forces, **310:**222
    substrata, **310:**221, 223
culture systems
    batch cultivation, **310:**443
    continuous flow systems,
        310:443–444
    overview, **310:**281–282, 284–285
dental applications, **310:**264
initial formation assays
    hydroxylapatite beads assay,
        310:505–506
    nitrocellulose overlay assay,
        310:504–505
    polystyrene microtiter plate assay,
        310:506–507
    surface plasmon resonance
        advantages, **310:**508–509
        data acquisition, **310:**509–510
        dissociation constant determination,
            310:510–511
        fimbrial binding affinity to host pro-
            teins, **310:**512–513
        materials, **310:**509
        principle, **310:**507–508
        software for data analysis,
            310:511–512
inocula for continuous flow cultures
    complex inocula, **310:**287
    defined complex inocula, **310:**287–288
    multispecies inoculation, **310:**286–287

media selection for growth, **310:**268–270,
    289–290
microscope flow cell studies of oral *Strep-
    tococcus*
    confocal microscopy
        growth analysis, **310:**328–329
        immunofluorescent staining,
            310:327–328
    culture, **310:**326–327
oxidative stress adaptation of biofilms
    diffusion limitation of oxygen,
        310:453–454
    oxidation–reduction potential assess-
        ment, **310:**455–457
    oxygen movement into biofilms, assays,
        310:454–455
pH of mouth and biofilm studies,
    310:281, 291–292, 441
plaque, **310:**264
plaque, stages in formation, **310:**513
Rototorque system culture
    generation time calculation, **310:**445
    inoculation, **310:**445–446
    media, **310:**446
    modifications of instrumentation,
        310:444
    pH control, **310:**445
    reporter gene analysis, **310:**457–460
saliva composition, **310:**269
stress adaptation, **310:**442
subgingival plaque, nutrient sources,
    310:270
Orcinol, ganglioside detection on high-per-
    formance thin-layer chromatography
    plates, **312:**123
Organelle motility
    chick embryo fibroblast cell-free system
        culture and harvest, **298:**355–356
        flow chamber assay, **298:**356
        fluorescence microscopy assay, **298:**358
        fractionation, **298:**356
        homogenization, **298:**356
        quantitative analysis, **298:**359–360
        solution preparation, **298:**354–355
        variability between preparations,
            298:360
        vesicle fractionation, **298:**358
    magnetic bead assay for microtubule–
        membrane interactions
        CLIP-170 identification, **298:**382

cytosol, preparation and manipulation,
    298:386–387
endosome preparation and manipula-
    tion, **298**:385–386
microtubule affinity matrix on mag-
    netic beads, preparation,
    298:383–385
organelle binding to microtubule
    beads, **298**:387–390
principle, **298**:382–383
melanosomes in cell-free *Xenopus* system
cyclic AMP effects, **298**:362
electron microscopy characterization,
    298:365
melanosome purification from cultured
    melanophores
    cell culture, **298**:362–363
    cell harvesting, **298**:363
    density gradient centrifugation,
        298:364, 372
    homogenization, **298**:363–364
    yield, **298**:364–365
motility assay
    axoneme preparation, **298**:368–369
    substrate preparation, **298**:369
    video-enhanced differential interfer-
        ence contrast microscopy,
        298:369–371
motor proteins and Western blotting,
    298:362, 367–368, 372
protein composition analysis by dena-
    turing gel electrophoresis,
    298:366–367
phagosome assay, **298**:380–381
protein components of motility,
    298:353–354
Reticulomyxa system
cell-free system
    cell culture, **298**:402–403
    light microscopy, **298**:403
    lysis, **298**:404
    network manipulations, **298**:407–408
    reactivated motility with ATP,
        298:404, 406–407
cytoplasmic streaming, **298**:400, 407
feeding network, **298**:400–401
microtubule bundles, splaying and
    bending, **298**:407
microtubule network structure,
    298:400–401

motor characterization, **298**:408
organelle movement along linear
    tracks, **298**:404, 406–407
surface transport of latex beads,
    298:408
video-enhanced differential interference
    contrast microscopy assay
coverslips, **298**:348
image acquisition and processing,
    298:349
instrumentation, **298**:439
Xenopus extracts
    direction of movement, **298**:351
    features, **298**:349–350
    quantitation of movement,
        298:350–351
    standard assay, **298**:350
*ori*P plasmid
applications
    gene expression analysis, **306**:309–310,
        312
    mutagen analysis, **306**:316
    overview, **306**:308–309
    promoter activity analysis, **306**:312–313
    selective advantage studies, **306**:313
    toxin assays, **306**:313, 315–316
drug resistance plasmids, **306**:310, 312
EBNA1
binding sites
    EBNA1, 321, 323
    genome, **306**:326
    *ori*P, **306**:317–318, 323
DNA linking, **306**:324–326
functional contributions, **306**:321,
    323–328
maintenance and segregation of plas-
    mids, **306**:326–327
requirement for plasmid replication,
    306:30, 308, 316–321, 327–328
HEK 293–Epstein–Barr virus expression
    system, **306**:31
replication
    long-term, **306**:318–320
    short-term, **306**:320–321
species specificity for replication, **306**:317
yeast artificial chromosomes, **306**:316
Ornithine transcarbamylase, conformational
    probing with osmotic stress measure-
    ment, **295**:58–59
Orotidylate decarboxylase

free energy profile, **308:**10–11
rate acceleration, **308:**10, 358
Osmotic stress
  conformational probing
    aspartate transcarbamylase
      allosteric regulation by nucleotide tri-
        phosphates, **295:**57–58
      mutant studies and physiological sig-
        nificance, **295:**61
      T to R transition, **295:**56–57
    assumptions, **295:**56
    ornithine transcarbamylase,
      295:58–59
  hydration changes in protein, mea-
    surement
    controls
      molecular crowding effects, **295:**51,
        53
      protein dissociation or aggregation,
        295:54–55
      solute size dependence, **295:**48–51
      solute types and protein interactions,
        295:47–48
      freezing point depression osmometry,
        295:45–46
    overview, **295:**42–43
    solute-induced osmotic pressure in
      buffer versus water, **295:**46–47,
      53–54
    vapor pressure osmometry, **295:**45
  structure-based calculations of hydration
    change, **295:**59–61
  temperature dependence of osmotic pres-
    sure, **295:**45–46
OTCase, *see* Ornithine transcarbamylase
Oubain/calcium-insensitive ATPase, amy-
    loid-β inhibition
  assay, **309:**741
  overview, **309:**739
Oxalate luminescence
  dioxetanedione reactions, **305:**27–28
  educational applications
    demonstrations, **305:**654–656
    fluorescers, **305:**655–656
    mechanisms of luminescence,
      305:656–657
    overview, **305:**652–653
  synthesis
    bis(2,4-dinitrophenyl)oxalate,
      305:653–654

bis-2,4,6-trichlorophenoxalate,
    305:654
  safety, **305:**653
  structural requirements, **305:**26–29
Oxidation–reduction potential, assessment
  in biofilms, **310:**455–457
Oxidative stress, *see also* Glutathione
    disulfide/glutathione ratio
  amyloid-β
    Alzheimer's disease, oxidative stress
      markers and pathogenesis,
      309:149–150, 747–750, 767–768
    cortical synaptosome membrane oxida-
      tion model, **309:**752–754
    creatine kinase inhibition and assay,
      309:748, 760
    electron paramagnetic resonance assay
      advantages, **309:**754–755
      spectral acquisition, **309:**757
      spin labels and labeling, **309:**755,
        757–758
      spin trapping, **309:**758
      stearic acid spin probes and lipid per-
        oxidation evaluation,
        309:765–766
    free fatty acid analysis, **309:**766–767
    glutamine synthetase inhibition and
      assay, **309:**748, 752, 759–760
    glutathione protection, **309:**737
    hippocampal neuronal culture oxida-
      tion, **309:**754
    4-hydroxynonenal–protein conjugates
      glucose transporter inhibition,
        309:740
      immunohistochemistry, **309:**738
      immunoprecipitation, **309:**737–738
      induction, **309:**734–735
      Western blot analysis, **309:**738, 766
    malondialdehyde assay, **309:**737
    neurotoxicity assays
      mitochondrial function assay,
        309:761
      morphometric cell counting assay,
        309:761
      overview, **309:**760–761
      Trypan blue exclusion assay, **309:**761
    overview of lipid peroxidation,
      309:734–735
    peptide assessment for study,
      309:751–752

reactive oxygen species detection with
  fluorescent probes, **309:**762
bioelectric effect role, **310:**666–667,
  669–670
carbonyl assay in proteins
  2,4-dinitrophenylhydrazine assays,
    309:138–139, 762–763
  histofluorescence, **309:**764–765
  immunohistochemistry, **309:**139–140
free radicals
  sources of reactive species, **309:**747
  types, **309:**746–747
immunocytochemistry
  advanced glycation end products,
    309:144–145, 149
  advanced lipoxidation end products,
    309:144–146, 149–150
  Alzheimer's disease samples,
    309:149–150
  antigen retrieval on tissue sections,
    309:143
  complications and artifacts,
    309:151–152
  heme oxygenase-1, **309:**148–150
  8-hydroxyguanosine, **309:**146–147
  imaging of optical density,
    309:150–151
  iron regulatory proteins, **309:**147–149
  peroxidase-coupled secondary anti-
    body, **309:**142–143
  tyrosine nitration assay, **309:**143–144
iron(II)/iron(III) as measure
  chelation and binding, **309:**142
  cytochemical detection, **309:**141
  oxidation, *in situ*, **309:**141–142
  rationale, **309:**140–141
limited diffusion of oxygen in biofilms,
  310:453–454
markers in brain cells, **309:**747–748
microelectrode measurement of oxygen
  gradient across biofilm, **310:**307
nervous system tissue preparation for
  assays, **309:**138
neurodegenerative disorder pathogenesis
  role, **309:**133–137, 747
oral biofilm adaptation
  diffusion limitation of oxygen,
    310:453–454
  oxidation–reduction potential assess-
    ment, **310:**455–457

oxygen movement into biofilms, assays,
  310:454–455
Pseudomonas aeruginosa biofilm adaptation, **310:**599–600
protein modifications, **309:**135
in situ models for neurodegenerative disorders, **309:**137–138
synergistic reactions, **309:**135–136
2′,5′-Oligoadenylate, *see* Ribonuclease L
8-Oxo-2′-deoxyguanosine
  dietary antioxidant effects on formation,
    300:182–184
  DNA oxidation marker, **300:**156, 174–175
  enzyme-linked immunosorbent assay,
    300:169–170
  high-performance liquid chromatography
    with electrochemical detection
    analysis of immunoaffinity column sam-
      ples, **300:**164–165
    artifacts, **300:**181
    calculations, **300:**163
    cells, DNA extraction and hydrolysis,
      300:178
    chromatography, **300:**162, 179–180
    DNA extraction
      chaotropic sodium iodide, **300:**158–
        159, 165
      chloroform/isoamyl alcohol extrac-
        tion, **300:**177
      phenol extraction, **300:**159–160,
        165
    DNA hydrolysis, **300:**160–161
    instrumentation, **300:**161–162, 179
    sensitivity, **300:**162–163
    tissue homogenization, **300:**157–158,
      177
    urine analysis by column switching,
      300:170–174
  immunoaffinity column purification from
    urine
    chromatography, **300:**164
    interpretation of values, **300:**165–166
    solid-phase extraction, **300:**163–164
  sources in urine, **300:**174–175
  urinary values, range, **300:**168–169, 182
8-Oxoguanine, *see also* DNA; 8-Oxo-2′-
  deoxyguanosine
  dietary antioxidant effects on formation,
    300:182–184
  DNA oxidation marker, **300:**156, 174–175

high-performance liquid chromatography
with electrochemical detection
analysis of immunoaffinity column samples, **300:**164–165
artifacts, **300:**181
calculations, **300:**163
cells, DNA extraction and hydrolysis,
300:178
chromatography, **300:**162, 179–180
DNA extraction
chaotropic sodium iodide, **300:**158–
159, 165
chloroform/isoamyl alcohol extraction, **300:**177
phenol extraction, **300:**159–160,
165
DNA hydrolysis, **300:**160–161
instrumentation, **300:**161–162, 179
sensitivity, **300:**162–163
tissue homogenization, **300:**157–158,
177
urine analysis by column switching,
300:170–174
immunoaffinity column purification from
urine
chromatography, **300:**164
interpretation of values, **300:**165–166
solid-phase extraction, **300:**163–164
urinary values, range, **300:**168–169
2-Oxohistidine
amino acid analysis, **300:**121, 123–124
high-performance liquid chromatography
with electrochemical detection,
300:120–121
hydrolysis of proteins, **300:**123
protein oxidation marker, **300:**120
structure, **300:**120–121
synthesis, **300:**121–122
3-Oxohexanoylhomoserine lactone
characterization, *see* Acylhomoserine lactone autoinducers
detection bioassay, **305:**292–293
stability, **305:**295, 300
synthesis
coupling with 1-ethyl-3-(3-dimethylami-
nopropyl)carbodiimide, **305:**305–
307, 309
silica gel chromatography, **305:**309
storage, **305:**309–310
3-Oxooctanoylhomoserine lactone

characterization, *see* Acylhomoserine lactone autoinducers
detection bioassay, **305:**294
enzymatic synthesis, **305:**301–302
Oxygen
concentration in saturated solution,
301:190–191
intracellular concentration, **301:**188
mitochondrial respiration
measurement, **301:**189–194
nitric oxide inhibition, **301:**188,
194–198
regulation of uptake, **301:**197–198
singlet state, *see* Singlet oxygen
triplet state, **319:**3
Oxygen radical absorbance capacity assay
hydroxyl radical assay
automated assay on Cobas Fara II,
299:60–61
manual assay, **299:**61
kinetics, **299:**51–52
peroxyl radical assay
automated assay on Cobas Fara II
data acquisition, **299:**54–56
reagents, **299:**54
comparison with other assays,
299:58–60
manual assay, **299:**56
reproducibility, **299:**56–57
sample preparation
animal tissues, **299:**53
biological fluids, **299:**52–53
foods, **299:**53
sensitivity of phycoerythrin, **299:**57
principle, **299:**51
transition metal assay, **299:**61–62
Oxytocin
antisense effects on adrenocorticotropic
hormone release, **314:**235–237
antisense targeting in hypothalamic nuclei, **314:**232–235
Ozone
antioxidant depletion in murine skin,
high-performance liquid chromatography analysis
advantages, **319:**546
ascorbate and urate analysis
equipment, **319:**539
operation, **319:**540
reagents, **319:**539

sample preparation, **319:**540
  standards, **319:**540
extraction, overview, **319:**538
glutathione analysis
  equipment, **319:**541
  operation, **319:**541
  reagents, **319:**540
  sample preparation, **319:**542
  standards, **319:**541–542
overview, **319:**537
ozone exposure, **319:**537
peak identification and quantification,
  319:544–545
tape stripping, **319:**537–538
troubleshooting, **319:**545
vitamin E analysis
  equipment, **319:**543
  operation, **319:**543
  reagents, **319:**542
  sample preparation, **319:**543–544
  standards, **319:**543
β scission pathways leading to aldehydes,
  319:573
biological targets, **319:**505, 546–547
Criegee ozonation, **319:**572
lipid peroxidation, **319:**4-hydroxy-2-
  nonenal adduct detection
  applications, **319:**563–564
  cytotoxicity, **319:**562–564
  enzyme-linked immunosorbent assay,
    319:569
  immunohistochemical analysis,
    319:569–570
  polyclonal antibody preparation
    amino acid–HNE adduct antibody
      characterization, **319:**565
    fluorescent lysine–HNE antibody
      preparation, **319:**566–567
    principle, **319:**564–565
    reductively-stabilized amino acid–
      HNE adduct antibody character-
      ization, **319:**565–566
  protein adducts, **319:**563, 567
  reduction of protein adducts
    sodium borohydride, **319:**567–568
    sodium cyanoborohydride, **319:**568
  Western blot analysis, **319:**568–569
lung effects
  cytokine-induced neutrophil chemoat-
    tractant induction

dexamethasone inhibition, **319:**560,
      562
    electrophoretic mobility shift assay,
      319:557–558, 560
    neutrophilia role, **319:**554–555
    Northern blot analysis, **319:**556–557
    reverse transcriptase polymerase
      chain reaction, **319:**556
  disease association, **319:**551–552
  inflammation, **319:**552–554
  nuclear factor-κB induction
    dexamethasone inhibition, **319:**560
    electrophoretic mobility shift assay,
      319:557–558, 560
    inflammation role, **319:**553–554
  penetration, **319:**571–572
  rat exposure, **319:**555–556
plant response assays
  Arabidopsis thaliana, ozone-sensitive
    mutant production
    advantages, **319:**532
    gene isolation, **319:**534
    genetic analysis, **319:**533
    mutagenization and screening,
      319:532
    Northern hybridization analysis,
      319:533–534
    resources, **319:**531
  chloroplast effects, **319:**521
  disease association, **319:**520–521
  ethylene emission, **319:**525–526
  oxidative burst, **319:**525
  pathogen attack effects, **319:**522
  pathogen treatment
    Pseudomonas syringae, **319:**535
    tobacco mosaic virus, **319:**535
  phytoalexin analysis, **319:**526
  phytotron exposure system,
    319:522–524
  plant dosing, **319:**524–525
  transcript induction
    complementary DNA libraries,
      319:527
    poly(A) RNA isolation, **319:**529–531
    RNA isolation from herbaceous
      plants, **319:**527
    RNA isolation from trees,
      319:528–529
pollutant levels, **319:**551
reactive absorption

measurement
  antioxidant comparison, **319:**510–511
  apparatus, **319:**506–507
  assumptions, **319:**508–509
  biomolecule concentration effects,
    319:509–510
  data analysis, **319:**507–509
  relative rate constants,
    319:510
  respiratory system application,
    319:510–512
modeling, **319:**505
signal transducers
  classes, **319:**570–571
  formation, **319:**571–574
  synthesis
    bis(1-hydroxyheptyl)peroxide,
      319:578–579, 581
    bis(1-hydroxynonyl)peroxide,
      319:578–579, 581
    glycerophospholipid hydroxyhydro-
      peroxides, **319:**574–575
    1-hydroxy-1-hydroperoxyalkanes,
      319:575–576
    materials, **319:**576
    1-palmitoyl-2-(1-hydroxy,1-
      hydroperoxynonanoyl)-*sn*-
      glycero-3-phosphocholine,
      319:578
    1-palmitoyl-2-[3-(8-octanoyl)-5-(1-
      octyl)-1,2,4-trioxolane]-*sn*-
      glycero-3-phosphocholine,
      319:577
    1-palmitoyl-2-(9-oxononanoyl)-*sn*-
      glycero-3-phosphocholine,
      319:577–578
    phospholipid aldehydes, **319:**574–575
    X-ray crystallography, **319:**581–582
singlet oxygen generation in plants
  ascorbic acid reaction, **319:**512–513,
    518–519
  chemiluminescence assay in leaves
    carrier gas effects on emission inten-
      sity, **319:**517–518, 520
    cell wall emission, **319:**516–517
    gas volume per unit surface area ef-
      fects, **319:**519–520
    instrumentation, **319:**513–514
    intensity of gas phase emission,
      319:516

intercellular extract measurements,
  319:515
kinetics, **319:**519
principle, **319:**513
spectral analysis, **319:**517
reactants in generation, **319:**514–515
stratum corneum effects, **319:**536
ultraviolet absorption, **319:**359, 536
uptake in plants, **319:**521
vitamin E reactions
  gas chromatography–mass spectrome-
    try analysis, **319:**550–551
  ozonation reaction, **319:**549–550
  products and mechanisms, **319:**547, 549

P

p21
  double-tagging assay, identification of
    binding domains
    controls, **303:**447
    cyclin binding, **303:**444–445, 447, 449
    cyclin-dependent kinase binding,
      303:444–445, 447, 449
    plasmids, **303:**447
  functions, **303:**444–445
  photodynamic therapy apoptosis role,
    319:352, 354
P22, *see also* RNA challenge phage system
  coat protein, native gel electrophoresis of
    aggregation intermediates, **309:**343
  tailspike protein intermediates
    native gel electrophoresis
      conformational isomer resolution,
        309:340–341
      quantitative analysis of aggregation,
        309:338, 340
      nondenaturing Western blotting,
        309:347–349
      resolution with various techniques,
        309:333–334
p23
  assays, **290:**429
  biological functions, **290:**421
  purification of recombinant human pro-
    tein from *Escherichia coli*
    anion-exchange chromatography,
      290:425
    cell growth and lysis, **290:**424–425

size-exclusion chromatography, **290:**426

p53
  photodynamic therapy apoptosis role,
    319:349, 352
  skin end point for ultraviolet A expo-
    sure, **319:**464
  squamous cell carcinoma mutations,
    319:355, 418

p60, *see* Hop/p60

P700, *see* Photosystem I; PsaC

p*K*, *see* Proton binding by proteins

PACT, *see* Photodynamic antimicrobial chemotherapy

PAGE, *see* Polyacrylamide gel electrophoresis

Pain, *see* Neuropathic pain

Pair correlation function, distribution analysis in confocal microscopy, **307:**218,
  220, 227

Paired helical filament, *see also* Alzheimer's
  disease; Tau
  electron microscopy, ultrastructure analy-
    sis, **309:**180–182
  immunoelectron microscopy of transglu-
    taminase, **309:**182–183
  isolation, **309:**179–180
  isopeptidase digestion, **309:**184–185
  purification, **309:**180

1-Palmitoyl-2-(1-hydroxy,1-hydroper-
  oxynonanoyl)-*sn*-glycero-3-phospho-
  choline, synthesis, **319:**578

1-Palmitoyl-2-[3-(8-octanoyl)-5-(1-octyl)-
  1,2,4-trioxolane]-*sn*-glycero-3-phos-
  phocholine, synthesis, **319:**577

1-Palmitoyl-2-(9-oxononanoyl)-*sn*-glycero-3-
  phosphocholine, synthesis, **319:**577–578

Papain, inhibition by peptidyl *N*-nitrosoani-
  lines,
  covalent modification, confirmation by nu-
    clear magnetic resonance, **301:**255,
    257
  kinetic data analysis, **301:**252, 255–256,
    258
  papain preparation and activity assay,
    301:252
  potency of individual compounds,
    301:256–257
  stability, **301:**256
  synthesis, **301:**250–251

Paper chromatography

ceramide glycanase assay, **311:**289
  gangliosides, **311:**93–94

Parallel plate flow chamber
  advantages and disadvantages, **310:**534
  biofilm eradication studies, **310:**530
  design, **310:**527–528
  initial microbial adhesion experiments
    design, **310:**528–529
  plasma coating effect on intial staphylo-
    coccal adhesion, **310:**530–531
  Pseudomonas aeruginosa detachment
    from contact lenses, **310:**532, 534
  Staphylococcus epidermidis adhesion and
    surface growth, **310:**531–532
  surface growth studies, **310:**529–530

Particle bombardment, nuclear transformation of *Chlamydomonas reinhardtii*,
  297:35–36

Partitioning, proteins in membranes
  binding, contrast with partitioning,
    295:65–66
  chemical potential of peptides in a phase,
    295:66–67
  coupling with folding, **295:**63–65
  fraction of partitioned peptide, calcula-
    tion, **295:**68, 71
  free energy of transfer from water to bi-
    layer, **295:**67
  location of partitioned peptides in bi-
    layer, **295:**69–70
  mole fraction partition coefficient
    calculation, **295:**67–68
    dependence on peptide concentration,
      295:68–69
    peptide solubility and aggregation in
      experimental determination,
      295:85–87
  separation of bound and free molecules
    in partition coefficient determination
    comparison with titration methods,
      295:70–71, 73
    equilibrium dialysis
      apparatus, **295:**74
      calculations, **295:**75
      high-performance liquid chromatog-
        raphy of peptides, **295:**74–75
    sucrose gradient centrifugation
      calculations, **295:**75
      high-performance liquid chromatog-
        raphy of peptides, **295:**77

speed of centrifugation, **295:**76–77
vesicle selection, **295:**73–74, 76
titration in partition coefficient determi-
    nation
    comparison with separation methods,
        295:70–71
    nonideality assessment, **295:**84–85
    plotting of data, **295:**71–73
    range of detectable partition coeffi-
        cients, **295:**85
    spectroscopic titration
        calculations, **295:**78–79
        circular dichroism, **295:**79, 82
        fluorescence titration, **295:**78–82, 87
        multistate equilibria in water and bi-
            layer, considerations, **295:**81–82
    titration calorimetry, **295:**83–84
    vesicle selection, **295:**84
Partitioning coefficient, biofilm nutrients,
    310:393
Patch-clamp, *see also* DRSCAN; Electro-
    physiology; Giant membrane patch
    clamp; Pressure clamp; Whole cell
    patch clamp
    antisense oligonucleotide loading
        advantages over antibody quenching,
            313:152–153
        $G\alpha_0$ knockdown studies, **313:**151–152
        membrane capacitance measurements,
            313:150–151
        overview, **313:**149
        time of loading, **313:**150
    archaea, **294:**520
    Bacillus subtilis, **294:**519–520
    biosensor, *see* Capillary electrophoresis
    Characeae, **294:**517
    charybdotoxin receptor reconstituted into
        bilayers, **294:**283–284
    Chlamydomas reinhardtii, **294:**517–518
    configurations
        cell-attached patch, **293:**221
        excised patches
            inside-out patches, **293:**221–222
            outside-out patches, **293:**222
        giant patches, **293:**222–223
        whole-cell recording, **293:**222
    cystic fibrosis transmembrane conduc-
        tance regulator, *see* Cystic fibrosis
        transmembrane conductance regu-
        lator

data analysis, *see* Hidden Markov model;
    Voltage clamp
Dictyostelium, **294:**516
Escherichia coli
    filament generation
        magnesium formation, **294:**462
        mutants, **294:**462
        snaking approach, **294:**460–461
        ultraviolet formation, **294:**462
    outer membrane, **294:**518
    protoplast generation, **294:**462–463
    reconstituted bacterial membranes,
        294:466–468
    reconstituted solubilized membrane
        proteins, **294:**468–469
    seal formation, **294:**463–466
    spheroplast formation, **294:**461–462
fibroblasts
    calexcitin studies
        data acquisition and analysis,
            293:198–201
        microinjection, **293:**197–198, 200
        purification, **293:**196–197
    culture, **293:**198
    inside-out patch preparation,
        293:198–199
    instrumentation, **293:**199–200
    ion channels, **293:**194–195
gating current measurement, **293:**338–340
GluR1 phosphorylation, whole-cell re-
    cording in functional analysis
    basal phosphorylation, **294:**368, 370
    intracellular patch perfusion solutions,
        294:367–368
    site-directed mutants, **294:**370
    stimulation of phosphorylation,
        294:366–368
M2 expressed in mouse erythroleukemia
    cells, whole cell patch-clamp analysis
    background permeability, **294:**497–498
    cell-to-cell variation, **294:**498–499
    combined measurements with fluores-
        cent pH indicators, **294:**505–506
    current/voltage relation, **294:**499–501
    ion selectivity, estimation from reverse
        potentials, **294:**501, 503–505
    overview, **294:**496
    perfusion, **294:**497
    single-channel parameter estimation,
        294:505

solution composition, **294:**496–497

Neurospora, **294:**515–516

noise

  electronic noise reduction

    amplifier, **293:**223, 225–228, 259

    capacity compensation, **293:**223

    whole cell compensations,
      293:223–224

  $1/f$ noise, **293:**224–227

  holder noise, **293:**228–229, 259

  patch configuration effects,
    293:221–223

  pipette noise

    dielectric noise and elastomer coat-
      ing, **293:**229, 231–239, 260

    distributed RC noise, **293:**231, 239–
      247, 249–251, 253–256, 260–261

    materials and geometry in minimiza-
      tion, **293:**218–219, 232–239,
      241–247, 249–251, 253–256,
      259–261, 265–266

    prediction, **293:**262–263

    R_e-C_p noise, **293:**231–232, 256–257,
      261–263

    seal noise, **293:**218, 220–221, 232,
      257–259

    thin film noise, **293:**231–232

  power spectral density, **293:**224, 226,
    228, 263–264

  total rms noise, **293:**225, 259

  whole-cell measurement noise,
    293:263–266

overview, **293:**79–80

Paramecium, **294:**508–511

planar bilayers, reconstituted channels,
  294:289–290, 337

plant cells

  comparison to animal cell patch clamp-
    ing, **294:**421–424

  exogenous second messenger adminis-
    tration

    caged calcium studies, **294:**437–440

    comparison of control and experi-
      mental cells, **294:**436

    repatching and pipette perfusion,
      294:436–437

  light and heat responses of protoplasts,
    294:423–424

  pipettes, **294:**422

  protoplast membrane exposure

enzymatic access, **294:**411–417

laser-assisted patch clamping,
  294:419

osmotic shock, **294:**417–418

seal formation, **294:**422–424

second messengers in ion channel regulation, **294:**424–425, 435

simultaneous calcium dye imaging

  calibration of fluorescence signal,
    294:431–432

  cameras for imaging, **294:**433–434

  compartmentalization of dye,
    294:430

  confocal laser scanning microscopy,
    294:427–428, 430–432

  dye selection, **294:**426–427, 434–435

  Indo-1 loading, **294:**430–431

  instrumentation, **294:**426, 428, 430,
    433–434

  photometry and imaging of ratio flu-
    orescence dyes, **294:**432–435

  pipette resistance, **294:**428

  solution compositions, **294:**421–422

potassium channel, heterologous expression in yeast, **293:**102–104

Saccharomyces cerevisiae, **294:**511–514

small cells

  interpretation of recordings,
    293:214–215

  neurons of *Caenorhabditis elegans*, *see*
    Whole cell patch clamp

Schizosaccharomyces pombe, **294:**514

Uromyces appendiculatus, **294:**514–515

Xenopus oocyte, *see Xenopus* oocyte

Patch cram, *Xenopus* oocyte, **293:**297

Paxillin

  blot overlay assay of binding proteins

    autoradiography, **298:**85

    buffer preparation, **298:**84

    denaturing polyacrylamide gel electro-
      phoresis, **298:**84–85

    incubation conditions, **298:**85

    iodine-125 labeling of paxillin, **298:**85

  domains, **298:**77–78

  phosphorylation

    assay, *in vivo*, **298:**88–90

    focal adhesion kinase assay, **298:**99

    signal transduction, **298:**77, 87

  precipitation assays

    binding proteins, **298:**85–86

kinase assays, **298:**86–87
purification from chicken gizzard
   ammonium sulfate precipitation, **298:**79
   anion-exchange chromatography,
      298:79
   buffer preparation, **298:**78
   hydroxylapatite chromatography,
      298:79
   immunoaffinity chromatography,
      298:79
   overview, **298:**77
   tissue homogenization, **298:**78–79
purification of recombinant fusion protein from *Escherichia coli*
   affinity purification, **298:**82
   cleavage of fusion protein, **298:**82–83
   expression and cell growth, **298:**81–84
   extraction, **298:**82
   materials, **298:**81
   optimization of expression, **298:**83
   overview, **298:**80–81
   vector construction, **298:**81
subcellular localization, **298:**77
PC, *see* Proprotein convertase
PCF, *see* Pair correlation function
PCNA, *see* Proliferating cell nuclear antigen
PCR, *see* Polymerase chain reaction
PDE, *see* Cyclic GMP phosphodiesterase
PDGF, *see* Platelet-derived growth factor
PDGFR, *see* Platelet-derived growth factor
   receptor
PDI, *see* Protein disulfide isomerase
PDMP, *see* 1-Phenyl-2-decanoylamino-3-
   morpholino-1-propanol
PDR subfamily, *see* Pleiotropic drug resis-
   tance subfamily
PDT, *see* Photodynamic therapy
PDT, *see* Photodynamic therapy
PDZ domain
   discovery, **303:**451
   mechanisms of protein–protein interac-
      tions, **303:**452
   peptides-on-plasmids library screening,
      see Peptides-on-plasmids library
      screening, neuronal nitric oxide syn-
      thase PDZ domains
PEG, *see* Polyethylene glycol
PEI, *see* Polyethyleneimine
Pentane

lipid peroxidation marker in breath,
   300:457
photoionization gas chromatography in
   breath
   column selection, **300:**458–459
   confounding variables, **300:**460–461
   principle, **300:**458
   sensitivity, **300:**461–462
   smokers versus nonsmokers,
      300:461–462
Pentylamine, transmission electron microscopy support treatment, **304:**221
Pepsin, antigen recovery in amyloid immunohistochemistry samples, **309:**18
Pepstatin, Gibbs energy of binding to aspartic proteases, **295:**114–116
Peptide antibody
   advantages and disadvantages,
      298:525–526
   affinity purification of antibody
      elution from beads, **298:**533
      peptide coupling to resin, **298:**531–533
      serum preparation, **298:**533–534
   immunoaffinity purification of cytoskele-
      tal proteins
      antibody binding to resin, **298:**534,
         536–537
      column preparation, **298:**536–537
      elution of peptides, **298:**536–538
      extraction of proteins, **298:**534, 536
      purity assessment of proteins, **298:**536,
         538–539
      septin complex from *Drosophila*,
         298:540–541
      γ-tubulin complex from *Xenopus* egg,
         298:541
      XKCM1, 539–540
   immunoprecipitation of cytoskeletal pro-
      teins, **298:**534
   keyhole limpet hemocyanin coupling of
      peptides for immunization,
      298:529–531
   peptide selection and parameters affect-
      ing antibody success, **298:**526–528
   thiol group assay of peptides,
      298:528–529
Peptide competition assay, rhodopsin ligands
   applications, **315:**453
   arrestin binding competition, **315:**454–455

peptide synthesis, **315:**453–454
Peptide mapping
  MDR1
    drug-binding site, **292:**305–306, 312
    phosphorylation site
      amino acid sequence of Lys-C pep-
        tide, **292:**334–335
      proteolysis following reduction and
        carboxamidomethylation,
        292:332–333
      reversed-phase high-performance liq-
        uid chromatography,
        292:333–335
      starting material requirements,
        292:328–330
      tryptic phosphopeptides, purification
        and sequences, **292:**336, 338–340
  Mdr1 phosphorylation sites, **292:**348–349,
    353–355
  P-glycoprotein, drug-binding site identifi-
    cation, **292:**305–306
  phosphopeptides of GluR1
    overview, **294:**357–358
    proteolytic digestion, **294:**358
    two-dimensional thin-layer chromatog-
      raphy, **294:**359–360
Peptide nucleic acid
  antisense studies
    bacteria, **313:**160–161
    controls, **313:**163
    mammalian cells, **313:**160
    size, **313:**163
    target selection, **313:**159–160
  applications, **313:**156–157
  melting temperature prediction for hy-
    brids, **313:**162
  specificity of binding, **313:**176
  structure, **313:**156, 158
  synthesis and handling,
    313:163–164
  triplex formation and gene inhibition,
    313:161–162
  uptake in cells
    conjugates, **313:**158
    lipofection, **313:**158
    structure relationship, **313:**158
Peptide-on-plasmids combinatorial library,
  rhodopsin ligands
  amplification of affinity purified plasmids,
    315:395–396

  binding to other G-protein-coupled recep-
    tors, **315:**404
  dark rhodopsin high-affinity sequences,
    315:397–398
  higher affinity peptide selection,
    315:398–399
  LacI lysate preparation, **315:**396–397
  library construction
    amplification of library, **315:**392–393
    electrocompetent cell preparation,
      315:392
    Escherichia coli strain, **315:**390, 392
    vector, **315:**390
  light rhodopsin high-affinity sequences,
    315:397–399
  maltose-binding protein fusion protein
    with clones
    crude lysate preparation, **315:**401
    enzyme-linked immunosorbent assay
      crude lysates, **315:**402–403
      overview, **315:**399, 402
      purified fusion protein, **315:**403–404
    purification of fusion proteins,
      315:401–402
    rationale, **315:**399, 402
    subcloning into vectors, **315:**399–400
  panning, **315:**393–395
  transducin Gα carboxyl terminal ligands,
    overview, **315:**388–390
Peptides-on-plasmids library screening, neuronal nitric oxide synthase PDZ domains
  affinity panning
    efficiency monitoring, **303:**461
    lysis of library, **303:**457
    principle, **303:**455
    reagents, **303:**455
    rounds of panning, **303:**455–458
  amplification of interacting plasmids
    DNA precipitation, **303:**458–459
    electroporation, **303:**459–460
    overview, **303:**458
    reagents, **303:**458
    rounds of amplification, **303:**460–461
  binding clone identification
    enzyme-linked immunosorbent assay,
      303:463, 465–466
    lysis of LacI clones, **303:**464–465
    overview, **303:**463
  complexity of library, **303:**453

enrichment determination, **303**:461–463
immobilization of panning proteins,
    303:454
LacI fusion proteins, **303**:452
principle, **303**:452
purification of panning proteins,
    303:453–454
sequence analysis and interacting protein
    search, **303**:466–467
Peptide synthesis
  caged tyrosine peptides, **291**:83–85
  μ-conotoxin GIIIA analogs, **294**:587–589
  gramicidin analogs
    acylation, **294**:531
    cleavage of peptide from resin,
        294:530–531
    N-Fmoc-amino acid synthesis,
        294:527–528
    N-formylamino acid preparation,
        294:526–527
    purification of product, **294**:531–532
    resin selection, **294**:528
    solid-phase synthesis, **294**:528–530
  scorpion venom peptide inhibitors,
      294:629–630
Peptidyl N-nitrosoanilines
  papain inhibition
    covalent modification, confirmation by
        nuclear magnetic resonance,
        301:255, 257
    kinetic data analysis, **301**:252, 255–256,
        258
    papain preparation and activity assay,
        301:252
    potency of individual compounds,
        301:256 257
    stability, **301**:256
    synthesis, **301**:250–251
Peptidylprolyl cis–trans-isomerase, see also
    FKBP12
  FKBP52 assay, **290**:429
  immunosuppressant binding and classifi-
      cation, **290**:84, 94
  plant organelle enzymes
    cyclosporin A binding assay
      chloroplast membrane preparation,
          290:91
      mitochondrial membrane prepara-
          tion, **290**:91
      radioactive tagging, **290**:91–92

FK506 binding assays
  immunoblot assay, **290**:96–98
  radioactive assay, **290**:95–96
organelle subfractionation
  chloroplast stroma and thylakoids,
      290:85–86
  mitochondria isolation and subfrac-
      tionation, **290**:88–90
  thylakoid grana and stroma mem-
      branes, **290**:86–88
organelle targeting, **290**:98–99
spectrophotometric activity assay,
    290:92, 94
types and characteristics, **290**:99–100
Perfused biofilm fermenter, see Biofilm fer-
    menter
Peripheral nerve injury, see also Peripheral
    nerve 3 protein
  neuropathophysiology, **314**:201–202
  voltage-gated sodium channels in neuro-
      pathic pain, **314**:202–205
Peripheral nerve 3 protein
  antisense studies in L5/L6 nerve-ligated
      rats
    clinical implications, **314**:212–213
    intrathecal administration, **314**:207–208
    nociception evaluation, **314**:208–209
    oligodeoxynucleotide design,
        314:206–207
    pain reduction, **314**:209–211
    specificity of effects, **314**:211–212
  neuropathic pain role, **314**:205–206
  tetrodotoxin resistance, **314**:204
Peripherin/rds
  domains involved in membrane fusion,
      316:75–76
  electroretinogram studies
    cone worse than rod disease, **316**:625
    equal rod and cone disease, **316**:624
    mutation studies, **316**:621–624
    rod worse than cone disease, **316**:624
  membrane fusion assays, see also Rod
      outer segment
    aggregation–adhesion analysis,
        316:84–85
    bilayer destabilization analysis,
        316:85–86
    characteristics of fusion with target
        membranes, **316**:74–75
    fusion competency assay of canine kid-

ney cells expressing mutant proteins
advantages, **316:**83
cell culture, **316:**80–81
pitfalls, **316:**83–84
probe labeling of membranes, **316:**82
resonance energy transfer-based
lipid mixing assay, **316:**81–83
peptide fragment assay
contents mixing assay, **316:**78–79
materials, **316:**78
peptide effects on fusion, **316:**79–80
pitfalls, **316:**80
quenching pair assay principle,
316:76
vesicle preparation, **316:**78
purification from bovine retina and recombination with vesicles,
316:69–71
mouse mutant phenotype, **316:**672
mutations in retinal dystrophies, **316:**671,
684
posttranslational modification, **316:**673
sequence homology between species,
316:672–673
transient transfection with rom-1 in COS-
1 cells
analysis of protein complexes
reagents, **316:**679, 681
retinal disease analysis, **316:**684, 687
velocity sedimentation analysis,
316:682–684
Western blot analysis, **316:**681–682
cell growth and maintenance,
316:674–675
expression vector construction for
wild-type and mutant proteins,
316:675–677
immunofluorescence assay of transfection efficiency, **316:**677–679
transfection, **316:**677–678
Permanganate footprinting
Drosophila salivary glands, genomic footprinting
buffers and reagents, **304:**488–489
hsp70 gene, **304:**481, 484
ligation-mediated polymerase chain reaction analysis, **304:**484, 487–488
naked DNA
formic acid treatment, **304:**483

permanganate treatment,
304:482–483
piperidine cleavage, **304:**483–484
overview, **304:**480–481
salivary gland
dissection, **304:**481
permanganate treatment, **304:**482
genomic footprinting of cell monolayers,
304:553–554
modification of DNA, **304:**616
permeability of cell membranes to potassium permanganate, **304:**616
polymerase chain reaction analysis, *see*
pfmdr1
copy number and drug sensitivity,
292:176–177
functional analysis by heterologous expression
Chinese hamster ovary cells,
292:177–178
functional complementation of Ste6p,
292:174, 178–180, 427
mutation analysis, **292:**178
role in *Plasmodium falciparum* drug resistance, **292:**174–177, 180–181
sequence homology with other P-glyco-
proteins, **292:**105, 109
subcellular localization of protein,
292:175, 179–181
substrate specificity, **292:**175–176
transcripts, **292:**180–181
Polymerase chain reaction
thymine reactivity, **304:**481
Peroxidase
amyloid immunohistochemistry detection
blocking of endogenous enzyme,
309:19, 37–38
primary versus secondary detection systems, **309:**21–22
substrates, **309:**22–23
horseradish enzyme, *see* Horseradish peroxidase
Peroxide, flow cytometry detection,
319:336
Peroxide chemiluminescence
analog studies, **305:**8–11
emission spectra characteristics, **305:**10
oxygen role in luminescence, **305:**4
quantum yields, **305:**8, 10
theory, **305:**8–11

Peroxiredoxin, *see also* Thiol-specific antioxidant
  assays
    coupled NADPH oxidation assay, **300:**223–224
    glutamine synthetase protection assay, **300:**223
    reagents, **300:**222–223
  functions, **300:**220–221
  purification
    Prx I, **300:**224
    Prx II, **300:**224–225
    Prx III, **300:**225–226
  sequence homology searching and classification, **300:**220–222
Peroxisome
  ATP-binding cassette proteins
    functions, **292:**769
    half-ATP-binding cassette proteins, **292:**758–759, 762, 768
    phylogenetic analysis, **292:**768, 770
    sequence alignment, **292:**760–761, 769, 771
  biogenesis
    disorders, *see also* Adrenoleukodystrophy
      classification, **292:**756–757
      Zellweger syndrome, **292:**757
    induction, **292:**754
    matrix proteins, **292:**754–755
    membrane proteins, **292:**755–756
  discovery, **292:**753–754
  Saccharomyces cerevisiae, see also PXA1; PXA2
    culture using oleic acid as sole carbon source, **292:**772–773
    induction of peroxisome biogenesis, **292:**773–774
    purification of peroxisomes
      fractionation on Nycodenz gradient, **292:**775
      gradient analysis, **292:**775–776
      high-density organelle isolation, **292:**774–775
      overview, **292:**773
    suitability for peroxisome studies, **292:**762–763
Peroxisome membrane protein-70
  biological function, **292:**758

homology with other ATP-binding cassette proteins, **292:**759–762
Peroxynitrite
  absorbance properties, **301:**343–344
  apoptosis induction, **301:**399–402
  benzoate hydroxylation assay, **301:**304–306
  biological effects, overview, **301:**288, 301, 333–334
  carbon dioxide effects on reactivity, *see* Carbon dioxide
  cell exposure in apoptosis studies, **301:**396–400
  cellular sources, **299:**210–211
  2,7-dichlorodihydrofluorescein assays of oxidation, **301:**371–373
  dihydrorhodamine 123 assays of oxidation, **301:**303–304, 371–373
  DNA damage, repair endonuclease assays
    cell-free supercoiled DNA, analysis of damage, **301:**314–315
    cells and reagents, **301:**313
    cellular DNA damage analysis, **301:**315–317
    damage profiles, **301:**317–318
    preparation of endonucleases, **301:**313
    principle, **301:**311–312
    types of endonucleases, **301:**311–312, 318
  formation from nitric oxide, **301:**13, 135, 152, 301, 333, 395, 425, 438, 455, 523
  generation from 3-morpholinosydnonimine, **301:**400–401
  hydroxycinnamate interactions
    m-coumaric acid, **299:**223, 234
    o-coumaric acid, **299:**222–223, 233–234
    p-coumaric acid, **299:**219–222, 233
    ferulic acid, **299:**223, 228, 234
    high-performance liquid chromatography analysis of products, **299:**214, 219, 222–223, 228
    mass spectrometry analysis of products, **299:**214–215, 221–222, 228
    reaction conditions, **299:**214
    spectral characterization of products, **299:**214, 219–221, 223, 228
  low-density lipoprotein oxidation

carotenoid and tocopherols
  analysis by high-performance liquid
    chromatography, **301:**322–323
  consumption and lipid peroxidation,
    301:331–332
  effects of peroxynitrite, **301:**324, 327,
    329, 331
  conjugated diene assay, **301:**323, 327
  isolation of lipoprotein, **301:**322
  protein modifications, **301:**321, 332
  thiobarbituric acid-reactive substances,
    301:323–324, 327, 329
mitochondria, response and effects
  calcium flux, **301:**385–386
  incubation conditions, **301:**382–383
  intactness, **301:**387
  membrane potential, **301:**383–384
  oxygen consumption, **301:**386–387
  pyridine nucleotide content,
    301:388–389
nitrate and nitrite contamination in preparations
  calibration curves, **301:**184
  effect of preparation age, **301:**182–183
  effect of synthesis method, **301:**185–187
  ultraviolet second-derivative spectros-
    copy, **301:**179–180, 184–185
nitric oxide donors from peroxynitrite
  detection of nitric oxide release
    electrochemical detection,
      301:294–296
    vasorelaxation bioassay, **301:**296–297
  synthesis, **301:**293–294, 298
perfusion of cardiac cells
  cell preparation, **301:**403
  effects
    cell motion, **301:**407–409
    intracellular pH, **301:**407–409
    membrane fluidity, **301:**407–409
  perfusion apparatus, **301:**404–406
  peroxynitrite synthesis, **301:**403–404
  pH adjustment, **301:**406, 409
polyhydroxylated compound modification, **301:**290–291
protectants
  assays
    α_1-antiproteinase inactivation protec-
      tion, **301:**338, 340–342
    tyrosine nitration inhibition,
      301:336–338

glutathione peroxidase, **301:**301–302,
  304–306, 309–311
selenium compounds, **301:**301–302,
  304–306, 309–311
protein modification, *see* Nitrotyrosine
reduction to nitrite, **301:**307, 344
stability in solution, **301:**178, 182, 291,
  333, 342, 396, 402–403, 438
stopped-flow spectroscopy
  bimolecular reactions and rate con-
    stants, **301:**350–352
  carbon dioxide reaction, **301:**351
  detection, **301:**345
  dimethyl sulfoxide oxidation,
    301:348–349
  pK_a determination, **301:**346–348
  protein modifications, **301:**350–352
  solutions, **301:**344–345
synthesis, **299:**212
  hydrogen peroxide reaction with
    nitrous acid, **301:**180–181, 292,
    322
  isoamyl nitrite reaction with hydrogen
    peroxide, **301:**180–181
  ozonation of azide, **301:**180
tocopherol reactions
  liposomes, **301:**486
  methanol solution, **301:**484–485
tyrosine nitration
  inhibitors
    catechins, **299:**215–216, 231–232
    hydroxycinnamates, **299:**216, 219,
      232–233
    scavenging mechanisms, **299:**230–235
  mechanism, **299:**211, 230
  product analysis, **299:**212, 215, 228
  scavenging assay
    calibration, **299:**213
    high-performance liquid chromatog-
      raphy analysis, **299:**213
    incubation conditions, **299:**212–213
    recovery, **299:**215, 228
unsaturated lipid reactions
  extraction and chromatographic analy-
    sis, **301:**460, 467–468
  incubation conditions, **301:**459–460
  synthesis and infusion of peroxynitrite,
    301:460–461, 467
Perrin–Weber equation, fluorescence anisotropy, **290:**315

Persistent pigment darkening, *see* Skin, ultraviolet protection
P-ethoxy oligonucleotide
  fluorescence localization in cells, **313:**376, 378
  hydrophobicity, **313:**372
  lipofection
    liposome incorporation, **313:**375
    separation of free and incorporated oligonucleotides, **313:**375–376
  pharmacokinetics
    calculations, **313:**386
    clearance, **313:**386
    extraction from whole blood and tissues, **313:**384–385
    organ distribution, **313:**386–387
    sample collection, **313:**385
  protein inhibition by liposomal delivery
    Bcl-2, **313:**378–379
    Crk1, 379, 382
    Grb2
      Philadelphia chromosome-positive leukemic cell inhibition, **313:**382–383
      protein expression analysis, **313:**379, 382
  radiolabeling with phosphorous-32, **313:**374
  safety studies in mice, **313:**384
  structure, **313:**373
Pf1 filamentous phage, nucleic acid alignment for dipolar coupling nuclear magnetic resonance
  alignment assay for phage preparations, **317:**227
  buffers, **317:**230, 232
  comparison with other phages, **317:**223
  concentration optimization, **317:**228–229
  flexibility and advantages of approach, **317:**222, 224–225, 239–240
  IRE-I RNA ^{1}H–^{15}N coupling, distance and angle information, **317:**233–237, 239
  line broadening, **317:**235
  mechanism of nucleic acid alignment, **317:**222–223
  phage preparation
    cell growth and harvesting, **317:**225
    homogeneity assay, **317:**226–227
    potassium bromide density gradient centrifugation, **317:**226
    precipitation of phage, **317:**225
    safety, **317:**229
    storage, **317:**226
  sample preparation, **317:**227–228
  stability of samples and recycling of components, **317:**232–233
  surface area of phage, **317:**223–224
  temperature dependence, **317:**222, 224–225, 230
  viscosity effects, **317:**235
  water alignment, **317:**230
PFGE, *see* Pulsed-field gel electrophoresis
Pgh1 gene, *see* pfmdr1
PGHS, *see* Prostaglandin endoperoxide H synthase
P-glycoprotein, *see also* MDR; *specific proteins*
  Arabidopsis thaliana, see ATPGP1; ATPGP2
  Caenorhabditis elegans, **292:**105, 768
  cell volume regulation via swelling-activated chloride channels
    channel activation effects, **292:**365
    channel activity measurement, **292:**361
    channel sensitivity to hypoosmotic solutions, P-glycoprotein effects, **292:**367
    electrophysiologic characterization, **292:**361, 363
    evidence of interaction, **292:**360–361
    mechanism of channel activity modulation, **292:**367–368
    pharmacologic characterization, **292:**363, 365
    phosphorylative modulation, **292:**369
    physiologic significance, **292:**368–370
    regulatory volume decrease, overview, **292:**359–360, 369
  Chinese hamster ATPase
    assays
      lactate dehydrogenase coupled assay, **292:**520
      malachite green assay, **292:**519–520
    kinetic parameters, **292:**520–521
    modification of essential residues, **292:**521
    overexpression in CR1R12 cells, **292:**515

plasma membrane preparation,
292:516–517
purification of P-glycoprotein
dye affinity chromatography,
292:518–519
reconstitution, **292:**519
solubilization, **292:**517–518
stimulators, **292:**521
suspension culture of CR1R12 cells,
292:516
vanadate trapping of nucleotide,
292:521–523
deletion in adrenoleukodystrophy,
292:111–112
domain organization, **292:**307–308, 318,
397, 415, 493, 504, 514
Drosophila, **292:**105
Entoamoeba histolytica, **292:**109
gene therapy, *see* Gene therapy, *MDR1*
genes
classification, **292:**103, 397
overview, **292:**290
half-molecules, **292:**111–112
heterologous expression systems, overview, **292:**398
human, *see* MDR1; *MDR1*
monoclonal antibodies
applications
detection of drug-resistant cells,
292:264
gene cloning, **292:**263
modulation of transport,
292:263–264
protein purification, **292:**263
selective killing of cells, **292:**264–265
Chinese hamster, **292:**260, 263
human, **292:**260–263
mouse, *see* mdr1, murine; mdr2, murine;
mdr3, murine
photoaffinity labeling in drug-binding site
identification
amino acid sequence analysis, **292:**312
assumptions, **292:**290, 308–309
chimeric protein studies, **292:**316–317
denaturing polyacrylamide gel electrophoresis, **292:**297–298, 310, 324
immunological mapping
high-resolution mapping,
292:314–315
low-resolution mapping, **292:**313–314

principle, **292:**312–313
immunoprecipitiation, **292:**297–298
inhibition with unlabeled drugs,
292:298–300
labeling reaction, **292:**296–297, 310,
322, 324
membrane preparations, **292:**311
peptide mapping, **292:**305–306, 312
probe selection, **292:**308–309
site-directed mutagenesis, **292:**304–305,
315–316
synthesis of analogs and labeling
colchicine, **292:**293–294, 300–301
daunorubicin, **292:**294, 301
dihydropyridine, **292:**294, 302, 304,
309–311
doxorubicin, **292:**294, 301
forskolin, **292:**309
phenothiazines, **292:**302–303
prazosin, **292:**305–306, 309, 313–314,
322, 324–325, 327
verapamil, **292:**294–296, 302
vinblastine, **292:**291, 293, 299–300,
304
vanadate trapping, effects on labeling,
292:325–328
Plasmodium falciparum, *see* pfmdr1
reconstitution, **292:**326–327
reversing agents of multidrug resistance,
mechanism and structure, **292:**300,
304
sequence homology between classes and
species, **292:**103, 105, 109, 111–112,
162–163, 290, 514
steps in drug transport, **292:**318
substrate specificity, **292:**259
tissue distribution and functions,
292:259–260
topology
alternative topologies, **292:**286–289
C terminus translocation reporters
criteria for reporter domains,
292:268, 283
fusion protein generation,
292:268–270
overview, **292:**267–268, 282–283
definition, **292:**265
determinants
characterization, **292:**270–271
types of sequences, **292:**266

expression systems for study
  microsomal membranes, **292:**272–
    273, 281
  rabbit reticulocyte lysate, **292:**272,
    274, 281
  RNA digestion in cell-free systems,
    292:273
  transcription *in vitro* using SP6 RNA
    polymerase, **292:**280
  wheat germ extract, **292:**281,
    286–287
  Xenopus laevis oocyte microinjec-
    tion, **292:**273–274
folding steps, **292:**266–267,
  285–288
mapping, **292:**277–278
membrane stripping with alkaline
  buffer, **292:**281–282
N-linked glycosylation detection,
  292:274–275, 277, 283–284
protease digestion assay, **292:**275–277,
  285
yeast homolog, *see* Ste6p
PGM, *see* Phosphoglucomutase
Pgp, *see* P-glycoprotein
PgpA
  gene amplification
    response to drugs, **292:**183
    detection of amplification
      agarose gel electrophoresis,
        292:184–186
      circular amplicon isolation,
        292:186–187
      DNA extraction, **292:**184
      Leishmania growth, **292:**184
      linear amplicon isolation,
        292:187–188
  gene disruption studies of resistance,
    292:190
  gene locus, **292:**183
  gene transfection studies of resistance,
    292:188–189
  metal specificity, **292:**182–183
  resistance mechanisms in *Leishmania*,
    292:190–192
  thiol conjugation and pump activity,
    292:190–191
pH
  acid adaptation of oral streptococci bio-
    films

acid tolerance response in plaque,
  310:446–447
alkali production assay, **310:**453
buffering capacity assays, **310:**449–450
killing resistance assay, **310:**447–448
pH change across biofilms, assessment,
  310:452–453
pH measurement within biofilms,
  310:451–452
proton movement into biofilms, assay,
  310:450–451
proton-translocating ATPase assay,
  310:448–449
Archaeoglobus fulgidus biofilm, induction
  by pH change, **310:**341–342
bioelectric effect, pH changes with electric current, **310:**665–666, 669
endosome monitoring with fluorescent
  dye, **302:**48
intracellular measurement with fluorescent dyes
  BCECF, vascular remodeling studies,
    307:257
  confocal microscopy dyes, **307:**127–129
  SNARF-1 loading of cells, **307:**128–
    129
microelectrode measurements across biofilm, **310:**307
modeling in steady-state biofilm systems,
  310:317–320
ratio confocal microscopy imaging with
  SNARF-1, **302:**349
sensor proteins, *see* Green fluorescent
  protein; Hisactophilin
Phage display
  BiP ligand biopanning
    amplification of eluted phages,
      290:402–403
    applications, **290:**385–386
    biotinylation of BiP, **290:**401–402
    materials and solutions, **290:**400–401
    panning reaction, **290:**402
    positive phage identification by colori-
      metric assay, **290:**403
    principle, **290:**400
    single-stranded phage DNA isolation,
      290:403–404
  protein–protein interaction assay,
    303:422–423, 452
  rhodopsin ligands

advantages in ligand screening, **315:**439–440

data analysis with arrestin sequences, **315:**448–451

data confirmation with arrestin sequences, **315:**451–452

insert preparation, **315:**440–442

library evaluation, **315:**442–443

M13 coat protein utilization, **315:**438–439

panning

  controls, **315:**445–446, 448

  DNA preparation and sequencing, **315:**447–448

  elution of bound phage, **315:**444–445, 447

  phage culture, **315:**446–447

  rounds, **315:**443–444, 447

  substrate coating on dishes, **315:**444

vectors, **315:**440

Phagocyte luminescence

activation regulation, **305:**591–592

basal and phorbol myristate acetate-stimulation assays of phagocyte function

  myeloperoxidase, **305:**611–613

  NADPH oxidase, **305:**610–611

chemiluminigenic substrates

  intensity of emission, **305:**598, 625

  rationale for use, **305:**597–598

luminol oxygenation assay

  classical peroxidase reaction, **305:**603–604

  myeloperoxidase-catalyzed oxygenation, **305:**602–603

  overview, **305:**601–602

  peroxynitrite-dependent luminescence, **305:**604–605

metabolic activation, **305:**592

multiple discriminant function analysis of blood luminescence, **305:**620–624

myeloperoxidase-catalyzed oxidations and oxygenations, **305:**595–596

NADPH oxidase and respiratory burst metabolism, **305:**593–595

NADPH oxidation measured with acridinium luminescence

  lucigenin reaction, **305:**598–599

  specificity for products, **305:**600–601

native luminescence and diagnostic limitations, **305:**597

opsonin receptor-dependent oxygenation assay

advantages over artificial stimulation assays, **305:**613

inflammatory activation ratios, **305:**615–617, 619–620

myeloperoxidase activity, **305:**613–614

opsonin receptor expression density effects, **305:**617, 619

oxygen species metabolism quantification by metabolite in neutrophils, **305:**628

oxygen spin quantum number and reactivity, **305:**592–593

quantitative analysis factors

  luminometer efficiency, **305:**625–626

  net luminescence efficiency, **305:**626

  overview, **305:**625

  phagocyte metabolic activity correlation with luminescence, **305:**626–629

  quantum yield, **305:**626

whole blood luminescence assay with lucigenin and luminol

  blood collection, **305:**607

  data analysis, **305:**609–610

  data collection, **305:**608–609

  overview, **305:**605–607

  subjects and conditions of testing, **305:**608–609

Phagocytosis, confocal microscopy quantification

bacteria, fluorochrome labeling, **307:**321–323

cell distribution of phagocytosis, **307:**319

distinguishing from endocytosis, **307:**319

equipment, **307:**323

external and internal particle separation, **307:**324, 328

fluorochrome selection, **307:**324–325

neutrophil phagocytosis of *Staphylococcus aureus*

  controls, **307:**327

  data analysis and interpretation, **307:**327–328

  modifications for other cell types, **307:**325–326

  neutrophil isolation, **307:**325–326

  phagocytosis conditions, **307:**326–327

  stopping of phagocytosis, **307:**327

opsonin enhancement and receptors,
  307:320
particle selection for assays, **307**:321–322
steps in phagocytosis, **307**:319–320
Phagosome
  lysosome fusion, **298**:373
  microtubule binding
    actin filament interactions, detection,
      298:381
    assay
      anchorage of microtubules,
        298:379–380
      buffer preparation, **298**:377–378
      cytosol preparation, **298**:377
      equipment, **298**:378
      fluorescence microscopy, **298**:379
      incubation conditions, **298**:378–379
      rhodamine microtubule preparation,
        298:377
    functions, **298**:373, 375
    membrane protein analysis, **298**:380
    screening for soluble binding factors,
      298:380
  motility assay, **298**:380–381
  preparation containing latex beads
    buffers, **298**:376
    fluorescent bead preparation,
      298:375–376
    homogenization, **298**:376
    internalization of beads, **298**:376
    materials, **298**:375
    sucrose gradient centrifugation,
      298:376–377
Phalloidin, *see* Cytoskeleton, confocal microscopy
Pharmacokinetics, antisense oligonucleotides, *see also* Oligonucleotide radiotherapy
  brain pharmacokinetics
    determinants, **314**:261
    fluorescence microscopy, **314**:273–274
    infusion system, **314**:272–273
    label selection, **314**:263–264
    oligonucleotide preparation, **314**:271
    phosphorothioate oligonucleotides,
      314:262–263
    radiolabeling of oligonucleotides,
      314:271–272
    recovery of oligonucleotides from
      brain, **314**:273

striatum, oligonucleotide infusion
  distribution, **314**:266
  rationale, **314**:268, 270
  stability, **314**:264–266
  uptake and intracellular localization,
    314:266–268
overview of cellular pharmacology,
  314:440–441
P-ethoxy oligonucleotides
  calculations, **313**:386
  clearance, **313**:386
  extraction from whole blood and tis-
    sues, **313**:384–385
  organ distribution, **313**:386–387
  sample collection, **313**:385
phosphorothioate oligodeoxynucleotides
  in mice, **314**:553–554
phosphorothioate oligodeoxynucleotides
  clearance, **313**:23–25, 55
  labeling, **313**:21
  oral bioavailability, **313**:25–26
  protein binding, **313**:23
  stability, **313**:21–22
  tissue distribution, **313**:23, 25–26
  uptake in cells, **313**:22
Raf-1, antisense knockdown in cultered
  artery smooth muscle cells
  cell culture, **314**:443–444
  cell viability assays, **314**:451
  dose–response analysis
    assay selection, **314**:444–445
    materials, **314**:445–447
    Northern blot analysis, **314**:447
    regression analysis, **314**:448
    Western blot analysis, **314**:448
  half-life
    oligodeoxynucleotides, **314**:452
    Raf-1, **314**:441, 452–453
  lipofection, **314**:442–444
  messenger RNA doubling time,
    314:452
  oligodeoxynucleotide design,
    314:442–443
  reversibility of inhibition, **314**:451–452
  time–response analysis
    assay selection, **314**:449
    parameters, **314**:448–449
    repeated application, **314**:449, 451
Phase-contrast microscopy
  Archaeoglobus fulgidus biofilm, **310**:342

plastic tab biofilms, **310:**102–103
plate biofilms, **310:**102
slide biofilms, **310:**101–102
Phase separation, *see* Lipid phase separation
Phase transition, *see* Liquid–crystalline phase transition
N-Phenacylthiazolium bromide, reversal of advanced glycation end product cross-links
  assays
    in vitro, **309:**171
    in vivo, **309:**171
  disaggregation of β-amyloid, **309:**160–161, 172
  overview, **309:**160
  synthesis, **309:**170
1,10-Phenanthroline-copper, *see* Copper phenanthroline
Phenolic antioxidants, *see also specific compounds*
  absorption and excretion in humans, **299:**92–95
  food additives, **299:**152
  gas chromatography–mass spectrometry of wine
    costs, **299:**151
    extraction and characterization, **299:**142–143
    gas chromatography conditions, **299:**141
    instrumentation, **299:**139, 141–142
    linearity, **299:**148
    mass spectrometry conditions, **299:**142
    materials, **299:**139
    peak identification, **299:**143–145
    precision, **299:**149–150
    recovery, **299:**148–149
    resolution, **299:**145, 147–148
    sample preparation, **299:**139
    sensitivity, **299:**149–150
    standards, **299:**139–140
  high-performance liquid chromatography
    chromatography conditions, **299:**103–105
    food sample preparation, **299:**100, 103
    retention times, **299:**105–106
    standards, **299:**100
    urine sample

    dietary manipulation, **299:**97, 100, 106
    preparation, **299:**103
  wine analysis
    absorbance characterization of peaks, **299:**119–120, 132–133
    anthocyanins, **299:**119–120, 135–136
    catechins, **299:**119
    difficulty of separations, **299:**113–114
    flavonols, **299:**121
    instrumentation, **299:**124, 131
    linearity, **299:**125, 127–129, 132
    low pH gradients, **299:**114, 116–117, 124–125, 131
    pH shift gradient, **299:**114–116
    polystyrene reversed-phase chromatography, **299:**117
    precision, **299:**129, 132
    procyanidins, **299:**137
    recovery, **299:**129, 132
    resolution, **299:**125, 132
    sample preparation, **299:**123
    sensitivity, **299:**122–123, 130, 132
    standards, **299:**118–119, 123–124, 130–131, 133, 135
  oxidation, pH dependence, **299:**153
  quinone formation and phenol regeneration, **299:**153
  radical scavenging activity, **299:**91–92
  total phenol analysis with Folin–Ciocalteu reagent
    advantages, **299:**177–178
    chemistry of reaction, **299:**160–161
    cinchonine precipitations, **299:**175–176
    comparison with other techniques, **299:**169–171
    flavonoid–nonflavonoid separations, **299:**170–173
    flow automation, **299:**156–158
    Folin–Denis reagent comparison, **299:**154–155
    incubation conditions and color development, **299:**155–156
    interference
      ascorbic acid, **299:**166–167
      nucleic acids, **299:**164
      oxidation, **299:**164
      proteins, **299:**164–165
      sugars, **299:**165–166
      sulfite, **299:**167–169

sulfur dioxide, **299:**167–168
molar absorptivity of specific phenols, **299:**161–163
phloroglucinol addition, **299:**173
preparation of reagent, **299:**155
sample preparation, **299:**159
standards and blanks, **299:**158–159
tannin separations, **299:**174–175
values for various wines, **299:**176–177
volume of reaction, **299:**156
1-Phenyl-2-decanoylamino-3-morpholino-1-propanol
  1-*O*-acylceramide synthase inhibitors, **311:**385–386
  cell growth inhibition assays, **311:**377–379
  ceramide accumulation mechanism, **311:**379–380
  ceramide glycanase inhibitors, **311:**387
  delivery into cultured cells
    bovine serum albumin complex, **311:**384
    liposomes, **311:**383–384
    Myrj 52 micelles, **311:**383
  diastereomers, **311:**374–375, 377, 378
  discovery, **311:**374–375
  functional group modification in homologs
    chain length effects, **311:**375, 379
    morpholino group substitutions, **311:**376, 378
    pyrrolidino analog, **311:**376–378
  glucosylceramide synthase inhibition, overview, **311:**105, 373
  metabolic effects
    ceramide levels, **311:**385–386
    glycosphingolipid levels, **311:**384–385
  1-phenyl-2-palmitoylamino-3-pyrrolidino-1-propanol homolog synthesis
    chiral chromatography, **311:**382
    diastereomers, **311:**381–382
    overview, **311:**380
    palmitoyl aminoacetophenone synthesis, **311:**380
    2-palmitoylamino-3-pyrrolidinopropiophenone, **311:**381
    physical characterization, **311:**382–383
  sphingomyelin synthase inhibitors, **311:**387
C-Phenylphosphoramidite, synthesis

2,3-*O*-isopropylidene-D-ribono-1,4-lactone as starting material, **317:**45–46
  overview, **317:**43
  2,3,5-tri-*O*-benzyl-D-ribono-1,4-lactone as starting material, **317:**43–45, 57–60
PHF, *see* Paired helical filament
PHGPX, *see* Phospholipid hydroperoxide glutathione peroxidase
Pholasin, enhancement of chemiluminescence, **305:**344
PHONTs, *see* Photolabile derivatives of oligonucleotides
Phosphate, caged
  caging groups, *see* 4-Coumarinylmethyl group; Desyl group; *p*-Hydroxyphenacyl group
  Fourier transform infrared spectroscopy
    band assignment by isotopic labeling, **291:**227–229
    difference spectra, **291:**226–229
    magnesium effects on spectra, **291:**229–231
  photochemical conversion, quantification, **291:**339
Phosphatidate phosphatase, sphingoid base inhibition
  assay
    detection, **312:**376
    incubation conditions, **312:**376
    phosphatidate, phosphorous-32 labeling, **312:**374–375
    unit definition, **312:**377
  enzyme preparation, **312:**374
  kinetic analysis
    dose–response, **312:**377–379
    inhibition constant determination, **312:**380
    phytosphingosine inhibition, **312:**379–380
    software, **312:**377
    sphinganine, **312:**379–380
    sphingosine, **312:**379
  overview, **312:**373–374
  sphingoid base preparation
    storage, **312:**375
    Triton X-100/lipid-mixed micelle preparation, **312:**375–376
  structural requirements for inhibition, **312:**377

Phosphatidylcholine, radiolabeling for sphingomyelin synthase assay, **311**:34–35

Phosphatidylcholine, translocation assay with Mdr2 in yeast secretory vesicles fluorescent phosphatidylcholine, **292**:391–392

principle, **292**:391–393

reaction conditions, **292**:395–396

secretory vesicle preparation with outer leaflet label, **292**:393, 395

Phosphatidylinositol, tritiated substrate preparation, **311**:123–124

Phosphatidylinositol 3-kinase, inhibitors, **298**:548

Phosphatidylserine-binding apoptosis assay, *see* Apoptosis

Phosphodiesterase, *see* Cyclic GMP phosphodiesterase

Phosphoglucomutase

difference Raman spectroscopy

symmetric stretch frequency of substrate phosphate, **308**:181, 198

vanadate spectroscopy, **308**:198

steps in catalysis, **308**:196, 198

Phospholiase A$_2$

all-*trans*-retinyl ester hydrolase inhibition studies, **316**:392–393

antisense knockdown, **314**:85

Phospholipase C, photodynamic therapy activation, **319**:346–348

Phospholipase D, chemiluminescence assay, **305**:385

Phospholipid exchange, giant patches, **293**:277–280

Phospholipid hydroperoxide glutathione peroxidase

assay by high-performance liquid chromatography with mercury cathode electrochemical detection

advantages over coupled assay, **300**:32

decay profile, **300**:30–32

lipid hydroperoxide substrate preparation, **300**:30

reaction mixture, **300**:30

coupled assay, **300**:29

Phospholipid peroxidation

detection by high-performance liquid chromatography with postcolumn chemiluminescence detection

cell culture, **300**:54

chemiluminescence detection, **300**:52–53

extraction, **300**:54–55

instrumentation, **300**:52

phosphate content assay, **300**:55–56

reagents and materials, **300**:53–54

standardization, **300**:56–57

phospholipids for analysis

fast-atom bombardment–mass spectrometry, **300**:64–67, 69

oxidizability measurement, **300**:65, 69

purification from crude soy lecithin, **300**:63–64

reversed-phase high-performance liquid chromatography, **300**:64–66

Phosphoprotein phosphatase

chemiluminescence assay

challenges, **305**:396–397

inhibitors, **305**:396–397, 401

luciferyl-*o*-phosphate assay of phosphotyrosine phosphatases, **305**:399–401

phosphoprotein phosphatase type 2B activity assay, **305**:397–398

membrane blotting, **305**:398–399

functions, **305**:391, 395–396

types, **305**:396

Phosphoramidate oligonucleotide

inhibition in *Xenopus* oocytes and embryos with antisense oligonucleotides An2, 426–427, 434–435

controls, **313**:427–428, 435–436

cyclins, **313**:427

efficacy, **313**:434–435

microinjection, **313**:432–433

RNA analysis, **313**:433

purification, **313**:430–432

ribonuclease cleavage of duplexes, **313**:423

specificity effects of modification, **313**:426–427

stability, **313**:424–426

structure of partially modified oligonucleotides, **313**:423

synthesis, **313**:428–430

Phosphoramidite, base-modified pyrimidine ribonucleoside synthesis

applications in hammerhead ribozyme studies, **317**:53–54

2'-O-tert-butyldimethylsilyl group introduction, 317:40, 55
5'-O-dimethoxytrityl group introduction, 317:40, 54
flash chromatography purification, 317:39–40
6-methylcytidine phosphoramidite, 317:49–51, 62–65
6-methyluridine phosphoramidite, 317:49–51, 62–64
oligoribonucleotide synthesis, purification, and characterization, 317:51–53
C-phenylphosphoramidite
  2,3-O-isopropylidene-D-ribono-1,4-lactone as starting material, 317:45–46
  overview, 317:43
  2,3,5-tri-O-benzyl-D-ribono-1,4-lactone as starting material, 317:43–45, 57–60
phosphitidylation, 317:55
2-pyridinone riboside phosphoramidite, 317:46–49, 60–62
4-pyridinone riboside phosphoramidite, 317:46–49, 60–62
riboabasic phosphoramidite, 317:41–43, 55–57
safety, 317:40, 43
solid-phase synthesis, 317:40–41
thin-layer chromatography monitoring of reactions, 317:39
5-Phosphoribosyl-1-pyrophosphate, isotopic labeling for kinetic isotope effect studies, 308:310
Phosphorothioate interference analysis, see also Nucleotide analog interference mapping
  metal-binding analysis, 317:77, 79
  principle, 317:77
  steric effects, 317:79
  transfer RNA aminoacylation analysis
    aminoacylation reaction, 317:87–88
    critical phosphate and 2'-hydroxyl position identification, 317:90–91
    iodine cleavage and cleavage pattern analysis, 317:89–90
    materials, 317:83–84
    overview, 317:82–83
    separation of charged and uncharged RNAs, 317:88–89

transcript
  dephosphorylation, 317:86–87
  nucleotide incorporation into RNA transcripts with mutant T7 RNA polymerase, 317:85–86
  phosphorous-32 labeling, 317:87
Phosphorothioate oligodeoxynucleotide
  advantages and limitations in antisense therapy, 313:37, 46, 96–97
  applications, 317:75
  association rates with RNA targets, 313:18
  chemical synthesis, 317:81–82
  degradation pathways, 313:269–270
  digoxigenin labeling
    intracellular distribution studies
      advantages over other techniques, 313:258–259
      cell fixation, 313:265
      confocal microscopy, 313:260–261, 265–266
      electron microscopy, 313:261–262, 264, 267–268
    synthesis
      DNA synthesizer setup, 313:264
      overview, 313:259
      purification, 313:265
  disadvantages in selection experiments, 313:73
  interference analysis, see Phosphorothioate interference analysis
  melting temperatures of hybridization, 313:18, 50, 56
  mercury-binding site engineering, 317:126
  minimally modified antisense oligonucleotides
    binding affinity, 313:278–279
    design criteria, 313:270
    efficacy in gene knockdown, 313:278, 281–284
    rationale, 313:269–270
    secondary modifications and ribonuclease H effects, 313:270–271
    specificity of effects, 313:280
    stability against nucleases, 313:273–276, 278
    synthesis, 313:272
  uptake by cells
    assay, 313:279
    concentration requirements without

uptake enhancers, **313**:280–282, 284

enhancers of uptake, **313**:279–280, 282–283

lipofection, **313**:283

sequence effects, **313**:279

nucleoside 5'-*O*-(1-thiotriphosphate) synthesis, **317**:82, 96–98

pharmacokinetics, *see* Pharmacokinetics, antisense oligonucleotides

protein binding

affinity, **313**:459

albumin, **313**:19, 23

classification of binding events, **313**:18–19

DNA polymerases, **313**:20

effects in antisense experiments, **313**:20

nucleases, **313**:19–20

overview of proteins, **313**:268–269

purification, *see* High-performance liquid chromatography

side effects, **313**:31

size requirements for antisense studies, **313**:26

stability, **313**:17, 21–22, 53, 148, 268, 361–362

stability, **314**:81, 91, 94, 123, 229, 249, 293, 314, 362, 480, 507–508, 539–541

stereochemistry

nucleotidyl transfer or hydrolysis reaction analysis, **317**:76–77

overview, **317**:75–76

structure, **314**:509, **317**:74–75

synthesis, **317**:76

therapeutic index, **313**:36–37, 54

toxicity, **314**:75, 161–162, 165–166, 194–195, 231, 262, 555–557

genotoxicity, **313**:33–34

humans, **313**:35–37

immunostimulation, **313**:32–33

mice, **313**:34

monkeys, **313**:34–35

rat, **313**:55

yield from solid-phase synthesis, **313**:203

Phosphorylation, ion channels

cystic fibrosis transmembrane conductance regulator, **294**:241, 243

glutamate receptors

extraction of labeled receptors, **294**:355–356

functional analysis by patch-clamp whole-cell recording

basal phosphorylation, **294**:368, 370

intracellular patch perfusion solutions, **294**:367–368

site-directed mutants, **294**:370

stimulation of phosphorylation, **294**:366–368

fusion proteins

phosphorylation *in vitro*, **294**:362–363

rationale for study, **294**:361–362

immunoprecipitation of labeled receptors, **294**:354, 356–357

investigation in native tissues, **294**:353–354

phosphoaminoacid analysis, **294**:360–361

phosphopetide mapping

overview, **294**:357–358

proteolytic digestion, **294**:358

two-dimensional thin-layer chromatography, **294**:359–360

radiolabeling in cell culture, **294**:354–355

site-directed mutagenesis of phosphorylation sites, **294**:363–364, 370

site-specific antibodies, generation and application, **294**:364–366

Phosphorylation, rhodopsin

arrestin binding to phosphorylated form, **315**:70–71

constitutively active opsin mutants as substrates, **315**:217–218

dephosphorylation, *see* Protein phosphatase type 2A, photoreceptors

gel assay of radiolabeled protein, **316**:271–272

inhibitors, *see* S-Modulin; Recoverin

kinase, *see* Rhodopsin kinase

kinetics, **316**:168

mutant studies in transgenic mice

DNA preparation for microinjection, **316**:171

founder screening, **316**:173–174

gene dosage control with knockout crosses, **316**:179–182, 184

overexpression and retinal degradation, **316**:178–179

rationale, **316**:168–169

reverse transcriptase–polymerase chain reaction for transgene quantification, **316:**176

single-cell recordings from mutant-containing photoreceptors, **316:**185

Southern blot analysis for sorting different lines from same founder, **316:**174

trafficking of mutants, **316:**184

transgene constructs, **316:**169–171

transgenic mouse generation, **316:**172–173

phototransduction role, **316:**167

preparation of phosphorylated rhodopsin anion-exchange chromatography, separation of forms, **315:**75–76

Chelex column chromatography, **315:**73–75

isoelectric focusing, **315:**71–72, 76

lectin affinity chromatography, **315:**72–73

phosphorylation conditions, **315:**71, 240

rod outer segment preparation, **315:**71

regulatory function, **316:**482

site identification

bovine rhodopsin

Asp-N digestion, **316:**484

mass spectrometry analysis of fragment phosphopeptides, **316:**487–488

overview, **316:**482–484

peptide separation and identification, **316:**484–486

phosphorylated protein preparation, **316:**484

subdigestion of C-terminal peptide, **316:**486–487

octopus rhodopsin

overview, **316:**488

peptide identification, **316:**492

phosphorylated protein preparation, **316:**488, 491

proteolysis with trypsin and thermolysin, **316:**491–492

overview of sites, **316:**167–168

stoichiometry, **316:**167

zebrafish larva eye, assay

incubation conditions and gel autoradiography, **316:**551–552

materials, **316:**551

principle, **316:**551

white light versus red light response, **316:**552

Phosphorylation/dephosphorylation, proteins

antibodies for phosphoproteins, **298:**558–559

criteria for protein function role, **298:**542

enrichment of cytoskeletal proteins for analysis, **298:**556

keratin intermediate filament

metabolic phosphorous-32 labeling, **298:**568–569

phosphorylation and disassembly, **298:**567–568

labeling of cytoskeletal proteins

in vitro, **298:**554–555

in vivo, **298:**551–553

phosphoamino acid analysis, **298:**557

phosphopeptide mapping, **298:**557

protein kinase inhibitors and activators, **298:**543, 547–550

protein phosphatase assay, **298:**555–556

protein phosphatase inhibitors, **298:**543–547

site of phosphorylation

determination, **298:**557–558

site-directed mutagenesis, **298:**558

transfection analysis, **298:**551

vasodilator-stimulated phosphoprotein

assays of phosphorylation, **298:**110–111

kinases, **298:**109–110

platelet inhibition, **298:**103

sites, **298:**103, 109–110

stimulators, **298:**103

vimentin

extraction of phosphoprotein, **298:**562

phosphopeptide antibody preparation, **298:**566–567

phosphorous-32 labeling

in vitro, **298:**562–563

in vivo, **298:**560–562

phosphorylation site determination, **298:**563, 565–566

Phosphotungstic acid, chromatin staining for transmission electron microscopy, **304:**227, 229

Photoacclimation, algae versus higher plants, **297:**221–222

Photoacoustic calorimetry
  apparatus, **295:**320
  cytochrome P450cam, carbon monoxide
    dissociation in site-directed mutants,
    295:325–327, 329–330
  myoglobin, carbon monoxide dissociation
    in site-directed mutants, **295:**321–325
  overview of techniques, **295:**316–317
  signal components and separation,
    295:317–319
Photoacoustic spectroscopy, *see* Laser-induced optoacoustic spectroscopy
Photoaffinity labeling, *see also* Ceramide;
  specific proteins
  carbene reagents, **311:**574
  carbohydrate-binding sites of glycosphin-
    golipid-binding proteins
    carbohydrate-binding domains,
      311:576–577
    monosaccharide-binding domains,
      311:579–581, 583
    oligosaccharide modification,
      311:577–579
    sialic acid derivatives, **311:**581, 583
  cyclic nucleotide-gated channels in rods
    applications with other proteins,
      315:771–772
    8-*p*-azidophenacylthio-cyclic GMP
      phosphorous-32 labeling,
        315:759–760
      purification, **315:**758–759
      synthesis, **315:**756–758
    cyanogen bromide cleavage of labeled
      subunits, **315:**761–762
    gel electrophoresis
      electroelution, **315:**761
      peptide fragments, **315:**762
      subunit separation, **315:**760–761
    labels
      labeling reaction, **315:**760
      selection, **315:**756–757
    partial purification of channels,
      315:760
    patch clamp recordings following cova-
      lent activation with 8-*p*-azidophe-
      nacylthio-cyclic GMP
      conductance states, **315:**770–771
      data acquisition, **315:**764–765, 768
      functional heterogeneity in binding
        sites, **315:**767

    ligand number assignment, **315:**768–
      770, 772
    macroscopic currents, **315:**764–765,
      767
    principle, **315:**764
    single-channel currents, **315:**768–771
    principle, **315:**756
    sequence analysis of labeled peptides,
      315:762
  dopamine transporter
    [^{125}I]DEEP, **296:**220–221, 225–227
    immunoprecipitation, **296:**221–222
    labeling conditions, **296:**220–221
    proteolysis
      gel purification and electroelution,
        296:223–225
      in situ proteolysis, **296:**225–227
    [^{125}I]RTI, **296:**220–221, 225, 227
  GTP–septin complexes, **298:**294–295
  ideal criteria for reagents, **311:**574
  lipid-binding sites of proteins
    lipid-binding domains, **311:**583–588
    sphingolipid-binding proteins,
      311:588–590
    temperature dependence, **311:**585
  nitrene reagents, **311:**574–575
  overview, **311:**573–574
  peptide analysis
    labile-substituted imidates, **311:**599–600
    residue identification, **311:**597, 599
    selective peptide cleavage, **311:**600
    separations, **311:**596–597
  photolysis conditions, **311:**595
  P-glycoprotein, drug-binding site identifi-
    cation
    assumptions, **292:**290
    denaturing polyacrylamide gel electro-
      phoresis, **292:**297–298
    immunoprecipitiation, **292:**297–298
    inhibition with unlabeled drugs,
      292:298–300
    labeling reactions, **292:**296–297
    peptide mapping, **292:**305–306
    site-directed mutagenesis, **292:**304–305
    synthesis of analogs
      colchicine, **292:**293–294, 300–301
      daunorubicin, **292:**294, 301
      dihydropyridine, **292:**294, 302, 304
      doxorubicin, **292:**294, 301
      phenothiazines, **292:**302–303

limitations compared to fluorescence photoactivation, **291**:63
Photobleaching, green fluorescent protein,
see also Fluorescence recovery after
photobleaching
reversibility, **302**:250, 252–254
solute diffusion analysis
complex phenomena, analysis, **302**:260
cytoplasm, **302**:254–255
diffusion coefficient, **302**:252
endoplasmic reticulum, **302**:257–258
membrane diffusion of fusion proteins,
302:259–260
mitochondria, **302**:255–256, 258–259
principle, **302**:241
quantitative analysis in organelles,
302:258–259
reversible versus irreversible photobleaching, analysis, **302**:253–254
Photobleaching energetics, rhodopsin
reaction scheme, **315**:156
enthalpy surface analysis, **315**:156
blue-shifted intermediate, **315**:156, 158
charge separation and energy storage,
315:158
limitation of vision at high light intensity,
315:159
measurement, *see* Cryogenic photocalorimetry; Laser-induced optoacoustic
spectroscopy
intermediates, **315**:348
Photobleaching recovery, photoreceptors
absorption spectra of pigments,
316:227–228
adaptation, definition, **316**:224
bleaching estimation
absorption cross-sections, **316**:228, 230
fraction of pigment bleached,
316:226–227
solution photosensitivity, **316**:233–234
specific optical density of rod outer segment, **316**:230–233
calcium role in bleaching adaptation,
316:249–252
cone electrophysiology, **316**:236, 238–239
enzyme assays with electrophysiology
calculations, **316**:242
cyclic GMP phosphodiesterase,
316:243, 245, 249
guanylyl cyclase, **316**:245, 247, 249

rapid solution changes, **316**:239–241
isolated rod electrophysiology, **316**:236,
239
light flash apparatus, **316**:226
photoreceptor preparation from tiger salamander, **316**:225–226
rationale for study, **316**:224–225
regeneration in retina, **316**:224
11-*cis*-retinal solution preparation
ethanol solution, **316**:235
exogenous application to isolated cells,
316:235–236
interphotoreceptor retinoid binding
protein solution, **316**:235
phospholipid vesicles, **316**:234–235
Photocalorimetry, *see* Cryogenic photocalorimetry
Photocleavable affinity tags
applications, **291**:137–138
design, **291**:138
photocleavable biotins
applications, **291**:152–154
criteria in design, **291**:152
leucine-enkephalin
avidin affinity purification, **291**:144–
145, 148–149
conjugation, **291**:144
fluorescamine assay, **291**:145
photocleavage kinetics, **291**:146–148
oligonucleotide labeling
affinity purification, **291**:145, 150–151
high-performance liquid chromatography, **291**:149–150
phosphorylation at 5'-end, **291**:153
photocleavage, **291**:145–146, 150–151
polyacrylamide gel electrophoresis
analysis, **291**:151–152
synthesis, **291**:145, 149, 153
structures, **291**:138–139
synthesis
materials, **291**:139–140
photocleavable biotin NHS carbonate, **291**:140–142
photocleavable biotin phosphoramidite, **291**:142–144, 153
Photodynamic therapy, *see also* Ultraviolet
radiation
antimicrobial chemotherapy, photosensitizers and mechanisms, **319**:197–198
apoptosis role

bc12 role, **319:**348–349
calcium signaling, **319:**346–348
caspase activation
    Bap31 fragments, **319:**350–351
    inhibitor studies, **319:**349–350
cell cycle mediation, **319:**351–353
ceramide generation, **319:**351
early event in tumor shrinkage,
    319:345–346
mechanisms, **319:**346
nitric oxide role, **319:**353
p21 role, **319:**352, 354
p53 role, **319:**349, 352
phospholipase C activation,
    319:346–348
retinoblastoma protein role,
    319:352–353
serine/threonine dephosphorylation,
    319:348
caries and periodontitis, **310:**275–276
cell damage mechanisms, overview,
    319:121–122
clinical applications, **319:**343–344
dissipation of singlet state energy,
    319:384
early response gene activation,
    319:355–356
glucose-regulated proteins in response,
    319:354–355
heat shock protein induction, **319:**357
historical background, **319:**343, 376
HOG1 activation, **319:**356–357
hydrogen peroxide formation and detec-
    tion, **319:**394–396
hydroxyl radical
    detection, **319:**395, 397
    formation, **319:**394
    reactivity, **319:**394–395
α_2-macroglobulin receptor/low-density li-
    poprotein receptor-related protein re-
    sponse, **319:**356
mitogen-activated protein kinase stimula-
    tion, **319:**358
mode of action, **319:**383–386
necrosis mechanism, **319:**346
nuclear factor-κB response, **319:**355
oxygen consumption rates, **319:**386
photochemical process, **319:**344, 383–384
photosensitizers, *see also* 5-Aminolevuli-
    nate; Aminopyropheophorbide

benzoporphyrin derivative monoacid
    ring A, **319:**381
electron-donating molecules, **319:**394
hematoporphyrin derivative, **319:**343
ideal criteria, **319:**379, 385
lutetium texaphrin, **319:**381
mono-L-aspartyl chlorin e6, 381
Photofrin II, **319:**378–379, 386
Photofrin limitations, **319:**344, 379
phthalocyanines, **319:**382–383, 385
tetra(*m*-hydroxyphenyl) chlorin,
    319:381
tin etiopurpurin, **319:**381
tumor uptake, **319:**377
principle, **310:**275, **319:**342–343, 377–378
protein tyrosine phosphorylation,
    319:357–358
singlet oxygen production
    detection, **319:**388–391, 393
    deuterium oxide effects, **319:**392–393
    diffusion rate, **319:**388
    mechanism, **319:**386–387
    quencher studies, **319:**391–392
stress-activated protein kinase activation,
    319:356–357
superoxide anion
    detection, **319:**395–396
    formation, **319:**394
    reactivity, **319:**394
tumor response mechanism, **319:**344–345
type I versus type II mechanisms
    cell type dependence, **319:**399–400
    lipophilicity effects, **319:**398
    macromolecular binding effects,
        319:398–399
    oxygen availability effects, **319:**397–
        398, 400
    polarity effects, **319:**398
Photoionization gas chromatography,
    breath hydrocarbons in lipid peroxida-
    tion measurement
column selection, **300:**458–459
confounding variables, **300:**460–461
pentane measurement, **300:**459–460
principle, **300:**458
sensitivity, **300:**461–462
smokers versus nonsmokers, **300:**461–462
Photolabeling, GroES with bisANS
binding conditions, **290:**320
conformational change analysis in

GroEL–GroES complex,
  290:322–323
fluorescent properties of fluorophore,
  290:318–319
hydrophobic binding site identification,
  290:318, 321–322
photolabeling conditions, **290:**320–321
Photolabile derivatives of oligonucleotides
  advantages in ribosome structure elucida-
    tion, **318:**119
  binding to ribosomal subunits, **318:**120,
    123–124
  controls, **318:**131–133
  criteria for target site-specific labeling,
    318:131–132
  design prospects, **318:**134–135
  distortion of ribosomal subunits,
    318:132–133
  identification of labeled nucleotides or
    limited ribosomal RNA sequences
    oligoDNA-directed ribonuclease H
      cleavage, **318:**125, 127
    overview, **318:**124
    reverse transcriptase primer extension
      assay, **318:**128–130
  identification of labeled proteins,
    318:129–131, 135–136
  photolysis conditions, **318:**124
  purification, **318:**122–123
  radiolabeling, **318:**123
  ribosomes, three-dimensional modeling,
    318:133–134
  synthesis
    amine functionality, **318:**121
    automated synthesis, **318:**121
    phosphorothioate functionality, **318:**121
    photolability introduction, **318:**121–122
  types, **318:**120, 122
Photolysis rate, calculation, **291:**131
Photomultiplier tube
  components, **305:**54
  design, **305:**55
  luminometer prospects, **305:**62–63
  overview of features, **305:**53
  photocathodes, **305:**55–57
  photon counting, **305:**65
  principle of detection, **305:**54–55, 64–65
  spectral sensitivity, **305:**65
Photon-counting microscopy, luciferase detection

absolute photon flux determination,
  310:160
background light levels, **310:**156
correlative microscopy, **310:**159–160
dark box, **310:**154–155
data acquisition, **310:**156–159
hardware, **310:**152–154
light leak testing, **310:**155–156
quantitative analysis, **310:**159
sensitivity, **310:**153
software, **310:**154
Vibrio fischeri cells, **310:**155–159
Photon energy, calculation, **305:**53
Photoprotecting compounds, *see* Caged
  compounds
PhotoSELEX, applications, **318:**99
Photosensitization, singlet oxygen production
  cell photosensitization
    linear dose–response, **319:**26
    photosensitizer efficiency comparisons,
      319:27–28
    rose bengal as photosensitizer,
      319:26–27
    subcellular localization of photosensitiz-
      ers, **319:**25–26
    uniform illumination of cells, **319:**26
  energy transfer efficiency, **319:**20–21
  light sources, **319:**21–22
  metal oxides, *see* Titanium dioxide; Zinc
    oxide
  photon absorption rate, **319:**20
  photosensitizers, **319:**21
  principle, **319:**20
  solution photosensitization
    1,3-diphenylisobenzofuran probe,
      319:22–24
    p-nitrosodimethylaniline bleaching,
      319:23, 25
    quantum yield measurement, **319:**23, 24
    substrate reaction modeling, **319:**24–25
Photosynthetic reaction center
  chlorophyll ligands in photosystem II re-
    action center, **297:**355–356
  crystallization for X-ray diffraction
    overview, **297:**58–59
    Rhodobacter sphaeroides reaction
      center
      amphiphile composition for type II
        crystals, **297:**72–73

crystal geometry and diffraction,
    297:61, 68–72
1,4-dioxane, role in crystallization,
    297:73
mutant proteins, **297**:75
orthorhombic crystals, **297**:68–
    70
pH, **297**:74
purity requirements, **297**:74
reproducibility, **297**:74
seeding, **297**:74–75
structure determination and refine-
    ment, **297**:76
subunit structure, **297**:62
temperature, **297**:73–74
tetragonal crystals, **297**:72
thin crystal preparation for spectros-
    copy, **297**:75
trigonal crystals, **297**:70–73
Rhodopseudomonas viridis reaction
    center
    crystal geometry and diffraction,
        297:61, 65
    crystallization by vapor diffusion,
        297:64–65
    structure determination and refine-
        ment, **297**:75–76
    subunit structure, **297**:62
mutagenesis by chloroplast transforma-
    tion of *Chlamydomonas reinhardtii*
    chimeric gene construction conferring
        antibiotic resistance, **297**:312–313,
        315–316
    complications, **297**:311
    gene gun transformation, **297**:311,
        317–318
    growth conditions, **297**:316–317
    growth media, **297**:316
    homologous recombination, **297**:311,
        315
    recipient strains, **297**:316
    selection of transformants, **297**:312,
        318–320
purification
    Rhodobacter sphaeroides reaction
        center
        anion-exchange chromatography,
            297:67–68
        cell growth and harvesting,
            297:65–66

molecular sieve chromatography,
    297:68
purification without sucrose gradient,
    297:67
solubilization, **297**:66
sonication, **297**:66
sucrose gradient centrifugation,
    297:66
Rhodopseudomonas viridis reaction
    center
    cell growth and harvesting,
        297:62–63
    molecular sieve chromatography,
        297:64
    solubilization, **297**:63
    sonication, **297**:63
    sucrose gradient centrifugation,
        297:63
Rhodobacter sphaeroides 2.4.1
    components, **297**:151–152
    mutagenesis techniques, **297**:155–156
    regulatory gene isolation
        complementation of regulatory mu-
            tants, **297**:156
        heterologous host gene expression,
            297:157
        mapping and cloning, **297**:158
        reverse genetics, **297**:159
        suppressor isolation, **297**:156–157
    regulatory mutant isolation
        cis-acting mutations, **297**:152–153
        decreased photosynthesis gene isola-
            tion using transcriptional fusions
            to *sacB*, **297**:154–155
        enrichment using *lacZ* fusions,
            297:153–154
        increased photosynthesis gene isola-
            tion using transcriptional fusions
            to *aph*, **297**:153
        spontaneous mutant isolation by pig-
            mentation and growth analysis,
            297:155
        trans-acting mutations, **297**:153
    transcriptional regulatory factors
        biochemical characterization,
            297:161–162
        interactions between regulatory fac-
            tors, **297**:162
        phenotypic characterization,
            297:159–160

protein–DNA interactions,
297:162–163
quantitation of photosynthetic expression, 297:160–161
sequence analysis, 297:159
structure–function analysis, 297:161
types and gene targets, 297:163–165
Synechocystis 6803 genes, see psbA;
psbB; psbD
Photosystem I
assays
electron acceptors functioning after F_X
cluster, flash photometry, 297:107–
108, 113, 115, 119
electron transfer efficiency of P700,
time-resolved optical absorption
spectroscopy, 297:105–107, 111,
113, 117, 119
electron transfer throughput by steady-
state $NADP^+$ reduction, 297:108–
109, 115–116, 119, 121, 123
oxygen evolution and uptake, 297:134
P700 estimation by absorbance changes
after oxidation or reduction,
297:135, 138
reductase assay, 297:135
components in cyanobacteria, 297:19, 124
pseudorevertants from Synechocystis
6803
genetic conformation for suppressor
mutations, 297:21–22
genetic transformation
competency dependence on growth
phase, 297:25
DNA preparation, 297:24–25
frequency, 297:23
growth conditions for primary mutants, 297:24
information attainable from system,
297:18
phenotypes of site-directed mutants,
297:19–20
phenotypic complementation in suppressor mutation locating, 297:22
polymerase chain reaction screening of
mutations, 297:21–22, 26
psaC mutants, pseudorevertant isolation, 297:20–21
secondary mutation at primary mutation site, 297:23

strains and growth conditions,
297:25–26
purification from Synechocystis 6803
anion-exchange chromatography,
297:126–127
overview of approaches, 297:127
solubilization, 297:126
sucrose gradient centrifugaton, 297:127
subunits, see PsaA; PsaB; PsaC; PsaD;
PsaE; PsaF; PsaI; PsaK; PsaL; PsaM
topographical characterization
chemical modification of exposed residues with biotin derivatives,
297:132–133
cross-linking of subunits, 297:133–134
limited proteolysis and gel electrophoresis, 297:130–131
trimer preparation from Synechocystis
6803, 125
Photosystem II
chloroplast-encoded genes
translational regulation in Chlamydomonas reinhardtii
affinity purification of factors interacting with 5'-untranslated region, 297:203–208
D1 labeling in vivo with carbon-14
acetate and gel electrophoresis
analysis, 297:196–198
messenger RNA preparation and
analysis, 297:198–203
nuclear mutant generation by glass
bead transformation,
297:193–195
screening of high-fluorescence mutants, 297:193, 195–196
types, 297:192
electron donors and acceptors, 297:339
singlet oxygen generation, 319:77–78, 80
spectroscopic characterization, see also
Plastoquinone redox state
chlorophyll fluorescence yield
charge recombination rate with 3-
(3',4'-dichlorophenyl)-1,1'-
dimethylurea, 297:344–346
factors affecting yield, 297:339–340
kinetic techniques, 297:340–341
mutant characterization, 297:341–346
secondary electron transfer rates,
297:342–344

electron nuclear double resonance,
   297:353–356
electron paramagnetic resonance,
   297:351–353, 356
electron spin echo envelope modulation, **297**:353–354, 356
Fourier transform infrared spectroscopy, **297**:357–358
Mössbauer spectroscopy, **297**:359
resonance Raman spectroscopy,
   297:358
thermoluminescence analysis of mutants, **297**:346–347
ultraviolet/visible absorbance spectroscopy characterization
   charge recombination rate analysis,
      297:349–350
   D1 mutant characterization,
      297:348–350
   flash detection, **297**:348
   manganese binding characterization,
      297:351
   net electron transport analysis,
      297:350–351
X-ray absorbance spectroscopy,
   297:358
Synechocystis 6803 studies, *see also* CP47
advantages as model system, **297**:166
chimeric gene introduction, **297**:303
combinatorial mutagenesis,
   297:303–305
deletion mutagenesis, **297**:295, 297, 299,
   301
domain deletion mutagenesis,
   297:302–303
genetic comparison with higher plants,
   297:299–300
genomic DNA isolation, **297**:169–170
growth conditions, **297**:167–169
homologous recombination,
   297:293–294
mapping of unknown genes influencing
   photosystem II function, **297**:307,
   309–310
pseudorevertant analysis, **297**:306–
   307
reaction center genes, *see psbA*; *psbD*
RNA isolation, **297**:170–171
segregation screening, **297**:297–298
site-directed mutagenesis, **297**:301–302

targeted random mutagenesis,
   297:305–306
transformation and vectors, **297**:171–
   172, 174–175, 294–295
Phycobilisome, *see* Complementary chromatic adaptation
Phycoerythrin, fluorescence assay of oxygen
   radical absorbing capacity, *see* Oxygen
   radical absorbance capacity assay
Phylogenetic analysis, neurotransmitter
   transporters
bootstrapping, **296**:433–434
DNAML program analysis, **296**:429, 433
DNAPARS analysis, **296**:434–435
evolutionary tree analysis, **296**:434–436
sequence acquisition and alignment,
   296:428–432
Phylogenetic analysis, rhodopsins
classification of visual pigments,
   315:312
wavelengths of maximal absorption, identification of critical amino acids
   ancestral pigment sequence elucida-
      tion, **315**:320–321
   phylogenetic tree construction
      neighbor-joining method, **315**:315,
         319–320
      programs, **315**:313–314
      unweighted pair-group method with
         arithmetic mean, **315**:314–315
   site-directed mutagenesis in functional
      testing
      absorbance spectroscopy, **315**:324
      applications, **315**:325
      bovine A292S, **315**:324
      cloning of genes, **315**:321–322
      mutagenesis, **315**:322–323
      purification, **315**:324
      regeneration, **315**:323–324
      RNA isolation, **315**:321
   wild-type pigment absorption, **315**:312,
      323–324
Phytoceramide, radiolabeling
amide labeling principles, **311**:503–504
extractions, **311**:502
flash column chromatography, **311**:501
materials, **311**:500–501
N-[1–³H]C₂-phytoceramide synthesis,
   311:505–506
purity analysis, **311**:502

radioactivity determination and detection, **311**:502

thin-layer chromatography, **311**:501

Phytochrome

light-regulated gene expression function in *Arabidopsis thaliana*

light sources and exposure schedules, **297**:183–184

luciferase reporter gene assays in transgenic plants

in vitro assay, **297**:189–190

in vivo imaging, **297**:190–191

Northern blot analysis of transcription

hybridization, **297**:185

overview, **297**:184–185

reagents, **297**:175–186

plant growth conditions, **297**:183

ribonuclease protection assay

digestion reaction, **297**:188

gel electrophoresis analysis, **297**:189

hybridization reaction, **297**:188

overview, **297**:186

probe synthesis, **297**:188

reagents, **297**:189

templates for RNA probe synthesis, **297**:186–187

photoconvertible forms, **297**:182

signal transduction, **297**:220–221

Phytoene desaturase

gene cloning and carotenoid biosynthesis, **297**:246–247, 251–252

inhibitors, **297**:262

Phytoene synthase, gene cloning and carotenoid biosynthesis, **297**:247–248, 249, 251

Phytosphingosine

phosphatidate phosphatase modulation, *see* Phosphatidate phosphatase, sphingoid base inhibition

radiolabeling

N-[1–³H-acetyl]phytosphingosine synthesis, **311**:505–506

amide labeling principles, **311**:503–504

extractions, **311**:502

flash column chromatography, **311**:501

materials, **311**:500–501

purity analysis, **311**:502

radioactivity determination and detection, **311**:502

thin-layer chromatography, **311**:501

yeast growth inhibition studies

liquid culture, **311**:323–324

positive selection markers, **311**:327

solid culture, **311**:324–325

strain background, **311**:323

synthetic media, **311**:325–326

PI, *see* Propidium iodide

PI 3K, *see* Phosphatidylinositol 3-kinase

Piceid

isomers, **299**:186

wine assays

content in wine, **299**:186

gas chromatography–mass spectrometry, **299**:186–187

high-performance liquid chromatography, **299**:187–188, 190

standards, **299**:188, 190

Pichia pastoris, *see* Yeast expression systems

Picosirius Red, connective tissue staining in optic nerve head sections, **307**:499–500

Piloty's acid, synthesis and nitroxyl release, **301**:214–215, 217, 286–287

Pine bark extract

inducible nitric oxide synthase, effects on activity, **301**:510, 512

medicinal uses, **301**:505

Pinopsin

absorption spectrum, **316**:290

function, **316**:279

gene cloning, **316**:279–282

immunohistochemical localization, **316**:282–283

overexpression in transfected mammalian cells

expression plasmid construction, **316**:284–285

human embryonic kidney cell growth, **316**:285

overview, **316**:283–284

transfection, **316**:287–288

purification from transfected cells

anion-exchange chromatography, **316**:288

cation-exchange chromatography, **316**:289

concentration determination, **316**:290

dialysis and centrifugation, **316**:289

extraction, **316**:288

nickel affinity chromatography of histi-
  dine-tagged protein, **316:**289
Pipette noise, *see* Patch clamp
PKA, *see* Protein kinase A
PKC, *see* Protein kinase C
PKG, *see* Protein kinase G
PKR, *see* Protein kinase R
PLA$_2$, *see* Phospholipase A$_2$
Planar lipid bilayer
  bilayer formation
    folding, **294:**291
    painting, **294:**291, 327–328
    polystyrene cups with holes, prepara-
      tion, **294:**326–327
  capacitance measurements, **294:**327–328
  channel insertion
    cross validation, **294:**296–297
    cystic fibrosis transmembrane conduc-
      tance regulator, **294:**301, 303
    examples of membrane preparations
      and reconstituted ion channels,
      294:295–296
    high-conductance calcium-activated po-
      tassium channels, **294:**299–300
    IRK1, 300
    mammalian cell lines as starting materi-
      als, **294:**300–301, 303
    nystatin/ergosterol-induced fusion
      channel density calculation,
        294:333–337
      enhancing vesicle delivery to bi-
        layers, **294:**329–330
      limitations, **294:**337–338
      lipid-to-protein ratio, **294:**332–333
      native fusigenic vesicle preparation,
        294:330–334
      osmotic gradient, **294:**328, 338
      overview, **294:**320–321
      stirring, **294:**329, 338
      vesicle lipids, **294:**321–322
      vesicle size, **294:**322–325
    rationale, **294:**287–288, 319
    recombinant channels, **294:**297–301,
      303
    Shaker, **294:**298, 301
    techniques, overview, **294:**291–292, 319
    Xenopus oocyte membranes,
      294:299–300
  microbial membrane vesicle incorpo-
    ration

Paramecium, **294:**510–511
Saccharomyces cerevisiae organelles,
  294:513–514
patch-clamp recording, **294:**289–290, 337,
  662–663
screening of membrane active compounds
  advantages, **294:**662–663
  amyloidogenic peptides, interactions
    with membranes, **294:**670–672
  antibiotic screening, **294:**672–673
  automation, **294:**673
  electrodes, **294:**669
  limitations, **294:**663–664
  lipids, **294:**668
  measuring system
    chamber cleaning, **294:**669
    overview of features, **294:**664
    perfusion chamber for solvent-free
      and native membrane vesicles,
      294:666–668
    perfusion setup for painted mem-
      branes, **294:**664–666
  porin channels, **294:**673
  recording equipment, **294:**669–
    670
  sensitivity, **294:**662
  single-channel reconstitution studies,
    294:670
  surfactant toxicity screening for microe-
    mulsions, **294:**673–674
  viral fusion analysis, **294:**674
stability, **294:**662–663
T-tubule membranes
  high-conductance calcium-activated po-
    tassium channels, **294:**292–293
  preparation of planar bilayers from rat
    muscle, **294:**293–295
  voltage-sensitive sodium channels,
    294:292–293
Planck–Benzinger thermal work function
  derivation, **295:**232–234
  temperature-invariant enthalpy
    DNA unwinding, **295:**255–256
    extrapolation, **295:**234–236, 238–239,
      241, 245
    hydrogen bonding
      carboxylic acid dimerization,
        295:264–269
      water, **295:**261–264

pH effects on ribonuclease A stability, **295**:247–248

protein unfolding applications, **295**:251, 253, 255–261

ribonuclease S′ system, peptide interactions, **295**:248–249, 251

self-associated protein systems, **295**:246–247

Planorbis giant dopamine neuron, *see* Neuron

Plasma membrane

aequorin targeting for calcium measurements with luciferase fusion protein, **305**:482, 494

fluorescent probes, **307**:121, 123

Plasma oxidizability, assays

Alzheimer's disease patients, **299**:46

coronary heart disease patients, **299**:46, 48

hyperlipidemic patients, **299**:46, 48

overview, **299**:36–37

photometric detection of conjugated dienes

comparison with other indices of lipid peroxidation, **299**:41, 43–44, 47

data acquisition, **299**:38–39

isolated low-density lipoprotein assay, **299**:44–45

kinetics, **299**:39–40, 48

oxidants, **299**:37–40

reproducibility, **299**:40–41

sample preparation, **299**:38, 40, 49

relationship to antioxidant and fatty acid composition, **299**:45, 47–48

Watanabe heritable hyperlipidemic rabbits, **299**:46–47

Plasmid

dendrimer complexes, **314**:36

limitations in antisense expression, **314**:33, 36

mammalian administration routes for protein expression

inhalation, **314**:35

intracerebroventricular injection, **314**:35

intramuscular injection, **314**:33–34

intravenous injection, **314**:34–35

vector components for antisense RNA expression, **314**:32

Plastocyanin

Coomassie stain analysis in native gels, **297**:276

spectroscopic analysis of soluble extracts, **297**:271–273

Western blot analysis, **297**:273–274

Plastoquinone redox state

algae model systems and culture, **297**:221–223

Chlamyodomonas mutants of photosynthetic electron transport, **297**:226–227

chlorophyll/cell measurements, **297**:230–231

fluorescence assays

photosystem II reaction center quantification, **297**:227

Q_A reoxidation kinetics, **297**:228

reaction centers with herbicides bound to Q_B pocket, quantification, **297**:228

redox state assay, **297**:229

oxidation by cyt b_6/f complex inhibitors, **297**:224–226, 229

reduction by 3-(3′,4′-dichlorophenyl)-1,1′-dimethylurea, **297**:223–224, 231

RNA analysis after herbicide treatment, **297**:231

sensing of light intensity, **297**:221

signal transduction analysis by gel retardation assay, **297**:232–234

Platelet-activating factor:lysophospholipid transacetylase

inhibitors, **311**:121–122

pH optimum, **311**:121

purification, **311**:122

reactions catalyzed, **311**:121–122

tissue distribution, **311**:121

Platelet-activating factor:sphingosine transacetylase

assay

incubation conditions, **311**:119

reagents, **311**:119

sample preparation, **311**:119

thin-layer chromatography, **311**:119–120

biological significance, **311**:118

inhibitors, **311**:121–122

kinetic parameters, **311**:120

pH optimum, **311**:121

reaction catalyzed, **311**:118

subcellular distribution, **311**:121
substrate specificity, **311**:120–121
tissue distribution, **311**:121
Platelet-derived growth factor, antisense
    knockdown in kidney, **314**:387
Platelet-derived growth factor receptor
    inhibitors, **319**:268–269
    radiation-induced signaling
        dephosphorylation assay *in vivo*,
            319:265
        overview, **319**:258
PLC, *see* Phospholipase C
Pleiotropic drug resistance subfamily, ATP-
    binding cassette transporters in *Sac-
    charomyces cerevisiae*, **292**:150–152
PMP70, *see* Peroxisome membrane protein-
    70
PMT, *see* Photomultiplier tube
PN3, *see* Peripheral nerve 3 protein
PNA, *see* Peptide nucleic acid
PNP, *see* Purine nucleoside phosphorylase
Podophyllotoxin
    intracellular concentration determination
        nonradiolabeled compounds and high-
            performance liquid chromatogra-
            phy, **298**:273–274
        radiolabeled compounds
            monolayer culture directly in ster-
                ilized scintillation vials, **298**:273
            monolayer or suspension cultures,
                298:272–273
    structure, **298**:256
    suppression of microtubule dynamics,
        298:264–265
    tubulin binding site, **298**:264
Point-spread function
    Airy disk, **307**:8, 12
    approximation, **307**:211–212
    collection pinhole point-spread function,
        307:9–10, 13
    confocal point-spread function, **307**:10
    full width at half-maximum intensity,
        307:13–14, 290, 304, 374, 502
    illumination source point-spread function,
        307:9–10, 13
    integrals, **307**:11
    lens aperture dependence, **307**:8–9
    measurement, **307**:114–115
    pattern, **307**:6, 8
    probability function, **307**:8, 10

Poisson–Boltzmann equation, *see also*
    Three-dimensional structure, molecular
    modeling
    derivation, **295**:367–368
    DNA–ion binding analysis
        charge distribution effects on competi-
            tive binding, **295**:376
        competition between two counterion
            species
            cylindrical geometry regime,
                295:374–375
            planar geometry regime, **295**:372–373
        competitive binding isotherm analysis
            competitive binding of several coun-
                terion species, **295**:378
            two counterion species, **295**:377–378
        counterion size effects on competitive
            binding, **295**:375–376
        single counterion species near highly
            charged surface
            cylindrical geometry regime,
                295:371–372
        Debye screening length, **295**:369
        physical parameters for DNA in wa-
            ter, **295**:369–370
        planar geometry regime, **295**:370–371
        reducing potential, **295**:368
Polarization microscopy, Congo red-stained
    amyloids, **309**:4–5, 7, 36–37, 68
Polarized epithelial cell, exogenous neuro-
    transmitter transporter expression
    cell culture, **296**:373–375
    cell lines, **296**:371–373
    cell surface biotinylation, **296**:376, 378
    immunofluorescence microscopy,
        296:376–378
    transfection, **296**:375–376
    transport assays, **296**:378, 380–381
Polyacrylamide gel electrophoresis, *see also*
    Electrophoretic mobility shift assay;
    Two-dimensional polyacrylamide gel
    electrophoresis; Western blot
    advantages in nucleoprotein analysis,
        304:312
    disulfide bond folding intermediates
        native gel electrophoresis, **290**:68
        sodium dodecyl sulfate gel electropho-
            resis, **290**:67–68
        urea polyacrylamide gel electrophore-
            sis, **290**:68–69

gel shift assay for RNA–peptide inter-
    action
  binding conditions, **318:**428
  competitor RNA, **318:**427
  concentration of ligands, **318:**428
  electrophoresis, **318:**430
  limitations, **318:**427
  materials, **318:**429
  purification of RNA, **318:**429–430
  quantificaion of RNA, **318:**430
  transcription, *in vitro*, **318:**429
guanosine 5'-monophosphoro-
    thioate–RNA
  [(β-acryloylamino)phenyl]mercuric
    chloride synthesis, **318:**231–232
  separation, **318:**232–234
histone acetyltransferase activity gel
    assay
  acetylation reaction, **304:**690
  controls, **304:**691
  electrophoresis, **304:**690
  gel preparation, **304:**689–690
  interpretation of results, **304:**691–692
  overview, **304:**689
  staining and fluorography, **304:**691
immunoprecipitation analysis, **293:**47, 552
modified denatured RNA fractionation,
    318:13–14
native gel electrophoresis of aggregation
    intermediates
  advantages, **309:**335
  α_1-antitrypsin, **309:**341
  carbonic anhydrase, **309:**343
  electrophoresis conditions, **309:**337
  P22 coat protein, **309:**343
  P22 tailspike protein
    conformational isomer resolution,
      309:340–341
    quantitative analysis of aggregation,
      309:338, 340
    sample denaturing and refolding/aggre-
      gation, **309:**335–336
  stacking gel, **309:**337
  stock solutions and buffers,
      309:336–337
native nucleosomes
  comparison to standard DNA gel elec-
    trophoresis, **304:**299–300
  cage effect in stabilization of com-
    plexes, **304:**300

casting of gels, **304:**300
electrophoresis conditions, **304:**300
staining, **304:**301
optimization, **304:**301
glycerol inclusion, **304:**302–303
particle mass separations, **304:**303
histone acetylation detection,
    304:303–304
effectiveness, **304:**304
nucleosome mobility assays
  interpretation of bands, **304:**311
  principle, **304:**309, 312
  two-dimensional gel electrophoresis,
    304:309–311
peptide quantification, **318:**426–427
positioned nucleosomes, principle of sepa-
    ration, **304:**304–307
preparative gel electrophoresis
  linker histones, **304:**13–14, 16–17
  nucleosome core particles, **304:**13–14,
    16–17
  positioned nucleosomes, **304:**307–308
principle, **309:**334–335
purification of suppressor transfer RNA,
    293:513–515
purified proteins, *see specific proteins*
replication-coupled assembly assay
  histone electrophoresis and transfer to
    membranes, **304:**84
  nucleosomal DNA, **304:**82–83
RNA, analysis of modification and cleav-
    age, *see specific probes*
small heat shock proteins
  oligomeric structure determination by
    nondenaturing pore-exclusion
    casting and running of gels,
      290:354–355
    principle, **290:**353
    solution preparation, **290:**354
  substrate binding assay with nondena-
    turing gels, **290:**361–363
subunit stoichiometry determination,
    293:26–27
Polyacrylamide substrate, cell culture
advantages, **298:**490–491
cell traction force measurement, **298:**490,
    496
collagen type I coating of polyacryl-
    amide, **298:**492–493
compliance measurement, **298:**494–495

deformation studies, **298**:496
flexibility effects on cells, **298**:489–490
flexibility measurement, **298**:494
glass surface activation, **298**:491–492
microscopy, **298**:496
polyacrylamide sheet preparation,
    298:492
thickness measurement, **298**:493–494
Polycomb, mapping of chromatin interac-
    tions, **304**:405, 417
Polyethylene glycol, folding enhancement
    of inclusion body proteins, **309**:227, 229
Polyethyleneimine, *see* Transfection, anti-
    sense oligonucleotides
Polyglutamine-containing protein aggre-
    gates
  dot-blot filter retardation assay
    aggregate isolation from COS-1 cells
        transfected with Huntington's dis-
        ease proteins, **309**:380, 382
    applications, **309**:385
    biotin/streptavidin–alkaline phospha-
        tase system for detection, **309**:381
    filter blotting and washing,
        309:380–381
    glutathione *S*-transferase–huntingtin fu-
        sion proteins
      aggregation potential by polygluta-
          mine tail size, **309**:381–382
      bacteria strains and media, **309**:378
      biotinylation tagging, **309**:385
      plasmids, **309**:377
      proteolytic cleavage, **309**:380, 382
      purification, **309**:379–380
      structures, **309**:377–378
    principle, **309**:376
  fibril characterization, overview of tech-
      niques, **309**:375–376
  neuronal intranuclear inclusions in neuro-
      degenerative diseases, **309**:375,
      687–689
  transgenic R6 mouse models of Hunting-
      ton's disease
    applications, **309**:689
    CAG repeats, **309**:687
    immunocytochemistry for neuronal in-
        tranuclear inclusions
      antibodies, **309**:689–690
      immunoelectron microscopy,
          309:692–694

light microscopy, **309**:692–694
reagents, **309**:690–692
sample preparation, **309**:692–693
onset of inclusions, **309**:687–688
pathogenesis, **309**:689
skeletal muscle inclusion detection
  applications, **309**:698–699
  dissection, **309**:699
  immunochemical staining and detec-
      tion, **309**:700–701
  reagents, **309**:699
  sectioning, **309**:700
Western blot filter retardation assay
    for neuronal intranuclear inclu-
    sions
  aggregate preparation, **309**:697–698
  immunoblotting, **309**:698
  nuclei isolation from brain,
      309:696–697
  reagents, **309**:694–696
Polylysine
  preparation for mounting of amyloid
      immunohistochemistry samples,
      309:17
  transmission electron microscopy support
      treatment, **304**:221
Polymerase chain reaction, *see also* Reverse
    transcriptase–polymerase chain reac-
    tion; RNA-arbitrarily primed polymer-
    ase chain reaction
  adeno-associated virus titering by quanti-
      tative polymerase chain reaction,
      316:756–757
  aequorin fuion protein construction,
      305:486, 488–490
  amplification of poorly represented
      species
    Long and Accurate polymerase chain
        reaction, **296**:509–511
    primer selection, **296**:512–513
  aquaporin homology cloning by degener-
      ate oligonucleotide polymerase chain
      reaction
    amplification reaction, **294**:565, 567
    gel electrophoresis analysis,
        294:565–566
    materials, **294**:562–565
    overview, **294**:562
    secondary amplification and DNA puri-
        fication, **294**:566–567

arbitrarily-primed technique for biofilm mutant analysis
  components of reactions, **310:**106
  first round, **310:**106–108
  overview, **310:**106
  second round, **310:**108
  transposon mutagenesis mutants, **310:**105
boranophosphate oligonucleotides, **313:**246–247
chimera contruction with overlap method, **293:**66, 68
combinatorial mutagenesis, photosystem II genes in *Synechocystis* 6803, 304
complementary DNA amplification from a small number of hematopoietic cells
  betaine and dimethyl sulfoxide improvement of efficiency, **303:**45–46, 50
  differential display analysis of products, **303:**50–54
  reaction conditions, **303:**50
complementary DNA selection
  complementary DNA fragments
    first step, **306:**86–87
    second step, **306:**87–88
  initial complementary DNA preparation, **306:**83
degenerate polymerase chain reaction in gene cloning
  overview, **292:**121–122
  probe design, **292:**164
  screening of libraries, **292:**165–166
deuteranomaly, green–red pigment hybrid characterization with long-range polymerase chain reaction amplification, **316:**663
differential display, **303:**244–246
direct selection of complementary DNA
  selected complementary DNA amplification, **303:**120
  starting complementary DNA pool amplification, **303:**117–118
DNA microarrays, target DNA preparation amplification, **303:**185
exon amplification with reverse transcription–polymerase chain reaction
  complementary DNA synthesis, **303:**97

electrophoretic analysis of products, **303:**97
large-insert trapping with cosmids
  background reduction, **303:**108
  electrophoretic analysis of products, **303:**109–110
  primers, **303:**107–109
  ribonuclease A treatmemt of products, **303:**108
  primers, **303:**95–96
exonuclease III mapping of chromatin, reiterative primer extension analysis
  amplification reaction, **304:**596–597
  end-labeling of oligonucleotides, **304:**596
  primer selection, **304:**595
  product analysis, **304:**597
  reagents, **304:**596
expression detection of ribozyme or substrate by reverse transcription, **306:**220–221
fluorescence *in situ* hybridization probe labeling, **304:**654
fluorescent differential display, **303:**301–302, 306–307
genotyping of mouse retinitis pigmentosa models
  cyclic GMP phosphodiesterase gene amplification, **316:**523
  overview, **316:**523
  rhodopsin gene amplification, **316:**524
genotyping of transgenic mice, **316:**716–718
green and green–red hybrid pigment gene expression in arrays, polymerase chain reaction/single-strand conformation polymorphism, **316:**668–670
high-density polymerase chain reaction membranes, **303:**215–216
HMG-D associated DNA, **304:**121
ligation-mediated polymerase chain reaction, *Drosophila* chromatin analysis, **304:**484, 487–488
ligation-mediated polymerase chain reaction, footprinting analysis
  direct primer labeling protocol
    amplification reaction, **304:**565
    direct labeling reaction, **304:**565–566
    end-labeling of primer, **304:**566

first primer extension, **304**:564
gel electrophoresis and band detection, **304**:566
ligation, **304**:564–565
overview, **304**:563–564
DNase I footprinting modifications, **304**:561, 563
guanine–adenine ligation-mediated polymerase chain reaction and dimethyl sulfate footprinting
amplification reaction, **304**:577
annealing reaction, **304**:577
applications, **304**:578, 580–584
guanine–adenine-specific cleavage reaction, **304**:576–577
labeling reaction, **304**:577–578
ligation reaction, **304**:577
overview, **304**:574
high-resolution mammalian nucleosome positioning, **304**:637–638
nucleosome boundary mapping at high resolution, **304**:625–626
optimization, **304**:624
overview, **304**:548–549, 623–624
selection of protocol, **304**:571
standard protocol
amplification reaction, **304**:558–559
electroblotting, **304**:559–560
first primer extension, **304**:557–558
gel electrophoresis of products, **304**:559
hybridization analysis, **304**:560
ligation, **304**:558
single-stranded probe preparation, **304**:560–561
transcription factor footprinting, **304**:622–623
mapping nonhistone chromatin-associated factors, **304**:402, 405, 412–414, 426–427, 430
mapping of unknown genes influencing photosystem II function, **297**:309
MDR1, random mutagenesis, **292**:478–479
microarray fluorescence-based transcription assay, **306**:7–9
mitochondria ATP-binding cassette genes disruption using polymerase chain reaction fragments, **292**:779–780, 785–786

search for genes, **292**:776–779
mitochondrial DNA, large-scale deletions in photoaging
advantages of detection, **319**:367
DNA extraction, **319**:368
nested polymerase chain reaction, **319**:369–370
primers, **319**:370
principle, **319**:368
quantitative analysis, **319**:371
short-cycle polymerase chain reaction, **319**:368–369
multiple primer mutagenesis, **294**:592–593
quantitative analysis with solid-phase capture of products and chemiluminescence detection
amplification cycle number determination, **305**:469
applications and advantages, **305**:466–467, 476
control assay, **305**:475
dioxetane substrates, **305**:467
principle, **305**:467–470
reagents and solutions, **305**:470–471
sensitivity, **305**:475
streptavidin-coated bead assay
detection, **305**:473
hybridization of fluorescein-labeled probe, **305**:473
overview, **305**:472–473
product capture, **305**:473
streptavidin-coated microplate assay
detection, **305**:472
hybridization of fluorescein-labeled probe, **305**:472
microplate coating, **305**:475
overview, **305**:471–472
product capture, **305**:472
streptavidin-coated paramagnetic particle assay
detection, **305**:474–475
hybridization of fluorescein-labeled probe, **305**:474
overview, **305**:474
product capture, **305**:474
radiolabeled DNA preparation for DNase I footprinting, **316**:606–608
red–green hybrid gene, polymerase chain reaction amplification and detection in females, **316**:660, 662

red/green pigment transcripts, ratio deter-
    mination in retina, **316:**667–668
representational difference analysis,
    306:72–73
restriction endonucleolytic analysis of dif-
    ferentially expressed sequences,
    303:261, 264, 267–269, 271–272, 275–
    276, 282–283, 286–289, 291–293
retinal pigment epithelium p65, amplifica-
    tion and single-strand conformation
    analysis
    labeled DNA, **316:**708, 710
    primers, **316:**709
    unlabeled DNA, **316:**710
retinitis pigmentosa mouse models, geno-
    typing
    DNA isolation from tail biopsy,
        316:522–523
    overview, **316:**521–522
    polymerase chain reaction amplifica-
        tion, **316:**523–524
    restriction digestion of amplified frag-
        ments, **316:**524
    sequencing, **316:**524–526
ribosomal RNA gene detection in pros-
    thetic joint biofilms
    amplification reaction, **310:**572–573
    primer design, **310:**572
RNA amplification, *see* Reverse tran-
    scriptase–polymerase chain reaction
screening of photosystem I mutations in
    Synechocystis 6803, 21–22, 26
sequencing by hybridization, amplifica-
    tion of plasmid inserts, **303:**169–170
site-directed mutagenesis, **293:**64, 97
Snorthern blot analysis, **306:**80
solid-phase differential display, hot start-
    mediated polymerase chain reaction,
    303:503
suppression subtractive hybridization
    selection of differentially expressed
        complementary DNAs,
        303:362–364
    subtraction efficiency analysis,
        303:364–365
    suppression polymerase chain reaction
        effect, **303:**350, 379
targeted display
    complementary DNA, **303:**400
    genomic DNA, **303:**400–401

terminal transferase-dependent polymer-
    ase chain reaction in footprinting
    analysis
    adapter analysis, **304:**569
    adapter ligation, **304:**570
    amplification reaction, **304:**570
    direct labeling of product, **304:**570
    first primer extension, **304:**567
    gel electrophoresis and analysis,
        304:570–571
    overview, **304:**548, 566–567
    tailing reaction, **304:**567, 569
transcript quantification, *in situ*,
    310:77–78
transposon insertion site identification
    with vectorette polymerase chain re-
    action, **303:**528–530
X-linked red–green photopigment genes,
    determination of number and ratio
    with competitive polymerase chain
    reaction
    amplification, **316:**658
    data analysis, **316:**655–656
    exons 2, **316:**4, and 5 analysis, **316:**659
    overview, **316:**655
    primers, **316:**657–658
    single-strand conformational polymor-
        phism, **316:**655, 658–659
Polyphenols, *see also* Flavonoids
    aortic ring relaxation assay
        antioxidant-independent effects of
            polyphenols, **301:**524–525
        aortic preparation and mounting,
            301:523
        endothelial nitric oxide generation,
            301:522, 525–526
    calcium flux in endothelial cells
        aortic endothelial cell culture, **301:**529
        fluorescence assay with Fura-2,
            301:530–532
    cyclic GMP accumulation as nitric oxide
        production assay, **301:**526–528
    electron paramagnetic resonance assay
        of nitric oxide production, **301:**528–
        529
Polyvinylidene difluoride film, *see* Micro-
    calorimetry, time-resolved studies of
    phototransduction
POR, *see* Protochlorophyllide oxidore-
    ductase

Pore occupancy, determination,
    293:389–390
Pore size, *see also* Ion channel permeabil-
    ity; Substituted-cysteine accesibility
    method; Three-dimensional structure,
    molecular modeling
  HOLE program in determination,
    293:605
  subunit stoichiometry and molecular mod-
    eling, **293:**605
Porphyrias, *see also* Erythropoietic proto-
    porphyria
  clinical manifestations, **319:**486–488
  diagnosis, **319:**486–487
  heme synthesis, **319:**485–486
  inflammation, **319:**487–488
  phototherapy
    matrix metalloproteinase induction,
      319:493
    melanin level manipulation, **319:**489
    psoralen–ultraviolet A therapy
      action spectra, **319:**491
      photosensitizers, **319:**491
      regimens, **319:**491–492
      risks, **319:**492–493
      sources, **319:**491
    ultraviolet B phototherapy
      dosing, **319:**490–491
      risks, **319:**490
      sources, **319:**490
    ultraviolet penetration in skin,
      319:489–490
  phototoxicity, immediate versus delayed,
    319:488–489
  porphyrin photochemistry, **319:**487
  treatment, **319:**485, 489
  types and enzyme defects, **319:**485–486
Position clamp, *see* Optical trap
Positional cloning, *see also* Genetically di-
    rected representational difference anal-
    ysis; Representational difference analy-
    sis; Snorthern blot analysis
  DNA preparation from mice
    large-cale preparation, **306:**71
    small-scale, **306:**69
  genetic crosses in mice for linkage analy-
    sis, **306:**67–69
  simple sequence length polymorphism
    genotyping, **306:**69, 71
Potassium channel, *see also* ATP-sensitive

potassium channel; G-protein-coupled
    inward rectifier potassium channel;
    Shaker-type potassium channel; Ykc1
Potassium channel, *see also* G-protein-cou-
    pled inward rectifier potassium chan-
    nel; Trk1; Trk2
  antibodies, voltage-gated channels
    autoimmunity, **294:**703
    mouse monoclonal channel antibodies,
      294:694, 702
    rabbit polyclonal antibodies against syn-
      thetic peptides, **294:**695–697, 702
  assembly
    overview, **293:**33
    Shaker-type potassium channel, *see*
      Shaker-type potassium channel
  classification, **294:**274, 286, 624
  κ-conotoxin PVIIA, voltage-gated chan-
    nel binding, **294:**620–621
  dendrotoxin blocking
    electrostatic interactions, **294:**656–658
    kinetic analysis, **294:**653–655
    mutation analysis of channel residue in-
      teractions, **294:**654–656
    voltage dependence, **294:**659–670
    Xenopus oocyte characterization,
      294:652–653, 657
  fluorescence recovery after photo-
    bleaching
    applications, **294:**49
    artifacts, corrections, and calibrations,
      294:56–58
    channel mobility measurement, **294:**53–
      54, 58
    confocal scanning laser microscopy
      measurements, **294:**54–56
    fluorescence resonance energy transfer in
      protein–protein interactions,
      294:52–53
    functional tagging in assembly studies,
      293:31
    fusion protein expression, **294:**51–52
    green fluorescent protein fusion protein
      construction, **294:**49–50
  heterologous expression in yeast mutants
    Akt1, 94, 103
    growth phenotype analysis, **293:**98–99
    IRK1, 94
    Kat1, 94, 98–99, 101–103
    media

ion selectivity selection, **293:**93,
   99–101
pH, **293:**93
potassium-limiting media in selection, **293:**92–93
potassium-permissive media,
   293:91–92
overview, **293:**89–91
patch-clamp studies, **293:**102–104
plasmid construction, **293:**91
random mutagenesis, **293:**97–98
sensitivity in functional analysis compared to electrophysiology,
   293:101–102
site-directed mutagenesis, **293:**95, 97
high-conductance calcium-activated channel, *see* Charybdotoxin receptor
immunoprecipitation of subunits from
   lysed cells
   antibody incubation, **293:**45–46
   fixed *Staphylococcus aureus* as matrix,
      293:43, 45
   gel electrophoresis, **293:**47
   imaging of gels
      autoradiography, **293:**47–48
      fluorography, **293:**48
      phosphorimaging, **293:**48–49
   precoating, **293:**44–45
   protein A agarose as matrix, **293:**43–
      44
   protein G agarose as matrix, **293:**44
inwardly-rectifying channel IRK1, reconstruction in planar bilayers, **294:**300
molecular dynamics simulated annealing
   with restrained molecular dynamics
   β-barrel models of pore, **293:**669–670,
      680–682
   H5 region mutagenesis, **293:**669
   α-helix bundle models of pore, **293:**671
   ranking ensembles of H5 models,
      293:671–672
scorpion venom peptide inhibitors, *see
   also* Charybdotoxin receptor
   applications, **294:**625
   binding assays, **294:**637–639
   classification, **294:**626
   expression of recombinant peptides in
      Escherichia coli
      extraction, **294:**631–632
      induction, **294:**631

purification of histidine-tagged proteins, **294:**632–633
transformation, **294:**630–631
vectors, **294:**630
peptide synthesis, **294:**629–630
purification
   assessment of purity, **294:**6290
   cation-exchange chromatography,
      294:627–628
   reversed-phase high-performance liq-
      uid chromatography, **294:**628
   storage, **294:**628–629
radiolabeling
   iodination, **294:**634–635
   tritiation by cysteine alkylation,
      294:635–637
sequence homology, **294:**626
specificity, **294:**626
structural overview, **294:**625
subunits
   expression in COS-1 cells
      cell cuture, **293:**33
      lysing cells in detergent, **293:**41–42
      metabolic radiolabeling
      pulse-chase labeling, **293:**39–41
      steady-state labeling, **293:**37–39
      transfection with calcium chloride,
         293:35, 37
   interaction forces, **293:**18
   types, **293:**32
tetramerization, **294:**47–48
transport vesicle isolation
   neuroblastoma cells, **294:**342–350
   squid giant axon, **294:**340–341
   velocity sedimentation on Ficoll gradi-
      ents, **294:**342–348
Potassium permanganate, *see* Permanganate
   footprinting
Power spectral density, patch clamp noise,
   293:224, 226, 228, 263–264
PP1, *see* Protein phosphatase-1
PP2A, *see* Protein phosphatase type 2A
PP2C, *see* Protein phosphatase type 2C,
   photoreceptors
PP7, *see* Protein phosphatase type 7, photoreceptors
PPDK, *see* Pyruvate phosphate dikinase
PPIase, *see* Peptidylprolyl *cis–trans*-iso-
   merase
PR, *see* Progesterone receptor

Prader–Willi syndrome, Southern blot analysis with chemiluminescence detection
  background on syndrome, **305:**460–462
  blotting, **305:**463
  dot blots, **305:**462–463
  hybridization and washes, **305:**463
  principle, **305:**462
  probe generation, **305:**462
  reagents and solutions, **305:**462
  signal generation and detection, **305:**463
Prazosin, photoaffinity labeling of P-glycoprotein with analogs, **292:**305–306, 309, 313–314, 322, 324–325, 327, 401–402, 404–406, 437–438
Premature infants, oxidative stress, **299:**341–342, 348
Prenylation
  electrospray ionization mass spectrometry, analysis of prenylated peptides and peptide methyl esters
    high-performance liquid chromatography–mass spectrometry, **316:**448–449, 451
    preparation of peptides, **316:**445–447
    rhodopsin kinase trypsinization, **316:**447–448
  examples of vision proteins, **316:**436, 452
  functions, **316:**452–453
  methyl iodide cleavage, **316:**437
  Raney nickel cleavage and analysis
    exogenous lipid removal by gel electrophoresis, **316:**439–440
    extraction of proteins
      cells, **316:**439–440
      gels, **316:**440–441
    gas chromatography–mass spectrometry analysis of nonradiolabeled prenyl groups, **316:**444–445
    quantification of radiolabeled proteins, **316:**440
    radioprenylation of proteins, **316:**438–439
    Raney nickel cleavage of prenyl groups, **316:**441–442
    reversed-phase high-performance liquid chromatography analysis, **316:**442–444
    Trypanosoma brucei proteins, **316:**439–444
  sequence motifs, **316:**436, 452

structure elucidation, overview, **316:**437–438
Pressure clamp
  construction tips, **294:**487–489
  electronic control, **294:**485, 487
  mechanical arrangement, **294:**483–485
  principle, **294:**482–483
  system volume minimization, **294:**484–485
  transducer/mixer chamber, **294:**485
  valves, **294:**483–484
Pressure myography, vascular remodeling, **307:**249–250, 261–262
Presynaptic terminal
  calcium assay with ratio fluorescence probes
    population measurements
      calibration of fluorescence signal, **294:**14–15
      data acquisition, **294:**13–14
      dye loading, **294:**12–13
    single terminal measurements in mossy fibers
      advantages, **294:**15–16
      calibration of fluorescence signal, **294:**19
      data acquisition, **294:**17–18
      dye loading, **294:**17
  diameter, **294:**3
Primer extension analysis
  detection of RNA modifications or cleavage
    chemical modification reactions, **318:**16
    gel electrophoresis, **318:**17
    primer annealing and extension, **318:**17
    primer selection, **318:**14, 16
  dimethyl sulfate probing *in vivo*
    extension reaction, **318:**485
    hybridization, **318:**484–485
    polyacrylamide gel electrophoresis, **318:**486
    primer labeling, **318:**484
  hydroxyl radical probing of S20 on 16S ribosomal RNA
    extension reaction, **318:**472–473
    hybridization, **318:**472
    interpretation of data, **318:**473–475
    overview, **318:**471
    polyacrylamide gel electrophoresis analysis, **318:**473
    reagents and buffers, **318:**471–472

photolabile derivatives of oligonucleotides, identification of labeled nucleotides, **318:**128–130

Primuline, sphingolipid staining on thin-
layer chromatography plates, **312:**78–79

PRINS, *see* Oligonucleotide-primed *in situ*
DNA synthesis

Prion protein, *see also* [*PSI*⁺]
insertion into planar lipid membranes,
294:672
isoforms, **309:**577–578
metabolic radiolabeling and immunoprecipitation from cultured cells,
309:126–127
monoclonal antibodies
fusion, **309:**110
immunization and peripheral immune
response, **309:**110, 113, 116
immunogen preparation, recombinant
bovine prion protein, **309:**108–110
screening of hybridomas
enzyme-linked immunofiltration
assay, **309:**110–112
enzyme-linked immunosorbent assay,
309:110–111
immunoprecipitation, **309:**112–113
nitrocellulose freezing of protein conformation, **309:**122
peptide libraries, **309:**112
Western blot analysis, **309:**112
specificity for native and disease prions
historical overview, **309:**106–108
PrP^C, **309:**116, 118
PrP^Sc, **309:**118–119
protease-resistant protein
conversion reaction and assays
animal bioassays, **309:**133
applications, **309:**127–128
autotemplate, **309:**127
guanidine hydrochloride cell-free
conversion reaction,
309:128–130
guanidine hydrochloride-free conversion reactions, **309:**130–132
limitations in sensitivity, **309:**133
tissue slices, *in situ* conversion reactions, **309:**132–133
proteinase K resistance, **309:**123–124
transmissible spongiform encephalopa-
thy pathogenesis, **309:**122–123

Western blot analysis
materials, **309:**125
prion isolation from cultured cells
for analysis, **309:**126
sample preparation, **309:**126
PrP^Sc formation, overview, **309:**106, 577
solid-state nuclear magnetic resonance,
309:557–558
synthetic peptides
Fourier transform infrared spectroscopy of isotopically labeled peptides
intermolecular interactions,
309:588–589
secondary structure analysis,
309:585–588
isotope labeling, **309:**580–581
ligation for prion protein preparation,
309:584–585
mass spectrometry characterization of
purity and isotope label,
309:581–584
purification, **309:**581
solid state nuclear magnetic resonance
of isotopically labeled peptides,
309:589, 591
solid-phase synthesis, **309:**579–580
structure, **309:**578–579

P RNA, *see* Ribonuclease P

Proanthocyanidin, *see* Tannin

Procyanidins, *see* Phenolic antioxidants

Profilin, **291:**6-nitroveratryloxycarbonyl chloride caged conjugate
functional analysis of caged protein,
291:108–110
preparation, **291:**108
recombinant protein purification,
291:107–108

Progesterone receptor
green fluorescent protein fusion protein,
302:77
heat shock protein 90 association,
290:419
subcellular distribution, **302:**74, 77

Prohormone
processing overview, **314:**103–104
proteases, *see* Proprotein convertase

Proliferating cell nuclear antigen, antisense
knockdown in kidney, **314:**388

Proline, *see* Nitrosoproline test

Proline racemase, free energy profile,
    308:11–13
Proline transporter, phylogenetic analysis,
    296:428–436
Propidium iodide
    Legionella pneumophila imaging with
        fluorescence microscopy,
        310:631–632
    membrane integrity assessment in bio-
        films with SYTO 9, **310:**173–174
    multicolor sdetection with green fluores-
        cent protein, **307:**615–616
    vascular remodeling studies, **307:**255–256,
        262–263
Proprotein convertase
    antisense inhibition of PC1 and PC2
        cell culture and transfection,
            314:107–108
        controls, **314:**115
        Northern blot analysis, **314:**108–109,
            112
        plasmid constrction, **314:**107
        procholecystokinin processing analysis,
            314:111–112, 115, 117
        radioimmunoassay of cholecystokinin,
            314:110
        rationale, **314:**106–107
        Western blot analysis, **314:**109, 112, 115
    autoactivation, **314:**106
    cleavage specificity, **314:**105
    hormone processing, **314:**105
    tissue distribution, **314:**105–106
    types, **314:**104
Prostaglandin endoperoxide H synthase
    activities, **319:**67–68
    chemiluminescence assay of arachidonic
        acid oxidation
        dimol emission, **319:**71, 75–77
        filters, **319:**70
        identification of singlet oxygen as emis-
            sion source, **319:**71
        inhibitors of chemiluminescence,
            319:74–75
        kinetics, **319:**72–73
        oxygen requirement, **319:**73
        peroxidase in singlet oxygen formation,
            319:75–76
        pH dependence, **319:**73
        single photon-counting apparatus,
            319:69–70

spectral analysis, **319:**73–74
suicide inactivation, **319:**68–69
Prosthetic joint biofilm
    culture for detection
        culture, **310:**569–570
        prosthesis sample processing,
            310:568–569
        sample collection, **310:**568
        tissue sample processing, **310:**569
    histological examination for detection,
        310:570
    immunological detection
        antibody production, **310:**574–575
        immunofluorescence microscopy,
            310:575–576
    infection rates, **310:**566–567, 576
    ribosomal RNA gene detection
        DNA extraction, **310:**571–572
        overview, **310:**571
        polymerase chain reaction
            amplification reaction, **310:**572–573
            primer design, **310:**572
        sequencing, **310:**574
        single-stranded conformational poly-
            morphism, **310:**573
Proteasome
    inhibitors
        boronic acids, **300:**350–357
        effect on bacterial cell wall-induced ar-
            thritis
            leukocyte isolation, **300:**358–359
            polyarthritis induction in rats,
                300:357–358
            prevention of arthritis by inhibitors,
                300:360–361
            residual 20S proteasome activity,
                assay, **300:**359–360
        effect on delayed-type hypersensitivity,
            300:361–362
        inhibition of nuclear factor-κB acti-
            vation
            electrophoretic mobility shift assay,
                300:351–352
            fluorescence immunoassay of cell ad-
                hesion molecule expression,
                300:354–356
            immunoassay of interleukin-2 expres-
                sion, **300:**356–357
            Western blot analysis of I-κB degra-
                dation, **300:**353–354

lactacystin, **300**:349–350
  peptide aldehydes, **300**:349
structure, **300**:347–348
ubiquitin pathway of degradation,
  300:347
Protection factor, *see* Low-barrier hydrogen
  bond
Protein A–green fluorescent protein fusion
  protein
  construction
    bacterial strains and plasmids,
      302:266–267
    plasmids, **302**:267, 269–270
  electrophoretic analysis, **302**:270–271
  hydropathy analysis, **302**:271
  purification, **302**:270
  Western blot applications
    advantages, **302**:274
    blotting, **302**:271
    conventional color detection,
      302:272–274
    enolase detection, **302**:272–274
    fluorescence detection, **302**:271–274
    overview, **302**:266
    sensitivity, **302**:273
Protein array
  antibody screening of membranes,
    303:231–232
  applications, **303**:230–231
  arraying and gridding, **303**:214–215
  complementary DNA library construc-
    tion, **303**:231
  high-density protein membranes, **303**:231
Proteinase K, antigen recovery in amyloid
  immunohistochemistry samples, **309**:18
Protein carbonyl
  brain tissue assay
    calculations, **300**:149
    protein concentration determination,
      300:150–151, 153–155
    range of observed carbonyl concentra-
      tions in clinical states, **300**:146–
      148, 151, 153, 156
    reaction conditions, **300**:148–150, 155
    reagents, **300**:147
    tissue sampling, **300**:148
  detection techniques, **300**:106
  enzyme-linked immunosorbent assay
    albumin standards, **300**:107
    antibody incubation conditions, **300**:108

    blanks, **300**:109
    calibration, **300**:108–109
    clinical applications, **300**:110
    nucleic acid interference, **300**:111
    principle, **300**:106–107
    reagents, **300**:107
    sample preparation, **300**:108
    sensitivity and reproducibility, **300**:109
    validation, **300**:109–110
  Western blot analysis, **300**:145
Protein disulfide isomerase
  assays
    bovine pancreatic trypsin inhibitor oxi-
      dation, **290**:52
    continuous ribonuclease refolding
      constraints, **290**:37–38
      data analysis, **290**:34–37
      incubation conditions, **290**:34
      principle, **290**:33–34
      substrate preparation, **290**:39
    insulin reduction, **290**:33, 52
    overview, **290**:32
    scrambled lysozyme refolding,
      290:51–52
    scrambled ribonuclease refolding,
      290:39, 51
  assisting *in vitro* protein folding,
    290:48–50
  catalytic activities, **290**:28, 41–42, 44
  chaperone/antichaperone activities,
    290:44–46
  concentration in endoplasmic reticulum,
    290:27, 44
  domains, **290**:27
  null mutants, **290**:47–48
  oligomerization, **290**:31–32
  properties of *Humicola insolens* enzyme
    physical properties, **290**:57–58
    sequence analysis, **290**:54
    stability, **290**:58–59
    substrate specificity, **290**:59
  properties of recombinant rat enzyme
    active sites, **290**:40–41
    catalytic efficiency, **290**:41
    chemical reactivity and redox proper-
      ties, **290**:42–43
    disulfide bonds, **290**:40
    peptide-binding site, **290**:43
    posttranslational modification, **290**:41
    reaction specificity, **290**:41–42

stability, **290:**46
purification of bovine liver enzyme,
    290:28
purification of *Humicola insolens* enzyme
    anion-exchange chromatography,
        290:53
    cloning and expression in *Bacillus brevis*, **290:**55
    extract preparation, **290:**52–53
    high-performance liquid chromatography, **290:**53
    lectin affinity chromatography, **290:**53
purification of recombinant rat enzyme
    from *Escherichia coli*
    anion-exchange chromatography
        column chromatography, **290:**30
        high-performance liquid chromatography, **290:**31
    bacterial strains and growth, **290:**29
    extraction, **290:**29–30
    zinc affinity chromatography,
        290:30–31
thioredoxin homology, **290:**39–40, 46–47,
    51
Protein energetics, model compounds,
    295:102
Protein folding, *see* Folding, proteins
Protein image hybridization
    cesium chloride gradient centrifugation,
        depletion of free protein,
        304:605–606
    cross-link types, **304:**610–611
    dimethylsulfate cross-linking reaction,
        304:604–605, 610
    DNA shearing, **304:**606–607
    exonuclease III trimming, **304:**607–608,
        611–612
    formaldehyde fixation, **304:**604–605
    hsp70 probing in *Drosophila*, **304:**611
    molecular weight estimation of cross-
        linked proteins, **304:**611
    overview, **304:**600–601
    phenol extraction of DNA–protein com-
        plexes, **304:**607–608
    protease inhibition, **304:**609
    two-dimensional gel electrophoresis
        electroblotting, **304:**609
        electrophoresis conditions, **304:**608–609
        probe hybridization, **304:**609
    ultraviolet cross-linking

cross-linking reaction, **304:**603–604
dose, **304:**601, 603
nuclei isolation, **304:**604
sources of light, **304:**610–611
Protein kinase A
    activation of cystic fibrosis transmem-
        brane conductance regulator,
        293:180–184
    activators, **298:**547, 549–550
    antisense knockdown of β_{II} isoform in re-
        nal glomerular cells
        antisense targeting overview, **313:**30–31
        controls, **314:**385, 387
        oligodeoxynucleotide design, **314:**384
        proliferation response, **314:**385
        rationale, **314:**383–384
        time course of knockdown, **314:**384–
            385
        transfection, **314:**385, 387
    caging at essential cysteine, **291:**129–130
    calexcitin as substrate, **293:**195
    ceramide modulation
        analog studies, **312:**363–364
        delivery to cells
            analogs, **312:**367
            binding protein considerations,
                312:367–368
            metabolism, **312:**368
        effective concentration, **312:**364–365
        overview, **312:**362
        short-chain ceramides, **312:**417–418
        verification, **312:**366
    classification of isozymes, **312:**361
    functions, **312:**361–362
    hydrolysis, **312:**110–114
    inhibitors, **298:**548, 550, **319:**269–270
    phosphatidylserine activation, **312:**361
    regulation of cystic fibrosis transmem-
        brane conductance regulator,
        293:182–184
    regulation of neurotransmitter trans-
        porter biosynthesis and surface ex-
        pression, **296:**368, 370
    solid-phase extraction, **312:**106–107
    sphingolipid modulation
        assays of responses
            activator competition analysis,
                312:371
            immune complex kinase assay,
                312:372–373

phorbol dibutyrate binding to intact
    cells, **312:**370–371
protein substrate phosphorylation
    assays, **312:**371–373
translocation assay, **312:**371
handling effects in studies, **312:**368–369
overview, **312:**362
sphingosine inhibition
    analog studies, **312:**363
    delivery to cells
        binding protein considerations,
            312:367–368
        bovine serum albumin conjugate
            preparation, **312:**370
        ethanolic sphingosine preparation,
            312:369–370
        metabolism, **312:**368
        toxicity prevention, **312:**367
    effective concentration, **312:**364–365
    overview, **312:**32, 362
    verification, **312:**366
thiophosphorylation of peptides, **291:**125
translocation mutant Chinese hamster
    ovary cell LY-A strain
    advantages of ovary cells, **312:**306–307
    defective step identification,
        312:316–317
    discovery, **312:**306
    GM3 analysis, **312:**314–316
    lysenin-resistant variant selection,
        312:308–310
    metabolic radiolabeling of lipids with
        serine
        cell culture and analysis,
            312:312–313
        classification of mutants, **312:**311–312
        materials, **312:**312
        rationale, **312:**310–311
    mutagenesis, **312:**307–308
    sphingomyelin analysis, **312:**314–316
Protein kinase G, inhibitors, **298:**548
Protein kinase R, targeting for ribonuclease
    L degradation
    antisense chimeras
        controls, **313:**533
        dose–response, **313:**532–533
        length of antisense sequences, **313:**533
    incubation conditions for inhibition
        assay, **313:**531–532
    messenger RNA preparation, **313:**531

Protein *S*-nitrosylation, *see* *S*-Nitrosylated
    protein
Protein overlay assay
    advantages and disadvantages in subunit
        interaction analysis, **293:**14
    principle, **293:**10–11
    Shaker-type potassium channel
        advantages and disadvantages,
            293:14
        probe-specific antibody assay, **293:**12
        radioactive labeling of ShB amino-ter-
            minal domain, **293:**10–11
        radioactive protein probe assay,
            293:11–12
Protein oxidation
    carbonyl group detection
        2,4-dinitrophenylhydrazine reaction,
            319:430
        generation of modifications,
            319:429–430
        tritiation, **319:**430
        Western blot analysis
            advantages, **319:**432
            controls, **319:**436
            dinitrophenylhydrazine derivatiza-
                tion, **319:**433
            electroblotting, **319:**434
            immunoassay, **319:**435
            materials, **319:**432–433, 435
            principles, **319:**431–432
            sample preparation, **319:**433
            staining of total protein bands,
                319:435
    detection methods, overview, **319:**429,
        431
    metal-catalyzed oxidation, **319:**428
    residues and modification types,
        319:428–430
Protein phosphatase, *see also* Ceramide-acti-
    vated protein phosphatase; *specific
    phosphatase types*
    regulation of cystic fibrosis transmem-
        brane conductance regulator,
        293:185–186
    type 1, inhibitors, **298:**544–546
    type-2A, inhibitors, **298:**544–546
Protein phosphatase type 2A, photore-
    ceptors
    activators and inhibitors
        overview, **315:**566–567

protamine activation mechanism, **315:**568–570
assay
incubation conditions, **315:**561
peptide dephosphorylation assay, **315:**561
phosphoopsin preparation, **315:**560
unit definition, **315:**561
classification of phosphatase types, **315:**570–571, 579–580
conserved sequences in humans, **315:**582–583
purification from bovine retina
anion-exchange chromatography, **315:**562
buffers, **315:**558–559
extraction, **315:**561–562
gel filtration, **315:**562, 564
heparin-agarose chromatography, **315:**562
overview, **315:**558–559
rod outer segment preparation, **315:**560
thiophosphatase *a* sepharose chromatography, **315:**564
rhodopsin recycling role, **315:**558, 568
substrate specificity, **315:**568
Western blot analysis of subunits, **315:**564–566
Protein phosphatase type 2C, photoreceptors
activation by fatty acids, **315:**578
assay
activity separation from other phosphatases, **315:**572–573
incubation conditions, **315:**571–572
product separation, **315:**572
radioactivity quantification, **315:**572
classification of phosphatase types, **315:**570–571, 579–580
enrichment from bovine retina, **315:**573–574
magnesium dependence, **315:**577–578
pH optima, **315:**577
physiological significance, **315:**578
purification
bovine retina as starting material, **315:**573, 575
gel electrophoresis, **315:**575–576
recombinant isozymes, **315:**574–575
storage, **315:**575

Protein phosphatase type 7, photoreceptors
assay
incubation conditions, **315:**588–589
inhibitor studies, **315:**589
p-nitrophenyl phosphate assay, **315:**589–590
phosphohistone substrate preparation, **315:**587–588
quantification of radioactivity, **315:**589
classification of phosphatase types, **315:**570–571, 579–580
conserved sequences in humans, **315:**582–583
divalent cation dependence, **315:**591–592
expression of human protein in baculovirus–insect cell system
complementary DNA production, **315:**581
identification and cloning of gene, **315:**581–582, 584
transfection and expression in Sf21 cells, **315:**586–587
vector construction, **315:**584–586
genomic DNA screening with polymerase chain reaction, **315:**592–593
homology between species, **315:**580
pH optimum, **315:**591
phosphatase inhibitor sensitivity, **315:**592
reverse transcription–polymerase chain reaction assessment of expression, **315:**592
substrate specificity, **315:**590
Protein–protein interaction
assays, *see* Double-tagging assay; Filter lift assay; Peptides-on-plasmids library screening; Phage display; RNA polymerase III-based two-hybrid system; Yeast two-hybrid system
mechanisms of physical interactions, **303:**451–452
modules
PDZ domain, *see* PDZ domain
types, **303:**451
Protein–protein interactions, *see also* Cross-linking; Dominant negative suppression; Gel-filtration chromatography; Immunoprecipitation; Protein overlay assay; Sucrose gradient centrifugation; Yeast two-hybrid system
functional tagging, **293:**27–29, 31

membrane dependence of subunit assembly, assays, **293:**24–25
stoichiometry determination, **293:**26–27
Protein shadow hybridization assay, phenanthroline–copper utilization, **304:**522
Protein tyrosine kinase, inhibitors, **298:**548
Protein tyrosine phosphatase, radiation-induced signaling
assays
immunoprecipitation and counting, **319:**267
incubation conditions, **319:**267
principle, **319:**266
substrate labeling and purification, **319:**266–267
targets, **319:**258–259
Proteoglycans
experimental murine AA amyloid content, **309:**707
immunolabeling, **309:**508
Proteome, analysis in biofilms
Bacillus subtilis biofilms
metabolic radiolabeling of proteins, **310:**434–435
phenotypic variability, **310:**434
two-dimensional polyacrylamide gel electrophoresis
autoradiogram analysis, **310:**435–436
database of maps, **310:**441
identification of proteins, **310:**436, 438
overview, **310:**433–434
running conditions, **310:**435
sample preparation, **310:**435
staining, **310:**435–436
heterogeneous cell populations, **310:**438–440
sequencing, **310:**434
Protochlorophyllide oxidoreductase
chlorophyll biosynthesis, overview, **297:**237–238
light-dependent isoforms, **297:**238
POR A-deficient *Arabidopsis* analysis, **297:**239–240
Proton binding by proteins, *see also* Bohr effect
absolute pK calculations
approaches, overview, **295:**176–177
continuum solvation models, **295:**176–178, 186–187

thermodynamic cycle for pK_a values, **295:**175–176
contributions to Gibbs energy of binding, **295:**108–109
flexibility of protein and protonation energetics, **295:**180–184
Gibbs energy of binding, proton linkage effects in determination, **295:**305–308
mean protonation calculation
flexible models, **295:**188–189
static models, **295:**187–188
microscopic pK calculations
Boltzmann-weighted sampling over configuration space, **295:**173
configurational partition function, **295:**171
free energy difference between states, **295:**172
long-range effects, **295:**173–174
number of contributing protonation states
calculation, **295:**171, 179
histidine tautomers, **295:**179–180
multiple site titration problem, **295:**178–179
relative electrostatic energies, estimating, **295:**184–187
Proton exchange, helix formation and amide proton exchange, **295:**7, 10–12, 15, 29
Proton transfer, *see also* Low-barrier hydrogen bond
carbonic anhydrase
His-64 in proton shuttling, **308:**286–287
Marcus parameters for proton transfer
Brønsted plots, **308:**291–292
His-64 proton transfer, **308:**291–293
His-67 proton transfer, **308:**293
human isozyme III as model system, **308:**287
imidazole and lutidine buffers, proton transfer, **308:**293–294
intrinsic kinetic barrier, **308:**295
kinetic isotope effects, **308:**292–293
mutagenesis and disruption of pK_a values, **308:**288–289
pH titration, **308:**290
rate measurements, **308:**289–290
work function dominance, **308:**290–291, 295–297

rate-limiting steps, **308:**285–286
stages in catalysis, **308:**286
Marcus rate theory
  Brønsted relations, **308:**280–282
  intrinsic barriers, **308:**276–279
  inverted region, **308:**280
  kinetic isotope effects, **308:**284–285
  overview, **308:**276–277
  variable transition-state principle,
    308:277–278
  work terms, **308:**282–284
Proton uptake, rhodopsin activation
  bacteriorhodopsin comparison, **315:**385
  overview, **315:**198, 377–378
  pH indicator dye assays
    advantages, **315:**378
    bromocresol purple, **315:**378
    calibration of dye, **315:**379, 381–383,
      387
    cresol red, **315:**378, 384
    detergent-solubilized rhodopsin prepa-
      ration, **315:**380
    general considerations, **315:**378–379
    instrumentation, **315:**379
    proton transfer results at different pH
      values, **315:**383–384
    pure pH signal determination, **315:**381
    sample preparation, **315:**380–381
  proton sources in activation, **315:**386
  proton transfer dynamics, **315:**384–385
Protoplast, *see* Transgenic plant
PrP^{Sc}, *see* Prion protein
PRPP, *see* 5-Phosphoribosyl-1-pyrophos-
  phate
PrtD, *see* Metalloprotease transporter
Prx I, *see* Peroxiredoxin
Prx II, *see* Peroxiredoxin
Prx III, *see* Peroxiredoxin
PsaA, *Synechocystis* 6803 gene, antibodies,
  and mutant strains, **297:**136, 138–139
PsaB, *Synechocystis* 6803 gene, antibodies,
  and mutant strains, **297:**136, 138–139
PsaC
  antibody preparation, **297:**127–128,
    138–139
  iron–sulfur clusters
    mutant characterization
      electron acceptors functioning after
        F_X cluster, flash photometry,
        297:107–108, 113, 115, 119

electron paramagnetic resonance
  characterization of structure,
    297:104–105, 111, 117, 121, 123
electron transfer efficiency of P700,
  time-resolved optical absorption
  spectroscopy, **297:**105–107, 111,
    113, 117, 119
electron transfer throughput by
  steady-state NADP$^+$ reduction,
  297:108–109, 115–116, 119, 121,
    123
structure, **297:**95–96
mutagenesis of *Synechocystis* iron–sulfur
  clusters
  in vitro mutagenesis
    2-mercaptoethanol ligand in C14D
      mutant, **297:**116–117, 121
    overexpression in *Escherichia coli*,
      297:100
    overview, **297:**96–98
    P700-F_X core preparation, **297:**98,
      100
    reconstitution of P700-F_A/F_B com-
      plexes, **297:**100–101
    site-directed mutagenesis, **297:**100
  in vivo mutagenesis
    overview, **297:**96–97, 101
    site-directed mutagenesis, **297:**103
    Synechocystis 6803 *psaC* deletion
      strain construction, **297:**101–103
    transformation and ghenetic segrega-
      tion, **297:**103–104
purification, dissociation from native complexes and ion-exchange chromatography, **297:**128
structure, **297:**95–96
Synechocystis 6803 gene and mutant
  strains, **297:**136
PsaD
  antibody preparation, **297:**127–128,
    138–139
  purification
    dissociation from native complexes and
      ion-exchange chromatography,
      297:128
    overexpressed recombinant protein
      from *Escherichia coli*
      cell growth and harvesting,
        297:128–129
      extraction, **297:**129

histidine tagging and nickel affinity
chromatography, **297:**129
ion-exchange chromatography,
297:129
Synechocystis 6803 gene and mutant
strains, **297:**136
PsaE
antibody preparation, **297:**127–128
purification
dissociation from native complexes and
ion-exchange chromatography,
297:128
overexpressed recombinant protein
from *Escherichia coli*
cell growth and harvesting,
297:128–129
extraction, **297:**129
histidine tagging and nickel affinity
chromatography, **297:**129
ion-exchange chromatography,
297:129
Synechocystis 6803 gene, antibodies, and
mutant strains, **297:**136, 138–139
PsaF, *Synechocystis* 6803 gene, antibodies,
and mutant strains, **297:**136, 138–139
PsaI, *Synechocystis* 6803 gene, antibodies,
and mutant strains, **297:**137–139
PsaK, *Synechocystis* 6803 genes, antibodies,
and mutant strains, **297:**137–139
PsaL, *Synechocystis* 6803 gene, antibodies,
and mutant strains, **297:**137–139
PsaM, *Synechocystis* 6803 gene, antibodies,
and mutant strains, **297:**137–139
psbA
mutagenesis by chloroplast transformation of *Chlamydomonas reinhardtii*
chimeric gene construction conferring
antibiotic resistance, **297:**312–313,
315–316
complications, **297:**311
gene gun transformation, **297:**311,
317–318
growth conditions, **297:**316–317
growth media, **297:**316
homologous recombination, **297:**311,
315
recipient strains, **297:**316
selection of transformants, **297:**312,
318–320
Synechocystis 6803

activation of *psbA1*, **297:**167, 180
DNA isolation for various applications
gel blot analysis, **297:**170
polymerase chain reaction analysis,
297:169–170
light-controlled transcription, analysis,
297:176–177
promoter characterization, **297:**177–178
pseudorevertant analysis of mutants,
297:180, 182
RNA isolation, **297:**170–171
site-directed mutagenesis, **297:**175–176
transcript stability, **297:**167, 178
transcription termination analysis,
297:178
transformation and mutant vectors,
297:171–172, 174–175
translational regulation in *Chlamydomonas reinhardtii*
affinity purification of factors interacting with 5′-untranslated region
extraction of crude cell lysate,
297:204–205
heparin affinity chromatography,
297:205–207
overview, **297:**203–204
RNA affinity chromatography,
297:207–208
D1 labeling *in vivo* with carbon-14 acetate and gel electrophoresis analysis, **297:**196–198
messenger RNA
Northern blot analysis, **297:**201, 203
polysomal RNA isolation,
297:199–200
polysome fractionation on continuous sucrose gradients,
297:201–203
total RNA isolation from polysomal
cell extract, **297:**201
nuclear mutant generation by glass
bead transformation, **297:**193–195
screening of high-fluorescence mutants,
297:193, 195–196
psbB
mutagenesis by chloroplast transformation of *Chlamydomonas reinhardtii*
chimeric gene construction conferring
antibiotic resistance, **297:**312–313,
315–316

complications, **297:**311
gene gun transformation, **297:**311,
    317–318
growth conditions, **297:**316–317
growth media, **297:**316
homologous recombination, **297:**311,
    315
recipient strains, **297:**316
selection of transformants, **297:**312,
    318–320
Synechocystis 6803, partial deletion strain
    construction, **297:**330–331
psbD, *Synechocystis* 6803, light-controlled
    transcription, analysis, **297:**176–177
PSD, *see* Power spectral density
Pseudomonas aeruginosa biofilm
alginate
    assay, **310:**582–584
    gene expression, **310:**74–76
antimicrobial paint additive studies
    efficacy, **310:**643–644
    extracellular polysaccharide evaluation,
        310:640–643
    modified Robbins device culture of bio-
        films, **310:**638–639
    overview, **310:**637
    substrata variation effects, **310:**641–
        643
    viable cell counting, **310:**639–641
architecture, **310:**117
bioelectric effect investigation
    antibiotic treatment, **310:**661, 663
    controls, **310:**664–665
    culture, **310:**658–659
    current dependence of effect,
        310:667–668
    electric circuit, **310:**663–664
    experimental chamber, **310:**660–661
    oxygen intermediates in mechanism,
        310:666–667, 669–670
    pH changes with electric current,
        310:665–666, 669
    temperature changes with electric cur-
        rent, **310:**666, 669
cellulose membrane filter
    culture, **310:**579
    penetration assay of antibacterial
        agents, **310:**585–586
CMC pouch infection model,
    310:591–593

contact lens detachment, parallel plate
    flow chamber studies, **310:**532, 534
green fluorescent protein imaging with
    confocal laser scanning microscopy
    advantages, **310:**47–48
    electroporation, **310:**48–50
    epifluorescence microscopy imaging,
        310:50–51
    filters, **310:**51
    plasmids, **310:**48
    quorum sensing mutants, **310:**51–52
hydrogen peroxide resistance
    catalase
        native gel assay, **310:**603–605
        spectrophotometric assay, **310:**602–
            603, 605
        types, **310:**600, 604
    culture, **310:**600–601
    planktonic cell resistance compared to
        biofilm, **310:**601–602, 604
    superoxide dismutase
        expression in biofilms, **310:**605–606
        native gel assay, **310:**603–604
        spectrophotometric assay,
            310:602–603
        types, **310:**604
once-flow-through continuous culture sys-
    tem, **310:**45–46
oxidative stress adaptation, **310:**599–600
primary adhesion to surfaces
    mechanisms, **310:**551–552
    surface evaluation
        antimicrobial surfaces, **310:**555–557
        cell growth and harvesting,
            310:552–553
        incubation conditions, **310:**553, 557
        scanning electron microscopy,
            310:554–555
        silicone surface, **310:**555–556
        statistical analysis, **310:**554
        viable cell counting, **310:**553–554
quorum sensing
    acylhomoserine lactone detection,
        310:119–120
    autoinducer detection
        lacZ fusion protein detection,
            310:120–122
        overview, **310:**119–120
        purification and detection,
            310:120–124

genes, **310:**44–45, 117–118
integrity of mutant biofilms, **310:**54–55, 118
lasR expression analysis using *lacZ* reporter system, **310:**125, 127
mutant generation
  autoinducer complementation, **310:**123–124
  cassettes, **310:**124
  selection of genes for mutation, **310:**123–124, 127
  role in biofilm development, **310:**44, 51–52, 124–125, 127
reporter genes, overview, **310:**125–128
Robbins device
  antibiotic resistance studies
    biofilm growth, **310:**213
    clinical importance, **310:**212–213
    enumeration of cells, **310:**213
    β-lactamase induction and quantification, **310:**213–216
    minimum inhibitory concentration determination, **310:**213–214, 216
  cystic fibrosis studies
    complement activation, **310:**211–212
    leukocyte response to biofilm, **310:**210–211
    pathogenesis of infection, **310:**210
  establishment of biofilms, **310:**209–210
  sample preparation and inoculation, **310:**47
  sodium dodecyl sulfate, biofilm integrity assay, **310:**54–55
Pseudorevertant
  definition, **297:**18
  photosystem I pseudorevertants from *Synechocystis* 6803
    genetic conformation for suppressor mutations, **297:**21–22
    genetic transformation
      competency dependence on growth phase, **297:**25
      DNA preparation, **297:**24–25
      frequency, **297:**23
      growth conditions for primary mutants, **297:**24
    information attainable from system, **297:**18
    phenotypes of site-directed mutants, **297:**19–20
  phenotypic complementation in suppressor mutation locating, **297:**22
  polymerase chain reaction screening of mutations, **297:**21–22, 26
  psaC mutants, pseudorevertant isolation, **297:**20–21
  secondary mutation at primary mutation site, **297:**23
  strains and growth conditions, **297:**25–26
  photosystem II pseudorevertants from *Synechocystis* 6803, 301–302, 306–307
  psbA pseudorevertants from *Synechocystis* 6803, characterization, **297:**180, 182
PSF, *see* Point-spread function
PSI, *see* Photosystem I
PSII, *see* Photosystem II
[*PSI+*]
  growth assay, **309:**655
  inheritance, **309:**649–651
  [*PIN+*] status, **309:**657
  reversibly curable nonsense suppression, **309:**654–656
  screening, **309:**650, 656
  Sup35p protein determinant
    differential centrifugation analysis
      cell growth, **309:**660
      centrifugation, **309:**661
      extract preparation, **309:**661
      overview, **309:**660
      Western blot analysis, **309:**662
    functions, **309:**651–652
    green fluorescent protein fusion protein
      applications, **309:**657–658
      induction of expression, **309:**658, 660
      plasmid constructiuon, **309:**658
    mutation studies, **309:**656–657
    polymerization studies of NM fragment
      8-anilino-1-naphthalenesulfonic acid assay of assembly, **309:**667
      Congo red assay, **309:**668–669
      electron microscopy analysis, **309:**671, 673
      limited proteolysis assay, **309:**669
      metabolic radiolabeling, **309:**671
      reaction conditions, **309:**666–667
      sedimentation analysis, **309:**669, 671

sodium dodecyl sulfate solubility
    assay, **309:**669
recombinant fragment expression and
    purification from *Escherichia coli*
  bacterial growth and induction,
      309:663
  chromatographic purification of non-
      tagged protein, **309:**664–665
  concentration and storage,
      309:665–666
  expression constructs, **309:**662–663
  histidine-tagged protein purification,
      309:664
  lysis of cells, **309:**663
  overview, **309:**662
  quantification, **309:**665
  structure analysis of function and solu-
      bility, **309:**652, 654
  suppression within single genetic back-
      ground, **309:**655
Psoralen, *see* Psoralen–ultraviolet A, por-
    phyria therapy; Site-specific psoralen;
    4,5′,8-Trimethylpsoralen
Psoralen–ultraviolet A, porphyria therapy
  action spectra, **319:**491
  photosensitizers, **319:**491
  regimens, **319:**491–492
  risks, **319:**492–493
  sources, **319:**491
PSR, *see* Picosirius Red
PTB, *see* *N*-Phenacylthiazolium bromide
PTK, *see* Protein tyrosine kinase
PTP, *see* Protein tyrosine phosphatase
Pull-down assay, ion channel associated pro-
    teins, **294:**373, 375–377
Pulse–chase, sphingosine metabolism
  advantages, **311:**657
  applications, **311:**657
  bulk separation of sphingolipids, **311:**661,
      663
  cell growth and incubations, **311:**659–660
  culture media analysis
    lipid extraction, **311:**666
    low molecular weight tritiated metabo-
        lite analysis, **311:**666
    rationale, **311:**665
    tritiated water determination, **311:**666
  delipidized pellet analysis, **311:**665
  extraction of sphingolipids, **311:**660
  identification of sphingolipids

β-*N*-acetylhexosaminidase treatment,
    311:668–669
  ceramide, **311:**680–681
  ceramide-1-phosphate, **311:**682
  chemical hydrolysis, **311:**667
  derivatizations, **311:**667
  diacylglycerol kinase treatment,
      311:670
  endoglycoceramidase treatment,
      311:669
  β-galactosidase treatment, **311:**667–668
  high-performance thin-layer chromatog-
      raphy, **311:**667
  isolation, **311:**666–667
  sialidase treatment, **311:**669
  sphingomyelinase treatment,
      311:669–670
  sphingosine, **311:**681–682
  sphingosine-1-phosphate, **311:**682
  materials, **311:**657–659
  overview, **311:**656–657
  pathway and trafficking route identifi-
      cation
    degradation pathways, **311:**673–674
    extralysosomal degradation processes,
        311:674–676
    ganglioside lactone biosynthesis,
        311:678–679
    glycosylation of tritiated gangliosides,
        311:677–678
    inhibition of enzymes, **311:**672–673
    manipulation of trafficking, **311:**672
    overview, **311:**670–672
    recycling processes, **311:**676–677
    signal transduction, **311:**679–682
  radioactivity determination, **311:**665
  sphingolipid solution preparation, **311:**659
  thin-layer chromatography, **311:**663
Pulsed-field gel electrophoresis, yeast artifi-
    cial chromosome preparation for micro-
    injection in transgenic mouse pro-
    duction
  agarase digestion, **306:**199–200
  agarose plug preparation, **306:**196–197
  overview, **306:**195
  preparative pulsed-field gel electrophore-
      sis, **306:**197–198
  second-dimension agarose gel electropho-
      resis for concentrating DNA, **306:**199
  shear avoidance, **306:**194

Pulse generator, electrophysiologic software modules, **293:**397–398

Pulse-labeling, chloroplast proteins in mutant maize phenotypes
buffers, **297:**51–52
in organello labeling, **297:**48–51
plant growth conditions, **297:**49
in vivo labeling, **297:**48–50

Purine nucleoside phosphorylase
commitment factors, **308:**341
functions and inhibition rationale, **308:**340
inhibitor design, **308:**349–350, 353–354, 394
isotope effects for arsenolysis, **308:**341
transition-state modeling, **308:**341–342

Puromycin, ribosome–polypeptide complex dissociation, **290:**7–8

PUVA, *see* Psoralen–ultraviolet A

PXA1
functions in yeast, **292:**153, 763, 766
gene cloning, **292:**763
sequence homology to other ATP-binding cassette proteins, **292:**133–134, 153, 764–766
site-directed mutagenesis, **292:**764–766

PXA2
functions in yeast, **292:**153, 767–768
heterodimerization with PXA1, 767
homology to human P-glycoproteins, **292:**133–134, 153

Pycnogenol, *see* Pine bark extract

N-(1-Pyrenyl)maleimide, high-performance liquid chromatography of thiol derivatives
derivatization reaction, **299:**259, 261–264
glutathione disulfide quantitation, **299:**259–260, 262
materials, **299:**260
metabolic modulation studies, **299:**265–266
sample preparation, **299:**264
standard curves, **299:**261–262

2-Pyridinone riboside phosphoramidite, synthesis, **317:**46–49, 60–62

4-Pyridinone riboside phosphoramidite, synthesis, **317:**46–49, 60–62

2-Pyridyldithioethylmethylamidodiethoxyethanemethylaminomethyltrimethyl psoralen, *see* Site-specific psoralen

Pyrophosphate, quantification with ATP sulfurylase and firefly luciferase ATP assay, **305:**367–368

L-*trans*-2,4-Pyrrolidine, glutamate transporter inhibition and transport, **296:**176, 180, 187–188

Pyruvate phosphate dikinase
evolution and efficiency, **308:**162
free energy profile construction
calculations, **308:**172–174
dissociation constant determinations, **308:**164–165
equilibrium constant determinations, **308:**167–168
kinetic model in simplification, **308:**163
physiological conditions, modeling, **308:**174, 176
rate constant determinations, **308:**160, 162, 164–166
time courses for single turnover reactions/simulations, **308:**168–172
kinetic mechanism, **308:**151–152, 162–163
rapid quench elucidation of chemical pathway
data acquisition, **308:**155
instrumentation, **308:**155
proposed pathways, **308:**151–152
radiolabeled compound preparation, **308:**154–155
single turnover experiment design, **308:**152–154
time courses for single turnover reactions
[β-^{32}P]ATP, phosphate and pyruvate, **308:**156–157, 168
[γ-^{32}P]ATP, divalent cation effects, **308:**160
[γ-^{32}P]ATP with and without phosphate, **308:**159–160
[γ-^{32}P]ATP, phosphate and pyruvate, **308:**157–159, 168
magnesium versus cobalt-activated enzyme, **308:**168
phosphoenol pyruvate, **308:**155–156
phosphoenol pyruvate, AMP and pyrophosphate, **308:**156, 168
simulations, **308:**169–172

Q

QLS, *see* Quasielastic light-scattering spectroscopy

QSAR, *see* Quantitative structure–activity relationship

Q-switching, production, **291:**195

Quantitative pressure injection, calcium introduction and membrane current measurement, **293:**375, 378–379

Quantitative structure–activity relationship
limitations, **308:**399
overview of inhibitor design, **308:**398
RNA, three-dimensional modeling
applications, **317:**457, 460, 470
hammerhead ribozyme analysis
base modifications for position 7 substitution, **317:**460–463, 466–468
limitations of crystallography, **317:**458–459
logical–combinatorial approach, **317:**461, 463
mathematical model generation, **317:**463–464
physical meaning of model, **317:**468–470
statistical significance and predictivity, **317:**464–466
overview, **317:**459–460

Quantum yield
caged carboxylates, **291:**43, 45–46
caged tyrosine peptides, **291:**85–87
calculation, **291:**131

Quasielastic light-scattering spectroscopy
advantages for protein assembly studies, **309:**430
Brownian motion and scattered light intensity fluctuation, **309:**433–434, 437–438
coherence area, **309:**435
correlation function
deconvolution, **309:**442
definition, **309:**435–436
determination, **309:**436–437
intensity correlation function relationship to field correlation function, **309:**436
data analysis
deconvolution, **309:**442
direct fit method, **309:**442

method of cumulants, **309:**442–443
polydispersity of samples, **309:**441–442
regularization, **309:**443–444
diffusion coefficient, determination, **309:**438–439
fibril growth studies
amyloid-β fibrillogenesis
concentration dependence of growth, **309:**452–453, 455
discrimination between fibrillogenesis and amorphous aggregation, **309:**455–456
Dutch mutant peptide nucleation and elongation, **309:**457
elongation rate measurement, **309:**451–453, 455
growth conditions and data collection, **309:**452
nucleation and early elongation, **309:**451
overview, **309:**450–451
thermodynamics, **309:**458–459
fibril–fibril interaction effects, **309:**446–447
flexibility effects, **309:**445–446
length effects, **309:**444–445
hydrodynamic radius, determination, **309:**439–440, 452–453
instrumentation
laser modes and stability, **309:**447–448
overview, **309:**437
temperature control, **309:**447
light scattering principle, **309:**431
purification of samples, **309:**448–450
signal origin, **309:**432–433
wavelength dependence on particle size, **309:**432–433

D-Quinolylluciferin, synthesis, **305:**191, 194

Quorum sensing, *see also* Acylhomoserine lactone autoinducers; *lux* control system, *Vibrio*
acylhomoserine lactone signaling, **310:**43, 118
definition, **310:**43, 117
Pseudomonas aeruginosa biofilms
acylhomoserine lactone detection, **310:**119–120
autoinducer detection
lacZ fusion protein detection, **310:**120–122

overview, **310:**119–120

purification and detection,
310:120–124

genes, **310:**44–45, 117–118

integrity of mutant biofilms, **310:**54–55,
118

lasR expression analysis using *lacZ* reporter system, **310:**125, 127

mutant generation

autoinducer complementation,
310:123–124

cassettes, **310:**124

selection of genes for mutation,
310:123–124, 127

role in biofilm development, **310:**44,
51–52, 124–125, 127

R

R17 phage coat, circular permutation analysis of RNA interactions, **317:**326–327

Rabbit reticulocyte lysate cell-free translation system

efficiency, **290:**3

P-glycoprotein translation, **292:**272, 274,
281

posttranslational processing, **293:**19

temperature optimum, **290:**3

Racemization, amyloid-β

aggregation rate effects of aspartyl racemization, **309:**202–203

aspartyl racemization assay, **309:**100–101,
105, 201, 204

nicotine effects on racemic peptide aggregation, **309:**200–203

seryl racemization assay, **309:**103–105

Radioiodination, *see* Iodination

Radiometric disk assay, *see* Biofouling,
membranes

Radiometry, *see* Ultraviolet radiation

Radiotherapy, *see* Oligonucleotide radiotherapy

c-Raf, antisense targeting, **313:**32

Raf-1, antisense knockdown pharmacokinetics in cultered artery smooth muscle
cells

cell culture, **314:**443–444

cell viability assays, **314:**451

dose–response analysis

assay selection, **314:**444–445

materials, **314:**445–447

Northern blot analysis, **314:**447

regression analysis, **314:**448

Western blot analysis, **314:**448

half-life

oligodeoxynucleotides, **314:**452

Raf-1, **314:**441, 452–453

lipofection, **314:**442–444

messenger RNA doubling time, **314:**452

oligideoxynucleotide design, **314:**442–443

reversibility of inhibition, **314:**451–452

time–response analysis

assay selection, **314:**449

parameters, **314:**448–449

repeated application, **314:**449, 451

Raman spectroscopy

caged versus uncaged GTP, **291:**235–237

difference spectroscopy

advantages in protein analysis,
308:178–179

dihydrofolate reductase, dihydrofolate
pK_a determination at active site,
308:193–195

inosine–uridine nucleoside hydrolase

electrostatic interactions and inhibitor design, **308:**200–201

transition-state structure, **308:**199

instrumentation, **308:**179, 181

isotope editing, **308:**178, 332

lactate dehydrogenase analysis

active site structure, **308:**187–188

stereospecific hydride transfer,
308:188–189

substrate binding energetics, **308:**188

transition-state stabilization,
308:189–191

ligand-bound proteins, **308:**177–178,
181

local oscillators, **308:**183–185

phosphoglucomutase

symmetric stretch frequency of substrate phosphate, **308:**181, 198

vanadate spectroscopy, **308:**198

principle, **308:**179

sensitivity, **308:**183

serine protease, acyl intermediate analysis, **308:**192–193

transition-state analog binding,
308:181–182

instrumentation for photolysis studies, **291**:225

interpretation tools
 Badger–Bauer relationships, **308**:183–185
 entropy determination, **308**:187
 ionization states, **308**:185–186
 phosphate–vanadate spectroscopy, **308**:186

liquid–crystalline phase transition in model bilayers
 C–C stretching mode region, **295**:480
 C–H stretching mode region, **295**:478–480
 data collection and analysis, **295**:477–478
 instrumentation, **295**:475–477
 mixed-chain-length phospholipid studies, **295**:490, 492–493
 sample preparation, **295**:475

protein structure studies, advantages and disadvantages, **309**:608
resolution, **308**:176–177, 181
spectral crowding in resonance spectroscopy, **308**:177

Ran-binding protein 2
glutathione *S*-transferase fusion protein with domain 4
 affinity purification, **315**:459–460
 cell lysis, **315**:459
 cleavage with thrombin, **315**:459–460
 expression in *Escherichia coli*, **315**:458–459
 rationale for construction, **315**:456–457
 vector design, **315**:456–458
screening for retinal binding proteins
 functional domain elucidation, **315**:468
 gel electrophoresis and analysis, **315**:462–463
 proteasome 19s regulatory complex subunit binding to cyclophilin-like domain, **315**:465–466
 pull-down assay, **315**:462
 Ran-GTPase identification
  gel electrophoresis and electroelution, **315**:464
  GTP dependence of binding, **315**:463
  sequencing, **315**:464
  Western blot analysis, **315**:465

retinal extract preparation, **315**:460–461
Raney nickel, prenyl group cleavage, *see* Prenylation
Rapamycin, dimerizer-regulated gene expression, **306**:280–281
Rapid filtration, neurotransmitter transporter release and uptake analysis, **296**:666
Rapid quench
advantages over steady-state measurements, **308**:149–150
dead time, **308**:150
free energy profile construction
 calculations, **308**:172–174
 dissociation constant determinations, **308**:164–165
 equilibrium constant determinations, **308**:167–168
 kinetic model in simplification, **308**:163
 physiological conditions, modeling, **308**:174, 176
 rate constant determinations, **308**:160, 162, 164–166
 time courses for single turnover reactions/simulations, **308**:168–172
kinetic mechanism, **308**:151–152, 162–163
nicotinate phosphoribosyltransferase studies, **308**:42
pyruvate phosphate dikinase, elucidation of chemical pathway
 data acquisition, **308**:155
 instrumentation, **308**:155
 proposed pathways, **308**:151–152
 radiolabeled compound preparation, **308**:154–155
 single turnover experiment design, **308**:152–154
 time courses for single turnover reactions
  [β-^{32}P]ATP, phosphate and pyruvate, **308**:156–157, 168
  [γ-^{32}P]ATP, divalent cation effects, **308**:160
  [γ-^{32}P]ATP with and without phosphate, **308**:159–160
  [γ-^{32}P]ATP, phosphate and pyruvate, **308**:157–159, 168
  magnesium versus cobalt-activated enzyme, **308**:168

phosphoenol pyruvate, **308:**155–156
phosphoenol pyruvate, AMP and py-
    rophosphate, **308:**156, 168
simulations, **308:**169–172
substrate channeling analysis
    thymidylate synthase–dihydrofolate re-
        ductase, **308:**135–137
    tryptophan synthase, **308:**117–118
Rappaport chamber, assembly and traction
    force assay, **298:**499–500
RAP-PCR, *see* RNA-arbitrarily primed
    polymerase chain reaction
RAR, *see* Retinoic acid receptor
Ras
    caged GTP studies in crystals
        caged GTP affinity, **291:**257, 259–260
        data collection, **291:**259
        flash lamp photolysis, **291:**95, 215–216,
            257, 261
        rapid freezing followed by data collec-
            tion, **291:**261
        Ras mutation for thermostability,
            291:260
        resolution of data, **291:**259–260, 262
    DNA sequence specificity of ultraviolet
        A damage with photosensitizer
        deuterium oxide effects, **319:**422–423
        folic acid as photosensitizer,
            319:422–423
        gel electrophoresis of DNA,
            319:420–422
        hematoporphyrin as photosensitizer,
            319:422–423
        irradiation conditions, **319:**419–420
        menadione as photosensitizer,
            319:422–423
        principle, **319:**418
        radiolabeling of DNA, **319:**419
        riboflavin as photosensitizer,
            319:421–423
        sequencing of DNA, **319:**420–421
        subcloning of *ras* fragments, **319:**418
        type I versus type II photosensitiza-
            tion, sequence effects, **319:**423,
            425–427
    Fourier transform infrared spectroscopy
        studies with caged GTP
        data collection, **291:**224–225
        difference spectra, **291:**243–245
    skin cancer mutations, **319:**418

Rate acceleration
    bimolecular enzymatic reactions,
        308:9–10
    orotidylate decarboxylase, **308:**10, 358
    unimolecular enzymatic reactions, **308:**6
Rate-limiting step
    commitments to catalysis, **308:**15–16,
        20
    energy barriers, **308:**17
    free energy profiling, **308:**17, 19
    isotope effects, **308:**15–16, 19
    least conductive step, **308:**17, 19
    stickiness of substrate dissociation, **308:**20
RatTGN38–green fluorescent protein fu-
    sion protein
    confocal laser scanning microscopy of
        membrane trafficking
        combining with immunofluorescence,
            302:9–10
        constitutive secretion assay, **302:**15–16
        data acquisition, **302:**6–7
        fluorescence recovery after photo-
            bleaching, **302:**7–9
        mounting, **302:**4, 6
        regulated secretion assay, **302:**19
    construct design, **302:**3–4
    expression in stably transfected cells,
        302:4
Raw3, *see Shaker*-type potassium channel
Rayleigh range, Gaussian beam optics,
    291:189
Rb, *see* Retinoblastoma protein
RCK4, *see Shaker*-type potassium channel
RDA, *see* Representational difference
    analysis
rDNA, *see* Ribosomal DNA
RDR, *see* Rotating disk reactor
Reactive nitrogen species, *see* Nitrosative
    stress
Reactive oxygen species, *see also* Hydroxyl
    radical; Oxidative stress; Superoxide
    anion radical
    classification, **299:**293
    disease association, **299:**3
    sources in cells, **299:**3–4
    total antioxidant power assays, *see* Chemi-
        luminescence; Ferric reducing/antiox-
        idant power assay; Oxygen radical
        absorbance capacity assay; Total per-
        oxyl radical-trapping potential assay;

Trolox equivalent antioxidant capacity assay

Reactive sulfur species, cellular sources, **300:**219

READS, *see* Restriction endonucleolytic analysis of differentially expressed sequences

RecA, ATP-driven molecular discrimination, **308:**47–48

Receptor–ligand internalization, confocal microscopy
advantages, **307:**340–341
applications, **307:**341
cell culture, **307:**346
chemokine receptors, **307:**385
flow chamber design and operation, **307:**341–344
internalization conditions and analysis
counterstaining of cells, **307:**348
immunoglobulin G internalization in endothelial cells, **307:**349–350
insulin and receptor internalization, **307:**345–346, 348–350
ligand–fluorescein conjugate preparation, **307:**345–346
membrane binding of ligands, analysis, **307:**346–347
microscope, **307:**344–345
neurotransmitter receptors, **307:**386
subcellular structure identification, **307:**350

Recombinant neurotransmitter transporter expression, *see* Baculovirus–insect cell expression system; Vaccinia virus–T7 RNA polymerase expression system

Recombinant protein expression systems, *see also* Adenovirus expression system; Baculovirus–insect cell expression system; Chinese Hamster ovary cell; COS cell; HEK 293–Epstein–Barr virus expression system; Murine erythroleukemic cells; Myeloma cell lines; Semliki Forest virus–Chinese hamster ovary cell expression system; *Xenopus* oocyte; Yeast expression systems
cell culture maintenance, **306:**19–20
comparison of expression systems, **306:**36–41
criteria in selection of system, **306:**41
overview, **306:**19–20

Recombinase-mediated cassette exchange
limitations, **306:**65–66
positive versus negative selection, **306:**64–65
reversibility, **306:**62
selectable markers, **306:**63–64
site-specific recombinases, **306:**62–63

Recoverin
affinity purification of rhodopsin kinase
advantages, **315:**405–406
column preparation, **315:**407–408
isolation of calcium-dependent recoverin binding proteins, **315:**408
recombinant kinase purification from baculovirus expression system, **315:**409–410
recombinant recoverin expression and purification from *Escherichia coli*, **315:**406–407
yield of kinase, **315:**408–409
calcium binding, **315:**404–405
conformational change studies
circular dichroism, **316:**113–114
hydrophobic site exposure, **316:**114
intrinsic fluorescence, **316:**113
flow dialysis analysis, **316:**110–112
calcium effects on protein–protein interactions, **316:**34
discovery, **316:**3
GRK1 surface plasmon resonance
binding kinetics analysis, **316:**30–31, 33
fragment binding analysis, **316:**39–40
homologs, **316:**122
immobilization for surface plasmon resonance studies, **316:**23–24
localization in retina, **316:**14
myristoyl switch
mechanism of action, **316:**122, 129
rhodopsin deactivation inhibition, **316:**123
N-terminal acylation, **315:**405
physical properties, **316:**13
purification
anion-exchange chromatography, **316:**11–12
bovine retina soluble protein preparation, **316:**10
hydrophobic affinity chromatography, **316:**10–11

recombinant protein from *Escherichia coli*, **316:**13

rhodopsin phosphorylation, inhibition, **316:**20

structure determination by nuclear magnetic resonance

  data acquisition and analysis, **316:**125–126

  isotopic labeling and purification of recombinant protein, **316:**124–125

  sample preparation, **316:**125

  three-dimensional structures

    calcium-induced conformational changes, **316:**128–129

    myristoylated protein with and without calcium, **316:**127

    unmyristoylated protein, **316:**126–127

Reflection, Gaussian beam optics, **291:**192–193

Reflection confocal microscopy

  electron microscopy samples, **307:**136–137, 147

  material science application, **307:**380

  scanning slit microscopes, **307:**231, 233

  tandem scanning confocal microscope

    cornea studies, **307:**541–542

    design overview, **307:**232–233, 398, 402–404, 541–542

    pulmonary acinar microcirculation, kinetics measurement

      erythrocyte behavior, **307:**408–409

      leukocyte behavior and adhesion molecule function, **307:**408, 411–415

      microscope design, **307:**404–407

      microvessel responses to physiological stimuli, **307:**415–418

      resolution of structures, **307:**409–410

      shear rates, **307:**410–411

      systemic microcirculation comparison, **307:**411–412

  thickness measurement of cornea

    backscatter estimation, **307:**244–245

    depth encoding system in measurement, **307:**236–238

    epithelial thickness, **307:**243

    image acquisition, **307:**234

    laser photoablation of central cornea, effects on thickness, **307:**244

objective lens design and control, **307:**235

overview, **307:**230–231

stromal thickness, **307:**243

through focusing approach, **307:**238–241, 243–245

tissue observation *in vivo*, **307:**231, 233–234

two-photon excitation microscopy, simultaneous detection, **307:**528–529

Refraction, Gaussian beam optics, **291:**194

Regeneration, rhodopsin, *see also* 11-*cis*-Retinal chromophore, rhodopsin

baculovirus–insect cell recombinant opsins, **315:**24–25

bovine opsin, **315:**241

bovine opsin expressed in *Pichia pastoris*, **315:**8, 11

early receptor current studies, rhodopsin regeneration, **315:**274, 286–287, 291

fragments of opsin, **315:**66

regeneration time

  assay with absorption spectroscopy, **315:**299–300, 302

  comparison between rhodopsin and cone pigments, **315:**294, 302

retinal analogs, **315:**237

Regulator of G-protein signaling, *see* RGS9

Regulatory volume decrease, *see* Chloride channel, swelling-activated

Rendering, *see* Cytotomography; Image data, confocal microscopy

Reporter gene, chemiluminescence detection, *see* Alkaline phosphatase; Bacterial bioluminescence; Firefly luciferase; β-Galactosidase; β-Glucuronidase, **305:**437–440

Representational difference analysis, *see also* Complementary DNA representational difference analysis; Genetically directed representational difference analysis

adaptors, changing on test amplicon, **306:**73–74

hybridization/amplification steps, **306:**74

overview and terminology, **306:**72

polymerase chain reaction amplification, **306:**72–73

Repressor, *see also* Biotin operon repressor; *c*I repressor

protein dimerization, **295:**425
thermodynamic binding measurements,
    295:425–426
Reserpine, vesicular monoamine trans-
    porter binding
  assay, **296:**153–154
  mutagenesis analysis, **296:**153
  proton gradient effects, **296:**153
Resonance energy transfer
  chemiluminescence resonance energy
      transfer, applications, **305:**498
  green fluorescent protein and mutants,
      305:209–210
  native microbial fluorescence proteins,
      305:165–166
Resonance Raman spectroscopy, analysis of
    photosystem II mutants, **297:**358
Resonant mirror spectroscopy, transducin
    activation of rhodopsin, assay,
    315:487–489
Resorcinol
  ganglioside detection on high-perfor-
      mance thin-layer chromatography
      plates, **312:**123
  sphingolipid staining on thin-layer chro-
      matography plates, **312:**78
Resorufin, caged compounds
  caging group selection, **291:**67
  practical considerations for use,
      291:67–69
  protein labeling, **291:**72–74
  synthesis, **291:**68–69
  types and structures, **291:**65–66
Respirometry
  instrumentation for high-resolution mea-
      surement, **301:**188–189
  mitochondrial oxygen consumption mea-
      surement, **301:**189–194
  nitric oxide inhibition of respiration,
      301:195–198
Restriction endonucleolytic analysis of dif-
    ferentially expressed sequences
  adapter
    assembly, **303:**266–267
    function, **303:**261
    ligation, **303:**267
  applications, **303:**262
  complementary DNA
    restriction digestion, **303:**266
    synthesis, **303:**264–266

fragment recovery from display gel,
    303:270–271
hidden differences, uncovering,
    303:269–270
neutrophils, profiling of gene expression
    with modified method
  activator effects
    gel display, **303:**294
    incubation with neutrophils,
        303:277–278
  advantages of neutrophil studies,
      303:273
  cloning of amplification products,
      303:290
  complementary DNA synthesis,
      303:275, 279–280, 291
  database comparisons, **303:**297
  fly adaptor
    ligation with digested complemen-
        tary DNA, **303:**275
    preparation, **303:**273–274
  gel analysis
    autoradiography, **303:**276, 285
    band recovery from dried gels,
        303:285–286
    data documentation, **303:**290
    materials, **303:**276
    polymerase chain reaction products,
        303:283–284
    sequencing, **303:**284–285, 289–290
  neutrophil isolation, **303:**274, 276–277
  oligonucleotides, **303:**273
  polymerase chain reaction
    fly adaptor-ligated complementary
        DNA, **303:**282–283, 291–293
    gel-eluted complementary DNA,
        303:286–287, 293
    materials, **303:**275–276
    primer FA-1, end labeling, **303:**275
    product purification, **303:**283
    sequencing of products, **303:**289–290,
        293
    unlabeled primers and complemen-
        tary DNA amplification,
        303:287–289
  reagents, **303:**274–276
  redundant band sources, **303:**294,
      296
  repetitive sequence bands, **303:**297
  reproducibility, **303:**294

restriction enzyme digestion, **303**:275, 280–281

RNA isolation, **303**:278–279, 291

oligonucleotides, **303**:264

overview, **303**:258–261

polymerase chain reaction, **303**:261, 264, 267–269, 271–272

reproducibility of gel patterns, **303**:261–262, 271

RNA isolation, **303**:262–264

spliced gene representations in gel, **303**:271–272

Restriction enzyme accessibility mapping

chromatin remodeling assays, *see* Chromatin accessibility complex; SWI/SNF

Drosophila chromatin accessibility, quantitative measurements

buffer preparation, **304**:495–496

digestion conditions, **304**:489–490, 493, 495

DNA purification, **304**:493–495

hsp26 gene, **304**:490, 492

larva collection, **304**:492–493

nuclei isolation

salivary glands, **304**:494–495

whole larvae, **304**:493

overview, **304**:489–490

nucleosome stability and dynamics, analysis

concentrations

substrate, **304**:295–296

restriction enzyme, **304**:296

contamination, free DNA, **304**:294–295

digestion conditions, **304**:291–292

DNA template design, **304**:290–291

equilibrium constants, **304**:293–294

kinetics of digestion, quantitative analysis, **304**:292–293

neutrality of probes, **304**:279

nucleosome core particle reconstitution and purification, **304**:291

polynucleosomal system analysis, **304**:296–298

rapid preequilibrium, testing, **304**:294

site exposure

cooperative binding by arbitrary proteins, modeling, **304**:288–290

definition, **304**:279

digestion kinetics, linking to simple

equilibrium-binding reactions, **304**:284–285

kinetic modeling, **304**:280, 282–283

spontaneous versus rapid equilibrium, **304**:283–284

structural model, **304**:283

system-specific, second-order effects, **304**:285, 287–288

specificity and sensitivity, **304**:278–279

stability controls during assay, **304**:295

steric hindrance at target site, **304**:278

yeast chromatin mapping

digestion conditions, **304**:376

nuclei preparation

cell growth and lysis, **304**:366–367

centrifugation, **304**:367–368

solution preparation, **304**:366

principle, **304**:365, 374

solution preparation, **304**:374

Southern blot analysis, **304**:376

Restriction enzyme–nick translation, DNA detection with confocal microscopy

CpG island identification, **307**:203

digestion, **307**:200

immunostaining, **307**:202

multiple fluorescence and reflectance detection, **307**:203–204, 206

nick translation, **307**:200

overview, **307**:199–200

Restriction site, mapping in genomic clones, **296**:504–507

Resveratrol

anticancer properties, **299**:185

antiinflammatory properties, **299**:185

antioxidant activity, **299**:184–185

food distribution, **299**:185–186

isomers, **299**:186

wine assays

content in wine, **299**:186

gas chromatography–mass spectrometry, **299**:186–187

high-performance liquid chromatography, **299**:187–188, 190

standards, **299**:188, 190

Reticulocyte lysate system

antisense knockdown assay

antisense oligonucleotide supplementation with ribonuclease H, **314**:299

gel electrophoresis of products,
      314:299, 301
kits, **314:**298

protein knockdown quantification,
      314:301
rationale, **314:**297–298
transcription/translation reactions,
      313:138–139, 148, **314:**298–299
vector design, **313:**137–138
plasmid selection, **314:**298
vesicular stomatitis virus translation ar-
      rest, antisense assay, **313:**199–200
Retina
avian, *see* Chicken retina
neurons, *in situ* hybridization
      cryosection hybridization for double-la-
            bel detection, **316:**584–587
      cryosection hybridization for single-la-
            bel detection
            applications, **316:**581–582
            fixation of tissue, **316:**582–583
            hybridization, **316:**583
            immunohistochemical detection,
                  316:584
            posthybridization, **316:**583–584
            prehybridization, **316:**583
      detection techniques, **316:**579, 584–585,
            589–590
      digoxigenin–RNA probe synthesis,
            316:580–581
      ribonuclease contamination prevention,
            316:580
      whole mounts
            fixation of tissue, **316:**587
            hybridization, **316:**588
            immunohistochemical detection,
                  316:588–589
            posthybridization, **316:**588
            prehybridization, **316:**587–588
transcription factors
      cloning and identification, *see* DNase I
            footprinting; Electrophoretic mobil-
            ity shift assay; Yeast one-hybrid
            assay
      overview, **316:**590–591
zebrafish, *see* Zebrafish retina
11-*cis*-Retinal
challenges of retinal quantification,
      316:313

high-performance liquid chromatography
      analysis, *see* Retinoids, high-perfor-
      mance liquid chromatography
      analysis
solution preparation for photoreceptor
      bleaching studies
      ethanol solution, **316:**235
      exogenous application to isolated cells,
            316:235–236
      interphotoreceptor retinoid binding
            protein solution, **316:**235
      phospholipid vesicles, **316:**234–235
11-*cis*-Retinal chromophore, rhodopsin
analog studies
      activation of constitutively active opsin
            mutants, **315:**264, 266
      binding site generalizations, **315:**220
      binding studies, **315:**228–229
      circular dichroism
            11,12-cyclopropylretinal, **315:**226–
                  228
            11,12-dihydroretinal binding,
                  315:222–223
            exciton-coupled spectroscopy,
                  315:222–228
            11-*cis*-locked seven-membered ring,
                  315:11, 12 dihydrorhodopsins,
                  315:223–226
            native rhodopsin spectra,
                  315:221–222
      conformational analysis of cyclopropyl-
            retinal analogs, **315:**229–230
      9-methyl group importance, **315:**264,
            266
      13-methyl group importance,
            315:230–233
      regeneration of opsins, **315:**237
      synthesis
            cyclopropyl retinal analogs, **315:**235
            Horner reaction, **315:**234–235
binding site
      geometry, **315:**179
      spectral tuning with site-directed muta-
            genesis, **315:**116–118
conformation analysis in photoisomeriza-
      tion, **315:**220–233
discovery, **315:**196
photobleaching energetics, *see* Photo-
      bleaching energetics, rhodopsin
photoisomerization reaction, overview,

315:12–13, 122–123, 148, 251, 269, 557
photoisomerization, sensing in transmembrane domains using glycine-121 mutants
phenylalanine-261 interactions mediated by retinal, 315:128–129
retinal C-9 methyl group interactions, analog studies, 315:126–127
substitution effects on activity, 315:124, 126
planarity in binding site, 315:220–222
regeneration, *see* Regeneration, rhodopsin
removal, 315:219, 236
reverse agonist activity, 315:233
Schiff base
absorption spectra of protonated form, 315:196–198
butylamine Schiff base complex preparation, 315:235–236
counterion protonation and thermal isomerization, 315:200–201
counterion tuning, 315:268
deprotonation, activation and noise reduction role, 315:160–161, 163, 198, 207, 269, 377
distribution in vertebrate visual pigments, 315:197–198
environment studies by Fourier transform infrared difference spectroscopy of rhodopsin mutants
Glu-113, mutation and role in stabilization/activation, 315:186–187, 189–190
Gly-90 mutants, 315:190–192
low-temperature studies, 315:191–192
water molecule interactions, 315:188–189
ethylamine or propylamine Schiff base complex preparation, 315:212
pK_a
bovine rhodopsin, 315:203, 207
chloride ion effects on gecko pigment, 315:204–205
counterion pK_a, 315:201
determination from absorbance spectra with model chromophores, 315:201, 203

environmental effects, 315:205–206
octopus rhodopsin, 315:205–206
P521, 315:204, 207
P531, 315:205
P568, 315:205
rationale for high value, 315:198–201, 207
thermal isomerization consequences and noise reduction, 315:199–201
proton exchange, *see* Proton uptake, rhodopsin activation
resonance Raman spectroscopy, 315:197
structure, 315:12, 91
thermal isomerization, 315:159–161
Retinal degeneration, *see* Peripherin/rds
Retinal pigment epithelium
microsomal enzyme preparations, *see* Isomerhydrolase; Lecithin:retinol acyltransferase; Retinal pigment epithelium retinal G-protein-coupled receptor; 11-*cis*-Retinol dehydrogenase; Retinyl ester hydrolase
visual cycle role, 316:324, 331, 359, 400, 565–566
Retinal pigment epithelium p65
function, 316:706, 723–724
gene regulation, 316:705
genotyping of mice with polymerase chain reaction, 316:716–718
knockout mouse
embryonic stem cell derivation, 316:714–716
generation, 316:716
phenotype, 316:722
retinoids, high-performance liquid chromatography analysis, 316:722
rhodopsin quantification and characterization, 316:722–723
mutation in disease
clinical phenotypes, 316:713–714
patient selection for analysis, 316:707
polymerase chain reaction/single-strand conformation analysis
labeled DNA, 316:708, 710
primers, 316:709
unlabeled DNA, 316:710
restriction digest analysis, 316:710
sequencing, 316:707–708, 710–711

types of mutations, **316:**706–707, 711–713

reverse transcriptase–polymerase chain reaction of gene expression in mice, **316:**718–719

sequence homology between species, **316:**705

Western blot analysis
  antibodies, **316:**719–720
  bovine retinal pigment epithelium membranes, **316:**720
  mouse eye, **316:**720–722

Retinal pigment epithelium retinal G-protein-coupled receptor
  absorption spectroscopy, **316:**417–418
  chromophore
    extraction by hydroxylamine derivatization, **316:**418–420
    high-performance liquid chromatography analysis of oximes, **316:**420–421
    identification, **316:**414
    irradiated protein chromophore, **316:**422
  homology
    rhodopsin similarity, **316:**413
    species comparison, **316:**414
  hydroxylamine reactivity, **316:**422
  purification
    extraction, **316:**416
    immunoaffinity chromatography, **316:**416–417
    retinal epithelium microsome preparation, **316:**414–415

Retinaldehyde dehydrogenase, phase partion assay
  enzyme preparation, **316:**363
  incubation conditions, **316:**364–365
  materials, **316:**361
  membrane preparation, **316:**363
  nicotinamide adenine dinucleotide tritiation, **316:**361
  pH, **316:**360
  principle, **316:**360

Retinitis pigmentosa
  animal models
    avian models, **316:**526–536
    cyclic GMP, genes affecting levels, **316:**521
    cyclic GMP phosphodiesterase rescue

      in mouse *rd* model, *see* Encapsidated adenovirus minichromosome
    genotyping of mouse models
      DNA isolation from tail biopsy, **316:**522–523
      overview, **316:**521–522
      polymerase chain reaction amplification, **316:**523–524
      restriction digestion of amplified fragments, **316:**524
      sequencing, **316:**524–526
    laboratory-generated models, **316:**517–519
    natural models, **316:**515, 517
    overview, **316:**515
  autosomal dominant retinitis pigmentosa
    paired-flash electroretinography in humans, **316:**222
    ribozyme targeting of mutant protein transcripts in rat, *see* Ribozyme
  rom-1 mutation, **316:**671–672
  retinal pigment epithelium p65 mutations, **316:**711–713
  rhodopsin mutations, **315:**79, **316:**618–619

Retinitis pigmentosa GTPase regulator
  mutation in X-linked retinitis pigmentosa type 3, **316:**688
  RCC1 homologous domain, **316:**688
  splice variant isolation, **316:**703–704
  tissue distribution analysis by Northern blot, **316:**703
  yeast two-hybrid system, detection of interacting retinal proteins
    advantages and disadvantages, **316:**689
    bait
      plasmid preparation, **316:**691–693
      plasmid, prescreen testing, **316:**695
      selection, **316:**691
    complementary cDNA library characteristics, **316:**690–691
    β-galactosidase filter lift assay, **316:**696–698
    host strain, **316:**694–695
    principle, **316:**688–689
    quantification of interaction, **316:**700–703
    target plasmid preparation, **316:**693–694
    transformation, **316:**695–696
    verification of interaction, **316:**698–700

Retinoblastoma protein, photodynamic therapy apoptosis role, **319:**352–353
Retinoic acid
absorbance properties, **299:**431–432
isomer separation
data analysis, **299:**437–438
extraction, **299:**433–435, 441
high-performance liquid chromatography, **299:**434–437, 441
resolution of peaks, **299:**438–440
sensitivity and reproducibility, **299:**440–441
standards, **299:**431, 433
receptors, **299:**430–431
Retinoic acid receptor, subcellular distribution, **302:**74
Retinoids, high-performance liquid chromatography analysis
absorption properties, **316:**314–315, 322, 324
calculation of results, **316:**319
challenges of retina retinoid studies, **316:**313
equipment, **316:**318
esterification
retinoic acid, **316:**317
retinol, **316:**316
extraction
reaction mixtures, **316:**314
tissues, **316:**317–318, 320–321
fluorescence detection, **316:**321–322
identification of peaks, **316:**321–322, 324
normal phase chromatography, **316:**318–319
oxime formation, **316:**317
reductions
retinal, **316:**314, 316
11-*cis*-retinal with tritiated sodium borohydride, **316:**316
retinal pigment epithelium retinal G-protein-coupled receptor chromophore
extraction by hydroxylamine derivatization, **316:**418–420
high-performance liquid chromatography analysis of oximes, **316:**420–421
irradiated protein chromophore, **316:**422
retinol oxidation to retinal, **316:**316

reversed-phase chromatography, **316:**318–320
saponification, **316:**316–317, 319
sensitivity, **316:**319
simultaneous assay of retinoid-utilizing enzymes in retina
calculations, **316:**339
extraction of retinoids, **316:**333–334
retinyl ester hydrolysis, **316:**337
separation of retinoids, **316:**334–335, 340
stability of compounds, **316:**319
standards, **316:**314
transgenic mouse mutants
constant illumination response, **316:**573
dissection of eye, **316:**567–568
extraction, **316:**568
flash response, **316:**572–573
individual variation of retinas, **316:**574
retinal pigment epithelium p65 knockout mouse analysis, **316:**722
zebrafish larva eye retinoid analysis
characterization of retinoids, **316:**556–557
chromatography, **316:**556
extraction, **316:**555–556
materials, **316:**555
principle, **316:**552, 555
Retinol, *see also* Vitamin A
absorbance properties, **299:**431–432
antioxidant activity, **299:**421
interactions with α-tocopherol in liposome oxidation prevention
chemicals and equipment, **299:**422
oxidation conditions
liposomes, **299:**422–423
retinal membrane oxidation, **299:**423–424
regeneration, **299:**421–422, 428, 430
retinal membrane protection, **299:**425, 427–429
soybean phosphatidylcholine liposome protection, **299:**424–425, 429
vitamin analysis with high-performance liquid chromatography, **299:**424
receptors, **299:**430–431
11-*cis*-Retinol
formation from all-*trans* retinyl esters, *see* Isomerhydrolase
high-performance liquid chromatography

analysis, *see* Retinoids, high-performance liquid chromatography analysis

11-*cis*-Retinol dehydrogenase
epitope ectopic site studies of topology in mutants, principle, **316:**347–348
glycosylation ectopic site studies of topology in mutants
endoglycosidase H analysis, **316:**355, 357
mutant construction, **316:**353–355
principle, **316:**346–347
high-performance liquid chromatography assay, **316:**365–366, 368, 370–371
hydropathy analysis, **316:**344–345
lumenal orientation of catalytic domain, **316:**358
phase partion assay
endogenous substrate assay, **316:**364
enzyme preparation, **316:**363
exogenous substrate assay, **316:**364
materials, **316:**361
membrane preparation, **316:**363
nicotinamide adenine dinucleotide tritiation, **316:**361
pH, **316:**360
principle, **316:**360
recombinant insect cell activity versus microsomes, **316:**367–368
protease protection analysis
principle of topology analysis, **316:**345–346
retinal pigment epithelium microsomes
microsome preparation, **316:**348–349
proteolysis reaction, **316:**349–350
Western blot analysis, **316:**350
Triton X-100 studies, **316:**356–357
in vitro-translated enzymes
comparison with retinal microsome enzyme topology, **316:**350, 357
microome preparation, **316:**352
proteinase K treatment, **316:**352–353
transcription, **316:**351
translation, **316:**351–352
visual cycle role, **316:**331, 344, 359, 371–372

Retinol isomerase, *see* Isomerhydrolase
Retinyl ester hydrolase, *see also specific enzymes*
assay with radiolabeled substrates

hydrolytic reaction confirmation, **316:**386
incubation conditions and quantification, **316:**385
overview, **316:**384–385
substrate synthesis, **316:**385
neutral, bile salt-independent activity in microsomes, **316:**385–386
11-*cis*-retinyl ester hydrolase
colocalization with 11-*cis*-retinyl esters in retinal pigment epithelium plasma membrane, **316:**393–395, 398–400
hydrolysis of 11-*cis*-retinyl esters *in situ*, **316:**395–397
isolation
hydrophobic interaction chromatography, **316:**398
Percoll gradient separation of S2 membranes, **316:**397–398
simultaneous detection with lecithin : retinol acyltransferase and retinol isomerase activities
all-*trans*-retinol preparation, **316:**333
cellular retinaldehyde-binding protein preparation and effects, **316:**332–333, 340–341
denaturing gel electrophoresis, **316:**339
high-performance liquid chromatography
calculations, **316:**339
extraction of retinoids, **316:**333–334
retinyl ester hydrolysis, **316:**337
separation of retinoids, **316:**334–335, 340
incubation conditions, **316:**333
kinetic effects of phosphates, ATP, and alcohols, **316:**341–343
materials, **316:**332
stability of proteins, **316:**339–340
ultraviolet treatment of retinal pigment epithelium microsomes, **316:**332, 340
substrate specificity of retinal pigment epithelium microsome enzymes
active site modification and inhibition analysis for different substrates, **316:**390–391
all-*trans*-retinyl ester hydrolase inhibitor studies

bis(*p*-nitrophenyl) phosphate,
316:391–392
phospholipase A$_2$, 316:392–393
fatty acyl group preference,
316:388–389
retinyl palmitate isomers, 316:386–388
substrate competition studies,
316:389–390
visual cycle role, 316:331, 384, 399–400
Retrovirus, *see also* Gene therapy; Hematopoietic stem cell, fluorescence selection of acid sphingomyelinase expression
antisense cyclin G$_1$ delivery, 314:37
gene therapy applications, 314:37–38
Retrovirus, green fluorescent protein expression
advantages in gene transfer, 302:329, 331, 358, 369
applications, 302:340–341, 358–359
cytotoxicity, 302:359
gene therapy, *see* Gene therapy
infection of target cells, 302:338–339
LEL vector
design and construction, 302:360–361
imaging of transduced tumors and tissue, 302:368
transduction of hematopoietic stem cells
cell culture, 302:366
clonogenic progenitor assays, 302:367
injection studies, 302:367–368
stem cell isolation, 302:366
transduction *ex vivo*, 302:367
transduction of human tumors in mice
A375 human tumor cell line transduction, 302:362–363
fluorescence-activated cell sorting, 302:366
fluorescence microscopy, 302:365–366
histology, 302:364
titer assays, 302:364–365
tumor cell suspension preparation, 302:364
tumor formation, 302:363–364
in vivo transduction, 302:364
transduction of murine embryos, *in vivo*, 302:368
virus production, 302:361–362

protein variants for mammalian expression, 302:330–332
stable virus-producing cell lines, selection, 302:337–338
titer determination
colony counting, 302:339, 364–365
green fluorescent protein expression by cell sorting, 302:340, 365
transduction
cell lines in culture, 302:336–337.362–363
efficiency assay, 302:339–340
transfection of vectors, 302:337
vector design and construction, 302:332–335, 340–341, 360–361
virus transduction, 302:335–336
Retzius P cell synapse
isolation of cells, 296:594–595
pairing of cells and culture, 296:595
serotonin transporter electrophysiology
external solutions, changing, 296:597–598
flash photolysis studies with DM-nitrophen
flash photolysis setup, 296:605–606
microinjection of DM-nitrophen, 296:604–605
photochemistry, 296:603–604
pre- and postsynaptic current isolation induced by flash photolysis, 296:602–603
two-electrode voltage clamp, 296:601
inward current, 296:597
patch pipette and microelectrode preparation, 296:606–607
potassium current, 296:597
simultaneous fluorescence imaging of 5,7-dihydroxytryptamine uptake with voltage clamping, 296:598–601
sodium dependent current, 296:595–596
solution preparation, 296:606–607
transporter-associated currents, 296:596
whole-cell patch clamp, 296:596
Rev
human immunodeficiency virus protein binding to REV, 318:424–425
peptide binding to RNA, *see* RNA–peptide interaction

RNA binding site of human immunodeficiency virus, modeling with comparative sequence analysis, **317:**506–507

translational repression for RNA binding screening, **318:**328–329

Reversed-phase chromatography, *see also* Reversed-phase chromatography high-performance liquid chromatography simultaneous determination of vitamin A, vitamin E, lycopene, and xanthophylls

calibration, **299:**352

extraction from plasma and serum, **299:**352–354, 356

instrumentation and chromatography conditions, **299:**350–352

peak identification, **299:**357–360

quality assurance, **299:**353, 356–357, 361–362

reagents and solutions, **299:**349

stability of samples, **299:**360–361

standards, **299:**350

Reversed-phase high-performance liquid chromatography

gangliosides, **312:**142–143

o-phthalaldehyde derivatives of sphingoid bases

free base measurement

applications, **312:**3

chromatography and detection, **312:**4–5

derivatization, **312:**4–5

lipid extraction, **312:**4, 6

standards, **312:**5

sphingoid base 1-phosphate measurement

chromatography and detection, **312:**7

derivatization, **312:**7

pH shifting, **312:**7–8

solid-phase extraction, **312:**6–7

sphingosine kinase assay, **312:**8–9

RNA–protein complexes, **318:**441–442

Reverse transcriptase–polymerase chain reaction

c-ABL knockdown analysis, **314:**435–436

cardiovascular ion channel transcript detection

RNA extracted from tissue, **293:**81–83

single-cell polymerase chain reaction, **293:**80–81, 84

cyclic GMP phosphodiesterase rescue evaluation in mouse *rd* model of retinitis pigmentosa, **316:**740

cytokine expression analysis in atopic dermatitis therapy with ultraviolet A radiation

β-actin standard, **319:**299–300

amplification cycles, **319:**298–299

primers, **319:**299

product purification and analysis, **319:**299–300

RNA extraction, **319:**298

cytokine-induced neutrophil chemoattractant, ozone induction in rat lung, **319:**556

differential reverse transcriptase polymerase chain reaction in keratinocytes, ultraviolet A-induced gene expression, **319:**285–286

fetal brain culture samples, **314:**255–257

β-globin splice variants, **313:**511–512, 518

lecithin:retinol acyltransferase transcripts, **316:**403–405

reverse ligation-mediated polymerase chain reaction of ribonuclease H cleavage reaction products

amplifications, **313:**134

ligation reaction, **313:**133

overview, **313:**131

reverse transcription, **313:**133–134

RNA linker synthesis, **313:**131–133

sequencing, **313:**134–135

ribozyme cleavage assay, **313:**499, **316:**774–775

transgene quantification in mice, **316:**176

Rex, binding element mapping by *in vitro* selection

aptamer analysis, **318:**213–214

binding assays, **318:**212

binding reactions, **318:**211

cellulose filter immobilization, **318:**211

cloning and sequence of selection products, **318:**212

competition experiments, **318:**212–213

polymerase chain reaction amplification, **318:**211

reagents, **318:**210

reverse transcription, **318:**211

rounds of selection, **318:**212

thermal equilibration, **318:**210–211

transcription, **318:**212

Reynolds number, characterization of flow conditions, **310:**225

RGR, *see* Retinal pigment epithelium retinal G-protein-coupled receptor

RGS9
enhancement of cyclic GMP phosphodiesterase, **315:**527
$G\alpha_t/G\alpha_{i1}$ chimera interactions, **315:**519–521, 523
$G\alpha_t/G\alpha_{i1}$ chimera switch II mutant interactions
glutathione *S*-transferase–RGS9 fusion protein pull-down assay, **315:**552–553
single-turnover GTPase assay, **315:**553–554
$G\beta_{5L}$ complex, **315:**526
GTPase, assay of regulatory effects, *see* Transducin
transducin conformation stabilization for GTP hydrolysis, **315:**526

Rhod-2, loading of mitochondria, **307:**453

Rhodamine 123
efflux assays
colony assay, **292:**246–247
MDR1 transgenic mice, **292:**584–586, 593
membrane potential assessment in disinfected biofilms, **310:**176–177

Rhodamine, caged compounds
applications, **291:**75, 77
caging group selection, **291:**67, 72
practical considerations for use, **291:**67, 69, 71–72
protein labeling, **291:**72–74
synthesis, **291:**69
thiol-reactive caged compound
actin labeling and cross-linking, **291:**170–171, 173–175
preparation, **291:**169–170
types and structures, **291:**66

Rhodamine-phalloidin
sliding filament motility assay, **298:**17
visualization of F-actin disruption by latrunculin-A, **298:**23–24

Rhodanese
assay in *Escherichia coli* cell-free translation system, **290:**25

reactivation assay with chaperonin 60, **290:**179–180

Rhodopsin kinase
affinity purification with recoverin
advantages, **315:**405–406
column preparation, **315:**407–408
isolation of calcium-dependent recoverin binding proteins, **315:**408
recombinant kinase from baculovirus expression system, **315:**409–410
recombinant recoverin expression and purification from *Escherichia coli*, **315:**406–407
yield of kinase, **315:**408–409
assays of rhodopsin phosphorylation, **315:**217–218, 237, 415–417
binding of opsin/all-*trans*-retinal complex, **315:**244, 249–250
kinetic light scattering of rhodopsin complexes
bimolecular rate constant, **315:**341–342
dissociation constant determination, **315:**339, 341
rhodopsin kinase preparation, **315:**335
overview of rhodopsin binding and activation, **315:**70, 411, 437–438
posttranslational modification, **315:**336
prenylated peptide analysis
high-performance liquid chromatography–mass spectrometry, **316:**448–449, 451
preparation of peptide standards, **316:**445–447
trypsinization, **316:**447–448
purification, **315:**217, 240–241
S-modulin cross-linking, **316:**19–20
sites of rhodopsin phosphorylation, **315:**557

Rhodopsin
absorption spectroscopy, *see* Absorption spectroscopy, rhodopsin
activation probability, **315:**268
apoprotein, *see* Opsin
chromophore, *see* All-*trans*-retinal; 11-*cis*-retinal chromophore, rhodopsin
concentration determination in transgenic mouse mutants, **316:**568
cone pigment comparison, *see* Cone visual pigment

crystallography, *see* Electron crystallography
cysteine scanning mutagenesis, *see* Split
receptor, mapping tertiary interactions in rhodopsin
early receptor current, *see* Early receptor
current, rhodopsin
electroretinogram studies of mutants,
316:618–621
fragment reconstitution studies
absorption spectroscopy of complexes,
315:66–68
chromophore formation, **315:**66
expression analysis by enzyme-linked
immunosorbent assay, **315:**65–66
gene fragments, vector construction,
315:60–62
glycosylation analysis, **315:**64
G-protein activation assays, **315:**68–69
immunoaffinity chromatography,
315:66
membrane integration assay, **315:**63–64
noncovalent interactions, **315:**69–70
rationale, **315:**59–60
trafficking, **315:**69
transient transfection in COS-1 cells,
315:62–63
Western blot analysis of expression,
315:63
genotyping of mouse retinitis pigmentosa
models, **316:**524
G-protein activation, *see* Transducin
intermediates, overview, **316:**425
lifetimes of intermediates, **315:**269
membrane topology, **315:**12, 116
meta intermediates
metarhodopsin-II decay rate and Glu-
122 role, **315:**294, 311–312
thermal behavior
absorption spectroscopy studies,
315:302–303, 305–308
cooling of samples, **315:**302–303
transducin activation assay, **315:**308–
309, 311
mutations in retinitis pigmentosa,
316:618–619
noise, origin and energetics, **315:**159–161,
163
peptide ligands, *see* Peptide competition
assay, rhodopsin ligands; Peptide-on-

plasmid combinatorial library, rhodopsin ligands; Phage display, rhodopsin ligands
phosphorylation, *see* Phosphorylation,
rhodopsin
photoaffinity labeling, *see* Photoaffinity labeling, rhodopsin
photobleaching energetics, *see* Photobleaching energetics, rhodopsin
photobleaching recovery, *see* Photobleaching recovery, photoreceptors
phylogenetic analysis, *see* Phylogenetic
analysis, rhodopsins
preparations
chicken protein
buffers, **315:**294, 296
concanavalin A affinity chromatography, **315:**296–297
ion-exchange chromatography,
315:296–298
transient transfection in HEK293S
cells, **315:**298–299
protein interaction studies
membrane preparation, **315:**413–415
transient transfection in HEK293
cells, **315:**411–413
proton exchange, *see* Proton uptake, rhodopsin activation
recombinant protein expression, *see* Baculovirus–insect cell recombinant expression systems; Yeast recombinant
expression systems
regeneration, *see* Regeneration, rhodopsin
retinal pigment epithelium p65 knockout
mouse characterization, **316:**722–723
site-directed mutagenesis, *see* Site-directed mutagenesis, rhodopsin
split receptor, *see* Split receptor, mapping
tertiary interactions in rhodopsin
stability of ground state, **315:**268
stable transfection, *see* Site-directed mutagenesis, rhodopsin
structure, *see* Electron crystallography;
Nuclear magnetic resonance
trafficking, *see* Membrane trafficking, rhodopsin
transient transfection, limitations, **315:**30
Xenopus transgenesis and mutant expression, **316:**64

Riboabasic phosphoramidite, synthesis, **317**:41–43, 55–57

Riboflavin, blue fluorescence protein ligand exchange, **305**:176–178

Ribonuclease, protein disulfide isomerase assays
 continuous ribonuclease refolding
  constraints, **290**:37–38
  data analysis, **290**:34–37
  incubation conditions, **290**:34
  principle, **290**:33–34
  substrate preparation, **290**:39
 scrambled ribonuclease refolding, **290**:39, 51

Ribonuclease A, pH effects on stability, **295**:247–248

Ribonuclease A mimics
 applications, **318**:159
 cleavage specificity, **318**:156, 158–159
 end labeling for RNA for cleavage, **318**:161–162
 imidazole buffer, RNA cleavage
  applications, **318**:153
  reaction conditions, **318**:161–162
  specificity, **318**:150–152
  suppression in double-stranded RNA, **318**:152–153
 imidazole conjugates, RNA cleavage
  minihelices derived from 3′ ends of viral RNAs, **318**:155–156, 163
  reaction conditions, **318**:162–164
  transfer RNA, **318**:153, 155, 158, 163–164
 mechanism of RNA hydrolysis, **318**:148–150
 polyacrylamide gel electrophoresis of cleavage products, **318**:164
 preparation, **318**:159–160
 rationale for development, **318**:147–148
 RNA preparation for cleavage, **318**:160–161
 types, **318**:150

Ribonuclease H
 binding of nucleic acids, **313**:16
 boranophosphate oligonucleotides, cleavage of hybrid duplexes
  assay
   incubation conditions, **313**:247–248
   product separation with high-perfor-

mance liquid chromatography, **313**:248
  endo/exo mechanisms in antisense effects, **313**:250–251
  melting temperature determination, **313**:249
  overview, **313**:229, 247
  thermostability of hybrids, effect on hydrolysis rate, **313**:249–250
 brain levels, **314**:87–88
 cellular distribution of activity, **313**:13
 cleavage of antisense DNA–RNA duplexes, **314**:77, 87, 181, 292, 314, 364
 functions in cell, **313**:13, **317**:133
 gene cloning, **313**:16
 length specificity for cleavage, **314**:230
 methylphosphonodiester antisense oligodeoxynucleotide cleavage assays
  Nonidet P-40-lysed cell assay
   advantages, **313**:118
   incubations and lysis, **313**:118
   Northern blot analysis, **313**:119–120
  reverse ligation-mediated polymerase chain reaction of cleavage reaction products
   amplifications, **313**:134
   ligation reaction, **313**:133
   overview, **313**:131
   reverse transcription, **313**:133–134
   RNA linker synthesis, **313**:131–133
   sequencing, **313**:134–135
  synthetic RNA assay with *Escherichia coli* ribonuclease H
   gel electrophoresis, **313**:108–109
   incubation conditions, **313**:108
  in vitro-transcribed RNA assay with cell extract ribonuclease H
   degradation analysis of chimeric oligonucleotides, **313**:117–118
   extract preparation, **313**:113–114
   formaldehyde gel electrophoresis and Northern blotting, **313**:116
   point mutation selectivity of chimeric oligonucleotides, **313**:113
   ribonuclease class identification, **313**:116–117
   RNA purification, **313**:115–116
  in vitro-transcribed RNA assay with *Escherichia coli* ribonuclease H

coupling transcription with cleavage
reaction, **313:**112–113
formaldehyde gel electrophoresis
and Northern blotting,
313:110–112
incubation conditions, **313:**109–110
open loop region identification in
RNA, **313:**112
transcription, **313:**109
modified nucleotide identification using
photolabile derivatives of oligonucleotides, **318:**125, 127
recognition elements, **313:**13–16
RNA–DNA duplex cleavage in antisense
inhibition, **313:**13–16, 26–27, 48, 95–
96, 147, 360, 442–443, 457–458
sensitivity to oligonucleotide modifications, **313:**270–271, 372–373, 422–423
site-specific RNA cleavage with 2'-O-
methyl RNA/DNA chimera
advantages, **317:**132–133
applications
detection of 2'-O-methyl RNA,
317:139
religated products, **317:**137–139
T7 RNA polymerase, transcript yield
increase, **317:**135–136
cleavage reaction, **317:**135
disadvantages, **317:**134
materials, **317:**134–135, 139
rationale, **317:**133
vesicular stomatitis virus translation arrest, antisense assay, **313:**200, 203
Ribonuclease L
dimerization, **313:**522
2',5'-linked oligoadenylates
activation of enzyme, **313:**522
antisense targeting principle, **313:**523
chimeric oligodeoxynucleotide, activation assay against generic substrate
chimeric oligodeoxynucleotide sequences, **313:**530
data analysis, **313:**529
incubation conditions, **313:**529
kinetics, **313:**528
sequence of substrate, **313:**528
substrates, overview, **313:**527
synthesis of substrate, **313:**528–529
protein kinase R targeting
controls, **313:**533

dose–response, **313:**532–533
incubation conditions for inhibition
assay, **313:**531–532
length of antisense sequences,
313:533
messenger RNA preparation,
313:531
synthetases, **313:**522–523
recombinant human enzyme
expression in baculovirus–insect cell
system, **313:**524
purification, **313:**524, 527
Ribonuclease L inhibitor subfamily, ATP-
binding cassette transporters in *Saccharomyces cerevisiae*, **292:**152–153
Ribonuclease P
advantages in RNA inhibition, **313:**456
catalysis, **313:**403–404
circular dichroism, magnesium-induced
folding transition measurements,
317:395–396, 398
circular permutation analysis
pre-transfer RNA interactions, **317:**327
quantitative analysis of cleavage,
317:327–328, 330
components, **318:**238–239
external guide sequence
cleavage assay of external guide sequence–RNA complex, **313:**449,
451
cloning into expression vectors
3/4 sequences, **313:**451–452
stem sequences, **313:**451
definition, **313:**444
design for antisense inhibition
3/4 sequences, **313:**447, 449
stem sequences, **313:**447
expression in *Escherichia coli* of stem
sequences
cell growth, **313:**452–453
plasmids, **313:**452
ribonuclease P-dependent inhibition,
313:453–454
strain, **313:**452
survival studies, **313:**453–455
expression in tissue culture,
313:455–456
external guide sequence
design from *in vitro* selection,
318:249–250

function, **318:**137, 238–239
structure, **313:**443–444
structure elucidation with photoaffinity
   cross-linking arylazides
   circularly permutated transcription tem-
     plates and RNAs, **318:**142
   cross-linking reaction, **318:**143–144
   handling and stability of photoagents,
     318:141–142
   photoagent incorporation
     3'-end attachment, **318:**140–141
     5'-end attachment, **318:**138, 140
   photochemistry, **318:**138
   phylogenetic–âcomparative data sets,
     318:145
   polyacrylamide gel electrophoresis anal-
     ysis, **318:**145
   three-dimensional modeling, **318:**145,
     147
substrates, *in vitro* selection
   buffers, **318:**250
   cleavage reaction in selection,
     318:245–246
   DNA pool
     amplification of double-stranded tem-
       plates, **318:**241, 243–244
     design, **318:**242
     reverse transcription, **318:**241, 243
   materials, **318:**240–242
   polyacrylamide gel electrophoresis anal-
     ysis, **318:**246–247
   primers, **318:**240–241
   reverse transcription–polymerase chain
     reaction, **318:**247
   selection progress assessment, **318:**249
   transcription, *in vitro*, **318:**241–242,
     244–245, 247
substrate specificity, **313:**443–444, **318:**239
target site selection for RNA inactivation, **313:**446–447
Ribonuclease protection assay
cardiovascular ion channel transcript detection, **293:**80, 83–84
light-activated gene transcription in
   Arabidopsis thaliana
   digestion reaction, **297:**188
   gel electrophoresis analysis, **297:**189
   hybridization reaction, **297:**188
   overview, **297:**186
   probe synthesis, **297:**188

reagents, **297:**189
   templates for RNA probe synthesis,
     297:186–187
Ribonuclease S system, peptide interactions, **295:**248–249, 251, 310–311
Ribonuclease T1
   inosine 3'-phosphodiester bond, specific
     cleavage of messenger RNA
   cellular RNA purification, **318:**50–51
   control RNA synthesis, **318:**49–50
   digestion reaction, **318:**54–55
   glyoxal reaction, **318:**54
   optimization, **318:**53–54
   oxidation of 3'-hydroxyl groups,
     318:51–52
   postcleavage processing
     cyclic phosphate removal, **318:**55
     glyoxal adduct removal, **318:**55–56
   principle, **318:**52–53
   ladder analysis of modified RNA, **318:**14,
     37, 39
Ribonucleic acid, *see* RNA
Ribonucleoprotein
antisense oligodeoxynucleotide studies
   controls, **313:**459
   design of oligodeoxynucleotides,
     313:458–459
   electrophoretic mobility shift assay for
     ribonucleoprotein formation
     binding conditions, **313:**462–463
     extract preparation, **313:**461
     neuron culture, **313:**460–461
     overview, **313:**459–460
     polyacrylamide gel electrophoresis,
       313:463
     radiolabeling of probe, **313:**461–462
     reagents, **313:**460
   fluorescence *in situ* hybridization assay
     for ribonucleoprotein formation
     and sorting
     antisense oligodeoxynucleotide incu-
       bation, **313:**465
     confocal microscopy, **313:**466
     fixation, **313:**465
     hybridization, **313:**466
     immunocytochemistry, **313:**466
     overview, **313:**464
     probe labeling, **313:**465
     reagents, **313:**464
   sorting of RNA, **313:**457

Ribonucleotide triphosphate reductase, isotopic labeling of substrates for kinetic isotope effect studies, **308**:309–310

Ribosomal DNA, hybridization with RNA probes, *see* Fluorescence *in situ* hybridization

Ribosomal RNA
  aminoglycoside complexes, structure analysis with nuclear magnetic resonance
  affinity parameters, determination, **317**:250–251
    aminoglycoside substituent effects, **317**:259–260
    biological prediction testing, **317**:258
    buffers for data collection, **317**:244
    chemical exchange, slow, intermediate, and fast, **317**:251–253
    chemical modification of RNA, **317**:242–243
    design of RNA, **317**:241
    drug design application, **317**:260
    labeling and synthesis of RNA, **317**:244–245
    mapping of antibiotic-binding site on RNA chemical shift, **317**:253–254
    mechanism of antibiotic action, **317**:240
    model comparison with ribosome, **317**:243
    mutant oligonucleotides, **317**:243–244, 258–259
    nuclear Overhauser enhancement spectroscopy, **317**:254–255
    prospects, **317**:261
    resonance assignments
      antibiotic, **317**:246–248
      exchangeable protons, **317**:245–246
      RNA protons, **317**:245
    structure
      calculations, **317**:255–256
      comparison with chemical modification data, **317**:256–257
      comparison with mutational data, **317**:257–258
    titration of antibiotic into RNA, **317**:249–250
  dimethyl sulfate probing *in vivo* of 18S rRNA, 489–490
  direct detection of modification or cleavage of 5S rRNA
    chemical modification

    diethyl pyrocarbonate, **318**:13
    dimethyl sulfate, **318**:12
    ethylnitrosourea, **318**:13
    hydrazine cleavage, **318**:13
    lead(II) acetate, **318**:12
    sodium borohydride cleavage, **318**:13
    denaturation and renaturtration, **318**:12
    end labeling, **318**:11–12
    polyacrylamide gel electrophoresis fractionation, **318**:13–14
  iron(II)–EDTA tethered complex for structure probing
    anticodon stem–loop analogs, **318**:186–188
    environment surrounding 5′-end of phenylalanine transfer RNA, **318**:185
    16S rRNA, **318**:183–184
  L11–23S rRNA interaction, *in vitro* selection from random RNA fragments
    advantages, **318**:252–253, 268
    antibiotic interference, **318**:255
    binding reaction, **318**:262
    binding sites, **318**:253–255
    cloning of fragments, **318**:265–266
    conservation of interaction between species, **318**:256
    cycles of selection, **318**:260–261
    minimal binding site identification, **318**:266–267
    nitrocellulose filter binding of complexes, **318**:261–262
    overview, **318**:253, 260
    polymerase chain reaction, **318**:260–261
    protein purification, **318**:256–257
    RNA pool
      construction, **318**:257–259
      diversity, **318**:259–260
    RNA:protein ratio, **318**:264
    selection progress, monitoring
      band-shift assay on native gels, **318**:262–264
      binding assays, **318**:262–263
    transcription, **318**:261
  loop E structure elucidation of 5S RNA, **318**:18–20
  primer extension analysis of 5S RNA modifications or cleavage
    chemical modification reactions, **318**:16
    gel electrophoresis, **318**:17
    primer annealing and extension, **318**:17

primer selection, **318:**14, 16
Ribosome
  assembly, **318:**251–252
  database of 30S structure, *see*
    RIBOWEB
  green fluorescent protein, ribosome syn-
    thesis reporter assay, **310:**42
  iron(II)–EDTA tethered hydroxyl radical
    probing
    advantages, **318:**461–462
    buffers, **318:**463–464
    equipment and supplies, **318:**464
    footprinting S20 on 16S ribosomal
      RNA, **318:**468
    hydroxyl radical probing conditions,
      318:470
    iron(II) derivatization of protein,
      318:467
    linker–iron(II) complex preparation,
      318:466–467
    nuclease contamination prevention,
      318:462
    overview, **318:**461
    primer extension analysis
      extension reaction, **318:**472–473
      hybridization, **318:**472
      interpretation of data, **318:**473–475
      overview, **318:**471
      polyacrylamide gel electrophoresis
        analysis, **318:**473
      reagents and buffers, **318:**471–472
    purification of iron(II)–S20–16S ribo-
      somal RNA complex, **318:**468–469
    reagents, **318:**462–463
    reconstitution of 30S subunits con-
      taining iron(II)-derivatized pro-
      tein, **318:**469–470
    RNA isolation following probing,
      318:470–471
    S20 mutation for cysteine substitution
      accessibility assay, **318:**465–466
      site selection, **318:**464–465
    S20–16S ribosomal RNA complex for-
      mation, **318:**467–468
  messenger RNA interaction assay, transla-
    tion-defective chloroplast mutants in
    maize
    buffer preparation, **297:**55
    extraction, **297:**53–54
    Northern blot analysis, **297:**54

  puromycin control, **297:**54–55
  sucrose gradient centrifugation, **297:**54
  number as bacteria growth measurement,
    310:42
  polypeptide complex isolation from cell-
    free translation systems
    size-exclusion chromatography, **290:**6–7
    sucrose density gradient centrifugation,
      290:7
  protein folding activity in *Escherichia
    coli*, **290:**18
  recombinant 30S ribosomal subunit prepa-
    ration
    cell lysis, **318:**450, 452
    fast protein liquid chromatography,
      318:452–454
    inclusion body solubilization,
      318:452–453
    nuclease contamination detection and
      avoidance, **318:**455
    overexpression in *Escherichia coli*,
      318:449–450
    physical properties of subunits, **318:**453
    S2 purification, **318:**454–455
  reconstitution of 30S ribosomal subunits
    using complete set of recombinant
    proteins
    applications, **318:**460
    buffers, **318:**447–449
    characterization of reconstituted sub-
      units, **318:**459–460
    equipment and supplies, **318:**449
    grouping of subunits for reconstitution,
      318:456–457
    microconcentration, **318:**459
    overview, **318:**446
    reagents, **318:**447
    reconstitution reactions, **318:**457–458
    sucrose density gradient purification,
      318:458–459
  structure
    Escherichia coli, **318:**251
    overview of elucidation, **318:**118–119
  three-dimensional cryoelectron mi-
    croscopy
    functional complexes, preparation
      antibiotic fixation, **317:**297–298
      A-site binding, PRE complex con-
        struction, **317:**300–301
      buffers, **317:**299–300

filter binding assay, **317:**301–302
gel filtration, **317:**303
Pi complex, **317:**300
PRE and POST states, overview,
    317:297–299
puromycin reaction, **317:**302
spun column isolation, **317:**303–304
stoichiometry determinations,
    317:304–305
sucrose cushion centrifugation,
    317:302–303
translocation and POST complex
    construction, **317:**301
hurdles in obtaining higher resolution,
    317:291
image processing
    contrast transfer function estimation,
        317:285
    micrograph selection and scanning,
        317:285
    particle picking and optical density
        normalization, **317:**286–287
    reconstruction of initial three-dimen-
        sional density map, **317:**287–288
    reconstruction strategies,
        317:283–284
    refinement, **317:**288–289
    visualization and interpretation of
        three-dimensional density map,
        317:289–291
isolation and purification of 70S ribo-
    somes
    bacteria growth, **317:**294
    crude 70S isolation, **317:**295
    materials and buffers, **317:**294
    overview and comparison of ap-
        proaches, **317:**277–279
    reassociated ribosome, advantages
        and preparation, **317:**293,
        296–297
    subunit isolation, **317:**295–296
    tight-coupled 70S isolation, **317:**295
low-dose electron microscopy, **317:**281
optical diffraction, **317:**282
overview, **317:**276–277
plunge-freezing, **317:**281
resolution, **317:**277, 291, 292–293
sample preparation, ligand complexes,
    317:279–281
specimen grid preparation, **317:**281

transfer RNA binding sites
    buffer composition, effects on bind-
        ing localization, **317:**306–309
    charging state of transfer RNA, ef-
        fects on binding localization,
        317:306–308
    E site role in protein synthesis,
        317:308–309
    localization, **317:**293, 305–306
    significance of additional sites,
        317:309
three-dimensional modeling using photo-
    labile derivatives of oligonucleotides,
    318:133–134
transfer RNA binding in *Escherichia coli*,
    radioactivity assays
    AcPhe-transfer RNA binding studies
        AcPhe-transfer RNA preparation,
            317:264–265
        artifacts distorting saturation curves
            of binding to poly(U)-pro-
            grammed ribosomes, **317:**265,
            267
        Phe-transfer RNA contamination, re-
            moval by enzymatic deacylation,
            317:269–270, 272–275
        Phe-transfer RNA contamination,
            stimulation of anomalous bind-
            ing, **317:**267, 269, 275
        radioactive component, vetting of
            specific activity, **317:**272–275
        ribosome preparations, **317:**275–276
        S100 enzymes free of RNA, prepara-
            tion, **317:**263–264
        site-specific binding to 70S ribo-
            somes, **317:**265
    exclusion principle for binding,
        317:262, 276
    peptide elongation reaction,
        317:261–262
    sites for binding, **317:**261
ultraviolet absorption and quantification,
    317:304
RIBOWEB
    access, **317:**483
    advantages of Web navigation,
        317:488–489
    extraction and analysis of data, **317:**483–
        484, 488
    ontology hierarchy

Data
Antibiotic-Resistance-Data, **317:**479–480, 482
attributes, **317:**476, 484–485
Cross-Linking-Data, **317:**485
Electron-Microscopic-Data, **317:**479
Hydroxy-Radical-Cleavage-Data, **317:**477–479
overview, **317:**475–476
instances versus classes, **317:**472–473
Physical-Thing
Molecular-ensemble, **317:**475
Molecular-part, **317:**475
Molecule, **317:**473, 475
Organism, **317:**473
Reference-Information, **317:**473, 489
software tools for design, **317:**482–483
prospects, **317:**490–491
rationale for development, **317:**470–472, 484
Ribozyme, *see also* Hairpin ribozyme; Hammerhead ribozyme; Leadzyme; Ribonuclease P; *Tetrahymena* group I intron
activity assays
small target cleavage, *in vitro*
cloning of targets, **313:**494–495
direct template production in target synthesis, **313:**495–496
group I intron ribozymes, **313:**497–498
hammerhead ribozymes, **313:**496–497
total RNA substrate
incubation conditions, **313:**498
reverse transcription-polymerase chain reaction analysis of products, **313:**499
ribonuclease protection assay, **313:**499–500
autosomal dominant retinitis pigmentosa, targeting of mutant protein transcripts in rat
challenges, **316:**764, 775–776
cleavage reactions, incubation conditions, **316:**771–772
cloning of ribozymes and targets, **316:**767–768
expression level requirements for ribozymes, **316:**776
materials, **316:**767

multiple turnover kinetic analysis of cleavage, **316:**772, 776
normal allele requirement for effective therapy, **316:**775
preparative gel electrophoresis of RNA, **316:**769–770
radiolabeling of RNA oligonucleotides, **316:**769
reverse transcription–polymerase chain reaction of products, **316:**774–775
RNA extraction from retina, **316:**770
selectivity of cleavage, **316:**775–776
single turnover kinetic analysis of cleavage, **316:**772–774
specific radioactivity determination for labeled RNAs, **316:**770–771
target site identification, **316:**765–766
transcription, *in vitro*, **316:**768–769
catalytic motifs
group I introns, **313:**402–403, 486–488
hairpin, **313:**404
hammerhead, **313:**404, 486–487
hepatitis D virus ribozyme, **313:**404
overview, **313:**402, 485–486
ribonuclease P, **313:**403–404
chemical synthesis, **313:**493–494
classification, **316:**761–762
delivery
constructs, **313:**500–501
transfection
lipofection, **313:**503
overview, **313:**417–418, 500
transferrin–polylysine conjugates, **313:**501–503
viral vectors
adenovirus, **313:**503–506
overview, **313:**415–417, 500
expression cassettes
auxiliary sequences, **313:**413–414
insulator sequences, **313:**412–413
promoter selection, **313:**409–410, 412
stability of expression, **313:**409
folding analysis, **295:**379, , 402
gene identification applications, **306:**225, 259
gene inactivation
advantages and limitations, **313:**401–402, 419–420
primer extension analysis of cleavage site, **313:**419

ribonuclease protection assay,
   313:418–419
strategies, **313**:442–443
target site accessibility assays
   cell extract assay, overview,
      313:405–406
   endogenous ribonuclease H diges-
      tion and analysis, **313**:406–408
   extract preparation, **313**:406
   library screening, **313**:405
   mapping techniques, **313**:446–447
   ribozyme testing, **313**:408–409
   in vivo assay, **313**:408
gene therapy
   advantages over antisense targeting,
      316:762
   applications and ribozyme develop-
      ment, **306**:207–208, 225, 245–247
   potential, **316**:762
   target specificity, **316**:763–764
secondary structures of hammerhead and
   hairpin ribozymes, **316**:763
small ribozyme features, **317**:149
structure and activity, **306**:246–247
transcription, *in vitro*
   group I introns, **313**:491
   hammerhead ribozyme, **313**:489, 491
   template
      cloning, **313**:489, 491
      direct production, **313**:491, 493
types, **306**:245–246
X-ray crystallography of small ribozymes
   chemical synthesis
      automated synthesis, **317**:151–152
      desalting, **317**:153
      overview, **317**:151
      reversed-phase high-performance liq-
         uid chromatography,
         317:152–153
   crystallization
      cations, **317**:155–158, 182
      precipitants, **317**:154–155, 182
      screening, **317**:154, 182–183
      crystal packing promotion, **317**:150–
         151, 181–182
   diffraction data acquisition
      cryoprotection, **317**:159–160, 184
      crystal requirements, **317**:158
      radiation sensitivity, **317**:158–159
      synthetic mother liquors, **317**:159

heavy atom derivatization, **317**:160–
      162, 183–184
   initial electron density maps, **317**:164–
      165, 184–185
   intermolecular contacts in crystal lat-
      tice, **317**:181–182
   Patterson methods for heavy atom posi-
      tions, **317**:162–164
   refinement, **317**:165–168, 185–186
   time-resolved crystallography
      catalysis initiation in crystals,
         317:187–188
      difference Fourier analysis,
         317:193–194
      extent of cleavage, analysis in crys-
         tals, **317**:196–198
      kinetic assays for cleavage in crys-
         tals, **317**:194–196
      Laue diffraction observation of tran-
         sient intermediates, **317**:192–193
      overview, **317**:186–187
      termination of catalysis in crystals,
         317:196
   trapping of reaction intermediates
      cryocrystallographic trapping,
         317:191–192
      kinetic bottleneck modifications,
         317:188–191
Ribulose-bisphosphate carboxylase, *see*
   Rubisco
Ricin, irreversible step and isotope effect,
   308:367–368
Rim protein, *see* ABCR
RLI subfamily, *see* Ribonuclease L inhibitor subfamily
RMCE, *see* Recombinase-mediated cassette
   exchange
RNA, *see also* Ribosomal RNA; Transfer
   RNA
   antibiotic complex, *see* Antibiotic–RNA
      complex
   folding, *see* Folding, RNA
   isolation, *see* RNA isolation
   isotopic labeling for kinetic isotope effect
      studies, **308**:316–318
   ligation, *see* RNA ligation
   mapping, *see* RNA mapping
   microinjection in *Xenopus laevis* oocyte,
      296:30, 46
   oligonucleotide, *see* RNA oligonucleotide

peptide interactions, *see* RNA–peptide interaction
protein cross-linking, *see* RNA–protein cross-linking
RNA cross-linking, *see* RNA–RNA cross-linking
size fractionation
  agarose gel electrophoresis
    GenePrep electrophoresis device, **296:**31, 33
    minigel electrophoresis, **296:**35–36
  sucrose density gradient centrifugation, **296:**33, 35
splicing, *see* RNA splicing
structure, *see* RNA structure
RNA-arbitrarily primed polymerase chain reaction
  advantages and limitations, **303:**311, 496
  applications, **303:**310
  cloning of amplified products, **303:**319
  confirmation of differential expression
    high-stringency reverse transcription–polymerase chain reaction, **303:**320–322
    low-stringency reverse transcription–polymerase chain reaction, **303:**319–321
    quantitation of polymerase chain reaction products, **303:**322
  differentially-amplified products, isolation and purification
    amplification of gel-eluted DNA, **303:**318–319
    Cot effect minimization, **303:**317, 324
    single-strand conformation polymorphism analysis, **303:**317–318
  equipment and reagents, **303:**312–313
  primer and polymerase selection, **303:**316, 323–324
  principle, **303:**311–312
  reduced complexity probe generation, **303:**381, 383–384, 390
  RNA fingerprinting, **303:**316–317
  RNA isolation, **303:**313, 323
  sequencing of amplified products, **303:**319
  Southern blot analysis of amplified products, **303:**319
  two arbitrary primers, rationale and advantages of use, **303:**322–323
RNA challenge phage system
  controls, **318:**347–348
  data analysis, **318:**347
  lysogenization frequencies, **318:**346–347
  materials
    media and reagents, **318:**336, 338
    plasmids, **318:**338–339
    primers, **318:**339–340
    strains and phages, **318:**337
  mutagenesis of RNA target site and phage isolation, **318:**348–349
  overview
    artificial RNA operator sites, **318:**335
    bacteriophage P22 life cycle, **318:**332–334
    phage modifications for system, **318:**334–335
  phage construction
    arc(Am) phage purification, **318:**343–345
    infection of cells, **318:**342
    large-scale production, **318:**345
    O$_{mnt}$-allele phage identification, **318:**342–343
    titering, **318:**345–346
    transformation of *Salmonella* recipient strains, **318:**341–342
  polymerase chain reaction/restriction fragment length polymorphism analysis, **318:**340–341
  protein lysate preparation from *Salmonella*, **318:**349–350
  radiolabeling of oligonucleotides, **318:**341
  Salmonella transformation and infection, **318:**346
  sequencing of DNA, **318:**341
  suppressor protein isolation, **318:**349
RNA folding, *see* Folding, RNA
RNA isolation
  differential display
    amplification of limited samples, **303:**240
    commercial kits for preparation, **303:**238
    DNase I treatment, **303:**239
    extraction, **303:**239
    fluorescent differential display, **303:**300, 305
    guanidine isothiocyanate cushion centrifugation, **303:**237–238
    ribonuclease inhibition, **303:**240

DNA microarray probe preparation,
  303:195
exon amplification
  large-insert trapping with cosmids,
    303:106–107
  vectors following transfection,
    303:94–95
low numbers of cells as starting material
  genomic DNA removal, **303:**333–334
  messenger RNA purification, **303:**334
  total RNA preparation, **303:**333
neutrophils, **303:**278–279, 291
restriction endonucleolytic analysis of dif-
    ferentially expressed sequences,
    303:262–264, 278–279, 291
RNA-arbitrarily primed polymerase
    chain reaction, **303:**313, 323
targeted display, **303:**398
Xenopus laevis oocyte expression cloning
  poly(A)⁺ RNA selection, **296:**25–26
  quantitation by ultraviolet spectros-
    copy, **296:**24–25
  total RNA, **296:**23–25
RNA ligation
  chemical ligation, **317:**110
  T4 DNA ligase joining of hybrid du-
    plexes
    circular RNA generation, **317:**113, 119
    concentration optimization
      DNA template, **317:**121
      enzyme, **317:**121
      RNA, **317:**121
    DNase treatment, **317:**123
    duplex specificity and mutation effects,
      317:122–123
    generic conditions for ligation, **317:**113,
      118–119
    lengths of substrates and templates,
      317:122
    multipart ligations, **317:**122
    principle and overview of splinted liga-
      tion, **317:**112–115
    radioactive labeling and ligation
      different tubes, **317:**113–115,
        120–121
      single tube, **317:**113–114, 120
    reagents, **317:**115, 117
    RNA substrate preparation,
      317:117–118
    side reactions, **317:**123

T4 RNA ligase, **317:**111–112
RNA mapping, antisense sequence se-
    lection
antisense oligonucleotide design from se-
    quencing data
  multiple oligonucleotides per accessible
    site, **314:**181–182
  primer extension reaction interretation,
    314:179, 181
libraries
  2′-methoxyribonucleotide residue incor-
    poration, **314:**170–171
  oligonucleotide length, **314:**170
  semirandom libraries, **314:**171
  synthesis and purification, **314:**172, 174
principle, **314:**169
rationale, **314:**169
sequencing of accessible sites
  nucleotide mixture preparation,
    314:177
  overview, **314:**177
  primer extension reactions,
    314:177–179
  sequencing reactions, **314:**178–179
transcript
  5′-end labeling, **314:**175
  mapping, **314:**176
  purification, **314:**174–175
  quantificxation, **314:**175
  structure comparison to *in vivo* tran-
    scripts, **314:**182–183
transcription, *in vitro*, **314:**174
validation, **314:**182
RNA oligonucleotide
  fluorescence studies, *see* Fluorescence
    spectroscopy
  heavy atom derivatization, *see* Heavy
    atom derivatives, RNA
  ligation, *see* RNA ligation
  phosphoramidite-modified oligoribo-
    nucleotide synthesis, purification,
    and characterization, **317:**51–53
  phosphorothioates, *see* Phosphorothioate
    oligoribonucleotide
  synthesis via 5′-silyl-2′-orthoester ap-
    proach
    N-acyl-5′-*O*-3′-*O*-
      tetraisopropyldisiloxanyl
      protecting group removal, **317:**6–7
    advantages, **317:**3–4, 17–18

anion-exchange high-performance liquid chromatography analysis,
317:15–16
2′-*O*-bis(2-acetoxyethoxy)methyl nucleoside synthesis, **317**:6
nucleotide addition cycles
overview, **317**:9–10
reactions and washes, **317**:12
reagent preparation, **317**:10–11
synthesizer instrumentation,
317:11–12
3′-*O*-phosphitylation, **317**:7–8
postsynthesis processing
cleavage and basic deprotection,
317:14–15
2′-*O*-deprotection, **317**:16–17
overview, **317**:13
phosphate deprotection, **317**:13–14
protecting groups, **317**:5
5′-*O*-silylation, **317**:7
solid supports, derivatization, **317**:8–9
solubility of products, **317**:4
yield, **317**:3, 17
synthesis with T7 RNA polymerase, *see*
T7 RNA polymerase
RNA–peptide interaction
absorbance melting curves
principles, **318**:435, 437
thermodynamic analysis, **318**:437–438
application to protein binding, **318**:423
chemical modification interference
overview, **318**:431–432
peptide affinity chromatography
coupling peptide to solid support,
318:432–434
fractionation of modified RNAs,
318:434
RNA modification, **318**:432
circular dichroism characterization of peptide secondary structure,
318:434–435
footprinting, **318**:431
gel shift assay for apparent dissociation
constant determination
binding conditions, **318**:428
competitor RNA, **318**:427
concentration of ligands, **318**:428
electrophoresis, **318**:430
limitations, **318**:427
materials, **318**:429

purification of RNA, **318**:429–430
quantificaion of RNA, **318**:430
transcription, *in vitro*, **318**:429
peptides
design
Rev, **318**:424–426
Tat, **318**:423–426
purification and handling, **318**:426
quantification
absorption, **318**:426
native polyacrylamide gel electrophoresis, **318**:426–427
secondary structure, **318**:425–426
RNA polymerase I, promoters in effector
RNA expression, **306**:248–249
RNA polymerase II, promoters in effector
RNA expression
messenger RNA, **306**:249–251
U1, 251–252
RNA polymerase III, promoters in effector
RNA expression
overview of types, **306**:252–253
type 2 promoters, **306**:253, 255, 257
type 3 promoters, **306**:257
RNA polymerase III-based two-hybrid
system
advantages over RNA polymerase II-
based system, **303**:413–414
5-fluoroorotic acid resistance test as reporter gene assay
cell culture, **303**:420–421
principle, **303**:420
principle, **303**:411–412
quantification of UAS$_G$–*SNR6* transcripts
Northern blot analysis, **303**:419–420
principle, **303**:417–418
RNA extraction, **303**:418–419
sensitivity, **303**:414
TFIIIC functions and utilization, **303**:412
three-hybrid system development,
303:422
transformation of yeast, **303**:416–417
troubleshooting, **303**:421–422
vectors, **303**:415
RNA–protein cross-linking, *see also* 5-Halo-
pyrimidines; Photolabile derivatives of
oligonucleotides; 4-Thiouridine
applications, **318**:22, 31–32
azide-substituted bases, **318**:88–89, 137
mass spectrometry analysis of cross-link

sites, *see* Matrix-assisted laser
      desorption/ionization mass spec-
      trometry
   methylene blue cross-linking
      advantages and disadvantages, **318:**32
      binding reaction, **318:**25–26
      cross-linking reaction, **318:**26
      dye titration, **318:**26–27
      gel electrophoresis, **318:**26
      immunoprecipitatyion, **318:**30
      labeled nucleotide selection, **318:**29
      light source and illumination duration,
         318:28
      materials and reagents, **318:**25
      mechanism and lack of base specificity,
         318:30–31
      overview, **318:**23–24
      purified complexes, cross-linking,
         318:29–30
      ribonuclease selection, **318:**29
      side reaction suppression, **318:**27–28
      thionine dye comparison, **318:**30
   overview of experiment, **318:**22
   photolabile derivatives of oligonucleo-
      tides, identification of labeled pro-
      teins, **318:**129–131, 135–136
   ultraviolet cross-linking of double-
      stranded RNA, **318:**22–23
RNA–protein fusion
   complementary DNA/messenger RNA–
      protein fusion, **318:**289–290
   fusion reaction
      cis versus *trans* fusion formation,
         318:285–287
      efficiency
         assay, **318:**283–285
         factors affecting, **318:**281
      ion effects, **318:**281–283
      kinetics, **318:**281
      modeling, **318:**287
      optimization of linker length and se-
         quence, **318:**285
   isolation
      oligo(dT) affinity chromatography,
         318:288
      overview, **318:**287–288
      thiopropyl-Sepharose chromatography,
         318:288–289
   ligation of messenger RNA and 3′-puro-
      mycin oligonucleotides, **318:**278–279

overview of fusion, **318:**269–270
   3′-puromycin oligonucleotide synthesis
      coupling to controlled pore glass,
         318:277
      linker sequences, **318:**277–278
      overview, **318:**274, 276
      phosphorylation, **318:**278
      N-trifluoroacetyl-5′-DMT-puromycin
         synthesis, **318:**276
      N-trifluoroacetyl-5′-DMT-2′-
         succinylpuromycin synthesis,
         318:276–277
      N-trifluoroacetylpuromycin synthesis,
         318:276
   translation, *in vitro*
      bacterial system, **318:**279
      reticulocyte lysate system, **318:**280–281
   in vitro selection
      advantages, **318:**268–269, 292–293
      DNA pool construction, **318:**271–273
      isolation of functional fusions with im-
         mobilized selection motif
         arginine-rich motif peptides, purifica-
            tion and fusions, **318:**290–292
         immunoprecipitation of myc peptide
            fusions, **318:**290
      overview, **318:**270–271
      polymerase chain reaction to generate
         enriched DNA pool, **318:**292
      transcription, **318:**273
RNA–protein interaction screening, *see*
      Antitermination, RNA-binding library
      screening; Frameshifting assay, RNA–
      protein interactions; Ribosomal RNA;
      RNA challenge phage system; Small
      nuclear ribonucleoprotein; Tat-fusion
      system; Translational repression assay
      procedure; Translational repression,
      RNA-binding protein screening; Yeast
      three-hybrid system
RNA–RNA cross-linking, *see also* Site-spe-
      cific psoralen; 4-Thiouridine
   azide-substituted bases, **318:**88–89, 137
   hammerhead ribozyme, disulfide cross-
      linking
      rationale, **318:**166–167
      alkaline hydrolysis analysis,
         318:172–173
      2′-amino nucleotide modification with
         isothiocyanates, **318:**167, 169

application to other ribozymes,
   318:174–175
catalytic activity of cross-linked ribozymes, **318:**173–174
cross-linking reaction, **318:**169
polyacrylamide gel electrophoresis analysis, **318:**169–173
radiolabeling of RNA, **318:**168
RNA synthesis, **318:**168
overview of approaches, **318:**104–105
ribonuclease P structure elucidation with
   photoaffinity cross-linking arylazides
circularly permutated transcription templates and RNAs, **318:**142
cross-linking reaction, **318:**143–144
handling and stability of photoagents,
   318:141–142
photoagent incorporation
   3′-end attachment, **318:**140–141
   5′-end attachment, **318:**138, 140
photochemistry, **318:**138
phylogenetic–âcomparative data sets,
   318:145
polyacrylamide gel electrophoresis analysis, **318:**145
three-dimensional modeling, **318:**145,
   147
RNase, *see* Ribonuclease
RNase H, *see* Ribonuclease H
RNase L, *see* Ribonuclease L
RNase P, *see* Ribonuclease P
RNase T1, *see* Ribonuclease T1
RNA splicing, 4-thiouridine site-specific
   cross-linking during splicing
antisense targeting, **313:**8–9, 506–510
P120 RNA, **318:**85
RNA–protein cross-link analysis,
   318:86–87
RNA–RNA cross-link analysis,
   318:86–87
thalassemia, *see* β-globin
ultraviolet irradiation, **318:**86
RNA structure, *see also* Branched RNA;
   Circular dichroism, RNA; Circular permutation analysis, RNA; Comparative
   sequence analysis; Folding, RNA; Kinetic oligonucleotide hybridization; Nuclear magnetic resonance, RNA; Phosphorothioate interference analysis;
   Quantitative structure–activity relation-

ship; Ribosomal RNA; RIBOWEB;
   Transient electric birefringence; X-ray
   crystallography, RNA
chemical probes, *see also specific probes*
applications of nucleotide modifications, **318:**7–8, 10, 17–21
equipment and reagents, **318:**11
identification of cleavage or modification sites, **318:**3, 11–14, 16–17
probe types
   base-specific probes, **318:**5–6
   chemical nucleases, **318:**6–7
   table, **318:**4
reaction conditions, **318:**9–10
cross-linking studies, *see* RNA–protein
   cross-linking; RNA–RNA cross-
   linking
ribonucleases for probing, **318:**9
Robbins device
applications, **310:**207–208
biofilm establishment, **310:**209–210
chemostat, **310:**256
conductance assays, biofilm culture,
   310:354–355
confocal microscopy, **310:**257–258
contamination, **310:**256
design, **310:**298
electron microscopy, **310:**257
flow control, **310:**263
leakage, **310:**256
media selection, **310:**254
modified device
   antimicrobial paint evaluation, **310:**639
   artificial urine model, **310:**562–563
   design, **310:**208, 255, 638–639
physiological studies of biofilms,
   310:258–259
Pseudomonas aeruginosa biofilm studies
antibiotic resistance studies
   biofilm growth, **310:**213
   clinical importance, **310:**212–213
   enumeration of cells, **310:**213
   β-lactamase induction and quantifi-
     cation, **310:**213–216
   minimum inhibitory concentration
     determination, **310:**213–214, 216
cystic fibrosis studies
   complement activation, **310:**211–212
   leukocyte response to biofilm,
     310:210–211

pathogenesis of infection, **310**:210
reproducibility, **310**:255–256
sessile sampler, **310**:208–209
Rod outer segment
ATPase activity, **315**:872–873
bleaching, **315**:236
cytoplasmic calcium, *see* Calcium flux
electron crystallography of rhodopsin,
    rod outer segment purification
  bovine, **315**:96
  frog, **315**:95
light activation avoidance in guanylate cy-
    clase assays, **315**:691–692, 710–711,
    718
membrane fusion, *see also* Peripherin/rds
  aggregation–adhesion analysis,
      316:84–85
  assay of peripherin/rds fragments
    caveats, **316**:80
    contents mixing assay, **316**:78–79
    materials, **316**:78
    peptide effects on fusion, **316**:79–80
    quenching pair assay principle,
        316:76
    vesicle preparation, **316**:78
  assay of target membrane fusions
    controls, **316**:75
    design principles, **316**:66
    disk membrane preparation, **316**:69
    disk rim-specific vesicle preparation,
        316:69
    fluorescence assay, **316**:71–72
    octadecylrhodamine B chloride label-
        ing, **316**:68–69
    overview, **316**:67
    peripherin/rds purification from bo-
        vine retina and recombination
        with vesicles, **316**:69–71
    plasma membrane preparation,
        316:67–68
  bilayer destabilization analysis,
      316:85–86
  characteristics of target model mem-
      brane fusions
    disk membranes, **316**:74
    disk rim-specific vesicles, **316**:74–75
    overview, **316**:73
    peripherin/rds recombined with vesi-
        cles, **316**:74–75
  fusion competency assay of canine kid-

ney cells expressing mutant
    peripherin/rds
  advantages, **316**:83
  cell culture, **316**:80–81
  pitfalls, **316**:83–84
  probe labeling of membranes, **316**:82
  resonance energy transfer-based
      lipid mixing assay, **316**:81–83
  fusion protein mediation, **316**:65
  physiological functions, **316**:65
  steps, **316**:84
phototransduction overview, **316**:188–189
preparation for assays
  cyclic GMP phosphodiesterase, pH
      assay, **315**:612–614
  cyclic GMP phosphodiesterase, single
      rod outer segment assay
    advantages, **315**:751–752
    amphibian rod outer segment prepa-
        ration, **315**:745–746
    calcium-dependent measurements,
        315:747–748, 750
    cone outer segment measurements,
        315:751
    inhibitor measurements, **315**:744
    overview, **315**:743–745
    pipette solutions, **315**:746
    washout minimization, **315**:751
  guanylate cyclase, **315**:695–696, 731
  guanylate cyclase, single rod outer seg-
      ment assay
    advantages, **315**:751–752
    amphibian rod outer segment prepa-
        ration, **315**:745–746
    calcium-dependent measurements,
        315:748, 750
    cone outer segment measurements,
        315:751
    overview, **315**:743, 745
    pipette solutions, **315**:746
    washout minimization, **315**:751
  sodium/calcium-potassium exchanger
    cation dependence of calcium up-
        take, **315**:855–856
    fluo-3 measurements and calibration,
        315:852–855
    preparation, **315**:849–850
    rhodopsin bleaching effects on cal-
        cium measurements, **315**:857
preparation for protein purification

antimicrobial concentration determi-
    nation, **310:**624
continuous treatment, **310:**627–628
overview, **310:**622–623
pulse treatment, **310:**626–627
antimicrobial testing
    oil field biofilms, **310:**622–628
    plating for viable cell counting,
        310:618–619
    sampling, **310:**617–618
    toilet bowl disinfection, **310:**620–622
    treatment protocol, **310:**617
apparatus, **310:**613–617
biofilm growth
    operation, **310:**617
    reagents, **310:**614–615
    repeatability, **310:**619–620
    setup, **310:**615–617
Rototorque
    design, **310:**298
    oral biofilm culture
        generation time calculation, **310:**445
        inoculation, **310:**445–446
        media, **310:**446
        modifications of instrumentation,
            310:444
        pH control, **310:**445
        reporter gene analysis, **310:**457–460
Rpd3, *see* Histone deacetylase
RPE, *see* Retinal pigment epithelium
RPE65, *see* Retinal pigment epithelium p65
RPGR, *see* Retinitis pigmentosa GTPase
    regulator
RRL, *see* Rabbit reticulocyte lysate
rRNA, *see* Ribosomal RNA
rRNA, *see* Ribosomal RNA
RS-20
    caged tyrosine peptide
        calmodulin binding analysis,
            291:90–93
        design of inactive caged peptide,
            291:88–90
        kinetics of photolysis, **291:**87–88
        spectral and photochemical properties,
            291:85–87
        synthesis, **291:**80–85
    structure and function, **291:**79–80
RSS, *see* Reactive sulfur species
RT–PCR, *see* Reverse transcriptase–
    polymerase chain reaction

RTC, *see* Tetrapotassium rubrene-2,3,8,9-
    tetracarboxylate
RTI-55, *see* 2β-Carbomethoxy-3β-(4-iodo-
    phenyl)tropane
RU-486-inducible gene expression
    combining inducible transactivator and re-
        porter into one vector, **306:**288
    GAL4–progesterone receptor–VP16
        fusion
        design for expression system, **306:**283,
            288–290
        liver targeting in transgenic mice,
            306:292, 294
        promoters, **306:**283, 285
        RU-486 control of gene expression,
            306:283–285, 292, 294
    inducible repressor construction,
        306:290–292
    progesterone receptor, truncation and li-
        gand binding, **306:**283, 288
    transient transfection of cell cultures,
        306:286–288
Rubidium efflux assay, ATP-sensitive potas-
    sium channels, **294:**452–453
Rubisco
    antisense RNA mutants
        developmental changes in expression,
            297:217–219
        feedback regulation analysis,
            297:219–220
        growth medium for selection, **297:**212
        photosynthesis rate control studies,
            297:216–217, 219–220
        promoters, **297:**211–212
        pulse-labeling and posttranscriptional
            control, **297:**213
        target gene considerations, **297:**210–211
        target sequence selection, **297:**211
        transcription analysis, **297:**212–214
        Western blotting, **297:**212–213
    bioassay for sphingolipid inhibition in
        Candida albicans, **311:**337–339
    DNA template availability in expression,
        297:218–219
    expression during leaf development,
        297:214–215, 217–219
    expression levels with chaperonin 60,
        290:153–155, 160–161
    inositol phosphoceramide synthase inhibi-
        tion, **311:**335, 348

mammalian mitochondrial chaperonin 60
  binding, **290:**204, 214
preparation
  fermentation, **311:**345–346
  isolation, **311:**346
reconstitution in chaperonin activity
  assays, **290:**184–185, 227–230
Rustmicin
  structure, **311:**337
  subunits and assembly, **297:**210
Ruthenium red, mast cell, activation assay
  advantages and limitations, **301:**26–27
  principle, **301:**23–24
  quantitative analysis, **301:**24, 26
Ruthenium red
  glycocalyx preservation for electron mi-
    croscopy
    glutaraldehyde–lysine fixation, **310:**196,
      203
    paraformaldehyde–glutaraldehyde–
      lysine fixation, **310:**198, 203
  structure, **310:**195
RVD, *see* Regulatory volume decrease

S

s26
  localization in retina, **316:**14
  physical properties, **316:**13
  purification
    anion-exchange chromatography,
      316:11–12
    frog retina soluble protein preparation,
      316:9–10
    hydrophobic affinity chromatography,
      316:10–11
    recombinant protein from *Escherichia
      coli,* **316:**13
S100, *see* Calcium-dependent guanylate cyclase activator protein
S100A11
  antibody production, **316:**109–110
  gel overlay assay of binding proteins,
    316:116–117
  immunohistochemistry, **316:**110
Salicylic acid, caged
  calcium response measurement with fluo-
    rescent probes, **291:**478, 481–482
  injection into plant cells, **291:**477–478
  photolysis conditions, **291:**478–480
  structure, **291:**475
  synthesis, **291:**475–477
Salicylic acid, hydroxyl radical assay
  dihydroxybenzoate product measurement
    by high-performance liquid chroma-
    tography, **300:**253, 299, 301–302
  electrochemical detection of other metab-
    olites, **300:**304
  guanidino compound effects on brain rad-
    ical generation, **300:**254–255
  interferences, **300:**303
  microdialysis perfusion application,
    300:302–303, 305
  oxidative injury assessment in brain tis-
    sue, **300:**305–307
  rat preparation for intact brain measure-
    ments, **300:**253
  reactivity of salicylic acid, **300:**299–302
  sensitivity, **300:**302
Saliva
  chromatography of proteins, **310:**503–504
  composition, **310:**269
SAP, *see* Serum amyloid P component
SAPK, *see* Stress-activated protein kinase
Saposins, *see also* Sphingolipid activator
    proteins
  sphingolipid translocation, **312:**576
SAPs, *see* Sphingolipid activator proteins
Sarcoplasmic reticulum/endoplasmic reticu-
    lum calcium ATPase
  calcium oscillation role, **291:**439
  2,5-di(*tert*-butyl)hydroquinone caged com-
    pound studies
    calcium studies with fluorescent indica-
      tors, **291:**435–437, 439
    cell loading, **291:**435
    synthesis, **291:**432
SAT, *see* Sialyltransferase
SBH, *see* Sequencing by hybridization
SC, *see* Stratum corneum
SCAM, *see* Substituted-cysteine accessibil-
    ity method
Scanning confocal laser microscopy, *see*
  Confocal laser scanning microscopy
Scanning electron microscopy, *see* Electron
  microscopy
Scanning transmission electron microscopy,
  see Electron microscopy
Scarlet protein, *Drosophila melanogaster*

eye pigmentation role, **292:**213–215
gene cloning, **292:**214
hydropathy plot, **292:**215, 218
sequence alignment with other transport proteins, **292:**215–217
site-directed mutagenesis, **292:**215, 218–219
White protein interactions, **292:**223
Scatchard plot, cooperativity of high mobility group protein binding to DNA, **304:**115–117
SCGE, *see* Single-cell gel electrophoresis
Schiff base, rhodopsin, *see* 11-*cis*-Retinal chromophore, rhodopsin
Scintillation proximity assay, neuronal nitric oxide synthase heme domain ligands
equilibrium binding assays, **301:**117–118
kinetics of binding, **301:**118, 121–122, 124–125
nitroarginine, **301:**117–122
principle, **301:**115–116
reagents, **301:**116
recombinant domain expression and purification, **301:**116
tetrahydrobiopterin, **301:**117–118, 121–124
SCOP2
function, **316:**279
gene cloning, **316:**279–282
Scorpion venom peptides, *see also* Charybdotoxin receptor
applications, **294:**625
binding assays, **294:**637–639
classification, **294:**626
expression of recombinant peptides in *Escherichia coli*
extraction, **294:**631–632
induction, **294:**631
purification of histidine-tagged proteins, **294:**632–633
transformation, **294:**630–631
vectors, **294:**630
peptide synthesis, **294:**629–630
potassium channel inhibition specificity, **294:**626
purification
assessment of purity, **294:**629
cation-exchange chromatography, **294:**627–628

reversed-phase high-performance liquid chromatography, **294:**628
storage, **294:**628–629
radiolabeling
iodination, **294:**634–635
tritiation by cysteine alkylation, **294:**635–637
sequence homology, **294:**626
structural overview, **294:**625
SDF-1, *see* Stromal cell-derived factor 1
SDK1, *see* Sphingosine-dependent kinase 1
SDR, *see* Short-chain dehydrogenase/reductase
SDS, *see* Sodium dodecyl sulfate
sec6-4, see Yeast expression systems
SecB
biological function, **290:**444–445
oligomeric analysis
data analysis, **290:**280–283
data collection, **290:**279–280
sample preparation, **290:**279
sedimentation velocity, **290:**284–285
peptide-binding assays
binding conditions, **290:**451
dissociation constants, **290:**450
fluorescence assay
calculations, **290:**452–453
competition assay, **290:**454
denaturant selection, **290:**453–454
enhancement of 1-anilinonaphthalene 8-sulfonate fluorescence, **290:**456
intrinsic tryptophan fluorescence of substrates, **290:**451–452
ligand binding site, protease protection and peptide mapping, **290:**457–459
proteolytic protection assay
ligand selection, **290:**454–455
proteinase K incubation conditions, **290:**455–456
size-exclusion chromatography of complexes, **290:**450
physical properties, **290:**449–450
purification of recombinant protein from *Escherichia coli*
anion-exchange chromatography, **290:**447–449
cell growth, **290:**445–446
cell lysis, **290:**446–447
size-exclusion chromatography, **290:**449

Secnidazole
    electrochemical reduction, **300:**314–315
    interaction with DNA, **300:**316–321
    structure, **300:**314
Secretion
    confocal microscopy measurement in rat
        salivary gland
            data acquisition, **307:**334–335
            dissection and cell dissociation,
                307:331–332
            exocytosis measurement, **307:**338–340
            fluorescent tracers, **307:**332–333
            perfusion, **307:**333
            secretory stimulation, **307:**335
            tight junctional permeability measure-
                ment, **307:**335, 337–338
        exocytosis and epithelial transport,
            307:329, 331
        overview of process, **307:**328–329
Secretory protein
    confocal laser scanning microscopy advan-
        tages over immunofluorescence mi-
        croscopy in study, **302:**3
    green fluorescent protein assays, *see* Hu-
        man chromogranin B–green fluores-
        cent protein fusion protein;
        RatTGN38–green fluorescent pro-
        tein fusion protein
Secretory vesicle, *Saccharomyces cerevisiae*
    advantages in membrane protein expres-
        sion, **292:**383–384, 396
    cystic fibrosis transmembrane regulator
        expression system, **292:**628
    electrochemical potential, **292:**384, 388
    invertase as marker, **292:**388
    morphological characterization, **292:**387
    murine P-glycoprotein expression
        applications, **292:**426
        drug transport studies
            buffers, **292:**389
            calculations, **292:**390
            cold stop solution, **292:**389–391
            filter assay, **292:**388–391
            incubation conditions, **292:**390–391
            isolation of secretory vesicles,
                292:386–387
        phosphatidylcholine translocation assay
            with Mdr2
            fluorescent phosphatidylcholine,
                292:391–392

    principle, **292:**391–393
    reaction conditions, **292:**395–396
    secretory vesicle preparation with
        outer leaflet label, **292:**393, 395
    transformation, **292:**385–386
    yeast strains and growth, **292:**384–385
Section thickness, correction between nomi-
    nal and real thicknesses, **307:**112
Sedimentation equilibrium
    principle, **290:**272, 274
    SecB, oligomeric analysis
        data analysis, **290:**280–283
        data collection, **290:**279–280
        sample preparation, **290:**279
Sedimentation velocity ultracentrifugation
    assay of nucleosome array saturation,
        304:31–32
    history of technique, **290:**270–271
    HMG14/17 reconstitution into chromatin,
        304:146–148
    instrumentation, **290:**271, 294
    quaternary structure determination
        advantages, **290:**278
        purity requirements for protein,
            290:278–279
    radial absorption scanning, **290:**272
    SecB, oligomeric analysis, **290:**284–285
    Stokes equation, **290:**275
    Svedberg equation, **290:**274–275
Selectin
    dual-radiolabeled monoclonal antibody
        technique for *in vivo* measurement
        advantages, **301:**15–16
        backflush procedure, **301:**18
        calculations, **301:**19–20
        E-selectin expression kinetics, **301:**21
        exsanguination, **301:**18
        injection of antibodies into mice,
            301:16–18
        P-selectin
            expression kinetics, **301:**20
            inflammatory mediator induction,
                301:21
        radioiodination of antibodies, **301:**16
        tissue preparation and analysis,
            301:18–19
    E-selectin, c-Jun N-terminal kinase anti-
        sense knockdown effects on tumor
        necrosis factor-α induction in endo-
        thelial cells

cell culture, **314:**366
flow cytometry analysis, **314:**369–370
JNK isoform specificity, **314:**377–378
knockdown efficacy, **314:**375, 377
oligonucleotide treatment, **314:**366–367
rationale for study, **314:**374–375
functions, **311:**548–549
ligands, *see* Sialyl Lewis X gangliosides;
Sulfo Lewis X gangliosides
types and expression, **301:**14, **311:**548
SELEX, *see In Vitro* selection, RNA aptamers
Semliki Forest virus–Chinese hamster
ovary cell expression system
affinity tagging of proteins, **293:**463–464
host cell culture, **293:**472–473
ion channel expression
electrophysiologic recording, **293:**479
glycosylation analysis, **293:**481–483
types expressed, **293:**459–460
optimization of expression, **293:**464–465,
478–479
posttranslational modification of proteins,
293:466
promoter selection, **293:**461, 463
protein expression level, **293:**459, 461
receptor assembly analysis, **293:**466
recombinant virus production
recombinant virion packaging, **293:**468
RNA transcription *in vitro*,
293:467–468
titration of virus, **293:**476–477
Senescence-accelerated mouse, *see also* Murine senile amyloid fibril
advantages over other species, **309:**679
aging characteristics
Gompertz function, **309:**676
grading score of senescence,
309:675–677
growth curve, **309:**675
life span, **309:**677
survivorship curve, **309:**676
genetic background of strains, **309:**679
history of development, **309:**674–675
pathobiological phenotypes, **309:**678
Senile plaque, *see* Alzheimer's disease
Septic shock, nitric oxide protection,
301:423
Septins
filament formation assay, **298:**287–288

filament length determination,
298:288–289
genes in yeast, **298:**279
GTP
exchange and hydrolysis, assay,
298:290–294
nucleotide content determination,
298:289–290
photocross-linking, **298:**294–295
purification of *Drosophila* complex
ammonium sulfate fractionation,
298:284–285
anion-exchange chromatography,
298:284, 286–287
cation-exchange chromatography,
298:284–286
high-speed supernatant preparation,
298:281–282
hydroxyapatite chromatography,
298:284, 287
immunoaffinity chromatography,
298:280–283, 540–541
overview of approaches, **298:**280–281
sequence analysis, **298:**279
Sequencing by hybridization
advantages, **303:**165–166, 168
applications, **303:**177–178
arrays
hybridization, **303:**172–173
preparation, **303:**170–172
types, **303:**168–169
clones
clustering, **303:**175–176
preparation, **303:**169
complementary DNA screening requirements, **303:**166–167, 177
data handling, **303:**173–174
differential expression of all genes,
303:176–177
hybridization signature comparison,
303:175
overview, **303:**167–168
polymerase chain reaction of plasmid inserts, **303:**169–170
principle, **303:**165–168
throughput, **303:**177–178
SERCA, *see* Sarcoplasmic reticulum/endo-
plasmic reticulum calcium ATPase
Serine palmitoyltransferase
assays

cell types, **311:**3–4
incubation conditions, **311:**5–6, 133
modifications, **311:**7–8
optimization, **311:**7, 134
principle, **311:**5, 133
radioactivity quantification,
    311:133–134
associated proteins, **311:**3, 5, 9
inhibitors, **311:**8, 335, 347–350
purification attempts, **311:**3
pyridoxal phosphate dependence, **311:**3, 7
reaction catalyzed, **311:**3
regulation, **311:**8–9
substrate specificity
    mammalian enzyme, **311:**6–7
    plant enzyme, **311:**134–135
temperature-sensitive Chinese hamster
    ovary cell mutants
    advantages of ovary cells, **312:**306–307
    LY-B strain
        defective step identification,
            312:316–317
        discovery, **312:**306
        GM3 analysis, **312:**314–316
        lysenin-resistant variant selection,
            312:308–310
        metabolic radiolabeling of lipids with
            serine, **312:**310–313
        mutagenesis, **312:**307–308
        sphingomyelin analysis, **312:**314–316
    SPB-1 cells, **312:**304–306
Serine protease, *see also* Chymotrypsin
    acyl intermediate analysis by difference
        Raman spectroscopy, **308:**192–193
    hydrogen bonds and energy for catalysis,
        308:23–24
    steps in catalysis, **308:**191–192
Serotonin, sample processing for confocal
    and electron microscopy, **307:**147, 149
Serotonin 2A receptor
    antisense targeting
        controls, **314:**82
        cortical culture system
            autoradiography of receptors,
                314:84–85
            duration of treatment, **314:**83
            phosphoinositide hydrolysis assay,
                314:83–84
            transfected cells, **314:**82–83
        intracerebroventricular infusion

dose, **314:**86
infusion system, **314:**85–86
receptor expression enhancement re-
    sponse and mechanism, **314:**86–
    87, 89
ribonuclease H requirement,
    314:87–88
oligonucleotide design, **314:**79–81
turnover in brain, **314:**86
Serotonin transporter, *see also* Vesicular
    monoamine transporter
baculovirus–insect cell expression system
    baculovirus production
        amplification, **296:**449
        overview, **296:**445–446
        plaque purification, **296:**447–448
        recombinant virus construction,
            296:447
        titration, **296:**448–449
    cell line maintenance, **296:**446
    infection, **296:**449–450
    inhibitor binding assays, **296:**453–455
    membrane preparation, **296:**450
    principle, **296:**444–445
    uptake assays, **296:**452–453
    Western blot analysis of neurotransmit-
        ter transporter expression,
        296:450–452
binding assays
    placenta brush border membrane vesi-
        cles, **296:**285–286
    transfected cells, **296:**351, 353
chimeric cross-species transporter
    binding assays, **296:**496
    construction approaches
        restriction enzyme-based chimeras,
            296:485–486
        restriction site-independent chime-
            ras, **296:**489–490, 492–493
        restriction site introduction and com-
            plementary DNA cassette swap-
            ping, **296:**486–489
    electrophysiological characterization,
        296:495
    expression analysis, **296:**496–497
    imipramine binding, **296:**481
    ion dependence of transport, assays,
        296:494
    pharmacological profiles, **296:**494–
        495

transmembrane domain functional
analysis
Drosophila–human transporter,
296:483
rat–human transporter, **296:**483, 497
transport assays, **296:**494
electrophysiology
current to flux ratios, **296:**591–592
flash photolysis studies with DM-nitrophen
flash photolysis setup, **296:**605–606
microinjection of DM-nitrophen,
296:604–605
photochemistry, **296:**603–604
pre- and postsynaptic current isolation induced by flash photolysis,
296:602–603
two-electrode voltage clamp, **296:**601
Retzius P cell synapse
external solutions, changing,
296:597–598
inward current, **296:**597
isolation of cells, **296:**594–595
pairing of cells and culture, **296:**595
patch pipette and microelectrode
preparation, **296:**606–607
potassium current, **296:**597
simultaneous fluorescence imaging of
5,7-dihydroxytryptamine uptake
with voltage clamping,
296:598–601
sodium dependent current,
296:595–596
solution preparation, **296:**606–607
transporter-associated currents,
296:596
whole-cell patch clamp, **296:**596
serotonin-induced inward currents,
296:591
transient expression in *Xenopus* oocytes, **296:**590
transporter cycles per channel opening,
296:592
gene cloning, **296:**426
human choriocarcinoma cell
applications, **296:**286
cell culture, **296:**287
vesicle preparation, **296:**288–289
intracerebral microdialysis and serotonin
detection

anesthesia, **296:**721–722
apparatus, **296:**726–727
contamination, **296:**723–724
guide cannula implantation, **296:**726
high-performance liquid chromatography with electrochemical detection
apparatus, **296:**728
mobile phase preparation, **296:**728
monoamine standard preparation,
296:729
peak identification, **296:**730
running conditions, **296:**730
perfusion solution, **296:**722, 727
presentation of data, **296:**725–726
principle, **296:**720–721
probes
calibration, **296:**727
placement verification, **296:**730
recovery performance, **296:**723
types, **296:**722–723
sample collection, **296:**727–728
uptake blockers
administration routes, **296:**725
monoamine effect, evaluation,
296:724
pharmacological profile, comparison from
different species, **296:**480–481, 497
phylogenetic analysis, **296:**428–436
placenta brush border membrane vesicle
preparation
advantages as starting tissue, **296:**278–
280, 289–290
centrifugation, **296:**282–283
dissection, **296:**281
federal regulations, **296:**278
flow chart, **296:**282
gross structure, **296:**280–281
homogenization, **296:**283
marker enzymes, **296:**284
storage, **296:**284
tissue collection, **296:**280
protein kinase C, regulation of biosynthesis and surface expression, **296:**368,
370
site-directed mutagenesis of conserved
aspartate in transmembrane domain
1, **296:**477–479, 498
transient transfection and cell culture,
296:350
uptake assays

human choriocarcinoma cells,
296:287–288
placenta brush border membrane vesicles, **296:**284–285
transfected cells, **296:**351
SERT, *see* Serotonin transporter
Serum amyloid A amyloidosis, *see* Experimental murine AA amyloid
Serum amyloid P component
binding to amyloid fibrils, assay,
309:45–46
electroimmunoassay, **309:**44–45
immunoelectron microscopy, **309:**508
immunohistochemical staining, **309:**40
radioiodination and detection of amyloid
A, **309:**713–714
removal from amyloid isolates, **309:**27–
28, 32–34
S fimbrial adhesin, gene activation during
agar growth, **310:**68
SGLT1, *see* Sodium/glucose cotransporter
Shaker-type potassium channel
adenovirus expression system
electrophysiologic studies, **293:**496–497
transfection tests using green fluorescent protein as reporter,
293:495–496
charybdotoxin binding, **294:**582, 584–586
chimeric channel analysis
chimera construction, **293:**13, 16
subunit interaction analysis, **293:**12–13,
17
vector, **293:**13
cut-open oocyte voltage clamp
gating current recording after internal
potassium removal, **293:**314
gating currents after analog compensation of linear resistive and capacity components, **293:**314
macropatch recording comparisons,
293:314, 316
tail currents in high external potassium, **293:**312–313
unsubtracting gating charge records,
293:313–314
gating current measurement
analog compensation of linear components, **293:**345–346
elementary gating event, **293:**348–350
expression systems, **293:**344–345

frequency domain recording, **293:**350,
352
inactivation and charge mobilization,
293:347–348
inactivation removed mutant,
293:345–347
inactivation
mechanism, **294:**640
peptides
concentration determination,
294:647–648
delivery of peptides, **294:**648
kinetics of blocking, **294:**640,
648–649
net charge effects, **294:**641
pore inner mouth probing,
294:644–646
purification, **294:**646
Raw3, 640–645
RCK4, 644
ShB, **294:**640–645
storage, **294:**647
tetraethylammonium slowing of inactivation, **294:**641
kinetic modeling, simultaneous data fitting, **293:**415
pore occupancy, **293:**390
protein overlay assay and subunit interaction analysis
advantages and disadvantages, **293:**14
principle, **293:**10–11
probe-specific antibody assay, **293:**12
radioactive labeling of ShB amino-terminal domain, **293:**10–11
radioactive protein probe assay,
293:11–12
PSD-95 association, coimmunoprecipitation assay, **294:**377–380
reconstruction in planar bilayers, **294:**298,
301
rise time parameters, **293:**410–413
site-directed mutagenesis, **293:**3–4
unidirectional flux measurement of *Xenopus* oocyte channels
efflux measurement, **293:**561–564
influx measurement
nonchannel flux correction, **293:**560
potassium-42 counting and specific
activity, **293:**559–560
reflux, **293:**560

tail current error, **293:**560–561
voltage clamp measurement,
    293:557–558
yeast two-hybrid system
    binding protein analysis, **293:**105–109
    subunit assembly analysis
        advantages and disadvantages,
            293:14
        filter assay of β-galactosidase, **293:**10
        fusion protein vector construction,
            293:6
        growth assays, **293:**8–9
        principle, **293:**4
        reporter genes, **293:**4, 8
        strains and reagents, **293:**5–6
        subcloning amino-terminal domains
            into fusion protein, **293:**6–7
        transformation, **293:**8
        yeast competent cell preparation,
            293:7–8
ShB
    concentration determination, **294:**647–
        648
    purification, **294:**646
    Shaker-type potassium channel blocking
        delivery of peptides, **294:**648
        kinetics of blocking, **294:**640, 648–649
        net charge effects, **294:**641
        pore inner mouth probing, **294:**644–646
        tetraethylammonium slowing of inacti-
            vation, **294:**641
    storage, **294:**647
Shear stress
    apparatus for endothelial cells,
        301:518–519
    atherosclerosis role, **301:**513–514
    calculation, **310:**226
    endothelial cell
        monolayer culture, **301:**515
        overview of shear stress effects,
            301:513
        transfection on glass plates, **301:**515–
            517, 519–520
    mitogen-activated protein kinase regu-
        lation
        assay by Western blot, **301:**520–521
        extracellular signal regulated kinase,
            p21Ras regulation, **301:**514,
            519–520
        Jun N-terminal kinase, **301:**514–515

nitric oxide release, measurement from
    sheared cells, **301:**519
oral biofilm model, **310:**222
sonication for determination in flow cul-
    ture, **310:**252
Short-chain dehydrogenase/reductase, *see*
    also specific enzymes
    activity of novel proteins, **316:**382–383
    complementary DNA cloning of retinal
        enzymes, **316:**376–377
    expressed sequence tags, database search-
        ing for cloning probes, **316:**373–376
    expression in baculovirus–insect cell sys-
        tem, **316:**380–382
    phase partition assay, **316:**368, 382
    sequence analysis
        homology of superfamily, **316:**372
        identification of family members,
            316:377–378
        phylogenetic analysis, **316:**378–379
    substrate specificity, **316:**371–372
    tissue distribution, Northern blot analy-
        sis, **316:**379–380
Sialic acid
    determination with periodic/thiobarbi-
        turic acid, **312:**351–352
    radiolabeled substrates
        sialidase assay, **312:**358
        sialyltransferase labeling, **312:**357
        tritium labeling by oxidation/reduction,
            312:357–358
Sialic acid, photoaffinity labels, **311:**581, 583
Sialic acid-dependent immunoglobulin fam-
    ily lectin
    ligands, **311:**560–561
    types, **311:**560
Sialidase, *see also* Ganglioside sialidases
    sphingolipid identification, **311:**669
Sialyl Lewis X gangliosides
    clinical applications, **311:**558–559
    fucose residue alterations, **311:**550–551
    selectin recognition, **311:**549–551
    structures, **311:**549–550
Sialyltransferase
    assay
        incubation conditions, **311:**87–88
        materials, **311:**85
        product identification and structural
            analysis, **311:**89–91
        product separations, **311:**88, 93–94

quantitative analysis, **311:**88–89
stock solutions, **311:**87
reaction catalyzed, **311:**85
SIGLEC, *see* Sialic acid-dependent immunoglobulin family lectin
Signal model, *see* Hidden Markov model
Signal sequence trap
invertase as reporter protein in yeast
accuracy in signal sequence protein
identification, **303:**476–478
applications, **303:**476–479
complementary DNA library construction, **303:**471–472
computational analysis of sequences,
303:475–476
plasmid
amplification by polymerase chain reaction, **303:**474–475
isolation, **303:**474
purification for sequencing, **303:**475
sequencing, **303:**475
rationale, **303:**469
selection for invertase secretion,
303:472–473
transformation of yeast, **303:**472–473
vector, **303:**470–471
stromal cell-derived factor 1 clones,
303:480–481
Tac tagging in COS cells
applications, **303:**494–495
complementary DNA library construction
DNA recovery from gels, **303:**486
first-strand synthesis and RNA hydrolysis, **303:**484–485
overview, **303:**481, 483
polymerase chain reaction of DNA
from gel blocks, **303:**486–487
restriction digestion of amplified
DNA, **303:**487–488
second-strand synthesis and size selection, **303:**485–486
criteria for positive clones, **303:**492–493
hydropathy analysis and database
searches, **303:**493–494
sib screening with anti-Tac immunostaining
cell surface immunostaining,
303:491–492
overview, **303:**483

pool plate preparation, **303:**489–490
smaller pool preparation for single
clone identification, **303:**492
transfection, **303:**490–491
vector
characteristics, **303:**481
insert ligation, **303:**488–489
preparation, **303:**488
transformation, **303:**489
Signal-to-noise ratio, confocal microscopy,
307:212, 215, 375
Silent information regulator 3, mapping of
chromatin interactions, **304:**02–403
Simian virus 40, large T antigen
functions in virus, **306:**297
neoplastic activity, **306:**299
protein interactions, **306:**297, 299
purification by immunoaffinity chromatography
antibody purification, **306:**300–302
baculovirus–insect cell expression
system
cell lines, **306:**299–300
growth media, **306:**300
infection of cells, **306:**303
lysate preparation, **306:**304, 306
chromatography conditions,
306:306–307
column preparation, **306:**302–303
structure and binding regions, **306:**297
Similarity measure, *see* Molecular surface
point similarity measures
Simulated annealing, *see* Three-dimensional
structure, molecular modeling
SIN-1, *see* 3-Morpholinosydnonimine
Single-cell gel electrophoresis
applications, **300:**184–185
DNA damage analysis
alkaline unwinding and electrophoresis, **300:**189–190
apoptotic DNA, **300:**193–194
cell lysis, **300:**189
cell types, **300:**185
comet patterns, evaluation and quantification, **300:**192–193
DNA repair enzyme treatment,
300:191
inducers of DNA damage, **300:**185–187
neutral assay, **300:**191–192
slide preparation, **300:**187–189

staining of slides, **300:**190–191
Single-channel measurements, gramicidin channels
  electrodes, **294:**542
  functional approach to structure identification by dimerization, **294:**545–547, 549–550
  membrane capacitance minimization, **294:**542–543
  purity requirements, **294:**543–544
  reconstitution, **294:**543
Single-photon detection, photoreceptors, *see* Rod outer segment
Single-photon photolysis, *see* Laser photolysis
Single-stranded conformation polymorphism
  green and green–red hybrid pigment gene expression in arrays, **316:**668–670
  red/green pigment, ratio determination in retina, **316:**667–668
  red pigment Ser180Ala polymorphism, **316:**659–660
  retinal pigment epithelium p65
    labeled DNA, **316:**708, 710
    primers, **316:**709
    unlabeled DNA, **316:**710
  ribosomal RNA gene detection in prosthetic joint biofilms, **310:**573
  X-linked red–green photopigment genes, **316:**655, 658–659
Singlet oxygen
  apoptosis induction, *see* Apoptosis
  biological targets, **319:**3–4
  catalase detection, *see* Catalase, singlet oxygen detection
  chemiluminescence, *see* Chemiluminescence, singlet oxygen
  cholesterol reporter, *see* Cholesterol
  diffusion in air and detection, **319:**112–113, 198
  erythropoietic protoporphyria, *see* Erythropoietic protoporphyria
  fluorescence assay, *see* DanePy
  gene expression activation, *see also* Mitogen-activated protein kinase; Nuclear factor-κB
    1,4-naphthalenedipropanoate effects, **319:**18–19

overview, **319:**18–19, 111, 130–131
  heat generation, **319:**59–60
  inorganic generators, **319:**4
  lifetimes and solvent effects, **319:**3, 14, 25, 86, 121
  peroxidase generation, **319:**59
  photosensitized production, *see* Photosensitization, singlet oxygen production
  quantum yield components, **319:**38–39
  quenchers, *see* Singlet oxygen quencher
  scavenging in phospholipid membranes
    carotenoid scavenging rates, **319:**104–106, 108–109
    factors in rate measurement, **319:**109–110
    liposome preparation, **319:**103
    photosensitized singlet oxygen generation, **319:**101–102
    rate constant calculation, **319:**102–103
    α-tocopherol scavenging rates, **319:**106–109
  sources in biological systems, **319:**3, 71, 85–86, 110–111, 144
  time-resolved detection
    applications, **319:**37
    laser-induced optoacoustic spectroscopy, **319:**43
    thermal lensing, **319:**42–43
    time-resolved near-infrared phosphorescence
      decay rate and quenching rate constant determination, **319:**47–49
      instrumentation, **319:**40–42, 49
      quantum yield measurement in homogeneous systems, **319:**44–47
      sample preparation and handling, **319:**43
      theory, **319:**38–40
      trace recording, **319:**43–44
  ultraviolet radiation production, *see* Ultraviolet radiation
  water-soluble carriers, *see* 1,3-Cyclohexadiene-1,4-diethanoate; 1,4-Dimethylnaphthalene, singlet oxygen carrier derivatives; 3-(4′-Methyl-1′-naphthyl)propionic acid
Singlet oxygen quencher, *see also specific compounds*
  amines, **319:**232–233
  amino acid-pyrrole *N*-conjugates, **319:**228

assay with monomol emission
  biological quenchers, **319:**224–226
  exogenous quenchers, **319:**225
  generation of singlet oxygen, **319:**222, 224
  instrumentation, **319:**223–224
  principle, **319:**223
indigoids, **319:**239–240
keratinocytes, ultraviolet A-induced gene expression effects, **319:**282–283
metal chelates, **319:**234–235, 237–239
mitogen-activated protein kinase, ultraviolet A induction effects of quenchers, **319:**132–138
nonaromatic heterocycles, **319:**239
olefins
  conjugated olefins and aromatic compounds, **319:**228, 230, 232
  nonconjugated olefins, **319:**227–228
phenols, **319:**233–234
physical versus chemical quenching, **319:**226
rates of quenching, **319:**227–228, 230, 232, 237–238
sulfides, **319:**232–233
SIR3, *see* Silent information regulator 3
Site-directed mutagenesis, *see also* Substituted-cysteine accessibility method
Brown protein, **292:**215, 218–219
buffer composition, **293:**68–69
chloroplast ATP synthase
  δ subunit, **297:**148
  ε subunit, **297:**146–147
CP47 mutagenesis in *Synechocystis* 6803
  advantages as model system, **297:**324–326
  applications and rationale, **297:**323–324
  characterization of mutants, **297:**336–337
  control strain K3, construction, **297:**329–330
  localized random mutagenesis, **297:**333–335
  mutant verification, **297:**333
  partial deletion strains, construction for *psbB*, **297:**330–331
  site-directed mutagenesis, **297:**331–332
  strains and maintenance, **297:**326–327
  vector construction, **297:**327–329

cyclic GMP phosphodiesterase Pγ subunit for autoinhibition assays
  cysteine insertion sites, **315:**636–638
  expression and isolation from *Escherichia coli*, **315:**640
  polymerase chain reaction site-directed mutagenesis, **315:**639–640
  vector design, **315:**637–639
D1, 226–227
folding enhancement of inclusion body proteins, **309:**232
Gα_t/Gα_{i1} chimeras
  alanine-scanning mutagenesis in switch II region, **315:**540
  cyclic GMP phosphodiesterase Pγ subunit binding assays, **315:**549–552
  GTPγS binding assay for folding analysis of mutants, **315:**548
  polymerase chain reaction-based mutagenesis, **315:**504–505, 542–544
  RGS interaction assays, **315:**552–554
  target selection and rationale, **315:**540, 542
  trypsin protection assay for folding analysis of mutants, **315:**545–546
  tryptophan fluorescence assay with AlF$_4^-$ for folding analysis of mutants, **315:**546–548
generation for pseudorevertant analysis, *see* Pseudorevertant
iron–sulfur clusters in PsaC, **297:**100, 103
M13 template mutagenesis
  mutant selection using *dut⁻*, *ung⁻* bacteria
    annealing reaction, **293:**59–60
    controls, **293:**60
    elongation reaction, **293:**60
    oligonucleotide phosphorylation, **293:**59
    overview, **293:**56–57
    uracil-containing DNA preparation, **293:**57, 59
  overview, **293:**55–56
MDR1
  applicability of MDR1 to mutagenesis studies, **292:**480–481
  assay for glycosylation-defective mutants, **292:**486–487
  ATPase, **292:**490, 492, 505

baculovirus–insect cell system expression, **292:**483–484
overview of mutagenesis, **292:**481–482
phosphorylation site, **292:**341–342
stable expression in mammalian cells, **292:**482–483, 505
transient expression in human embryonic kidney cells
 advantages, **292:**484
 ATPase assay of reconstituted protein, **292:**489–490, 492
 purification of histidine-tagged mutants, **292:**487–489
 transfection, **292:**484–485
Mdr1 phosphorylation site, **292:**355, 357–358
Mdr3, 417–418
oligonucleotides
 GC content, **293:**70
 length, **293:**69–70
 procurement, **293:**55
 restriction site inclusion, **293:**70
photoaffinity labeling sites, **292:**304–305, 315–316
photosystem II genes in *Synechocystis* 6803, 301–302
plasmid-based mutagenesis
 annealing reaction, **293:**63
 competent cell preparation, **293:**70–71
 controls, **293:**63
 double-stranded DNA template preparation, **293:**62–63
 elongation reaction, **293:**63
 oligonucleotide phosphorylation, **293:**62
 overview, **293:**60–62
 plasmid minipreparation, **293:**69
polymerase chain reaction-based mutagenesis, **293:**64, 97
protein structure model testing
 ab initio modeling, **293:**603–604
 homology modeling, **293:**642–643
PrtD, **292:**76
psbA of *Synechocystis* 6803, 175–176
PXA1, 764–766
reaction center mutagenesis by transformation of *Chlamydomonas reinhardtii* chloroplasts
 chimeric gene construction conferring

antibiotic resistance, **297:**312–313, 315–316
complications, **297:**311
gene gun transformation, **297:**311, 317–318
growth conditions, **297:**316–317
growth media, **297:**316
homologous recombination, **297:**311, 315
recipient strains, **297:**316
selection of transformants, **297:**312, 318–320
Rhodobacter cytochromes, **297:**93–94
rhodopsin, *see* Site-directed mutagenesis, rhodopsin
Scarlet protein, **292:**215, 218–219
Shaker-type potassium channel, **293:**3–4
spectroscopic characterization of photosystem II mutants
chlorophyll fluorescence yield
 charge recombination rate with 3-(3′,4′-dichlorophenyl)-1,1′-dimethylurea, **297:**344–346
 factors affecting yield, **297:**339–340
 kinetic techniques, **297:**340–341
 mutant characterization, **297:**341–346
 secondary electron transfer rates, **297:**342–344
electron nuclear double resonance, **297:**353–356
electron paramagnetic resonance, **297:**351–353, 356
electron spin echo envelope modulation, **297:**353–354, 356
Fourier transform infrared spectroscopy, **297:**357–358
Mössbauer spectroscopy, **297:**359
resonance Raman spectroscopy, **297:**358
thermoluminescence analysis of mutants, **297:**346–347
ultraviolet/visible absorbance spectroscopy characterization
 charge recombination rate analysis, **297:**349–350
 D1 mutant characterization, **297:**348–350
 flash detection, **297:**348
 manganese binding characterization, **297:**351

net electron transport analysis,
    297:350–351
X-ray absorbance spectroscopy,
    297:358
Ste6p, **292**:210–211
value in structure–function analysis,
    293:54
White protein, **292**:215, 218–219, 224
Site-directed mutagenesis, rhodopsin
chromophore binding site and spectral
    tuning, **315**:116–118
chromophore photoisomerization, sensing
    in transmembrane domains
    absorption spectroscopy analysis,
        315:123–124
    glycine-121 interactions
        phenylalanine-261 interactions medi-
            ated by retinal, **315**:128–129
        retinal C-9 methyl group interac-
            tions, analog studies,
            315:126–127
        substitution effects on activity,
            315:124, 126
    overview, **315**:122–123
constitutively active opsin mutants, *see*
    Opsin
Fourier transform infrared difference
    spectroscopy of rhodopsin mutants
    active state characterization
        aspartate/glutamate mutants, **315**:184
        dark state, **315**:183
        H211N mutant, **315**:184–185
        lipid vesicle versus detergent spectra,
            315:183–184
        metarhodopsin-II, **315**:181, 183
    attenuated total reflection sampling
        principle, **315**:180, 192–193
        proton exchange studies,
            315:193–195
    band assignment, **315**:195
    molecular changes leading to activa-
        tion, **315**:185–186
    sample preparation, **315**:180–181
    Schiff base environment studies
        Glu-113, mutation and role in
            stabilization/activation, **315**:186–
            187, 189–190
        Gly-90 mutants, **315**:190–192
        low-temperature studies,
            315:191–192

water molecule interactions,
    315:188–189
human gene mutagenesis
    absorption spectroscopy of mutants,
        315:50–51, 57
    immunoaffinity chromatography,
        315:49–50
    immunocytochemistry, subcellular local-
        ization and quantification,
        315:38–39
    materials, **315**:31–32
    plasmid preparation, **315**:33–34, 36
    primer
        annealing, **315**:34
        design and sequences, **315**:32–33
        phosphorylation, **315**:33
        selection primer, **315**:34, 36
    slot immunoblot assay for quantifica-
        tion, **315**:39, 44–45, 47–48
    vectors, **315**:32, 36
hydrogen-bonding network of water mole-
    cules in mutants, **315**:129
stable transfection of human mutants in
    HEK293S cells
    advantages, **315**:30–31
    calcium phosphate transfection,
        315:36–37
    expression levels, mutant dependence,
        315:45, 48–49, 55
    fluorescent probing of expression,
        315:56–57
    large-scale culture, **315**:58
    materials, **315**:31–32
    sampling frequency for expression anal-
        ysis, **315**:56
    selection of clones, **315**:37–38, 51, 55
    supertransfection and secondary drug
        selection, **315**:45, 48–49, 55–57
    vectors, **315**:37, 51
tryptophans in transmembrane segment
    absorption spectroscopy
        data collection, **315**:119–120
        difference spectroscopy analysis of
            mutants, **315**:120, 122
    conservation in G-protein-coupled re-
        ceptors, **315**:118
    targeting of specific tryptophans,
        315:118–119
    wavelengths of maximal absorption, iden-
        tification of critical amino acids

ancestral pigment sequence elucida-
    tion, **315:**320–321
phylogenetic tree construction
    neighbor-joining method, **315:**315,
        319–320
    programs, **315:**313–314
    unweighted pair-group method with
        arithmetic mean, **315:**314–315
site-directed mutagenesis for functional
    testing
    absorbance spectroscopy, **315:**324
    applications, **315:**325
    bovine A292S, **315:**324
    cloning of genes, **315:**321–322
    mutagenesis, **315:**322–323
    purification, **315:**324
    regeneration, **315:**323–324
    RNA isolation, **315:**321
wild-type pigment absorption, **315:**312,
    323–324
Site-specific psoralen
applications, **318:**118
RNA–RNA cross-linking
    advantages, **318:**105–106
    analysis
        polyacrylamide gel electrophoresis,
            318:115, 117
        reverse transcription primer arrest
            assay, **318:**117
    DNA sequence design for targeting re-
        actions, **318:**109
    DNA–site-specific psoralen
        chromatography, **318:**110–111
        hybridization to RNA, **318:**111, 113
        synthesis on 3′-terminus of DNA,
            318:111
        synthesis, **318:**109–110
    irradiation conditions for cross-linking,
        318:114–115
    overview, **318:**105–106
    RNA–site-specific psoralen
        azidophenacylbromide, **318:**114
        irradiation conditions for formation,
            318:113
        purification, **318:**113–114
    structures, **318:**106, 108
    synthesis, **318:**106, 108
In Situ hybridization, *see also* Fluorescence
    in situ hybridization
    ABCR localization in eye, **315:**883–884

chemiluminescence detection
    alkaline phosphatase detection,
        305:578–579, 583–584
    applications, **305:**585–586
    controls, **305:**582
    dual hybridization reaction,
        305:584–586
    horseradish peroxidase detection,
        305:579, 583–584
    hybridization reaction, **305:**583
    imaging, **305:**579–580, 590
    probes, **305:**581–582
    quantitative analysis, **305:**580–581,
        585
    sensitivity, **305:**578, 585
    specimen preparation, **305:**581
heat shock protein 70 antisense oligonu-
    cleotide localization
    administration in rat brain,
        314:239–240
    frozen sections
        preparation, **314:**240
        processing, **314:**241–242, 247
    hybridization mix, **314:**241
    localization
        fluorescent probe detection,
            314:243–244
        radiolabeled probes, **314:**243
    oligonucleotide design, **314:**239
    radiolabeling of probe, **314:**240–241
    stability analysis, **314:**245
    subcellular localization, **314:**246–247
    vibratome sections
        preparation, **314:**240
        processing, **314:**242–243, 247
luciferin-binding protein from *Gonyaulax
    polyedra*, **305:**275–276
neuronal tissue
    coating of microslides, **293:**156
    complementary DNA probe hybrid-
        ization
        autoradiography, **293:**159–160
        controls, **293:**160
        design of probe, **293:**157
        hybridization and washing condi-
            tions, **293:**158–159
        prehybridization, **293:**158
        probe synthesis and radiolabeling,
            293:157–158
        thionin staining, **293:**160

complementary RNA probe hybrid-
    ization
    autoradiography and staining,
        293:165
    controls, **293:**165
    design of probe, **293:**160–161
    hybridization and washing condi-
        tions, **293:**162–165
    prehybridization, **293:**162
    probe synthesis and radiolabeling,
        293:161–162
    template construction, **293:**161
    observation of developed sections,
        293:165
    solution preparation, **293:**156
    tissue preparation, **293:**156–157
principle, **305:**577–578
retinal neurons
    cryosection hybridization for double-la-
        bel detection, **316:**584–587
    cryosection hybridization for single-la-
        bel detection
        applications, **316:**581–582
        fixation of tissue, **316:**582–583
        hybridization, **316:**583
        immunohistochemical detection,
            316:584
        posthybridization, **316:**583–584
        prehybridization, **316:**583
    detection techniques, **316:**579, 584–585,
        589–590
    digoxigenin–RNA probe synthesis,
        316:580–581
    ribonuclease contamination prevention,
        316:580
    whole mounts
        fixation of tissue, **316:**587
        hybridization, **316:**588
        immunohistochemical detection,
            316:588–589
        posthybridization, **316:**588
        prehybridization, **316:**587–588
in Situ transcription
    principle, **303:**3–4
    single cell analysis of messenger RNA
        with *in situ* transcription/antisense
        RNA amplification
        amplification efficiency, **303:**16–17
        applications and rationale, **303:**13–15
        complementary DNA synthesis, **303:**16

fixation and embedding, **303:**15
immunohistochemical labeling,
    303:15–16
Size-exclusion chromatography, *see* Gel fil-
    tration
SKFF-89976-A, γ-aminobutyric acid trans-
    porter inhibition, **296:**171, 174
Skin
    compartmentalization of antioxidants,
        300:428
    iodometric hydroperoxide assay of oxida-
        tion status
        aging effects, **300:**435–436
        electrochemical detection, **300:**435, 437
        γ-irradiation effects, **300:**436
        principle, **300:**434
    low molecular weight antioxidants, deter-
        mination
        aging effects, **300:**432
        electrochemical detection, **300:**429–430,
            433–434
        extraction procedure, **300:**429–431, 437
        principle, **300:**429–430
        secretion of antioxidants and quantifi-
            cation, **300:**431–432
    photon emission detection
        age effects on oxidative status, **300:**425,
            428
        antioxidants, effects of topical applica-
            tion, **300:**426–427
        control unit, **300:**421
        data processing unit, **300:**423
        horny layer moisture content effects,
            300:425
        light excitation unit, **300:**423
        photomultiplier, **300:**421–423
        photosensitizer effects, **300:**424
        rubbing effects, **300:**425
        ultraviolet-A induction, **300:**423–424
        ultraviolet filters, effects of topical ap-
            plication, **300:**424–425
    α-tocopherol analysis in stratum corneum
        gradient and physiological relevance,
            300:418–419
        high-performance liquid chromatogra-
            phy, **300:**416
        lipid extraction, **300:**415–416
        quantitative analysis, **300:**416–418
        tape stripping, **300:**414–415
Skin, oxidation product measurement

ethanol-soluble sample preparation, **319:**613

high-performance liquid chromatography squalene peroxide analysis, **319:**616–617, 619–621

    topical antioxidant analysis in tape-stripped samples, **319:**618, 621–622

lipid hydroperoxides, colorimetric analysis, **319:**615–616

oxidation product generation, **319:**619

tape stripping, **319:**613, 615

Skin, two-photon excitation microscopy imaging of NAD(P)H metabolism

correlative microscopy, **307:**515

data acquisition, **307:**531–532

multiphoton process, experimental characterization, **307:**515

validation, **307:**519

wavelengths, **307:**517

Skin, ultraviolet protection

antioxidant defenses, **319:**466

chlorpromazine induced phototoxicity in skin equivalents, sunscreen assay applications, **319:**456

    cell viability assay, **319:**45

    chlorpromazine dose–response, **319:**455

    Episkin preparation, **319:**454

    formulas for testing, **319:**454–456

    incubation, **319:**455

    interleukin-1α assay, **319:**455

    principle, **319:**453–454

    ultraviolet A radiation source, **319:**454

delayed-type hypersensitivity assay of immunosuppression

    action spectra, **319:**462, 464

    experimental groups, **319:**460

    hypersensitivity test, **319:**460, 462

    recall antigens, **319:**460

    statistical analysis, **319:**462

    sunscreen testing on human skin, **319:**460–462

end points for ultraviolet A exposure, **319:**464

immediate pigment darkening, ultraviolet A assay

    source definition and efficacy spectrum, **319:**451

    sunscreen evaluation, **319:**452–453

    time course, **319:**450

ultraviolet A protection factor, **319:**451–452

lipid peroxidation assay in skin equivalents

    Episkin preparation, **319:**457

    K assay, **319:**458

    sunscreen evaluation, **319:**458–459

    ultraviolet A source, **319:**457–458

persistent pigment darkening, ultraviolet A exposure measurement, **319:**449–450

topical antioxidant screening

overview, **319:**466

prescreening, *in vitro*

    cell culture, **319:**467

    coenzyme Q10, 469–471

    α-glucosyl rutin, **319:**469–471

    irradiation, **319:**467

    mitochondria transmembrane potential, **319:**467–468

    phosphotyrosine assay, **319:**467–468

    statistics, **319:**468–469

    thiol assay, **319:**467–468

    viability assays, **319:**468

    vitamin E, **319:**469

ex vivo/in vitro screening

    biopsy and cell preparation, **319:**473–474

    coenzyme Q10, 474–475

    α-glucosyl rutin, **319:**474

    mitochondria transmembrane potential, **319:**473

    thiol assay, **319:**473

in vivo screening

    electron spin resonance, **319:**478

    erythema assay, **319:**477

    α-glucosyl rutin, **319:**476–477

    polymorphous light eruption prophylaxis, **319:**476–477

    ultraweak photon emission, **319:**475–478

    vitamin E, **319:**476

transmission measurements of sunscreen materials

advantages and limitations of assay, **319:**448–449

carrier matrix, **319:**446–447

light sources, **319:**446–447

monochromatic protection factor calculation, **319:**447–448

principle, **319:**447

sun protection factor calculation, **319:**447–448

ultraviolet penetration, **319:**359, 366, 445, 489–490

Slide, biofilm growth, **310:**254–255

Slot immunoblot, quantification of human opsin mutants, **315:**39, 44–45, 47–48

Small heat shock proteins, *see also specific proteins*

classification in plants, **290:**350

expression in recombinant *Escherichia coli*, **290:**350–352

oligomeric structure determination by nondenaturing pore-exclusion poly-acrylamide gel electrophoresis, **290:**352–355

Small nuclear ribonucleoprotein

conservation between yeast and mammals, **318:**385

functions, **318:**385

SmD3, associated peptide analysis

affinity chromatography

nickel affinity chromatography, **318:**392

α-Py antibody column, **318:**390, 392

affinity tagging

construction of epitope-tagged gene, **318:**388–389

rationale, **318:**387

tag selection, **318:**387

vectors, **318:**395, 397–398

glycerol gradient separation of spliceosomal small nuclear ribonucleoproteins, **318:**392–393

mass spectrometry microsequencing analysis of U4/U6U5 proteins, **318:**395

transformation of knockout strain, **318:**389

yeast growth and extraction, **318:**389–390

types in yeast, **318:**386

SmD3, *see* Small nuclear ribonucleoprotein

SNAP, *see* S-Nitroso-N-acetylpenicillamine

SNAP-25, *see* Synaptosomal-associated protein-25

SNARF-1

compartmentation of ester-loaded fluorophore, **302:**350

loading of cells, **307:**128–129

ratio confocal microscopy imaging of pH, **302:**349

SNF, *see* SWI/SNF

Sniffer-patch detector

applications, **294:**191

spatial resolution, **294:**191

SNOC, *see* S-Nitrosocysteine

SNOG, *see* S-Nitrosoglutathione

Snorthern blot analysis

adaptor ligation, **306:**79–80

complementary DNA

digestion, **306:**79

preparation, **306:**78–79

messenger RNA preparation, **306:**78

overview, **306:**76, 78

polymerase chain reaction, **306:**80

screening in complementary DNA selection, **306:**88

sensitivity, **306:**78

Southern blotting, **306:**80–81

SNQ2, functions in *Saccharomyces cerevisiae*, **292:**151

SNR, *see* Signal-to-noise ratio

snRNP, *see* Small nuclear ribonucleoprotein

SOD, *see* Superoxide dismutase

Sodium/bicarbonate cotransporter, expression cloning in *Xenopus laevis* oocyte, **296:**21

Sodium borohydride, direct detection of RNA modification, **318:**13

Sodium/calcium-exchanger

antisense knockdown

cell culture

arterial myocyte cells, **314:**318–319

cardiac myocytes, **314:**461

colon carcinoma, **314:**461

neonatal cardiac myocytes, **314:**317–318

Sf9 insect cells, **314:**461–462

controls, **314:**316–317, 456–457, 459, 471, 473, 476

functional assays of effects

flash photolysis of caged calcium, **314:**467, 469–471, 476

reverse mode measurements, **314:**467, 476

half-life of exchanger, **314:**322

immunoprecipitation, **314:**320–321

knockdown efficacy of oligonucleotides, **314:**322–323
oligodeoxynucleotide design
length, **314:**455–456
sequences, **314:**315–316, 455–456
targeting region on messenger RNA,
314:457, 459
phosphorothioate oligodeoxynucleotide
synthesis and purification,
314:315–316, 459–461
pulse–chase radiolabeling of proteins,
314:319–320
specificity of effects, **314:**475–476
time course and dose response, **314:**471
toxicity assays
cardiac myocytes, **314:**475
colon carcinoma cells, **314:**464–466
uptake in cells, **314:**462–463
Western blot analysis, **314:**321
calcium assay
fluorescent dye assays, **314:**321–322,
466
solutions, **314:**466
DM-nitrophen caged calcium studies
electrophysiological measurements,
291:302–306
photolysis kinetics, **291:**301
rate constant determinations,
291:305–306
regulatory site studies, **291:**301–302,
304–305
inhibitor development, **314:**454
mechanism of exchange, **314:**455
stoichiometry of translocation, **291:**303,
314:454–455, 469
Sodium/calcium-potassium exchanger, photoreceptors
inhibitors, **315:**848
ion selectivity, **315:**849
phototransduction role, **315:**831
purification from bovine rod outer segments with cyclic nucleotide-gated
channel
anion-exchange chromatography,
315:836–837
calmodulin affinity chromatography,
315:837–838
dye affinity chromatography, **315:**837
gel electrophoresis and Western blot
analysis, **315:**841–843, 846–847

immunoaffinity methods
antibodies, **315:**838–839
immunoaffinity chromatography,
315:840
immunoprecipitation, **315:**839–840
ricin-agarose chromatography, **315:**838
rod outer segment
isolation, **315:**833–835
solubilization, **315:**835–836
reconstitution into liposomes
buffers, **315:**844
calcium flux measurements with
Arsenazo III, **315:**844–846
incubation conditions, **315:**844
spectrofluorometric assay
calcium dye loading, **315:**850–851
cell culture, **315:**850
cell transfection studies
A23187-induced fluorescence,
315:859
calcium measurements, **315:**858–859
cell lines, **315:**857–858
gramicidin effects, **315:**860
noise reduction, **315:**860, 862
single-cell imaging, **315:**862–863
transfection efficiency, **315:**860
distinguishing activity from sodium/cal-
cium exchanger by cation dependence, **315:**849, 855–856
instrumentation
single cell measurements,
315:851–852
suspension preparations, **315:**851
rod outer segment
cation dependence of calcium uptake, **315:**855–856
fluo-3 measurements and calibration,
315:852–855
preparation, **315:**849–850
rhodopsin bleaching effects on calcium measurements, **315:**857
solutions, **315:**849
stoichiometry of transport, **315:**848
structure, **315:**833, 846–847
tissue distribution of isoforms,
315:847–848
Sodium channel, *see also* Peripheral nerve
3 protein
antibodies, voltage-gated channels,
294:703

blocking peptides, *see also* μ-Conotoxin
conotoxin binding
  arginine interactions in sodium chan-
    nel binding, **294:**618–619
  δ-conotoxin inhibition of sodium
    channel inactivation, **294:**614,
    617, 619–620
  μO-conotoxin, **294:**620
  electrostatic compliance analysis, inter-
    residue distance measurement,
    294:597–598
  mechanism, **294:**581, 603–604
  mutant cycle analysis
    applications, **294:**596–597
    coupling coefficient calculation,
      294:595
    interaction energy, **294:**596
    principle, **294:**594–595
  stoichiometry of binding, **294:**580–581
  types, **294:**580
classification of voltage-gated channels,
  314:203–205
ligands delaying inactivation, **294:**614,
  617
outer vestibule modeling
  chemical modification probing, **294:**599
  effective dielectric constant determina-
    tion, **294:**598–599
  secondary structure prediction, **294:**599
pain-relief drug blocking, **314:**203
pore region, **294:**578–580, 584, 590, 602,
  604–605
selectivity and permeation factor analysis,
  294:602–603
transmembrane topology, **294:**575–577
transport vesicle isolation
  neuroblastoma cells, **294:**342–350
  squid giant axon, **294:**340–341
  velocity sedimentation on Ficoll gradi-
    ents, **294:**342–348
voltage-gated channels in neuropathic
  pain, **314:**202–203
voltage-sensitive channels in T-tubule
  membranes, **294:**292–293
voltage sensor, **294:**577, 601–602
Sodium dodecyl sulfate
  biofilm integrity assay, **310:**54–55, 118
  Sup35p NM fragment assembly assay us-
    ing solubility, **309:**669
Sodium/glucose cotransporter, expression

  cloning in *Xenopus laevis* oocyte,
    296:19
Sodium Green, loading of cells, **307:**127
Sodium nitroprusside
  cell toxicity, **301:**394
  effects on oxidative stress, **301:**415–418
  nitric oxide delivery in lipid peroxidation
    studies, **301:**445
Sodium, potassium-ATPase
  amyloid-β inhibition
    assay, **309:**741
    overview, **309:**739
  electrophysiological measurements with
    caged ATP
    bilayer experiments, **291:**291–295
    binding of caged compounds, **291:**293
    patch clamp studies, **291:**295–298
    photolysis conditions, **291:**291, 293
    time-resolved measurements, **291:**293–
      294, 297–298
  ion specificity, **291:**292–293
  physiological function, **291:**290
  reaction mechanism, **291:**290–291
Sodium trioxodinitrate, *see* Angeli's salt
Solid-phase differential display
  complementary DNA synthesis,
    303:502–503
  hematopoietic stem cells, **303:**496, 500,
    502
  hot start-mediated polymerase chain reac-
    tion, **303:**503
  messenger RNA capture, **303:**502
  overview, **303:**498–500
  probe-coated bead preparation,
    303:500–502
  product isolation from differential dis-
    play, **303:**503–504
  reproducibility, **303:**500
  RNA lysate preparation, **303:**502
  sensitivity, **303:**500
Solid-phase extraction, sphingolipid purification
  advantages, **312:**101–102
  aminopropyl chromatography, **312:**103,
    106
  ceramides, **312:**106–107
  loading capacity, **312:**109–110
  materials, **312:**102
  recovery, **312:**110
  sample sources, **312:**108–109

solvent systems, **312:**103, 105–106
sphingoid bases
  hydrolysis conditions
    ceramides, **312:**110–114
    cerebrosides, **312:**112
  resolution, **312:**107–108
  serum and urine samples, **312:**114
  weak cation-exchange chromatography,
    312:103, 106
Solid-phase nucleus
  applications, **304:**64, 75–76
  components, **304:**63–64
  DNA replication
    assays, **304:**70–72
    characteristics, **304:**64, 68, 70
    restriction digest analysis, **304:**70–71
  DNA template preparation, **304:**65–66
  extract preparation from *Xenopus* eggs,
    304:66
  α-fetoprotein gene transcription
    chromatin assembly, **304:**66–68
    coupling with replication, **304:**72,
      74–75
    gene constructs, **304:**65
    hepatoma factor effects, **304:**75
    methylation of gene, **304:**70
    repressed template transcription,
      304:74–75
Solid-state nuclear magnetic resonance, *see*
    Nuclear magnetic resonance, solid-state
Solute diffusion, *see* Photobleaching, green
    fluorescent protein
Solvent accessibility, calculation for pro-
    teins, **295:**110–111, 297–301
Sorbarod biofilm fermenter, *see* Biofilm fer-
    menter
Southern blot
  chemiluminescence detection
    advantages, **305:**451–452, 463
    clinical applications, **305:**452–454, 456
    Prader–Willi syndrome or Angelman
      syndrome detection
      background on syndromes,
        305:460–462
      blotting, **305:**463
      dot blots, **305:**462–463
      hybridization and washes, **305:**463
      principle, **305:**462
      probe generation, **305:**462
      reagents and solutions, **305:**462

signal generation and detection,
        305:463
    T-cell receptor, gene rearrangement
      assay
      blotting, **305:**458–460
      hybridization and washes, **305:**458
      overview, **305:**456–457
      probe generation, **305:**458–459
      reagents and solutions, **305:**457–458
      signal generation and detection,
        305:458–459
  chromatin remodeling during nucleotide
    excision repair, **304:**345
  confirmation of positive genomic clones,
    296:502–504
  cross-linked DNA, **304:**507–510,643–645
  indirect end-labeling and low-resolution
    mammalian nucleosome positioning,
    304:634–635
  micrococcal nuclease and yeast mono-
    nucleosome analysis, **304:**395–396
  principle, **305:**450–451
  restriction enzyme accessibility mapping
    of chromatin, **304:**376
  screening of *MDR1* transgenic mice,
    292:577–578
  transgene analysis, **306:**201
  transgenic mouse, sorting different lines
    from same founder, **316:**174
  X-linked red–green photopigment genes,
    determination of number and ratio,
    316:655
  yeast nucleosome positioning assay,
    304:395–396
Sp2/0 cells, *see* Myeloma cell lines
SP-10, green fluorescent protein as pro-
    moter reporter
  applications, **302:**283–284
  microinjection
    transgene purification, **302:**275–276
    transgenic mouse generation, **302:**277
  overview, **302:**272–273, 283
  reporter gene constructs, **302:**273–275
  spermatogenesis, stages of expression
    cycle of the seminiferous epithelium,
      302:278–279
    patterns, **302:**282–283
    phases, **302:**278
    transillumination-assisted microdissec-
      tion, **302:**279–282

tissue specificity of expression,
  302:277–278
SPA, *see* Scintillation proximity assay
SPE, *see* Solid-phase extraction
Spectroradiometry, *see* Ultraviolet radiation
Sperm, *see* Centrifuge microscope
Spherical aberration
  cell volume measurement error, **307:**287–
    288, 290
  overview, **307:**113–114
Sphinganine, *see also* Dihydrosphingosine
  chemical synthesis
    column chromatography, **311:**521–522
    (2*S*)-3-ketosphinganine hydrochloride
      synthesis, **311:**531–532
    materials, **311:**519
    overview, **311:**421–423, 519, 523–525
    purity analysis, **311:**522
    thin-layer chromatography,
      311:520–522
  hydroxylase, *see* Dihydrosphingosine 4–
    hydroxylase
  phosphatidate phosphatase modulation,
    see Phosphatidate phosphatase, sphin-
    goid base inhibition
  radiolabeled ganglioside preparation,
    311:654–655
  radiolabeling
    [4,5–³H]D-*erythro*-dihydrosphingosine
      synthesis
      D-*erythro*-4, **311:**5-dehydrosphingo-
        sine as starting material,
        311:43–44
      principle, **311:**487–488, 519
      D-*erythro*-sphingosine as starting ma-
        terial, **311:**42–43
    extractions, **311:**487
    flash column chromatography, **311:**486
    purity analysis, **311:**487
    radioactivity determination and detec-
      tion, **311:**487
    thin-layer chromatography,
      311:485–487
    D-*erythro*-sphinganine, tritiated com-
      pound synthesis, **311:**148–149
    [3–³H]sphinganine synthesis
      3-ketodihydro-C₂-ceramide as start-
        ing material, **311:**495
      3-keto-sphinganine as starting mate-
        rial, **311:**495

principle, **311:**491–492
stereochemistry, **311:**244
tandem mass spectrometry, **312:**38
Sphinganine *N*-acyltransferase, *see* Acyl-
  CoA : sphinganine *N*-acyltransferase
Sphinganine kinase
  assay of plant enzyme, **311:**146–147
  substrate specificity of plant enzyme,
    311:147
Sphinganine-1-phosphate, radiolabeling
  enzymatic synthesis, overview, **311:**513,
    515
  [4,5–³H]sphinganine-1-phosphate synthe-
    sis, **311:**517–518
Sphinganine-1-phosphate phosphatase, *see*
  Dihydrosphingosine-1-phosphate phosphatase
Sphingofungin, serine palmitoyltransferase
  inhibition, **311:**335, 350
Sphingolipid, *see also specific lipids*
  binding protein labeling
    affinity labeling, **311:**571–573
    cross-linking, **311:**570–571
    overview, **311:**569–570
    photoaffinity labeling, *see* Photoaffinity
      labeling
  biosynthesis mutants, *see* Serine palmi-
    toyltransferase; Yeast sphingolipid
    biosynthesis mutant
  biosynthetic pathways, overview,
    311:303–305, 480, 569
  fungus, synthetic pathways, **311:**335
  inhibitors
    bioassay for sphingolipid inhibition in
      Candida albicans, **311:**337–339
    isolation using bioassay guidance,
      311:340–341
  isolation, *see* Sphingolipid isolation
  phase separation, *see* Lipid phase sepa-
    ration
  plants
    metabolism overview, **311:**132
    types, **311:**130–131
  pulse–chase studies of metabolism
    advantages, **311:**657
    applications, **311:**657
    bulk separation of sphingolipids,
      311:661, 663
    cell growth and incubations,
      311:659–660

culture media analysis
  lipid extraction, **311:**666
  low molecular weight tritiated metab-
    olite analysis, **311:**666
  rationale, **311:**665
  tritiated water determination,
    311:666
delipidized pellet analysis, **311:**665
extraction of sphingolipids, **311:**660
identification of sphingolipids
  β-N-acetylhexosaminidase treatment,
    311:668–669
  ceramide, **311:**680–681
  ceramide-1-phosphate, **311:**682
  chemical hydrolysis, **311:**667
  derivatizations, **311:**667
  diacylglycerol kinase treatment,
    311:670
  endoglycoceramidase treatment,
    311:669
  β-galactosidase treatment,
    311:667–668
  high-performance thin-layer chroma-
    tography, **311:**667
  isolation, **311:**666–667
  sialidase treatment, **311:**669
  sphingomyelinase treatment,
    311:669–670
  sphingosine, **311:**681–682
  sphingosine-1-phosphate, **311:**682
materials, **311:**657–659
overview, **311:**656–657
pathway and trafficking route identification
  degradation pathways, **311:**673–674
  extralysosomal degradation pro-
    cesses, **311:**674–676
  ganglioside lactone biosynthesis,
    311:678–679
  glycosylation of tritiated ganglio-
    sides, **311:**677–678
  inhibition of enzymes, **311:**672–673
  manipulation of trafficking, **311:**672
  overview, **311:**670–672
  recycling processes, **311:**676–677
  signal transduction, **311:**679–682
radioactivity determination, **311:**665
sphingolipid solution preparation,
  311:659
thin-layer chromatography, **311:**663

radiolabeling of backbone, overview,
  311:480–481, 483–485, 497–500
trafficking, see BODIPY-labeled sphingolipids; GM1; Translocation, sphingolipids
yeast as model system
  advantages, **311:**319
  cloning of genes, **311:**328–331
  media preparation, **311:**321, 323
  metabolism overview, **311:**319, 321
  phytosphingosine growth inhibition
    studies
    liquid culture, **311:**323–324
    positive selection markers, **311:**327
    solid culture, **311:**324–325
    strain background, **311:**323
    synthetic media, **311:**325–326
  sphingolipid metabolites, sources and
    preparations, **311:**321
  thin-layer chromatography of sphingo-
    lipids, **311:**327–328
Sphingolipid activator proteins, see also
  GM2 activator protein
  genes, **311:**255
  glycosphingolipid uptake in deficient fi-
    broblasts, serine radiolabeling,
    311:274–275
  mechanism of action, **311:**255, 257
  processing, **311:**255, 258–259
  purification, **311:**263
  sphingolipid activator protein B
    defects, **311:**258, 260
    globotriaosylceramide degradation
      assay with α-galactosidase A
      incubation conditions, **311:**264–265
      liposome preparation, **311:**264
      materials, **311:**263–264
      product separation and quantifica-
        tion, **311:**265
    GM1 degradation with β-galactosidase
      incubation conditions, **311:**266
      materials, **311:**266
      principle, **311:**265–266
      product separation and quantifica-
        tion, **311:**267
    sulfatide degradation assay with arylsul-
      fatase A, **311:**267
  sphingolipid activator protein C
    defects, **311:**259
    galactosylceramide degradation assay

with β-galactosylceramidase,
311:269
glucosylceramide degradation assay
with glucocerebrosidase
incubation conditions, **311:**268–269
liposome preparation, **311:**268–269
materials, **311:**267–268
product separation and quantification, **311:**269
sphingolipid activator protein D
acid ceramidase requirement, **311:**201–
202, 259
ceramide degradation assay with acid
ceramidase
incubation conditions, **311:**270–271
liposome preparation, **311:**270–
271
overview, **311:**270
defects, **311:**269–270
separation from acid ceramidase,
311:206
sulfatide loading in sphingolipid activator
protein-deficient fibroblasts,
311:273–274
Sphingolipid ceramide N-deacylase
ω-aminoceramide synthesis and purification, **311:**687–688
assay
incubation conditions, **311:**299
principle, **311:**298–299
substrate preparation, **311:**299
thin-layer chromatography, **311:**299
unit definition, **311:**300, 684
biological significance, **311:**297–298
condensation reaction in radiocompound
synthesis
[14C]ceramide synthesis and purification, **311:**684–685
[14C]galactosylceramide synthesis and
purification, **311:**685–687
[14C]globotetoraosylceramide synthesis
and purification, **311:**685–687
[14C]GM1a synthesis and purification,
311:685–687
materials, **311:**683
overview, **311:**682–684
specificity of reaction, **311:**684
[14C]sulfatide synthesis and purification, **311:**685–687
ion effects, **311:**300

lysosphingolipid preparation in biphasic
system, **311:**302–303
organic solvent effects, **311:**302
pH optimum, **311:**300
purification, **311:**298, 300
reaction catalyzed, **311:**298, 682
substrate specificity, **311:**300
Sphingolipid isolation
anion-exchange chromatography
chromatography, **312:**70–71
column preparation, **312:**70
resin preparation, **312:**70
sample preparation, **312:**70
separation of anionic and neutral sphingolipids, **312:**69
denatured protein removal,
312:67
desalting
overview of techniques, **312:**67–68
reversed-phase chromatography,
312:68–69
extraction, **312:**66–67
lipid contaminant removal, **312:**67
solid-phase extraction, see Solid-phase extraction, sphingolipid purification
structural overview, **312:**64–65
thin-layer chromatography
applications, **312:**71–72
autoradiography, **312:**75–77
development of chromatograms,
312:72–73
sample application, **312:**72
solvent systems, **312:**73–75
staining, **312:**78–79
Sphingolipid methods, compilation from
Methods in Enzymology
inhibitors of metabolism, **312:**586–587
isolation and analysis, **312:**589–591
metabolism, **312:**583–586, 591
protein interactions and cellular targets,
312:591–592
synthesis, **312:**587–589
transport and trafficking, **312:**592
Sphingomyelin
analog synthesis, **311:**400
assay in caveolin-enriched membranes
bacterial sphingomyelinase assay,
311:192–193
thin-layer chromatography, **311:**193
assay of in situ synthesis

choline incorporation radioassay,
  311:40
fluorescence assay with NBD-ceramide,
  311:39–40
serine radiolabeling assay, **311:**40–41
biological functions, **311:**31
biosynthesis, **311:**31–33
BODIPY-sphingomyelin, *see also* Hematopoietic stem cell, fluorescence selection of acid sphingomyelinase expression
  preparation, **312:**332
  pulse–chase labeling, **312:**332–333
chemical synthesis
  demethylated sphingomyelin synthesis,
    311:533–534
  lysosphingomyelin synthesis, **311:**533
  N-palmitoylsphingomyelin synthesis
    3-*O*-benzylceramide preparation,
      311:539–541, 544–546
    3-*O*-benzylsphingomyelin prepara-
      tion and conversion,
      311:546–547
    materials, **311:**543
    overview, **311:**536–539, 541–543
    phosphatidylating agents,
      311:541–542
    thin-layer chromatography, **311:**543
  principles, **311:**532–533
  problems in total synthesis,
    311:536–539
delayed ion extraction matrix-assisted desorption ionization time-of-flight
  mass spectrometry, **312:**92, 99
fluorescence labeling, **312:**297
radiolabeling
  [^{14}C-choline]lysosphingomyelin synthe-
    sis, **311:**512–513
  [^{14}C-choline]sphingomyelin synthesis,
    311:509, 511–512
  extractions, **311:**502
  N-[1–^{14}C-hexanoyl]D-*erythro*-
    sphingosylphosphorylcholine
    synthesis, **311:**508
  flash column chromatography, **311:**501
  materials, **311:**500–501
  principles, **311:**503–504, 509
  purity analysis, **311:**502
  radioactivity determination and detec-
    tion, **311:**502

thin-layer chromatography, **311:**501
tandem mass spectrometry, **312:**42
Sphingomyelinase, *see also* Acid sphingomyelinase; Alkaline sphingomyelinase;
  Neutral sphingomyelinase
activators, **311:**164
assay with fluorescent sphingolipids
  lysosomal acid enzyme, **312:**299–300
  neutral magnesium-dependent enzyme,
    312:300
hematopoietic stem cell, fluorescence selection of acid sphingomyelinase expression
  applications, **312:**330–331, 337–338
  BODIPY-sphingomyelin
    preparation, **312:**332
    pulse–chase labeling, **312:**332–333
  cell preparation, **312:**331
  c-Kit immunolabeling, **312:**332, 335
  colony-forming units–spleen assay,
    312:334, 336
  enzyme assays, **312:**333–334
  enzyme deficiency in Niemann-Pick dis-
    ease, **312:**331, 337–338
  fluorescence-activated cell sorting,
    312:333–335
  lineage depletion, **312:**331–332
  retrovirus
    amphotropic versus ecotropic vi-
      ruses, **312:**336
    transduction, **312:**333
reaction catalyzed, **311:**176
sphingolipid identification, **311:**669–670
sphingomyelinase D toxicology,
  311:234–235
substrates for assays, **311:**176
translocation assay of sphingolipids,
  312:570–571
types, **311:**164
Sphingomyelin synthase
BODIPY-labeled ceramide assay on replica colonies
  applications, **312:**530
  incubation on polyester disks,
    312:530–531
  lipid transfer from disk to thin-layer
    chromatography plate, **312:**531
  thin-layer chromatography, **312:**531,
    533
fluorescence assay with NBD-ceramide

incubation conditions, **311:**37
  optimization, **311:**38
  quantification, **311:**37–38
  substrate, **311:**37
inhibitors, **311:**42, 387
kinetic mechanism, **311:**32
membrane association, **311:**31–32
pH optimum, **311:**32
phosphorylcholine donors, **311:**31
radiolabeled substrate assay
  substrate delivery to membranes
    bovine serum albumin, **311:**35–36
    lipid transfer protein, **311:**35
    small uniamellar vesicles, **311:**36–37
  substrate preparation, **311:**34–35
in situ activity assays, *see* Sphingomyelin
tissue distribution, **311:**32–33
Sphingosine
  caged compounds, *see* Caged sphingo-
    lipids
  chemical synthesis of derivatives
    aldol reactions
      chiral auxiliaries, **311:**431–433
      protected glycines, **311:**430
      2-nitroethanol, **311:**428–430
      titaniumm enolate of iminoglycinate,
        311:433
      L-valineglycine, **311:**430–431
    1-amino derivatives
      2-amino-1-(4-morpholinyl)-4-
        octadecene-3-ol, **311:**455
      2-amino-1-(1-pyrrolidinyl)-4-
        octadecene-3-ol, **311:**455–456
      2-azido-1-(methylsulfonyloxy)-4-
        octadecene-3-ol, **311:**454
      2-azido-1-(4-morpholinyl)-4-
        octadecene-3-ol, **311:**454–455
      2-azido-1-(1-pyrrolidinyl)-4-
        octadecene-3-ol, **311:**455
      3-benzoyloxy-1,2-diazido-4-
        octadecene, **311:**456–457
      3-benzoyloxy-1,2-di-*O*-
        trifluoroacetylamino-4-
        octadecene, **311:**457
      1,2-diamino-4-octadecene-3-ol,
        311:457
      1,2-diazido-4-octadecene-3-ol,
        311:456
    overview, **311:**445–446
    azidosphingosine, **311:**445, 448

D-*erythro*-*N*-Boc-sphingosine synthesis,
  311:526–527
column chromatography, **311:**521–522
(*R*,*R*)-1,2-divinylethylene glycol as starting material, **311:**426
diastereoselective iodocyclization of
  trichloroacetimidates, **311:**434–
  436
Diels-Alder cycloaddition of *N*-sulfi-
  nylcarbamates, **311:**436
diethyl tartate derivatives as starting
  material, **311:**426, 428, 437–438
D-galactose as starting material,
  311:413–417, 441–442
D-glucosamine as starting material,
  311:408–412
D-glucose as starting material,
  311:407–408
heptadec-2-en-1-ol as starting material,
  311:421–422
Horner-Wittig reactions, **311:**433–434
intramolecular opening of epoxides,
  311:422–423
D-isoascorbic acid as starting material,
  311:419–421
(2*S*)-3-keto-*N*-Boc-sphingosine synthesis, **311:**528–531
(2*S*)-3-ketosphingosine hydrochloride
  synthesis, **311:**531
Koskinen and Koskinen synthesis
  α,β-unsaturated enone formation,
    311:462–464
  analog synthesis, **311:**471–474
  antiselective reduction, **311:**466–469
  deprotection, **311:**341
  1,1-dimethylethyl 4-[(dimethoxyphos-
    phinyl)acetyl]-2,2-dimethyl-3-
    oxazolidinecarboxylate,
    311:474–475
  1,1-dimethylethyl 4-[(2(*E*)-hexade-
    cenyl-1-oxo)-2,2-dimethyl-3-
    oxazolidinecarboxylate, **311:**475
  1,1-dimethylethyl-4-(1-(*R*)-hydroxy-
    2-(*E*)-hexadecenyl)-2,2-dimethyl-
    3-oxazolidinecarboxylate,
    311:478–479
  1,1-dimethylethyl-4-(1-(*S*)-hydroxy-2-
    (*E*)-hexadecenyl)-2,2-dimethyl-
    3-oxazolidinecarboxylate,
    311:479

1,1-dimethylethyl-4-{1-(R)-hydroxy-2-(E)-[7-(1-pyrenyl)]-heptenyl}-2,2-dimethyl-3-oxazolidinecarboxylate, 311:479

1,1-dimethylethyl-4-{1-oxo-2-(E)-[7-(1-pyrenyl)-nonenyl]}-2,2-dimethyl-3-oxazolidinecarboxylate, 311:478

enone reduction, overview, 311:464–465

ethyl 7-(1-pyrenyl)heptanoate, 311:476–477

ethyl 7-(1-pyrenyl)-hept-6-enoate, 311:476

ethyl 6-triphenylphosphine hexanoate bromide, 311:475–476

Horner-Wadsworth-Emmons reaction, 311:458–459, 463–464, 472

β-ketophosphonate formation, 311:459–461, 472

materials, 311:473–474

overview, 311:458–459

7-(1-pyrenyl)-heptanal, 311:477–478

7-(1-pyrenyl)-heptanol, 311:477

D-serine as starting material, 311:472–473

stereochemical assignments by nuclear magnetic resonance, 311:470

syn selective reduction, 311:465–466

yield, 311:471

materials, 311:519

1-O-methyl derivative, see Ceramide

4-methyl-cis-sphingosine synthesis

2-azido-1,3-O-benzylidene-4-methyl-4-octadecene-1,3-diol, 311:453

3,5-O-benzylidene-3,5-dihydroxy-4-(methylsulfonyloxy)-2-pentanone, 311:452

1,3-O-benzylidene-4-methyl-2-(methylsulfonyloxy)-4-octadecene-1,3-diol, 311:452–453

1,3-O-benzylidene-2-(methylsulfonyloxy)pentane-1,3,4-triol, 311:451–452

1,3-O-benzylidenepentane-1,2,3,4-tetrol, 311:450–451

4-methyl-cis-azidosphingosine, 311:453

4-methyl-cis-sphingosine, 311:453

Nugent and Hudlicky synthesis, 311:436–437

overview, 311:391–392, 519, 523–525

photoaffinity labels, 311:594–595

purity analysis, 311:522

ribonolactone starting material, 311:418–419

L-serine as starting material
biomimetic approach, 311:402–404

blocking groups, 311:394

Boutin and Rapoport synthesis, 311:400–401

deprotection of oxazolidine-carboxylate intermediates, 311:397

double bond generation between C-4 and C-5, 311:399

Garner and Parks synthesis and modifications, 311:392–399

Koskinen and Koskinen synthesis, 311:458–479

nucleophile addition to Garner aldehyde, 311:395–399

Polt synthesis, 311:402

Tkaczuk and Thornton synthesis and modifications, 311:399–400

Sharpless asymmetric epoxidations, 311:421–426

solid-supported sphingosines, 311:438–439

(2S,3R)-D-erythro-sphingosine synthesis, 311:527

C₁₈-sphingosine
(2S,3R,4E)-2-azido-1,3-benzylidene-4-octadecene-1,3,4-diol, 311:448

azidosphingosine conversion to sphingosine, 311:448–449

(2R,3R,4E)-1,3-O-benzylidene-4-octadecene-1,3,4-triol, 311:447–448

2,4-O-benzylidene-D-threose, 311:447

thin-layer chromatography, 311:520–522

titanium diisopropoxide diazide in stereosynthesis, 311:425–426

D-xylose as starting material, 311:404–406

fluorescent substrates, see BODIPY-labeled sphingolipids; Lissamine rhodamine-labeled sphingolipids

identification in pulse–chase studies of
signaling, **311:**681–682
oxidation of double bond in glycosphingolipids for serine and ceramide acid
precursors for glyconjugate synthesis
gel electrophoresis analysis of glycopro-
teins, **312:**478–479
glycoconjugate binding assays
neoglycoprotein conjugates,
312:486–487
neohydrocarbon conjugates,
312:485–486
glycoconjugate synthesis
bovine serum albumin conjugation
and isolation, **312:**483–484
neohydrocarbon conjugate preparation, **312:**484–485
glycosyl aldehyde versus carboxylic
acid products, **312:**474
materials, **312:**476–478
oxidation reactions and workup,
312:481–482
peracetylated glycolipid synthesis and
purification, **312:**479
N-protected deacyl glycosphingolipid
synthesis
N-tert-butoxycarbonyl derivatives,
312:480–481
deprotection following oxidation,
312:482–483
N-trifluoroacetyl derivatives, **312:**480
rationale, **312:**473–474
thin-layer chromatography, **312:**478
phosphatidate phosphatase modulation,
see Phosphatidate phosphatase, sphingoid base inhibition
protein kinase C modulation, *see* Protein
kinase C
radiolabeling
N-[1–³H-acetyl]-D-*erythro*-sphingosine
synthesis, **311:**505
amide labeling principles, **311:**503–504
[3–³H]L-*threo*-azidosphingosine synthesis, **311:**449–450
carbon-14, **311:**444
extractions, **311:**487, 502
flash column chromatography, **311:**486,
501
N-[1–¹⁴C-hexanoyl]-D-*erythro*-sphin-
gosine synthesis, **311:**506–507

materials, **311:**500–501
N-palmitoyl-D-*erythro*-sphingosine synthesis, **311:**507–508
purity analysis, **311:**487, 502
radioactivity determination and detection, **311:**487, 502
[3–³H]sphingosine synthesis
3-keto-*N-tert*-butoxycarbonyl-
sphingosine as starting material,
311:493–494
3-keto-C₂-ceramide as starting material, **311:**494–495
3-keto-sphingosine as starting material, **311:**492–493
principle, **311:**491–492
thin-layer chromatography, **311:**485–
487, 501
tritium, **311:**443–444
signal transduction role, **311:**234, 441
stereochemistry
biological function, **312:**217
circular dichroism analysis, *see* Circular
dichroism, sphingolipids
high-performance liquid chromatography analysis, **312:**227
stereoisomers, **311:**392
structure, **311:**391
tandem mass spectrometry, **312:**38
tritium labeling for ganglioside sialidase
assay
assay, **312:**356
C-3 labeling, **312:**354–355
double-bond labeling, **312:**355–356
Sphingosine-dependent kinase 1
activity assays
simplified radioimmunoassay,
312:386–387
standard assay, **312:**385–386
14-3-3 as substrate
effects of phosphorylation, **312:**381
phosphorylation site, **312:**381
isolation
anion-exchange chromatography,
312:385
buffers, **312:**382–383
cell culture and stimulation with 12-*O*-
tetradecanoylphorbol 13-acetate,
312:382
extract preparation, **312:**383
sphingolipid specificity, **312:**381

subcellular localization, **312:**382
Sphingosine kinase
  activators, **311:**216–219
  assay
    cell lysate preparation, **311:**220–221
    incubation conditions, **311:**221–222
    materials, **311:**220
    overview, **311:**219
    product identification and characteriza-
      tion, **311:**222–223
    reversed-phase high-performance liq-
      uid chromatography of *o*-phthalal-
      dehyde derivatives, **312:**8–9
    substrate preparation, **311:**221
    thin-layer chromatography, **311:**222
  biological significance, **311:**215–218
  genes in yeast, **311:**218
  inhibitors, **312:**10
  LCB4 and *LCB5* cloning in yeast,
    311:329–330
  reaction catalyzed, **311:**215
  sphingosine 1-phosphate measurement,
    see Sphingosine 1-phosphate
  subcellular localization, **311:**245
  substrate specificity, **311:**219
  tissue distribution, **311:**218
  transfection studies, **311:**218–219
Sphingosine 1-phosphate
  assays
    enzymatic assay
      lipid extraction, **312:**12, 15
      phospholipid normalization, **312:**13
      radiolabel quantification, **312:**13
      rat tissue and cultured cell levels,
        312:15–16
      reaction conditions, **312:**12–13
      sphingosine interference testing,
        312:14–15
      sphingosine kinase preparation,
        312:11–12
      standard curve, **312:**13–14
    mass spectrometry, **312:**10–11
    reversed-phase high-performance liq-
      uid chromatography of *o*-phthalal-
      dehyde derivative
      chromatography and detection,
        312:4–5
      derivatization, **312:**4–5
      lipid extraction, **312:**4, 6
  chemical synthesis, **311:**519

functions, **311:**233–234, 568, 589, **312:**9,
  401
identification in pulse–chase studies of
  signaling, **311:**682
phosphorous-32 radiolabeling,
  312:402–403
radiolabeling
  enzymatic synthesis, overview, **311:**513,
    515
  N-[^{3}H-palmitoyl]D-*erythro*-sphingosine-
    1-phosphate synthesis, **311:**516
  D-*erythro*-[^{32}P]sphingosine-1-phosphate
    synthesis, **311:**238–239, 516–517
receptors
  binding assay, **312:**403–404
  Edg receptors, **312:**401–402
  factors affecting binding
    cells versus membranes, **312:**405–406
    ligand concentration, **312:**406
    nonspecific binding, **312:**404
    receptor expression, **312:**406
    temperature, **312:**405
  overview of types, **312:**10, 401–402
  signal transduction, **312:**401–402
  tissue distribution, **312:**402
regulation of levels, **312:**9–10
tandem mass spectrometry, **312:**39
Sphingosine 1-phosphate lyase
  assay
    incubation conditions, **311:**251
    materials, **311:**250–251
    overview, **311:**249–250
    substrate synthesis, **311:**252–254
    thin-layer chromatography, **311:**252
  biological significance, **311:**245
  inhibitors, **311:**248–249
  reaction catalyzed, **311:**245–246
  sequence analysis, **311:**246–247
  stereospecificity, **311:**247–248
  substrate specificity, **311:**247–248
  tissue distribution, **311:**246
Sphingosine 1-phosphate phosphatase, *see*
  also Dihydrosphingosine-1-phosphate
  phosphatase
  assay
    intact cell assay, **311:**241–242
    substrate synthesis, **311:**238–239
    in vitro assay, **311:**240
  biological significance, **311:**233–234
  requirements for catalysis, **311:**244

structure, **311:**236
substrate specificities, **311:**236–237
types in mammals, **311:**235–236
Sphingosine transacetylase, *see* Platelet-activating factor:sphingosine transacetylase
Sphingosylphosphorylcholine, *see* Sphingomyelin
Spin-label electron spin resonance
    instrumentation, **294:**61–62
    ion channel applications, **294:**59–60, 65–66, 70
    labeling of lipids and proteins, **294:**60–63
    lipid–protein interactions
        difference spectroscopy, **294:**63–64
        number of lipid assciation sites on protein, **294:**64
        stoichiometry determination, **294:**64–66
    sample cells, **294:**62
    saturation and relaxation enhancements, **294:**70–75
    saturation transfer measurements, **294:**66–70
    sensitivity, **294:**59
Splinted ligation, *see* RNA ligation
Split receptor, mapping tertiary interactions in rhodopsin
    cysteine scanning mutagenesis
        absorption spectroscopy of double mutants, **315:**139
        number of mutants, **315:**132
        transducin activation by double mutants, **315:**139–140
    disulfide bond formation in cysteine mutants
        detection with Western blot analysis with concanavalin A, **315:**132, 138
        induction of formation, **315:**132–134, 137–138, 140
        kinetics of formation, **315:**140–143
        native cysteine cross-linking, **315:**141
    fragment design and structure, **315:**130–132
    interpretation of results, **315:**142
    principles and overview, **315:**130, 134–135
    receptor preparation
        constructs, **315:**135
        immunoaffinity chromatography, **315:**136–137
        reconstitution, **315:**136

transfection and expression in COS-1 cells, **315:**135–136
SPR, *see* Surface plasmon resonance
SPT, *see* Serine palmitoyltransferase
Squalene, peroxide analysis by high-performance liquid chromatography, **319:**616–617, 619–621
Src, regulation of cystic fibrosis transmembrane conductance regulator, **293:**184–185
Src kinase, chemiluminescence immunoassay
    detection, **305:**415
    dioxetane substrate characteristics, **305:**411–412
    overview of kinase assays, **305:**410–411
    peptide capture and kinase reaction, **305:**415–416
    principle, **305:**412–413
    reagents, **305:**413–414
    sensitivity, **305:**415–416
    solutions, **305:**414–415
    troubleshooting, **305:**415
SSCP, *see* Single-stranded conformational polymorphism
SSH, *see* Suppression subtractive hybridization
SSP, *see* Site-specific psoralen
Stannic chloride, effect on Chesapeake Bay sediment bacteria, **310:**638
Staphylococcal nuclease
    denaturation studies with automated titrating fluorometer
        chemical denaturation
            ammonium sulfate titrations, **295:**162–163
            apparent equilibrium constant calculation, **295:**152–153
            background correction, **295:**155–156
            denaturant delivery, **295:**156–157, 159
            free energy calculations, **295:**153
            reversibility testing, **295:**159–161
            sample preparation, **295:**154–155
            schedule of titrant addition, **295:**151–152, 157–159
        data processing, **295:**152
        instrumentation, **295:**150–151
        pH denaturation, **295:**166–167
        thermal denaturation

apparent equilibrium constant calculation, **295:**153–154
reversibility testing, **295:**165
sample preparation, **295:**163–165
van't Hoff enthalpy calculation, **295:**154
fluorescence quenching
  acrylamide quenching, **295:**169
  potassium iodide quenching, **295:**167–169
  Stern–Volmer plots, **295:**168–169
Staphylococcus epidermidis, adhesion and surface growth studies with parallel plate flow chamber, **310:**531–532
Stargardt disease, ABCR mutations and pathogenesis, **315:**864, 879–880, 896–897
Stathmin
  antisense knockdown studies in cultured neurons
    dose–response and time course of inhibition, **314:**158
    functional effects in PC12 cells, **314:**158, 160
    oligonucleotide design, **314:**155, 158
    oligonucleotide uptake, **314:**158
  neuronal differentiation role, **314:**149–150
  phosphorylation, **314:**155
Ste6p
  a-factor export role, **292:**156, 193
  assays
    halo assays of a-factor export
      patch halo assays, **292:**201
      principle, **292:**198
      sensitivity, **292:**198–199
      spot assays, **292:**199–201
    immunoprecipitation of a-factor, **292:**201–202
    mating assay of a-factor export
      filter assay, **292:**196–197
      overview, **292:**156, 195–196
      patch mating assay, **292:**197–198
      plate assay, **292:**197
  epitope tagging, **292:**203–204
  functional replacement
    chimeric molecules, **292:**212
    mouse Mdr3, 134–135, 157, 159, 194, 211, 382, 398, 421, 423–425
    pfmdr1, **292:**174, 178–180, 211

half-life, **292:**203
half-molecule reconstitution studies, **292:**209–210
heterologous expression in *Escherichia coli* and topology studies, **292:**379, 381
immunofluorescence assay in subcellular localization, **292:**206–208
immunoprecipitation, **292:**205–206
pulse-chase metabolic labeling, **292:**204–205
sequence homology with P-glycoproteins, **292:**109
site-directed and random mutagenesis analysis, **292:**210–211
Western blot analysis, **292:**208–209
STEM, *see* Scanning transmission electron microscopy
Stern–Volmer plot, fluorescence quenching experiments, **295:**168–169
Sterol dehydrogenase, phase partion assay
  applications, **316:**368, 382
  enzyme preparation, **316:**363
  incubation conditions, **316:**365
  materials, **316:**361
  membrane preparation, **316:**363
  nicotinamide adenine dinucleotide tritiation, **316:**361
  pH, **316:**360
  principle, **316:**360
Stimulated emission, theory, **291:**179–182
Stochastic dynamics, *see* Three-dimensional structure, molecular modeling
Stopped-flow spectroscopy
  advantages over steady-state measurements, **308:**149–150
  dead time, **308:**150
  peroxynitrite reactions
    bimolecular reactions and rate constants, **301:**350–352
    carbon dioxide reaction, **301:**351–352, 356–358
    detection, **301:**345
    dimethyl sulfoxide oxidation, **301:**348–349
    pK_a determination, **301:**346–348
    protein modifications, **301:**350–352
    solutions, **301:**344–345
  substrate channeling analysis

thymidylate synthase–dihydrofolate re-
    ductase, **308:**137–139
tryptophan synthase, **308:**117–118
Stratum corneum
  ceramides, *see* Stratum corneum, infrared
    spectroscopy of ceramides
  structure, **300:**413–414
  α-tocopherol analysis
    gradient and physiological relevance,
        300:418–419
    high-performance liquid chromatogra-
        phy, **300:**416
    lipid extraction, **300:**415–416
    quantitative analysis, **300:**416–418
    tape stripping, **300:**414–415
Stratum corneum, infrared spectroscopy of
    ceramides
  ceramide types, **312:**228–229
  horizontal attenuated total reflection,
      312:231–232
  sample preparation, **312:**231–232
  spectral regions sensitive to structure and
      organization
    CD_2 rocking modes, **312:**241–243
    methylene scissoring modes,
        312:236–238
    methylene stretching modes,
        312:233–236
    methylene wagging modes,
        312:238–241
    overview, **312:**232–234
    polar region vibrations, **312:**243–244
    transmission spectroscopy, **312:**231–
        232
Streptococcus biofilm, *see also* Endocarditis;
    Oral biofilm
  cellulose membrane filter
    culture, **310:**579
    penetration assay of antibacterial
        agents, **310:**585–586
  glucosyltransferases of mutans strepto-
      cocci
    assay, **310:**516–517
    mutagenesis in *Streptococcus mutans*
        insertional inactivation of *gtfB* and
            gtfC, **310:**521–522
        rationale, **310:**519
        reconstruction of mutants,
            310:522–523
        transformation, **310:**520

plaque formation role, **310:**513–514,
    516
purification of cell-associated enzyme
    anion-exchange chromatography,
        310:518
    cell growth, **310:**518
    extraction, **310:**518
    hydroxylapatite chromatography,
        310:519
    types, **310:**517–518
hexose assay, **310:**582–584
hydrophobicity assay, **310:**514–515
PA surface antigen, hydrophobicity,
    310:514
salivary glycoprotein adsorption of mu-
    tans streptococci
    assay, **310:**515–516
    saliva preparation, **310:**515
sucrose-dependent adhesion to glass sur-
    faces, **310:**517
tissue cage infection model
    animals, **310:**589
    antibiotic susceptibility assay,
        310:590–591
    apparatus, **310:**589
    infection induction, **310:**590–591
Stress-activated protein kinase
  inhibitors, **319:**270
  photodynamic therapy activation,
      319:356–357
Stromal cell-derived factor 1, signal se-
    quence trapoing, **303:**479–480
Subcellular localization
  antiaffinity-tagged antibody application,
      303:510–511
  messenger RNA, **303:**12–13
Substituted-cysteine accessibility method
  acetylcholine receptor, **293:**124, 130, 133,
      135–139, 143–144
  assumptions, **293:**123
  assumptions, **296:**341
  classification of substituted residues,
      293:123–134
  comparison with site-directed mutagene-
      sis in structure–function analysis,
      296:346
  cysteine modification
    functional assay, **296:**333–334
    reagents, **296:**333
  cystic fibrosis transmembrane conduc-

tance regulator, **293:**124, 130, 135–
136, 145
dopamine D$_2$ receptor, **296:**331, 335–337,
342–345
dopamine transporter
binding assays, **296:**338–339, 341–342
cysteine modification with methane thiosulfonate reagents, **296:**339–340
kinetics of modification, **296:**340
native cysteine residues, identification
and replacement, **296:**335–337
protection from modification,
296:340–341
secondary structure determination,
296:344–345
site-directed mutagenesis and functional effects, **296:**338, 343–344
stable transfection, **296:**338
transient transfection, **296:**338
transmembrane topology, **296:**345
uptake assays, **296:**339
endogenous cysteine reactivity, **293:**134
expression of target proteins, **293:**133
functional integrity of channels
assay, **293:**133–134
preservation in mutation, **293:**124
GABA$_A$, **293:**124, 130, 135–137, 139,
144–145
mutagenesis, **293:**134–135
principle, **296:**330–333
rate constant analysis of sulfhydryl group
reactivity
accessibility analysis, **293:**137–138
channel blocker binding sites, locating,
293:139
channel gates, locating, **293:**138–139
charge-selectivity filter, locating,
293:144–145
electrostatic potential dependence,
293:139–143
factors affecting rate constants, **293:**137
measurement, **293:**147
membrane potential dependence,
293:143–144
screening conditions, **293:**135
secondary structure studies, **293:**136–137
sulfhydryl-specific reagents
assay of free sulfhydryls, **293:**127
hydrolysis of thiosulfonates,
293:127–129

membrane permeability of thiosulfonates, **293:**130–131
nonthiosulfonate reagents, **293:**132
reactions with 2-mercaptoethanol, rate
constants, **293:**129–130
size and channel accessibility, **293:**130,
135–136
synthesis of thiosulfonates, **293:**125–
127, 146
voltage-dependent conformational
changes, analysis
modification rate measurement,
293:147
state-dependent accessibility, **293:**154
state-dependent movement of fluores-
cently tagged cysteine residues,
293:154
voltage-dependent accessibility,
293:147–150, 152–154
Substrata
biocompatibility of implantable medical
devices, **310:**534–535
biofouling studies, substrata preparation
cellulose acetate thin films on internal
reflection elements and glass coupons, **310:**473–475, 492
cellulose acetate, **310:**469–471
polyamide thin-film composite, **310:**471,
473
rationale, **310:**469
continuous flow systems for oral biofilm
culture, **310:**221, 223, 283, 289
types for biofilm growth, **310:**249
X-ray photoelectron microscopy analysis
chemically-modified substrata, **310:**385
conditioned substrata, **310:**385–387
native substrata, **310:**384–385
Substrate channeling
carbamoyl phosphate synthase
catalytic activities and subunits,
308:140–142
kinetic studies, **308:**142
putative channel, **308:**141
rationale, **308:**140
criteria for establishing
steady-state assays, **308:**113–115
structural analysis, **308:**116
transient kinetic assays, **308:**115–116
cyclosporin synthetase, **308:**143
diffusion comparison, **308:**113

electrostatic channeling, **308:**133–135

enzyme complexes, general features, **308:**144–145

FK506 synthetase, **308:**143–144

isotope dilution assays, **308:**114–115

malate dehydrogenase–citrate synthase, **308:**142–143

physiological relevance, **308:**111–112

thymidylate synthase–dihydrofolate reductase, *see* Thymidylate synthase–dihydrofolate reductase, substrate channeling

transient time approximation, **308:**113–114

tryptophan synthase, *see* Tryptophan synthase, substrate channeling

Substrate metabolism rate, biofilms, **310:**393, 419

Subtilisin, organic solvent transfer of crystals, **308:**205

Subtractive hybridization, *see also* Complementary DNA representational difference analysis; Suppression subtractive hybridization

Subunit dissociation constant, determination

hemoglobins

bovine hemoglobin, **306:**127

hemoglobin A, **306:**123–124

hemoglobin F, **306:**124–125

hemoglobin F$_1$, **306:**125

hemoglobin S, **306:**123–124

preparation, **306:**114–115

recombinant hemoglobins with substitutions in $\alpha_1\beta_2$ interface, **306:**126

recombinant sickle hemoglobins, **306:**125–126

small zone gel filtration on Superose-12

analysis time, **306:**118

calculations, **306:**121–122

care of column, **306:**118–119

elution conditions, **306:**117

elution profiles and curve analysis, **306:**117, 119–122

overview, **306:**114

peak concentration analysis, **306:**117–118

sample dilution and peak broadening, **306:**127–129

standards and calibration, **306:**115–117

systematic errors, **306:**122

Subunit interactions, *see* Assembly, ion channels; Protein–protein interactions

Sucrose density gradient centrifugation

advantages and disadvantages in subunit interaction analysis, **293:**15

cell-free translation system products, **290:**5–7

GLYT2 hydrodynamic characterization, **296:**9–12

heat shock protein 18.1, **290:**357

molecular weight calculations, **296:**10–12

partial specific volume calculation, **296:**9, 12

partition coefficient determination, proteins in membranes

calculations, **295:**75

high-performance liquid chromatography of peptides, **295:**77

speed of centrifugation, **295:**76–77

principle, **290:**272, 274

RNA fractionation, **296:**33, 35

sedimentation coefficient calculation, **296:**10–11

symbionin, **290:**183

Sugar–SNAP, *see* S-Nitroso-N-acetylpenicillamine

Sulfarlem, *see* Anethole dithiolthione

Sulfatidase, *see* Arylsulfatase A

Sulfatide, *see also* Sulfo Lewis X gangliosides

biological significance, **311:**94–95

2-O-fucosyl sulfatide

selectin binding assays, **311:**557–558

synthesis, **311:**558

radiolabeling, **311:**101

selectin binding, overview, **311:**549, 552–553, 555

[^{14}C]sulfatide, enzymatic synthesis and purification, **311:**685–687

tissue distribution, **311:**94

urine assay with high-performance liquid chromatography

alkaline hydrolysis, **311:**96, 98

desulfation, **311:**98

lipid extraction, **311:**96

metachromic leukodystrophy diagnosis, **311:**100–101

perbenzoylation, **311:**98–99

reagents, **311:**96

separation and analysis of perbenzoyl-
ated glycolipids, **311:**99–100
sulfatide isolation by anion-exchange
chromatography, **311:**98
Sulfo Lewis X gangliosides, *see also* Sulfatide
clinical applications, **311:**559–560
derivatives containing ceramide or 2-
(tetradecyl)hexadecyl residue
overview, **311:**553
selectin binding assays, **311:**555–558
structure analysis, **311:**555
selectin binding, overview, **311:**549,
552–553
Sulfonylurea receptor, *see also* ATP-sensitive potassium channel
deglycosylation assay, **292:**741
immunoprecipitation, **292:**741
insulin secretion role, **292:**132, 154, 732
iodoglyburide
photoaffinity labeling
membrane preparations, **292:**737, 739
specificity, **292:**738
whole cells, **292:**737–738
synthesis of radiolabeled iodine-125
compound, **292:**733–734
proteolytic degradation assay, **292:**741
purification from cultured cells
cell culture and isolation, **292:**736
digitonin solubilization, **292:**739–740
dye affinity chromatography, **292:**740
lectin affinity chromatography, **292:**740
membrane preparation and storage,
292:738–739
polyacrylamide gel electrophoresis
analytical, **292:**737–738
preparative, **292:**740
purification of histidine-tagged protein
from COSm6 cells
cell culture and transfection, **292:**743
membrane preparation, **292:**744
nickel affinity chromatography, **292:**744
rubidium efflux assays, **292:**743–744
vector construction, **292:**743
sequence homology with other ATP-binding cassette transporters,
292:109–110
Sun protection factor, *see* Skin, ultraviolet
protection
Sunscreen, *see* Skin, ultraviolet protection

Sup35p, *see* [*PSI*⁺]
Superoxide
anion generation by nitric oxide synthase
adrenochrome assay, **301:**75–76
electron acceptor-stimulated superoxide generation, **301:**175–177
electron spin resonance detection
advantages, **301:**74, 170
loop gap resonator measurements,
301:172–173
neuronal nitric oxide synthase
assays, **301:**173–177
spin trapping reagents, **301:**171–172
specificity of production by isoforms,
301:75–76
electron spin resonance assay of anion
radical scavenging activity
ascorbic acid contribution, **299:**28, 32,
34
calibration, **299:**32
electron spin resonance measurements,
299:29–30
reaction conditions, **299:**28–31
sample preparation, **299:**29
Fenton reaction, **301:**437, 471
ferricytochrome *c* assay, **301:**169–170
generation, **301:**439
nitric oxide protection against cell damage, **301:**420–421, 426–427
platelet production
assay, **301:**67, 69–70
effects, **301:**67
Superoxide dismutase, *see also* Hydrogen
peroxide
chemiluminescence assay, **305:**384
copper,zinc-superoxide dismutase, effects
on *S*-nitrosoglutathione decomposition
assay, **301:**222–224
glutathione effects, **301:**224–226
physiological implications, **301:**226–227
tyrosine nitration of manganese enzyme
in renal allograft rejection, **301:**136,
141, 144
Suppression subtractive hybridization
applications
complementary DNA identification
preferentially expressed in two or
more tissues, **303:**373–375
tissue-specific complementary DNA,

generation and cloning, **303:**369,
371
tissue-specific gene identification on Y
chromosome, **303:**371, 373, 379
differential screening
approaches, **303:**366, 377
background reduction, **303:**376–377
complementary DNA dot blots of polymerase chain reaction products,
303:367
complementary DNA probe preparation, **303:**368
hybridization with subtracted and re-
verse-subtracted complementary
DNA probes, **303:**368–369
insert amplification, **303:**366–367
efficiency, **303:**375–376
limitations, **303:**378–379
materials, **303:**354–356
oligonucleotides, **303:**354–355
overview, **303:**350–351, 353–354
subtracted library preparation
adapter ligation and efficiency test,
303:358–361
complementary DNA synthesis,
303:356–357
hybridizations
first, **303:**361
second, **303:**361–362
polymerase chain reaction
selection of differentially expressed
complementary DNAs,
303:362–364
subtraction efficiency analysis,
303:364–365
RNA preamplification, **303:**377–378
*Rsa*I digestion, **303:**357, 378
subcloning of amplified products,
303:365–366
suppression polymerase chain reaction effect, **303:**350, 379
Suppressor mutant, *see* Yeast sphingolipid
biosynthesis mutant
Suppressor transfer RNA, unnatural amino
acid incorporation
acylation of transfer RNAs, **293:**519–521
amino acid incorporation, assessment of
fidelity, **293:**524–525
background current assessment, **293:**524
controls, **293:**525–526

design of suppressor transfer RNA,
293:506–507
electrophysiologic analysis, **293:**523
heteromultimeric protein expression,
293:525
ion channel types in study, **293:**523, 529
materials, **293:**505–506
microinjection of messenger and transfer
RNAs into *Xenopus* oocytes, **293:**523
overview, **293:**504
purification of suppressor transfer RNA
modifications of Qiagen total RNA kit,
293:515–516
polyacrylamide gel electrophoresis,
293:513–515
reacylation of transfer RNAs, **293:**524
suppression testing with in vitro translation, **293:**526–529
synthesis of suppressor transfer RNA
annealing reaction, **293:**510
buffers, **293:**507–509
ligation, **293:**510
oligonucleotide phosphorylation,
293:509–510
transcription, *in vitro*, **293:**511–513
unnatural amino acid synthesis, **293:**516–
517, 519
SUR, *see* Sulfonylurea receptor
SUR1 *see* ATP-sensitive potassium channel
Surface plasmon resonance
applications, overview, **309:**386
biosensor, *see* Surface plasma resonance
biosensor
calmodulin binding to cyclic nucleotide-
gated channels
buffers and materials, **315:**787
controls, **315:**791
immobilization by amine coupling,
315:789
immobilization by ligand thiol coupling, **315:**788
kinetic data interpretation, **315:**794, 796
mass transfer limitations, **315:**786
overview of experimental design,
315:788
principle, **315:**786–787
qualitative analysis, **315:**791–792
quantitative analysis, **315:**793–794
recording of titration series,
315:789–791

Scatchard plot analysis, **315**:796–797
sensor chip geometry, **315**:791
fibril elongation assay with BIACORE
    system
  advantages, **309**:387, 401–402
  amyloid-β fibril preparation, **309**:390
  association phase time, varying,
    309:397–398
  association time, **309**:392
  background binding, **309**:390–391
  binding responses, assessment, **309**:394
  contact time and peptide concentra-
    tions, varying, **309**:398–399
  data analysis, **309**:400–401
  dissociation phase, normalizing,
    309:398
  experimental conditions, varying,
    309:400
  fibril decay monitoring, **309**:395–396
  flat surface data, **309**:399–400
  immobilization of fibrils, **309**:391–392
  limitations, **309**:388
  model of fibril elongation, overview,
    309:395
  principle, **309**:388, 390
  reference surfaces, **309**:392
  reproducibility of binding experiments,
    309:392, 394
glycolipid receptor analysis
  applications, **312**:205–206, 216
  cholera toxin B pentamer, **312**:211–
    214, 216
  dissociation constant determination,
    312:214–216
  glycolipid immobilization on sensor
    chips
    capture of liposomes on chips,
      312:208–210
    liposome fusion to alkane thiol
      monolayers, **312**:210–211
    overview of approaches, **312**:207
    preparation of liposomes,
      312:207–208
  heat-labile enterotoxin, **312**:211–212
  kinetic analysis, **312**:214–216
  monoclonal antibody binding, **312**:212
  standards, **312**:212–213
GRK1 binding kinetics analysis with re-
    coverin or calmodulin, **316**:30–31, 33
instrumentation, **312**:206–207

ligand immobilization
  amine group coupling, **316**:21–22
  biotinylated protein coupling to immo-
    bilized streptavidin, **316**:24–25,
    27–28
  histidine-tagged proteins, **316**:28
  preservation of activity, **316**:29
  sulfhydryl group coupling, **316**:22
oral biofilm formation assay
  advantages, **310**:508–509
  data acquisition, **310**:509–510
  dissociation constant determination,
    310:510–511
  fimbrial binding affinity to host pro-
    teins, **310**:512–513
  materials, **310**:509
  software for data analysis, **310**:511–512
overview, **316**:20–21
principle, **310**:507–508, **312**:206–207
sensogram analysis, **316**:29
small molecule effects on protein–protein
    interactions in visual transduction
    cascade, **316**:33–34, 36, 39
structure–function analysis of mutant pro-
    teins, **316**:29–30
transducin activation of rhodopsin, assay
  overview, **315**:482–483
  patterns of rhodopsin membranes,
    315:485, 487
  self-assembled monolayers, **315**:483,
    485, 487
  supported lipid bilayers containing rho-
    dopsin, **315**:483–485
Surface plasma resonance biosensor
  advantages of biosensors in kinetic stud-
    ies, **295**:268–269
  binding capacity determination, **295**:276
  binding stoichiometry analysis,
    295:290–291
  commercial sources, **309**:387
  complexities in data analysis,
    295:269–270
  configurations of binding assays
    capturing, **295**:274
    direct coupling, **295**:273–274
    ligand immobilization, **295**:273
    reference surface, **295**:274–275
  corroboration of experiments,
    295:292–293
  data analysis

CLAMP data analysis program,
    295:279–281, 288
  global analysis, **295:**279
  integration, **295:**278–279
  linear transformation, **295:**277–278
  model fitting to data, **295:**281–285
  normalized plots, **295:**281
  quality of fit, assessment, **295:**283–285
erythropoietin ligand–receptor binding ki-
    netics
  analyte concentrations, **295:**276
  capturing of ligand, **295:**274
  data collection, **295:**277
  equilibrium dissociation constant,
    295:292
  flow rate, **295:**275–276
  mass transport coefficient, **295:**291–292
  model fitting, **295:**282–286
  overview of system, **295:**270, 293
  sample injection volume, **295:**277
nonspecific binding to biosensor, **295:**272
parameter confidence statistics
  linear approximation, **295:**286–287
  Monte Carlo analysis, **295:**287–288
  profiling, **295:**288–290
reaction models, experimental testing,
    295:285–286
sample requirements
  aggregation state characterization,
    295:272
  purity, **295:**271–272
Surfactant, *see* Biosurfactant
Surlactin, *see* Biosurfactant
SV40, *see* Simian virus 40
Svedberg equation, analytical ultracentrifu-
    gation, **290:**274–275
Swinholide A
  actin binding
    assays, **298:**30
    mechanism, **298:**26–27
  actin dimer formation assay, **298:**30–31
  applications, **298:**28–32
  effects on actin organization in living
    cells, **298:**27–28
  F-actin depletion *in vitro*, **298:**28–29,
    31–32
  solubility, **298:**30
  structure, **298:**26
Swinnex biofilm fermenter, *see* Biofilm fer-
    menter

SWI/SNF
  ATPase assays
    activated charcoal binding of phospho-
      rous-32, **304:**729–730
    kinetic parameter determination,
      304:731
    thin-layer chromatography assay
      chromatgraphy, **304:**730–731
      incubation conditions, **304:**730
      materials and solutions, **304:**730
  nucleosome array remodeling assay by re-
    striction enzyme coupling
    applications, **304:**741
    glycerol effects on restriction enzyme
      activity, **304:**735–736
    incubation conditions, **304:**739–740
    nucleosome array reconstitution
      dialysis conditions, **304:**733
      DNA template, **304:**732
      histone octamer preparation,
        304:732–733
      nucleosome saturation, determina-
        tion, **304:**733–734
    overview, **304:**731
    *Sal*I cleavage kinetics, **304:**738–739
    salt effects
      nucleosome array conformation and
        self-association, **304:**736–738
      restriction enzyme activity,
        304:735–736
  nucleosome disruption, overview, **304:**726
  purification of histidine-tagged protein
    DNA cellulose chromatography,
      304:728–729
    glycerol gradient centrifugation,
      304:729
    ion-exchange chromatography, **304:**729
    materials, **304:**727–728
    nickel affinity chromatography, **304:**728
    overview, **304:**727
    yeast growth and lysis, **304:**728
    yield, **304:**729
  subunits, **304:**726
Symbionin
  autophosphorylation
    assays
      site identification, **290:**189–190
      in vitro, **290:**189
      in vivo, **290:**187–189
    heat shock response, **290:**181

chaperonin activity assays
  Rubisco reconstitution, **290**:184–
    185
  self-reconstitution, **290**:185–186
gene, **290**:181
homology with GroEL, **290**:181
oligomerization, **290**:184
phosphotransferase activity
  assay, **290**:190–191
  GTP production, **290**:181
  transphosphorylation to other endosym-
    biotic proteins, **290**:192–193
purification from pea aphid
  ammonium sulfate fractionation,
    290:182–183
  homogenization and extraction,
    290:182
  size-exclusion chromatography, **290**:184
  storage, **290**:184
  sucrose gradient centrifugation, **290**:183
synthesis by isolated endosymbionts
  assay, **290**:187
  endosymbiont isolation from pea
    aphid, **290**:186–187
Synaptosomal-associated protein-25
  antisense knockdown studies in cultured
    neurons
    cortical neuron culture, **314**:85
    dose schedule, **314**:152
    dose–response and time course of inhi-
      bition, **314**:152
    functional effects
      PC12 cells, **314**:154–155
      primary neurons, **314**:152, 154
    oligonucleotides
      design, **314**:150–151
      uptake, **314**:151–152
  antisense targeting, **313**:29
  neuronal differentiation role,
    314:149–150
  vesicle fusion role, **314**:150, 154
Synaptosome, neurotransmitter uptake in
  homogenized versus intact tissue,
  296:690–692
Synchrotron X-ray footprinting
  applications in RNA folding, **317**:354–355
  beam flux calculation, **317**:355–358
  beamline set-up, **295**:393
  dose–response of cleavage, **295**:382–383,
    317:358

factors affecting nucleic acid cleavage,
    295:393–394
gel electrophoresis and analysis, **295**:400–
    401, **317**:367–368
hydroxyl radical, advantages in foot-
    printing, **317**:354
kinetics of cleavage, **295**:382, 386–388
magnesium-initiated RNA folding,
    295:399–400
manual mixing experiments, **295**:395–396
  electronic shutter, **317**:360
  reaction conditions, **317**:364–365
mechanism of nucleic acid cleavage,
    295:381–382, 386
millisecond exposure shutter,
    295:388–389
prospects, **317**:368
quenching of radical, **295**:387
radiolysis of water, **295**:382, 384–385
rapid-mixing footprinting
  device and protocol, **295**:389–392,
    396–398
  instrument set-up, **317**:361–363, 365
  magnesium initiation and data collec-
    tion, **317**:365–366
RNA preparation, **317**:363–364
sample preparation, **295**:394–395
sample storage, **317**:360
solutions, **317**:363
sources of radiation, **295**:388
spatial resolution, **317**:368
steady-state hydroxyl radical concentra-
    tion, **317**:358
temporal resolution, **317**:352, 354, 368
Tetrahymena ribozyme, **295**:402, **317**:354–
    355, 358–360, 364–365, 368
SYR2, cloning in yeast, **311**:329
SYTO 9, membrane integrity assessment in
  biofilms with propidium iodide,
  310:173–174

                            T

T4 DNA ligase, RNA ligation in hybrid duplexes
  circular RNA generation, **317**:113, 119
  concentration optimization
    DNA template, **317**:121
    enzyme, **317**:121

RNA, **317:**121
DNase treatment, **317:**123
duplex specificity and mutation effects, **317:**122–123
engineering of metal-binding sites in large RNAs, **317:**128
generic conditions for ligation, **317:**113, 118–119
lengths of substrates and templates, **317:**122
multipart ligations, **317:**122
principle and overview of splinted ligation, **317:**112–115
radioactive labeling and ligation
  different tubes, **317:**113–115, 120–121
  single tube, **317:**113–114, 120
reagents, **317:**115, 117
RNA substrate preparation, **317:**117–118
side reactions, **317:**123
T4 RNA ligase
engineering of metal-binding sites in large RNAs, **317:**128–129
synthesis of circular RNA for permutation analysis, **317:**315–317
in vitro ligations, **317:**111–112
T7 RNA polymerase
nucleoside analog incorporation into RNA with Y639F mutant enzyme
  bulky 2′ substituent handling, **317:**73–74
enzyme
  characteristics, **317:**65–66
  expression in *Escherichia coli*, **317:**66–67
  purification, **317:**67–68
  storage, **317:**68
noncanonical substrates, **317:**73–74
templates
  gapped templates, **317:**72–73
  initially transcribed sequence, **317:**72
  nicked templates, **317:**72–73
  partially single-stranded templates, **317:**72–73
Tetrahymena group I intron analysis, phosphorothioate nucleotide incorporation into RNA transcripts
  normalization of incorporation efficiency, **317:**99–100
  nucleotides and reactions, **317:**98–99
transcription buffers

acetate, **317:**71
composition optimization, **317:**68–69
detergents, **317:**71
manganese, **317:**71–72
polyamine stimulation, **317:**69–71
pyrophosphatase inclusion, **317:**71
transfer RNA aminoacylation analysis, phosphorothioate nucleotide incorporation into RNA transcripts, **317:**85–86
TAR RNA of human immunodeficiency virus-2, synthesis from isotopically-labeled ribonucleotides for nuclear magnetic resonance studies, **317:**36
transcript yield increase using ribonuclease H, **317:**135–136
Tac, *see* Signal sequence trap
Tandem mass spectrometry, *see* Mass spectrometry
Tandem scanning confocal microscope, *see* Reflection confocal microscopy
Tannin
extraction and purification from tissue, **299:**180
gel-filtration chromatography, **299:**179
proanthocyanidin classification and structures, **299:**178–179
procyanidin separation by high-performance liquid chromatography, **299:**137
separations in Folin–Ciocalteu reagent total phenol analysis, **299:**174–175
size separation by high-performance liquid chromatography
  degree of polymerization, analysis, **299:**182, 184
  gradient, **299:**182
  peak identification, **299:**180–181
  sample preparation, **299:**180–181
T antigen, *see* Simian virus 40
TAP, *see* Transporters associated with antigen processing
Tape stripping, stratum corneum α-tocopherol analysis, **300:**414–415
TAR RNA, human immunodeficiency virus-2
structure determination with nuclear magnetic resonance
  heteronuclear single quantum coherence, **317:**37–38

nuclear Overhauser enhancement spectroscopy, **317**:37

synthesis from isotopically-labeled ribonucleotides using T7 RNA polymerase, **317**:36

TAR, *see* Human immunodeficiency virus

Targeted display

agarose gel electrophoresis of amplification products, **303**:401

applications, **303**:394–395

cell culture, **303**:398

cloning, confirmation, and sequencing of differentially expressed complementary DNAs, **303**:401, 403

genome coverage, **303**:406–407

genomic DNA isolation, **303**:399

goals in differential gene expression analysis, **303**:393

optimization, **303**:403

polymerase chain reaction

complementary DNA, **303**:400

genomic DNA, **303**:400–401

primers

experimental determination of performance, **303**:406

hit rate, **303**:394–398

length, **303**:395–396

sequence, **303**:396

theoretical performance

basic helix–loop–helix primer set, **303**:403–405

human genome primer set, **303**:405

principle, **303**:392–394

prospects and limitations, **303**:407–408

RNA

DNase I treatment, **303**:399

isolation, **303**:398

reverse transcription for complementary DNA synthesis, **303**:399–400

Targeted random mutagenesis, photosystem II genes in *Synechocystis* 6803, 301–302, 305–306

Tat, *see also* Human immunodeficiency virus; Tat-fusion system

bovine immunodeficiency virus protein binding to TAR, **318**:423–424

peptide binding to RNA, *see* RNA–peptide interaction

Tat-fusion system

applications, **318**:362–363

electroporation, **318**:362

fluorescence-activated cell sorting, **318**:360–361

fusion protein design, **318**:353–355

green fluorescent protein and reporter cell line generation, **318**:356–358

library delivery into reporter cells

overview, **318**:358

protoplast fusion, **318**:359–360

protoplast preparation, **318**:359

overview of RNA-binding library screening, **318**:350–353

plasmid recovery, **318**:360–361

plasmids and reporters, **318**:351, 353, 355–356

Tau, *see also* Alzheimer's disease; Paired helical filament

dephosphorylation of preparations, **309**:88–89

functions with microtubules, **309**:81–82

isoforms, **309**:81–83

paired helical filament-tau

disease pathogenesis role, **309**:81–83

purification

brain tissue selection, **309**:83–84

guanidine isothiocyanate extraction, **309**:85

homogenization of tissue, **309**:84

Sarkosyl extraction, **309**:84

sucrose density gradient centrifugation, **309**:84–85

posttranslational modifications, **309**:82

purification of normal protein

biopsy tissue, **309**:86

fetal protein, **309**:88

large biopsy samples as starting material, **309**:86–87

small biopsy samples and autopsy tissue as starting material, **309**:87–88

Taxol

cancer treatment, **298**:238, 267

cell-type specific differences in microtubule response, **298**:275–276

cellular uptake and loss, **298**:270–271

glioma therapy, **313**:371

intracellular concentration determination

nonradiolabeled compounds and high-performance liquid chromatography, **298**:273–274

radiolabeled compounds

monolayer culture directly in sterilized scintillation vials, **298:**273

monolayer or suspension cultures, **298:**272–273

intracellular concentration during initial incubation, **298:**271

microtubule affinity, **298:**268

microtubule bundling, **298:**268

microtubule stabilization, **298:**239, 267–269

photoaffinity labeling of microtubules
direct photolabeling, **298:**239
^{3}H-3′-(*p*-azidobenzamido)taxol
labeling reaction, **298:**241, 243–245
site of photincorporation, **298:**245–246, 250–251
structure, **298:**240
^{3}H-2-(*m*-azidobenzoyl)taxol
digestion of β-tubulin, **298:**242–243, 246, 248
labeling reaction, **298:**241, 243–245
peptide analysis, **298:**242
site of photincorporation, **298:**246, 248–251
structure, **298:**240
high-performance electrophoresis chromatography of labeled β-tubulin, **298:**241–242, 246
microtubule assembly assay, **298:**240
β-tubulin isolation and formic acid digestion, **298:**241, 245–246

solubility, **298:**268

structure, **298:**256

suppression of microtubule treadmilling and dynamic instability, **298:**269–270

synergistic effects with other antimitotic drugs, **298:**276

T cell apoptosis, *see* Apoptosis

T cell receptor, gene rearrangement assay with chemiluminescence
hybridization and washes, **305:**458
overview, **305:**456–457
probe generation, **305:**458–459
reagents and solutions, **305:**457–458
signal generation and detection, **305:**458–459
Southern blotting, **305:**458–460

TCP-1, *see* Cytosolic chaperonin

TD, *see* Targeted display

Tea, high-performance liquid chromatography of catechins
chromatography conditions, **299:**109, 111, 203–204, 206
comparison of green teas, **299:**112–113, 203
extraction, **299:**204
instrumentation, **299:**108–109
precision, **299:**204, 206
quantitative analysis, **299:**111–112
sample extraction, **299:**109
stability of samples, **299:**206
standards, **299:**109, 204

TEAC assay, *see* Trolox equivalent antioxidant capacity assay

TEB, *see* Transient electric birefringence

Teeth, *see* Dentin; Enamel

TEM, *see* Transmission electron microscopy

TEMP, *see* 2,2,6,6-Tetramethyl-4-piperidone

Temperature effects
activation energy, **308:**13–15
confocal microscopy
acetoxymethyl ester dyes
cleavage of dyes, **307:**22–23
loading of dyes, **307:**448–449
focus, **307:**25–26

Temperature-invariant enthalpy, *see* Enthalpy

Temperature jumps, neurotransmitter transporter release and uptake analysis, **296:**665–666

Tenascin, sample mounting for confocal microscopy, **307:**85, 87

Tepy, singlet oxygen detection in plants
leaf infiltration, **319:**81
trapping, **319:**78–79

Terminal transferase-dependent polymerase chain reaction, *see* Polymerase chain reaction

Tetrabenazine, vesicular monoamine transporter binding
distinguishing of isoforms, **296:**87–88
α-dihydrotetrabenazine, **296:**74–76
α-[*O-methyl*-^{3}H]dihydrotetrabenazine, **296:**74–75, 87–88, 92–93
inhibition analysis, **296:**154–155
single-cell visualization of radioactive derivative binding, **296:**92

Tetracycline, derepression of green fluores-

cent protein expression *in planta*,
302:324–325, 327–328
7,8-*cis*-Tetradecenoylhomoserine lactone,
structure elucidation, 305:297, 299
Tetrahydrobiopterin, scintillation proximity
assay of binding to neuronal nitric oxide synthase heme domain, 301:117–
118, 121–124
Tetrahymena group I intron
kinetic oligonucleotide hybridization analysis of folding
gel electrophoresis, 317:341–342
incubation conditions and quenching,
317:340–341
kinetic folding assays, 317:346–348, 350
optimization of quenching reaction,
317:342, 344–346
overview, 317:331–334
reagents, 317:337–338, 340
thermodynamic analysis, 317:350–351
in vitro selection, 317:351–352
nucleotide analog interference mapping
materials and reagents, 317:93, 95
nucleotide incorporation into RNA
transcripts with mutant T7 RNA
polymerase
normalization of incorporation efficiency, 317:99–100
nucleotides and reactions, 317:98–99
phosphorothioate-tagged nucleotide triphosphate synthesis, 317:96–98
quantitative analysis, 317:102, 104
RNA selection assay, 317:100–102
synchrotron X-ray footprinting, folding
analysis, 317:354–355, 358–360, 364–
365, 368
Tetrahymena thermophila, see Lipoic acid
Tetrakis(dimethylamino)ethylene, educational demonstrations of luminescence,
305:646–647
2,2,6,6-Tetramethyl-4-piperidone
photodynamic therapy singlet oxygen detection, 319:393
singlet oxygen detection in plants
buffers, 319:78–79
electron paramagnetic resonance spectra, 319:79
hydroxylamine conversion prevention,
319:79–80
leaf infiltration, 319:81

photoinhibitory treatment, 319:80–81
reaction conditions, 319:79
trapping, 319:78–79
Tetramethylrhodamine, electrical potential
imaging, 302:346–347, 353–354
Tetrapotassium rubrene-2,3,8,9-tetracarbox-
ylate, singlet oxygen trapping, 319:50
TEV, *see* Two-electrode voltage clamp
TFO, *see* Triplex-forming oligonucleotide
TGF-*β*, *see* Transforming growth factor-*β*
TGN, *see* Trans-Golgi network
Thalassemia, *see* *β*-globin
Therapeutic proteins, aggregation
consequences for patient, 309:236–
237
infrared spectroscopy characterization,
309:254–255
stresses and prevention
freeze-drying, 309:245–251
freezing, 309:244–245
interfacial stresses, 309:251–254
long-term storage in aqueous solution,
309:241–242
overview, 309:237–238
pathways for aggregation, 309:242–244
thermal stress, 309:238–241
Thermal stress, protein aggregation
heat treatment of therapeutic proteins,
309:238
mechanisms, 309:238
stabilizers
mechanism of action, 309:239–241
types, 309:239
Thermodynamic cycle analysis
allosteric effects on second ligand binding, 294:132–133
double-mutant analysis, 294:133–135
principle, 294:132
Thermoluminescence, analysis of photosystem II mutants, 297:346–347
Thermozymocidin, *see* Myriocin
Thin-layer chromatography, *see also* High-
performance thin-layer chromatography
acylhomoserine lactone autoinducer identification, 305:300
carotenoid photobleaching products,
299:418
ceramidase assay, 311:197
ceramide

analysis in glycosylphosphatidylinositol anchors, **312:**514–515

derivatives, **311:**485–487, 501, 520–522, 603

ceramide galactosyltransferase products, **311:**62–63

ceramide kinase assay, **311:**211

ceramide phosphate in enzymatic ceramide determination, **312:**28–30

ceramide synthase assay, **311:**17, 136–137

cholesterol oxidation products, **319:**91–93, 100

dehydrosphingosine, **311:**520–522

detergent-insoluble membrane fraction lipids from mouse melanoma B16 cells, **312:**491, 493–494

dihydroceramide, **311:**485–487

dihydroceramide desaturase assay chromatography, **311:**27

sheet preparation, **311:**25

dihydrosphingomyelin, **311:**543

dihydrosphingosine 4–hydroxylase assay, **311:**12

fumonisins, **311:**371–372

glucosylceramide synthase assay, **311:**52, 306

glucosylthioceramide, **311:**603

glycosphingolipids, **312:**49, 54, 146

lactosylceramide synthase assay, **311:**76

lactosylthioceramide, **311:**603

phosphoaminoacid analysis, **294:**360–361

phosphopeptide mapping, **294:**359–360

phytoceramide, **311:**501

phytosphingosine, **311:**501

platelet-activating factor:sphingosine transacetylase assay, **311:**119–120

short-chain ceramides, **312:**416

sphinganine and derivatives, **311:**485–487, 520–522

sphingolipid ceramide N-deacylase, **311:**299

sphingolipid isolation applications, **312:**71–72

autoradiography, **312:**75–77

development of chromatograms, **312:**72–73

sample application, **312:**72

solvent systems, **312:**73–75

staining, **312:**78–79

sphingomyelin, **311:**193, 501, 543

sphingomyelin synthase, BODIPY-labeled ceramide assay on replica colonies, **312:**531, 533

sphingosine and derivatives, **311:**501, 520–522

sphingosine kinase assay, **311:**222

sphingosine metabolites, **311:**663, 667

sphingosine-1-phosphate lyase assay, **311:**252

verotoxin overlay assay of glycolipid binding advantages, **312:**460–461

overview, **312:**460

platelet Gb$_3$ binding, **312:**463

specificity assay, **312:**461–462

yeast sphingolipids, **311:**327–328

Thioctic acid, *see* Lipoic acid

Thioflavin S, amyloid staining, **309:**12, 276

Thioflavin T advantages and disadvantages of amyloid staining, **309:**11–12, 475–476

agitation-stimulated fibrillogenesis monitoring advantages, **309:**209–210, 212

controls, **309:**210

fluorescence detection, **309:**211–212

sampling versus *in situ* monitoring, **309:**216

stock solution preparation, **309:**210

amyloid-β fibrils extension assay, **309:**481–482, 487–488

nucleation assay, **309:**480–481, 486–488

quantitative assay comparison with other techniques, **309:**283–284

fibril formation assay, **309:**281–283

fluorescence spectra, **309:**276–277

materials, **309:**279–280

precautions, **309:**283

sensitivity, **309:**280–281

specificity, **309:**276, 279

spectral shift mechanism, **309:**278–279

standard curve, **309:**282

flow cytometric characterization of amyloid fibrils with thioflavin T and Congo red criteria for fibril detection, **309:**463

data acquisition, **309:**462

immunoglobulin light chain amy-

loidosis, amyloid burden determi-
    nation, **309:**464–466
instrument settings, **309:**461
overview, **309:**460–461
staining of sample, **309:**462–463
kinetics of fibril formation, fluorescence
    assay
    assumptions, **309:**311
    binding conditions, **309:**310–311
    first-order kinetic model of extension,
        309:313, 315–317
    initial rate measurement, **309:**316
    polymerization reaction
        composition, **309:**312–313
        initiation and termination, **309:**313
        temperature of reaction, **309:**313
    wavelengths for detection, **309:**310
pitfalls in use, **309:**12–13
structure, **309:**11, 275
Thiol-specific antioxidant
    function, **300:**219–220
    reduction of hydrogen peroxide, **300:**220
    sequence homology searching,
        300:219–220
Thiol trapping, *see* Disulfide bond for-
    mation
Thionine, RNA–protein cross-linking,
    318:30
Thiophosphoryl group, caged protein conju-
    gates
    kemptide modification with 2-nitrobenzyl
        group, **291:**126
    modification in presence of cysteinyl resi-
        dues, **291:**119
    selective reactivity in 2-nitrobenzyl modi-
        fication, **291:**126–127
    thiophosphorylation of peptides, **291:**125
Thioredoxin reductase
    assays
        cell lysate assay, **300:**237–238
        Ellman's reagent assay, **300:**232–233
        insulin reduction assay, **300:**233,
            237–238
        NADPH oxidase assay, **300:**235
        unit definition, **300:**233
    genes and cloning, **300:**228–229
    inhibitors
        assays in analysis, **300:**236–237
        types, **300:**235–236
    isoenzymes, **300:**238–239

purification of mammalian enzyme
    absorption spectra, **300:**232
    chromatography, **300:**231–232
    homogenization, **300:**231
    kinetic parameters from calf thymus,
        300:232
    tissue sources, **300:**230
reaction catalyzed, **300:**226
selenocysteine in mammalian protein and
    function, **300:**229–230, 233–234
sequence homology searching, **300:**229
structure of *Escherichia coli* protein,
    300:227
subtrate specificity, **300:**234
Thioredoxin superfamily, chaperones,
    290:46–47
Thiosulfonates, sulfhydryl reactivity in sub-
    stituted-cysteine accesibility method
    assay of free sulfhydryls, **293:**127
    hydrolysis, **293:**127–129
    membrane permeability, **293:**130–131
    rate constant analysis of sulfhydryl group
        reactivity
        accesibility analysis, **293:**137–138
        channel blocker binding sites, locating,
            293:139
        channel gates, locating, **293:**138–139
        charge-selectivity filter, locating,
            293:144–145
        electrostatic potential dependence,
            293:139–143
        factors affecting rate constants, **293:**137
        membrane potential dependence,
            293:143–144
    reactions with 2-mercaptoethanol, rate
        constants, **293:**129–130
    size and channel accessibility, **293:**130,
        135–136
    synthesis, **293:**125–127
4-Thiouridine
    RNA–protein cross-linking
        advantages, **318:**71
        disadvantages, **318:**22–23
        site-specific cross-linking during pre-mes-
            senger RNA splicing
            P120 RNA, **318:**85
            RNA–protein cross-link analysis,
                318:86–87
            RNA–RNA cross-link analysis,
                318:86–87

ultraviolet irradiation, **318:**86
site-specific incorporation into RNA
applications, **318:**85–88
cellularly-derived RNA incorporation,
318:85
ligation of half-RNAs with bridging
deoxyoligonucleotide and T4
DNA ligase, **318:**83–84
overview of approaches, **318:**71–74
preparation of 5′- and 3′-half RNAs
phage RNA polymerase transcription, **318:**75–78
principles, **318:**73, 75
site-specific ribonuclease H cleavage
directed by 2′-O-methyl RNA–
DNA chimeras, **318:**78–80
4-thiouridine 5′,3′-diphosphate
ligation to 3′ end of 5′-half RNA,
318:82–83
synthesis, **318:**80–82
structure, **318:**72
Three-dimensional cryoelectron microscopy, *see* Electron microscopy
Three-dimensional structure, molecular
modeling
ab initio modeling
building of models, **293:**602–603
site-directed mutagenesis studies,
293:603–604
binding energy transduction to ion channel opening, **293:**616–619
channel representation
explicit representation
data files in analysis, **293:**707–708
overview, **293:**706–707
software programs, **293:**707–709
rigid wall, **293:**705–706
consensus sequence alignment, **293:**596–
598, 622–624
continuum calculations
conductance prediction, **293:**687–688,
690–691
current–voltage curves, **293:**688–689,
692–693
electric field calculations
boundary charge method,
293:718–719
Poisson–Boltzmann equation,
293:685–687, 693, 717–719
limitations, **293:**692

overview, **293:**684–685
pore radius profile, **293:**689–690
databases of known structures
folds, **293:**595–596
proteins, **293:**593–595
electrophysiologic data, relating to structural models
dose–response studies, **293:**634–636
ligand analog data, **293:**637–640
related channel data, **293:**640–642
energy minimization of molecular models
final minimization, **293:**630–631, 634
initial minimization
adjustment of initial models,
293:629–630, 633
evaluation, **293:**628–629, 632–633
template construction, **293:**627–628
secondary minimization, **293:**630,
633–634
ensembles, overview, **293:**698–699
ergodic theorem, **293:**697
free-energy calculation
particle insertion, **293:**720
thermodynamic integration,
293:720–721
gramicidin, **293:**694–695, 721
homology modeling, *see also* Cyclic nucleotide-activated channel
conserved residue placement, **293:**624
deletions in alignment, **293:**623–624
fragment-based approach, **293:**599–600
insertions in alignment, **293:**623–624,
631
limitations, **293:**642
loop regions, initial construction,
293:624–626
reference protein identification,
293:622–623, 631
single-step approach, **293:**600–602
site-directed mutagenesis in model testing, **293:**642–643
importance of modelng, **293:**647–648
ion intermolecular potentials, **293:**705
ligand docking in molecular modeling
cyclic nucleotide-activated channel,
293:626, 637–640
hot spot determination, **293:**607–610,
613–615
refinement of glutamate receptor binding sites by alignment

S1, **293:**610–611

S2, **293:**611

S3, **293:**611

S4, **293:**612

S5, **293:**612

S6, **293:**612

module combination methods, **293:**615–616, 619

molecular dynamics simulated annealing with restrained molecular dynamics

acetylcholine receptor, **293:**656, 658–659, 722

alamethicin, **293:**659–663, 689–692, 722

boundary conditions, **293:**713–715

dynamic behavior simulation, **293:**651–652, 699, 711–712

electrostatic energy calculation, **293:**651, 700–702

empirical energy function, **293:**650, 652

force calculation, **293:**712

integration of motion equations, **293:**712–713

ion modeling with water in pore, **293:**679–682

limitations in ion channel modeling, **293:**684

lipid bilayer modeling, **293:**682–684

LS peptide, **293:**666–668

M2, 664–666, 668

pore-forming domains, **293:**649, 655

potassium channel modeling

β-barrel models of pore, **293:**669–670, 680–682

H5 region mutagenesis, **293:**669

α-helix bundle models of pore, **293:**671

ranking ensembles of H5 models, **293:**671–672

programs, **293:**655

restraints, **293:**652–653

sampling, **293:**713

simulated annealing, **293:**653–654, 710–711

structures for restraints

channel-forming peptides, **293:**659–663

low-resolution structure availability, **293:**656, 658–659

transmembrane segments, **293:**663–668

temperature dependence, **293:**652

in vacuo simulations, **293:**673

van der Waals component, **293:**651, 700

water modeling

effects on pore structure, **293:**675–676

energy minimization in positioning, **293:**674–675

goals, **293:**673–674

NB protein pore, **293:**676–679

solvation of model, **293:**674

Monte Carlo simulation

boundary conditions, **293:**713–715

moving of molecules, **293:**709–710

overview, **293:**699

sampling of phase space, **293:**710–711

overview of simulation steps, **293:**722–723

pH setting, **293:**627

potential energy sources, **293:**700–709

pressure calculation, **293:**721

programs for sequence scanning, resources, **293:**593–597, 621–622, 644–646

stochastic dynamics simulation, **293:**699, 715–716

subunit stoichiometry and pore size, **293:**605

water intermolecular potentials

forces, **293:**700–701

point charge models, **293:**701–703

polarizable models, **293:**703–705

water modeling in pore, **293:**626–627, 631, 636, 696–697

Three-hybrid system, *see* Yeast three-hybrid system

Thrombin receptor, signaling inhibition by amyloid-β, **309:**744

Thymidine kinase, ribonuclease P targeting of herpes simplex virus, **313:**455–456

Thymidine triphosphate, caged compound synthesis, **291:**489

Thymidylate synthase–dihydrofolate reductase, substrate channeling

catalytic activities, **308:**132–133

domains, **308:**133

electrostatic channeling, **308:**133–135

transient kinetic analysis

pulse chase, **308:**139

rapid quench, **308:**135–137

stopped-flow fluorescence, **308:**137–139

Thyroid receptor
  green fluorescent protein fusion protein,
    302:78
  subcellular distribution, **302:**74, 78
TIM, *see* Triose phosphate isomerase
Time-resolved near-infrared phosphores-
    cence, singlet oxygen detection
  decay rate and quenching rate constant
    determination, **319:**47–49
  instrumentation, **319:**40–42, 49
  quantum yield measurement in homoge-
    neous systems, **319:**44–47
  sample preparation and handling, **319:**43
  theory, **319:**38–40
  trace recording, **319:**43–44
Tinidazole
  electrochemical reduction, **300:**314–315
  interaction with DNA, **300:**316–321
  structure, **300:**314
Tissue cage, *see Streptococcus* biofilm
Tissue-type plasminogen activator, refold-
    ing of recombinant protein from inclu-
    sion bodies, **309:**222, 225–226, 229
Titanium dioxide
  cosmetic applications, **319:**29, 36
  oxidation product analysis
    butylated hydroxytoluene, **319:**30–32,
      35–36
    electron paramagnetic resonance,
      319:31, 35
    gas chromatography/mass spectrome-
      try, **319:**31, 33–34
    light source, **319:**30
    materials, **319:**30
    methyl oleate, **319:**30–35
    2,2,6,6-tetramethyl-4-piperidone,
      319:30, 35
    uric acid, **319:**30–32
  photochemical generation of active oxy-
    gen species, **319:**29
Titin, *see* Myocyte
Titration calorimetry
  Hill plot, **295:**363
  Kirchoff's law and temperature depen-
    dence of enthalpy change, **295:**91
  Omega instrument, **295:**350
  partition coefficient determination, pro-
    teins in membranes, **295:**83–84
  temperature dependence of equilibrium
    constant

    determination by isothermal calo-
      rimetry
    affinity calculation over experimental
      temperature range, **295:**98–99
    binding enthalpy change measure-
      ment, **295:**92–93
    self-association status of reactants
      and products, defining,
      295:96–97
    temperature increase for tight-bind-
      ing complexes, **295:**97–98
    temperature range and effects on na-
      tive conformations, **295:**93–96
  Gibbs–Helmholtz equation, **295:**89–90
  van't Hoff equation
    derivation, **295:**90–91
    ligand binding equilibria of biologi-
      cal macromolecules, **295:**91–92,
      98–99
  Wiseman isotherm in isothermal titration
    calorimetry
  derivation
    complex cases, **295:**355–356
    general multisite interactive case,
      295:360
    single-site cases, **295:**351–354
    two independent site cases, **295:**360
    two-site interactive case, **295:**360–363
  limitations to interpretation,
    295:363–364
  overview, **295:**333–334, 350–351
  plotting
    axes, **295:**356–357
    n as a measure of $[M_T]$, 358–360
    $n = 1$ intercept, **295:**358
    Y intercept, **295:**357–358
TLC, *see* Thin-layer chromatography
TMAE, *see* Tetrakis(dimethylamino)eth-
  ylene
TNF-α, *see* Tumor necrosis factor-α
Tocopherol, *see also* α-Tocopherol; Vitamin
  E
  low-density lipoprotein oxidation
    analysis by high-performance liquid
      chromatography, **301:**322–323
    consumption and lipid peroxidation,
      301:331–332, 448–449
    effects of peroxynitrite, **301:**324, 327,
      329, 331
    oxidation by reactive nitrogen species

chemistry of nitration, **301:**489
gas chromatography–mass spectrometry analysis of 5-nitro-γ-tocopherol, **301:**487, 489
high-performance liquid chromatography analysis, **301:**486–487
3-morpholinosydnonimine reactions, **301:**485–486
peroxynitrite reactions
    liposomes, **301:**486
    methanol solution, **301:**484–485
susceptibility of tocopherol types, **301:**489–490
types and structures, **301:**319–321
α-Tocopherol, *see also* Vitamin E
gas chromatography–mass spectrometry
    calibration curve preparation, **299:**311, 317–318
    extraction and derivatization, **299:**311–312
    instrumentation, **299:**312
    oxidation product analysis from rat liver, **299:**314–317
    selected ion monitoring, **299:**313–314
high-performance liquid chromatography with electrochemical detection, **299:**309–310
interactions with retinol in liposome oxidation prevention
    chemicals and equipment, **299:**422
    oxidation conditions
        liposomes, **299:**422–423
        retinal membrane oxidation, **299:**423–424
    regeneration, **299:**421–422, 428, 430
    retina membrane protection, **299:**425, 427–429
    soybean phosphatidylcholine liposome protection, **299:**424–425, 429
    vitamin analysis with high-performance liquid chromatography, **299:**424
peroxyl radical trapping and products, **299:**309
prooxidant activity in low-density lipoprotein
    antioxidant versus prooxidant activity, **299:**362–364, 373, 375
    assay
        high-performance liquid chromatography, **299:**368–370

reagents, **299:**364–365
    dose response of lipid oxidizability, **299:**372–373
    inverse deuterium kinetic isotope effect, **299:**371–372
    lipoprotein preparation
        deuterium-labeled low-density lipoprotein, **299:**367–368
        native and tocopherol-enriched low-density lipoprotein, **299:**365–366
        tocopherol-depleted low-density lipoprotein, **299:**366–367
        tocopherol-replenished low-density lipoprotein, **299:**367
    mechanism, **299:**362–364
    phase transfer activity effects, **299:**370–371
synthesis of deuterium-labeled compound, **299:**310–311
Topology, *see* Transmembrane topology
Total antioxidant power assays, *see* Chemiluminescence; Ferric reducing/antioxidant power assay; Oxygen radical absorbance capacity assay; Total peroxyl radical-trapping potential assay; Trolox equivalent antioxidant capacity assay
Total internal reflection microscopy, green fluorescent protein, **302:**152–153
Total peroxyl radical-trapping potential assay
    chemiluminescence enhancement
        accuracy assessment, **299:**9
        contributions of individual components, **299:**10–14
        effects on results
            age, **299:**11–12
            antioxidant supplementation of diet, **299:**10
            disease states, **299:**12, 14
            fasting, **299:**9–10
            gender, **299:**11–12
            metabolic rate, **299:**14
            pH of assay buffer, **299:**13
            smoking, **299:**10
        low-density lipoprotein assay, **299:**8–9
        plasma assay, **299:**7–8
        principle, **299:**6–7
        stoichiometric peroxyl radical-scavenging factors, **299:**10
        total antioxidant reactivity, **299:**12–13

electrode stability, **299:**50
α-Toxin, photoreceptor permeabilization
  cell death analysis, **316:**277
  confocal immunofluorescence localization
    antibody preparation, **316:**272–273
    bovine retina sectioning and staining,
      316:273
    data collection, **316:**273, 275
    neurobiotin distribution following per-
      meabilization, **316:**277–278
    toxin distribution, **316:**277
  expression and purification of recombi-
    nant toxin, **316:**271
  mutant pore blocking with zinc, **316:**271,
    275
  permeabilization mechanism and fea-
    tures, **316:**270
  rationale, **316:**269–270
  rhodopsin phosphorylation assay of per-
    meabilization, **316:**271–272, 275
tPA, *see* Tissue-type plasminogen activator
TR, *see* Thyroid receptor
Traction force assay
  calibration of silicone films
    overview, **298:**505–506
    pinch experiment with needle pair,
      298:508–509, 511
    single needle calibration, **298:**506, 508
  chamber assembly, **298:**501
  collagen gels as substrates, **298:**520–521
  compliance of silicone films, optimization
    to accomodate different cells,
    298:512–513
  definition of traction force, **298:**497
  design of flexible substrata, **298:**498
  imaging
    bead centroid sampling, **298:**515, 517
    film drift compensation, **298:**513
    image analysis, **298:**513, 515, 517,
      519–520
    mesh construction for generating trac-
      tion map image, **298:**517, 519–520
    microscopy, **298:**512
    resolution in traction map, **298:**520
  nonwrinkling silicone assay
    film preparation, **298:**503, 505
    overview, **298:**497
    Rappaport chamber assay, **298:**499–500
  polyacrylamide substrates, **298:**490, 496
  silicone oil properties, **298:**506, 520

  wrinkling silicone assay
    film preparation, **298:**501
    flow chamber assay, **298:**499
    overview, **298:**497
    simple assay, **298:**498–499
Traffic ATPase, *see* ATP-binding cassette
    transporter
Transcription, green fluorescent protein as
    reporter
  flow cytometry assay
    applications, **302:**206
    cytometer setup, **302:**200
    data analysis, **302:**202–205
    DNA constructs, **302:**201
    overview, **302:**199–200
    sensitivity, **302:**206
    transfectants and controls in experimen-
      tal design, **302:**201–202
  NF-κB reporter assay
    advantages, **302:**32–33
    destabilized protein for rapid turnover,
      302:33, 37
    high-throughput drug screening, **302:**38
    principle, **302:**33
    tumor necrosis factor-α induction
      dose response, **302:**35–36
      time course, **302:**34–35
  plants, *see also* Transgenic plant
    advantages of green fluorescent protein
      as reporter, **302:**316, 328
    autofluorescence interference, **302:**317
    derepression of expression by tetracy-
      cline *in planta*, **302:**324–325,
      327–328
    detection systems, **302:**316–317
    fluorescence microscopy and imaging,
      302:319–321, 325
    β-glucuronidase reporter, **302:**309, 312,
      316, 318
    luciferase reporter, **302:**316
    pathogen studies, **302:**322–323
    promoter activity reduction, analysis,
      302:321–324
    quantitative analysis in real time,
      302:320–321
    root systems, **302:**321
    transformation marker, **302:**317–319
  testis-specific promoters in transgenic
    mice
    applications, **302:**283–284

microinjection
    transgene purification, **302:**275–276
    transgenic mouse generation, **302:**277
    overview, **302:**272–273, 283
    reporter gene constructs, **302:**273–275
    spermatogenesis, stages of expression
        cycle of the seminiferous epithelium,
            302:278–279
        patterns, **302:**282–283
        phases, **302:**278
        transillumination-assistedmicrodissec-
            tion, **302:**279–282
    tissue specificity of expression,
        302:277–278
transfection assays, *see* Transfection
Transcription assay, *see* Microarray, DNA
Transducin
    abundance in rod outer segment, **316:**253
    activation assays of rhodopsins
        absorption spectroscopy assays
            extra metarhodopsin-II, **315:**476, 478
            photoregeneration from metarhodop-
                sin-II state, **315:**479
        bovine opsin expressed in *Pichia pas-
            toris*, **315:**9
        comparison of intrinsic biophysical
            assays, **315:**471–472
        constitutively active opsin mutants,
            transducin activation assay
            filter binding assay, **315:**213–214
            pH dependence, **315:**208, 214–215
            transducin purification, **315:**213, 241
        constitutively active opsin mutants,
            transducin activation assay with
            all-*trans*-retinal
            filter-binding assay, **315:**255–256
            intrinsic tryptophan fluorescence
                assay, **315:**256
        cysteine double mutants, **315:**139–140
        intrinsic tryptophan fluorescence of
            transducin in assay, **315:**118–119,
            241, 243–245, 247–249, 473–476
        meta intermediate thermal behavior,
            comparison of rod and cone pig-
            ments, **315:**308–309, 311
        near-infrared light scattering assay,
            315:479–482
        opsin/all-*trans*-retinal complex activa-
            tion by transducin, **315:**241, 243–
            245, 247–249

reconstituted fragments of opsin,
    315:68–69
resonant mirror spectroscopy,
    315:487–489
surface plasmon resonance assay
    overview, **315:**482–483
    patterns of rhodopsin membranes,
        315:485, 487
    self-assembled monolayers, **315:**483,
        485, 487
    supported lipid bilayers containing
        rhodopsin, **315:**483–485
expression of $G_t\alpha$ in baculovirus–insect
    cell system, **315:**539
fluorescent labels as transducin conforma-
    tion probes in rhodopsin activation
    acrylodan probing of amino-terminal
        conformational changes,
        315:499–501
    cysteine-substituted mutants for label-
        ing, **315:**500–502
    functional activity of labeled proteins,
        315:495–497
    $G\alpha_t/G\alpha_{i1}$ chimera construction in *Esch-
        erichia coli*, **315:**491
    high-performance liquid chromatogra-
        phy of labeled species, **315:**493
    lucifer yellow
        carboxyl-terminus conformational
            changes, **315:**498–499
        labeling reaction, **315:**491–492
        sites of labeling, **315:**500
        stoichiometry of labeling,
            315:492–495
        switch II region conformational
            changes, **315:**497–498
    principle, **315:**490–491
    proteolytic analysis of labeled peaks,
        315:495
$G\alpha_t/G\alpha_{i1}$ chimeras
    applications, **315:**523–524, 554
    concentration determination,
        315:508–509
    construction in *Escherichia coli*,
        315:491, 503–507
    expression, **315:**507–508, 544
    functional assays
        conformational change induction by
            AlF_4^-, intrinsic tryptophan fluo-
            rescence, **315:**512–513

GTPase assay, **315:**512
GTP binding stimulation by rhodopsin, **315:**511–512
overview, **315:**510–511
GDP dissociation with GTPγS binding
binding reactions, **315:**514
GDP binding to Gα subunits, **315:**513, 517
intrinsic fluorescence assay, **315:**513–514
spontaneous release rates, **315:**514, 516–517
homology of subunits, **315:**503
low-solubility chimeras, expression and purification, **315:**508
purification
anion-exchange chromatography, **315:**507–508, 545
nickel chromatography of histidine-tagged proteins, **315:**507–508, 544–545
receptor interactions
binding to inhibitory subunit of phosphodiesterase, **315:**517–519
overview, **315:**517
RGS9 interactions, **315:**519–521, 523
site-directed mutagenesis
alanine-scanning mutagenesis in switch II region, **315:**540
cyclic GMP phosphodiesterase Pγ subunit binding assays, **315:**549–552
GTPγS binding assay for folding analysis of mutants, **315:**548
polymerase chain reaction-based mutagenesis, **315:**504–505, 542–544
RGS interaction assays, **315:**552–554
target selection and rationale, **315:**540, 542
trypsin protection assay for folding analysis of mutants, **315:**545–546
tryptophan fluorescence assay with AlF$_4^-$ for folding analysis of mutants, **315:**546–548
solubility, **315:**509–510
Gα$_t$ peptide binding to rhodopsin, *see* Peptide-on-plasmid combinatorial library, rhodopsin ligands
GTPase acceleration, *see* Cyclic GMP phosphodiesterase; RGS9

GTPase assays
multiple-turnover approach, **315:**527–528
single -turnover approach
controls, **315:**529, 533–534
data analysis, **315:**530, 532, 537
detection techniques, **315:**528
equipment, **315:**535–537
GTPase regulators, assay of effects, **315:**530–531, 534–535, 553–554
initialization by GTP binding to G$_t$α, **315:**528–531, 536, 538
magnesium in binding reaction, **315:**531
overview, **315:**528
preactivation assay, **315:**537–538
quenching, **315:**532
rhodopsin complex formation, **315:**529–530
salt concentration, **315:**531
separation of radiolabeled products, **315:**532–533, 536–538
solutions, **315:**536–538
standard assay, **315:**535–537
time scale of reactions, **315:**531–532
GTPase cycle in phototransduction, **315:**524–526
immobilization for surface plasmon resonance studies, **316:**25, 27–28
isoprenyl/methyl function studies of βγ
ADP ribosylation by pertussis toxin, **316:**459–460, 478–481
effector enzyme activation assays, **316:**459, 463, 465
farnesylated cysteine analog probing of binding specificity, **316:**462–464
gel filtration analysis of subunit interactions, **316:**481
GTP exchange assay, **316:**459, 462
GTPγS binding to α catalyzed by meta-rhodopsin II, **316:**476–478
liposome binding assay, **316:**460–461
protein preparation for assays
defarnesylated βγ, **316:**457
demethylated βγ, **316:**457–459, 467, 469
mutant βγ expression and purification in baculovirus–insect cell system, **316:**471–474
reversed-phase high-performance liq-

uid chromatography analysis of
  γ, **316:**469, 471, 474
rhodopsin, **316:**456
rod outer segment preparation,
  316:467–468
transducin, **316:**456, 468–469
rhodopsin binding, **316:**462–463
sites of modification, **316:**466–467
stripped rod outer segment binding
  assay, **316:**475–476
kinetic light scattering studies of
  association/dissociation with rhodopsin
dissociation signal
  kinetic analysis, **315:**346–347
  overview, **315:**344–345
light-independent binding, **315:**343
modeling of binding reaction,
  315:343–344
nucleotide affinity, **315:**342–343
origin of signal, **315:**329–330
phosphodiesterase signal
  kinetic analysis, **315:**346–347
  overview, **315:**345–346
  phosphodiesterase preparation,
    315:335
rates and dissociation constant determination, **315:**344
sample preparation, **315:**334–335
limited proteolysis of G-proteins as conformational probes
rationale, **315:**364–365
antibody preparation, **315:**369, 372
GDP binding effects, **315:**367
GTP binding effects, **315:**365–367
incubation conditions
  protein binding, **315:**372–374
  proteolysis, **315:**374
materials, **315:**369, 371–372
protected cleavage sites, **315:**376
proteolytic fragment identification
  reversed-phase high-performance liq-
    uid chromatography, **315:**367,
    369
  Western blot analysis, **315:**367, 369,
    375–376
rod outer segment preparation, **315:**371
transducin preparation, **315:**371
tryptic cleavage sites in Gα and Gβγ,
  315:365

V8 protease clevage sites, **315:**367, 376
vesicle preparation, **315:**371
posttranslational modification, **315:**336
promoter analysis of α subunit in transiently transfected *Xenopus* embryos,
  316:57
rhodopsin activation
  conformational changes with activa-
    tion, **315:**178–179, 364, 490
  overview, **315:**70, 342–343, 471–473
signal amplification, **315:**422
subunits, functions and effectors,
  316:454–456
time-resolved low-temperature absorption spectroscopy of rhodopsin intermediates in activation
  absorption maxima of intermediates,
    315:361–362
  applications and overview, **315:**350–351
  collected spectra, **315:**353, 355
  experimental setup, **315:**351–352
  GTP analog studies, **315:**352, 359, 361,
    363
  intermediates interacting with trans-
    ducin, **315:**348–349, 358–359,
    361–363
  photoconverted pigment, quantifica-
    tion, **315:**352–353
  rhodopsin purification, **315:**349
  spectral analysis, **315:**355, 357–358
  transducin purification, **315:**349–350
visual cycle role, **316:**453–454
Xenopus oocyte expression
  advantages, **316:**41, 49
  α subunit expression, **316:**47–49
  immunoprecipitation of radiolabeled
    proteins, **316:**45–46
  materials and solutions, **316:**41–43
  membrane preparation, **316:**45
  nucleic acid injection, **316:**43–44
  oocyte preparation, **316:**43
  voltage clamp recording, **316:**44–45
zebrafish larva eye, activation assays
  GTPγS *in situ* binding assay
    incubation conditions and radioactiv-
      ity detection, **316:**542
    materials, **316:**541
    pob mutant analysis, **316:**542, 544
    principle, **316:**540–541
    tissue preparation, **316:**541–542

GTPγS *in vitro* binding assay
  incubation conditions and radioactivity detection, **316:**544
  materials, **316:**544
  principle, **316:**544
  white light versus red light response, **316:**544–545
Transfection
  antisense oligonucleotides, *see also* viral vectors
    cationic liposome delivery, **314:**293, 324–325, 344–347, 385, 387, 407–408, 442–444, 485, 487–489
    polyethyleneimine transfection of neurons
      cation:DNA ratio, **314:**295–297
      efficacy, **314:**294–295
      electrophysiology study compatibility, **314:**293–294
      incubation conditions, **314:**297
      stock solutions, **314:**296–297
      toxicity, **314:**295
    renal tubular cells, **314:**380–381
    transfected neuron identification, **314:**306–307
  efficiency assay with green fluorescence protein and flow cytometry
    applications, **302:**206
    cytometer setup, **302:**200
    data analysis, **302:**202–205
    DNA constructs, **302:**201
    overview, **302:**199–200
    sensitivity, **302:**206
    transfectants and controls in experimental design, **302:**201–202
  electroporation of mammalian cells for flow cytometry, **302:**190–192
  green fluorescent protein fusion protein expression in stably versus transiently transfected cells, **302:**4, 192–193
  lipofection
    advantages and disadvantages, **313:**346, 362, 373
    angiotensinogen antisense oligodeoxynucleotides, **313:**55–56
    application with non-nucleic acid drugs, **313:**322
    biodegradable pH-sensitive surfactants

to facilitate endosome release, **313:**351–352
c-Myc antisense oligonucleotides, cationic lipid delivery
  efficacy, **313:**395–396
  incubation conditions, **313:**391
  Lipofectamine versus Lipofectin, **313:**391, 395–396
  optimization, **313:**391–392, 394
  toxicity, **313:**392, 394
C5-propyne-containing oligophosphorothioate oligonucleotide, **313:**88–90
cationic liposome lipoplexes, **313:**323, 348–350
direct chemical conjugation, **313:**323
efficacy, **313:**155
encapsulation
  active encapsulation with cationic lipids, **313:**330
  anionic liposomes, **313:**347–348
  delivery to inflammation sites and tumors, **313:**338–340
  detergent dialysis, **313:**328–329
  dry lipid hydration, **313:**324–325, 328
  efficacy in intercellular adhesion molecule-1 inhibition, **313:**340
  efficiency of methods, **313:**347
  lipophilic backbone modification and encapsulation, **313:**329–330, 346
  literature review, **313:**326–327
  mechanisms of therapy enhancement, **313:**333–334
  minimal volume entrapment, **313:**329
  pharmacokinetics, **313:**336–338
  principle, **313:**324
  reverse-phase hydration, **313:**328
  stability assays, **313:**334, 336
  stabilized antisense-lipid particles, **313:**331, 333
  toxicity, **313:**340–341
minimally modified antisense oligonucleotides, **313:**282–283
miniosmotic pump in glioma model, **313:**367, 370
P-ethoxy oligonucleotides
  liposome incorporation, **313:**375
  separation of free and incorporated oligonucleotides, **313:**375–376
peptide nucleic acid conjugates, **313:**158

phosphorothioate oligoribonucleotides
    in thalassemic splice mutant correc-
    tion, **313:**510–511
  umbrella amphiphiles, **313:**350–351
myocytes
  calcium phosphate transfection,
    302:182–183
  efficiency, **302:**178
  lipofection, **302:**180, 182
  microinjection, **302:**183
naked DNA, **313:**55
neurons
  calcium phosphate transfection,
    302:221
  cerebellar granule neurons, primary
    culture
    dissection, **302:**215
    polylysine coating of culture sur-
      faces, **302:**214–215
    supplements, **302:**216
    trypsinization/trituration, **302:**215–
      216, 218
  green fluorescent protein vector analy-
    sis of transfectants
    advantages, **302:**222–225
    constructs, **302:**218–221
    glutamate receptor subunits, **302:**224
    inositol triphosphate 3-kinase,
      302:223–224
  lipofection, **302:**221–222
  stable versus transient transfection,
    302:213
polyethyleneimine comparison with lipofection
  agarose gel analysis of carrier complex
    formation, **313:**63–65
  cell culture, **313:**58–59
  confocal microscopy of nuclear localiza-
    tion, **313:**59–60, 65, 67
  efficacy of H-Ras depletion, **313:**71–72
  flow cytometry analysis, **313:**61, 67, 69
  fluorescence quenching by carriers,
    313:62, 67
  Northern blot analysis, **313:**60–63
  oligonucleotide synthesis, **313:**57–58
  stability analysis of modified oligonucle-
    otides, **313:**61–62, 69–72
  transfectant preparation, **313:**58
ribozyme gene delivery
  lipofection, **313:**503

overview, **313:**417–418, 500
transferrin–polylysine conjugates,
  313:501–503
transfectant screening with green fluorescent protein
  advantages, **302:**207, 212
  applications, **302:**212
  bidirectional vector system, **302:**207–
    210, 212
  cloning, **302:**210
  fluorescence detection, **302:**208
  luciferase coexpression and assay,
    302:211–212
  materials and reagents, **302:**208
  stable transfection and selection,
    302:210
  transient transfection assays of antisense
    inhibition, **313:**90–93
Transfer RNA, *see also* Suppressor transfer
  RNA, unnatural amino acid incorporation
  aminoacylation analysis with phosphoro-
    thioate interference
    aminoacylation reaction, **317:**87–88
    critical phosphate and 2'-hydroxyl posi-
      tion identification, **317:**90–91
    iodine cleavage and cleavage pattern
      analysis, **317:**89–90
    materials, **317:**83–84
    overview, **317:**82–83
    separation of charged and uncharged
      RNAs, **317:**88–89
    transcript
      dephosphorylation, **317:**86–87
      nucleotide incorporation into RNA
        transcripts with mutant T7 RNA
        polymerase, **317:**85–86
      phosphorous-32 labeling, **317:**87
  circular permutation analysis of tRNA^Phe
    folding, **317:**326
  cleavage by imidazole conjugates,
    318:153, 155, 158, 163–164
  Levitt's pair modeling with comparative
    sequence analysis, **317:**505–506
  ribosome binding in *Escherichia coli*, ra-
    dioactivity assay
    AcPhe-transfer RNA binding studies
      AcPhe-transfer RNA preparation,
        317:264–265
      artifacts distorting saturation curves

percutaneous absorption,
   314:497–498
scar formation effects, **314:**498–499
wounded skin, oligonucleotide distribution, **314:**496–497
functions and clinical effects, **314:**493–494
Transgenic frog, *see Xenopus* transgenesis
Transgenic mouse
  construction of green fluorescent protein
    transgenic mice through embryonic
    stem cells
    chimeric mice generation, **302:**245–246
    electroporation, **302:**243–244
    gene integration marker, advantages,
      302:234–235
    germ line transmission, **302:**246
    identification of expressing cells
      fluorescence, **302:**244–245
      Northern blot analysis, **302:**246–247
    vector construction, **302:**242–243, 249
  development of technique, **306:**187
  fluorescence microscopy of green fluores-
    cent protein expression
    filters, **302:**247
    fresh tissue, **302:**247–248
    histological analysis, **302:**248–249
  GC-E knockout mouse
    electroretinogram recordings,
      316:560–564
    generation, **316:**559
    histological analysis, **316:**560–561
    single-cell electrophysiology, **316:**560–
      561, 563–564
  genotyping of retinitis pigmentosa
    models
    DNA isolation from tail biopsy,
      316:522–523
    overview, **316:**521–522
    polymerase chain reaction amplifica-
      tion, **316:**523–524
    restriction digestion of amplified frag-
      ments, **316:**524
    sequencing, **316:**524–526
  green fluorescent protein expression in
    preimplantation embryos
    embryo collection, **302:**236
    gene integration marker, advantages,
      302:234–235
    microinjection
      buffer, **302:**239

DNA fragment, preparation for injec-
    tion, **302:**238–239
holding pipette preparation, **302:**237
injection chamber, **302:**238
injection pipette preparation, **302:**237
instrumentation, **302:**239
technique, **302:**239–240
vector for expression, **302:**237,
    249
monitoring of expression, **302:**240
selective production of transgenic mice,
    302:240–242
superovulation, **302:**235–236
homologous recombination
  cis-regulatory DNA element analysis,
    306:43–45
  selectable marker removal with recom-
    binases, **306:**53–55
MDR1, see Transgenic mouse, *MDR1*
opsin phosphorylation-site mutant ex-
    pression
  rationale, **316:**168–169
  constant illumination protocols
    excised eyes, **316:**569
    living mice, **316:**569
    retinoid responses, **316:**573
  DNA preparation for microinjection,
    316:171
  flash protocols
    excised eyes, **316:**569
    living mice, **316:**568
    retinoid responses, **316:**572–573
  founder screening, **316:**173–174
  gene dosage control with knockout
    crosses, **316:**179–182, 184
  housing of animals, **316:**567
  overexpression and retinal degrada-
    tion, **316:**178–179
overview, **316:**565–567
reporter genes, overview, **302:**233–234
retinal pigment epithelium p65 knockout
    mouse
  embryonic stem cell derivation,
    316:714–716
  generation, **316:**716
  phenotype, **316:**722
  retinoids, high-performance liquid chro-
    matography analysis, **316:**722
  rhodopsin quantification and character-
    ization, **316:**722–723

RU-486-inducible gene expression in liver, **306:**292, 294

testis-specific promoters, green fluorescent protein as reporter

applications, **302:**283–284

microinjection

transgene purification, **302:**275–276

transgenic mouse generation, **302:**277

overview, **302:**272–273, 283

reporter gene constructs, **302:**273–275

spermatogenesis, stages of expression phases, **302:**278

cycle of the seminiferous epithelium, **302:**278–279

transillumination-assistedmicrodissection analysis, **302:**279–282

patterns, **302:**282–283

tissue specificity of expression, **302:**277–278

visual cycle mutants

mouse generation, **316:**172–173

photoreceptor membrane preparation, **316:**569, 571

retinoid analysis by high-performance liquid chromatography

dissection of eye, **316:**567–568

extraction, **316:**568

individual variation of retinas, **316:**574

reverse transcriptase–polymerase chain reaction for transgene quantification, **316:**176

rhodopsin concentration determination, **316:**568

single-cell recordings from mutant-containing photoreceptors, **316:**185

Southern blot analysis for sorting different lines from same founder, **316:**174

trafficking of mutants, **316:**184

transgene constructs, **316:**169–171

Western blot analysis of protein expression, **316:**574–575

yeast artificial chromosome expression, *see* Yeast artificial chromosome

Transgenic mouse, *MDR1*

actin promoter and stability of expression, **292:**573–575

Eμ transgenic mice production, overview, **292:**572, 575–576

expression assays

immunofluorescence and fluorescence-activated cell sorting, **292:**580–581, 583–584

Northern blot analysis of RNA, **292:**580

ribonuclease protection assay, **292:**581–582

Western blot analysis, **292:**584

functional assays

drug resistance measurements in white blood cells, **292:**589–593

drug sources, preparation, and doses, **292:**586, 588–589

rhodamine-123 efflux, **292:**584–586, 593

homozygous transgenic mouse identification, **292:**578–580

protection against toxic substrates, **292:**546, 572, 589–593

reversal agent testing, **292:**588, 593

screening of founder mice, **292:**577–578

strain selection, **292:**576

tissue expression and targeting, **292:**574, 582–583

Transgenic plant, green fluorescent protein expression

advantages as transgene markers, **302:**310

cytotoxicity, **302:**312–313

flow cytometry

autofluorescence, **302:**310, 312

lasers for excitation, **302:**315

mutant green fluorescent protein applications, **302:**314–315

protoplast and subcellular organelle analysis

instrument settings, **302:**302–303

transfected protoplasts, **302:**305–306

transgenic plant nuclei, **302:**307–309

transgenic plant protoplasts, **302:**307

setting up

Coulter Elite, **302:**301–302

Cytomation MoFlo, **302:**302

sorting setup

Coulter Elite, **302:**303–304

Cytomation MoFlo, **302:**304–305

growth and maintenance of plants, **302:**299

nuclei

preparation, **302:**299

protein targeting, **302:**311–313

plasmids, **302**:298–299
preparation of plants, **302**:299
protoplast
 preparation, **302**:300, 310
 transfection, **302**:301
Transglutaminase, isopeptide bond formation
activity assay, **309**:174–175
colocalization with insoluble protein complexes, **309**:178–179
cross-linking of proteins, **309**:172–173
detection of isopeptide bonds
 immunohistochemistry, **309**:176
 overview, **309**:174
immunoelectron microscopy with paired
 helical filaments and tangles,
 309:182–183
immunohistochemistry, **309**:176, 178–179
isopeptidase cleavage of bonds
 assay, **309**:173
 digestion of paired helical filaments
  and tangles, **309**:184–185
 mechanism of cleavage in assay system, **309**:183–184
levels in brain, **309**:173–175
substrate specificity, **309**:174
Western blot analysis, **309**:176–178
Trans-Golgi network, *see* Membrane trafficking, rhodopsin
Transient electric birefringence
data analysis
 birefringence response conversion,
  317:447
 Levenberg–Marquardt analysis of decay curves, **317**:447–448
 normalization of field-free decay curve,
  317:447
 phased τ ratio approach, analysis of
  nonhelix element flexibility,
  317:450, 452
 τ ratio approach, analysis of birefringence decay times, **317**:448–450
instrumentation
 cell, **317**:444
 detector, **317**:445
 high-voltage pulse generator, **317**:444
 laser source, **317**:444
 oscilloscope, **317**:445
 overview, **317**:443–444
 temperature control, **317**:448

principle, **317**:441–443
RNA experiments
 applications, **317**:440
 branched RNA analysis, **317**:392
 nonhelix element flexibility, **317**:450,
  452
 overview, **317**:440–441
 protein–RNA interactions,
  317:452–453
RNA design
 central placement in RNA helix of
  region of interest, **317**:445
 length of helix, **317**:445–445
Transient time, approximation in substrate
 channeling, **308**:113–114
Transient transfection, *see* Acetylcholine receptor; COS cell; Transfection
Transillumination-assisted microdissection,
 testis-specific promoter analysis with
 green fluorescent protein as reporter,
 302:279–282
Transition-state
binding
 affinity, **308**:303–304, 357–358
 energetics, **308**:304–305, 358, 399–400
catalytic theory, **308**:303, 399–400
difference Raman spectroscopy
 inosine–uridine nucleoside hydrolase,
  transition-state structure, **308**:199
 lactate dehydrogenase, transition-state
  stabilization, **308**:189–191
 transition-state analog binding analysis,
  308:181–182
$G_{i\alpha1}$, GTP hydrolysis, **308**:72, 81–85,
  90–91
kinetic isotope effect analysis
 S-adenosylmethionine synthetase,
  308:343
 AMP deaminase
  catalytic mechanism, **308**:333, 335
  inhibitor design, **308**:348–349, 394
  isotope effects and transition-state
   modeling, **308**:335–336
  molecular electrostatic potential surface of transition-state,
   308:347–348
 binding isotope effects, **308**:330–332
 bond energy/bond order vibrational
  analysis of transition-state
  structure

hydrolysis application, **308:**385–
    388, 395–396
  overview, **308:**368, 370–371, 382–383
  reaction space and interpolation phi-
    losophy, **308:**383–384
  reference structures, **308:**387–392
lactate dehydrogenase, transition-state sta-
    bilization, **308:**189–191
lifetime, **308:**302–303
Marcus rate theory, variable transition-
    state principle, **308:**277–278
molecular surface point similarity mea-
    sures in binding energy modeling
  accuracy, **308:**409
  adenosine deaminase, **308:**404–405,
    407, 409
  AMP nucleosidase, **308:**405, 407, 409
  cytidine deaminase, **308:**405–407, 409
  indications for use, **308:**425
  overview, **308:**401–404
  quantum mechanical descriptions of
    molecules, **308:**400–401
  similarity as a function of α,
    308:402–403
neural networks in binding energy calcu-
    lation
  advantages, **308:**410, 425–426
  architecture of network, **308:**412–414
  back propagation, **308:**414
  cytidine deaminase analysis, **308:**414–
    415, 424, 426
  hidden layer magnitudes and inhibitors
    design, **308:**423–424
  inosine–uridine nucleoside hydrolase
    analysis
    accuracy, **308:**419, 424
    binding sites, **308:**416
    orientation of ligand binding,
      308:417
    substrate specificity, **308:**416–417
    training of network, **308:**418
    transition-state structures and affini-
      ties, **308:**418–419, 422–424
  inputs and mapping, **308:**411–412
  learning, **308:**410, 414
  output layer, **308:**413
  overview of biological applications,
    308:410–411
  pattern classification, **308:**425
  weights and biases, **308:**413–414

triosephosphate isomerase, active site in-
    teractions with transition-state,
    308:272–273
Translation, *see* Cell-free translation system
Translational repression assay
  advantages, **318:**383–384
  cloning of RNA-binding proteins
    applications, **318:**383
    cell growth and enrichment, **318:**381
    selection of clones, **318:**381, 383
    transformation of cells, **318:**379
  components of system, **318:**375–377, 383
  green fluorescence protein as reporter,
    318:375–376
  plasmids, **318:**377
  principles, **318:**374–377
  prospects, **318:**384
  RNA–protein interaction analysis
    binding affinity range, **318:**377, 384
    flow cytometry analysis, **318:**378–379,
      384
    plasmids, **318:**377–378
    suitability of RNA-binding sequences,
      318:378
    yeast transformation and growth,
      318:378
Translational repression, RNA-binding pro-
    tein screening in bacteria
  applications, **318:**329, 331–332
  competent cells
    preparation, **318:**317–318
    transformation, **318:**318
  controls, **318:**321
  dissociation constant estimation, **318:**327
  β-galactosidase reporter assays
    colony color assay, **318:**318–319
    solution assay, **318:**319–321
  mathematical modeling, **318:**311–312
  optimization, **318:**318–319
  plasmids
    reporter, **318:**315–317
    RNA-binding protein, **318:**313–315
  principle, **318:**309–310
  RNA-binding protein characterization
    binding assay, **318:**325
    mutagenesis, **318:**323–325
    nucleolin, **318:**329
    Rev, **318:**328–329
    suppressor variant identification,
      318:325–327

U1A, **318:**331
RNA targets
    characterization by random mutagene-
        sis, **318:**321–323
    size, **318:**310–311
strains of *Escherichia coli*, **318:**312–313,
    321
translation efficiency determinants in
    Escherichia coli, **318:**310
troubleshooting
    low expression of binding proteins,
        318:327–328
    RNA structure, failure to form,
        318:328
    target definition, **318:**328
Translocation, sphingolipids
    analog assays
        fluorescence measurements on-line,
            312:578
        short-chain lipids, **312:**576
        short-chain sphingolipids added to out-
            side of cells or organelles,
            312:576–578
        short-chain spin-labeled sphingolipids,
            312:578–579
    calculations
        distribution, **312:**565–566
        mobility, **312:**566–567
    chemical modification assay
        oxidation of glycolipids
            sodium periodate, **312:**568–569
            TEMPO, **312:**569
        principle, **312:**567
        reagents, **312:**567–568
    criteria for assays, **312:**564–565
    enzymatic modification assay
        N-deacylase, **312:**572–573
        endoglycoceramidase, **312:**572
        galactose oxidase, **312:**573
        glucocerebrosidase, **312:**572
        neuraminidase, **312:**572
        principle, **312:**569
        sphingomyelinase, **312:**570–571
    glycosphingolipid functions, **312:**562–563
    kinetic measurements, **312:**565
    lipid exchange assay
        nonspecific transfer protein, **312:**575
        principle, **312:**574–575
        saposins, **312:**576
        transfer proteins, **312:**575

noncovalent protein binding assay
    antibodies, **312:**574
    cholera toxin, **312:**574
    lectins, **312:**574
    principle, **312:**573
pools and sorting, **312:**563
transporters, **312:**564
Transmembrane topology
    ATP-binding proteins, **292:**8–9
    CP47, 321–322
    cysteine modification with membrane im-
        permeant and membrane permeant–
        biotin reagents
        limitations, **296:**329–330
        maleimide compound modifications
            binding reactions, **296:**323
            biotin label detection, **296:**322
            cell lysis, **296:**323–324
            immunoprecipitation, **296:**322, 324
            Western blotting, **296:**322, 324
            mapping principle, **296:**320, 322, 329
        methane thiosulfate compound modifi-
            cations
            binding reactions, **296:**327
            cell lysis, **296:**326–328
            compound types, **296:**325–326
            recovery of biotinylated proteins,
                296:328
            Western blotting, **296:**326, 328–329
        protein engineering, **296:**321–322
    definition, **292:**265
    fusion proteins in transmembrane topol-
        ogy analysis, **296:**293–294
    GAT-1 epitope tagging in analysis
        carbonate extraction, **296:**301–302
        expression in *Xenopus* oocytes,
            296:299–300
        glycosylation analysis, **296:**302
        immunoprecipitation of fusion pro-
            teins, **296:**300–301
        plasmid construction, **296:**296–298
        principle, **296:**294–295
        protease protection assay, **296:**294–296,
            300, 302–305, 307
        RNA transcription, **296:**298
    heterologous expression of P-glycopro-
        teins in *Escherichia coli*
        Mdr1, 378–379
        Ste6p, **292:**379
    hydropathy analysis, **296:**295–296

immunofluorescence staining analysis, **293:**542–544

metalloprotease transporter, *Erwinia chrysanthemi*, **292:**75

molecular dynamics simulated annealing with restrained molecular dynamics, **293:**663–668

molecular modeling, **293:**591–593, 597

MscL analysis using alkaline phosphatase as reporter gene, **294:**477–478

overview of mapping techniques, **296:**318–319

P-glycoprotein

   alternative topologies, **292:**286–289

   C terminus translocation reporters

      criteria for reporter domains, **292:**268, 283

      fusion protein generation, **292:**268–270

      overview, **292:**267–268, 282–283

   determinants

      characterization, **292:**270–271

      types of sequences, **292:**266

   expression systems for study

      microsomal membranes, **292:**272–273, 281

      rabbit reticulocyte lysate, **292:**272, 274, 281

      RNA digestion in cell-free systems, **292:**273

      transcription *in vitro* using SP6 RNA polymerase, **292:**280

      wheat germ extract, **292:**281, 286–287

      Xenopus laevis oocyte microinjection, **292:**273–274

   folding steps, **292:**266–267, 285–288

   mapping, **292:**277–278

   membrane stripping with alkaline buffer, **292:**281–282

   N-linked glycosylation detection, **292:**274–275, 277, 283–284

   protease digestion assay, **292:**275–277, 285

photosystem I

   chemical modification of exposed residues with biotin derivatives, **297:**132–133

   cross-linking of subunits, **297:**133–134

limited proteolysis and gel electrophoresis, **297:**130–131

sodium channels, **294:**575–577

substituted-cysteine accessibility method analysis, **296:**345

Transmission electron microscopy, chromatin and DNA–protein complexes

direct mounting, **304:**216, 218

fixation

   artifacts, **304:**225–226

   dimethyl suberimidate, **304:**224

   functions, **304:**222

   glutaraldehyde and formaldehyde, **304:**223–224

   optimization, **304:**224–225

freeze-fracture, **304:**218

negative staining

   advantages, **304:**226

   ammonium molybdate, **304:**229

   aurothioglucose, **304:**229

   DNA, **304:**226–227

   methylamine tungstate, **304:**229

   phosphotungstic acid, **304:**227, 229

   reaction conditions, **304:**227

   resolution, **304:**226

   surfactant utilization, **304:**230

   uranyl acetate, **304:**227

support preparation

   carbon supports, **304:**218–220

   treatment to facilitate sample adherence

      Alcian blue, **304:**221–222

      amylamine, **304:**221

      ethidium bromide, **304:**222

      glow charging, **304:**220

      ions, **304:**220

      pentylamine, **304:**221

      polylysine, **304:**221

surface spreading

   Kleinschmidt method, **304:**214–215

   Miller method, **304:**215–216

Transporter, comparison to pores and channels, **296:**660–662

Transport current, *see* Electrophysiology, neurotransmitter transporters

Transporters associated with antigen processing

   components of major histocompatibility peptide transport system, **292:**111

gene inactivation by herpes simplex virus, **292:**118, 120

sequence homology of TAP proteins with P-glycoprotein, **292:**111

subunit structure, **292:**745

transport assay

competition experiments, **292:**749, 751–753

incubation conditions, **292:**747, 748, 752

melanoma cell line WM98.1, 749

peptide proteolysis and glycosylation, **292:**752–753

precipitation of reporter peptide, **292:**748–749

solubilization of cells, **292:**747–748

streptolysin O permeabilization of cells, **292:**746–748

Transposon mutagenesis

advantages over null mutation studies, **303:**512

antibiotic selection of transposition, **297:**13

arbitrarily-primed polymerase chain reaction for biofilm mutant analysis

components of reactions, **310:**106

first round, **310:**106–108

overview, **310:**105–106

second round, **310:**108

colony purification following selection, **297:**13

conjugal transposon transfer from *Escherichia coli*, **297:**4, 8, 10–13

conjugative plamid tagging, **310:**39–40

Fremyella diplosiphon, **297:**5, 9

HAT tag insertion

effect on protein function, **303:**531

hemagglutinin protein insertion, **303:**515–516

immunofluorescence microscopy detection, **303:**524–525

mini-transposon constructs, **303:**514–515

yeast carrier generation, **303:**523–524

insertion sequences, **297:**5, 9

insertion site identification

plasmid rescue, **303:**526–528

vectorette polymerase chain reaction, **303:**528–530

marine bacteria

green fluorescent protein fusions, **310:**76

lacZ fusions, **310:**70

mutagenized genomic libraries, preparing and transforming DNA, **303:**525–526

Mutator element in maize chloroplast mutation, **297:**41–42, 56–57

Nostoc 7120, **297:**4–5, 7–10, 13–17

Nostoc ellipsosporum, **297:**9, 14

Nostoc punctiforme, **297:**3–5, 8, 10–12, 14–17

reconstruction of mutation in heterocyst-forming cyanobacteria

electroporation of *Escherichia coli*, **297:**15

genomic DNA isolation and digestion, **297:**15

ligation of recovered plasmids, **297:**16

purpose, **297:**14

sucrose sensitivity screening, **297:**16–17

reporter gene systems, **310:**57–58

Rhodobacter cytochromes, **297:**91–92

Rhodobacter sphaeroides 2.4.1 photosynthesis genes, **297:**156

screening for mutant phenotypes, **303:**526

segregation of mutant and wild-type chromosomes, **297:**14

shuttle mutagenesis

mating of yeast, **303:**519

mini-transposon derivatives, characteristics, **303:**514–516

overview, **303:**513–514, 517, 520

plasmid isolation, **303:**519, 521

reagents, **303:**516–519

transformation of bacteria, **303:**517

symbiotic interaction in screening, **297:**3–4

Tn5

derivatives

antibiotic resistance gene replacement, **297:**7

mutation stabilization, **297:**8–9

recovery with flanking genomic sequences, **297:**6–7

transcription-only fusions, **297:**7–8

transposition frequency, **297:**7

efficiency of transposition, **297:**5–6

type II restriction endonuclease activity in screening, **297:**9–10

yeast transformants

assays, overview, **303:**521
  green fluorescent protein in screening,
    303:522
  insert characterization, **303:**523
  selection, **303:**521–522
Transthyretin amyloid
  formation mechanism, **309:**592
  hydrogen exchange measured with elec-
      trospray mass spectrometry
    exchange properties of wild-type and
        variant proteins, **309:**639, 641
    monomer analysis, **309:**641, 643
    preparation of samples, **309:**637
  monoclonal antibody probing of confor-
      mation
    antibody production
      immunization of mice, **309:**596
      screening of hybridomas,
          309:596–597
    applications, **309:**604–605
    binding to extracted fibrils, **309:**603
    enzyme-linked immunosorbent assay
      cryptic epitope demonstration,
          309:600–601
      direct assay, **309:**595
      sandwich-type assay, **309:**596
    immunoprecipitation, **309:**602–603
    mutants of protein for aggregation,
        309:593–595
    phage display of epitopes
      DNase I digestion, **309:**598
      fractionation of DNA fragments,
          309:598–599
      fusion gene cloning, **309:**599
      overview, **309:**595
      screening, **309:**599–600
    precautions, **309:**604
    rationale, **309:**592–593
    strategy overview, **309:**593
  subunit structure, **309:**592
  X-ray fiber diffraction, **309:**529–530
TRAP assays, *see* Total peroxyl radical-trap-
    ping potential assay; Translational re-
    pression assay
Trehalose, stabilization of reverse tran-
    scriptase, **303:**19, 22
Triboluminescence, educational demonstra-
    tions, **305:**658–659
Tributyltin, antimicrobial activity as paint
    additive

efficacy, **310:**643–644
  extracellular polysaccharide evaluation,
      310:640–643
  modified Robbins device culture of bio-
      films, **310:**638–639
  overview, **310:**637
  substrata variation effects, **310:**641–
      643
  viable cell counting, **310:**639–641
TRiC, *see* Cytosolic chaperonin
7-Trifluoromethyl-4-chloro-*N*-methylquino-
    lium, colorimetric assay of thiols
  absorbance properties, **299:**277, 279–281
  glutathione-specific assay
    incubation conditions, **299:**284–286
    instrumentation, **299:**283
    interferences, **299:**281–283
    reaction mechanism, **299:**280, 286
    reagents, **299:**283
    reproducibility, **299:**281
    sample preparation, **299:**283–284
    sensitivity, **299:**281, 286
  reaction with thiols, **299:**277–278
  total mercaptan assay, **299:**277–280,
      284
N-(*p*-Trifluoromethyldiazinirinyl)phe-
    nylethyl-2-[^{35}S]-2-thioacetyl-D-*erythro*-
    C$_{18}$-sphingosine, synthesis, **311:**594–595
N-{2-[4-(3-Trifluoromethyl-3*H*-diazirine-3-
    yl)phenyl]ethylsulfanyl-acetyl}[^{3}H]lyso-
    GM1, synthesis, **311:**592–593
Trihexyphenidyl, inhibition of cocaine bind-
    ing to dopamine transporter, **296:**461–
    462, 464
4,5',8-Trimethylpsoralen
  linker region mapping between nucleo-
      somes, **304:**618–619
  nucleosome positioning by indirect end-la-
      beling, **304:**636
  permeability of cell membranes, **304:**639
  supercoiling assay
    applications, **304:**639
    calculations, **304:**641
    cell culture, **304:**641–642
    cross-linking reaction, **304:**642
    DNA topological domain determi-
        nation
      calculations, **304:**647–650
      principle, **304:**639–640
    extraction of DNA, **304:**642–643

nicking of DNA, **304:**642
principle, **304:**640
single-strand breaks, measurement
    eukaryotic cells, **304:**645
    specific chromosomal loci, **304:**646
Southern blot analysis of cross-linked
    DNA, **304:**643–645
N,N′,N″-Tripropylamidothionophosphate
antioxidant activity mechanism,
    299:294–295
peroxyl radical reactivity, **299:**299
reactivity
    hydrogen peroxide decomposition,
        299:296
    lipid hydroperoxide reduction,
        299:297–300
    sodium hypochlorite, **299:**296–297
synthesis, **299:**296
Triosephosphate isomerase
active site interactions with transition-
    state, **308:**272–273
catalytic problem description,
    308:248–250
entropic barrier, lowering by enzyme,
    308:272
hydrogen bonds, strengthening accompa-
    nying charge rearrangement
    basis for effect, **308:**251–254
    catalytic contribution, quantitative esti-
        mation, **308:**254–257, 261, 266–268
    dielectric environment lowering by en-
        zyme, **308:**261, 264–266, 274
    Hine equation and rate enhancement
        calculation, **308:**257, 274–276
    linear-free energy analysis of intramo-
        lecular hydrogen bond strength,
        308:261–263, 266–268
    multiple hydrogen bond analysis,
        308:269–270
    overview, **308:**250–251
refolding of recombinant protein from in-
    clusion bodies, **309:**226
Triplex-forming oligonucleotide
advantages in gene expression inhibition,
    313:94–95
binding assays
    cross-linking, **313:**93–94
    DNA footprinting, **313:**93
delivery into cells, **313:**88–90
design

controls, **313:**79
criteria, **313:**74
modifications for stability, **313:**76–79
target sequence selection, **313:**74–76
electrophoretic mobility shift assays of
    binding
    electrophoresis conditions, **313:**82–83
    incubations, **313:**80, 82
    principle, **313:**79–80
    restriction endonucleases, **313:**80–81
    self-association structures, **313:**83
    target concentration variation, **313:**83
inhibition assays
    overview, **313:**90
    transient transfection assays with re-
        porter genes, **313:**90–93
principle of gene inhibition, **313:**144
secondary structure analysis with ultravio-
    let melting
    data acquisition, **313:**84–85
    principle, **313:**83–84
    self-structure formation, **313:**86–88
    thermodynamic equilibrium, **313:**85–86
Trk1
deletion mutants
    heterologous potassium channel expres-
        sion, **293:**90–91, 93–95
    selection by growth on potassium-lim-
        iting media, **293:**91–93
    potassium transport in yeast, **293:**89–90
Trk2
deletion mutants
    heterologous potassium channel expres-
        sion, **293:**90–91, 93–95
    selection by growth on potassium-lim-
        iting media, **293:**91–93
    potassium transport in yeast, **293:**89–90
tRNA, *see* Transfer RNA
Trolox equivalent antioxidant capacity
    assay
comparison with oxygen radical ab-
    sorbance capacity assay, **299:**58–60
decolorization assay, **299:**384, 389
extraction and analysis of tomato,
    299:383–385, 387, 389
materials, **299:**381
preparation of preformed 2,2′-azinobis(3-
    ethylbenzothiazoline-6-sulfonic acid)
    radical cation, **299:**381
principle, **299:**50–51, 380

standards, preparation, **299**:381–383
Trypan blue exclusion assay, amyloid-β neurotoxicity, **309**:761
Trypsin
antigen recovery in amyloid immunohistochemistry samples, **309**:18
proteolysis of polypeptides in cell-free
translation systems, **290**:10–11
Tryptophan intrinsic fluorescence
GroEL
application in folding studies, **290**:101
contaminating tryptophan fluorescence
emission and purity, **290**:115–116,
136–137, 139–140
peptide-binding assays, fluorescence of
substrates, **290**:451–452
redox equilibrium determination in disulfide bond formation, **290**:72–73
Tryptophan synthase, substrate channeling
alternative substrate studies, **308**:121
aminoacrylate and α reaction activation,
308:121–122, 128
hydrophobic tunnel
conformational changes, **308**:126–127
crystal structure, **308**:116, 126–127
mutation studies, **308**:122–124
mutation studies
β Cys-170, 122–124
Glu-109, 122
Lys-87, **308**:122
rapid quench analysis, **308**:117–118
solid-state nuclear magnetic resonance
studies
advantages over solution-state studies,
308:128
aminoacrylate intermediate,
308:128–132
serine binding site, **308**:132
stopped-flow analysis, **308**:117–118
subunits
communication modeling
tests of model, **308**:121–124
transient kinetic experiments,
308:118–121
reactions catalyzed, **308**:116
X-ray crystallography
available structures, **308**:124–125
flexible loop region of α subunit,
308:125–126
hydrophobic tunnel, **308**:116, 126–127

TSA, *see* Thiol-specific antioxidant
Tubulin, *see also* Cytoskeleton, confocal microscopy; Microtubule; Tubulin polymerization
caged C2CF fluorescein derivative
applications, **298**:125–126
concentration and labeling stoichiometry determination, **298**:129–132
marking experiments, **298**:132
microinjection, **298**:132
photoactivation advantages over photobleaching studies, **298**:125
photobleaching susceptibility,
298:126
polymerization competency assay,
298:132
preparation
labeling procedure, **298**:127–129
reagents and buffers, **298**:127
spacer arm, **298**:126
GTP-binding site of β-tubulin, **298**:251
immunoaffinity purification of γ-tubulin
complex from *Xenopus* egg, **298**:541
isotypes and antimitotic drug response,
298:275–276
prokaryotic homolog, *see* FtsZ
γ tubulin ring complex
electron microscopy, **298**:224–225
microtubule cosedimentation assay,
298:225–226
microtubule end-binding assay,
298:226–227
microtubule nucleation assay, **298**:225
minus-end blocking assay, **298**:227–228
purification from *Xenopus* eggs
ammonium sulfate fractionation,
298:220
anion-exchange chromatography,
298:223
cation-exchange chromatography,
298:219–220, 222–223
extract preparation, **298**:218–219
gel filtration chromatography,
298:219–220, 223
immunoaffinity chromatography,
298:219, 222
sucrose gradient centrifugation,
298:223
yield, **298**:222
caged fluorescein-labeled tubulin

applications in transport and microtubule dynamics, **291:**348, 353, 356

electron microscopy following photoactivation, **291:**352–353, 355–356

microinjection into neurons, **291:**348, 350

neuron culture for microinjection, **291:**349–350

photoactivation, **291:**350–351

preparation, **291:**349

video microscopy, **291:**351

microtubule preparation, **291:**335

preparation from bovine brain, **291:**335

Tubulin polymerization

minimal mechanism, **308:**93

self-assembly models

ATP role

bidirectional polymerization with hydrolysis at time of monomer addition, **308:**99–101

delayed hydrolysis after monomer addition, **308:**101–102

difficulty of study, **308:**110–111

overview, **308:**94

timing control of assembly and disassembly, **308:**102–103, 110

tubulinGTP cap, **308:**105–106

unidirectional polymerization with hydrolysis at time of monomer addition, **308:**97–99

condensation equilibrium model, **308:**94–96

elongation without nucleotide hydrolysis, **308:**96–97

exchange-limited polymerization zone model

depolymerization zone, **308:**109

nucleation, **308:**108–109

oligoproline sequences, **308:**107

rationale for model, **308:**106–107

rates of elongation, **308:**106–107

Tumor necrosis factor-α

cell adhesion induction, assay, **300:**400

c-Jun N-terminal kinase activation, characterization, **300:**342–344

NF-κB induction with green fluorescent protein as reporter

advantages of assay, **302:**32–33

destabilized protein for rapid turnover, **302:**33, 37

dose response, **302:**35–36

high-throughput drug screening, **302:**38

principle, **302:**33

time course, **302:**34–35

Turbidity, *see* Light scattering

Turnover number, upper limit, **308:**3

Twitching motility, assay, **310:**98–99

Two-body interaction coefficient, *see* Molecular crowding

Two-dimensional crystallization, membrane proteins

aquaporin I, *in vitro* reconstitution

dialysis, **294:**148–150

lipids for reconstitution, **294:**147–148

gap junctions

membrane isolation from enriched recombinant preparations, **294:**139–142

in situ crystallization, **294:**144, 147

lipid monolayer crystallization, **294:**150–153

monitoring by negative stain electron microscopy, **294:**143

structural analysis, *see* Electron cryomicroscopy, structural analysis of membrane proteins

variables, **294:**142–143

Two-dimensional dwell-time analysis, single-channel gating

applications, **293:**438–439

component dependencies

calculation, **293:**442–444

extracting useful information, **293:**444–446, 454–456

plots, **293:**441–442, 456

similarity and extended compound states, **293:**446–448, 454–456

data collection for analysis, **293:**440

histogram, **293:**440–442

limitations in analysis, **293:**451–454

mechanism testing, **293:**456

microscopic reversibility testing, **293:**448–451

theory, **293:**439–440

Two-dimensional polyacrylamide gel electrophoresis, *see also* Polyacrylamide gel electrophoresis

apparatus and isoelectric focusing tubes, **316:**495–496

Bacillus subtilis biofilms

autoradiogram analysis, **310:**435–436
database of maps, **310:**441
identification of proteins, **310:**436, 438
overview, **310:**433–434
running conditions, **310:**435
sample preparation, **310:**435
staining, **310:**435–436
bovine retina protein patterns, **316:**500,
    502
buffers and solutions, **316:**496–497
genome database complementation,
    316:493–494
high mobility group protein interactions
    with nucleosome core particle
antibody preparation and immunofrac-
    tionation, **304:**141–142
electrophoresis conditions, **304:**142
overview, **304:**134, 140–141
isoelectric focusing gel electrophoresis,
    316:498–500
mass spectrometry for peptide mass fin-
    gerprinting
applications, **316:**494–495, 510–511
in-gel digestion
    protease selection, **316:**502–503
    trypsin digestion, **316:**503–504
matrix-assisted laser desorption ioniza-
    tion–time of flight analysis
    criteria for identification using MS-
        Fit, **316:**508–509
    database searching, **316:**505–506
    equipment, **316:**505
    large fragment fingerprinting,
        316:509–510
    principle, **316:**504–505
    spectra acquisition, **316:**505
prospects, **316:**510–511
methylation of DNA and cross-linking to
    histones
DNA versus protein gels, **304:**265, 267
electrophoresis conditions, **304:**269–271
interpretation of bands, **304:**273, 275
linker histone binding on reconstituted
    nucleosomes, **304:**529
peptide resolution on gels, **304:**272–273
resolution in first dimension, **304:**267
short preliminary fractionation for sepa-
    ration time estimation,
    304:267–269
overview, **316:**492–493

protein image hybridization
electroblotting, **304:**609
electrophoresis conditions, **304:**608–609
probe hybridization, **304:**609
radical-induced protein–DNA cross-link-
    ing, **304:**520
sample preparation, **316:**497–498
sodium dodecyl sulfate–polyacrylamide
    gel electrophoresis, **316:**500
supercoiling assay, **304:**177, 179
Two-electrode voltage clamp
data analysis, *see* Hidden Markov model;
    Voltage clamp
prediction of nonideal data, **293:**412–413
rise time parameters in data analysis,
    293:410–412
Xenopus oocyte
frequency response improvement,
    293:291–292
instrumentation, **293:**286–287
intracellular electrodes, **293:**287–289
noise minimization, **293:**292
overview, **293:**285–287, 319
performance testing, **293:**293–294
solution switching, **293:**292–293
voltage clamp adjustment, **293:**289–291
Two-hybrid system, *see* RNA polymerase
    III-based two-hybrid system; Yeast
    two-hybrid system
Two-photon excitation microscopy, *see also*
    Multiphoton excitation microscopy
absorption cross section, **307:**522
advantages and limitations, **307:**15–16,
    18, 469, 513, 516
detectors, **307:**525–527
excitation wavelength, **307:**353–354
green fluorescent protein, **302:**152
instrumentation overview, **307:**524–528
Jablonski diagram of excitation,
    307:513–514
lasers
pulse picker to minimize damage,
    307:532, 534–536
sources, **307:**469, 522–523, 526
lifetime fluorescence imaging, **307:**528,
    532
localization of excitation, **307:**354
nicotinamide adenine dinucleotide(phos-
    phate) imaging
cornea imaging, **307:**517, 519

data analysis and quantification,
    307:359
instrumentation for confocal micros-
    copy, **307:**355–357
linearity of detection, **307:**358–359
muscle metabolism imaging,
    307:365–368
pancreatic β cells, glucose response
    analysis
    immunostaining of enzymes and hor-
        mones, **307:**362–365
    islet cell studies, **307:**360–362
    principle, **307:**360
photobleaching correction, **307:**359
skin metabolism imaging
    correlative microscopy, **307:**515
    data acquisition, **307:**531–532
    multiphoton process, experimental
        characterization, **307:**515
    validation, **307:**519
    wavelengths, **307:**517
    viability controls, **307:**358
principle, **307:**353–354
pulse compensation, **307:**524
reflected light confocal microcopy, simul-
    taneous detection, **307:**528–529
resolution, **307:**354
video rate data acquisition, **307:**529–530
wavelength selection, **307:**522
Two-photon photolysis
caged calcium
    apparatus, **291:**361–365
    applications, **291:**356–357, 380
    calcium yield determination, **291:**358,
        369–374
    temporal behavior of calcium concen-
        tration distribution, **291:**374–379
    uncaging action cross section determi-
        nation, **291:**365–369
development, **291:**176, 178
laser systems, **291:**178–179, 201–202,
    357–358, 361–365
pulse duration, **291:**178
spatial resolution, **291:**359–361
Tyrosine, *see also* m-Tyrosine; o-Tyrosine
chlorination, *see* 3-Chlorotyrosine
nitration, *see* Nitrotyrosine
m-Tyrosine
atherosclerotic lesion content, **300:**143
formation from phenylalanine by hy-

droxyl radical, **300:**125, 141, **301:**471–
    472, 480
gas chromatography–mass spectrometry
    amino acid analysis, **301:**473–474
    arterial tissue analysis, **301:**481–483
    derivatization with N-methyl-N-(*tert*-bu-
        tyldimethylsilyl)trifluoroacetamide
        and spectrometry, **301:**474–475,
        477
    protein analysis, **301:**477, 480–481
isotope dilution negative-ion electron cap-
    ture mass spectrometry
    advantages in protein oxidation analy-
        sis, **300:**127–1
    derivatization, **300:**133
    gas chromatography, **300:**133–134
    internal standard synthesis,
        300:128–129
    mass spectrometry settings,
        300:134
    protein and tissue hydrolysis,
        300:130–132
    selected ion monitoring in quantifica-
        tion, **300:**134–136, 138
    solid-phase extraction, **300:**132
    tissue collection and preparation,
        300:129–130
metal ion role in formation, **300:**141
o-Tyrosine
atherosclerotic lesion content, **300:**143
formation from phenylalanine by hy-
    droxyl radical, **300:**125, 141, **301:**471–
    472, 480
gas chromatography–mass spectrometry
    amino acid analysis, **301:**473–474
    arterial tissue analysis, **301:**481–483
    derivatization with N-methyl-N-(*tert*-bu-
        tyldimethylsilyl)trifluoroacetamide
        and spectrometry, **301:**474–475,
        477
    protein analysis, **301:**477, 480–481
isotope dilution negative-ion electron cap-
    ture mass spectrometry
    advantages in protein oxidation analy-
        sis, **300:**127–128
    derivatization, **300:**133
    gas chromatography, **300:**133–134
    internal standard synthesis,
        300:128–129
    mass spectrometry settings, **300:**134

protein and tissue hydrolysis,
  300:130–132
selected ion monitoring in quantification, **300**:134–136, 138
solid-phase extraction, **300**:132
tissue collection and preparation,
  300:129–130
metal ion role in formation, **300**:141
o,o'-Tyrosine
atherosclerotic lesion content, **300**:143
formation from tyrosine radicals in proteins, **300**:125, 140
isotope dilution negative-ion electron capture mass spectrometry
  advantages in protein oxidation analy-
    sis, **300**:127–128
  derivatization, **300**:133
  gas chromatography, **300**:133–134
  internal standard synthesis,
    300:128–129
  mass spectrometry settings, **300**:134
  protein and tissue hydrolysis,
    300:130–132
  selected ion monitoring in quantifica-
    tion, **300**:134–136, 138
  solid-phase extraction, **300**:132
  tissue collection and preparation,
    300:129–130
myeloperoxidase role in formation,
  300:140
Tyrosine, caged peptides
applications, **291**:78–79, 94
design of inactive caged peptides,
  291:88–90, 94
extinction coefficient determination,
  291:85
kinetics of photolysis, **291**:87–88
quantum yield determination, **291**:85–87
synthesis
  2-bromo-α-carboxyl-2-nitrophenyl
    methyl ester, **291**:80–82
  coupling reaction, **291**:82
  deprotection, **291**:83
  peptide synthesis, **291**:83–85
  protecting of tyrosine, **291**:82
thiophosphoryltyrosine modification,
  291:133
Tyrosine, nitration, *see* Nitrotyrosine; Peroxynitrite
Tyrosine, oxidation in proteins, *see* 3-Chlo-

rotyrosine; 3-Nitrotyrosine; *m*-Tyro-
  sine; *o*-Tyrosine; *o,o'*-Tyrosine

U

U1A
  RNA complex structure refinement with
    residual dipolar couplings, **317**:218
  RNA-dependent protein–protein inter-
    face analysis with nuclear magnetic
    spectroscopy, **317**:201, 206–208
  translational repression for RNA binding
    screening, **318**:331
U2, dimethyl sulfate probing *in vivo*,
  318:486, 489
U3, dimethyl sulfate probing *in vivo*,
  318:490, 492
Ubiquinol-10, loss in low-density lipoprotein oxidation, **300**:20–22
Ubiquinone, nitric oxide inhibition,
  301:197
Ugp system, *see* Uptake glycerol phosphate
  system
UL13, antisense targeting of initiation codon, **313**:9–10
Ultracentrifugation, *see* Sedimentation equilibrium; Sedimentation velocity; Sucrose density gradient centrifugation
Ultraviolet cross-linking, *see* Cross-linking
Ultraviolet footprinting
  genomic footprinting, reaction conditions,
    304:556–557
  polymerase chain reaction analysis, *see*
    Polymerase chain reaction
  principle, **304**:617
  protein image hybridization, *see* Protein
    image hybridization
Ultraviolet radiation
  apoptosis induction, *see* Apoptosis
  atopic dermatitis therapy with ultraviolet
    A radiation
  cytokine expression analysis with re-
    verse transcriptase polymerase
    chain reaction
  β-actin standard, **319**:299–300
  amplification cycles, **319**:298–
    299
  primers, **319**:299

product purification and analysis,
      319:299–300
RNA extraction, **319:**298
irradiation technique, **319:**297–298,
      305–306
overview, **319:**296–297, 302–303
T cell apoptosis
   cell isolation, **319:**303, 305
   deuterium oxide effects, **319:**304
   DNA laddering assay, **319:**306
   double-staining technique for detec-
      tion, **319:**308
   immunofluorescence microscopy,
      319:301
   irradiation device and dosimetry,
      319:304
   irradiation studies, *in vitro*,
      319:303–309
   mitochondria transmembrane poten-
      tial changes, **319:**307
   morphology, **319:**300–301
   phosphatidylserine assay, **319:**307
   singlet oxygen generator and
      quencher effects, **319:**304–
      305
   TUNEL assay, **319:**306–308
   ultraviolet B effects, **319:**302
carcinogenesis
   action spectrum of murine squamous
      cell carcinoma, **319:**364–355
   mechanisms, **319:**331, 360
   p53 mutations, **319:**355
   papillomas, **319:**365
   tumor initiation and promotion,
      319:364
cellular responses, **319:**363–364
cholesterol oxidation, *see* Cholesterol
classification and radiation sources,
      319:259–260, 280, 359, 417
DNA damage, *see* DNA; Mitochondrial
      DNA
dosimetry of ultraviolet A
   cosine laws, **319:**246
   inverse square law, **319:**246–247
   objectives, **319:**245
   radiometric terms and units,
      319:245–246
   surface irradiation, **319:**247
   time of exposure, **319:**246
   volume irradiation, **319:**247–248

genotoxicity, **319:**153–154, 161–162,
      360–362
histological analysis of skin effects,
      319:612
keratinocyte, ultraviolet A-induced gene
      expression
   AP-2, **319:**280, 288, 290
   cell culture, **319:**281
   deuterium oxide studies, **319:**284
   differential reverse transcriptase poly-
      merase chain reaction, messenger
      RNA expression analysis,
      319:285–286
   electrophoretic mobility shift assay for
      transcription factor identification,
      319:288–290
   fluorescence-activated cell sorting, pro-
      tein expression analysis, **319:**285
   irradiation, **319:**281–282
   materials, **319:**284–285
   mimicing with singlet oxygen genera-
      tors, **319:**284
   reporter gene assays, **319:**286–288
   singlet oxygen quencher effects,
      319:282–283
signal transduction induction, *see also*
      Epidermal growth factor receptor;
      Mitogen-activated protein kinase; Nu-
      clear factor-κB; Platelet-derived
      growth factor receptor
   DNA damage, transcriptional arrest as
      signal generator, **319:**257–258
   dominant-negative mutant studies,
      319:270–272
   kinase inhibitors, **319:**267–270
   protein tyrosine phosphatases as tar-
      gets, **319:**258–259
   targets and sensors, overview, **319:**256–
      257, 465–466
   threshold hypothesis, **319:**259
   translational elongation, block as signal
      generator, **319:**258
singlet oxygen role in ultraviolet A radia-
      tion effects
   deuterium oxide studies, **319:**292
   iron role
      calcein assay, **319:**295
      chelator studies, **319:**294
      iron regulatory protein measure-
         ments, **319:**295

lipid peroxidation, **319:**294, 296
mimicing with singlet oxygen generators, **319:**293–294
modifiers of other active oxygen intermediates, **319:**293
overview, **319:**144, 290–291
singlet oxygen quencher effects,
319:292–293
skin protection studies, *see* Skin, ultraviolet protection
solar ultraviolet A levels, **319:**446
spectroradiometry
calibration of instruments, **319:**250
commercial instruments, **319:**250–251
detector, **319:**250
error sources, **319:**250
input optics, **319:**249
monochromator, **319:**250
spectral power distribution of sources,
319:248–249
T cell lymphoma therapy, **319:**308
ultraviolet A radiometers
angular response, **319:**251, 253
calibration, **319:**253–255
spectral sensitivity, **319:**251
stability, **319:**255
Ultraviolet/visible spectroscopy
aromatic amino acid absorption, **309:**609
Beer–Lambert law, **309:**607
protein structure studies
advantages and disadvantages, **309:**608
aggregation prevention and correction,
309:613–614
baseline determination, **309:**613
concentration of protein, **309:**613
cuvettes and volume, **309:**612
difference spectroscopy, **309:**610–611
slit width, **309:**613
solvent perturbation studies,
309:611–612
spectra interpretation, **309:**609–610
transmittance relationship to absorption,
309:609
Ultraweak photon emission
comparison to other luminescence processes, **300:**419–420
detectors, **300:**420
skin detection
age effects on oxidative status, **300:**425,
428

antioxidants, effects of topical application, **300:**426–427
control unit, **300:**421
data processing unit, **300:**423
horny layer moisture content effects,
300:425
light excitation unit, **300:**423
photomultiplier, **300:**421–423
photosensitizer effects, **300:**424
rubbing effects, **300:**425
ultraviolet-A induction, **300:**423–424
ultraviolet filters, effects of topical application, **300:**424–425
sources, **300:**421
topical antioxidant screening,
319:475–478
unc17, see Vesicular acetylcholine transporter
Unc104, *see* Kinesins
UPE, *see* Ultraweak photon emission
Uptake glycerol phosphate system
appearance in suppressor strains,
292:40–41
assay of transport, **292:**42–43
ATP-binding protein
functional exchange with MalK,
292:49–50
regulation of *pho* regulon, **292:**50–51
genes and regulation, **292:**43–45
ligand-binding protein
binding assays, **292:**47–49
purification, **292:**47
phosphate scavenger role, **292:**45
phosphoglycerol diester phosphodiesterase activity, **292:**42
proteins
UgpA, **292:**46
UgpB, **292:**45–49
UgpC, **292:**46, 49–51
UgpE, **292:**46
substrates, **292:**41–42
Uracil phosphoribosyltransferase
assay, **317:**27
purification of recombinant enzyme from
Escherichia coli, **317:**23–25
Uranyl, cleavage mapping of RNA
structure
cleavage reaction, **318:**45
comparison with iron(II) reaction, **318:**47
end labeling of RNA, **318:**45

materials, **318:**44–45

metal ion binding site probing, **318:**45–47

overview, **318:**43–44

polyacrylamide gel electrophoresis analysis, **318:**45

Uranyl acetate, chromatin staining for transmission electron microscopy, **304:**227

Uranyl cation-binding assay, membrane surface charge, **310:**481–482

Urease, quantification of urea with firefly luciferase ATP assay, **305:**366

Urethral catheter biofilm, *see* Bladder catheter biofilm

Uric acid

high-performance liquid chromatography analysis of ozone depletion in skin

equipment, **319:**539

operation, **319:**540

reagents, **319:**539

sample preparation, **319:**540

standards, **319:**540

metal oxide reaction product analysis, **319:**30–32

Uridine-5′-triphosphate

isotopic labeling using glycolysis and pentose phosphate pathway enzymes

enzyme preparation and assays, **317:**23–28

monitoring of reaction mixtures, **317:**32

overview, **317:**19–23

preparation from glucose, **317:**29–32

purification from reaction pools, **317:**33–35

phosphorothioate-tagged nucleotide triphosphate analogs for nucleotide analog interference mapping, **317:**108–109

Urine, *see* Bladder catheter biofilm

Uronic acid, *see* Exopolysaccharide, biofilms

UTP, *see* Uridine-5′-triphosphate

U-tube

advantages as concentration jump system, **294:**180, 187, 189

equipment, **294:**183, 185

exchange solutions

flow rate, **294:**180

time required for complete exchange, **294:**181–182

volume, **294:**180

mounting, **294:**183

operation, **294:**186–187

preparation, **294:**185

principle of function, **294:**181–183

testing, **294:**185–186

troubleshooting, **294:**187

UvrA, evolutionary relationship to ATP-binding cassette transporter, **292:**115

V

Vaccinia virus

cesium chloride gradient centrifugation of forms, **307:**593

confocal microscopy assays

advantages, **307:**591, 606–607

antibody blocking of virus binding, **307:**594–595, 599–602

binding assay, **307:**596–597

cell growth and virus production, **307:**594, 606

data acquisition and analysis, **307:**596

entry assay, **307:**603–604

extracellular enveloped virus receptor identification, **307:**599–602

indirect immunofluorescent staining, **307:**595

intracellular mature virus receptor identification, **307:**599–602

lysosome staining, **307:**595–596, 604

particle/plaque-forming unit ratio determination, **307:**598–599

plasma membrane staining, **307:**595

weak base effects on entry, **307:**604–606

structure and forms, **307:**592–593

Vaccinia virus–T7 RNA polymerase expression system

Hela cell maintenance, **296:**438–439

infection, **296:**440–441

optimization of expression, **296:**441–442

overview, **296:**437–438

transport assay of dopamine transporter, **296:**442

virus preparation, **296:**439–440

VAChT, *see* Vesicular acetylcholine transporter

VAI, *see* 3-Oxohexanoylhomoserine lactone

Valinomycin, Mdr3 ATPase stimulation, **292:**409

Vanadate
MDR1 ATPase inhibition, **292:**438–441
Mdr3 ATPase inhibition, **292:**410–411
trapping of ATPase
Chinese hamster P-glycoprotein, **292:**521–523
effects on P-glycoprotein photoaffinity labeling, **292:**325–328

van't Hoff enthalpy
derivation of equation, **295:**90–91
fluorescence titration in determination, **295:**154
ligand binding equilibria of biological macromolecules, **295:**91–92, 98–99
liquid–crystalline phase transition in model bilayers, **295:**474–475

Vascular cast, *see* Gelatin, vascular casts with fluorochrome labeling

Vascular cell adhesion molecule-1
dual-radiolabeled monoclonal antibody technique for *in vivo* measurement
advantages, **301:**15–16
backflush procedure, **301:**18
calculations, **301:**19–20
expression kinetics, **301:**21–22
exsanguination, **301:**18
injection of antibodies into mice, **301:**16–18
radioiodination of antibodies, **301:**16
tissue preparation and analysis, **301:**18–19
expression on endothelial cels, **301:**15
ligands, **301:**14

Vascular kinetics, *see* Microcirculatory kinetics

Vascular remodeling
cell culture studies, **307:**251
confocal microscopy
advantages for study, **307:**251–252
fluorophores
cytoplasmic dyes, **307:**257, 259
extracellular dyes, **307:**259
nuclear dyes, **307:**253, 255–257
selection factors, **307:**252
iterative multilevel thresholding and segmentation, **307:**271
limitations
diffraction of light, **307:**270

intensity attenuation with depth, **307:**270
object fusion, **307:**269–270
object intensity heterogeneity, **307:**270
segmentation problems, overview, **307:**269
object classification and measurement, **307:**271–272
pressure myography combination studies
advantages and limitations, **307:**272–273
fixation, **307:**262
image quantification, **307:**265–267
microscopy conditions and image capturing, **307:**263–265
parameter calculation for vessel characterization, **307:**266–267
pressurization, **307:**261–262
rationale, **307:**259–261
semiautomated quantification of three-dimensional objects and volumes, **307:**267, 269
staining with nuclear dyes, **307:**262–263, 272
definition, **307:**246–247
developmental changes, **307:**246
histology analysis, **307:**250–251
hypertension, **307:**246
noninvasive *in vivo* measurements, **307:**248
pressure myography, **307:**249–250
wire myography, **307:**248–249

Vasodilator-stimulated phosphoprotein
blot overlay assay of binding proteins
applications, **298:**111–112
phosphorous-32 labeling of vasodilator-stimulated phosphoprotein, **298:**112–113
cellular distribution, **298:**104
functions, **298:**104–105
immunofluorescence, **298:**110
immunoprecipitation, **298:**110–111
phosphorylation
assays, **298:**110–111
kinases, **298:**109–110
platelet inhibition, **298:**103
sites, **298:**103, 109–110
stimulators, **298:**103

purification
  human platelet protein, **298**:105–106
  porcine platelet protein, **298**:106–107
  recombinant protein from baculovirus–
    insect cell expression system
    baculovirus production, **298**:107
    immunoaffinity chromatography,
      298:108–109
  sequence analysis, **298**:103–104
  solid-phase binding assays, **298**:113
  Western blot analysis, **298**:109–111
Vasopressin
  antisense effects on adrenocorticotropic
    hormone release, **314**:235–237
  antisense targeting in hypothalamic nu-
    clei, **314**:232–235
Vasopressin receptor, messenger RNA
  structure and antisense targeting,
  314:228–229
VASP, *see* Vasodilator-stimulated phosphoprotein
VCAM-1, *see* Vascular cell adhesion mole-
  cule-1
Verapamil
  MDR1 ATPase stimulation, **292**:439–440
  Mdr3 ATPase stimulation, **292**:408–409
  photoaffinity labeling of P-glycoprotein
    with analogs, **292**:294–296, 302
Verotoxin
  Gb$_4$ binding, **312**:459
  neoglycoconjugate binding assays
    neoglycoprotein conjugates, **312**:487
    neohydrocarbon conjugates,
      312:485–486
  receptor enzyme-linked immunosorbent
    assay of glycolipid binding
    antibody incubations and color develop-
      ment, **312**:466
    binding conditions, **312**:465
    blocking, **312**:465
    microtiter plate preparation,
      312:465–466
    overview, **312**:463–465
  retrograde intracellular glycolipid traf-
    ficking
    fluorescein-tagged toxin
      frozen tissue section staining,
        312:471–472
      intracellular staining, **312**:471
      labeling, **312**:470–471

overview, **312**:459, 466, 468–469
visualization of toxin-bound glycolipid,
  312:469–470
  thin-layer chromatography overlay assay
    of glycolipid binding
    advantages, **312**:460–461
    overview, **312**:460
    platelet Gb$_3$ binding, **312**:463
    specificity assay, **312**:461–462
Vesicle sizing, **294**:322–325
Vesicular acetylcholine transporter
  Caenorhabditis elegans
    discovery, **296**:93–94
    functional assays
      behavioral assays, **296**:537–538
      lethality rescue, **296**:538–539
    mutation phenotypes, **296**:532
    unc-17 gene, **296**:531–532
  human protein expression in PC-12 cells
    acetylcholine transport, **296**:94–96, 98–
      99, 146
    antibody production, **296**:96
    generation of stable cell lines,
      296:96–97
    immunohistochemistry, **296**:97–98
    lipofection, **296**:97
    postnuclear homogenate preparation,
      296:98
  photoaffinity labeling
    [^{3}H]azidoacetylcholine
      denaturing gel electrophoresis and
        autofluorography, **296**:113
      photolysis conditions, **296**:110–113
      reversibility of binding, **296**:107–109
      structure, **296**:101
      synthesis, **296**:101–102
    [^{3}H]azidoaminobenzovesamicol
      denaturing gel electrophoresis and
        autofluorography, **296**:114–116
      photolysis conditions, **296**:113–114
      reversibility of binding, **296**:109
      structure, **296**:101
      synthesis, **296**:102–104
  Torpedo protein
    characteristics, **296**:94
    isolation of vesicles, **296**:99–100,
      104–106
    ligand binding or transport assays,
      296:107
  vesamicol inhibition, **296**:100, 155

Vesicular γ-aminobutyric acid transporter,
see also GAT-1
Caenorhabditis elegans
  functional assay, 296:541
  mutation phenotypes, 296:533
  unc-47 gene, 296:532
proton gradient in mechanism,
  296:116–117
reconstitution
  cholate solubilization, 296:120
  dilution technique, 296:120–122
  gel filtration chromatography technique, 296:121–122
  phospholipid preparation, 296:119
salt gradient driving of uptake,
  296:122–123
synaptic vesicle isolation, 296:117–118
uptake assay and properties in synaptic
  vesicles, 296:119–120, 123–124
Vesicular glutamate transport system
chloride stimulation, 296:140–141
proton gradient in mechanism, 296:126,
  140–142
purification on glycerol gradients
  ATPase activity, 296:135, 137
  centrifugation, 296:135
  glutamate uptake activity, 296:135, 137
  reagents, 296:134–135
  reconstitution, 296:135
reconstitution
  coreconstitution with endogenous and
    exogenous proton pumps, 296:137,
    139–144
  purified proteins, 296:135
  synaptic vesicles, 296:132–134
role in glutamate neurotransmission,
  296:125–126
solubilization, 296:128–129, 131–132
synaptic vesicles, large-scale preparation,
  296:127–128
Vesicular monoamine transporter
assays in functional identification
  distinguishing of isoforms, 296:87–88
  inhibitors of human transporters,
    296:86
  kinetic parameters, 296:85
  permeabilized cell assay
    cell culture and collagen coating,
      296:89–90
    detergents, 296:88, 91–92

  intracellular medium, 296:88–89, 91
  single-cell visualization of functional
    activity, 296:92–93
  transfection of monkey kidney fibroblasts, 296:85–86, 90–91
  vaccinia virus infection of monkey
    kidney fibroblasts, 296:90–91
assays of uptake
  background determination,
    296:150–151
  bioenergetic analysis, 296:151–152
  incubation conditions and quantitative
    analysis, 296:150–151
  membrane preparation, 296:149
  permeabilized cell assays, 296:149–150
  transient expression in COS1 cells,
    296:148–149
binding sites, 296:73, 144
bovine transporter
  chromaffin ghost preparation, 296:145
  isoelectric point, 296:65, 67
  kinetic properties, 296:65
  purification
    histidine-tagged protein, 296:67,
      69–70
    native protein, 296:65
    photolabeled protein, 296:79–81, 83
  reconstitution, 296:64–65
  size, 296:66–67
Caenorhabditis elegans
  cat-1 gene, 296:532–533
  functional assays
    food sensation assay, 296:539–540
    male mating efficiency, 296:540–541
    mutation phenotypes, 296:532–533
energetics of transport, 296:73, 144–145,
  147–148, 151–152
flux reversal assay with radiolabeled serotonin, 296:156–157
glyoxylic acid-induced fluorescence assay
  of dopamine transport, 296:146–147
immunohistochemistry, 296:97–98
isoforms, 296:67, 87–88
ketanserin binding
  distinguishing from 5-HT2$_A$ receptor
    binding, 296:76
  photoaffinity labeling with 7-azido[8-
    [125]I]iodoketanserin
    purification of labeled protein,
      296:79–81, 83

quantitative analysis, **296:**77–79
reaction conditions, **296:**77, 79
synthesis, **296:**77
rat transporter, purification of histidine-
tagged protein
VMAT1, 72
VMAT2, 67, 69, 71–72
reserpine binding
assay, **296:**153–154
mutagenesis analysis, **296:**153
proton gradient effects, **296:**153
substrate specificity, **296:**73, 87
tetrabenazine and derivatives, binding
α-dihydrotetrabenazine, **296:**74–76
α-[*O-methyl*-³H]dihydrotetrabenazine,
296:74–75
inhibition analysis, **296:**154–155
VMAT2
purification of histidine-tagged protein
metal affinity chromatography,
296:158–160
rat transporter, **296:**67, 69, 71–72
screening for transient expression
with *N*-methyl-4-phenylpyridin-
ium, **296:**158
solubilization, **296:**158–159
reconstitution into proteoliposomes,
296:160–161
transport assay of reconstituted protein, **296:**161–162
Vesicular stomatitis virus
host range, **313:**189
protein synthesis, **313:**190–191
safety in handling, **313:**189–190
serotypes, **313:**189
stock suspension preparation,
313:192–193
titration of stocks
limit dilution, **313:**193–195
plaque assay, **313:**195–196
translation arrest, antisense assay
applications, **313:**200–201, 203
gel electrophoresis of proteins, **313:**200
messenger RNA preparation
infection, **313:**196–197
poly(A) RNA purification,
313:198–199
RNA extraction, **313:**197–198
overview, **313:**190–193
prehybridization reaction, **313:**199

reticulocyte lysate incubations,
313:199–200
ribonuclease H mechanism delineation,
313:200, 203
VGCC, *see* Voltage-gated calcium channel
Vibrio bioluminescence, *see* Bacterial bioluminescence
Video-enhanced differential interference
contrast microscopy
alignment of microscope, **298:**321,
323–324
image recording and data storage,
298:324–325, 327
kinectin vesicle motility assay
antibody inhibition, **298:**195–196
cell culture, **298:**192–193
data analysis, **298:**195
flow chamber incubation, **298:**194–195
vesicle preparation, **298:**193–194
video-enhanced differential interference contrast microscopy,
298:194–195
melanosome motility assay, **298:**369–371
molluscan thick filament motility assay,
298:428, 433–435
motility assays
coverslips, **298:**348
image acquisition and processing,
298:349
instrumentation, **298:**439
organelle movement in *Xenopus* extracts
direction of movement, **298:**351
features, **298:**349–350
quantitation of movement,
298:350–351
standard assay, **298:**350
network formation assay, **298:**351, 357
schematic diagram, **298:**321
spatial resolution, **298:**317–319
temporal resolution, **298:**319
yeast
preparation for imaging, **298:**319–320
spindle and nuclear cycle kinetics analysis, **298:**329
Vimentin, *see also* Cytoskeleton, confocal
microscopy
extraction of phosphoprotein, **298:**562
phosphopeptide antibody preparation,
298:566–567

phosphorous-32 labeling
  in vitro, **298:**562–563
  in vivo, **298:**560–562
phosphorylation site determination,
    298:563, 565–566
Vinblastine
  cellular uptake and loss, **298:**259–260
  drug concentration effects, **298:**252–253,
    256–257
  intracellular concentration determination
    nonradiolabeled compounds and high-
        performance liquid chromatogra-
        phy, **298:**273–274
    radiolabeled compounds
      monolayer culture directly in ster-
          ilized scintillation vials, **298:**273
      monolayer or suspension cultures,
          298:272–273
  Mdr3 ATPase stimulation, **292:**409
  microtubule binding
    affinity, **298:**257–258
    binding site, **298:**251–252
    effects in living cells, **298:**258–261
    suppression of treadmilling and dy-
        namic instability, **298:**258
  photoaffinity labeling of P-glycoprotein
      with analogs, **292:**291, 293, 299–300,
      304
  radiolabeled drug accumulation assays of
      MDR1
    intact cells, **292:**471–472
    reconstituted proteoliposomes,
        292:500–503
  structure, **298:**256
  tubulin binding effects, **298:**257–258
Violaxanthin, biosynthesis, **297:**256–257
Viral vectors, *see also* specific vectors
  adeno-associated virus, *see* Adeno-associ-
      ated virus
  adenovirus, **314:**38–39
  ideal criteria, **314:**36–37
  retrovirus
    angiotensin II type 1 receptor, anti-
        sense targeting
      complementary DNA preparation,
          314:582–583
      concentrating of vectors, **314:**588
      infection of target cells, **314:**588
      titration of virus, **314:**587–588
      transduction, *in vivo*, **314:**588, 590

viral particle preparation, **314:**583–
    585, 587
  overview, **314:**37–38
Viridiofungins
  bioassay for sphingolipid inhibition in
      Candida albicans, **311:**337–339
  preparation
    fermentation, **311:**344–345
    isolation, **311:**345
  serine palmitoyltransferase inhibition,
      311:335, 347, 350
  structure, **311:**337
Virus inactivation
  1,4-dimethylnaphthalene singlet oxygen
      carrier derivatives
    cultivation of virus, **319:**200–201
    cytotoxicity assay, **319:**202–203
    N,N'-di(2,3-dihydroxypropyl)-1,4-
        naphthalenedipropanamide
        efficacy, **319:**204–207
    enveloped virus inactivation,
        319:203–206
    exposure conditions, **319:**201
    nonenveloped virus inactivation,
        319:206
    overview, **319:**16–17, 199
    survival assays, **319:**201–202
  fresh frozen plasma
    photodynamic inactivation with methyl-
        ene blue
      absorption, **319:**208
      photoproducts, **319:**212
      principle, **319:**208
      production, **319:**212–216
      tolerability, **319:**208, 212
      virus sensitivity, **319:**208–209, 216
    screening, **319:**207
Virus infection, *see* Electron microscopy;
    Vaccinia virus
Vitamin A, *see also* Carotenoids; Retinoic
    acid; Retinol
  absorbance properties, **299:**431–432
  antioxidant activity, **299:**421
  dietary supplementation, rationale,
      300:167
  receptors, **299:**430–431
  simultaneous determination with vitamin
      E, lycopene, and xanthophylls with
      reversed-phase high-performance liq-
      uid chromatography

calibration, **299:**352

extraction from plasma and serum, **299:**352–354, 356

instrumentation and chromatography conditions, **299:**350–352

peak identification, **299:**357–360

quality assurance, **299:**353, 356–357, 361–362

reagents and solutions, **299:**349

stability of samples, **299:**360–361

standards, **299:**350

Vitamin C, *see* Ascorbic acid

Vitamin E, *see also* α-Tocopherol

C11-BODIPY, multiwell screening of topical effectiveness, **319:**608–609

components and structures, **299:**318–319, 330

dietary supplementation, rationale, **300:**167

free radical neutralization, **300:**78

half-life of radical, **300:**78

high-performance liquid chromatography

diol column chromatography, **299:**320, 323–325

fluorescence detection, **299:**325

materials, **299:**321

milk replacer diet effects in piglet tissues, **299:**328–329

oils and diets, extraction, **299:**322

separation of components, **299:**323–324

simultaneous determination with coenzyme Q

animal maintenance, **299:**334–335

chromatography conditions, **299:**337–338, 340, 344, 347

electrochemical detection, **299:**333–334, 342, 344–345

extraction, **299:**331–332

materials, **299:**334

plasma sample preparation, **299:**344

recovery and reproducibility, **299:**340–341, 345

saponification, **299:**332–333

standards, **299:**336–337, 342–344, 347

storage of samples, **299:**332

simultaneous determination with vitamin A, lycopene, and xanthophylls with reversed-phase high-performance liquid chromatography

calibration, **299:**352

extraction from plasma and serum, **299:**352–354, 356

instrumentation and chromatography conditions, **299:**350–352

peak identification, **299:**357–360

quality assurance, **299:**353, 356–357, 361–362

reagents and solutions, **299:**349

stability of samples, **299:**360–361

standards, **299:**350

standards, **299:**322–323

tissue pulverization and extraction, **299:**321–322, 329

triacylglycerol influence on quantitation, **299:**325–329

high-performance liquid chromatography analysis of ozone depletion in skin

equipment, **319:**543

operation, **319:**543

reagents, **319:**542

sample preparation, **319:**543–544

standards, **319:**543

loss in low-density lipoprotein oxidation, **300:**21–22

low-density lipoprotein oxidation protection, **300:**44

ozone reactions

gas chromatography–mass spectrometry analysis, **319:**550–551

ozonation reaction, **319:**549–550

products and mechanisms, **319:**547, 549

regeneration by ascorbic acid

hepatocyte assay system from ascorbate-deficient rats

animal maintenance, **300:**80–81

ascorbate determinations, **300:**83–84

tert-butyl hydroperoxide protection studies, **300:**85, 88

isolation and culture of hepatocytes, **300:**81–82

tocopherol determinations, **300:**83

vitamin content of cells, **300:**79–80

vitamin supplementation, **300:**82–83

mechanism, **300:**79

saponification in extraction, **299:**318–320

singlet oxygen scavenging rates in phospholipid membranes, **319:**106–109

stratum corneum analysis

gradient and physiological relevance, **300:**418–419

high-performance liquid chromatography, **300:**416
lipid extraction, **300:**415–416
quantitative analysis, **300:**416–418
tape stripping, **300:**414–415
In Vitro selection, RNA aptamers, *see also* Antibiotic–RNA complex; Bradykinin; Rex; Ribonuclease P; Ribosomal RNA; Yop51* tyrosine phosphatase
ATP-binding aptamers, **318:**269
binding equilibrium, **318:**204
buffer conditions for selection, **318:**203–204
competitor inclusion in initial binding reaction, **318:**205
constant sequence utilization, **318:**196
ligand types, **318:**202–203
modified ribonucleotide pools, **318:**195–196
negative selection, **318:**205
pool synthesis
initial DNA pool
complexity assessment, **318:**197
large-scale amplification, **318:**200–201
primer extension, **318:**199–200
purification, **318:**197
synthesis, **318:**197
overview, **318:**196
transcription, **318:**201–202
principle, **318:**193
randomization of pool
complete randomization, **318:**193–195
length of random sequence region, **318:**195
partial randomization, **318:**193–194
segmental randomization, **318:**194–195
RNA : ligand ratio, **318:**204
stringency of selection, **318:**203–205
In Vitro selection, RNA–protein fusion, *see* RNA–protein fusion
In Vitro translation, *see* Rabbit reticulocyte lysate; Wheat germ agglutinin system
In Vivo confocal neuroimaging, *see* Neuron
VMAT, *see* Vesicular monoamine transporter
V_{max}, *see* Maximum velocity
Voltage clamp, *see also* Cut-open oocyte voltage clamp; Electrophysiology;

Patch clamp; Two-electrode voltage clamp
data analysis, *see also* Hidden Markov model; Markov model; Two-dimensional dwell-time analysis
kinetic modeling
equilibrium distribution, **293:**397
Markov models, **293:**395–397
overview, **293:**394–395
model-independent analysis, **293:**393–394
on-line versus off-line analysis, **293:**392–393
PulseSim program, **293:**399–400, 414
simultaneous data fitting
Monte Carlo simulation, **293:**416–417
multiple populations of channels, **293:**416
nonstationary noise analysis, **293:**417–419
single-channel simulation, **293:**417
tense-relaxed activation model, **293:**415–416
storage of data, **293:**397–399
leak correction
approaches, **293:**400–401
leak template design, **293:**401–402
linear off-line correction, **293:**402–403
nonlinear off-line correction, **293:**404–406
low-pass filtering
causal Bessel filter simulation, **293:**408–410
filter parameter determination, **293:**407–408
Gaussian filter with linear phase, **293:**406–407
validation of parameters, **293:**411–412
volume change effects in cardiac myocytes, **307:**293–294
Voltage clamp biosensor, *see* Capillary electrophoresis
Voltage-gated calcium channel
antibodies
autoimmunity, **294:**702
channel antibodies, **294:**690–692, 702
neurotoxin antibodies, **294:**693, 702
ω-conotoxin binding, **294:**614, 618
drug screening with fluorescence assay of calcium

activity of various drugs, **294:**27, 30–32
advantages of assay, **294:**46–47
limitations of assay, **294:**45–46
pharmacologic validation, **294:**26–27
stable cell line expression of recombinant channels, **294:**25–26
Volume exclusion, *see* Molecular crowding
Volume investigation, *see* Cell volume, confocal microscopy measurement
VSV, *see* Vesicular stomatitis virus
VT, *see* Verotoxin

W

Waist position, Gaussian beam optics, **291:**188, 190
Waist radius magnification, Gaussian beam optics, **291:**189
Water, modeling in protein structure
ligand binding modeling, **293:**626–627
molecular dynamics simulated annealing with restrained molecular dynamics
effects on pore structure, **293:**675–676
energy minimization in positioning, **293:**674–675
goals, **293:**673–674
ion modeling with water in pore, **293:**679–682
NB protein pore, **293:**676–679
solvation of model, **293:**674
pore modeling, **293:**631, 636, 696–697
Water, temperature-invariant enthalpy of hydrogen bonding, **295:**261–264
Wave polarization, Gaussian beam optics, **291:**190–191
Western blot
ABCR, **315:**868–869
affinity-tagged antigen analysis, **303:**508–509
androgen receptor–green fluorescent protein fusion protein, **302:**123–124
antisense knockdown of proteins, analysis
$G\alpha$ subunits, **314:**15, 17, 101
c-Jun N-terminal kinase, **314:**368
leukemia cells, **313:**126–128
c-Myc, **314:**125–127, 560–561, 563–564, 571–572
nerve growth factor, **314:**162–163, 165

neuronal nitric oxide synthase, **314:**394, 397–398
opioid receptors, **314:**13–15
proprotein convertases, **314:**109, 112, 115
Raf-1, **314:**448
sodium–calcium exchanger, **314:**321
aquaporin I expressed in *Xenopus* oocytes, **294:**570, 572
Arp2/3 complex, **298:**43–44
arrestin, **315:**431–432
ATPGP1
extraction of protein, **292:**171
gel electrophoresis and blotting, **292:**171
immunostaining, **292:**171–172
transgenic plant expression analysis, **292:**173
wild-type plant expression analysis, **292:**172–173
calcium-dependent guanylate cyclase activator protein, **315:**734, 737–739
cationic amino acid transporters, **301:**89–91
cell surface biotinylated neurotransmitter transporters, **296:**314–315
centrosome proteins, **298:**237–238
chemiluminescence detection with 1,2 dioxetane substrates
advantages, **305:**427
alkaline phosphatase-conjugated secondary antibody detection, **305:**421
antiphosphospecific primary antibody, **305:**421–422
applications, **305:**418
biotinylated secondary antibody/protein detection, **305:**422–424
blot preparation, **305:**418–419
dual detection with alkaline phosphatase and β-galactosidase-conjugated secondary antibodies, **305:**424–425
β-galactosidase-conjugated secondary antibody detection, **305:**424
imaging, **305:**426
overview, **305:**417–418
reagents, **305:**419–420
solutions, **305:**420
stripping of blots, **305:**425–426
troubleshooting, **305:**426–427

chloroplast mutant phenotypes in maize
  detection of bands, **297:**46
  extraction of proteins, **297:**43–44
  immunostaining, **297:**46
  overview, **297:**42–43
  plant growth conditions, **297:**43
  polyacrylamide gel electrophoresis,
    297:45
  quantification of extracts, **297:**44–45
  transfer to membranes, **297:**45
constitutively active opsin mutant expression, **315:**212
cyclic GMP phosphodiesterase, transgene
  expression, **316:**740–741
cyclic nucleotide-gated channel, **315:**841–
  843, 846–847
cysteine-modified neurotransmitter transporters, **296:**322, 324, 326, 328–329
cystic fibrosis transmembrane regulator,
  292:641–643, 645, 650, 655–657
cytochrome c_6, **297:**273–274
cytoplasmic linker protein-170, 202–203
dimethylsulfate cross-linking in replicating chromatin, **304:**532–533
dynein, **298:**367–368, 372
epidermal growth factor receptor phosphorylation assay, **319:**261–262
excitatory amino acid receptors
  blotting with semidry electrophoretic
    transfer, **294:**406, 409
  immunodetection, **294:**406–407, 410
  receptor activation studies *in vivo*,
    294:407–408
  sample preparation from mouse brain,
    294:405
  solution preparation, **294:**404–405
  troubleshooting, **294:**410
filter retardation assay for neuronal intranuclear inclusions
  aggregate preparation, **309:**697–698
  immunoblotting, **309:**698
  nuclei isolation from brain,
    309:696–697
  reagents, **309:**694–696
focal adhesion kinase, **298:**97–98
focal adhesion kinase-related nonkinase,
  298:98
Gcn5 phosphorylation, **304:**706
β-globin, splice variant analysis,
  313:512–514

glucocorticoid receptor–green fluorescent
  protein fusion protein, **302:**79
glucosylceramide synthase, **311:**53
glutamine transporters, antisense oligodeoxynucleotide inhibition analysis,
  296:523–526
GLYT2 glycoprotein, **296:**14
green fluorescent protein mutant degradation, **302:**441, 443–444
HMG-D and chromatin interactions
  blocking and probing of membranes,
    304:133
  Drosophila embryo collection, **304:**129
  gel electrophoresis, **304:**131
  membrane transfer, **304:**131–132
  micrococcal nuclease digestion of chro-
    matin, **304:**130–131
  nuclei isolation, **304:**130
  principle, **304:**126–127
  rationale, **304:**125–126, 129
HMG-I(Y)
  antibody production, **304:**167–169
  incubations and washings, **304:**169
4-hydroxy-2-nonenal adduct detection,
  319:568–569
4-hydroxynonenal–protein conjugates,
  309:738, 766
I-κB
  degradation, **300:**353–354
  induction by nitric oxide, **300:**375–379
inositol 1,4,5-triphosphate receptor–green
  fluorescence protein fusion protein,
  302:230
kinesin, **298:**367–368, 372
lecithin:retinol acyltransferase, **316:**410
M2 expression in mouse erythroleukemia
  cells, **294:**495–496
MDR1
  multidrug-resistant KB-V1 cell protein,
    292:497–499
  rodent cell expression, **292:**507–508
  transgenic mouse expression, **292:**584
  vaccinia virus–T7 RNA polymerase sys-
    tem protein, **292:**463–464
mitogen-activated protein kinase phosphorylation, **319:**133–134
multidrug-resistance associated protein
  expression, **292:**603–604
neurotransmitter antibody specificity testing, **296:**394–395, 397–398, 417–420

neurotransmitter transporter expression in baculovirus–insect cell expression system, **296**:450–452

nondenaturing blotting of aggregation intermediates
  antibodies
    incubations, **309**:346–347
    preparation, **309**:345
  applications, **309**:349
  comparison to denaturing blotting, **309**:343–345
  controls, **309**:345
  detection, **309**:347
  gel electrophoresis, **309**:346
  membranes, **309**:345
  P22 tailspike protein, **309**:347–349
  principle, **309**:343
  stock solutions and buffers, **309**:345–346
  transfer to membrane, **309**:346

norepinephrine transporter, **296**:354–357

nuclear factor-κB
  degradation and phosphorylation, **319**:277–278
  electrophoretic mobility shift assays, **319**:598

nuclear translocation of p65 in NF-κB activation, **300**:369

opsin fragments, analysis of expression, **315**:63

peripherin/rds-rom-1 complex analysis, **316**:681–682

plastocyanin, **297**:273–274

prion protein
  monoclonal antibodies, **309**:112
  protease-resistant protein blotting materials, **309**:125
    prion isolation from cultured cells for analysis, **309**:126
    sample preparation, **309**:126

protein A–green fluorescent protein fusion protein applications
  advantages, **302**:274
  blotting, **302**:271
  conventional color detection, **302**:272–274
  enolase detection, **302**:272–274
  fluorescence detection, **302**:271–274
  overview, **302**:266
  sensitivity, **302**:273

protein carbonyl group detection
  advantages, **319**:432
  controls, **319**:436
  dinitrophenylhydrazine derivatization, **319**:433
  electroblotting, **319**:434
  immunoassay, **319**:435
  materials, **319**:432–433, 435
  principles, **319**:431–432
  sample preparation, **319**:433
  staining of total protein bands, **319**:435

protein phosphatase type 2A subunits, **315**:564–566

Ran-GTPase, **315**:465

retinal pigment epithelium p65
  antibodies, **316**:719–720
  bovine retinal pigment epithelium membranes, **316**:720
  mouse eye, **316**:720–722

11-*cis*-retinol dehydrogenase, **316**:350

rhodopsin
  quantification of transient transfection in HEK293 cells, **315**:413–415
  trafficking assay, **315**:86

Rubisco in antisense RNA mutants, **297**:212–213

sodium/calcium-potassium exchanger, **315**:841–843, 846–847

Ste6p, **292**:208–209

target protein inhibition by antisense RNA, **313**:472–473

transducin fragments, **315**:367, 369, 375–376

transglutaminase, **309**:176–178

tyrosine nitrated proteins, **301**:137, 139, 307–309

vision mutant protein expression analysis in transgenic mice, **316**:574–575

yeast two-hybrid system, confirmation of protein–protein interactions, **303**:436

Wheat germ agglutinin system, posttranslational processing, **293**:19

Wheat germ extract, cell-free translation of P-glycoprotein, **292**:281, 286–287

White protein, *Drosophila melanogaster*
  ATP-binding domain, **292**:219–220
  Brown protein interaction sites, **292**:221–223
  eye pigmentation role, **292**:213–215
  gene cloning, **292**:214

ascorbic acid, **299:**166–167
nucleic acids, **299:**164
oxidation, **299:**164
proteins, **299:**164–165
sugars, **299:**165–166
sulfite, **299:**167–169
sulfur dioxide, **299:**167–168
molar absorptivity of specific phenols,
    299:161–163
phloroglucinol addition, **299:**173
preparation of reagent, **299:**155
sample preparation, **299:**159
standards and blanks, **299:**158–159
tannin separations, **299:**174–175
values for various wines, **299:**176–177
volume of reaction, **299:**156
Wire myography, vascular remodeling,
    307:248–249
Wiseman isotherm, *see* Titration calo-
    rimetry
Wye base, intrinsic fluorescence, **317:**419

X

Xenopus embryo
advantages for photoreceptor studies,
    316:50–51
retinal development, **316:**50
transient transfection
    embryo preparation, **316:**54–55
    lipofection, **316:**56
    luciferase reporter assay, **316:**56–57
    materials and solutions, **316:**52–54
    principle, **316:**51–52
    promoters, analysis of *cis*-acting ele-
        ments, **316:**57
    trypsinization, **316:**55–56
Xenopus melanophore, *see* Melanopsin
Xenopus oocyte
antisense inhibition of proteins
    advantages of system, **314:**303
    dose and timing, **314:**304–305
    microinjection, **314:**304
    preparation and maintenance of oo-
        cytes, **314:**303–304
    protein inhibition analysis, **314:**305
aquaporin I expression
    complementary DNA expression con-
        struct preparation, **294:**568–569

complementary RNA
    injection, **294:**570
    preparation, **294:**569–570
materials, **294:**568
osmotic water permeability measure-
    ment, **294:**570
Western blot analysis, **294:**570,
    572
caged inositol 1,4,5-triphosphate studies
    of calcium release
caffeine studies, **291:**393, 395
dose–response relationship,
    291:391–392
Ins*P*$_4$ agonist studies, **291:**393
microinjection, **291:**383–384
chronoamperometry, dopamine clearance
    measurement, **296:**718–719
cut-open oocyte voltage clamp
    advantages and disadvantages, **293:**297,
        318–320, 340
    agar bridges, **293:**305–306, 321–322
    amplifier, **293:**304–305, 321
    calcium channel
        calcium chelator removal of inactiva-
            tion, **293:**317–318
        current measurement after removing
            contaminating chloride currents,
            293:316–317
        G protein inhibition, GTP perfusion
            analysis, **293:**330
        speed of intracellular perfusion,
            293:328–329
        whole-cell clamping comparison,
            293:326–328
    capacity current linearity, **293:**311–312
    chamber, **293:**302, 304, 321
    clamp speed, **293:**308–310
    impaling and voltage clamping,
        293:325–326
    intracellular perfusion, **293:**320–322,
        325, 328–329
    mounting, **293:**304, 322, 325
    potassium channel
        gating current recording after inter-
            nal potassium removal, **293:**314
        gating currents after analog compen-
            sation of linear resistive and ca-
            pacity components, **293:**314
        macropatch recording comparisons,
            293:314, 316

tail currents in high external potassium, **293:**312–313
unsubtracting gating charge records, **293:**313–314
saponin permeabilization, **293:**306–308, 320
solution preparation, **293:**301–302
voltage inhomogeneity in upper domus, **293:**310–311
cystic fibrosis transmembrane regulator expression after RNA microinjection, **292:**623–625
detection of expressed ion channels
immunofluorescence staining
background fluorescence minimization, **293:**540–541
confocal microscopy, **293:**547–548
epitope tagging of proteins and staining, **293:**538–540
frozen sections, **293:**544–547
monoclonal versus polyclonal antibodies, **293:**537–538
oocyte preparation, **293:**539
topology staining, **293:**542–544
immunoprecipitation
cytosolic versus membrane proteins, **293:**548–549
homogenization, **293:**550
incubation conditions, **293:**551–552
membrane isolation by sucrose gradient centrifugation, **293:**551
polyacrylamide gel electrophoresis, **293:**552
reconstitution in lipid bilayers
membrane isolation, **293:**554
overview, **293:**552–553
reconstitution reaction, **293:**554–555
solution preparation, **293:**553
voltage clamping, **293:**536–537
electrophysiology analysis of neurotransmitter transporters
cut-open oocyte technique, **296:**572
expression analysis, **296:**552–554
simultaneous fluorescence microscopy, **296:**572–573, 576–578
two-microelectrode oocyte voltage clamp, **296:**572–573, 576
expression cloning
assay development, **296:**29–30

blocking of native channels or transporters, **296:**23, 48
complementary DNA discoveries
channels, enzymes, and transporters, **296:**18, 20
receptors, **296:**19, 21
complementary DNA
functional screening, **296:**46–47
hybrid-depletion screening approach, **296:**49
library plating and DNA isolation from filters, **296:**42–44
ligation into plasmid vector, **296:**41
size selection by agarose gel electrophoresis, **296:**40–41
synthesis, **296:**36, 39
transfection of library in *Escherichia coli*, **296:**41–42
complementary RNA synthesis, **296:**44, 46
functional complementation, **296:**47–48
limitations, **296:**48–49
oocyte behavior during protein expression, **296:**30
microinjection
plasmid DNA, **296:**50
RNA, **296:**30, 46
RNA isolation
poly(A)+ RNA selection, **296:**25–26
quantitation by ultraviolet spectroscopy, **296:**24–25
size fractionation, **296:**31, 33, 35–36
total RNA, **296:**23–25
source organism and tissue selection, **296:**19, 21, 48
fluorescence microscopy of expressed neurotransmitter transporters
charge-coupled device detection, **296:**569, 571–572, 575
fluorescence labeling, **296:**573–574
ion measurements with fluorescent dyes
applications, **296:**575–576
loading of dye, **296:**576
simultaneous electrophysiological measurements, **296:**572–573, 576–578
optical measurements, **296:**567, 569–570, 574–575

photodiode measurements, **296:**570–571, 575
GAT-1 expression, **296:**299–300
glycine receptor, transient expression
 applications of expression system, **294:**263–264
 injection capillary preparation, **294:**268
 injection of cytoplasm, **294:**269
 injection of nucleus, **294:**269–270
 messenger RNA isolation from neural tissue, **294:**264
 preparation of oocytes, **294:**266–267
 transcription of complementary DNA *in vitro*, **294:**264–266
 vector linearization, **294:**265
heterologous protein expression
 distribution of ion channels, **293:**298
 follicular cell layer removal, **293:**285, 534
 incubation of oocytes, **293:**284–285, 301, 534–535
 messenger RNA
  microinjection, **293:**283–284, 535–536
  preparation, **293:**281–283, 300–301, 533–534
 nonsense codon suppression and unnatural amino acid incorporation
  acylation of transfer RNAs, **293:**519–521
  amino acid incorporation, assessment of fidelity, **293:**524–525
  background current assessment, **293:**524
  controls, **293:**525–526
  electrophysiologic analysis, **293:**523
  heteromultimeric protein expression, **293:**525
  ion channel types in study, **293:**523, 529
  materials, **293:**505–506
  messenger RNA generation, **293:**521–522
  microinjection of messenger and transfer RNAs, **293:**523
  overview, **293:**504, 564–565
  reacylation of transfer RNAs, **293:**524
  suppression testing with *in vitro* translation, **293:**526–529

suppressor transfer RNA design, **293:**506–507
suppressor transfer RNA synthesis and purification, **293:**507–516
unnatural amino acid synthesis, **293:**516–517, 519
seasonal variation of expression, **293:**298
vectors for high-level expression, **293:**530, 532–533
inositol 1,4,5-triphosphate receptor–green fluorescence protein fusion protein expression
 confocal microscopy 228–229
 expression levels, **302:**230–231
 fixation of oocytes 229
 microinjection 227–228
 microsome preparation 229–230
 plasmid, **302:**227
 subcellular localization, **302:**226, 232–233
 Western blot analysis, **302:**230
isolation
 collagenase treatment, **296:**28–29
 dissociation, **296:**28
 maintenance, **296:**29
 ovarectomy, **296:**27–28
 solutions and equipment, **296:**26–27
 sorting, **296:**29
membrane insertion into planar bilayers, **294:**299–300
microinjection of P-glycoprotein messenger RNA, **292:**273–274
native expression of ion channel subunits, **293:**19
neurotransmitter receptor studies with caged compounds, **291:**470–472
patch clamp
 giant membrane patches, **293:**297–298
 inside-out patches, **293:**296
 macropatch recording, **293:**294–295, 299–300
 omega patches, **293:**296–297
 outside-out patches, **293:**296
 patch cramming, **293:**297
 pipette fabrication, **293:**294–295
 seal formation, **293:**295–296
 simultaneous recording with two-electrode voltage clamp, **293:**299
 vitelline envelope removal, **293:**294

phototransduction protein expression
  advantages, **316:**41, 49
  immunoprecipitation of radiolabeled
      proteins, **316:**45–46
  materials and solutions, **316:**41–43
  membrane preparation, **316:**45
  nucleic acid injection, **316:**43–44
  oocyte preparation, **316:**43
  opsin
      glycosylation forms, **316:**46–47
      Limulus opsin, **316:**49
      reconstitution, **316:**49
  transducin α subunit, **316:**47–49
  voltage clamp recording, **316:**44–45
sectioning
  coating slides with bovine serum albu-
      min for microscopy, **293:**546–547
  cryostat sectioning, **293:**545–546
  freezing in optimal cutting temperature
      compound, **293:**544–545
size, **296:**17–18
two-electrode voltage clamp
  detection for capillary electrophoresis,
      294:197–199
  frequency response improvement,
      293:291–292
  instrumentation, **293:**286–287
  intracellular electrodes, **293:**287–289
  noise minimization, **293:**292
  overview, **293:**285–287, 319
  performance testing, **293:**293–294
  solution switching, **293:**292–293
  voltage clamp adjustment, **293:**289–291
unidirectional flux measurement of ex-
      pressed potassium channels
  efflux measurement, **293:**561–564
  influx measurement
      nonchannel flux correction, **293:**560
      potassium-42 counting and specific
          activity, **293:**559–560
      reflux, **293:**560
      tail current error, **293:**560–561
      voltage clamp measurement,
          293:557–558
Xenopus oocyte/embryo
  phosphoramidate-containing oligonucleo-
      tides, antisense inhibition
  An2, 426–427, 434–435
  controls, **313:**427–428, 435–436
  cyclins, **313:**427

efficacy, **313:**434–435
microinjection, **313:**432–433
RNA analysis, **313:**433
RNA depletion in oocytes with antisense
      oligonucleotides
  applications, **313:**436–437, 442
  function monitoring, **313:**441
  microinjection
      depleted RNA sequences, **313:**437,
          441–442
      equipment, **313:**438
      technique, **313:**440–441
  oligonucleotide design and synthesis,
      313:439
  oocyte isolation, **313:**440
  reversibility, **313:**437
  RNA synthesis *in vitro*, **313:**439–
      440
  transcription, **313:**421, 434–435, 437
Xenopus transgenesis
  efficiency, **316:**64
  egg quality, **316:**58
  embryo selection, **316:**63–64
  interphase egg extract, **316:**60–61
  materials and solutions, **316:**59–60
  nuclear transplantation, **316:**62–63
  overview, **316:**58–59
  restriction enzyme-mediated integration
      reaction, **316:**62
  rhodopsin mutant expression, **316:**64
  sperm nuclei preparation, **316:**61–62
Xenopus melanophore, melanopsin
      transgene expression
  materials, **316:**296–297
  nuclear transplantation, **316:**298–299
  overview, **316:**294–296
  restriction-enzyme-mediated integra-
      tion, **316:**298
  sperm and egg preparation,
      316:297–298
  transgene incorporation analysis,
      316:302
  video microscopy assay of light re-
      sponse, **316:**304–305
X-linked chronic granulomatous disease,
      gene therapy, *see* Gene therapy,
      MDR1
X-ray absorbance spectroscopy, analysis of
      photosystem II mutants, **297:**358
X-ray crystallography

$G_{i\alpha 1}$ with ligands, GTP hydrolysis mechanism
  crystallization conditions, **308**:74–75
  data collection, **308**:74, 76
  data sets, **308**:73–74
  domains, **308**:76
  GDP complex, **308**:86–88
  GDPP$_i$ ternary complex trapping, **308**:85–86, 91
  ground state enzyme–substrate complex, **308**:78, 80–81, 90
  hydrogen bonding interactions, **308**:76–78
  intermediates, forcing within crystals, **308**:88–90
  regulators of G-protein signaling, mechanism of action, **308**:92
  transition-state as modeled by GDPAlF$_4^-$Mg^{2+}, **308**:81–85, 90–91
membrane proteins
  classification of crystal types, **297**:58–59
  cocrystallization with antibodies, **297**:76–77
  difficulty in crystallization, **297**:57–58
  type II crystals
    amphiphile addition, **297**:59
    crystallization techniques, **297**:60–61
    examples, **297**:60, 360
    multilateral interactions, **297**:59–60
organic solvent trapping of intermediates
  chymotrypsin, *see* Chymotrypsin
  cross-linking of crystals, **308**:204
  lysozyme, **308**:205
  side chain rearrangements, **308**:204
  stability of crystals, **308**:205–206
  substrate addition, **308**:206–207
  subtilisin, **308**:205
  transfer technique, **308**:205
  water structure changes on protein surface, **308**:204
photolysis studies, *see* X-ray crystallography, photolysis studies
photosynthetic reaction center from *Rhodobacter sphaeroides*
  crystal geometry and diffraction, **297**:61, 68–72
  crystallization by vapor diffusion
    amphiphile composition for type II crystals, **297**:72–73

    1,4-dioxane, role in crystallization, **297**:73
    mutant proteins, **297**:75
    orthorhombic crystals, **297**:68–70
    pH, **297**:74
    purity requirements, **297**:74
    reproducibility, **297**:74
    seeding, **297**:74–75
    temperature, **297**:73–74
    tetragonal crystals, **297**:72
    thin crystal preparation for spectroscopy, **297**:75
    trigonal crystals, **297**:70–73
  purification
    anion-exchange chromatography, **297**:67–68
    cell growth and harvesting, **297**:65–66
    molecular sieve chromatography, **297**:68
    purification without sucrose gradient, **297**:67
    solubilization, **297**:66
    sonication, **297**:66
    sucrose gradient centrifugation, **297**:66
  structure determination and refinement, **297**:76
  subunit structure, **297**:62
photosynthetic reaction center from *Rhodopseudomonas viridis*
  crystal geometry and diffraction, **297**:61, 65
  crystallization by vapor diffusion, **297**:64–65
  purification
    cell growth and harvesting, **297**:62–63
    molecular sieve chromatography, **297**:64
    solubilization, **297**:63
    sonication, **297**:63
    sucrose gradient centrifugation, **297**:63
  structure determination and refinement, **297**:75–76
  subunit structure, **297**:62
RNA, *see* X-ray crystallography, RNA
substrate–product equilibrium in intermediate trapping, **308**:202–203

tryptophan synthase, substrate channeling studies
  available structures, **308:**124–125
  flexible loop region of α subunit, **308:**125–126
  hydrophobic tunnel, **308:**116, 126–127
  ultrafast data collection of enzyme intermediates, **308:**201
X-ray crystallography, photolysis studies
  caged enzymes, **291:**255, 266
  concentration requirements for caged substrates, **291:**252–254
  crystal size, **291:**262
  glycogen phosphorylase, **291:**252–253
  heat damage of crystals by flash, **291:**216–218, 258
  irreversibility of photolysis, **291:**262–264
  measuring compounds released by photolysis, **291:**219, 253, 256–257
  muscle fibers, **291:**253–254
  photolysis yield considerations, **291:**254–255, 258–259
  Ras GTPase studies in crystals
    caged GTP affinity, **291:**257, 259–260
    data collection, **291:**259
    flash lamp photolysis, **291:**215–216, 257, 261
    rapid freezing followed by data collection, **291:**261
    Ras mutation for thermostability, **291:**260
    resolution of data, **291:**259–260, 262
  time-resolved measurements with Laue crystallography, **291:**251, 255–256, 259, 264, 277
X-ray crystallography, RNA
  heavy atom derivatization, *see* Heavy atom derivatives, RNA
  large RNA structures
    cryostabilization, **317:**172–173
    crystallization module introduction into solvent-exposed regions, **317:**170–171
    crystal requirements, **317:**171–172
    intermolecular contacts in crystal lattice, **317:**169–170
    map interpretation and model building, **317:**176–177
    nonisomorphism of crystals, **317:**173

    phasing by heavy atom substitution, **317:**173–175, 179–180
    radiation sensitivity of crystals, **317:**172
    solvent electron density peak interpretation, **317:**177–179
  small ribozymes
    chemical synthesis
      automated synthesis, **317:**151–152
      desalting, **317:**153
      overview, **317:**151
      reversed-phase high-performance liquid chromatography, **317:**152–153
    crystallization
      cations, **317:**155–158, 182
      precipitants, **317:**154–155, 182
      screening, **317:**154, 182–183
    crystal packing promotion, **317:**150–151, 181–182
    diffraction data acquisition
      cryoprotection, **317:**159–160, 184
      crystal requirements, **317:**158
      radiation sensitivity, **317:**158–159
      synthetic mother liquors, **317:**159
    hammerhead ribozyme, **317:**458–459
    heavy atom derivatization, **317:**160–162, 183–184
    initial electron density maps, **317:**164–165, 184–185
    intermolecular contacts in crystal lattice, **317:**181–182
    Patterson methods for heavy atom positions, **317:**162–164
    refinement, **317:**165–168, 185–186
    time-resolved crystallography
      catalysis initiation in crystals, **317:**187–188
      difference Fourier analysis, **317:**193–194
      extent of cleavage, analysis in crystals, **317:**196–198
      kinetic assays for cleavage in crystals, **317:**194–196
      Laue diffraction observation of transient intermediates, **317:**192–193
      overview, **317:**186–187
      termination of catalysis in crystals, **317:**196
    trapping of reaction intermediates

cryocrystallographic trapping, **317:**191–192, 468–469
kinetic bottleneck modifications, **317:**188–191

X-ray diffraction, liquid–crystalline phase transition in model bilayers
bilayer partial thickness estimation, **295:**486
data collection, **295:**483–484
instrumentation, **295:**482–483
low-angle scattering analysis, **295:**484
mixed-chain-length phospholipid studies, **295:**490–491
sample preparation, **295:**480–482
wide-angle scattering analysis, **295:**484–487

X-ray diffraction spectroscopy, urinary tract device encrustation, **310:**561

X-ray fiber diffraction, amyloid fibrils
α-helical structure, **309:**528
alignment of fibrils
magnetic alignment, **309:**532
stretch frame, **309:**531–532
amyloid-β fibrils from senile plaques, **309:**528–529
β strand structure, **309:**526–528
calcitonin fibrils, **309:**528
core structure of protofilaments, **309:**529–530
data acquisition, **309:**532–534
insulin fibrils, **309:**528
interpretation of diffraction data, **309:**534–535
light chain amyloid, **309:**526
lipid reflections, **309:**530
ordered specimens, **309:**535–536
sample preparation
synthetic peptide fibrils, **309:**530–531
tissue materials, **309:**530
serum amyloid A, **309:**526–527
structural spacing, **309:**526
transthyretin amyloid, **309:**529–530

X-ray footprinting, *see* Synchrotron X-ray footprinting

X-ray microanalysis
biofilm sample preparation, **310:**307–308
catheter biofilm formation in bladder model, **310:**500–501

X-ray photoelectron microscopy

adhesion analysis of *Azospirillum brasilense*
extracellular protein effects, **310:**382–383
free cell analysis, **310:**381
surface chemical composition analysis, **310:**381–382
tetracycline effects, **310:**382
atomic force microscopy of samples, **310:**387, 389
biofilm surface analysis, **310:**388
dehydration techniques, **310:**389
depth resolution, **310:**376–377
functional groups
cell growth analysis, **310:**379–380
identification from spectra, **310:**377–378
microorganism constituent analysis, **310:**379–380
peak position, chemical environment effects, **310:**377–378
principles, **310:**377
substrata analysis
chemically-modified substrata, **310:**385
conditioned substrata, **310:**385–387
native substrata, **310:**384–385
suspended cell surface analysis, **310:**387–388

Xanthine oxidase
chemiluminescence assay, **305:**385
nitric oxide effects on toxicity, **301:**418–420, 426

Xanthommatin
Drosophila melanogaster eye color role, **292:**213
transport, *see* Scarlet protein; White protein

Xanthophylls, simultaneous determination with vitamin E, lycopene, and vitamin A with reversed-phase high-performance liquid chromatography
calibration, **299:**352
extraction from plasma and serum, **299:**352–354, 356
instrumentation and chromatography conditions, **299:**350–352
peak identification, **299:**357–360
quality assurance, **299:**353, 356–357, 361–362
reagents and solutions, **299:**349

stability of samples, **299:**360–361
standards, **299:**350
XKCM1, *see* Kinesin
XPS, *see* X-ray photoelectron microscopy

Y

y+, *see* L-arginine transport
YAC, *see* Yeast artificial chromosome
YCF1
functions in yeast, **292:**154
multidrug-resistance protein, functional homology, **292:**154–155
subcellular localization, **292:**155
Yeast artificial chromosome, transgenic mouse expression
applications, **306:**203
embryonic stem cell transfection, **306:**193–194
identification and structural analysis of transgenics
fiber fluorescent *in situ* hybridization, **306:**202
long-range restriction enzyme mapping, **306:**202
overview, **306:**201
RecA-assisted restriction endonuclease cleavage, **306:**201–202
*Sfi*I fragmentation, **306:**202
Southern blot analysis, **306:**201
large DNA sequence maintenance, **306:**188–189
mutagenesis and segregation, **306:**192–193
optimization of vectors, **306:**189, 191–192
pronuclear microinjection
DNA purification
agarase digestion, **306:**199–200
agarose plug preparation for preparative pulsed-field gel electrophoresis, **306:**196–197
overview, **306:**195
pulsed-field gel electrophoresis, **306:**197–198
second-dimension agarose gel electrophoresis for concentrating DNA, **306:**199
shear avoidance, **306:**194
efficiency of transgenesis, **306:**195
materials and reagents, **306:**195–196
microinjection protocol, **306:**200–201
Yeast expression systems, *see also* Copper-regulated expression vectors
advantages over other systems, **306:**169–170
codon bias, **306:**185–186
5′-untranslated region, effect on expression, **306:**182–183
folding and sorting of proteins, **306:**183–185
Pichia pastoris, expression of recombinant proteins
alcohol oxidase promoter, **306:**154–156
applications, **306:**169
cell biology, **306:**155
cell culture
density, **306:**162–163, 166
fermentor and monitoring, **306:**163, 166–168
inoculum, **306:**166–167
media, **306:**163, 166
methanol feed, **306:**167–168
comparative cell composition, **306:**154–155
electroporation, **306:**164
glyceraldehyde-3-phosphate promoter, **306:**168
host strains, **306:**158–160
levels of expression, **306:**154
screening for recombinant protein expression
dot-blot hybridization, **306:**160–161
G418 resistance, **306:**164–165
high throughput screening, **306:**161
immunodetection, **306:**161, 165–166
shake flask evaluation, **306:**161–162
transposon determinant selection, **306:**160
transformation, **306:**160
vectors, **306:**156, 158–159
promoters
classes, **306:**145
induction/repression times, **306:**145–146
sec6-4 mutant, expression of recombinant proteins
advantages of system, **306:**181–182, 186
applications, **306:**171–172

aquaporin expression
  activation energy for water transport, **306:**180–181
  promoters, **306:**172
  proteoliposome reconstitution, **306:**180–181
  shrinkage assay of function, **306:**177–180
  vesicle characteristics, **306:**177
  water permeability calculation, **306:**180–181
  growth conditions, **306:**173–175, 183
  plasmid construction, **306:**173
  temperature sensitivity, **306:**170
  vesicle
    electron microscopy, **306:**177
    isolation, **306:**175–176
    trafficking, **306:**170–171
  yield and plasmid loss, **306:**183
Yeast one-hybrid assay
  advantages and limitations, **316:**592–593
  kits, **316:**593
  principle, **316:**591–592
  retinal transcription factor cloning
    bait plasmid construction
      cloning target sequences into reporter vectors, **316:**593–594
      target DNA sequence selection, **316:**593
    complementary DNA library from retina, construction and amplification, **316:**594–596
    dual-reporter selection
      false-positive minimization, **316:**596
      separate reporter strains and diploid analysis, **316:**597–598
      single dual-reporter strain, **316:**597
    positive clone characterization, **316:**598–600
Yeast recombinant expression systems, bovine opsin expression in *Pichia pastoris*
  absorption spectroscopy characterization, **315:**8
  advantages of system, **315:**10
  chromophore formation, **315:**8, 11
  expression vector construction, **315:**4–5, 10
  G-protein activation, **315:**9
  glycosylation analysis, **315:**7, 10
  large-scale culture, **315:**5–6

media, **315:**4
  purification
    buffers, **315:**6–7
    immunoaffinity chromatography, **315:**8
  subcellular localization, **315:**7
  transformation by electroporation, **315:**5
  yield, **315:**5, 11
Yeast sphingolipid biosynthesis mutants
  applications, **312:**329–330
  characterization of suppressor mutants
    cloning and disrupting wild-type *TSC* genes, **312:**325–327
    complementation grouping of *TSC* mutants, **312:**325
    sphingolipid biosynthesis defects, **312:**327–329
    temperature sensitivity due to same mutation causing suppression, **312:**325
  genes, **312:**320–321
  genetic selection principles, **312:**320, 322
  selection of suppressor mutants, **312:**324
  sphingolipid biosynthesis overview, **312:**318–320
  suppressors of calcium-sensitive phenotype of *csg2* null mutants
    SCS screen, **312:**322–323
    TSC screen, **312:**323–324
Yeast three-hybrid system
  applications, **306:**113, **318:**399
  comparison with other techniques for RNA–protein interactions, **306:**93–94
  components, **318:**400–401
  hybrid RNA
    MS2 sites and signal strength, **306:**99–100
    secondary structure, **306:**99
    size, **306:**99, 101
    uridine runs, **306:**100–101
  hybrid RNAs
    copy numbers, **318:**405
    runs of U's, **318:**406
    secondary structure, **318:**404–405
    size of RNA element, **318:**406
  known RNA-binding protein, finding RNA partner
    activation domain plasmid introduction, **306:**111

binding specificity determination,
  306:112
false positive elimination, 306:111–112
overview, 306:110
RNA library construction, 306:111
second reporter gene, screening for activation, 306:111
sequencing of RNAs, 306:112
known RNA–protein interactor testing,
  306:101–103
known RNA sequence, finding protein
  partner
  3-aminotriazole in initial selection,
    306:104
  binding specificity determination using
    mutant and control RNAs,
    306:108–109
  false positives, elimination by colony
    color, 306:104–105
  functional tests, 306:110
  β-galactosidase assays
    filter assay, 306:105–106
    liquid assay, 306:106–107
  positive complementary DNA identifi-
    cation, 306:109–110
  protein libraries, 306:104
  RNA plasmid transformation, 306:103
  selection stringency, 306:103
  URA3 counterselection, 306:107
plasmid vectors, 318:402–403
  pACTII/CAN, 306:98–99
  pIII/MS2-1 and pIII/MS2-2, 306:97
  pIIIA/MS2-1 and pIII/MS2-2, 306:97
principle, 306:94–96, 318:399–401
prospects, 318:401–402, 418–419
protein partner, finding for known RNA
  sequence
  3-aminotriazole in initial selection,
    318:409
  binding specificity determination
    fraction of true positives,
      318:413–414
    functional screening, 318:414–415
    mating assay, 318:413
    RNAs and controls, 318:413
  colony color assay for elimination of
    false-positives, 318:409–410
  complementary DNA library transfor-
    mation, 318:409
  β-galactosidase assays

filter assay, 318:411
  solution assay, 318:411–412
  positive complementary DNA identifi-
    cation, 318:414–415
  RNA plasmid transformation, 318:409
  stringency of selection, 318:408
  URA3 counterselection, 318:412–413
RNA partner, finding for known binding
  protein
  binding specificity determination,
    318:418
  false-positive elimination, 318:417
  fraction of true positives, 318:418
  overview, 318:415–416
  protein-independent activators, 318:417
  reporter gene activation screening,
    318:417
  RNA library construction, 318:416–417
  sequencing of RNAs, 318:418
  transformations, 318:416
strains of yeast, 306:101, 318:406–407
success rates, 306:97–97
testing known or suspected interactors,
  318:407
Yeast two-hybrid system, see also RNA
  polymerase III-based two-hybrid
  system
advantages and disadvantages, 293:14,
  119
bait
  constructs, 293:108–109
  self-activation, 293:110–111
Caenorhabditis elegans protein interactions, 296:544, 546
confirmation of positives
  bait and fish gene switching, 303:435
  binding assays, in vitro, 303:436
  cyclohexamide screening of bait-nega-
    tive cells, 303:433–434
  double-tagging assay, 303:450
  fish plasmid isolation and selection,
    303:434–435
  rare positives, gene types, 303:435
  tryptophan media screening of bait-neg-
    ative cells, 303:434
  Western blot analysis, 303:436
β-galactosidase assays, 303:431–433
ion channel associated protein identification, 294:371, 373
kinase screen, 303:437

library screen quality, assessment, **293:**117–119

media preparation, **293:**120–122

N-methyl-D-aspartate receptor subunit interactions, **293:**105–109

plasmid recovery, **293:**116–117

plasmids, **303:**425–428, 436–437

principle of RNA polymerase II-based systems, **303:**411, 423–424

principle, **293:**4, 22, 106–107, 111

reagents, **303:**428–429

reporter genes

β-galactosidase filter assay, **293:**10, 114–116, 122

types, **293:**4, 8, 22, 107

retinal protein interactions, *see* Yeast two-hybrid system, detection of interacting retinal proteins

reverse two-hybrid screen, **303:**437

Shaker-type potassium channel

binding protein analysis, **293:**105–109

subunit interaction analysis

fusion protein vector construction, **293:**6

growth assays, **293:**8–9

strains and reagents, **293:**5–6

subcloning amino-terminal domains into fusion protein, **293:**6–7

transformation of yeast, **293:**7–8, 112–114, 122, **303:**429–431

triple-hybrid assay, **303:**437

troubleshooting, **303:**436–437

vectors and reporter strains, **293:**6, 107–108

yeast strains, **303:**424–425

Yeast two-hybrid system, detection of interacting retinal proteins

advantages and disadvantages, **316:**689

principle, **316:**688–689

retinitis pigmentosa GTPase regulator as bait

bait

plasmid preparation, **316:**691–693

plasmid, prescreen testing, **316:**695

selection of region, **316:**691

complementary cDNA library characteristics, **316:**690–691

β-galactosidase filter lift assay, **316:**696–698

host strain, **316:**694–695

quantification of interaction, **316:**700–703

target plasmid preparation, **316:**693–694

transformation, **316:**695–696

verification of interaction, **316:**698–700

YEF3, *see* Elongation factor-3

Yellow fluorescence protein

absorption spectroscopy, **305:**166, 174

bioluminescence activity, **305:**180

fluorescence spectra, **305:**176

green fluorescent protein variant, **305:**204–206

ligand exchange, **305:**170–172

purification from *Vibrio fischeri*, **305:**172

resonance energy transfer, **305:**165–166

stability, **305:**173

structure, **305:**165

YFP, *see* Yellow fluorescence protein

Ykc1, forward genetics analysis, **294:**523–524

y⁺L, *see* L-arginine transport

Yop51* tyrosine phosphatase, RNA aptamer identification by *in vitro* selection

aptamer analysis, **318:**213–214

binding assays, **318:**208–210

binding reactions, **318:**207

cellulose filter immobilization, **318:**207

cloning and sequence of selection products, **318:**208–209

polymerase chain reaction amplification, **318:**207–208

reagents, **318:**206

reverse transcription, **318:**207

rounds of selection, **318:**208

thermal equilibration, **318:**207

transcription, **318:**208

YOR1, function in yeast, **292:**155

YSR2, *see* Dihydrosphingosine-1-phosphate phosphatase

YSR3, *see* Dihydrosphingosine-1-phosphate phosphatase

Z

Zeaxanthin, *see also* Carotenoids

biosynthesis, **297:**256–257

oxidative metabolites, **299:**460–461

photooxidation protection of retina,
299:457–460
stereoisomer separations
dicarbamate diastereomers, preparation, **299:**463–466
extraction, **299:**462
high-performance liquid chromatography, **299:**466–467
retinal distribution, **299:**467
separation from lutein, **299:**462–463
serum and tissue sampling, **299:**462
serum distribution, **299:**467
types of isomers, **299:**461
Zebrafish retina
advantages as model system, **316:**536–537
cyclic GMP phosphodiesterase assay
incubation conditions and product separation, **316:**546–547
light sensitivity determination, **316:**547
materials, **316:**546
principle, **316:**546
time course of activation, **316:**547, 549
guanylyl cyclase assay
calcium dependence, **316:**550–551
incubation conditions and product separation, **316:**550
materials, **316:**549–550
principle, **316:**549
homogenization of larval eye,
316:539–540
larval stocks and culture conditions,
316:538
nrb model, **316:**537
retinoid analysis by high-performance liquid chromatography
characterization of retinoids,
316:556–557
chromatography, **316:**556
extraction, **316:**555–556
materials, **316:**555
principle, **316:**552, 555
rhodopsin and cone opsin phosphorylation assay
incubation conditions and gel autoradiography, **316:**551–552
materials, **316:**551
principle, **316:**551
white light versus red light response,
316:552
transducin activation assays

GTPγS *in situ* binding assay
incubation conditions and radioactivity detection, **316:**542
materials, **316:**541
pob mutant analysis, **316:**542, 544
principle, **316:**540–541
tissue preparation, **316:**541–542
GTPγS *in vitro* binding assay
incubation conditions and radioactivity detection, **316:**544
materials, **316:**544
principle, **316:**544
white light versus red light response,
316:544–545<hsi>317
Zeste, ultraviolet cross-linking of protein–
DNA complexes *in vivo*, **304:**498, 510–
512, 515
Zinc
functions in metalloproteins, **301:**126
release by nitric oxide
cell types and response, **301:**131–
132
nitrosothiol synthesis addition to cell
cultures, **301:**127–128, 130
NONOate addition to cell cultures,
301:128–129
Zinquin fluorescence assays, **301:**129–
130, 132
Zinc affinity chromatography, protein disulfide isomerase, **290:**30–31
Zinc finger transcription factors, inhibition
by nitric oxide, **301:**126, 132–135,
210–211
Zinc oxide
cosmetic applications, **319:**29, 36
oxidation product analysis
butylated hydroxytoluene, **319:**30–32,
35–36
electron paramagnetic resonance,
319:31, 35
gas chromatography–mass spectrometry, **319:**31, 33–34
light source, **319:**30
materials, **319:**30
methyl oleate, **319:**30–35
2,2,6,6-tetramethyl-4-piperidone,
319:30, 35
uric acid, **319:**30–32
photochemical generation of active oxygen species, **319:**29

Zoo blotting, evolutionary sequence conser-
    vation analysis of genes, **303**:84,
    145
Z ring, *see* FtsZ
Zyxin
    antibodies, **298**:63–64
    binding proteins
        assays, overview, **298**:76
        types, **298**:72
    blot overlay assay of binding proteins
        advantages, **298**:75
        autoradiography, **298**:75
        denaturing gel electrophoresis and blot-
            ting, **298**:75
        incubation conditions, **298**:75
        reagents and buffers, **298**:72–73
        zyxin radiolabeling
            iodine-125, 74
            phosphorous-32, **298**:73–75
    domains, **298**:62, 65–66
    functions, **298**:76
    gene locus, **298**:66–67
    isoelectric point, **298**:67
    phosphorylation, **298**:67

polyacrylamide gel electrophoresis,
    298:67
purification from chicken gizzard
    anion-exchange chromatography,
        298:70
    buffer preparation, **298**:68–69
    hydroxylapatite chromatography,
        298:70–71
    overview, **298**:67–68
    phenyl–Sepharose chromatography,
        298:70
    recombinant protein expression sys-
        tems, **298**:71–72
    storage, **298**:71
    tissue extraction, **298**:69–70
    yield, **298**:71
purification from human platelets, **298**:71
quaternary structure, **298**:67
related proteins, **298**:66
sequence homology between species,
    298:64–66
subcellular distribution, **298**:62–63, 76
tissue distribution, **298**:64
zinc binding, **298**:67

Contributor Index

Boldface numerals indicate volume number

A

Abdulaev, Najmoutin G., **315,** 3, 59; **316,** 87
Abe, Akira, **311,** 42, 105, 373
Abeygunawardana, Chitrananda, **308,** 219
Abogadie, Fe C., **314,** 136
Ackers, Gary K., **295,** 190, 450
Acland, Gregory M., **316,** 777
Acquotti, Domenico, **312,** 247
Acworth, Ian N., **300,** 297
Adcock, Ian M., **319,** 551
Adelman, John P., **293,** 53
Aebischer, Claude-Pierre, **299,** 348
Aggarwal, Bharat B., **300,** 339; **319,** 585
Agmon, Vered, **312,** 293
Agrawal, Rajendra K., **317,** 276, 292
Agre, Peter, **294,** 550
Aguilar-Bryan, Lydia, **292,** 732; **294,** 445
Aguini, Nadia, **299,** 276
Ahearn, Donald G., **310,** 551
Ahmad, Nihal, **319,** 342
Ahn, Jinhi, **296,** 370; **315,** 864
Akabas, Myles H., **293,** 123
Akaike, Akinori, **293,** 319
Akerboom, Theodorus P. M., **301,** 145
Akhtar, Muhammad, **315,** 557
Akong, Michael, **294,** 20
Al-Abed, Yousef, **309,** 152
Alahari, Suresh, **313,** 342
Albert, Arlene D., **315,** 107
Alberts, Bruce, **298,** 218
Alderton, Wendy K., **301,** 114
Aleksandrov, Andrei A., **292,** 616
Alexander, J. Steven, **301,** 3
Alexander, Rebecca W., **318,** 118
Alfieri, Jennifer A., **304,** 35
Alho, Hannu, **299,** 3
Alkon, Daniel L., **293,** 194
Allan, Andrew C., **291,** 474
Allan, Viki J., **298,** 339
Allen, Cynthia M., **319,** 376
Allen, Margaret, **294,** 649
Allen, Robert C., **305,** 591
Allewell, Norma M., **295,** 42
Allikmets, Rando, **292,** 116
Allin, Christoph, **291,** 223
Allis, C. David, **304,** 675
Allison, David G., **310,** 232

Almouzni, Geneviève, **304,** 333
Al-Shawi, Marwan K., **292,** 514
Altman, Russ B., **317,** 470
Altman, Sidney, **313,** 442
Amano, Atsuo, **310,** 501
Amara, Susan G., **296,** 307, 318, 436, 466
Ambudkar, Suresh V., **292,** 318, 492, 504
Ames, Bruce N., **299,** 83; **300,** 70, 156
Ames, James B., **316,** 121
Anand, Vibha, **316,** 777
Andersen, Jens Bo, **310,** 20
Andersen, Olaf S., **294,** 208, 525
Andersen, Roxanna N., **310,** 322
Anderson, Brian B., **296,** 675
Anderson, Karen S., **308,** 111
Anderson, Michael T., **302,** 296
Anderson, Neil, **310,** 566
Andersson, Carol R., **305,** 527
Andersson, Karin, **309,** 591
Andrade, Joseph D., **305,** 660
Andrews, John H., **307,** 607
Andriambeloson, Emile, **301,** 522
Andriantsitohaina, Ramaroson, **301,** 522
Androlewicz, Matthew J., **292,** 745
Annunziato, Anthony T., **304,** 76
Anstey, Nicholas M., **301,** 49
Aoki, Takashi, **302,** 264
Aparicio, Jennifer G., **315,** 673
Apel, Klaus, **297,** 237
Appelt, Denah M., **309,** 172
Applebury, Meredithe L., **315,** 673
Aragón, Carmen, **296,** 3
Archer, Trevor K., **304,** 584
Ares, Manuel, Jr., **318,** 479
Argentieri, Dennis C., **311,** 168
Ariga, Toshio, **312,** 115
Armstrong, Gregory, **297,** 237
Arnér, Elias S. J., **300,** 226
Arnosti, C., **310,** 403
Arribas, Silvia M., **307,** 246
Arshavsky, Vadim Y., **315,** 524
Artemyev, Nikolai O., **315,** 539, 635
Arts, Ilja C. W., **299,** 202
Asami, Osamu, **290,** 50
Asensi, Miguel, **299,** 267
Assmann, Sarah M., **294,** 410
Aston, Christopher, **303,** 55
Astriab, Anna, **313,** 342

507

Atkinson, Howard J., **302,** 316
Attwell, David, **296,** 608, 617
Aubry, Jean-Marie, **319,** 3, 50, 197
Avdonin, Vladimir, **293,** 724
Ayar, Ahmet, **312,** 387
Ayoob, Joseph C., **302,** 171
Ayscough, Kathryn, **298,** 18
Azem, Abdussalam, **290,** 253

B

Babcock, George F., **307,** 319
Bacikova, Dagmar, **312,** 317
Bacot, Karen, **290,** 218
Bada, Michael A., **317,** 470
Badminton, Michael N., **305,** 479
Baehr, Wolfgang, **316,** 515
Baillie, George S., **310,** 644
Bajjalieh, Sandra, **311,** 207
Baker, Brian M., **295,** 294
Baker, Elliott R., **299,** 287
Baker, Herman, **299,** 287
Baldwin, Michael A., **309,** 576
Baldwin, Robert L., **295,** 1
Baldwin, Thomas O., **290,** 1; **305,** 135, 157, 180
Balin, Brian J., **309,** 172
Ballmaier, Daniel, **301,** 311
Bamberg, E., **291,** 289
Banting, George, **302,** 3
Bao, Yijia, **314,** 168
Barak, Larry S., **302,** 153
Baraldini, Mario, **305,** 120, 577
Barbeito, Luis, **301,** 393
Barbone, Arminda G., **311,** 168
Bardwell, James C. A., **290,** 59
Barenholz, Yechezkel, **299,** 293
Barkan, Alice, **297,** 38
Barker, Eric L., **296,** 475
Barlev, Nickolai A., **304,** 696
Barlow, Keith D., **311,** 215
Barlow, Robert B., **316,** 41
Barnes, Stephen, **301,** 454
Barrick, Jeffrey E., **318,** 268
Bartelsen, Oliver, **311,** 149
Barthel, Linda K., **316,** 579
Barton, Jacqueline K., **319,** 165
Barton, Michelle Craig, **304,** 63
Bartsch, Helmut, **301,** 40
Baschong, Werner, **307,** 173

Baskaran, Namadev, **303,** 45, 272
Baskin, Ronald J., **298,** 413
Bast, Aalt, **301,** 490
Basu, Manju, **311,** 287; **312,** 196
Basu, Shib S., **312,** 196
Basu, Subhash C., **311,** 287; **312,** 196
Bat, Olga, **304,** 639
Batchelor, Robert, **311,** 207
Bates, Gilliam P., **309,** 687
Batni, Suchitra, **316,** 50
Bätz, Michael, **301,** 532
Baudard, Marion, **292,** 538
Bavykin, Sergei G., **304,** 516
Bayer, Peter, **317,** 198
Bayley, Hagan, **291,** 117
Beal, M. Flint, **300,** 297
Beale, Michael H., **291,** 474
Bear, Christine E., **294,** 227
Beattie, Ruth, **314,** 290
Beaudet, Lucille, **292,** 397, 414
Beauwens, Renaud, **294,** 304
Beavo, Joseph A., **315,** 597
Becker, Peter B., **304,** 742
Beckerle, Mary C., **298,** 62
Beckett, Dorothy, **295,** 424
Beckman, Joseph S., **301,** 342, 373, 393
Beckman, Kenneth B., **300,** 156
Bednar, Jan, **304,** 191
Begenisich, Ted, **293,** 383, 556
Beger, Carmela, **306,** 207
Behlke, Joachim, **290,** 269
Beigelman, Leonid, **317,** 39
Beinfeld, Margery C., **314,** 103
Beisiegel, Ulrike, **299,** 35
Beit-Yannai, Elie, **300,** 285
Béjà, Oded, **292,** 370
Belasco, Joel G., **318,** 309
Belikov, S., **304,** 600
Bellido, David, **313,** 257
Benazzi, Louise, **295,** 208
Bence, K. K., **302,** 213
Bender, M. A., **306,** 42
Bendig, Jürgen, **291,** 415
Benedek, George B., **309,** 429
Benimetskaya, Lyuba, **313,** 287
Benjamin, Nigel, **301,** 92
Bennett, Jean, **316,** 777
Bennett, Kristine A., **318,** 332
Benovic, Jeffrey L., **315,** 422
Benzie, Iris F. F., **299,** 15

Bergamaschi, Gaetano, **314,** 429
Berger, Shelley L., **304,** 696
Bergmeier, Wolfgang, **290,** 203
Berlett, Barbara S., **300,** 239
Bermudez, Luiz E., **302,** 285
Berneburg, Mark, **319,** 366
Berova, Nina, **312,** 217; **315,** 219
Beroza, Paul, **295,** 170
Berrios, Miguel, **307,** 55
Berry, Elliot M., **300,** 285
Berthold, Fritz, **305,** 62
Berti, Paul J., **308,** 355
Bertrand, Edouard, **318,** 493
Betts, Scott D., **309,** 333
Betz, Heinrich, **294,** 260
Beutner, Stefan, **319,** 226
Bezanilla, Francisco, **293,** 300, 331; **296,** 566
Bhardwaj, Gabu, **292,** 594
Bhattacharjee, Hiranmoy, **292,** 82
Bhoola, Kanti D., **307,** 368
Bibi, Eitan, **292,** 370
Bickmore, Wendy, **304,** 650
Bieberich, Erhard, **312,** 339
Bielawska, Alicja E., **311,** 480, 499, 518; **312,** 22, 420
Biere, Anja Leona, **309,** 386
Bierfreund, Uwe, **311,** 255
Bieri, Christoph, **315,** 471
Biessen, E. A. L., **314,** 324
Biggin, Mark D., **304,** 496
Bijsterbosch, M. K., **314,** 324
Billups, Brian, **296,** 617
Bingham, James B., **298,** 171
Bingham, Scott E., **297,** 310
Birch, David G., **316,** 202
Birch, Mike, **307,** 496
Birge, Robert R., **315,** 143
Björk, Ulla, **304, 313**
Bkaily, Ghassan, **307,** 119
Blaha, Gregor, **317,** 292
Blair, L. A. C., **302,** 213
Blakely, Randy D., **296,** 347, 475
Blatt, Thomas, **319,** 465
Blatter, Lothar A., **307,** 274
Blinks, John R., **305,** 223
Block, Steven M., **298,** 460
Bloedorn, Britta, **319,** 226
Blomberg-Wirschell, Maureen, **298,** 228
Blomhoff, Rune, **299,** 430
Blond, Sylvie, **290,** 384

Bloodsworth, Allison, **301,** 454
Bloom, Kerry, **298,** 317
Bloomfield, Victor A., **295,** 364
Blough, Neil V., **300,** 202
Blount, Paul, **294,** 458
Blum, H. E., **314,** 524
Blumsom, Jennifer K., **302,** 316
Bode, Ann M., **299,** 77
Bodennec, Jacques, **312,** 101
Boesze-Battaglia, Kathleen, **316,** 65
Bogdanov, Mikhail B., **300,** 297
Böhmer, Frank D., **319,** 255
Böhnke, Matthias, **307,** 536
Bok, Dean, **316,** 400
Bokman, Anne M., **294,** 92
Bold, Richard, **310,** 463
Bolzan, Rachel M., **301,** 178
Bond, Chris T., **293,** 53
Bond, Tamara D., **292,** 359
Bondar, Vladimir S., **305,** 223
Bone, Richard A., **299,** 457
Bönisch, Heinz, **296,** 259
Bonnet, Jean-Jacques, **296,** 203
Bonte, Edgar J., **304,** 742
Boonaert, Christophe J. P., **310,** 375
Boos, Winfried, **292,** 40
Borazjani, Roya N., **310,** 551
Borghetti, Alberico, **307,** 340
Borhan, Babak, **316,** 423
Borisy, Gary G., **298,** 570
Borst, J. Gerard G., **293,** 352
Bose, Suchira, **290,** 409, 418
Bost, Frédéric, **314,** 342
Boudvillain, Marc, **317,** 140
Bourinet, Emmanuel, **293,** 71
Bourne, Louise C., **299,** 91
Bova, Michael P., **290,** 365
Boveris, Alberto, **301,** 188
Bowden, George H. W., **310,** 216
Bowser, David N., **307,** 441
Boyd, Beth, **312,** 459
Bradbury, E. Morton, **304,** 264
Bradshaw, David J., **310,** 279
Branchini, Bruce R., **305,** 188
Brand, Stephen, **300,** 345
Braslavsky, Silvia E., **319,** 37
Braunheim, Benjamin B., **308,** 398
Brazil, Bill T., **290,** 318
Bredtved, Berit K., **310,** 353
Brenowitz, Michael, **295,** 379; **317,** 353

Bretschneider, Frank, **294,** 180
Brett, Ana Maria Oliveira, **300,** 314
Brichon, Gérard, **312,** 101
Bricker, Terry M., **297,** 320
Bridges, Richard J., **296,** 175
Brindley, David N., **311,** 233
Bristol, Lynn A., **296,** 514
Britton, Dustin H. O., **314,** 168
Briviba, Karlis, **301,** 301, 311; **319,** 3, 130, 222
Broach, James R., **304,** 414
Broaddus, William C., **314,** 121
Brockerhoff, Susan E., **316,** 536
Brodsky, Jeffrey L., **306,** 169
Bronson, J. Darin, **316,** 515
Bronstein, Irena, **305,** 370, 410, 417, 428, 466
Brotchie, Daniel, **307,** 496
Browdy, Angeleah, **311,** 123
Brown, Alastair J. H., **303,** 392
Brown, David A., **314,** 136
Brown, Deborah A., **312,** 272
Brown, Edward B., **291,** 356
Brown, Michael C., **298,** 77
Brown, Patrick O., **303,** 179
Brown, R. G., **310,** 629
Brown, R. Lane, **315,** 755
Brownell, James E., **304,** 675
Brözel, Volker S., **310,** 3
Brueggemann, Lioubov, **315,** 268
Bruick, Richard K., **297,** 192
Brunden, Kurt R., **309,** 723
Brunel, Christine, **318,** 3
Brunner, Josef, **312,** 429
Bruns, Dieter, **296,** 593
Bryan, Joseph, **294,** 445
Bubb, Michael R., **298,** 26
Bucala, Richard, **309,** 152
Buchner, Johannes, **290,** 323, 339, 409, 418
Buck, Kari Johnson, **296,** 466
Buckler, Alan J., **303,** 83
Buckley, Noel J., **314,** 136
Buddecke, Eckhart, **313,** 268
Buehler, Christof, **307,** 513
Buell, Gary, **293,** 459
Bügl, Hans, **290,** 418
Buijs, Jos, **309,** 402
Bukhtiyarov, Yuri, **318,** 118
Bunin, Melissa A., **296,** 689
Burdelski, Martin, **299,** 341
Burke, John M., **317,** 409
Burke, Julian F., **303,** 392

Burkhardt, Janis K., **298,** 373
Burkhardt, Nils, **317,** 261
Burmester, Christoph, **291,** 251
Burne, Robert A., **310,** 441
Burr, Jeanne A., **299,** 309
Burston, Steven G., **290,** 141
Burz, David S., **295,** 450
Bush, Valerie J., **305,** 402
Bushnev, Anatoliy S., **311,** 535
Buss, I. Hendrikje, **300,** 106
Busscher, Henk J., **310,** 426, 523
Büssow, Konrad, **303,** 205
Bustin, Michael, **304,** 133
Butterfield, D. Allan, **309,** 746
Buxser, Stephen E., **300,** 256
Buzby, Jeffrey S., **313,** 388

C

Cadenas, Enrique, **301,** 188; **319,** 67
Caderas, Georg, **306,** 352
Cadet, Jean, **319,** 143
Cahill, Dolores J., **303,** 205
Cai, Yuping A., **297,** 3
Caligan, Thomas B., **312,** 3
Callamaras, Nick, **291,** 380; **307,** 152
Callender, Robert, **308,** 176
Cammack, Richard, **301,** 279
Campbell, Anthony K., **305,** 479
Campbell, Lisa, **305,** 499
Cantalupo, Paul, **306,** 297
Cao, Guohua, **299,** 50
Cao, Yongwei, **296,** 551
Capitani, Silvano, **307,** 190
Caplan, Michael J., **296,** 370
Cardarelli, Carol, **292,** 248, 492
Carleton, Alan, **313,** 143
Carman, George M., **312,** 373
Carninci, Piero, **303,** 19
Caron, Marc G., **302,** 153
Carpenter, Barry K., **291,** 30
Carpenter, John F., **309,** 236
Carroll, F. Ivy, **296,** 248
Carruthers, Lenny M., **304,** 19
Carter, Barrie J., **292,** 717
Cartwright, Iain L., **304,** 462
Cascio, Wayne E., **302,** 341
Case, David A., **295,** 170
Cashikar, Anil G., **309,** 649

Castanotto, Daniela, **313,** 401
Cate, Jamie H., **317,** 169
Catsicas, Marina, **314,** 148
Catsicas, Stefan, **314,** 148
Caughey, Bryon, **309,** 122
Caulfield, Malcolm P., **314,** 136
Cavanagh, H. Dwight, **307,** 230
Cazzola, Mario, **314,** 429
Cedergren, Robert, **317,** 457
Celander, Daniel W., **318,** 332
Cepus, Valentin, **291,** 223
Cha, Albert, **296, 566**
Chae, Ho Zoon, **300,** 219
Chafin, David R., **304,** 231
Chalfant, Charles E., **312,** 420
Chamberlin, A. Richard, **296,** 175
Chambers, Timothy C., **292,** 328
Chance, Mark R., **295,** 379; **317,** 353
Chang, Byeong S., **309,** 236
Chang, Chung-Yu, **291,** 117
Chang, Louise, **304,** 76
Chang, Xiu-Bao, **292, 616**
Changela, Anita, **304,** 99
Chardon, Alain, **319,** 445
Chartrand, Pascal, **318,** 493
Chatterjee, Subroto, **311,** 73
Chatturvedi, Prasson, **311,** 94
Chaturvedi, Madan M., **319,** 585
Chaudhry, F. A., **296,** 388
Chaudière, Jean, **299,** 276
Chen, Ching-Kang, **315,** 404
Chen, Hsiu-Hua, **304,** 548
Chen, Jeannie, **316,** 167
Chen, Jiunn-Liang, **318,** 136
Chen, Lily, **318,** 423
Chen, Nian-Hang, **296,** 248, 719
Chen, Pu, **316,** 413
Chen, S., **290,** 242
Chen, Shiming, **316,** 590
Cheney, Richard E., **298,** 3
Cheng, Seng Hing, **292,** 697
Chervenak, Robert, **302,** 189, 199
Chevalier, Mathieu, **290,** 384
Cheynier, Véronique, **299,** 178
Chigorno, Vanna, **311,** 639
Chitnis, Parag R., **297,** 124
Chitnis, Vaishali P., **297,** 124
Chittock, R. S., **291,** 245
Chittock, R. S., **291,** 483
Chizhmakov, I. V., **294,** 490

Choi, Byung-Min, **314,** 493
Chou, Chuan-Chu, **314,** 411
Christensen, Bjarke Bak, **310,** 20
Christodoulou, Danae, **301,** 201
Chu, Vi T., **317,** 140
Chudziak, F., **294,** 445
Chumley, Phillip H., **301,** 454
Chun, Paul W., **295,** 227
Chung, Hun-Taeg, **314,** 493
Chung, Kian Fan, **319,** 551
Chung, Namjin, **311,** 319
Chung, Shin-Ho, **293,** 420
Church, Daniel F., **300,** 63
Church, Deanna M., **303,** 83
Churchill, Mair E. A., **304,** 99
Cideciyan, Artur V., **316,** 611
Cinti, Caterina, **307,** 190
Cioffi, Catherine L., **314,** 363
Clark, A. Clay, **290,** 100; **305,** 157
Clark, David J., **304,** 35
Clark, Janet A., **296,** 293
Clark, Matthew D., **303,** 205
Clarke, David M., **292,** 480
Clarke, Steven, **309,** 89
Clayton, Alison L., **304,** 533
Cleland, W. Wallace, **308,** 3
Clement, John P., IV, **292,** 732; **294,** 445
Closs, Ellen I., **301,** 78
Coffin, Deborah, **301,** 201
Cohen, Amybeth, **297,** 192
Cohen, Michael F., **297,** 3
Colbert, Maureen, **303,** 468
Cole, Douglas L., **313,** 203
Cole, Susan P. C., **292,** 594
Coleman, David E., **308,** 70
Colflesh, David E., **307,** 55
Colin, Fernand, **294,** 304
Collins, Brian E., **312,** 438
Collins-Racie, Lisa A., **303,** 468
Colón, Wilfredo, **309,** 605
Commentz, Jens, **299,** 341
Compton, Duane A., **298,** 331
Conley, Edward C., **294,** 704
Conlon, Kimberly A., **307,** 55
Conner, Elaine M., **300,** 345
Conzelmann, Andreas, **312,** 506
Cook, John C., **301,** 201, 413
Cook, Richard G., **304,** 675
Cook, Terry A., **315,** 597
Cooney, Joseph J., **310,** 637

Cooper, Conan B., **315,** 847
Cooperman, Barry S., **318,** 118
Corera, Amadou T., **296,** 203
Cornwall, M. Carter, **316,** 224
Costa, Lidia E., **301,** 188
Costa, Maria, **317,** 491
Costerton, J. William, **310,** 656
Cote, Rick H., **315,** 646
Coury, Larry A., **306,** 169
Cowan, Christopher W., **315,** 524
Cowan, Nicholas J., **290,** 230
Cox, Guy, **307,** 29
Crabtree, Jean E., **305,** 577
Craig, Peter J., **314,** 290
Crane-Robinson, Colyn, **304,** 533
Crawford, Isabelle T., **292,** 652
Cremesti, Aida E., **311,** 123
Crino, Peter B., **303,** 3
Cronin, Stephen R., **302,** 58
Crooke, Stanley T., **313,** 3
Croop, James M., **292,** 101
Cross, Carroll E., **301,** 471
Crothers, Donald M., **317,** 132
Crouch, Rosalie K., **315,** 219
Crow, John P., **301,** 151, 367
Crowe, Alison J., **304,** 63
Crowley, Jan R., **300,** 88, 124
Cruz, Jeffrey A., **297,** 139
Cruz, Lourdes J., **294,** 605
Cruz, Maria L., **300,** 314
Cryderman, Diane E., **304,** 462
Cueto, Rafael, **301,** 178
Cullen, Daniel, **307,** 607
Culver, Gloria M., **318,** 446, 461
Cummings, Kellie B., **304,** 376
Curfman, Christopher, **311,** 391
Curran, Martin D., **310,** 566
Curto, Monica, **314,** 247

D

D'Avino, Pier Paolo, **306,** 129
Dabiri, Guissou A., **302,** 171
Dacklin, Ingrid, **309,** 591
Dagan, Arie, **312,** 293, 330
Dagle, John M., **313,** 420
Daldal, Fevzi, **297,** 81
Dale, David C., **305,** 591
Daly, Craig J., **307,** 246

Danbolt, N. C., **296,** 388
Danek, Krisstina, **296,** 649
Dani, John A., **294,** 3
Daniels, Gwynn M., **296,** 307
Danos, Olivier, **306,** 225
Dantzig, Jody A., **291,** 307
Dar, Dalit E., **296,** 456
Darley-Usmar, Victor M., **301,** 288, 454
Darrah, Charlotte, **313,** 485
Dastgheib, Sara, **312,** 196
Davidson, Amy L., **292,** 20
Davidson, Norman, **293,** 483
Davies, Stephen W., **309,** 687
Davis, Jennifer L., **306,** 225
Davis, Richard I., **310,** 566
Davis, Ronald W., **306,** 3
Davis, Trisha N., **302,** 87
Dayrell, Mariza, **314,** 136
de Antonio, Isabel, **314,** 3
de Beer, Frederick C., **309,** 701
de Bernardez Clark, Eliana, **309,** 217
de Graaf, David, **292,** 225
de Gruijl, Frank R., **319,** 359
De Kievit, Teresa R., **310,** 117
De Leo, T., **300,** 245
de Waard, Michel, **314,** 290
Deamond, Scott F., **313,** 297
Dean, Michael, **292,** 116
Dean, Nicholas M., **314,** 342
DeAngelis, Barbara, **299,** 287
Declercq, Lieve, **319,** 612
Deeb, Samir S., **316,** 651
Deeley, Roger G., **292,** 594
DeFelice, Louis J., **296,** 578
DeGraff, William, **301,** 413
DeGrip, Willem J., **315,** 12
Dehnes, Y., **296,** 388
Delepelaire, Philippe, **292,** 67
Delmas, Patrick, **314,** 136
Demaimay, Rémi, **309,** 122
Dembo, Micah, **298,** 497
den Dunnen, Johan T., **303,** 100
Deng, Hua, **308,** 176
Denicola, Ana, **301,** 353
Deprez, Eric, **295,** 316
Deras, Michael L., **317,** 353
Deretic, Dusanka, **315,** 77
Desai, Arshad, **298,** 125, 279
Deshmukh, Ranjit R., **313,** 203
Desmaisons, David, **313,** 143

Destree, Antonio, **300,** 345
Detwiler, Peter B., **316,** 133
Deutsch, Carol, **293,** 17
Devine, Jerry H., **305,** 279
Dewilde, Anny, **319,** 197
Dey, Saibal, **292,** 318
Dheur, Sonia, **313,** 56
Di Mascio, Paolo, **319,** 3
Di Meo, S., **300,** 245
Di Primo, Carmelo, **295,** 316
Diamantopoulos, Georgios S., **298,** 197
Diamond, Deborah L., **290,** 444
Dias, Rui M. B., **300,** 194
Diatchenko, Luda, **303,** 349
Diaz-Torres, Maria R., **310,** 433
Dibdin, George H., **310,** 296
Dickinson, Shelly D., **296,** 708
Dickson, Ramona, **290,** 218
Dickson, Robert C., **311,** 3
Diffey, Brian L., **319,** 245
Dijkwel, Paul P., **297,** 182
Diner, Bruce A., **297,** 337
Ding, Linlin, **290,** 365
Ding, Tomas T., **309,** 510
Dinur, Tama, **312,** 293, 330
Dizhoor, Alexander M., **315,** 708
Do Régo, Jean-Claude, **296,** 203
Dobrowsky, Rick T., **311,** 184
Doherty, Patrick, **309,** 687
Dominguez, Michel, **312,** 523
Dong, Beihua, **313,** 522
Doudna, Jennifer A., **317,** 169
Dougherty, Dennis A., **293,** 504
Douglas, L. Julia, **310,** 644
Douki, Thierry, **319,** 143
Dow, Linda K., **304,** 99
Doxsey, Stephen J., **298,** 228
Doyle, Michael L., **295,** 88
Drenser, Kimberly A., **316,** 761
Drmanac, Radoje, **303,** 165
Drmanac, Snezana, **303,** 165
Drummen, G. P. C., **319,** 603
Druzina, Zhanna, **318,** 118
Duan, Rui-Dong, **311,** 276
Dubaquié, Yves, **290,** 193
Dube, Michael G., **314,** 184
Duckett, McKeough, **303,** 468
Dudler, Robert, **292,** 162
Dudley, Samuel C., Jr., **294,** 575
Duerr, Janet S., **296,** 529

Duff, Robert J., **313,** 297
Dufrêne, Yves F., **310,** 375
Dugas, Tammy R., **300,** 63
Dumoulin, Antoine, **306,** 113
Dunaway-Mariano, Debra, **308,** 149
Dunlap, Walter C., **319,** 29
Dunn, Teresa M., **312,** 317
Duong, Tommy, **302,** 32
Durnford, Dion G., **297,** 220
Dürrenberger, Markus, **307,** 173
Dzeja, Claudia, **291,** 415; **315,** 797

E

Earles, Cynthia, **296,** 660
Eberhard, Anatol, **305,** 301
Ebert, Thomas A., **305,** 402
Eberwine, James, **303,** 3
Ebrey, Thomas G., **315,** 196
Echegoyen, Luis, **301,** 235
Eckstein, Fritz, **317,** 74
Edgar, Rotem, **292,** 370
Edmiston, Charles E., Jr., **310,** 194
Edsall, Lisa, **312,** 9
Edwards, Robert H., **296,** 144
Eertmoed, Alison L., **293,** 564
Egan, Marie E., **292,** 664
Egger, Marcel, **314,** 454
Ehrengruber, Markus U., **293,** 483
Ehrnsperger, Monika, **290,** 339
Eickhof, Holger, **309,** 375
Eisen, Michael B., **303,** 179
Eisenstein, Edward, **290,** 119
Eiserich, Jason P., **301,** 454, 471
Elgin, Sarah C. R., **304,** 462
Eliceiri, George L., **313,** 436
Elkins, James G., **310,** 599
Ellington, Andrew D., **318,** 193
Elliott, Peter, **300,** 345
Embury, Paula, **309,** 723
Engelmann, Ralf, **307,** 563
Ennas, Maria Grazia, **314,** 247
Epe, Bernd, **301,** 311; **319,** 436
Epel, David, **291,** 278
Eppihimer, Michael J., **301,** 14
Eraso, Jesus M., **297,** 151
Ercal, Nuran, **299,** 258
Erickson, Harold P., **298,** 305
Erickson, Jeffrey D., **296,** 84

Eriksson, Jan, **297,** 166
Eriksson, John E., **298,** 542
Eriksson, Ulf, **316,** 344
Eritja, Ramon, **313,** 257
Erlich, Shai, **312,** 330
Ernst, Dieter, **319,** 520
Ernst, Oliver P., **315,** 471
Escoubas, Jean-Michel, **297,** 220
Esler, William P., **309,** 350
Estévez, Alvaro G., **301,** 373, 393
Euteneuer, Ursula, **298,** 400
Evans, Cheryl, **303,** 468
Evans, Gregory L., **292,** 572
Evans, Pat, **301,** 333
Evans, Patricia, **300,** 145
Evans, W. Howard, **305,** 479
Everatt, Louise, **313,** 485
Ewart, Gary D., **292,** 213
Ewing, Andrew G., **296,** 675

F

Fahmy, Karim, **315,** 178
Fain, Gordon L., **316,** 146, 224
Falch, Erik, **296,** 165
Falkowski, Paul G., **297,** 220
Fanberstein, David, **300,** 428
Fang, Xingwang, **317,** 393
Fang, Yu, **302,** 207
Fantini, Jacques, **311,** 626; **312,** 495
Farber, Debora B., **315,** 617; **316,** 724
Farber, Gregory K., **308,** 201
Farber, John L., **300,** 78
Farfouri, Sharon, **300,** 428
Fariss, Robert N., **316,** 269
Farre, Cecilia, **294,** 189
Fassel, Theresa A., **310,** 194
Favre, Isabelle, **294,** 287
Fay, Fredric S., **291,** 78
Fedorov, Alexey N., **290,** 1
Felden, Brice, **318,** 147
Feltz, Anne, **314,** 290
Fendler, K., **291,** 289
Fenton, Wayne A., **290,** 141
Ferguson, Stephen S. G., **302,** 153
Fernández-Piñas, F., **305,** 513
Ferreira, Paulo A., **315,** 455; **316,** 688
Ferrone, Frank, **309,** 256
Field, Christine M., **298,** 279, 525

Field, Stanley, **318,** 399
Fields, Stanley, **306,** 93
Fiering, Steven, **306,** 42
Fillmore, Helen L., **314,** 121
Finckh, Barbara, **299,** 341
Findeis, Mark A., **309,** 476
Finer, Jeffrey T., **298,** 436
Fink, Anthony L., **309,** 559
Finn, J. Patrick, III, **296,** 144
Fischl, Anthony S., **311,** 123
Fisher, Harvey R., **295,** 331, 350
Fisher, Mark T., **290,** 119
Fishman, Harvey A., **294,** 189
Flaus, Andrew, **304,** 251
Flory, Mark R., **302,** 87
Flotte, Terence R., **292,** 717
Fluiter, K., **314,** 324
Fong, Henry K. W., **316,** 413
Fontana, Margherita, **307,** 482
Forte, Pablo, **301,** 92
Fortun, Susana, **310,** 656
Fouchard, Robert C., **299,** 318
Fouts, Derrick E., **318,** 332
Fragoso, Gilberto, **304,** 626
François, Jean-Christophe, **313,** 74
Frank, Joachim, **317,** 276, 292
Frank, Ludmila A., **305,** 223
Frankel, Alan D., **318,** 297, 350, 423
Frankel, E. N., **299,** 190
Fraser, Paul E., **309,** 526
Frederick, Jeanne, **316,** 515
Freedman, Jane E., **301,** 61
Freeman, Bruce A., **301,** 353, 454
Frei, Balz, **300,** 17, 51
Freire, Ernesto, **295,** 100
French, Robert J., **294,** 575
Friddle, Carl, **303,** 512
Frieden, Carl, **290,** 100
Frings, Stephan, **291,** 415; **315,** 797
Frisby, Dennis L., **296,** 529
Fritzsch, Günter, **297,** 57
Frølund, Bente F., **296,** 165
Fronczek, Frank R., **319,** 570
Fthenakis, Christina, **319,** 612
Fukada, Yoshitaka, **316,** 278, 465
Fukuoka, Takashi, **310,** 577
Fukuzawa, Kenji, **319,** 101
Funakoshi, Takeshi, **291,** 348
Fung, David C., **298,** 114
Furuichi, Teiichi, **302,** 225

Furuta, Toshiaki, **291,** 50
Futerman, Anthony H., **312,** 179, 187

G

Gaber, Richard F., **293,** 89
Gable, Ken, **312,** 317
Gabriel, Daniela, **291,** 95
Gabriel, Manal M., **310,** 551
Gaestel, Matthias, **290,** 339
Gage, Peter W., **293,** 420
Galbraith, David W., **302,** 296
Galley, Kevin, **294,** 227
Galli, Aurelio, **296,** 578
Gallivan, Justin P., **293,** 504
Gallo, Gloria, **309,** 69
Ganapathy, Vadivel, **296,** 278
Ganz, Michael B., **314,** 378
Garami, Elizabeth, **294,** 227
Garber, Mitchell E., **306,** 352
Garbers, David L., **316,** 558
Garcia, Maria L., **294,** 274, 624
Garcia-de-la Asuncion, Jose, **299,** 267
Gardiner, Katheleen, **303,** 144
Garwin, Gregory G., **316,** 313, 359, 565
Garzón, Javier, **314,** 3
Gasnier, Bruno, **296,** 73
Gasser, Susan M., **304,** 663
Gatt, Shimon, **312,** 293, 330
Gatti, Rita, **307,** 340
Gaut, Joseph P., **300,** 88
Gavigan, Stacey A., **295,** 403
Gavin, Igor M., **304,** 516
Gazula, Valeswara-Rao, **311,** 184
Gazzola, Gian Carlo, **307,** 340
Gdula, David A., **304,** 757
Gee, K., **291,** 63
Gee, Kyle R., **291,** 30, 403
Gee, Ricardo L., **316,** 103
Geiger, Peter G., **300,** 23
Geisse, Sabine, **306,** 19
Gejyo, Fumitake, **309,** 305
Gelb, Michael H., **316,** 436
Gelfand, Vladimir I., **298,** 361
Gelmini, S., **305,** 557
Genka, Chokoh, **301,** 402
George, Christopher H., **305,** 479
Georget, Virginie, **302,** 121
Georgiou, George, **309,** 48

Geraghty, F. M., **294,** 490
Gerdes, Hans-Hermann, **302,** 11
Gerhardt, Greg A., **296,** 708
Gerisch, Günther, **302,** 51
Germann, Ursula A., **292,** 427
Germino, F. Joseph, **303,** 422
Gerwert, Klaus, **291,** 223
Ghafourifar, Pedram, **301,** 381
Ghebramedhin, Haile, **297,** 166
Ghomashchi, Farideh, **316,** 436
Ghosh, Chandramallika, **313,** 135
Giacomoni, Paolo U., **319,** 612
Giangiacomo, Kathleen M., **294,** 274
Gibson, Bruce G., **305,** 164
Giegé, Richard, **318,** 147
Gilbert, H. F., **290,** 26
Gilbert, Peter, **310,** 232
Gilchrist, Annette, **315,** 388
Giles, Richard V., **313,** 95
Gillard, Samantha, **314,** 290
Gillies, George T., **314,** 121
Gilmor, Michelle L., **296,** 407
Gilmour, David S., **304,** 462
Girotti, Albert W., **300,** 23, 85
Girzadas, Mark, **311,** 287
Givens, Richard S., **291,** 1
Givskov, Michael, **310,** 20
Giwercman, Birgit, **310,** 207
Glascott, Peter A., Jr., **300,** 78
Glavy, J. S., **292,** 342
Go, Young-Mi, **301,** 513
Godar, Dianne E., **319,** 309
Goeldner, Maurice, **291,** 265
Goeres, Darla, **310,** 608
Gohil, Kishorchandra, **300,** 402
Golbeck, John H., **297,** 95
Goldberg, Andrew F. X., **316,** 671
Goldberg, David M., **299,** 122, 137
Golden, Barbara L., **317,** 124
Golden, Susan S., **305,** 527
Goldenberg, Sarah, **292,** 248
Golden-Fleet, Margaret, **303,** 468
Goldin, Alan L., **293,** 529
Goldman, Yale E., **291,** 307
Goldsteins, Gundars, **309,** 591
Goloubinoff, Pierre, **290,** 203, 253
Golstein, Philippe E., **294,** 304
Gomelsky, Mark, **297,** 151
Gomez, Christina M., **299,** 457
González, Daniel G., **305,** 212, 672

Gonzalez, G., **294**, 445
González-Cabezas, Carlos, **307**, 482
Goodman, M. B., **293**, 201
Goody, Roger S., **291**, 251
Gopalakrishnan, Vidya, **306**, 328
Gorczyca, Wojciech A., **315**, 689
Gordon, Sharona E., **315**, 772
Gorman, Linda, **313**, 506
Gorman, Sean P., **310**, 558, 566
Gorn, Vladimir V., **313**, 164
Gorovits, Boris M., **290**, 313
Goss, Steven P. A., **301**, 444
Goto, Tetsuhisa, **299**, 107
Gotta, Monica, **304**, 663
Gottenbos, Bovrt, **310**, 523
Gottesman, Michael M., **292**, 248, 318, 441, 456, 474, 523, 546, 557
Gould-Fogerite, Susan, **314**, 411
Goulian, M., **294**, 208
Gow, Andrew, **301**, 367
Gracz, Lawrence M., **296**, 99
Grallert, Holger, **290**, 323
Granger, D. Neil, **301**, 14, 424
Granger, Donald L., **301**, 49
Granovsky, Alexey E., **315**, 635
Grant, Caroline E., **292**, 594
Grassucci, Robert A., **317**, 276
Gratton, Enrico, **307**, 513
Gray, Richard, **294**, 3
Gray-Keller, Mark P., **316**, 133
Greathouse, Denise V., **294**, 525
Green, Carla B., **316**, 291
Green, Michael E., **293**, 694
Green, Mitchell, **296**, 660
Green, Vicki A., **305**, 135, 180
Green, William N., **293**, 564
Greenberg, E. Peter, **305**, 288; **310**, 43
Greenberger, Lee M., **292**, 307
Gregory, Philip D., **304**, 365
Gremo, Fulvia, **314**, 247
Grenningloh, Gabriele, **314**, 148
Grether-Beck, Susanne, **319**, 280
Grewer, Christof, **291**, 443
Grierson, Ian, **307**, 496
Griffith, Jack, **304**, 214
Grilley, Michelle M., **311**, 9
Grisham, Matthew B., **300**, 345; **301**, 201, 220, 413, 437
Grondin, Katherine, **292**, 182
Gros, Philippe, **292**, 397, 414

Gross, Jeffrey W., **308**, 28
Gross, Michael, **290**, 296
Gross, Paul R., **306**, 225
Grossman, Arthur R., **297**, 279
Groudine, Mark, **306**, 42
Grubmeyer, Charles Timmis, **308**, 28
Grunstein, Michael, **304**, 399
Grunwald, Maria E., **315**, 817
Guerrier-Takada, Cecilia, **313**, 442
Guggino, William B., **292**, 664
Guillas, Isabelle, **312**, 506
Gundersen, Thomas E., **299**, 430
Gunshin, Hiromi, **296**, 17
Guo, Zhengmao, **301**, 249
Gurevich, Vsevolod V., **315**, 422
Gusev, Pavel A., **293**, 194
Gutierrez, Peter L., **300**, 202
Guttoh, David S., **292**, 745
Gutz, Ivano G. R., **300**, 314
Guyot, Sylvain, **299**, 178

H

Hackos, David H., **315**, 797
Haeberli, Peter, **317**, 39
Haenen, Guido R. M. M., **301**, 490
Haeseleer, Françoise, **316**, 359, 372
Hagen, Volker, **291**, 415; **315**, 797
Hager, Gordon L., **302**, 73; **304**, 626
Hagerman, Paul J., **317**, 440
Haimeur, Anass, **292**, 182
Haimo, Leah, **298**, 389
Hakomori, Sen-Itiroh, **312**, 381, 447, 488
Haley, Jane E., **314**, 136
Hall, Jason A., **292**, 3, 20
Hallaway, Brenda J., **305**, 390, 402
Hallböök, Finn, **314**, 148
Halliwell, Barry, **300**, 145, 333, 471; **319**, 401
Ham, Amy J. L., **299**, 309
Hamada, Shigeyuki, **310**, 501, 513
Hamaguchi, Akikazu, **312**, 381
Hamel, Christian P., **316**, 705
Hamill, Owen P., **294**, 482
Hamilton, Marty, **310**, 608
Hamm, Heidi E., **315**, 363, 388, 490, 502
Hammache, Djilai, **312**, 495
Hammer, Mara, **305**, 660
Hampton, Randolph Y., **302**, 58
Han, E.-K., **292**, 342

Han, May, **315,** 116, 251
Han, Yung Jin, **298,** 427
Hanada, Kentaro, **312,** 304
Hanakam, Frank, **302,** 51
Handa, Kazuko, **312,** 447, 488
Handelman, Garry J., **300,** 43
Hanner, Markus, **294,** 274, 624
Hannun, Yusuf A., **311,** 156, 164, 480, 499, 518; **312,** 22, 361, 407, 420
Hanrahan, John W., **293,** 169
Hansberg, Wilhelm, **319,** 110
Hansen, Jeffrey C., **304,** 19
Hansen, Mark R., **317,** 220
Hanson, Paul, **317,** 220
Hanzelka, Brian L., **305,** 288
Hao, Wenshan, **316,** 413
Harada, Kazuo, **318,** 297
Harasym, Troy O., **313,** 322
Hardesty, Boyd, **290,** 18
Hardy, Simon J. S., **290,** 444
Hargrave, Paul A., **315,** 70, 91, 437
Harker, Mark, **297,** 244
Härmälä-Braskén, Ann-Sofi, **298,** 542
Harper, James D., **309,** 510
Harris, Guy H., **311,** 335
Harris, Thomas K., **308,** 219
Hart, Ian K., **294,** 677
Hartung, K., **291,** 289
Hartzell, Patricia L., **310,** 335
Hashimoto, Hisashi, **307,** 84
Hassett, Daniel J., **310,** 599
Hassler, Daniel F., **311,** 176
Hathaway, David R., **300,** 456
Hauser, Hansjörg, **318,** 363
Hausladen, Alfred, **300,** 389
Hauswirth, William W., **316,** 743, 761
Hay, A. J., **294,** 490
Hayashi, Takaaki, **316,** 651
Hayashizaki, Yoshihide, **303,** 19
Hayes, Jeffrey J., **304,** 231
Hazen, Stanley L., **300,** 88, 124
He, Kaizhang, **313,** 226
He, Mai, **314,** 411
Heagle, Amy B., **317,** 276
Hebb, Matthew O., **314,** 261
Hebbes, Tim R., **304,** 533
Hecht, Andreas, **304,** 399
Heck, Martin, **315,** 329
Hediger, Matthias A., **296,** 17
Heersink, Joanna, **310,** 608

Heidecker, Manfred, **291,** 95
Heilig, Markus, **314,** 261
Heim, Roger, **302,** 408
Heinecke, Jay W., **300,** 88, 124
Heinemann, Christine, **310,** 426
Heinemann, Stefan H., **293,** 391
Heinrich, Michael, **312,** 429
Heinrich, Ulrike, **319,** 494
Heiser, Volker, **309,** 375
Helbock, Harold J., **300,** 70, 156
Hell, Johannes W., **296,** 116
Helmchen, Fritjof, **293,** 352
Hennesthal, Christian, **293,** 391
Henry, Jean-Pierre, **296,** 73
Hensler, Julie G., **314,** 76
Hensley, Preston, **295,** 88
Hentze, Matthias W., **318,** 374
Herick, Klaus, **305,** 62
Hermann, Thomas, **318,** 33
Herr, John C., **302,** 272
Herrlich, Peter, **319,** 255
Herschlag, Daniel, **308,** 246
Herzberg, Mark C., **310,** 109
Herzenberg, Leonard A., **302,** 296
Hess, George P., **291,** 30, 443
Hess, Stephen D., **294,** 20
Heumann, Hermann, **318,** 33
Hideg, Éva, **319,** 77
Hideg, Kálmán, **319,** 77
Higgins, Christopher F., **292,** 359
Higuchi, Hajime, **301,** 424
Higuchi, Hideo, **291,** 307
Higuchi, Keiichi, **309,** 674
Hilgemann, Donald W., **293,** 267
Hiort, Catharina, **303,** 55
Hirabayashi, Yoshio, **311,** 303
Hirai, Masana, **290,** 50
Hirama, Tomoko, **312,** 205
Hiramatsu, Midori, **300,** 252
Hirayoshi, Kazunori, **304,** 351
Hirokawa, Nobutaka, **291,** 348
Hirschberg, Joseph, **297,** 244
Hlady, Vladimir, **309,** 402
Ho, Siew Peng, **314,** 168
Hoa, Gaston Hui Bon, **295,** 316
Hobbs, Carl, **309,** 687
Hoesch, Robert E., **291,** 431
Hoffman, Robert M., **302,** 20
Hoffmann, Thomas, **319,** 226
Hofmann, Klaus Peter, **315,** 238, 329, 377, 471

Hogg, Neil, **301,** 169, 444
Hogg, Penny, **307,** 496
Hoglen, Niel C., **301,** 483
Høiby, Niels, **310,** 207
Holian, Andrij, **319,** 562
Hollman, Peter C. H., **299,** 202
Holmgren, Arne, **300,** 226
Holt, Jeffrey T., **314,** 499
Hong, Anita, **309,** 189
Honjo, Tasuku, **303,** 479
Honkanen, Richard E., **315,** 579
Hood, Donald C., **316,** 202
Hope, Michael J., **313,** 322
Hopkins, William F., **294,** 649
Hoppe, Udo, **300,** 419
Horiuchi, Motohiro, **309,** 122
Horn, Richard, **293,** 145
Horowitz, Paul M., **290,** 313, 318
Horsburgh, Brian C., **306,** 337
Horwich, Arthur L., **290,** 141
Horwitz, Joseph, **290,** 365
Horwitz, Susan Band, **292,** 342; **298,** 238
Hörz, Wolfram, **304,** 365
Hoshi, Toshinori, **293,** 724
Hoskins, Brian C., **310,** 248
Hosokawa, Masanori, **309,** 674
Hou, Yongchun, **301,** 242
Howard, C. Vyvyan, **307,** 496
Howell, David N., **307,** 573
Howells, Anthony J., **292,** 213
Hristova, Kalina, **295,** 62
Hrncic, Rudi, **309,** 69
Hrycyna, Christine A., **292,** 456
Hsu, Fong F., **300,** 88, 124
Hu, James C., **305,** 499
Huan, Ling-Jun, **294,** 227
Huang, Chiao-Chian, **302,** 32, 207, 438
Huang, Qing-Ling, **290,** 365
Huang, Xizhong, **315,** 579
Hubank, Michael, **303,** 325
Hubinette, Maria M., **306,** 337
Hübner, Christoph, **299,** 341
Hudson, Alan L., **314,** 61
Huganir, Richard L., **294,** 353
Hughes, Jeff, **313,** 342
Hughes, John M. X., **318,** 479
Hughes, Martin N., **301,** 279
Hunter, John C., **314,** 201
Hurley, James B., **315,** 404, 708; **316,** 536

Hutchings, Catherine, **303,** 392
Hutner, Seymour H., **299,** 287

I

Ichikawa, Shinichi, **311,** 303
Igel, A. Haller, **318,** 479
Iglewski, Barbara H., **310,** 117
Ikami, Takao, **311,** 547
Ikebe, Mitsuo, **291,** 78
Ikura, Mitsuhiko, **316,** 121
Illarionov, Boris A., **305,** 223
Illarionova, Victoria A., **305,** 223
Imai, Hiroo, **315,** 293, 347
Imai, Noritaka, **319,** 29
Indyk, Lawrence, **295,** 350
Inglefield, Jon R., **307,** 469
Inoue, Masayasu, **319,** 29
Inoue, Sadayuki, **309,** 496
Inouye, Masayori, **313,** 467
Inouye, Satoshi, **302,** 444
Isambert, Marie-Françoise, **296,** 73
Ischiropoulos, Harry, **301,** 367
Ishida, Hideharu, **311,** 547
Ishida, Hideyuki, **301,** 402
Ishida, Kenneth, **310,** 463
Ishii, Takahiro M., **293,** 53
Ishikawa, Dai, **312,** 145, 157
Ishikawa, Hajime, **290,** 181
Ishikawa, Hiroshi, **307,** 84
Ito, Kimiko, **319,** 417
Ito, Makoto, **311,** 297, 682
Ito, Takashi, **303,** 298
Ivaldi, Serge, **312,** 495
Iversen, Patrick, **313,** 135
Iwabuchi, Kazuhisa, **312,** 381, 488
Iwamura, Michiko, **291,** 50
Izu, Hiroyuki, **311,** 297

J

Jackson, Alisa C., **311,** 168
Jacob, Robert F., **309,** 285
Jacobs, Kenneth A., **303,** 468
Jacobson, Ingemar, **294,** 189
Jacobson, Ken, **298,** 497
Jacques, Danielle, **307,** 119

Jahn, Reinhard, **296,** 116
Jain, Chaitanya, **318,** 309
Jakob, Ursula, **290,** 323, 409
Jakoi, Emma R., **313,** 456
Jang, Geeng-Fu, **316,** 359
Jankowsky, Eckhard, **317,** 140
Janssen, Yvonne M. W., **300,** 363
Jansson, Christer, **297,** 166
Jarchau, Thomas, **298,** 103
Jardemark, Kent, **294,** 189
Jasek, Mark C., **293,** 483
Jasinska, Renata, **311,** 233
Javitch, Jonathan A., **296,** 331
Jedlitschky, Gabriele, **292,** 607
Jekunen, Antti P., **314,** 506
Jenkins, W. Terry, **306,** 113
Jenner, Andrew, **301,** 471; **319,** 401
Jennissen, Herbert P., **309,** 402
Jensen, Roy A., **314,** 499
Jester, James V., **307,** 230
Ji, Ming, **316,** 50
Ji, Yanbin, **301,** 145
Jia, Yanlin, **293,** 169
Jiang, Xin, **302,** 438
Jin, Ping, **297,** 124
Jo, Hanjoong, **301,** 513
Johansen, Charlotte, **310,** 353
Johnson, Edwin C., **294,** 20
Johnson, Kenneth H., **309,** 3
Johnston, Jeffrey P., **315,** 673
Jones, David S., **310,** 558
Jones, G. J., **316,** 224
Jones, Helen E., **305,** 479
Jones, Katherine A., **306,** 352
Jones, Peter L., **304,** 715
Jordan, Mary Ann, **298,** 252
Jordan, Ryan N., **310,** 393
Joseph, Simpson, **318,** 175
Jothi, Sumana, **319,** 536
Jourd'heuil, David, **301,** 201, 220, 437
Juhaszova, Magdalena, **314,** 313
Juliano, R. L., **313,** 342
Jun, Chang-Duk, **314,** 493
Jung, Andreas H., **291,** 1
Jung, Barbara, **303,** 309
Jung, Karl-Heinz, **311,** 441
Jungalwala, Firoze B., **311,** 94
Jungmann, Holger, **319,** 494
Justice, Jay B., Jr., **296,** 649

K

Kacharmina, Janet Estee, **303,** 3
Kaczorowski, Gregory J., **294,** 274, 624
Kaes, Christoph, **311,** 82
Kaether, Christoph, **302,** 11
Kagan, Bruce L., **294,** 661
Kagayama, Manabu, **307,** 79
Kain, Steven R., **302,** 32, 38, 207, 329, 358, 438
Kairemo, Kalevi J. A., **314,** 506
Kajino, Tsutomu, **290,** 50
Kakuev, Dmitri L., **316,** 87
Kálai, Tamás, **319,** 77
Kallenbach, Neville R., **295,** 26
Kallio, Marko, **302,** 272
Kallury, Krishna M. R., **299,** 318
Kalra, Pushpa S., **314,** 184
Kalra, Satya P., **314,** 184
Kalyanaraman, B., **301,** 169, 444
Kamaid, Andres, **301,** 393
Kambayashi, Yasuhiro, **319,** 216
Kamboj, Sunjeev, **294,** 353
Kamisaka, Yasushi, **311,** 50
Kanai, Yoshikatsu, **296,** 17
Kaneko, Shuji, **293,** 319
Kang, David, **301,** 220
Kang, Sang Won, **300,** 219
Kang, Shin-Hong, **313,** 506
Kangasjärvi, Jaakko, **319,** 520
Kannagi, Reiji, **312,** 160
Kanofsky, Jeffrey R., **319,** 59, 505, 512
Kantor, David B., **293,** 483
Kao, Joseph P. Y., **291,** 431
Kaplan, Batia, **309,** 69
Kaplan, Samuel, **297,** 151
Kapurniotu, Aphrodite, **309,** 152
Karin, Michael, **319,** 273
Karle, Jesper, **314,** 20
Karlin, Arthur, **293,** 123
Karnaukhova, Elena, **315,** 219
Karoui, Hakim, **301,** 169
Karpeisky, Alexander, **317,** 39
Karpen, Jeffrey W., **315,** 755
Karpov, V., **304,** 600
Kartner, Norbert, **292,** 616, 629
Kashina, Anna, **298,** 133
Katayama, Haruko, **313,** 359
Katovich, Michael J., **314,** 581
Katsuki, Motoya, **302,** 233

Katzhendler, Yehoshua, **299,** 293
Kaufmann, Raimund, **299,** 390
Kaupp, U. Benjamin, **291,** 415; **315,** 797
Kavanaugh, Michael, **296,** 566
Kawabata, Shigetada, **310,** 513
Kawamura, Akira, **312,** 217
Kawamura, Satoru, **316,** 3
Kawanishi, Shosuke, **319,** 331, 417
Kazantsev, Alexei V., **318,** 136
Kazmerzak, Karen, **310,** 322
Ke, An, **297,** 124
Keaney, John F., Jr., **300,** 17, 51, 61
Kefalov, V. J., **316,** 224
Kehoe, David M., **297,** 279
Kelleher, Joseph F., **298,** 42
Kelleher, Kerry, **303,** 468
Kelly, Patrick, **311,** 287
Kendall, Jonathan M., **305,** 479
Kendrick, Brent S., **309,** 236
Kennedy, Matthew J., **316,** 536
Keppler, Dietrich, **292,** 607
Kessel, Martin, **290,** 203
Kettle, Anthony J., **300,** 111
Keyhani, Reza, **319,** 465
Khan, Saleem A., **306,** 328
Khanna, Savita, **299,** 239
Kharazmi, Arsalan, **310,** 207
Khokhar, Santosh, **299,** 202
Khurana, Ritu, **309,** 559
Kielbassa, Christopher, **319,** 436
Kilpatrick, Michael W., **313,** 485
Kim, Ki Hean, **307,** 513
Kim, Sungmee, **301,** 201
Kim, Tom S. Y., **315,** 831
Kimata, Yukio, **302,** 373
Kindle, Karen L., **297,** 27
Kindy, Mark S., **309,** 701
King, Alastair J., **315,** 557
King, Jonathan, **309,** 333
King, LaShaunda, **290,** 384
King, Linda A., **302,** 394
King, S. Bruce, **301,** 211
King, Scott A., **292,** 686
King, Stephen J., **298,** 171
Kingston, David G. I., **298,** 238
Kirk, Marion, **301,** 454
Kirkegaard, Robin D., **310,** 224
Kirsch, Dieter, **299,** 390
Kiser, Gretchen L., **292,** 616

Kishikawa, Katsuya, **312,** 420
Kisilevsky, Robert, **309,** 496
Kiso, Makoto, **311,** 547
Kissner, Reinhard, **301,** 342
Kita, Katsuhiro, **311,** 297, 682
Klaassen, Corné H. W., **315,** 12
Kladde, Michael P., **304,** 431
Kliger, David S., **315,** 164
Klimuk, Sandra K., **313,** 322
Klinger, Alexandra L., **295,** 190
Kloser, Andrew W., **292,** 616
Klotz, Lars-Oliver, **301,** 301; **319,** 130
Klumpp, Susanne, **315,** 570
Klunk, William E., **309,** 285
Knaus, Hans-Günther, **294,** 274, 624
Knebel, Axel, **319,** 255
Knobel, Susan M., **307,** 351
Knoell, Tom, **310,** 463
Knoll, Friederike, **311,** 568
Knox, Barry E., **316,** 41, 50
Ko, Samuel, **307,** 20
Ko, Young Hee, **292,** 675
Kobuchi, Hirotsugu, **301,** 504
Koch, Hans P., **296,** 175
Koch, Hans-Georg, **297,** 81
Koch, Karl-Wilhelm, **315,** 785
Koch, Katherine S., **302,** 264
Koch, Tad H., **318,** 88
Kocher, Hans P., **306,** 19
Kochevar, Irene E., **319,** 20
Kocsis-Angle, June, **309,** 723
Koeppe, Roger E., II, **294,** 208, 525
Koga, Tetsufumi, **310,** 577
Kohen, Ron, **299,** 293; **300,** 285, 428
Kohlschütter, Alfried, **299,** 341
Kohno, Kenji, **302,** 373
Kohno, Masahiro, **299,** 28
Kojima, Naoya, **312,** 447
Kolb-Bachofen, V., **301,** 126
Kole, Ryszard, **313,** 506
Kolenbrander, Paul E., **310,** 322
Kollmus, Heike, **318,** 363
Kolter, Roberto, **310,** 91
Kolter, Thomas, **311,** 255, 568
Komatsu, Makiko, **300,** 252
Komori, Naoka, **316,** 492
Komura, Juon-ichiro, **304,** 548
Konaka, Ryusei, **319,** 29
Kone, Zie, **293,** 169

Kong, S. K., **307**, 20
Kontush, Anatol, **299**, 35, 341
Koonce, Michael P., **298**, 400
Kooy, Neil W., **301**, 367
Koppal, Tanuja, **309**, 746
Koppenol, Willem H., **301**, 342
Korber, Darren R., **310**, 3
Korenbrot, Juan I., **315**, 797
Korth, Carsten, **309**, 106
Korth, Hans-Gert, **301**, 532
Korytowski, Witold, **300**, 23; **319**, 85
Kosek, Anne B., **317**, 92
Koskinen, Ari M. P., **311**, 458
Koskinen, Päivi M., **311**, 458
Kostka, Peter, **301**, 227
Koutalos, Yiannis, **315**, 742
Kowal, Anthony S., **309**, 649
Kraemer, Brian, **306**, 93; **318**, 399
Kramer, Gisela, **290**, 18
Kramer, John K. G., **299**, 318
Kramer, Phillip R., **304**, 639
Krasnoperova, Natalija V., **316**, 167
Kreis, Thomas E., **298**, 197, 381
Kresge, A. Jerry, **308**, 276
Kricka, Larry J., **305**, 333, 370
Krieg, Adam J., **304**, 99
Krishna, Murali, **301**, 413
Kriz, Ronald, **303**, 468
Krogsgaard-Larsen, Povl, **296**, 165
Kröncke, K.-D., **301**, 126
Krönke, Martin, **312**, 429
Krüger, Martin, **306**, 207
Krutmann, Jean, **319**, 280, 296, 302, 366
Krzymañska-Olejnik, Edyta, **291**, 135
Kubes, Paul, **301**, 22
Kubitscheck, Ulrich, **307**, 207
Kudlicki, Wieslaw, **290**, 18
Kues, Thorsten, **307**, 207
Kuhn, J., **305**, 543
Kumar, Ashok, **300**, 339
Kumar, Janardan, **298**, 185
Kumar-Singh, Rajendra, **316**, 724
Kung, Ching, **294**, 458, 507
Kuo, Yu-Min, **309**, 58
Kupzig, Sabine, **302**, 3
Kurita, Toyohisa, **311**, 297
Kuroda, Masayuki, **292**, 82
Kurose, Iwao, **301**, 424
Kusakabe, Moriaki, **307**, 84

Kutyavin, Igor, **313**, 164
Kvam, Egil, **300**, 322

L

Labarca, Cesar G., **293**, 504
Labbé, Simon, **306**, 145
Lacoste, Jérome, **313**, 74
Lacoste, Judith, **298**, 89
Lacroix, Laurent, **313**, 74
Ladisch, Stephan, **312**, 135
Ladokhin, Alexey S., **295**, 62
Laethem, Ronald M., **311**, 176
Lai, Josephine, **314**, 201
Lakshman, M. R., **299**, 441
Lam, Derek, **316**, 436
Lambert, Regis C., **314**, 290
Lamuela-Raventós, Rosa M., **299**, 152, 184
Landrum, John T., **299**, 457
Landt, Stephen G., **318**, 350
Langebartels, Christian, **319**, 520
Lansmann, Stefanie, **311**, 149, 201
Lanzrein, Markus, **293**, 483
LaPaglia, Christopher, **310**, 335
Lapham, Jon, **317**, 132
Laporte, Stephane A., **302**, 153
Laroche, Thierry, **304**, 663
Laroux, F. Stephen, **301**, 220
Larsson, Magnus, **303**, 495
La-Scalea, Mauro A., **300**, 314
Lau, Yun-Fai Chris, **303**, 349
Laube, Bodo, **294**, 260
Laue, Thomas M., **295**, 494
LaVallie, Edward R., **303**, 468
Lawrence, John R., **310**, 131, 145
Le Roux, Erwan, **299**, 178
Lebedeva, Nadya V., **305**, 527
Lebleu, Bernard, **313**, 189
Leclerc, Fabrice, **317**, 457
Lee, C. Y., **307**, 20
Lee, Caroline G. L., **292**, 557
Lee, Garrett J., **290**, 350
Lee, Hon Cheung, **291**, 403
Lee, Hyeonmoo, **297**, 310
Lee, John, **305**, 87, 164
Lee, Kyu-Min, **304**, 231
Lee, Lihsueh, **311**, 373
Lee, Nancy R., **316**, 526

Lee, San San, **302,** 3
Lee, Ten-Ching, **311,** 117
Lee, Virginia M.-Y., **309,** 81
Lee, W. Theodore, **290,** 154
Leeuwenburgh, Christiaan, **300,** 124
Leffert, Hyam L., **302,** 264
Leganés, F., **305,** 513
Légaré, Danielle, **292,** 182
Lehner, Martin, **299,** 267
Lehr, Hans-Anton, **300,** 462
Lehrach, Hans, **303,** 205; **309,** 375
Lehre, K. P., **296,** 388
Leibach, Frederick H., **296,** 278
Leier, Inka, **292,** 607
Leighton, Barbara H., **296,** 318
Leighton, Jonathan, **292,** 776
Leinonen, Janne, **299,** 3
Lejeune, Philippe, **310,** 56
Lelong, Isabelle H., **292,** 492
Lem, Janis, **316,** 167
Lemasters, John J., **302,** 341
Lerner, Leonid E., **315,** 617
Lester, Henry A., **293,** 483, 504; **296,** 551
Lester, Robert L., **311,** 3
Levey, Allan I., **296,** 407
LeVine, Harry, III, **309,** 274, 467
Levine, Mark, **299,** 65
Levine, Rodney L., **300,** 120
Levis, Richard A., **293,** 218
Levitan, Edwin S., **294,** 47
Levy, John P., **302,** 329, 358, 424
Levy, L. M., **296,** 388
Lewin, Alfred S., **316,** 743, 761
Lewin, Gudrun, **300,** 437
Lewis, James W., **315,** 164
Lewis, Scott M., **296,** 125
Lewisch, Sandra A., **300,** 120
Leyh, Thomas S., **308,** 48
Li, Anli, **315,** 388
Li, Beibei, **300,** 202
Li, Canhui, **294,** 227
Li, Liming, **305,** 249
Li, Min, **293,** 3; **303,** 451
Li, Nanxin, **319,** 273
Li, Qiao, **304,** 313
Li, Ruixiang, **312,** 135
Li, Xianqiang, **302,** 32, 207, 438
Li, Zhixiong, **311,** 287; **312,** 196
Liang, Earvin, **313,** 342
Liao, James K., **300,** 374

LiCata, Vince J., **295,** 42
Licht, Thomas, **292,** 248, 546
Liebler, Daniel C., **299,** 309; **301,** 483; **319,** 546
Lill, Holger, **296,** 425
Lillard, Sheri J., **294,** 189
Lilley, David M. J., **317,** 368
Lim, Chun Ren, **302,** 373
Lin, Steven W., **315,** 116
Lin, Zhicheng, **296,** 456
Lindquist, Susan L., **290,** 430; **309,** 649
Ling, Victor, **292,** 51
Lingwood, Clifford A., **312,** 459, 473
Link, Charles J., Jr., **302,** 329, 358, 424
Linke, Thomes, **311,** 201
Linsdell, Paul, **293,** 169
Liotta, Dennis C., **311,** 391, 535
Liour, Shyh-Shyurng, **312,** 339
Lipp, Peter, **314,** 454
Lis, John T., **304,** 351
Lisle, John T., **310,** 166
Liu, Betty, **305,** 428
Liu, Bin, **311,** 156, 164
Liu, Fenyong, **318,** 238
Liu, Jiankang, **300,** 70
Liu, Meng, **303,** 45
Liu, Qing-Rong, **296,** 52
Liu, Rihe, **318,** 268
Liu, Yongsheng, **311,** 123
Liu, Zhi-Ren, **318,** 22
Livingstone-Zatchej, Magdalena, **304,** 447
Livrea, Maria A., **299,** 421
Lledias, Fernando, **319,** 110
Lledo, Pierre-Marie, **313,** 143
Llorente, Belsis, **317,** 457
Lloret, Ana, **299,** 267
Lockery, Shawn R., **293,** 201
Loewy, Ariel G., **309,** 172
Loft, Steffen, **300,** 166
Logie, Colin, **304,** 726
Lomakin, Aleksey, **309,** 429
London, Erwin, **312,** 272
Long, Roy H., **318,** 493
Loo, Tip W., **292,** 480
López-Corcuera, Beatriz, **296,** 3
Lorimer, George H., **290,** 135, 203
Lory, Philippe, **293,** 71
Lou, Jihong, **315,** 219
Loukin, Stephen H., **294,** 507
Lovett, Michael, **303,** 111
Lowe, Peter N., **301,** 114

Lowenson, Jonathan D., **309,** 89
Lu, Chin-Chih, **293,** 267
Lu, Chunlin, **298,** 305
Lu, Di, **314,** 581
Luberto, Chiara, **312,** 407
Lucas, V. J., **310,** 629
Lucotti, Claudia, **314,** 429
Luger, Karolin, **304,** 3
Lui, P. Y., **307,** 20
Lukhtanov, Eugene A., **313,** 164
Lukyanov, Sergey, **303,** 349
Luna, Elizabeth J., **298,** 32
Lundbæk, J. A., **294,** 208
Lundeberg, Joakim, **303,** 495
Lundgren, Erik, **309,** 591
Lundin, Arne, **305,** 346
Luo, Jiexin, **293,** 169
Luque, Irene, **295,** 100
Lurtz, Monica M., **294,** 117
Lutkenhaus, Joe, **298,** 296
Lybarger, Lonnie, **302,** 189, 199
Lykkesfeldt, Jens, **299,** 83
Lynch, Daniel V., **311,** 130
Lynch, Stephen R., **317,** 240
Lyras, Leonidas, **300,** 145

M

Ma, Amy, **298,** 89
Ma, Jing-Tyan, **302,** 38
Ma, Ju-Fang, **310,** 599
Ma, Kan, **309,** 189
MacDonald, Raynard, **310,** 3
MacKenzie, C. Roger, **312,** 205
Mackey, David, **306,** 308
MacMillan-Crow, Lee Ann, **301,** 135
Maer, A. M., **294,** 208
Maes, Daniel, **319,** 612
Mager, Sela, **296,** 551
Maggio, John E., **309,** 350
Magleby, Karl L., **293,** 437
Maguire, Albert M., **316,** 777
Maiorino, Matilde, **300,** 33
Maitra, Ratan K., **313,** 522
Majors, Paul D., **310,** 248
Makhov, Alexander M., **304,** 214
Maland, Mathew C., **301,** 513
Mallavarapu, A., **291,** 63
Maloney, Peter C., **292,** 652

Mammen, Andrew L., **294,** 353
Mandala, Suzanne M., **311,** 335
Mandinova, Anna, **307,** 173
Mangiarini, Laura, **309,** 687
Mani, Shobana S., **316,** 50
Maniak, Markus, **302,** 43
Maniotis, Despina, **313,** 485
Mann, Giovanni E., **301,** 78
Manning, James M., **306,** 113
Manning, Lois R., **306,** 113
Manning, Mark C., **309,** 236
Mantilla, Carlos B., **307,** 296
Mantyh, Patrick W., **309,** 350
Manz, Werner, **310,** 79
Manzoli, Francesco A., **307,** 190
Mao, Cungui, **311,** 223
Maraldi, Nadir M., **307,** 190
Marenus, Ken, **319,** 612
Maresca, Marc, **312,** 495
Maretzki, Dieter, **315,** 238
Margulis, Alexander, **315,** 730
Margulis, Michael, **291,** 431
Marks, David L., **311,** 50
Markwardt, Fritz, **294,** 180
Marquis, Robert E., **310,** 441
Marriott, Gerard, **291,** 95, 155
Marsh, Derek, **294,** 59
Marsh, Philip D., **310,** 279
Marshall, J., **302,** 213
Marshall, John, **292,** 697
Marshall, Kristin A., **318,** 193
Marsolier, Marie-Claude, **303,** 411
Martásek, Pavel, **301,** 70, 169
Martin, Chris S., **305,** 428, 466
Martin, Hans-Dieter, **319,** 226
Martin, Katherine J., **303,** 234
Martinac, Boris, **294,** 458
Martinac, Boris, **294,** 507
Mashima, Ryuichi, **319,** 29
Mason, Jeffrey T., **295,** 468
Mason, R. Preston, **309,** 285
Massire, Christian, **317,** 491
Masters, Barry R., **307,** 513, 536
Masters, Bettie Sue Siler, **301,** 169
Masters, Betty Sue S., **301,** 70
Mata, Nathan L., **316,** 384
Mathai, John C., **294,** 550
Matheson, Julie M., **302,** 225
Mathews, Ceri J., **293,** 169
Mathews-Roth, Micheline M., **319,** 479

Mathieu-Daudé, Françoise, **303,** 309, 380
Matroule, Jean-Yves, **319,** 119
Matsuda, Takahiko, **316,** 465
Matsumoto, Hiroyuki, **316,** 492
Matsuno, Akira, **313,** 359
Matthews, Hugh R., **316,** 146, 224
Mattoo, Autar K., **290,** 84
Mattson, David L., **314,** 389
Mattson, Mark P., **309,** 733
Matulic-Adamic, Jasenka, **317,** 39
Maulet, Yves, **314,** 290
Mayfield, Stephen, **297,** 192
Maylie, James, **293,** 53
Mayne, Lynne V., **303,** 392
Mayr, Christian, **290,** 409
Mazo, Ilya A., **302,** 329, 358
Mazzoni, Maria R., **315,** 363, 490
McBride, Don W., Jr., **294,** 482
McCabe, Douglas R., **300,** 297
McCapra, Frank, **305,** 3, 633
McCarthy, Nina E. M., **315,** 557
McCarty, Richard E., **297,** 139
McClelland, Michael, **303,** 309, 380
McConnell, Patrick, **294,** 92
McCord, Jeffrey D., **299,** 113
McCoy, John M., **303,** 468
McCray, James A., **291,** 175
McDermott, Timothy R., **310,** 599
McDowell, J. Hugh, **315,** 70
McFadden, Bruce A., **290,** 147
McFeters, Gordon A., **310,** 166
McGill, Andrea D., **301,** 235
McGoldrick, Carol, **298,** 353
McGrath, John C., **307,** 246
McIntosh, J. Michael, **294,** 605
McIntosh, Lee, **297,** 18
McKay, David B., **317,** 149
McKay, Robert, **314,** 342
McKeever, MaryAnn, **319,** 612
McLean, Mark A., **295,** 316
McLean, Robert J. C., **310,** 248
McLeod, Bruce R., **310,** 656
McMahon, Timothy J., **301,** 99
McManus, Owen B., **294,** 274
McNally, Frank, **298,** 206
Meeks, John C., **297,** 3
Megidish, Tamar, **312,** 381
Mehta, Amit D., **298,** 436
Mei, Wei Ping, **300,** 419
Meineke, Petra, **301,** 532

Meisenheimer, Kristen M., **318,** 88
Meisenheimer, Poncho L., **318,** 88
Meldrum, R. A., **291,** 483
Melikian, Haley E., **296,** 347
Melnik, Svetlana, **304,** 516
Meluh, Pamela B., **304,** 414
Mendelsohn, Richard, **312,** 228
Mendez, Ana, **316,** 167
Menger, Michael D., **300,** 462
Mennerick, Steven, **296,** 632
Merabet, Eddine K., **295,** 450
Merberg, David, **303,** 468
Merbs, Shannath, **316,** 626
Merchant, Sabeeha, **297,** 263
Mercola, Dan, **314,** 342
Meredith, Filmore I., **311,** 361
Mergny, Jean-Louis, **313,** 74
Merickel, Andrew, **296,** 144
Merrill, Alfred H., Jr., **311,** 15, 194; **312,** 3, 361, 583
Mesáros, Stefan, **301,** 160
Metzger, Thomas, **296,** 456
Meyer, Christoph K., **315,** 377
Meyer, David, **298,** 133
Michaelis, Susan, **292,** 130, 193
Michalowski, Susan, **304,** 214
Michel, Christoph, **311,** 22
Michel, François, **317,** 491
Michikawa, Takayuki, **302,** 225
Mikhailov, Andrey, **298,** 542
Mikoshiba, Katsuhiko, **302,** 225
Milbradt, Rainer, **299,** 330
Mildvan, Albert S., **308,** 219
Miles, Allen M., **301,** 220, 437
Milesi, David, **313,** 164
Mileusnic, Radmila, **314,** 213
Miller, Brian S., **310,** 433
Miller, Gary W., **296,** 407
Miller, R. Timothy, **301,** 70
Miller, Sara E., **307,** 573
Miller, W. Todd, **291,** 117
Miller, William C., **301,** 49
Millstein, Jack H., **310,** 335
Milne, Eric, **301,** 92
Milstien, Sheldon, **312,** 9
Min, Hongtao, **305,** 527
Minton, Allen P., **295,** 127
Miranda, Neil, **301,** 249
Miranda, Silvia R. P., **312,** 330
Mirochnitchenko, Oleg, **313,** 467

Mirzabekov, Tajib A., **294,** 661
Mitchell, James B., **301,** 413
Mitchison, Timothy J., **291,** 63; **298,** 52, 125, 218, 279, 525
Mitra, Alok K., **294,** 135
Mitsutake, Susumu, **311,** 682
Mittag, Maria, **305,** 258
Mittelman, Marc W., **310,** 534
Miyawaki, Atsushi, **302,** 225
Miyazaki, Chie, **290,** 50
Mizzen, Craig A., **304,** 675
Möbius, Wiebke, **312,** 534
Moczydlowski, Edward, **294,** 287
Moe, Paul C., **294,** 458
Moggs, Jonathan G., **304,** 333
Mohler, Emile R., III, **300,** 456
Mohr, Detlef, **299,** 362
Mohr, Harald, **319,** 207
Mohuczy, Dagmara, **314,** 32
Molday, Laurie L., **294,** 246
Molday, Robert S., **294,** 246; **315,** 831, 864; **316,** 671
Molin, Søren, **310,** 20
Molineaux, Susan M., **309,** 476
Møllegaard, Niels Erik, **318,** 43
Møller, Simon G., **302,** 316
Møller, Søren, **310,** 353
Mommone, Tom, **319,** 612
Momose, Kazutaka, **307,** 425
Monaghan, Erin, **312,** 317
Monia, Brett P., **314,** 363
Moody, Thomas P., **295,** 494
Moore, David J., **312,** 228
Moore, Linda L., **299,** 457
Moore, Melissa J., **317,** 109
Moradpour, D., **314,** 524
Morcos, Paul A., **313,** 174
Mori, Akitane, **299,** 28
Morioka, Mizue, **290,** 181
Morita, Akimichi, **319,** 302
Morris, Nicola S., **310,** 494
Morrow, Jason D., **300,** 3
Morse, Daniel P., **318,** 48
Morse, David, **305,** 258
Morse, Randall H., **304,** 376
Mortensen, Alan, **299,** 408
Morton, Thomas A., **295,** 268
Moscho, Alexander, **294,** 189
Mosier, Jennifer, **305,** 410, 417, 428
Moskovitz, Jackob, **300,** 239

Moskowitz, Neal K., **303,** 422
Moslehi, Jahan J., **309,** 649
Mowri, Hiro-omi, **300,** 17
Moyal, Dominique, **319,** 445
Mueller, Dianne M., **300,** 124
Mukherjee, Amit, **298,** 296
Mukhopadhyay, Asok, **319,** 585
Mukhtar, Hasan, **319,** 342
Muldoon, Rebecca R., **302,** 329, 358, 424
Mullins, R. Dyche, **298,** 42
Mund, Thomas, **298,** 103
Mundis, Dmitri, **318,** 104
Muradov, Khakim G., **315,** 635
Mural, Richard J., **303,** 77
Murphy, Charles L., **309,** 69, 204
Murphy, Kenneth P., **295,** 294
Murray, James B., **317,** 180
Murrell-Lagnado, Ruth D., **294,** 640
Musiani, Monica, **305,** 120, 577
Mussini, Isabella, **314,** 247
Muth, Theodore R., **296,** 370
Müthing, Johannes, **312,** 45
Muto, Akira, **302,** 225
Muzyczka, Nicholas, **316,** 743
Myers, Fiona A., **304,** 533
Myllykallio, Hannu, **297,** 81
Mylvaganam, Murugesapillai, **312,** 473
Myszka, David G., **295,** 268; **309,** 386

N

Nagasawa, Herbert T., **301,** 211
Nagashima, Tadashi, **313,** 359
Nagel, G., **291,** 289
Nagiec, M. Marek, **311,** 3
Naiki, Hironobu, **309,** 305
Naito, Mikihiko, **292,** 258
Nakagawa, Ichiro, **310,** 501
Nakamura, Kenji, **302,** 233
Nakamura, Robert L., **293,** 89
Nakamura, Tomoyuki, **303,** 479
Nakanishi, Koji, **312,** 217; **315,** 219; **316,** 423
Nakano, Minoru, **319,** 216
Nakao, Kazuki, **302,** 233
Nakazawa, Hiroe, **301,** 402
Nardello, Véronique, **319,** 50
Nargeot, Joël, **293,** 71
Nasi, Enrico, **316,** 41
Nathans, Jeremy, **315,** 879; **316,** 626

Natochin, Michael, **315,** 539
Natowicz, Marvin R., **311,** 94
Nawrocki, Joseph P., **315,** 70
Nelson, Daniel A., **292,** 732
Nelson, Nathan, **296,** 52, 425
Nelson, Thomas J., **293,** 194
Neri, Luca M., **307,** 190
Nettleton, Ewan J., **309,** 633
Neu, Thomas R., **310,** 131, 145
Neumann, Inga D., **314,** 223
Newburg, David S., **311,** 94
Newburger, Peter E., **303,** 272
Newman, Dianne K., **310,** 91
Niblack, Brett, **291,** 117
Nicolas, Jean-Claude, **302,** 121
Nielsen, Alex Toftgaard, **310,** 20
Nielsen, C., **294,** 208
Nielsen, Ellen Holm, **309,** 491
Nielsen, Mogens, **314,** 20
Nielsen, Peter E., **313,** 156; **318,** 43
Niepel, Tanja, **310,** 3
Nierhaus, Knud H., **317,** 261, 292; **318,** 251
Niethammer, Martin, **293,** 104
Nieves, E., **292,** 342
Niggli, Ernst, **314,** 454
Nijbroek, Gaby L., **292,** 193
Nikaido, Hiroshi, **292,** 3, 20
Nikolova-Karakashian, Mariana, **311,** 31, 194
Nilsson, Åke, **311,** 276
Nishijima, Masahiro, **312,** 304
Nissen, Mark S., **304,** 155
Nixon, James R., **310,** 566
Noda, Yasuko, **299,** 28
Noland, Brian W., **305,** 157
Noller, Harry F., **318,** 175, 446, 461
Nonell, Santi, **319,** 37
Northrop, Dexter B., **308,** 3
Nourooz-Zadeh, Jaffar, **300,** 13, 58
Nowak, Mark W., **293,** 504
Nowak, Thaddeus S., Jr., **314,** 238
Núñez, Megan E., **319,** 165
Nutikka, Anita, **312,** 459
Nybo, Mads, **309,** 491

O

O'Donnell, Valerie B., **301,** 454
O'Kane, Dennis J., **305,** 87, 390, 402, 450

O'Malley, Bert W., **306,** 281
O'Toole, George A., **310,** 91
Obeid, Lina M., **311,** 223, 319; **312,** 361
Ochiya, Takahiro, **314,** 401
Odeberg, Jacob, **303,** 495
Oegema, Karen, **298,** 279, 525
Oesch, Bruno, **309,** 106
Offensperger, S., **314,** 524
Offensperger, Wolf-Bernhard, **314,** 524
Offord, James, **294,** 92
Ogden, David C., **294,** 490
Ogita, Kiyokazu, **294,** 385
Ohata, Hisayuki, **302,** 341; **307,** 425
Ohguro, Hiroshi, **316,** 482
Ohshima, Hiroshi, **301,** 40
Ohyama, Tsuyoshi, **315,** 797
Oikawa, Shinji, **319,** 331
Okano, Toshiyuki, **316,** 278
Okayama, Naotsuka, **301,** 3
Olejnik, Jerzy, **291,** 135
Olesen, Corinne E. M., **305,** 410, 417, 428
Oliver, Tim, **298,** 497
Olivera, Ana, **311,** 215
Olivera, Baldomero M., **294,** 605
Olmsted, J. B., **302,** 103
Olofsson, Anders, **309,** 591
Olson, Keith R., **302,** 103
Oprian, Daniel D., **315,** 130
Orkin, Stuart H., **304,** 572
Orlandini, Guido, **307,** 340
Orr, George A., **292,** 342; **298,** 238
Orthofer, Rudolf, **299,** 152
Ortoleva-Donnelly, Lori, **317,** 92
Orwar, Owe, **294,** 189
Osada, Kyoichi, **319,** 188
Osawa, Shoji, **315,** 411
Ossipov, Michael H., **314,** 201
Ossipow, Vincent, **304,** 757
Oswald, Robert E., **293,** 589
Ottl, Johannes, **291,** 95, 155
Otto-Bruc, Annie E., **316,** 269
Ou, Joyce, **312,** 3
Ouellette, Marc, **292,** 182

P

Pace, Norman R., **318,** 136
Pacelli, Roberto, **301,** 413

Packer, Lester, **299,** 28, 239, 247, 300, 330; **300,** 395, 402, 413; **301,** 504
Pae, Hyun-Ock, **314,** 493
Pagano, Richard E., **311,** 50; **312,** 523
Palczewski, Krzysztof, **316,** 330, 359, 372, 436
Pallardó, Federico V., **299,** 267
Palm, Gottfried J., **302,** 378
Palmer, Robert J., Jr., **310,** 20, 152, 160, 224, 322
Palombella, Vito J., **300,** 345
Pan, Peng, **291,** 117
Pan, Tao, **317,** 313
Pan, Ying-Xian, **314,** 51
Panagiotidis, Cynthia H., **292,** 30
Pannala, Ananth Sekher, **299,** 207; **301,** 319
Panopoulou, Georgia D., **303,** 205
Panten, U., **294,** 445
Pap, Eward H. W., **319,** 603
Papadopoulou, Barbara, **292,** 182
Papineni, Rao V. L., **294,** 117
Paraskeva, Efrosyni, **318,** 374
Pardee, Arthur B., **303,** 234
Pardi, Arthur, **317,** 220
Parimoo, Satish, **303,** 127
Parish, Craig A., **316,** 451
Park, Chan-Ho, **291,** 1
Park, Heonyong, **301,** 513
Park, Juliet K. J., **301,** 227
Parker, Amy, **302,** 285
Parker, George A., **314,** 411
Parker, Ian, **291,** 380; **307,** 152
Parsek, Matthew R., **305,** 288; **310,** 43
Parsons, J. Thomas, **298,** 89
Parsons, Stanley M., **296,** 99
Partridge, L. Donald, **293,** 371
Pasini, Patrizia, **305,** 120, 577
Pastan, Ira, **292,** 248, 318, 441, 456, 474, 523, 546, 557
Pasternak, Gavril W., **314,** 51
Pastorino, Antonio M., **300,** 33
Patel, Rakesh P., **301,** 288
Patrick, Sheila, **310,** 566
Paul, Pascal, **311,** 50
Pazzagli, M., **305,** 557
Pedersen, Peter L., **292,** 675
Pedersen, Steen E., **294,** 117
Peled-Zehavi, Hadas, **318,** 297
Pelham, Robert J., Jr., **298,** 489

Pelle, Ed, **319,** 612
Pellegrini, Nicoletta, **299,** 379
Pellieux, Corinne, **319,** 197
Pelluffo, Hugo, **301,** 393
Penczek, Pawel, **317,** 276
Peng, Bihai, **314,** 411
Peng, Ling, **291,** 265
Pennings, Sari, **304,** 298
Pepperberg, David R., **316,** 202
Perini, Laura T., **295,** 403
Perrella, Michele, **295,** 208
Perry, David K., **312,** 22
Perry, George, **309,** 133
Persson, Hakan, **309,** 591
Peters, Katherine, **312,** 3
Peters, Reiner, **307,** 207
Peterson, Craig L., **304,** 726
Peterson, Kenneth R., **306,** 186
Petroll, W. Matthew, **307,** 230
Petrou, Steven, **307,** 441
Petushkov, Valentin N., **305,** 164
Peyman, Anusch, **313,** 268
Pfefferli, Martine, **312,** 506
Pfeifer, Gerd P., **304,** 548
Phadke, Nikhil D., **290,** 59
Phillips, M. Ian, **313,** 46; **314,** 32
Phipps, Donald, **310,** 178
Phylactou, Leonidas A., **313,** 485
Pien, Christine, **300,** 345
Pierce, Daniel W., **298,** 154
Pierlot, Christel, **319,** 3, 197
Piéroni, Gérard, **312,** 495
Pietrini, Grazia, **296,** 370
Piette, Jacques, **319,** 119
Pile, Lori A., **304,** 462
Pinto, Ashwin, **294,** 677
Pinzani, P., **305,** 557
Pipas, James M., **306,** 297
Piperakis, Stylianos M., **300,** 184
Pirkl, Franziska, **290,** 418
Piston, David W., **307,** 351
Pitts, Betsey, **310,** 608
Piulats, Jaume, **313,** 257
Plattner, Ronald D., **311,** 348
Player, Mark R., **313,** 522
Podda, Maurizio, **299,** 330
Poderoso, Juan J., **301,** 188
Poh-Fitzpatrick, Maureen B., **319,** 485
Pohlentz, Gottfried, **311,** 82

Polach, K. J., **304,** 278
Polans, Arthur S., **316,** 103
Pollard, Thomas D., **298,** 42
Pollock, Jamie, **312,** 387
Pollock, Roy, **306,** 263
Popov, Igor, **300,** 437
Porreca, Frank, **314,** 201
Porter, Ken, **313,** 226
Portoukalian, Jacques, **312,** 101
Porwol, Torsten, **307,** 108
Possee, Robert D., **302,** 394
Post, J. A., **319,** 603
Postnikov, Yuri V., **304,** 133
Poston, J. Michael, **300,** 239
Potapova, O., **314,** 342
Pothier, Pierre, **307,** 119
Pouget, Jean-Pierre, **319,** 143
Poulsen, Henrik Enghusen, **300,** 166
Povlock, Sue L., **296,** 436
Pozdnyakov, Nikolay, **315,** 730
Prabhu, Sujit S., **314,** 121
Prakash, Y. S., **307,** 296
Prasad, Puttur D., **296,** 278
Prashar, Yatindra, **303,** 258
Prasil, Ondrej, **297,** 220
Pratt, Leslie A., **310,** 91
Preobrazhenskaya, O., **304,** 600
Preston, Gregory M., **294,** 550
Price, Steven F., **299,** 113
Prigent-Combaret, Claire, **310,** 56
Prior, Ronald L., **299,** 50
Pritchard, Kirkwood A., Jr., **301,** 169
Provencio, Ignacio, **316,** 291
Providence, Lyndon L., **294,** 525
Pruss, Dmitry, **304,** 516
Pryor, William A., **301,** 178; **319,** 570
Puglisi, Joseph D., **317,** 240
Pulvermüller, Alexander, **315,** 329
Purich, Daniel L., **308,** 93
Putnam-Evans, Cindy, **297,** 320
Pyle, Anna Marie, **317,** 140

Q

Qian, Ting, **302,** 341
Qian, Yan, **296,** 347
Query, Charles C., **317,** 109
Quinn, Jeanette M., **297,** 263

Quintero, E., **310,** 403
Quivey, Robert G., Jr., **310,** 441

R

Radford, Kathryn, **293,** 459
Radford, Sheena E., **290,** 296
Radi, Rafael, **301,** 353
Rae, James L., **293,** 218
Raffen, Rosemarie, **309,** 318
Raggers, René J., **312,** 562
Raidoo, Deshandra M., **307,** 368
Rait, Vladimir, **313,** 226
Raizada, Mohan K., **314,** 581
Rajavel, Mathumathi, **308,** 28
Rajski, Scott R., **319,** 165
Ralston, Corie Y., **317,** 353
Ramachandra, Muralidhara, **292,** 318, 441,
    456
Ramage, Gordon, **310,** 566
Ramamoorthy, Sammanda, **296,** 347
Raman, Dayanidhi, **315,** 411
Ramanathan, Ragulan, **290,** 100
Ramirez, Johnny, **301,** 242
Ramjeesingh, Mohabir, **294,** 227
Rand, James B., **296,** 529
Randall, Linda L., **290,** 444
Rando, Robert R., **316,** 324, 451
Randolph, Theodore W., **309,** 236
Rao, Manjunath N., **299,** 441
Rao, Srinivasa, **298,** 238
Rapp, Gert, **291,** 202
Rasmussen, Jane E., **300,** 124
Raub, Thomas J., **300,** 256
Ravanet, Jean-Luc, **319,** 143
Raveché, Elizabeth S., **314,** 411
Ray, Jennifer A., **316,** 384
Raymond, Gregory, **309,** 122
Raymond, Pamela A., **316,** 579
Re, Roberta, **299,** 379
Reaves, Phyllis Y., **314,** 581
Rechsteiner, Thomas J., **304,** 3
Recht, Michael I., **317,** 240
Reddi, P. Prabhakara, **302,** 272
Reddy, Prasad, **290,** 119
Redmond, Robert W., **319,** 20
Redmond, T. Michael, **316,** 705
Ree, Anne Hansen, **303,** 495

Reed, Michael W., **313,** 164
Reeves, Raymond, **304,** 155
Rehman, Almas, **319,** 401
Reid, Gregor, **310,** 426
Reily, Michael D., **294,** 92
Reinhard, Matthias, **298,** 103
Reith, Maarten E. A., **296,** 248, 719
Remington, S. James, **305,** 196
Resnekov, Orna, **302,** 136
Rhee, Sue Goo, **300,** 219
Ribeiro, Angela R., **305,** 479
Riboni, Laura, **311,** 656
Ricci, Andrea R., **304,** 584
Rice-Evans, Catherine A., **299,** 91, 207, 379; **301,** 319
Richard, Sylvain, **293,** 71
Richmond, Timothy J., **304,** 3, 251
Richter, Christoph, **301,** 381
Rickard, Janet E., **298,** 197
Ridge, Kevin D., **315,** 3, 59; **316,** 87
Ridgeway, Theresa M., **295,** 494
Ridgway, Harry F., **310,** 178, 463
Ridnour, Lisa A., **299,** 258
Rie, Gousei, **307,** 425
Rieke, Fred, **316,** 186
Rigaud, Jacques, **299,** 178
Riggs, Arthur D., **304,** 548
Rijken, P. J., **319,** 603
Riley, Ronald T., **311,** 348
Rines, Daniel R., **298,** 133
Riordan, John R., **292,** 616, 629
Ripamonti, Marilena, **295,** 208
Ritchie, David M., **311,** 168
Rivera, Victor M., **306,** 263
Robbins, Ian, **313,** 189
Roberts, A. D. G., **310,** 629
Roberts, L. Jackson, II, **300,** 3
Roberts, Neil, **307,** 496
Roberts, Richard W., **318,** 268
Robinson, Carol V., **290,** 296; **309,** 633
Robinson, Emma S. J., **314,** 61
Robinson, Michael B., **296,** 189
Robinson, Phyllis R., **315,** 207
Robinson, Susan W., **316,** 558
Robinson-Benion, Cheryl, **314,** 499
Roby, Clinton, **313,** 297
Roda, Aldo, **305,** 120, 577
Rodermel, Steven, **297,** 209
Rodriguez, Grisel, **310,** 178, 463

Roeder, G. Shirleen, **303,** 512
Roenneberg, Till, **305,** 104
Roepman, Ronald, **316,** 688
Rogers, Gary A., **296,** 99
Rogers, Stephen L., **298,** 361
Roher, Alex Eugene, **309,** 58, 89
Rohl, Carol A., **295,** 1
Röhrdanz, Beatrix, **317,** 261
Rollag, Mark D., **316,** 291
Roman, Linda J., **301,** 70
Romano, Lisa, **294,** 410
Romby, Pascale, **318,** 3
Romero, Michael F., **296,** 17
Romert, Anna, **316,** 344
Ronda, Nicoletta, **307,** 340
Roninson, Igor B., **292,** 225
Rook, Fred, **297,** 182
Roper, M. A., **310,** 629
Rose, Richard C., **299,** 77
Roseman, A. M., **290,** 242
Rosen, Barry P., **292,** 82
Rosenfeld, Michael G., **306,** 67
Rosenfeld, Stuart A., **306,** 154
Røsok, Øystein, **303,** 495
Rospert, Sabine, **290,** 193
Rossi, David, **296,** 617
Rossi, Francis M., **291,** 431
Rossi, John J., **313,** 401
Ross-Macdonald, Petra, **303,** 512
Rosti, Vittorio, **314,** 429
Rothberg, Brad S., **293,** 437
Rothschild, Kenneth J., **291,** 135
Rothstein, Jeffrey D., **296,** 514
Rouxhet, Paul G., **310,** 375
Rouzina, Ioulia, **295,** 364
Roy, Sashwati, **299,** 239, 247; **300,** 395, 402
Royall, James A., **301,** 367
Rudnick, Gary, **296,** 233
Rudolph, Rainer, **309,** 217
Ruetz, Stephan, **292,** 382
Ruiz, Alberto, **316,** 400
Ruiz, MariaLuisa, **315,** 755
Rump, E. T., **314,** 324
Rumsey, Steven C., **299,** 65
Russo, Rosaria, **295,** 208
Ruth, Adam, **292,** 225
Ryan, Colleen A., **304,** 76
Ryan, Michael P., **304,** 376

Ryder, Sean P., **317,** 92
Rye, Hays S., **290,** 141
Ryte, Antonina, **313,** 268
Ryter, Stefan, **300,** 322

S

Saari, John C., **316,** 313, 359, 536, 565
Sabel, Bernhard A., **307,** 563
Sachs, Kristina, **315,** 238
Sáenz-Robles, M. Teresa, **306,** 297
Safa, Ahmad R., **292,** 289
Safarik, Jana, **310,** 463
Saggau, Peter, **294,** 3
Sagné, Corinne, **296,** 73
Sahlgren, Cecilia, **298,** 542
Saibil, H. R., **290,** 242
Saimi, Yoshiro, **294,** 507
Saison-Behmoaras, E. Tula, **313,** 56
Saito, Hirihisa, **301,** 424
Sakaki, Yoshiyuki, **303,** 298
Sakmar, Thomas P., **315,** 116, 178, 251
Sala-Newby, Graciela B., **305,** 479
Salgo, Maria G., **319,** 570
Salih, Gaza, **297,** 166
Salmon, E. D., **298,** 317
Samii, Jason M., **300,** 51
Sánchez-Blázquez, Pilar, **314,** 3
Sand, Wolfgang, **310,** 361
Sandaltzopoulos, Raphael, **304,** 757
Sandermann, Heinrich, Jr., **319,** 520
Sandhoff, Konrad, **311,** 82, 149, 201, 255, 568
Sangari, Felix J., **302,** 285
Sanger, Jean M., **302,** 171
Sanger, Joseph W., **302,** 171
Sanghvi, Yogesh S., **313,** 203
Sansom, Mark S. P., **293,** 647
Santi, Spartaco, **307,** 190
Sarantis, Monique, **296,** 608
Sarasin, Alain, **319,** 153
Sargueil, Bruno, **318,** 22
Sasano, Yasuyuki, **307,** 79
Sastre, Juan, **299,** 267
Satchwell, Michael F., **315,** 30
Sathasivam, Kirupa, **309,** 687
Sato, Kohji, **293,** 155
Satoh, Masamichi, **293,** 319
Satpaev, Daulet K., **316,** 20
Sattsangi, Saroj, **294,** 339

Sauermann, Gerhard, **300,** 419; **319,** 465
Sawada, Geri, **300,** 256
Sawin, K. E., **291,** 63
Sayers, Lee G., **302,** 225
Sayre, Lawrence M., **309,** 133
Scalzitti, Joanne M., **314,** 76
Scaringe, Stephen A., **317,** 3
Schaefer, Amy L., **305,** 288
Schaefer, Hans, **319,** 445
Schatz, David G., **303,** 325
Schatz, Gottfried, **290,** 193
Schatz, Peter, **303,** 451
Scheel, Jochen, **298,** 381
Scheidig, Axel, **291,** 251
Schena, Mark, **306,** 3
Schenk, James O., **296,** 660
Scherr, Michaela, **313,** 401
Schertler, Gebhard F. X., **315,** 91
Scherzinger, Eberhard, **309,** 375, 687
Scheule, Ronald K., **292,** 697
Schick, Diana, **316,** 688
Schieltz, David, **316,** 436
Schierle, Joseph, **299,** 348
Schineller, Jeffrey B., **305,** 301
Schirmer, Eric C., **290,** 430
Schliwa, Manfred, **298,** 400
Schlueter, Charisse, **316,** 50
Schmalhofer, William A., **294,** 274
Schmeichel, Karen L., **298,** 62
Schmidt, Annette, **313,** 268
Schmidt, Richard R., **311,** 441
Schnaar, Ronald L., **312,** 438
Schneider, Carolyn A., **304,** 76
Schnetkamp, Paul P. M., **315,** 718, 847
Scholey, Jonathan M., **298,** 133
Scholten, Jeffrey D., **309,** 467
Schönfeld, Hans-Joachim, **290,** 269
Schramm, Vern L., **308,** 301
Schroeder, Renée, **318,** 214
Schroer, Trina A., **298,** 171
Schuchman, Edward H., **312,** 330
Schüep, Willy, **299,** 348
Schuldiner, Shimon, **296,** 64
Schulze, Heike, **311,** 22
Schumacher, Christoph, **314,** 440
Schuman, Erin M., **293,** 483
Schütze, Stefan, **312,** 429
Schwaller, Beat, **314,** 454
Schwanstecher, C., **294,** 445
Schwanstecher, M., **294,** 445

Schwartz, David C., **303,** 55
Schwartz, Steven D., **308,** 398
Schwartz-Bloom, Rochelle D., **307, 469**
Schwarz, Andreas, **312,** 179, 187
Schwarz, Elisabeth, **309,** 217
Schwarzmann, Günter, **311,** 601; **312,** 534
Schwehm, Jeffery M., **295,** 150
Schweizer, Matthias, **301,** 381
Schwiebert, Erik M., **292,** 664
Sclavi, Bianca, **295,** 379; **317,** 353
Scopes, Robert K., **290,** 161
Scott, Lincoln G., **317,** 18
Scott, Roderick H., **312,** 387
Scott, Sean-Patrick, **293,** 620
Scott, William G., **317,** 180
Scully, Michael S., **314,** 168
Seagren, E., **310,** 403
Seal, Rebecca P., **296,** 318
Seale, Jeffrey W., **290,** 318
Segawa, Akihisa, **307,** 328
Seibert, Fabian S., **292,** 616
Selke, Dagmar, **315,** 570
Sellers, James R., **298,** 427
Semple, Sean C., **313,** 322
Semple-Rowland, Susan L., **316,** 526
Sen, Chandan K., **299,** 239, 247, **300,** 363, 395, 402
Senear, Donald F., **295,** 403
Sener, Abdullah, **294,** 304
SenGupta, Dhruba, **306,** 93; **318,** 399
Senior, Alan E., **292,** 514
Sentenac, André, **303,** 411
Seregina, Tatiana, **302,** 424
Sergueev, Dmitri, **313,** 226, 342
Sergueeva, Zinaida, **313,** 226
Serio, Tricia R., **309,** 649
Serpell, Louise C., **309,** 526
Serrano, Silvia H. P., **300,** 314
Seshadri, Sangita, **309,** 559
Setzu, Mariella, **314,** 247
Sevanian, Alex, **319,** 188
Severt, W. L., **313,** 456
Shacter, Emily, **319,** 428
Shadel, Gerald S., **305,** 279
Shan, Shu-ou, **308,** 246
Shani, Noam, **292,** 753
Shao, Haiyan, **309,** 189
Sharma, Mukul M., **310,** 248
Sharman, Wesley M., **319,** 376
Sharpe, Lindsay T., **316, 626**

Shaw, Barbara Ramsay, **313,** 226, 342
Shaw, Lynn C., **316, 761**
Shaw, Sid, **298,** 317
Shayman, James A., **311,** 42, 105, 373
Shea, Thomas, **301,** 70
Sheahan, Laura, **306,** 328
Shear, Jason B., **294,** 189
Shearman, Mark S., **309, 716**
Sheehan, Amy, **303,** 512
Sheetz, Michael P., **298,** 185, 353
Sheikh, Saifuddin, **306,** 328
Shelton, Jeffrey, **305,** 527
Shelton, Valerie M., **317,** 393
Sheng, Morgan, **293,** 104, **294,** 371
Sheng, ZuFang, **293,** 17
Shepard, Harvey K., **295,** 494
Sheps, Jonathan A., **292,** 51
Shichida, Yoshinori, **315,** 293, 347
Shigenaga, Mark K., **301,** 27
Shih, Theodore M., **293,** 529
Shobana, Sundaram, **294,** 525
Shoshani, Tzipora, **292,** 474
Shu, Liming, **311,** 373
Shukla, Pragati, **315,** 268
Shuman, Howard A., **292,** 30
Shwaery, Glenn T., **300,** 17, 51
Sidler, Michael, **292,** 162
Siebert, Friedrich, **315,** 178
Siebert, Paul D., **303,** 349
Sieck, Gary C., **307,** 296
Siegele, Deborah A., **305,** 499
Siekierka, John J., **290,** 74
Sierakowska, Halina, **313,** 506
Sies, Helmut, **299,** 390; **301,** 145, 301, 311; **319,** 3, 67, 130, 222, 494
Siewe, Ruth M., **305,** 62
Sigurdsson, Snorri Th., **318,** 165
Sil'nikov, Vladimir N., **318,** 147
Silberstein, Anatoly Y., **294,** 661
Sillence, Dan J., **312,** 562
Silverman, David N., **308,** 276
Silverman, Robert H., **313,** 522
Silverman, Scott K., **293,** 504
Sima, Paul D., **319,** 505, 512
Simmons, Andrew D., **303,** 111
Simmons, Robert B., **310,** 551
Simões-Wuest, A. Paula, **314,** 477
Simon, Andras, **316,** 344
Simpson, Robert T., **304,** 431
Sinclair, James F., **305,** 152

Sinden, Richard R., **304,** 639
Singer, Robert H., **318,** 493
Singh, Surinder, **299,** 207; **301,** 319
Singleton, Vernon L., **299,** 152
Sitaramayya, Ari, **315,** 730
Sittler, Annie, **309,** 375
Skach, William R., **292,** 265
Skiba, Nikolai P., **315,** 490, 502
Skibbens, Bob, **298,** 317
Skibsted, Leif H., **299,** 408
Slaughter, Robert, **294,** 624
Slepak, Vladlen Z., **316,** 20
Sligar, Stepher G., **295,** 316
Slodzinski, Martin K., **314,** 313
Smeekens, Sjef C. M., **297,** 182
Smeeton, Allister H., **293,** 589
Smith, Alan E., **292,** 697
Smith, Barbara L., **294,** 550
Smith, Christopher W. J., **318,** 22
Smith, Colin A., **318,** 297, 423
Smith, Drew, **318,** 229
Smith, E. R., **312,** 361
Smith, Gary K., **311,** 176
Smith, Geoffrey L., **307,** 591
Smith, Janet B., **314,** 537
Smith, Lorna M., **301,** 92
Smith, Mark A., **309,** 133
Smith, Raymond D., **293,** 529
Smith, Virginia F., **290,** 444
Smith, W. Clay, **315,** 437
Snyder, Michael, **303,** 512
Snyman, Celia J., **307,** 368
So, Peter T. C., **307,** 513
Sogos, Valeria, **314,** 247
Sohn, Dong-Hwan, **314,** 493
Soleas, George J., **299,** 122, 137
Solomon, Alan, **309,** 69, 204, 460
Sommer, Wolfgang, **314,** 261
Sonnino, Sandro, **311,** 639; **312,** 247
Sornson, Mark W., **306,** 67
Sorscher, Eric J., **292,** 686
Sosnick, Tobin R., **317,** 393
Souquet, Jean-Marc, **299,** 178
Sousa, Rui, **317,** 65
Southwick, Frederick S., **308,** 93
Souto, Maria L., **316,** 423
Spahn, Christian M. T., **317,** 261, 292; **318,** 251
Spaulding, Vikki, **303,** 468
Spear, Nathan, **301,** 393

Spear, Russell N., **307,** 607
Spector, Ilan, **298,** 26
Speed, Margaret, **309,** 333
Spek, Erik J., **295,** 26
Spiecker, Martin, **300,** 374
Spiegel, Sarah, **311,** 215; **312,** 9, 401
Spiess, Eberhard, **307,** 108
Spiller, David G., **313,** 95
Spitz, Douglas R., **299,** 258
Sprang, Stephen R., **308,** 70
Sprong, Hein, **311,** 59
Spudich, James A., **298,** 436
Squadrito, Giuseppe L., **301,** 178; **319,** 570
Squarzoni, Stefano, **307,** 190
Sreekumar, R., **291,** 78
Stäb, Franz, **300,** 419; **319,** 465
Stadtman, Earl R., **300,** 239
Stafford, Grace A., **304,** 376
Ståhl, Stefan, **303,** 495
Stahl, Wilhelm, **299,** 390; **319,** 494
Stamler, Jonathan S., **300,** 389; **301,** 99
Stampe, Per, **293,** 556
Standifer, Kelly M., **314,** 90
Stanley, Philip E., **305,** 47, 96
Staple, Julie K., **314,** 148
Stary, Anne, **319,** 153
Stauderman, Kenneth A., **294,** 20
Stecher, Hartmut, **316,** 330
Stefani, Enrico, **293,** 300, 331
Steffan, Rüdiger, **293,** 391
Stein, C. A., **313,** 287
Stein, David, **313,** 135
Steitz, Erica A., **318,** 332
Stelzl, Ulrich, **317,** 292; **318,** 251
Stephenson, F. Anne, **294,** 677
Sternberg, Claus, **310,** 20
Stevens, Fred J., **309,** 318
Stevens, Scott W., **318,** 385
Stewart, Philip S., **310,** 166, 656
Stickler, David J., **310,** 494
Stimson, Evelyn R., **309,** 350
Stites, Wesley E., **295,** 150
Stocker, Roland, **299,** 362
Stockman, Andrew, **316,** 626
Stoclet, Jean Claude, **301,** 522
Stookey, George K., **307,** 482
Stover, Jen, **303,** 468
Strain, J. J., **299,** 15

Stratford, Michael R. L., **301,** 259
Strauss, Erich C., **304,** 572
Streit, Peter, **309,** 106
Stricker, Nicole L., **303,** 451
Strobel, Scott A., **317,** 92
Strohmaier, Anja-Rose, **307,** 108
Stronach, Beth E., **298,** 62
Struthers, Mary, **315,** 130
Stryer, Lubert, **316,** 121
Stühmer, Walter, **293,** 280
Stuppia, Liborio, **307,** 190
Sturman, Paul, **310,** 608
Subramanyam, Y. V. B. K., **303,** 45, 272
Suetterlin, Rosmarie, **307,** 173
Sueyoshi, Noriyuki, **311,** 297
Sugano, Sumio, **302,** 233
Sugden, Bill, **306,** 308
Sugden, David, **316,** 291
Sugimoto, Yoshikazu, **292,** 474, 523
Sugiyama, Eiko, **312,** 80
Sukharev, Sergei I., **294,** 458
Sullards, M. Cameron, **312,** 32
Sullivan, Jack M., **315,** 30, 268
Sullivan, Michael, **295,** 379; **317,** 353
Sultan, Charles, **302,** 121
Sulzer, David, **296,** 675
Summers, Jack, **313,** 226
Sun, Hui, **315,** 879
Sun, Jun, **297,** 124
Sun, Xuejun, **307,** 135
Sun, Ye-Ming, **294,** 287
Sunde, Margaret, **309,** 526
Sustmann, Reiner, **301,** 532
Sutcliffe, Michael J., **293,** 589
Suter, Bernhard, **304,** 447
Svehag, Sven-Erik, **309,** 491
Sverdrup, Francis, **306,** 328
Svitkina, Tatyana M., **298,** 570
Swandulla, Dieter, **293,** 371
Swanson, Richard, **316,** 41
Sweedler, David, **317,** 39
Swezey, Robert R., **291,** 278
Swiatek, Gavin C., **305,** 672
Swindell, Charles S., **298,** 238
Swingle, Mark R., **315,** 579
Szatkowski, Marek, **296,** 617
Szerencsei, Robert T., **315,** 847

Szostak, Jack W., **318,** 268
Szulc, Zdzislaw, **311,** 480, 518
Szweda, Luke I., **319,** 562
Szweda, Pamela A., **319,** 562

T

Tabita, F. Robert, **290,** 154
Tachibanaki, Shuji, **315,** 347
Tada-Oikawa, Saeko, **319,** 331
Taglicht, Daniel, **292,** 130
Taguchi, Hideki, **290,** 169
Takada, Tatsuyuki, **302,** 233
Takahashi, Michiko, **296,** 608
Takeda, Toshio, **309,** 674
Takemoto, Jon Y., **311,** 9
Taketomi,Tamotsu, **312,** 80
Taki, Takao, **312,** 145, 157
Tally, Jon, **295,** 331
Tamura, Akira, **313,** 359
Tan, Qiang, **315,** 219
Tan, Ruoying, **318,** 350
Tanaka, Jacqueline C., **293,** 620
Tang, Cha-Min, **291,** 431
Tang, Ruey-Jing, **310,** 637
Tang, Xiaoping, **314,** 32
Tani, Motohiro, **311,** 682
Tao, Lin, **310,** 109
Tari, Ana M., **313,** 372
Tarrasón, Gemma, **313,** 257
Tashiro, Kei, **303,** 479
Tassiou, Aspasia M., **300,** 184
Tate, Christopher G., **296,** 443
Tatsuzawa, Hidetaka, **319,** 216
Taylor, Fletcher B., Jr., **305,** 591
Taylor, Michael R., **316,** 536
Taylor, Walter, **305,** 104
Telo, João P., **300,** 194
Tempel, Bruce L., **294,** 649
Tenhunen, Mikko, **314,** 506
Tennent, Glenys A., **309,** 26
Tenner, Kathleen S., **305,** 450
Teplow, David B., **309,** 429
Tepper, Annemiek D., **312,** 16
Terada, Masaaki, **314,** 401
Terakita, Akihisa, **315,** 293, 347
Terouanne, Béatrice, **302,** 121
Terstappen, Georg C., **312,** 187

Tesoriere, Luisa, **299,** 421
Tettamanti, Guido, **311,** 639, 656
Thatcher, Nicola M., **312,** 387
Theriot, Julie A., **291,** 63, **298,** 114
Thiede, Bernd, **318,** 438
Thiele, Dennis J., **306,** 145
Thiele, Jens J., **300,** 413; **319,** 536
Thom, Stephen R., **301,** 367
Thoma, C., **314,** 524
Thoma, Fritz, **304,** 447
Thomas, Brian C., **318,** 136
Thomas, Carole J., **302,** 394
Thomas, James A., **301,** 145
Thomas, Michael D., **305,** 135, 315
Thomas, Tarita O., **315,** 490, 502
Thompson, Debra A., **316,** 41
Thompson, James D., **306,** 241
Thompson, John A., **301,** 135
Thorne, Alan W., **304,** 533
Thummel, Carl S., **306,** 129
Tidd, David M., **313,** 95
Tint, Irina S., **298,** 361
Tirosh, Oren, **299,** 293; **300,** 285, 428
Tischmeyer, Wolfgang, **314,** 275
Todd, Matthew J., **290,** 135
Tohyama, Masaya, **293,** 155
Toivola, Diana M., **298,** 542
Tolbert, Thomas J., **317,** 18
Tonkinson, John, **313,** 287
Topping, Traci B., **290,** 444
Toro, Ligia, **293,** 529
Torrence, Paul F., **313,** 522
Torres-Ruiz, Jose A., **290,** 147
Toschi, Nicola, **314,** 223
Traber, Katrina, **299,** 300
Traber, Maret G., **299,** 330
Tran, Ken, **316,** 724
Trang, Phong, **318,** 238
Treiber, Daniel K., **317,** 330
Tremethick, David John, **304,** 50
Trenkle, Thomas, **303,** 309, 380
Trewavas, Anthony J., **291,** 474
Triana-Alonso, Francisco J., **317,** 261
Trimmer, James S., **293,** 32
Trojanowski, John, **309,** 81
Trollinger, Donna R., **302,** 341
Tronnier, Hagen, **319,** 494
Troullier, Agnes, **291,** 223
True, Heather L., **318,** 332
Truscott, Kaye, **290,** 161

Ts'o, Paul O. P., **313,** 297
Tsai, Sophia Y., **306,** 281
Tse, Christin, **304,** 19
Tsin, Andrew T. C., **316,** 384
Tsinoremas, Nicholas F., **305,** 527
Tsuji, Frederick I., **302,** 444
Tsujimoto, Gozoh, **302,** 233
Tsukiyama, Toshio, **304,** 757
Tsuruo, Takashi, **292,** 258, 523
Tufaro, Frank, **306,** 337
Tunney, Michael M., **310,** 558, 566
Turk, John, **300,** 124
Turnacioglu, Kenan K., **302,** 171
Turner, Christopher E., **298,** 77
Tyrrell, Rex M., **300,** 322; **319,** 290

U

Udaka, Shigezo, **290,** 50
Ueda, Tetsufumi, **296,** 125
Uhl, George, **296,** 456
Uhlmann, Eugen, **313,** 268
Ujike, Yosuke, **307,** 425
Ulbrich, Carola, **291,** 223
Ulitzur, S., **305,** 543
Umesono, Kazuhiko, **302,** 444
Unger, Vinzenz M., **294,** 135
Uppu, Rao M., **301,** 178
Urbatsch, Ina L., **292,** 397, 514
Ursini, Fulvio, **300,** 33
Urwin, Peter E., **302,** 316
Usachenko, Sergei I., **304,** 264

V

Vajkoczy, Peter, **300,** 462
Valax, Pascal, **309,** 48
Vale, Ronald D., **298,** 154
Valle, David, **292,** 753
Vallejo, Yolanda F., **293,** 564
Valverde, Miguel A., **292,** 359
van Berkel, T. J. C., **314,** 324
van Blitterswijk, Wim J., **312,** 16
Van Brocklyn, James R., **312,** 401
van der Mei, Henny C., **310,** 523
van der Sluijs, Peter, **311,** 59
van der Vliet, Albert, **301,** 471

van Echten-Deckert, Gerhild, **311,** 22; **312,** 64
Van Epps, Heather A., **316,** 536
van Ginkel, Paul R., **316,** 103
Van Hooser, J. Preston, **316,** 269, 565
Van Horn, Gerald, **306,** 328
van Lier, Johan E., **319,** 376
van Meer, Gerrit, **311,** 59; **312,** 562
van Tilburg, Anita, **305,** 315
Van Veldhoven, Paul P., **311,** 244
Vandenbergh, David J., **296,** 498
Vanderplasschen, Alain, **307,** 591
Vanin, A. F., **301,** 269
Vann, Lewis, **312,** 9
Varadarajan, Sridhar, **309,** 746
Varani, Gabriele, **317,** 198
Varani, Luca, **317,** 198
Varga-Weisz, Patrick D., **304,** 742
Varney, Mark A., **294,** 20
Varoqui, Helene, **296,** 84
Vásquez-Vivar, Jeanette, **301,** 169
Vass, Imre, **319,** 77
Vaughan, Roxanne A., **296,** 219
Veliçelebi, Gönül, **294,** 20
Velraeds, Martine, **310,** 426
Venditti, P., **300,** 245
Venema, Dini P., **299,** 202
Verkman, Alan S., **302,** 250
Vermaak, Danielle, **304,** 715
Vermaas, Wim F. J., **297,** 293
Viani, Paola, **311,** 656
Vieira, Abel J. S. C., **300,** 194
Viera, Liliana, **301,** 373
Vierling, Elizabeth, **290,** 350
Vietsch, H., **314,** 324
Viitanen, Paul V., **290,** 203, 218
Vilaró, Senén, **313,** 257
Villazana, Elia T., **316,** 384
Viña, José, **299,** 267
Vincent, Angela C., **294,** 677
Vincent, Jean-Didier, **313,** 143
Virgili, Fabio, **301,** 504
Vishnuvardhan, Daesety, **314,** 103
Visscher, Koen, **298,** 460
Visser, Antonie J. W. G., **305,** 164
Visvardis, Evangelos-E., **300,** 184
Vladimirov, Serguei N., **318,** 118
Vlassov, Valentin V., **318,** 147
Vodovotz, Yoram, **301,** 201, 413
Vogel, Horst, **315,** 471
Vogt, Thomas, **303,** 309

Volkman, Sarah, **292,** 174
Vollmar, Brigitte, **300,** 462
Volsen, Stephen G., **314,** 290
von Coburg, Alexander, **312,** 534
von Rège, Henry, **310,** 361
von Weizsäcker, F., **314,** 524
Vörtler, L. Claus S., **317,** 74
Voss, Kenneth J., **305,** 53
Vought, Bryan W., **315,** 143
Voyta, John C., **305,** 370, 410, 417, 428, 466
Vuong, T. Minh, **316,** 253
Vysotski, Eugene S., **305,** 223

W

Wade, Paul A., **304,** 715
Waggoner, David W., **311,** 233
Walczak, Claire E., **298,** 525
Walke, Stefan, **290,** 339
Walker, Diane, **310,** 608
Walker, James T., **310,** 629
Walker, Jeffery W., **291,** 78
Walker, Kerfoot P., III, **304,** 19
Walker, Teresa M., **316,** 103
Wall, Jonathan, **309,** 204, 460
Wallace, Scot T., **318,** 214
Wallrath, Lori L., **304,** 462
Walter, Nils G., **317,** 409
Walter, Ulrich, **298,** 103
Wang, Elaine, **311,** 15; **312,** 3
Wang, Hongwei, **314,** 581
Wang, Jianqiang, **301,** 235, 242
Wang, Jun, **309,** 81; **318,** 238
Wang, Peng George, **301,** 235, 242, 249
Wang, Ruo, **318,** 118
Wang, Suming, **302,** 424
Wang, Yanchun, **294,** 227
Wang, Yaohui, **299,** 65
Wang, Yaolin, **306,** 281
Wang, Ye, **292,** 745
Wang, Yu-li, **298,** 489
Wanker, Erich E., **309,** 375, 687
Ward, Jane L., **291,** 474
Ward, William W., **305,** 212, 672
Warren, René, **315,** 831
Watabe, Hiroyuki, **302,** 264
Watanabe, Rikio, **312,** 523
Waterhouse, Andrew L., **299,** 113, 184
Watnick, Paula I., **310,** 91

Watson, Gregory M. F., **290,** 154
Wayment, Hollie, **296,** 660
Weaver, Valerie B., **310,** 91
Webb, Chris D., **302,** 136
Webb, Robert H., **307,** 3
Webb, Tom, **290,** 218
Webb, Watt W., **291,** 356
Webber, Andrew N., **297,** 310
Weber, Christine, **299,** 330
Weber, Janet A., **304,** 462
Weber, Jörg F. W., **291,** 1
Weber, Stefan U., **319,** 536
Weber, Thomas, **312,** 429
Wecker, Matt, **318,** 229
Wedekind, Joseph E., **317,** 149
Weeks, Daniel L., **313,** 420
Wei, Ping, **306,** 352
Weikl, Tina, **290,** 418
Weinberg, J. Brice, **301,** 49
Weiner, Ronald, **310,** 403
Weiss, Celeste, **290,** 203, 253
Weiss, Deborah T., **309,** 69
Weiss, Ellen R., **315,** 411
Weissman, Jonathan S., **290,** 141
Weissman, Sherman M., **303,** 127, 258, 272
Welch, Matthew D., **298,** 52
Weller, Dwight, **313,** 135
Wells, Sandra E., **318,** 479
Welsh, John, **303,** 309, 380
Wemmer, David E., **309,** 536
Wensel, Theodore G., **315,** 524
Westermark, Gunilla T., **309,** 3
Westermark, Per, **309,** 3
Westhof, Eric, **317,** 491
Whalen, Patrick O., **316,** 761
Wharton, C. W., **291,** 245, 483
Wheatley, Christine, **312,** 523
White, C. Roger, **301,** 288
White, David C., **310,** 152, 224
White, Stephen H., **295,** 62
Whiteley, Marvin, **310,** 248
Whiteman, Matthew, **301,** 333
Whitten, Mark E., **316,** 436
Wickel, Marc, **312,** 429
Wickens, Marvin, **306,** 93; **318,** 399
Wickstrom, Eric, **314,** 537
Widom, J., **304,** 278
Wiederrecht, Gregory, **290,** 74
Wightman, R. Mark, **296,** 689

Wiklund, Ronney, **297,** 166
Wilkinson, Nicola, **302,** 394
Williams, David Alan, **307,** 441
Williams, Shirley A., **313,** 388
Williamson, James R., **317,** 18, 330
Williamson, Mark J., **303,** 468
Wilson, David M., **305,** 402
Wilson, Leslie, **298,** 252
Wilson, Michael, **310,** 264
Wilson, Timothy J., **295,** 494
Wilson-Kubalek, Elizabeth M., **312,** 515
Wimley, William C., **295,** 62
Wimpenny, Julian, **310,** 296
Winderickx, Joris, **316,** 651
Wingerath, Thomas, **299,** 390
Wink, David A., **301,** 201, 220, 413
Winoto-Morbach, Supandi, **312,** 429
Winslow, Robert M., **306,** 113
Winston, Anette, **316,** 324
Winterbourn, Christine C., **300,** 106
Winters, Carole, **310,** 494
Winters, Roger A., **299,** 258
Winzeler, Elizabeth A., **306,** 3
Wirth, Dyann, **292,** 174
Wirtz, K. W. A., **319,** 603
Witting, Paul K., **299,** 362
Wittmann-Liebold, Brigitte, **318,** 438
Wlodawer, Alexander, **302,** 378
Wo, Z. Galen, **293,** 589
Wolber, Rainer, **319,** 465
Wolbring, Gregor, **315,** 718
Wolfaardt, Gideon M., **310,** 3
Wolffe, Alan P., **304,** 715
Wolfson, M., **292,** 342
Wolk, C. Peter, **297,** 3; **305,** 513
Wollenzien, Paul, **318,** 104
Wonderlin, William F., **294,** 339
Wong, Henry, **300,** 345
Wong, Mei Lie, **298,** 218, 279
Wong-Staal, Flossie, **306,** 207
Wood, Stephen J., **309,** 386
Woodbury, Dixon J., **294,** 319
Woodcock, Christopher L., **304,** 191
Woodson, Sarah, **295,** 379; **317,** 353
Workman, Jerry L., **304,** 696
Wrange, Örjan, **304,** 313
Wu, Carl, **304,** 757
Wu, Jituo, **297,** 320
Wu, Rosemary, **310,** 322

Wu, Wen-I., **312,** 373
Wu-Pong, Susanna, **314,** 121
Wyszynski, Michael, **294,** 371

X

Xia, Xiao-ming, **293,** 53
Xu, Cen, **296,** 248
Xu, Jia, **293,** 3
Xu, Jim, **311,** 233
Xu, M., **304,** 431
Xu, Xiaolian, **312,** 272
Xu, Youfeng, **293,** 483

Y

Yadan, Jean-Claude, **299,** 276
Yahi, Nouara, **312,** 495
Yaida, Yutaka, **314,** 238
Yamada, Yukio, **290,** 50
Yamaguchi, Kazuhiro, **307,** 394
Yamaguchi, Tomohiko, **316,** 651
Yamamoto, Masayuki, **307,** 425
Yamamoto, Yorihiro, **319,** 29
Yamashita, Clyde K., **316,** 724
Yan, Weiming, **316,** 761
Yang, C.-P. H., **292,** 342
Yang, Chii-Shen, **315,** 490
Yang, Jing, **309,** 189
Yang, Lynda J.-S., **312,** 438
Yang, Min, **299,** 379
Yang, Yifan, **301,** 235
Yarrow, Justin, **305,** 527
Yasuda, Hiroshi, **310,** 577
Yates, John R., III, **316,** 436
Yates, Stephen L., **309,** 723
Yatin, Servet M., **309,** 746
Yau, King-Wai, **315,** 742, 817
Ye, Yao Zu, **301,** 373
Yeager, Mark, **294,** 135
Yeagle, Philip L., **315,** 107
Yeh, Elaine, **298,** 317
Yelin, Rodrigo, **296,** 64
Yeliseev, Alexei A., **297,** 151
Yeo, Helen C., **300,** 70
Yohn, Christopher B., **297,** 192

Yokoyama, Hidekatsu, **300,** 275
Yokoyama, Kohei, **316,** 436
Yokoyama, Shozo, **315,** 312
Yoneda, Yukio, **294,** 385
Yoo, Hoon, **313,** 342
Yoo, Ji-Chang, **314,** 493
Yoshida, Akiko, **315,** 730
Yoshida, Kenichi, **302,** 233
Yoshida, Masasuke, **290,** 169
Yoshida, Yuko, **299,** 107
Yu, Jianping, **297,** 18
Yu, Liuning, **304,** 376
Yu, Robert K., **312,** 115, 339
Yu, Yi-Tao, **318,** 71
Yufu, Katsumi, **314,** 238

Z

Zack, Donald J., **316,** 590
Zagorski, Michael G., **309,** 189
Zahniser, Nancy R., **296,** 708
Zamburlini, Adriana, **300,** 33
Zander, Thomas, **290,** 59
Zangemeister-Wittke, Uwe, **314,** 477
Zare, Richard N., **294,** 189
Zaret, Kenneth S., **304,** 612
Zehavi, Uri, **312,** 387
Zeidel, Mark L., **306,** 169
Zeigler, Miriam M., **305,** 135
Zeilhofer, Hanns Ulrich, **293,** 371
Zeilstra-Ryalls, Jill H., **297,** 151
Zeikowicz, Abraham, **300,** 428
Zelver, Nick, **310,** 608
Zeng, Hong, **309,** 189
Zenkova, Marina A., **318,** 147
Zennaro, Lucio, **300,** 33
Zerangue, Noa, **296,** 566
Zerr, Patricia, **293,** 53
Zhang, Beilin, **306,** 93; **318,** 399
Zhang, Fang, **292,** 51
Zhang, Jian-Ting, **292,** 279
Zhang, Jiaping, **292,** 492
Zhang, Jie, **302,** 153
Zhang, Shudong, **292,** 474
Zhang, Tong, **290,** 18
Zhang, Y. Clare, **313,** 46
Zhao, Xiaoning, **302,** 32, 438
Zheng, Yixian, **298,** 218, 525

ISBN 0-12-182221-4

90038

9 780121 822217